The Human Body
in Health & Disease

Text and Illustration Team

Frederic H. Martini received his Ph.D. from Cornell University in comparative and functional anatomy. He has broad interests in vertebrate biology, with special expertise in anatomy, physiology, histology, and embryology. Dr. Martini's publications include journal articles, technical reports, contributed chapters, magazine articles, and a book on the biology and geology of tropical islands. He is the author of *Fundamentals of Anatomy and Physiology* (4e, 1998); he has coauthored *Human Anatomy* (3e, 2000) with Profs. Michael J. Timmons and Michael P. McKinley and *Essentials of Anatomy and Physiology* (2e, 2000) with Prof. Edwin F. Bartholomew. Dr. Martini has been involved in teaching undergraduate courses in anatomy and physiology (comparative and/or human) since 1970. He is currently on the faculty of University of Hawaii and remains affiliated with the Shoals Marine Laboratory (SML), a joint venture between Cornell University and the University of New Hampshire. Dr. Martini is involved with all aspects of the A & P learning system that accompanies each of his textbooks. He also maintains an active research program on vertebrate physiology, anatomy, and ecology. Dr. Martini is a member of the Human Anatomy and Physiology Society, the American Physiological Society, and the American Association of Anatomists. He is also a member of the National Association of Biology Teachers, the Society for College Science Teachers, the Society for Integrative and Comparative Biology, the Western Society of Naturalists, and the International Society of Vertebrate Morphologists.

Edwin F. Bartholomew received his undergraduate degree from Bowling Green State University in Ohio and his M.S. from the University of Hawaii. His interests range widely, from human anatomy and physiology to the marine environment and the "backyard" aquaculture of escargots and ornamental fish. He has taught human anatomy and physiology at both the secondary and undergraduate levels, including a wide variety of other science courses (from botany to zoology) at Maui Community College. He is presently teaching at historic Lahainaluna High School, the oldest school west of the Rockies. Mr. Bartholomew has written journal and magazine articles, a weekly newspaper column, and, with Dr. Martini, coauthored *Essentials of Anatomy and Physiology* (2e, Prentice Hall, 2000) and *Structure and Function of the Human Body* (Prentice Hall, 1999). He is a member of the Human Anatomy and Physiology Society, the National Association of Biology Teachers, and the American Association for the Advancement of Science.

Dr. Kathleen Welch received her M.D. from the University of Seattle and did her residency at the University of North Carolina in Chapel Hill. For two years, she served as Director of Maternal and Child Health at the LBJ Tropical Medical Center in American Samoa before taking a position in Family Practice at the Kaiser Permanente Clinic in Lahaina, Hawaii. She has been in private practice since 1987. Dr. Welch is a Fellow of the American Academy of Family Practice. She is a member of the Hawaii Medical Association and the Human Anatomy and Physiology Society. Drs. Martini and Welch were married in 1979; they have one child, "P. K.," born in January 1995.

Dr. William C. Ober (art coordinator and illustrator) received his undergraduate degree from Washington and Lee University and his M.D. from the the University of Virginia in Charlottesville. While in medical school he also studied in the Department of Art as Applied to Medicine at Johns Hopkins University. After graduation Dr. Ober completed a residency in family practice, and is currently on the faculty of the University of Virginia as an Instructor in the Division of Sports Medicine. He is also part of the Core Faculty at Shoals Marine Laboratory, (Cornell University), where he teaches biological illustration in the summer program. Dr. Ober now devotes his full attention to medical and scientific illustration.

Claire W. Garrison, R.N. (illustrator) practiced pediatric and obstetric nursing for nearly 20 years before turning to medical illustration as a full-time career. Following a five-year apprenticeship, she has worked as Dr. Ober's associate since 1986. Ms. Garrison is also a Core Faculty member at Shoals.

Texts illustrated by Dr. Ober and Ms. Garrison have received national recognition and awards from the Association of Medical Illustrators (Award of Excellence), American Institute of Graphics Arts (Certificate of Excellence), Chicago Book Clinic (Award for Art and Design), Printing Industries of America (Award of Excellence), and Bookbuilders West. They are also recipients of the Art Directors Award.

The Human Body in Health & Disease

Frederic H. Martini, Ph.D.
Edwin F. Bartholomew, M.S.
Kathleen Welch, M.D.

with

William C. Ober, M.D.
Art coordinator and illustrator

Claire W. Garrison, R.N.
Illustrator

Prentice Hall, Upper Saddle River, New Jersey 07458

Library of Congress Cataloging-in-Publication Data
Martini, Frederic.
   The human body in health & disease/Frederic H. Martini, Edwin F. Bartholomew,
Kathleen Welch; with William C. Ober, art coordinator and illustrator; Claire W.
Garrison, illustrator. --1st ed.
      p.   cm.
   ISBN 0-13-856816-2
   1. Human physiology. 2. Human anatomy. 3. Physiology, Pathological. I. Title:
Human body in helath and disease. II. Bartholomew, Edwin F. III. Welch, Kathleen
IV. Title

QP36.M423 2000
612—dc21
                                                99-058973

Senior Editor: Halee Dinsey
Project Manager and Development Editor: Don O'Neal
Editorial Assistant: Leslie Anderson
Production Editor: Shari Toron
Assistant Vice President of Production & Manufacturing: David W. Riccardi
Executive Managing Editor: Kathleen Schiaparelli
Director, Creative Services: Paul Belfanti
Art Director: Heather Scott
Assistant to Art Director: John Christiana
Interior Design: Douglas & Gayle, Judith A. Matz-Coniglio
Cover Designer: Elizabeth Nemeth
Cover Image: ©Image Bank, Terje Rakke
Art Manager: Gus Vibal
Manufacturing Manager: Trudy Pisciotti
Buyer: Michael Bell
Page Layout: Karen Noferi, Joanne Del Ben
Editor in Chief: Paul F. Corey
Editor in Chief, Development: Ray Mullaney
Editor in Chief, Development: Carol Trueheart
Executive Marketing Manager for Biology: Jennifer Welchans
Senior Marketing Manager: Martha McDonald
Marketing Assistant: Andrew Gilfillan
Special Projects Manager: Barbara A. Murray
Supplements Production Editor: Meaghan Forbes
Illustrators: William C. Ober, M.D.; Claire Garrison, R.N.

Printed in the United States of America
10 9 8

ISBN 0-13-856816-2

Prentice-Hall International (UK) Limited, *London*
Prentice-Hall of Australia Pty. Limited, *Sydney*
Prentice-Hall Canada Inc., *Toronto*
Prentice-Hall Hispanoamericana, S.A., *Mexico*
Prentice-Hall of India Private Limited, *New Delhi*
Prentice-Hall of Japan, Inc., *Tokyo*
Pearson Education Asia Pte. Ltd
Editora Prentice-Hall do Brasil, Ltda., *Rio de Janeiro*

To the Student

You're beginning an exciting journey into the study of human anatomy (*structure*) and physiology (*function*). You'll be learning how the body works both when the body is healthy and when it is suffering from disease. In addition, you'll be introduced to clinical applications that demonstrate how this information relates to the real world. This material might seem foreign at first, that's quite understandable. However, as you read through the next several pages you'll see a variety of features that we've designed for you to overcome that feeling, and help you understand and master this new material.

CHAPTER OUTLINE AND CHAPTER OBJECTIVES

Each chapter opens with an outline that provides an overview of the concepts. The outline shows you where you are going and gives you the ability to turn to a particular section for specific information.

The list of objectives gives you the important concepts you should learn from the chapter.

Emphasizing Concepts

SYSTEM BRIEF

The System Brief provides you with an overview of the important functions of a particular system. Briefs are given each time a new system is introduced.

CONCEPT LINKS

The chain-link icon provides a quick visual signal that tells you where to look to review background material important to your understanding of new material.

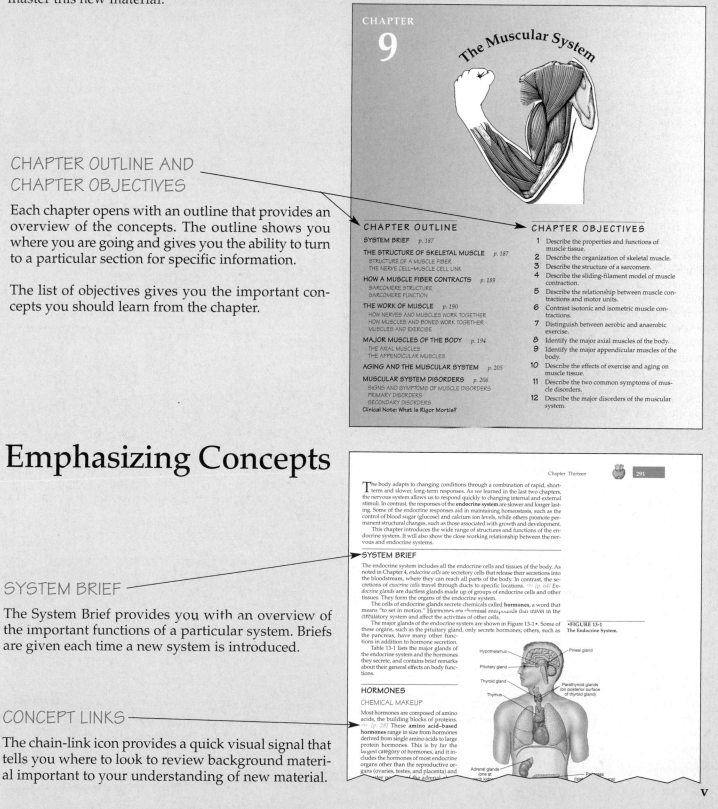

CHAPTER 9
The Muscular System

CHAPTER OUTLINE

SYSTEM BRIEF   *p. 187*

THE STRUCTURE OF SKELETAL MUSCLE   *p. 187*
STRUCTURE OF A MUSCLE FIBER
THE NERVE CELL–MUSCLE CELL LINK

HOW A MUSCLE FIBER CONTRACTS   *p. 189*
SARCOMERE STRUCTURE
SARCOMERE FUNCTION

THE WORK OF MUSCLE   *p. 190*
HOW NERVES AND MUSCLES WORK TOGETHER
HOW MUSCLES AND BONES WORK TOGETHER
MUSCLES AND EXERCISE

MAJOR MUSCLES OF THE BODY   *p. 194*
THE AXIAL MUSCLES
THE APPENDICULAR MUSCLES

AGING AND THE MUSCULAR SYSTEM   *p. 205*

MUSCULAR SYSTEM DISORDERS   *p. 206*
SIGNS AND SYMPTOMS OF MUSCLE DISORDERS
PRIMARY DISORDERS
SECONDARY DISORDERS
Clinical Note: What Is Rigor Mortis?

CHAPTER OBJECTIVES

1 Describe the properties and functions of muscle tissue.
2 Describe the organization of skeletal muscle.
3 Describe the structure of a sarcomere.
4 Describe the sliding-filament model of muscle contraction.
5 Describe the relationship between muscle contractions and motor units.
6 Contrast isotonic and isometric muscle contractions.
7 Distinguish between aerobic and anaerobic exercise.
8 Identify the major axial muscles of the body.
9 Identify the major appendicular muscles of the body.
10 Describe the effects of exercise and aging on muscle tissue.
11 Describe the two common symptoms of muscle disorders.
12 Describe the major disorders of the muscular system.

Chapter Thirteen   **291**

The body adapts to changing conditions through a combination of rapid, short-term and slower, long-term responses. As we learned in the last two chapters, the nervous system allows us to respond quickly to changing internal and external stimuli. In contrast, the responses of the **endocrine system** are slower and longer lasting. Some of the endocrine responses aid in maintaining homeostasis, such as the control of blood sugar (glucose) and calcium ion levels, while others promote permanent structural changes, such as those associated with growth and development.

This chapter introduces the wide range of structures and functions of the endocrine system. It will also show the close working relationship between the nervous and endocrine systems.

SYSTEM BRIEF

The endocrine system includes all the endocrine cells and tissues of the body. As noted in Chapter 4, *endocrine cells* are secretory cells that release their secretions into the bloodstream, where they can reach all parts of the body. In contrast, the secretions of *exocrine cells* travel through ducts to specific locations. ∞ *[p. 64] Endocrine glands* are ductless glands made up of groups of endocrine cells and other tissues. They form the organs of the endocrine system.

The cells of endocrine glands secrete chemicals called **hormones**, a word that means "to set in motion." Hormones are chemical compounds that travel in the circulatory system and affect the activities of other cells.

The major glands of the endocrine system are shown in Figure 13-1 • Some of these organs, such as the pituitary gland, only secrete hormones; others, such as the pancreas, have many other functions in addition to hormone secretion.

Table 13-1 lists the major glands of the endocrine system and the hormones they secrete, and contains brief remarks about their general effects on body functions.

HORMONES

CHEMICAL MAKEUP

Most hormones are composed of amino acids, the building blocks of proteins. ∞ *[p. 28]* These **amino acid–based hormones** range in size from hormones derived from single amino acids to large protein hormones. This is by far the largest category of hormones, and it includes the hormones of most endocrine organs other than the reproductive organs (ovaries, testes, and placenta) and

•FIGURE 13-1
The Endocrine System.

Hypothalamus
Pituitary gland
Thyroid gland
Thymus
Pineal gland
Parathyroid glands (on posterior surface of thyroid gland)
Adrenal glands (one at each kidney)

v

Understanding Diseases and Clinical Applications

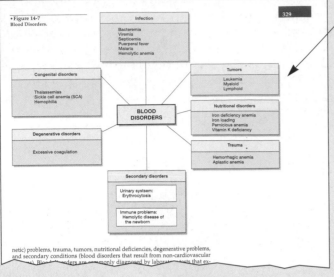

CONCEPT MAPS OF DISEASES

These illustrations organize the diseases affecting each system by the underlying causes. Although different diseases may affect different organ systems, there are recurring patterns. An awareness of those patterns will help you understand the underlying links between diseases, no matter which body system you're studying.

CLINICAL NOTES

Clinical or health-related topics of particular importance are presented in boxes set off from the main text. Some boxes provide information about the use of anatomy and physiology in the diagnosis and treatment of disease.

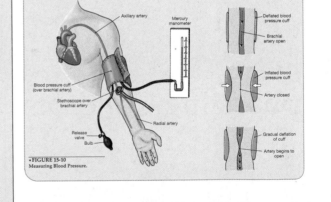

TABLES OF DIAGNOSTIC PROCEDURES AND LABORATORY TESTS

These tables provide an excellent resource for information about specific diagnostic procedures and laboratory tests used for the diagnosis of diseases.

Visualizing Body Structures, Functions, and Disorders

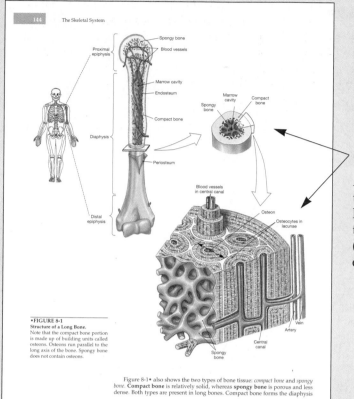

•FIGURE 8-1
Structure of a Long Bone.
Note that the compact bone portion is made up of building units called osteons. Osteons run parallel to the long axis of the bone. Spongy bone does not contain osteons.

Figure 8-1• also shows the two types of bone tissue: *compact bone* and *spongy bone*. **Compact bone** is relatively solid, whereas **spongy bone** is porous and less dense. Both types are present in long bones. Compact bone forms the diaphysis

OUTSTANDING ANATOMY ART

Anatomy is a visual subject, and without effective illustrations it's often difficult to grasp three-dimensional relationships. This book uses several helpful illustration techniques. These include: macro-to-micro drawings (shown here), and clearly labeled anatomical art large enough so that you can easily see the important features.

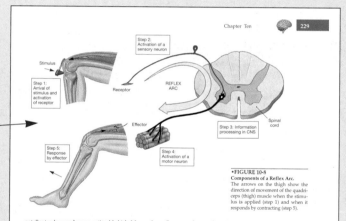

•FIGURE 10-8
Components of a Reflex Arc.
The arrows on the thigh show the direction of movement of the quadriceps (thigh) muscle when the stimulus is applied (step 1) and when it responds by contracting (step 5).

patellar tendon produces a noticeable kick. Many other reflexes may have at least one interneuron placed between the sensory neuron and the motor neuron. Due to the larger number of synapses, these reflexes have a longer delay between a stimulus and response; however, they can produce far more complicated responses. In general, the larger the number of interneurons involved, the more complicated the response. For example, reflexes involved with maintaining bal

CLEAR PHYSIOLOGY

Physiology can seem complex and abstract unless you keep the "big picture" in mind. In this text, physiological processes are made easier to understand through stepped-out diagrams and flow charts.

•FIGURE 11-8
Meningitis.
The brain of a patient who died of *Streptococcus pneumoniae* meningitis. Note the accumulation of pus from the infection.

Bacteria. Examples of infections of the brain and central nervous system caused by bacteria include meningitis and *brain abscesses*. Different forms of meningitis strike different age groups and are caused by different bacteria. For example, meningitis caused by *Haemophilus influenzae* (hē-MOF-il-us in-flū-EN-zi) occurs most often during the first year of life and rarely affects children over 5 years old. Untreated children who recover from this infection often have some degree of mental retardation. Vaccination has dramatically reduced the incidence of *Haemophilus* meningitis. *Neisseria meningitidis* (ni-SE-rē-a me-NIN-ji-ti-dis) infects individuals between 5–40 years, and is called *meningococcal* meningitis. These bacteria infect the upper throat, pass into the blood, and are carried to the meninges. If they also spread to all parts of the body, death can occur within hours. During World War II, this was the leading cause of death by infectious disease of U.S. armed forces. Treatment involves antibiotics. Adults over the age of 40 (and occasionally children) may contract *Streptococcus* meningitis, which is caused by *Streptococcus pneumoniae* (noo-MŌ-nē-i). Bacteria that initially infect the lungs, sinuses, and ears can pass through the blood to the meninges (Figure 11-8•).

The most common clinical assessment involves checking for a "stiff neck" by asking the patient to touch the chin to the chest. Meningitis affecting the cervical portion of the spinal cord results in a marked increase in the muscle tone, or tension, of the extensor muscles of the neck. So many motor units become activated that voluntary or involuntary flexion of the neck becomes painfully difficult, if not impossible.

An **abscess** is a localized collection of pus lying within an enclosed tissue space. A **brain abscess** may be caused by different types of bacteria. They reach the brain through wounds to the head or by blood from other infected areas. As they grow, they compress the brain, affecting the functions of the compressed regions.

Viruses. Examples of viral infections of the brain and central nervous system include viral meningitis, *rabies*, and different varieties of encephalitis. **Viral meningitis** is caused by different types of viruses, including the *mumps* virus. In contrast to untreated bacterial meningitis, which is usually fatal, viral meningitis is usually not fatal.

Rabies is an acute disease of the central nervous system. The rabies virus infects and kills mammals worldwide. Not all mammals are susceptible to rabies,

DISORDERS IN CONTEXT

Examples of a wide array of body disorders are discussed and illustrated with clinical photographs.

Study Aids

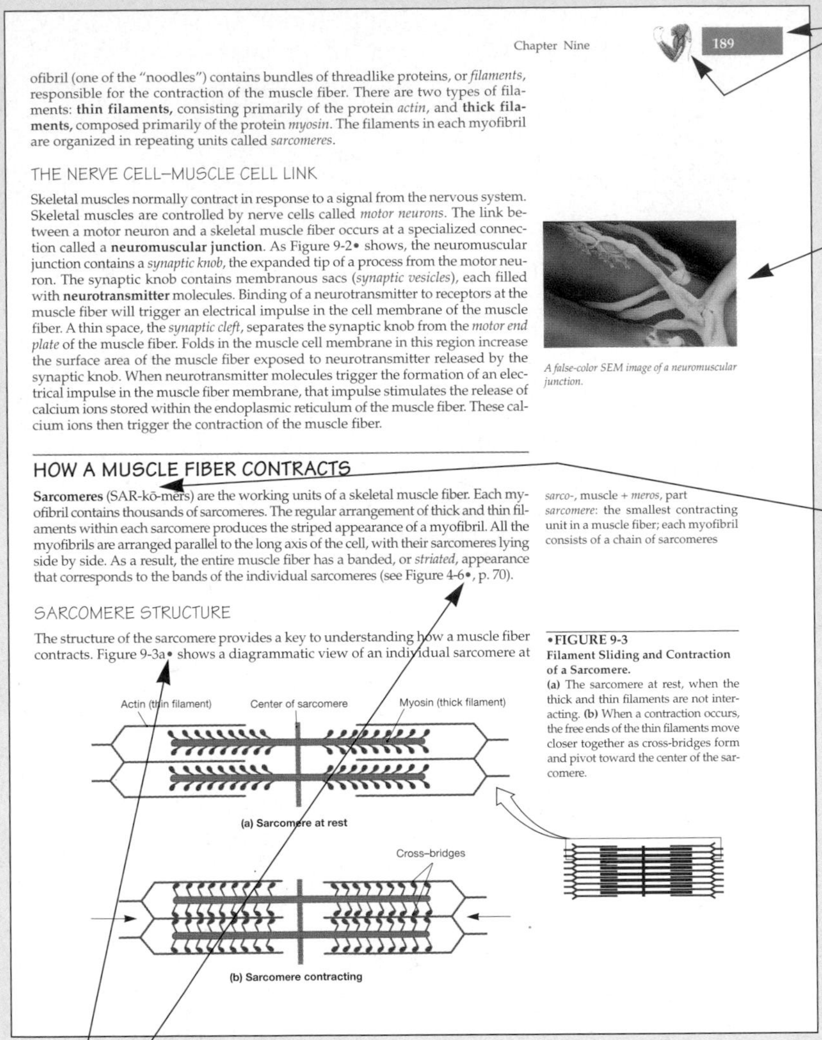

NAVIGATION AIDS

NAVIGATION AIDS

An icon and color-coded thumb-tab help you find your place in the book.

A VISUAL THEME

Numbered figures are discussed in detail in the narrative. Unnumbered images that provide additional perspective are placed in the margins.

KEY WORDS

Important terms are highlighted in bold type and often include their pronunciation. Many of these key words are also listed at the end of the chapter for easy review.

FIGURE REFERENCE DOTS

Red dots are placed next to figure numbers in the text. These dots serve as place markers, making it easier to return to your spot in the text after you've studied an illustration.

The sample textbook page shown reads:

ofibril (one of the "noodles") contains bundles of threadlike proteins, or *filaments*, responsible for the contraction of the muscle fiber. There are two types of filaments: **thin filaments**, consisting primarily of the protein *actin*, and **thick filaments**, composed primarily of the protein *myosin*. The filaments in each myofibril are organized in repeating units called *sarcomeres*.

THE NERVE CELL–MUSCLE CELL LINK

Skeletal muscles normally contract in response to a signal from the nervous system. Skeletal muscles are controlled by nerve cells called *motor neurons*. The link between a motor neuron and a skeletal muscle fiber occurs at a specialized connection called a **neuromuscular junction**. As Figure 9-2• shows, the neuromuscular junction contains a *synaptic knob*, the expanded tip of a process from the motor neuron. The synaptic knob contains membranous sacs (*synaptic vesicles*), each filled with **neurotransmitter** molecules. Binding of a neurotransmitter to receptors at the muscle fiber will trigger an electrical impulse in the cell membrane of the muscle fiber. A thin space, the *synaptic cleft*, separates the synaptic knob from the *motor end plate* of the muscle fiber. Folds in the muscle cell membrane in this region increase the surface area of the muscle fiber exposed to neurotransmitter released by the synaptic knob. When neurotransmitter molecules trigger the formation of an electrical impulse in the muscle fiber membrane, that impulse stimulates the release of calcium ions stored within the endoplasmic reticulum of the muscle fiber. These calcium ions then trigger the contraction of the muscle fiber.

A false-color SEM image of a neuromuscular junction.

HOW A MUSCLE FIBER CONTRACTS

Sarcomeres (SAR-kō-mers) are the working units of a skeletal muscle fiber. Each myofibril contains thousands of sarcomeres. The regular arrangement of thick and thin filaments within each sarcomere produces the striped appearance of a myofibril. All the myofibrils are arranged parallel to the long axis of the cell, with their sarcomeres lying side by side. As a result, the entire muscle fiber has a banded, or *striated*, appearance that corresponds to the bands of the individual sarcomeres (see Figure 4-6•, p. 70).

sarco-, muscle + *meros*, part
sarcomere: the smallest contracting unit in a muscle fiber; each myofibril consists of a chain of sarcomeres

SARCOMERE STRUCTURE

The structure of the sarcomere provides a key to understanding how a muscle fiber contracts. Figure 9-3a• shows a diagrammatic view of an individual sarcomere at

•FIGURE 9-3
Filament Sliding and Contraction of a Sarcomere.
(a) The sarcomere at rest, when the thick and thin filaments are not interacting. **(b)** When a contraction occurs, the free ends of the thin filaments move closer together as cross-bridges form and pivot toward the center of the sarcomere.

Actin (thin filament)   Center of sarcomere   Myosin (thick filament)

(a) Sarcomere at rest

Cross-bridges

(b) Sarcomere contracting

Study Aids

✔ If the activity of osteoclasts exceeds the activity of osteoblasts in a bone, how will the mass of the bone be affected?

Unlike compact bone, spongy bone has no osteons. It consists of an open network of interconnecting calcified rods or plates. These open areas of spongy bone provide space for red marrow or yellow marrow.

BONE DEVELOPMENT AND MAINTENANCE

A person's height and general body proportions are determined by his or her skeleton. Bone formation is called **ossification**. Ossification begins about 6 weeks after fertilization, when the embryo is approximately 12 mm (1/2 in.) long. Bone formation and growth continue through adolescence, and parts of the skeleton usually do not stop growing until ages 18 to 25.

BONE GROWTH

endo, inside + *chondros*, cartilage
endochondral ossification: bone formation within an existing cartilage

Before bone formation begins, an embryo has a cartilaginous skeleton. As development proceeds, many of these cartilages are replaced by bones. This process is called **endochondral ossification** (en-dō-KON-dral). Endochondral ossification begins as cartilage cells within a cartilaginous model break down and are then replaced by osteoblasts that begin producing bone matrix. As more cartilage breaks down, the number of osteoblasts increases, and more and more of the cartilage is replaced by bone. Steps in the growth and ossification of a limb bone are diagrammed in Figure 8-2•. Bone formation starts at the shaft surface and then begins within the shaft before it occurs in the ends of the bone.

This process continues until the cartilaginous model has been almost completely replaced. However, until growth is completed, zones of cartilage remain

•FIGURE 8-2
Endochondral Ossification.

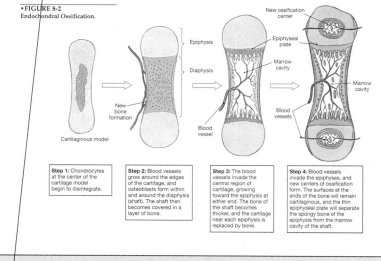

Cartilaginous model

Epiphysis

Diaphysis

New bone formation

Blood vessel

New ossification center

Epiphyseal plate

Marrow cavity

Blood vessels

Marrow cavity

Step 1: Chondrocytes at the center of the cartilage model begin to disintegrate.

Step 2: Blood vessels grow around the edges of the cartilage, and osteoblasts form within and around the diaphysis (shaft). The shaft then becomes covered in a layer of bone.

Step 3: The blood vessels invade the central region of cartilage, growing toward the epiphysis at either end. The bone of the shaft becomes thicker, and the cartilage near each epiphysis is replaced by bone.

Step 4: Blood vessels invade the epiphyses, and new centers of ossification form. The surfaces at the ends of the bone will remain cartilaginous, and the thin epiphyseal plate will separate the spongy bone of the epiphysis from the marrow cavity of the shaft.

CONCEPT CHECK QUESTIONS

These questions, located at the end of major sections in the narrative, will help you "check" your understanding of the basic concepts you've just covered. Answers are provided in the chapter review at the end of each chapter.

VOCABULARY DEVELOPMENT

At various points throughout each chapter there are notes in the margin that point out important word roots and their meaning. These word roots form the basis of the vocabulary in the chapter.

Reviewing Concepts

KEY WORDS

Key words from the chapter are defined and the pronunciations revisited.

SELECTED CLINICAL TERMS

Clinical terms are defined, and pronunciations often provided, for easy reference.

STUDY OUTLINE

This outline provides a detailed summary of all sections in the chapter including page references and all corresponding figure and table numbers.

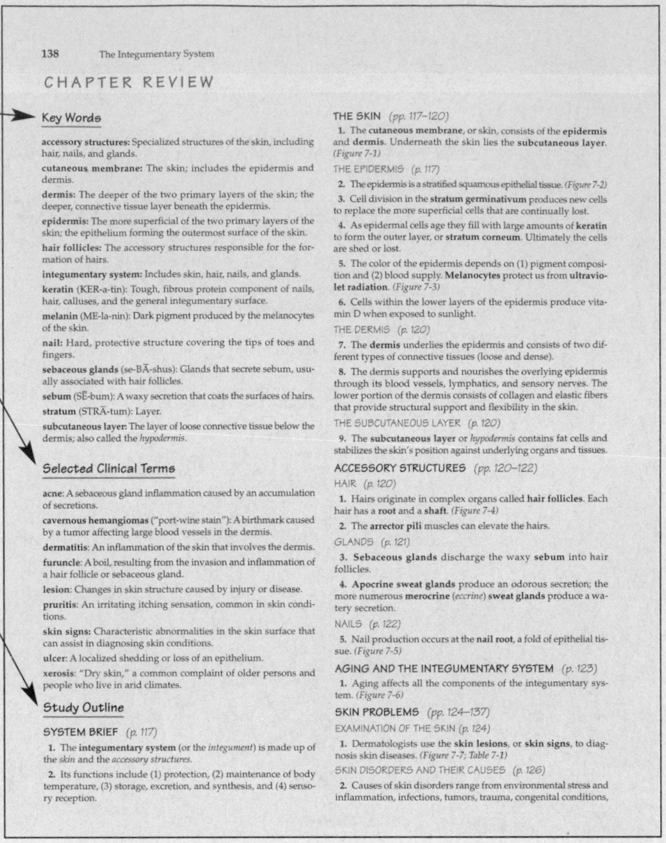

Sample page (right top)

CHAPTER REVIEW

Key Words

accessory structures: Specialized structures of the skin, including hair, nails, and glands.

cutaneous membrane: The skin; includes the epidermis and dermis.

dermis: The deeper of the two primary layers of the skin; the deeper, connective tissue layer beneath the epidermis.

epidermis: The more superficial of the two primary layers of the skin; the epithelium forming the outermost surface of the skin.

hair follicles: The accessory structures responsible for the formation of hair.

integumentary system: Includes skin, hair, nails, and glands.

keratin (KER-a-tin): Tough, fibrous protein component of nails, hair, calluses, and the general integumentary surface.

melanin (ME-la-nin): Dark pigment produced by the melanocytes of the skin.

nail: Hard, protective structure covering the tips of toes and fingers.

sebaceous glands (se-BĀ-shus): Glands that secrete sebum, usually associated with hair follicles.

sebum (SĒ-bum): A waxy secretion that coats the surfaces of hairs.

stratum (STRĀ-tum): Layer.

subcutaneous layer: The layer of loose connective tissue below the dermis; also called the *hypodermis*.

Selected Clinical Terms

acne: A sebaceous gland inflammation caused by an accumulation of secretions.

cavernous hemangiomas ("port-wine stain"): A birthmark caused by a tumor affecting large blood vessels in the dermis.

dermatitis: An inflammation of the skin that involves the dermis.

furuncle: A boil, resulting from the invasion and inflammation of a hair follicle or sebaceous gland.

lesion: Changes in skin structure caused by injury or disease.

pruritis: An irritating itching sensation, common in skin conditions.

skin signs: Characteristic abnormalities in the skin surface that can assist in diagnosing skin conditions.

ulcer: A localized shedding or loss of an epithelium.

xerosis: "Dry skin," a common complaint of older persons and people who live in arid climates.

Study Outline

SYSTEM BRIEF (p. 117)

1. The **integumentary system** (or the *integument*) is made up of the *skin* and the *accessory structures.*

2. Its functions include (1) protection, (2) maintenance of body temperature, (3) storage, excretion, and synthesis, and (4) sensory reception.

THE SKIN (pp. 117–120)

1. The **cutaneous membrane**, or skin, consists of the **epidermis** and **dermis**. Underneath the skin lies the **subcutaneous layer**. *(Figure 7-1)*

THE EPIDERMIS (p. 117)

2. The epidermis is a stratified squamous epithelial tissue. *(Figure 7-2)*

3. Cell division in the **stratum germinativum** produces new cells to replace the more superficial cells that are continually lost.

4. As epidermal cells age they fill with large amounts of **keratin** to form the outer layer, or **stratum corneum**. Ultimately the cells are shed or lost.

5. The color of the epidermis depends on (1) pigment composition and (2) blood supply. **Melanocytes** protect us from **ultraviolet radiation**. *(Figure 7-3)*

6. Cells within the lower layers of the epidermis produce vitamin D when exposed to sunlight.

THE DERMIS (p. 120)

7. The **dermis** underlies the epidermis and consists of two different types of connective tissues (loose and dense).

8. The dermis supports and nourishes the overlying epidermis through its blood vessels, lymphatics, and sensory nerves. The lower portion of the dermis consists of collagen and elastic fibers that provide structural support and flexibility in the skin.

THE SUBCUTANEOUS LAYER (p. 120)

9. The **subcutaneous layer** or *hypodermis* contains fat cells and stabilizes the skin's position against underlying organs and tissues.

ACCESSORY STRUCTURES (pp. 120–122)

HAIR (p. 120)

1. Hairs originate in complex organs called **hair follicles**. Each hair has a **root** and a **shaft**. *(Figure 7-4)*

2. The **arrector pili** muscles can elevate the hairs.

GLANDS (p. 121)

3. **Sebaceous glands** discharge the waxy **sebum** into hair follicles.

4. **Apocrine sweat glands** produce an odorous secretion; the more numerous **merocrine** (*eccrine*) **sweat glands** produce a watery secretion.

NAILS (p. 122)

5. Nail production occurs at the **nail root**, a fold of epithelial tissue. *(Figure 7-5)*

AGING AND THE INTEGUMENTARY SYSTEM (p. 123)

1. Aging affects all the components of the integumentary system. *(Figure 7-6)*

SKIN PROBLEMS (pp. 124–137)

EXAMINATION OF THE SKIN (p. 124)

1. Dermatologists use the **skin lesions**, or **skin signs**, to diagnosis skin diseases. *(Figure 7-7; Table 7-1)*

SKIN DISORDERS AND THEIR CAUSES (p. 126)

2. Causes of skin disorders range from environmental stress and inflammation, infections, tumors, trauma, congenital conditions,

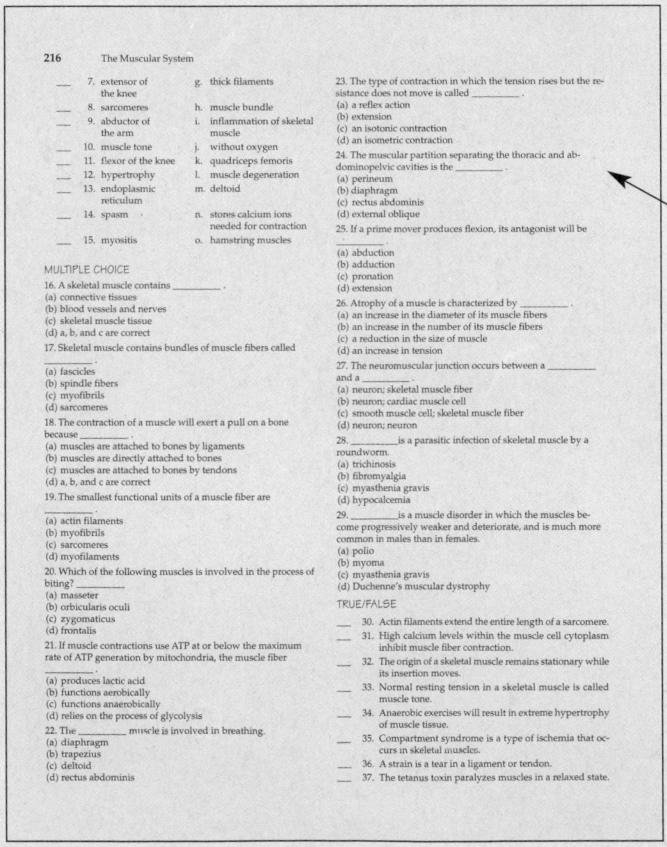

Sample page (left middle)

___ 7. extensor of the knee
___ 8. sarcomeres
___ 9. abductor of the arm
___ 10. muscle tone
___ 11. flexor of the knee
___ 12. hypertrophy
___ 13. endoplasmic reticulum
___ 14. spasm
___ 15. myositis

g. thick filaments
h. muscle bundle
i. inflammation of skeletal muscle
j. without oxygen
k. quadriceps femoris
l. muscle degeneration
m. deltoid
n. stores calcium ions needed for contraction
o. hamstring muscles

MULTIPLE CHOICE

16. A skeletal muscle contains _____ .
(a) connective tissues
(b) blood vessels and nerves
(c) skeletal muscle tissue
(d) a, b, and c are correct

17. Skeletal muscle contains bundles of muscle fibers called _____
(a) fascicles
(b) spindle fibers
(c) myofibrils
(d) sarcomeres

18. The contraction of a muscle will exert a pull on a bone because _____
(a) muscles are attached to bones by ligaments
(b) muscles are directly attached to bones
(c) muscles are attached to bones by tendons
(d) a, b, and c are correct

19. The smallest functional units of a muscle fiber are _____
(a) actin filaments
(b) myofibrils
(c) sarcomeres
(d) myofilaments

20. Which of the following muscles is involved in the process of biting? _____
(a) masseter
(b) orbicularis oculi
(c) zygomaticus
(d) frontalis

21. If muscle contractions use ATP at or below the maximum rate of ATP generation by mitochondria, the muscle fiber _____
(a) produces lactic acid
(b) functions aerobically
(c) functions anaerobically
(d) relies on the process of glycolysis

22. The _____ muscle is involved in breathing.
(a) diaphragm
(b) trapezius
(c) deltoid
(d) rectus abdominis

23. The type of contraction in which the tension rises but the resistance does not move is called _____ .
(a) a reflex action
(b) extension
(c) an isotonic contraction
(d) an isometric contraction

24. The muscular partition separating the thoracic and abdominopelvic cavities is the _____ .
(a) perineum
(b) diaphragm
(c) rectus abdominis
(d) external oblique

25. If a prime mover produces flexion, its antagonist will be _____
(a) abduction
(b) adduction
(c) pronation
(d) extension

26. Atrophy of a muscle is characterized by _____
(a) an increase in the diameter of its muscle fibers
(b) an increase in the number of its muscle fibers
(c) a reduction in the size of muscle
(d) an increase in tension

27. The neuromuscular junction occurs between a _____ and a _____
(a) neuron; skeletal muscle fiber
(b) neuron; cardiac muscle cell
(c) smooth muscle cell; skeletal muscle fiber
(d) neuron; neuron

28. _____ is a parasitic infection of skeletal muscle by a roundworm.
(a) trichinosis
(b) fibromyalgia
(c) myasthenia gravis
(d) hypocalcemia

29. _____ is a muscle disorder in which the muscles become progressively weaker and deteriorate, and is much more common in males than in females.
(a) polio
(b) myoma
(c) myasthenia gravis
(d) Duchenne's muscular dystrophy

TRUE/FALSE

___ 30. Actin filaments extend the entire length of a sarcomere.
___ 31. High calcium levels within the muscle cell cytoplasm inhibit muscle fiber contraction.
___ 32. The origin of a skeletal muscle remains stationary while its insertion moves.
___ 33. Normal resting tension in a skeletal muscle is called muscle tone.
___ 34. Anaerobic exercises will result in extreme hypertrophy of muscle tissue.
___ 35. Compartment syndrome is a type of ischemia that occurs in skeletal muscles.
___ 36. A strain is a tear in a ligament or tendon.
___ 37. The tetanus toxin paralyzes muscles in a relaxed state.

REVIEW QUESTIONS

A variety of questions are provided to test your recall of the chapter's basic information and terminology, and to let you develop your powers of reasoning and analysis by applying chapter material to real-world and clinical applications. The types of questions include: **Matching, Multiple Choice, True/False, Short Essay,** and **Applications.**

Sample page (bottom right)

SHORT ESSAY

38. List three functions of skeletal muscle.
39. Describe the three characteristic properties of all muscle tissue.
40. What is the functional difference between the axial musculature and the appendicular musculature?
41. The muscles of the spine include many dorsal extensors but few ventral flexors. Why?
42. What are the effects of aging on the muscular system?
43. What is carpal tunnel syndrome?
44. How are muscle contractions affected by conditions of hypercalcemia and hypocalcemia?

APPLICATIONS

45. While lifting his 8-year-old daughter, Jim experiences a sharp pain in his groin. He can feel a small bulge in his inguinal region. a. What is your diagnosis? b. Describe possible causes of this problem. c. Would you expect it to be more common in men or women, or equally common in both? Explain your answer.

46. Mr. West, 57 years old, has noticed that his eyelids sag, sometimes blocking his vision. He also has occasional difficulty with chewing and swallowing his food. Explain a possible cause for these symptoms.

ANSWERS TO CONCEPT CHECK QUESTIONS

For easy reference, the concept check questions are answered at the end of each chapter.

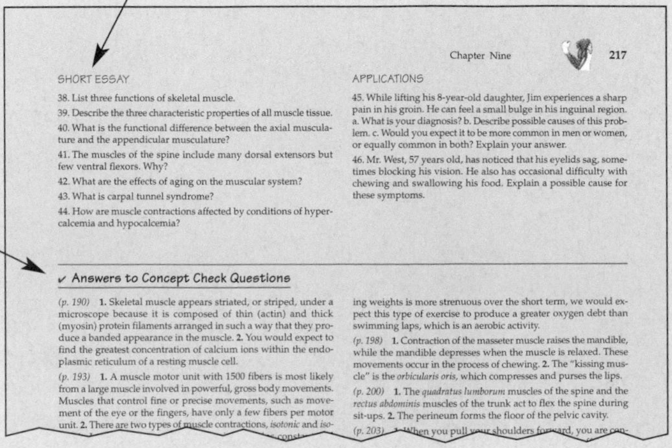

Sample page (bottom)

✓ Answers to Concept Check Questions

(p. 190) 1. Skeletal muscle appears striated, or striped, under a microscope because it is composed of thin (actin) and thick (myosin) protein filaments arranged in such a way that they produce a banded appearance in the muscle. 2. You would expect to find the greatest concentration of calcium ions within the endoplasmic reticulum of a resting muscle cell.

(p. 193) 1. A muscle motor unit with 1500 fibers is most likely from a large muscle involved in powerful, gross body movements. Muscles that control fine or precise movements, such as movement of the eye or the fingers, have only a few fibers per motor unit. 2. There are two types of muscle contractions, isotonic and iso-

ing weights is more strenuous over the short term, we would expect this type of exercise to produce a greater oxygen debt than swimming laps, which is an aerobic activity.

(p. 198) 1. Contraction of the masseter muscle raises the mandible, while the mandible depresses when the muscle is relaxed. These movements occur in the process of chewing. 2. The "kissing muscle" is the *orbicularis oris*, which compresses and purses the lips.

(p. 200) 1. The *quadratus lumborum* muscles of the spine and the *rectus abdominis* muscles of the trunk act to flex the spine during sit-ups. 2. The perineum forms the floor of the pelvic cavity.

(p. 203) 1. When you pull your shoulders forward, you are con-

Brief Contents

Contents

CHAPTER 9

The Muscular System 186

CHAPTER 10

The Nervous System 1: Neurons and the Spinal Cord 218

CHAPTER 11

The Nervous System 2: The Brain and Cranial Nerves 240

CHAPTER 12

The Senses 266

CHAPTER 13

The Endocrine System 290

CHAPTER 14

The Blood 316

CHAPTER 15

The Heart and Circulation 340

CHAPTER 16

The Lymphatic System and Immunity 380

CHAPTER 17

The Respiratory System 404

CHAPTER 18

The Digestive System 434

CHAPTER 19

Nutrition and Metabolism 466

CHAPTER 20

The Urinary System and Body Fluids 486

CHAPTER 21

The Reproductive System 518

CHAPTER 22

Development and Inheritance 550

APPENDIX I

Weights and Measures 576

APPENDIX II

Normal Physiological Values 579

APPENDIX III

Foreign Word Roots, Prefixes, Suffixes, and Combining Forms 581

Glossary/Index GI-1

Preface

The Human Body in Health and Disease is an invitation to discover more about yourself and how your body works, both when it is healthy and when it is suffering from disease. It is intended for anyone interested in gaining a basic understanding of the human body, and seeing how that information can be used to diagnose and treat various diseases . You don't need a background in biology or science to follow the concepts presented here. We've designed this book to make it easy for you to build a foundation of basic knowledge (What structure is that? How does it work? What happens if it doesn't work?). We've also organized the material to provide a framework for understanding related information obtained outside of the classroom.

There are three general themes in *The Human Body in Health and Disease.* The first is that the human body functions as an integrated and coordinated unit. Coordination exists at all levels, from single cells, the smallest living units in the body, to organs like the brain, heart, and lungs. The second theme, which is closely related to the first, is that the components of the body work together to maintain a stable internal environment. These two themes are introduced in Chapter 1 and reinforced in all subsequent chapters. The third theme is that diseases fall into a relatively limited number of categories. Those general categories remain valid no matter which body system is considered. For example, diseases resulting from infection or tumors may affect any system. An awareness of these common patterns will help you organize new information and make it possible for you to make predictions about the cause, effects, and treatment of specific diseases. This third theme is introduced in the chapter on human diseases (Chapter 6) and it forms the organizing principle behind the Disorders sections in Chapters 7-22.

GENERAL FEATURES OF THIS BOOK

No two people use books the same way, and everyone has a different learning style. We have organized each chapter to help you learn the important material, whatever your particular learning style. The three key features are (1) an emphasis on concepts and concept organization, (2) clearly organized and integrated illustrations, and (3) extensive review materials and self-tests.

AN EMPHASIS ON CONCEPTS

- *Learning Objectives*: Each chapter begins with a short list of learning objectives. These objectives focus attention on the key concepts presented in the chapter text.

- *System Brief*: Each of the chapters that deals with a major body system begins with an overview of that system's functions. After this quick summary of what the system does, we spend the rest of the chapter discussing the hows and the whys.

- *The Use of Analogies*: Whenever possible, the basic functions of the human body are related to familiar physical principles or events in everyday life. This helps to create a mental picture that makes it easier to follow abstract concepts.

- *Clinical Notes*: Most chapters include at least one boxed clinical note. These boxes, found near the relevant narrative, provide useful insights into the relevance or application of important concepts.

- *Discussions of the Effects of Aging on Body Systems*: These discussions summarize the structural and functional changes associated with aging and relate them to normal anatomy and physiology. An understanding of the aging process is becoming increasingly important because the proportion of the population over age 65 is increasing dramatically.

- *Disorders of Each Body System*: These discussions begin with an overview of the signs and symptoms of the diseases for that body system followed by information about diagnostic procedures. Concept maps organize the diseases according to their underlying causes. The diseases for each body system may change, but these recurring causes of disease will help you understand the underlying links between diseases, no matter which body system you're studying.

- *An Emphasis on Vocabulary Development*: Important terms are highlighted in the page margins near their first appearance in the narrative. The word roots are shown, and once you are familiar with them you will be able to understand the meaning of most new terms before you find their formal definitions.

- *Concept Checkpoints*: A few questions are placed near the end of each major section in a chapter. These questions are a quick way to check your reading comprehension. The answers are located at the end of the chapter; if you find you've made an error, you can reread the appropriate section before continuing through the chapter.

- *Cross-referencing*: A concept link icon (∞ *[p. 000]*) and page reference will be found wherever the development of a new concept builds on material presented earlier in the text.

- *Thumb-tabs:* Color-coded thumb-tabs are associated with icons representing specific systems. This combination makes it easy for you to find a particular section within the book.

INTEGRATED AND COMPREHENSIVE ILLUSTRATIONS

- *One View, One Vision*: The art program and the text evolved together, and the layout helps you correlate the information provided by the text and the illustrations. All of the illustrations were done by the same two illustrators, so the color usage and presentation style is consistent throughout the book.

- *Integrating Structural Relationships at All Levels*: You are most familiar (and probably most comfortable) when dealing with individuals, organ systems, or organs—things that can easily be seen. You are probably much less comfortable when dealing with events at the molecular or cellular level. *The Human Body in Health and Disease* includes keystone figures that bridge the gap between the immediate, large-scale world and the unfamiliar microscopic world of cells and tissues.

- *Figure Dots*: Each figure callout in the text is followed by a red dot (•) that refers the reader to the red dot that precedes the figure captions. The dot in the text provides a convenient placemark for the reader, making it easy to return to the narrative after studying the figure.

EXTENSIVE CHAPTER REVIEW MATERIAL

Each chapter ends with an extensive Chapter Review that will help you study, apply, and integrate new material into the general framework of the course. Each Chapter Review contains the following elements:

- *Key Words* and *Selected Clinical Terms*: The most important key terms or words in the chapter are listed in this section, along with their definitions.

- *Study Outline*: The Study Outline reviews the major concepts and topics in their order of presentation in the text. Relevant page numbers are indicated for major headings, and related key terms are boldfaced. For ease of reference, the related figure and table numbers are indicated as appropriate.

- *Review Questions*: The basic review questions are intended to test the understanding and recall of basic concepts and related terminology. The *Short Essay* questions encourages you to combine and relate the basic concepts of the chapter and to promote critical thinking skills. The *Applications* questions require you to synthesize and apply concepts to real-world problems.

OTHER USEFUL FEATURES

The appendices contain material that most students and instructors will use at some time in the course.

- *Appendix I* reviews the systems of weights and measures used in the text. You should review this material while reading Chapter 1, because this will prevent confusion and distress later in the text.
- *Appendix II* contains a list of foreign word roots, prefixes, suffixes, and combining forms.
- *Glossary/Index*: The glossary provides pronunciations and definitions of important terms.

ANCILLARIES

The ancillary package has been carefully crafted and integrated with the textbook to meet the needs of the instructor and the student.

For The Instructor

- *Instructor's Manual and Test Item File* (by Steven Bassett, Southeast Community College). Complete with teaching strategies that are linked to the chapter opening learning objectives, this unique ancillary also contains answers for the textbook's end-of-chapter questions. To simplify the task of developing lectures, this ancillary also includes a topic outline for each chapter. Presented in an easy-to-read grid format, the outline relates key vocabulary terms, illustrations, and transparency acetates to each topic from the text. The test item file includes over 2,000 questions. It is filled with multiple choice and fill-in-the-blank questions that are linked to the textbook's learning objectives, as well as questions that will challenge students to synthesize two or more objectives. (0-13-017265-0)
- *Prentice Hall Custom Test*. Offering complete question editing capabilities, this electronic version of the Test Item File can be launched from either Windows or a Macintosh platform. (Windows: 0-13-018895-6, Macintosh: 0-13-018896-4)
- *Transparency Acetates*. This set of over 200 full-color transparencies includes key illustrations from the textbook. To simplify their integration into lectures, each transparency has been linked to its topic in the Instructor's Manual's topic outline. (0-13-017267-7)
- *Companion Website* (**www.prenhall.com/martini**). Beyond an invaluable resource for students, the Companion Website offers instructors unique tools and support to make it easy to integrate Prentice Hall's on-line resources into the course. These tools include the Syllabus Manager which allows instructors to construct an on-line syllabus tailored to assignments and events for a particular class.

For The Student

- *Study Guide*. (by Steven Bassett, Southeast Community College). Designed to help students master the textbook material, this study guide is organized around the textbook's learning objectives. A variety of exercises, including matching questions, concept maps, and fill-in-the-blank narratives, promote this mastery. Problem-solving skills are developed using clinical concept questions. (0-13-017266-9)

- *Companion Website.* **(www.prenhall.com/martini)**. An exciting Companion Website has been developed specifically for this text. In addition to multiple-choice, matching, and short-answer questions, this site's self-grading quizzes offer exercises in critical thinking and labeling. Numerous interesting, related Websites are referenced and annotated in the Destinations sections, and the NetSearch offers students a convenient gateway to hundreds of other sites of interest.

- *Science on the Internet: A Student's Guide.* This hands-on supplement brings you up to speed on what the Internet is and how to navigate it. (0-13-021308-x)

- *The New York Times "Themes of the Times" Program.* Prentice Hall's unique alliance with *The New York Times* enhances your access to current, relevant information and applications. Articles are selected by the text authors and are compiled into a free supplement that helps you make the connection between your classroom and the outside world.

ACKNOWLEDGMENTS

This textbook represents a group effort that involved the authors and many other people whose efforts are deeply appreciated. Foremost on the list stand the faculty and reviewers whose advice, comments, and collective wisdom helped to shape the text into its final form. Their interest in the subject, their concern for the accuracy and method of presentation, and their experience with students of widely varying abilities and backgrounds made the review process an educational experience. To these individuals, who carefully recorded their comments, opinions, and sources, we express our sincere thanks and best wishes.

The following individuals devoted large amounts of time reviewing drafts of the sections dealing with the normal anatomy and physiology of the body:

Susan Baldi, *Santa Rosa Junior College*
Barbara Barger, *Clarion County Correction Center*
Steven Bassett, *Southeast Community College*
Judith Bell, *Midstate College*
Mitzie Bryant, *St. Louis Board of Education—Practical Nursing*
Peggy Guichard, *City College of San Francisco*
Rosemarie Harris, *University of Mississippi Medical Center*
Roger Johnson, *University of Mississippi Medical Center*
Karen Jones, *Pitt Community College*
Mary Jordan, *St. Louis Board of Education—Practical Nursing*
Leanna Konechne, *Pima Medical Institute—Respiratory Therapy Program*
Jack Lazarre, *University of Phoenix*
Judi Lindsley, *Lourdes College*
Mary Jane Lofton, *Belmont Technical College*
Evelyn Moffett, *Jones County Junior College*
Cheryl Perugini, *Porter and Chester Institute*
Mary Rahr, *Northeast Wisconsin Technical College*
Barbara Ramutkowski, *Pima Medical Institute—Medical Assistant Program*
Wayne Seifert, *Brookhaven College*
Charles Seiger, *Atlantic Community College*
Janet Sesser, *High Tech Institute, Inc.*
Ann Smith, *Joliet Junior College*
Kim Smith, *Shelton State Community College*
Connie Vinton-Schoepske, *Iowa State University*
Jeff Walmsley, *Lorain County Community College*
Pam White, *Northwest Mississippi College*

Our gratitude is also extended to the faculty members and students at campuses across the country (and out of the country) whose suggestions and comments stimulated the decision to write *The Human Body in Health and Disease.*

A text has two components: narrative and visual. In preparing the narrative, we were ably assisted by Don O'Neal, Development Editor and Project Manager, and Shari Toron, Production Editor, who managed to keep text and art moving in the proper directions at the appropriate times. Virtually without exception, reviewers stressed the importance of accurate, integrated, and visually attractive illustrations in aiding the students to understand essential material. The illustrations were produced and assembled by the team of Bill Ober, M.D. and Claire Garrison, RN.

The authors wish to express their appreciation to the editors and support staff at Prentice Hall who made the entire project possible and who kept the text, art, and production programs on schedule and in relative harmony. Special thanks are due to Ray Mullaney, Editor in Chief, College Book Editorial Development; to Tim Bozik, President of Engineering, Science and Mathematics, and Paul Corey, Editor-in-Chief for the Sciences, for their support of the project; and to Halee Dinsey, Senior Editor for Applied Biology, for coordinating everyone's efforts. We also thank David Riccardi, Assistant Vice President of Production & Manufacturing, and Kathleen Shiaparelli, Executive Managing Editor, for their support.

Additional thanks are due the following individuals for their important contributions to this project: David K. Brake, Laura Edwards, Deena Cloud, Steven Bassett, Charles Seiger, Byron Smith, and Damien Hill.

Any errors or oversights within this text are strictly those of the authors, and not the reviewers, artists, or editors. In an effort to improve future editions, we ask that readers with pertinent information, suggestions, or comments concerning the organization or content of this textbook send their remarks to us care of Halee Dinsey, Senior Editor for Applied Biology, Prentice Hall, Inc., 1 Lake Street, Upper Saddle River, NJ 07458. You may also reach us directly, using the email addresses below. Any and all comments and suggestions will be deeply appreciated, and carefully considered in the preparation of the second edition.

Frederic Martini
Haiku, Hawaii
martini@maui.net

Edwin F. Bartholomew
Lahaina, Hawaii
edbarth@maui.net

Kathleen Welch, MD
Makawao, Hawaii
Kwelch@maui.net

Dedication:
To PK, Ivy, and Kate.

Illustration Credits

Chapter 1 01-05a,b,c Custom Medical Stock Photo, Inc.; 01-07a CNRI/Photo Researchers, Inc.; 01-07b Photo Researchers, Inc.

Chapter 2 02-10b Photo Researchers, Inc.; 02-10c, Monte S. Buchsbaum, M.D., Mount Sinai School of Medicine, New York, NY; page 33 Roger Tully/Tony Stone Images

Chapter 3 page 45 Michal Heron

Chapter 6 06-05a Eric Grave/Science Source/Photo Researchers, Inc.; 06-05b Michael Abbey/Visuals Unlimited; 06-05c Michael Abbey/Visuals Unlimited, 06-05d Arthur M. Siegelman/Visuals Unlimited; 06-06a, L. West/Photo Researchers, Inc. 06-06b, Cath Wadforth/Science Photo Library/Photo Researchers, Inc.; 06-06c Runk/Schoenberger/Grant Heilman Photography, Inc.; 06-06d John Shaw/Tom Stack & Associates; 06-06e A.M. Siegelman/Visuals Unlimited; page 101 David M. Phillips/Visuals Unlimited; page 104 J. Forsdyke/Science Photo Library/Photo Researchers, Inc.; page 110 Richard Hutchings/Photo Researchers, Inc.

Chapter 7 07-03 Pearson Education/PH College; 07-06 D.Falconer/PhotoDisc, Inc.; 07-09b D. Yeske/Visuals Unlimited; 07-09a Frederic H. Martini; 07-09c CNRI/Phototake NYC; 07-10a Leonard Morse, M.D./Medical Images Inc.; 07-10b Courtesy of Charles J. Kirkpatrick, M.D., President, Innovative Therapeutics, Inc; 07-11a Courtesy of Elizabeth A. Abel, M.D., from the Leonard C. Winograd Memorial Slide Collection, Stanford University School of Medicine. ; 07-11b Courtesy of Elizabeth A. Abel, M.D., from the Leonard C. Winograd Memorial Slide Collection, Stanford University School of Medicine; 07-15 BioPhoto Associates/Science Source/Photo Researchers, Inc.; page 128 Courtesy of Dr. Hikka Helovuo, K. Kakkarainen, and K. Pannio. Oral Microbiol. Immuno. 8:75–79, (1993); page 128 Kenneth E. Greer/Visuals Unlimited; page 137 Corbis Digital Stock

Chapter 8 08-17cL, Ralph T. Hutchings; 08-17cR, Ralph T. Hutchings; 08-17d, Ralph T. Hutchings; 08-19a Ralph T. Hutchings; 08-22aB Harold Chen, M.D.; 08-22aT Harold Chen, M.D. 08-22b Grantpix/Monkmeyer Press; 08-23a Science Photo Library/Custom Medical Stock Photo, Inc.; 08-23b National Medical Slide/Custom Medical Stock Photo, Inc.; 08-23c Princess Margaret Rose Orthopaedic Hospital/Science Photo Library/Photo Researchers, Inc.; 08-24a Southern Illinois University/Visuals Unlimited; 08-24b Grace Moore/Medichrome/The Stock Shop, Inc.; 08-24c Southern Illinois University/Peter Arnold, Inc.; 08-24d Scott Camazine/Photo Researchers, Inc.; 08-24e Patricia Barber, RBP/Custom Medical Stock Photo, Inc.; 08-24f Southern Illinois University/Visuals Unlimited; 08-24g Southern Illinois University/Visuals Unlimited; 08-24h Custom Medical Stock Photo, Inc.; 08-24i Project Masters, Inc./The Bergman Collection; 08-26a Smith & Nephew, Inc.; 08-26b Smith & Nephew, Inc.; 08-26c Smith & Nephew, Inc.; 08-27a SIU Biomed Comm/Custom Medical Stock Photo, Inc.; 08-27b Courtesy of Dr. Eugene C. Wasson, III and staff of Maui Radiology Consultants, Maui Memorial Hospital; 08-27c Courtesy of Dr. Eugene C. Wasson, III and staff of Maui Radiology Consultants, Maui Memorial Hospital; Page 156 Armed Forces Institute of Pathology, Neg. No. 95-166; page 160 Courtesy of Alfonso Perello' Roso, CID: 22: 1113-4, (1996); page 177 Prof. P. Motta, Dept. of Anatomy, University "La Sapienza," Rome/Science Photo Library/Photo Researchers, Inc.; page 177 Prof. P. Motta, Dept. of Anatomy, University "La Sapienza," Rome/Science Photo Library/Photo Researchers, Inc.

Chapter 9 page 189 Don Fawcett/Science Source/Photo Researchers, Inc. page 191 NASA/Johnson Space Center; page 209 CNRI/Photo Researchers, Inc.; Page 210 T.J. Beveridge/Biological Photo Service

Chapter 10 10-10a Barts Medical Library/Phototake NYC; 10-11a Kenneth E. Greer, M.D., University of Virginia Health Sciences Center; 10-11b Courtesy M. Hogeweg, Ophthalmology Consultant. From Trop. Doctor Supplement, 1:51-21 (1992); Page 234 Richard Ellis/Agence France Presse/Corbis

Chapter 11 11-08 From Golden J. A., and D. N. Louis, NEJM. 331:34 (1994). Massachusetts General Hospital and Harvard Medical School, Boston, MA; 11-09 Science VU/Visuals Unlimited; 11-10 LSHTM/Tony Stone Images; 11-11 Visuals Unlimited; Page 247 Marcus E. Raichle, M.D., Washington University School of Medicine; Page 260 Monte S. Buchsbaum, M.D., Mount Sinai School of Medicine, New York, NY.

Chapter 12 12-14 National Medical Slide/Custom Medical Stock Photo, Inc.; 12-15 Courtesy, Director-General and Programme Manager Prevention of Blindness, World Health Organization; page 283 Richmond International, Inc.; page 285 Jon Meyer/Custom Medical Stock Photo, Inc.

Chapter 13 13-07 Project Masters, Inc./The Bergman Collection; 13-08 John Paul Kay/Peter Arnold, Inc.; 13-09 Project Masters, Inc./The Bergman Collection; 13-10 Custom Medical Stock Photo, Inc.; 13-11 Biophoto Associates/Science Source/Photo Researchers, Inc.; page 298 Beckman/Custom Medical Stock Photo, Inc.

Chapter 14 14-02a Ed Reschke/Peter Arnold, Inc.; 14-03 Custom Medical Stock Photo, Inc.; 14-09a Stanley Flegler/Visuals Unlimited; 14-09b Stanley Flegler/Visuals Unlimited; page 320 David Scharf/Peter Arnold, Inc.

Chapter 15 15-20a B & B Photos/Custom Medical Stock Photo, Inc.; 15-20b William Ober/Visuals Unlimited; 15-21a Biopho to Associates/Photo Researchers, Inc.; 15-21b SIU/Custom Medical Stock Photo, Inc.; 15-21c Medtronic, Inc.; 15-23 Peter Arnold, Inc., page 351 Professor P. M. Motta, A Caggiati, and G. Macchiarelli/Science Photo Library/Photo Researchers, Inc.

Chapter 16 16-10 Ken Greer/Visuals Unlimited; 16-11 National Medical Slide/Custom Medical Stock Photo, Inc.; 16-12 From Cote, J. Internat, J. Dermatol. 30: 500-501, 1991; 16-13 Ken Greer/Visuals Unlimited; page 387 J.G. Hirsch/J.G. Hirsch; page 391 Ken Greer/Visuals Unlimited; page 397 Centers for Disease Control and Prevention (CDC).

Chapter 17 17-12 Biophoto Associates/Photo Researchers, Inc.; 17-13 Science VU/Visuals Unlimited; 17-14a Biophoto Associates/Science Source/Photo Researchers, Inc.; 17-14b Biophoto Associates/Science Source/Photo Researchers, Inc.; page 412 AP/Wide World Photos; page 424 David M. Philips/Visuals Unlimited

Chapter 18 18-10a Dr. R. Gottsegen/Peter Arnold, Inc.; 18-10b Science VU/Max Listgarten/Visuals Unlimited 18-11 Centers for Disease Control and Prevention (CDC); 18-12 Kenneth E. Greer/Visuals Unlimited; 18-13a Dennis E. Feely, Stanley L. Erlandsen, and David G. Case; 18-13b Stanley L. Erlandsen, University of Minnesota Medical School; 18-14 Bruce Iverson/Science Photo Library/Photo Researchers, Inc.

Chapter 19 19-06 United Nations; page 481 Don King/The Image Bank; page 481 Robert Semeniuk/The Stock Market

Chapter 20 20-09 Photo Researchers, Inc.; 20-11 Peter Arnold, Inc.; page 491 Ralph T. Hutchings

Chapter 21 21-15 Courtesy of U.S. Public Health Service; 21-16 CNRI/Phototake NYC; 21-17 Kenneth E. Greer/Visuals Unlimited

Chapter 22 22-08 CNRI/Science Photo Library/Photo Researchers, Inc.; 22-10a CNRI/Science Photo Library/Photo Researchers, Inc.; 22-10b Lawrence Migdale/Pix; page 552 Francis Leroy, Biocosmos/Science Photo Library/Custom Medical Stock Photo, Inc.; page 552 Photo Lennart Nilsson/Albert Bonniers Forlag; page 552 Photo Lennart Nilsson/Albert Bonniers Forlag; page 569 Photo Researchers, Inc.

Frontmatter

Page ii Frederic H. Martini; page ii Edwin Bartholomew; page ii Frederic H. Martin; page ii Medical and Scientific Illustrations; page vii From Golden J. A. and D. N. Louis, NEJM. 331:34 (1994). Massachusetts General Hospital and Harvard Medical School, Boston, MA; page viii Don Fawcett/Science Source/Photo Researchers, Inc.

CHAPTER

1

Introduction to the Human Body

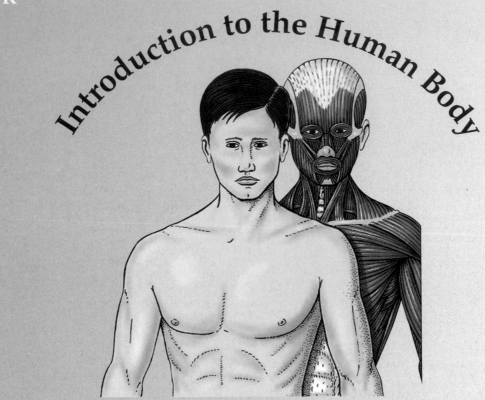

CHAPTER OUTLINE

CHAPTER OBJECTIVES

1 Define the terms *anatomy* and *physiology*.

2 Identify the major levels of organization in humans and other living organisms.

3 Explain the importance of homeostasis.

4 Describe how positive and negative feedback are involved in homeostatic regulation.

5 Use anatomical terms to describe body regions, body sections, and relative positions.

6 Identify the major body cavities and their subdivisions.

7 Distinguish between *visceral* and *parietal* portions of *serous membranes*.

8 Describe the four common techniques used in a physical examination.

The world around us contains an enormous variety of living organisms, but none tends to fascinate us as much as ourselves and our bodies. The scientific study of living organisms, including humans, is called **biology**. This text describes two important aspects of human biology: body structure, or *anatomy*, and body function, or *physiology*. As you read through this book, you will learn for yourself why the human body has often been called the "incredible machine."

THE SCIENCES OF ANATOMY AND PHYSIOLOGY

The words *anatomy* and *physiology* originated in ancient Greece. **Anatomy** (a-NAT-o-mē), which comes from Greek words that mean "to cut open," is the study of the body's internal and external structures and the physical relationships among body parts. **Physiology** (fiz-ē-OL-o-jē), the study of body function, examines the physical and chemical processes that keep us alive and healthy. As you will see throughout your study of anatomy and physiology, structure and function are closely linked. An understanding of anatomy gives clues to likely functions, and physiology can be explained only in terms of the parts of the body involved.

The link between structure and function is always present but not always understood. The anatomy of the heart, for example, was clearly described in the fifteenth century. Almost 200 years passed, however, before the pumping action of the heart was demonstrated. On the other hand, many important cell functions were known decades before people first used microscopes to study cellular structure.

LEVELS OF ORGANIZATION IN THE BODY

One distinct difference between living and nonliving things is the complexity of their structural organization: living things are much more complex. The human body contains several *levels of organization*. Structures in the simpler levels combine to create structures in the more complex levels. Understanding how the functions of one level affect other levels allows us to predict how the the body will respond to external or internal changes. Look now at Figure 1-1●, which shows the various levels of organization in the human body, using the cardiovascular system as an example. The following list describes the structures from simplest to most complex:

- *Molecules*: As you'll discover in the next chapter, *atoms* are the smallest stable units of matter, and atoms combine to form *molecules*. Some molecules have only two atoms, but others consist of many dozens of atoms that form complex shapes. Even at this simplest level, the specialized shape of a molecule often determines its function. This is the *molecular level* of organization.

- *Cells*: *Cells* are the smallest living units in the body. Each cell contains numerous *organelles*, which in turn are composed of many different molecules. Heart muscle cells contain large numbers of special molecules that give these cells the ability to contract, or shorten.

- *Tissues*: A *tissue* is composed of a group of similar cells working together to perform a specific function. The muscle cells of the heart make up one form of *muscle tissue*, whose function is to contract and thereby cause movement. In this case, the movement causes the heart to pump blood. (The four tissue types will be discussed in Chapter 4.)

- *Organs:* An *organ* consists of several different tissues that work together to perform specific functions. For example, the heart is a hollow, three-dimensional organ. Its walls contain layers of muscle and other tissues.

- *Organ System*: Each time the heart contracts, it pushes blood into a network of blood vessels. The heart, blood, and blood vessels form the *cardiovascu-*

•FIGURE 1-1
Levels of Organization.
Interacting atoms form molecules that combine to form cells, such as heart muscle cells. Groups of cells combine to form tissues with specific functions, such as heart muscle. Two or more tissues combine to form an organ, such as the heart. The heart is one component of the cardiovascular system, which also includes the blood and blood vessels. All the organ systems combine to create an organism, a living human being.

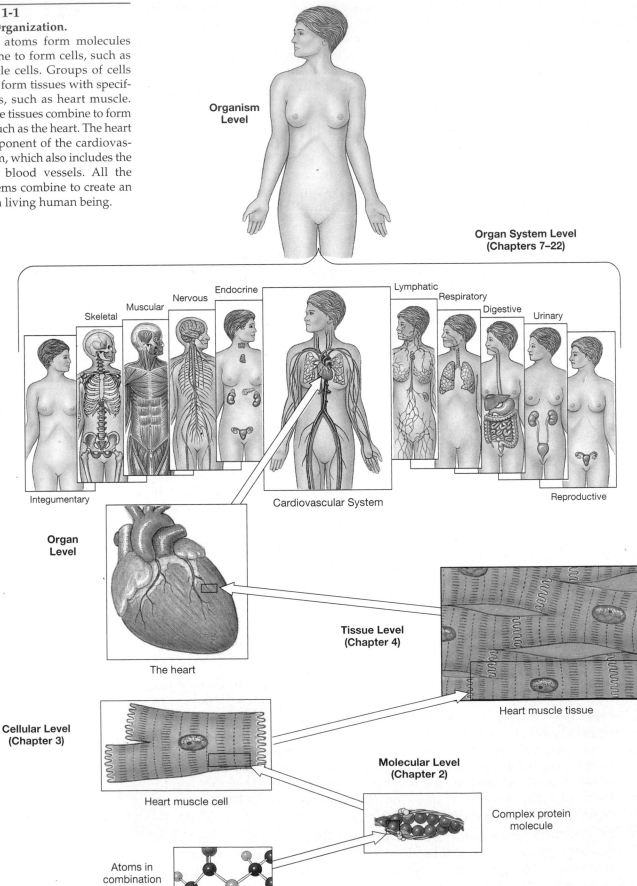

Organism
Level

Organ System Level
(Chapters 7–22)

Skeletal   Muscular   Nervous   Endocrine   Lymphatic   Respiratory   Digestive   Urinary

Integumentary   Cardiovascular System   Reproductive

Organ
Level

The heart

Tissue Level
(Chapter 4)

Heart muscle tissue

Cellular Level
(Chapter 3)

Heart muscle cell

Molecular Level
(Chapter 2)

Complex protein
molecule

Atoms in
combination

lar system, an example of an *organ system*. Organ systems are also sometimes called *body systems*.

- *Organism*: All 11 organ systems of the body work together to maintain the highest level of organization, that of the individual *organism*—in this case, a human being.

Each level of structural organization is totally dependent on the others. For example, damage at the cellular, tissue, or organ level may affect an entire organ system. A molecular change in heart muscle cells can cause abnormal contractions or even stop the heartbeat. Physical damage to the heart muscle tissue, as in a chest wound, can make the heart ineffective—even when most of the heart muscle cells are structurally intact. Abnormal structure at the organ level can make the heart an ineffective pump even if the muscle cells and muscle tissue are perfectly normal.

Finally, something that affects an organ system will sooner or later affect all its components. The heart, for example, cannot pump blood effectively if the organism has experienced massive blood loss. If the heart cannot pump and blood cannot flow, the tissues will begin to break down as cells suffocate or starve. These changes will not be restricted to the cardiovascular system. All the cells, tissues, and organs in the body will be damaged.

✔ What is the smallest living unit in the human body?

✔ What is the difference between an organ and a tissue?

HOMEOSTASIS

Organ systems in the human body are packed together in a relatively small space. The cells, tissues, organs, and organ systems exist together in a shared environment like people in a large city. Just as city dwellers breathe the city air and eat food from local restaurants, cells absorb oxygen and nutrients from their environment. All living cells are in contact with blood or some other body fluid. Within this fluid environment, factors such as temperature, volume, salt levels, or acidity (pH) must remain within a relatively narrow range for the body to function properly. If body fluids become abnormal, cells may become injured or die.

Although very small changes do occur regularly in the fluid environment of the cells, our bodies have many ways of preventing drastic changes. **Homeostasis** (hō-mē-ō-STĀ-sis), derived from the root words *homeo* (unchanging) and *stasis* (standing), refers to the relatively constant internal environment of the body. Our body's systems normally maintain homeostasis regardless of our ongoing activities. But when mechanisms that control homeostasis fail, organ systems begin to malfunction, and the individual then experiences symptoms of illness, or **disease**.

Do you use a thermostat in your house or apartment to control room temperature? If so, you are probably already familiar with the basic principles of homeostatic control, although you may not realize it. A thermostat monitors room temperature, and its function is to keep room temperature within a degree or two of the set point you select. In the summertime, the thermostat does this by controlling an air conditioner. The principle is simple: the thermostat turns on the air conditioner if the room becomes warmer than the set point, and turns it off when the temperature reaches the set point again.

The mechanisms that control homeostasis are activated in response to a *stimulus*. A **stimulus** is anything in the surroundings that produces a *response*—that is, a change in the activities of cells, tissues, organs, or organ systems. In the thermostat analogy, the stimulus is an increase in room temperature. The response produced is the activation of the air conditioner. The air conditioner cools the room and removes the stimulus. This type of homeostatic control is called *negative feedback*.

NEGATIVE FEEDBACK

In **negative feedback**, the response reverses the effect of a stimulus. Most homeostatic control in the body involves negative feedback. One such example is the

•FIGURE 1-2
Negative Feedback.
In negative feedback, a stimulus produces a response that eliminates the original stimulus. Body temperature is regulated by a control center in the brain that functions as a thermostat with a set point of 37° C. If body temperature climbs above 37.2° C, heat loss is increased through enhanced blood flow to the skin and increased sweating.

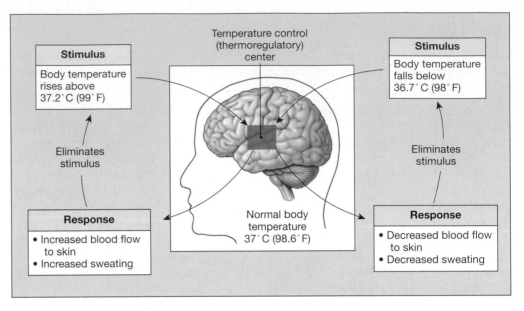

therme, heat
thermoregulation: the control of body temperature

control of body temperature, a process called *thermoregulation* (ther-mō-reg-ye-LĀ-shun). Thermoregulation balances the amount of heat lost with the amount of heat produced. In the human body, heat is lost mostly through the skin, whereas body heat is produced mainly by the skeletal muscles.

The cells of the the body's "thermostat," the *thermoregulatory center*, are located in the brain (Figure 1-2•). The thermoregulatory center has a set point near 37° C (98.6° F). When your body temperature rises above 37.2° C (99° F), the brain cells of the thermoregulatory center send signals that result in the widening, or *dilation* (dī-LĀ-shun), of the blood vessels in the skin, and this increases blood flow to the body surface. Your skin then acts like a radiator, losing heat to the environment. In addition, sweat glands are stimulated, so you begin to perspire. The evaporation of sweat helps carry heat away from the skin surface. When your body temperature returns to normal, the thermoregulatory center reduces blood flow to the skin surface, and sweat gland activity decreases.

POSITIVE FEEDBACK

Not all homeostatic mechanisms work to reverse the effects of a stimulus. In **positive feedback**, the initial stimulus produces a response that *magnifies* the stimulus. For example, suppose that the thermostat in your home was accidentally wired so that when the temperature rose it would turn on the heater, rather than the air conditioner. In that case, the initial stimulus (rising room temperature) would cause a response (heater turning on) that would exaggerate the stimulus, rather than reverse it. The room temperature would continue to rise until some external factor turned off the thermostat or unplugged the heater (or the house burned down).

Remember that negative feedback provides long-term control that results in relatively unchanging internal conditions. Positive feedback, on the other hand, controls less frequent or unusual (but necessary) changes that, once begun, need to be completed quickly. For example, the process of birth can be dangerous to a baby if it continues for too long a time. The primary stimulus that begins labor and delivery is the stretching of the womb, or *uterus*, by the growing fetus. This stimulates a control center in the brain to release a chemical that stimulates muscle contractions in the wall of the uterus. These contractions begin moving the fetus toward the birth canal. This movement causes more uterine stretching and, in turn, additional stimulation of the control center. Each time the control center responds, the uterine muscles produce even more movement and stretching. This kind of cycle, a *positive feedback loop*, can be broken only by some external force or process; in this instance, the delivery of the newborn infant eliminates all stretching of the uterus. Homeostasis is then restored.

✔ Why is homeostasis important to human beings?

✔ How does negative feedback keep internal body conditions within an acceptable range?

✔ What is the difference between negative feedback and positive feedback?

LEARNING ANATOMY

Early anatomists faced a serious communication problem when describing various parts of the human body: the ordinary terms used to describe the body were too vague. For example, stating that a bump is "on the back" does not give very exact information about its location. So anatomists began to create maps of the human body, using prominent anatomical structures as landmarks and reporting distances in centimeters or inches. These early anatomists spoke Latin or Greek, and many of the names they gave to anatomical structures are still used today.

ANATOMICAL POSITION AND DIRECTIONAL TERMS

Even precise terminology about the body would be confusing without a standard body position to serve as a reference. (Are your eyes above your nose? What if you're standing on your head?) The standard reference position of the body is known as the **anatomical position**. In this position, the arms are at the sides with the palms facing forward (Figure 1-3a●). A person is said to be **supine** (soo-PĪN) when lying faceup, and **prone** when lying facedown.

Table 1-1 and Figure 1-3● give the most important directional terms and examples of their use. Note that the terms *anterior* and *ventral* both refer to the front of the body. Likewise, *posterior* and *dorsal* are terms that refer to the back of the human body. Remember, "left" and "right" always refer to the left and right sides of the subject, not of the observer.

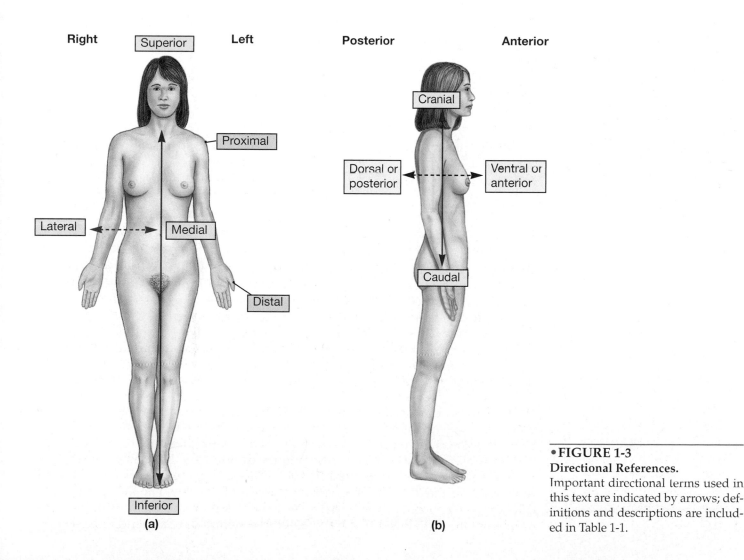

●FIGURE 1-3
Directional References.
Important directional terms used in this text are indicated by arrows; definitions and descriptions are included in Table 1-1.

Table 1-1	Directional Terms (see Figure 1-3)	
Term	**Region or Reference**	**Example**
Anterior	The front; before	The navel is on the *anterior* (ventral) surface of the trunk
Ventral	The belly side (equivalent to anterior when referring to human body)	
Posterior	The back; behind	The shoulder blade is located *posterior* (dorsal) to the rib cage
Dorsal	The back (equivalent to posterior when referring to human body)	The *dorsal* body cavity encloses the brain and spinal cord
Cranial or cephalic	The head	The *cranial*, or *cephalic*, border of the pelvis is *superior* to the thigh
Superior	Above; at a higher level (in human body, toward the head)	
Caudal	The tail (coccyx in humans)	The hips are *caudal* to the waist
Inferior	Below; at a lower level	The knees are *inferior* to the hips
Medial	Toward the body's longitudinal axis	The *medial* surfaces of the thighs may be in contact; moving medially from the arm across the chest surface brings you to the sternum
Lateral	Away from the body's longitudinal axis	The thigh articulates with the *lateral* surface of the pelvis; moving laterally from the nose brings you to the eyes
Proximal	Toward an attached base	The thigh is *proximal* to the foot; moving proximally from the wrist brings you to the elbow
Distal	Away from an attached base	The fingers are *distal* to the wrist; moving distally from the elbow brings you to the wrist
Superficial	At, near, or relatively close to the body surface	The skin is *superficial* to underlying structures
Deep	Farther from the body surface	The bone of the thigh is *deep* to the surrounding skeletal muscles

REGIONS OF THE HUMAN BODY

To avoid confusion, anatomists and healthcare professionals use very specific names for body structures and for the general area, or region, around each. For example, rather than say "arm," we can say *brachium* to refer specifically to the upper arm, or *brachial* (the adjective form) to refer to something in the region of the upper arm. These regional names are listed in Table 1-2 and, as adjectives, in Figure 1-4•. Mastering these terms will be helpful as you learn about the various organs and systems of the body. For example, the term *cranium* refers to the skull, and later chapters will discuss the *cranial nerves*, the *cranial arteries*, and so forth.

Because of the many organs lying within the lower part of the trunk of the body, two types of more detailed surface maps of this area, the *abdominopelvic region*, have been developed. The first approach separates this region into **abdominopelvic quadrants**: four segments divided by imaginary lines that intersect at the *umbilicus* (navel) (Figure 1-5a•). This method provides useful references for the description of aches, pains, and injuries. The location of pain can help a healthcare provider determine the possible cause. For example, tenderness in the right lower quadrant (RLQ) is a symptom of appendicitis, where-

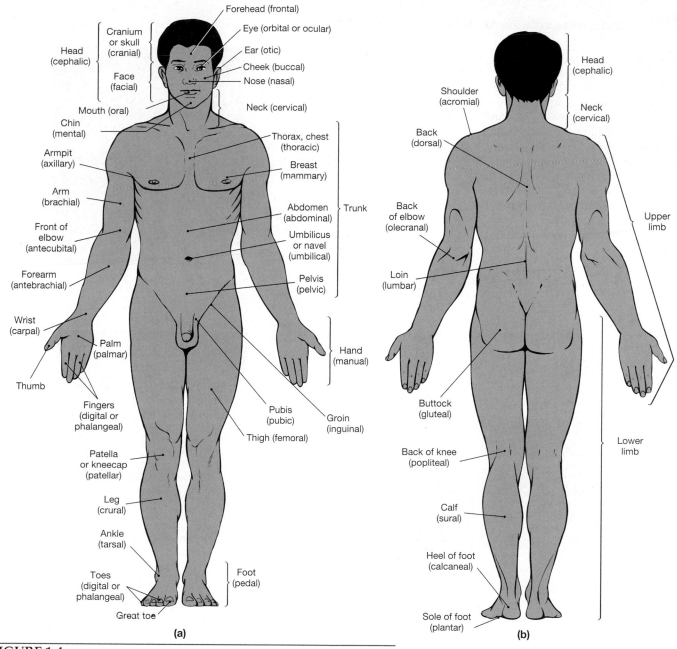

•FIGURE 1-4

Anatomical Landmarks.

The common names of anterior and posterior body landmarks are listed first; the anatomical adjectives (in parentheses) follow.

Table 1-2	Regions of the Human Body (see Figure 1-4)		
Structure	**Area**	**Structure**	**Area**
Cephalon (head)	Cephalic region	Axilla (armpit)	Axillary region
Cervicis (neck)	Cervical region	Brachium (arm)	Brachial region
Thoracis (chest)	Thoracic region	Antebrachium (forearm)	Antebrachial region
Abdomen	Abdominal region	Manus (hand)	Manual region
Pelvis	Pelvic region	Thigh	Femoral region
Loin (lower back)	Lumbar region	Leg	Crural region
Buttock	Gluteal region	Calf	Sural region
Pubis (anterior pelvis)	Pubic region	Pes (foot)	Pedal region
Groin	Inguinal region		

•FIGURE 1-5
Abdominopelvic Quadrants and Regions.
(a) Abdominopelvic quadrants divide the area into four sections. These terms, or their abbreviations, are most often used in clinical discussions. **(b)** More precise regional descriptions are provided by reference to the appropriate abdominopelvic region. **(c)** Quadrants or regions are useful because there is a known relationship between superficial anatomical landmarks and underlying organs.

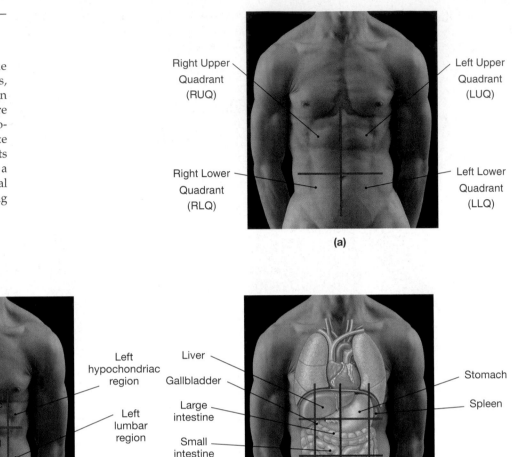

as tenderness in the right upper quadrant (RUQ) may indicate gallbladder or liver problems.

The other approach, which is even more precise, recognizes nine **abdominopelvic regions** (Figure 1-5b•). Figure 1-5c• shows the relationships among quadrants, regions, and internal organs.

BODY SECTIONS

Many times, the only way to see the inner structure of a body part is to cut a *section* through it. Any slice through a three-dimensional object can be described with reference to three **sectional planes**. These planes, described next, are shown in Figure 1-6• and reviewed in Table 1-3.

1. **Transverse plane.** The **transverse plane** lies at right angles to the long axis (head-to-foot) of the body, dividing it into **superior** and **inferior** sections. A cut in this plane is called a **horizontal section**, a *transverse section*, or a *cross section*.

2. **Frontal plane.** The **frontal plane**, or **coronal plane**, parallels the long axis of the body. The frontal plane extends from side to side, dividing the body into **anterior** and **posterior** sections.

3. **Sagittal plane.** The **sagittal plane** also parallels the long axis of the body, but it divides the body into *left* and *right* sections. A cut that passes along the

midline and divides the body into equal left and right halves is a *midsagittal section*. Because they parallel the long axis of the body, frontal and sagittal sections are also often called *longitudinal sections*.

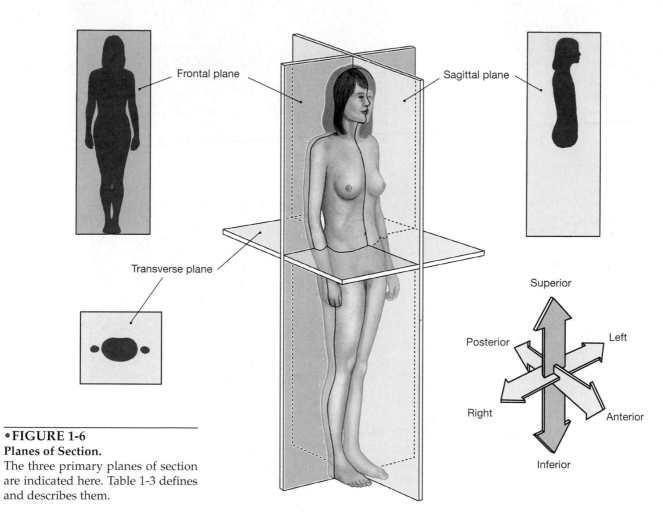

Frontal plane

Sagittal plane

Transverse plane

Superior

Posterior   Left

Right   Anterior

Inferior

•FIGURE 1-6
Planes of Section.
The three primary planes of section are indicated here. Table 1-3 defines and describes them.

Table 1-3	Terms That Indicate Planes of Sections (see Figure 1-6)	
Orientation of Plane	**Adjective**	**Description**
Parallel to long axis	Sagittal	A *sagittal* section separates right and left portions
	Midsagittal	In a *midsagittal* section, the plane passes through the midline, dividing the body in half and separating right and left sides
	Frontal or coronal	A *frontal*, or *coronal*, section separates anterior and posterior portions of the body; *coronal* usually refers to sections passing through the skull
Perpendicular to long axis	Transverse or horizontal	A *transverse*, or *horizontal*, section separates superior and inferior portions of the body

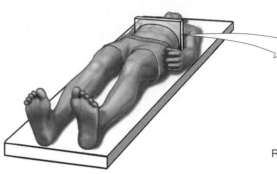

Liver   Stomach   Spleen

Rib

Right kidney   Vertebra   **(a)**   Left kidney

Liver   Stomach

Right kidney   Vertebra   **(b)**   Left kidney

•FIGURE 1-7
Scanning Techniques.
(a) A color-enhanced CT scan of the abdomen. **CT** (**C**omputerized **T**omography), formerly called **CAT** (**C**omputerized **A**xial **T**omography), uses computers to reconstruct sectional views. CT scans show three-dimensional relationships and soft tissue structure more clearly than do standard X-rays. **(b)** A color-enhanced **MRI** (**M**agnetic **R**esonance **I**maging) scan of the same region. Details of soft tissue structure are usually much more clearer than in CT scans.

An understanding of sectional views has become increasingly important since the development of electronic imaging techniques that enable us to see inside the living body without resorting to surgery. Figure 1-7• presents examples of CT and MRI scans of a transverse section through the abdomen.

BODY CAVITIES AND BODY LININGS

Many vital organs of the body are suspended within internal chambers called *body cavities*. These cavities have two essential functions: (1) They cushion delicate organs, such as the brain and spinal cord, from the thumps and bumps that occur during walking, jumping, and running; and (2) they permit changes in the size and shape of internal organs. Because the lungs, heart, stomach, intestines, urinary bladder, and many other organs are situated within body cavities, they can expand and contract without disrupting the activities of nearby organs.

The two main body cavities are the **dorsal body cavity**, which surrounds the brain and spinal cord, and the much larger **ventral body cavity**, which surrounds the organs of the respiratory, cardiovascular, digestive, urinary, and reproductive systems.

The dorsal body cavity (Figure 1-8a•) is a fluid-filled space that contains the brain and spinal cord and is surrounded by the **cranium** (the bones of the skull) and the **vertebrae** (the bones of the spine). The dorsal body cavity is subdivided into the *cranial cavity*, which encloses the brain, and the *spinal cavity*, which surrounds the spinal cord.

diaphragma, a partition wall
diaphragm: a muscular sheet that separates the thoracic cavity from the abdominopelvic cavity

The **diaphragm** (DĪ-a-fram), a flat muscular sheet, divides the ventral body cavity into an upper (superior) **thoracic cavity**, enclosed by the chest wall, and a lower (inferior) **abdominopelvic cavity** (Figure 1-8a•). The abdominopelvic cavity has two continuous subdivisions: the upper **abdominal cavity** and the lower **pelvic cavity**.

The surfaces of organs and their respective body cavities are covered by shiny, slippery linings called *serous membranes*. The *parietal* portion of a serous membrane

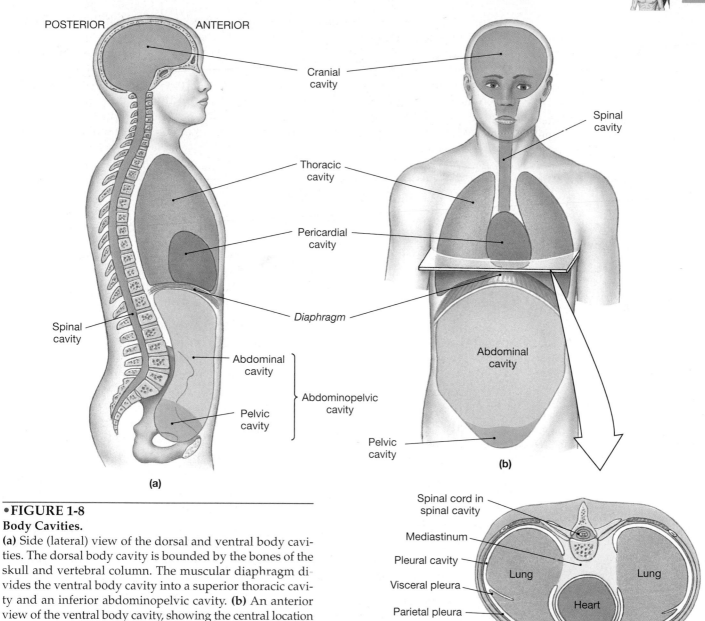

•FIGURE 1-8
Body Cavities.
(a) Side (lateral) view of the dorsal and ventral body cavities. The dorsal body cavity is bounded by the bones of the skull and vertebral column. The muscular diaphragm divides the ventral body cavity into a superior thoracic cavity and an inferior abdominopelvic cavity. **(b)** An anterior view of the ventral body cavity, showing the central location of the pericardial cavity within the chest (thoracic) cavity. **(c)** This cross (transverse) section shows how the mediastinum divides the thoracic cavity into two pleural cavities.

forms the outer wall of the body cavity. The *visceral* portion covers the surfaces of internal organs, or **viscera** (VIS-e-ra), where they project into the body cavity. The space between these opposing portions of a serous membrane is very small, but the space contains a thin layer of lubricating fluid. Many visceral organs undergo periodic changes in size and shape. For example, the lungs inflate and deflate with each breath, and the volume of the heart changes during each heartbeat. A covering of serous membrane prevents friction of adjacent organs and between the organs and the body wall.

The thoracic (chest) cavity contains the heart and lungs. The heart projects into a space known as the **pericardial cavity**. The serous membrane lining this space is called the *pericardium*, a term that means "around the heart." Each lung is surrounded by a **pleural cavity**. The serous membrane lining the pleural cavities is called the *pleura* (PLOO-ra). The region between the two pleural cavities contains a large mass of tissue known as the *mediastinum* (mē-dē-as-TĪ-num or mē-dē-AS-ti-num) (Figure 1-8c•). The mediastinum contains the

peri, around + *kardia*, heart
pericardial cavity: a cavity that surrounds the heart

✔ Is the elbow proximal or distal to the wrist?

✔ What type of section would divide the body into right and left portions?

✔ If a surgeon makes an incision just inferior to the diaphragm, what body cavity will be opened?

thymus, trachea, esophagus, the large arteries and veins attached to the heart, and the pericardial cavity.

Most of the abdominopelvic cavity and the organs it contains are covered by a serous membrane called the *peritoneum* (pe-ri-tō-NĒ-um). This lined portion of the abdominopelvic cavity is known as the *peritoneal* (per-i-tō-NĒ-al) *cavity*. Organs such as the stomach, small intestine, and portions of the large intestine are suspended within the peritoneal cavity by double sheets of peritoneum, called *mesenteries* (MES-en-ter-ēs). Mesenteries provide support and stability while permitting limited movement.

THE PHYSICAL EXAMINATION

Physical examinations, or *physical exams*, are done for a host of reasons during our lives. For example, they are used to follow the relative rates of growth and development of babies and children, or to check the general levels of health later in life. Physical exams also provide a vital part of the diagnosis of illness or disease.

The common techniques used in a physical examination include *inspection* (vision), *palpation* (touch), *percussion* (tapping), and *auscultation* (listening).

- **Inspection** is careful observation. A general inspection involves examining body proportions, posture, and patterns of movements. Local inspection is the examination of sites or regions of suspected disease. Of the four components of the physical exam, inspection is often the most important because it provides a large amount of useful information. Many diagnostic conclusions can be made on the basis of inspection alone; most skin conditions, for example, are identified in this way. A number of endocrine (glandular) problems and metabolic disorders can produce subtle changes in body proportions that can be detected by the trained eye.

- In **palpation** the clinician uses hands and fingers to feel the body. This procedure provides information on skin texture and temperature, the presence of abnormal tissue masses, the pattern of the pulse, and the location of tender spots. Once again, the procedure relies on an understanding of normal anatomy. A small, soft, lumpy mass in one spot is a salivary gland; in another location it could be a tumor. A tender spot is important in diagnosis only if the observer knows what organs lie beneath it.

- **Percussion** is tapping with the fingers or hand to obtain information about the densities of underlying tissues. For example, the chest normally produces a hollow sound, because the lungs are filled with air. That sound changes in pneumonia, when the lungs contain large amounts of fluid. Of course, to get the clearest chest percussions, the fingers must be placed in the right spots.

- **Auscultation** (aws-kul-TĀ-shun) is listening to body sounds, often using a stethoscope. This technique is particularly useful for checking the condition of the lungs during breathing. The wheezing sound heard in asthma is caused by constriction of the airways, and pneumonia produces a gurgling sound, indicating that fluid has accumulated in the lungs. Auscultation is also important in diagnosing heart conditions. Many cardiac problems affect the sound of the heartbeat or produce abnormal swirling sounds during blood flow.

✔ What component of the physical examination often provides the most information?

✔ The measurements of which body functions are referred to as vital signs?

Every examination also includes measurements of certain vital body functions, including body temperature, blood pressure, respiratory rate, and heart (pulse) rate. The results, called **vital signs**, are recorded on the patient's chart. Each of these values can vary over a normal range that differs according to the age, sex, and general condition of the individual. Table 1-4 indicates the normal ranges for vital signs in infants, children, and adults.

Table 1-4	Normal Range of Values for Resting Individuals by Age Group		
Vital Sign	**Infant** *(3 months)*	**Child** *(10 years)*	**Adult**
Blood pressure (mm/Hg)	90/50	125/60	95/60 to 140/90
Respiratory rate (per minute)	30–50	18–30	8–18
Pulse rate (per minute)	70–170	70–110	50–95

CHAPTER REVIEW

Key Words

anatomy (a-NAT-o-mē): The study of the structures of the body.

anatomical position: The standard reference position of the body; the body viewed from the anterior surface with the palms facing forward.

frontal plane: A sectional plane that divides the body into anterior and posterior portions.

homeostasis (hō-mē-ō-STĀ-sis): A relatively stable or constant internal environment.

negative feedback: Correcting process in which the body's response reverses the effect of a stimulus and restores homeostasis.

organ: A combination of different tissues that perform complex functions.

physiology (fiz-ē-OL-o-jē): The study of how the body functions.

positive feedback: Mechanism that increases a deviation from normal limits following an initial stimulus.

sagittal plane: Sectional plane that divides the body into left and right portions.

serous membrane: Slippery, delicate tissue sheet, consisting of *parietal* and *visceral* portions, that lines body cavities and covers internal organs to prevent friction.

tissue: A collection of specialized cells that perform a specific function.

transverse plane: Sectional plane that divides the body into superior and inferior portions.

viscera: Organs in the ventral body cavity.

Selected Clinical Terms

abdominopelvic quadrants: One of four divisions of the anterior abdominal surface.

abdominopelvic regions: One of nine divisions of the anterior abdominal surface.

auscultation: Listening to body sounds using a stethoscope.

CT, CAT (computerized [axial] tomography): An imaging technique that uses X-rays and a computer to reconstruct the body's three-dimensional structure.

disease: A malfunction of organs or organ systems resulting from the body's failure to maintain homeostasis.

inspection: The careful observation of a person's appearance and actions.

MRI (magnetic resonance imaging): An imaging technique that employs a magnetic field, radio waves, and a computer to portray subtle structural details with greater precision than a CT scan.

palpation: Using the hands and fingers to feel the body as part of a physical exam.

percussion: Tapping with the fingers or hand to obtain information about the densities of underlying tissues.

X-rays: High-energy radiation that can penetrate living tissues.

Study Outline

THE SCIENCES OF ANATOMY AND PHYSIOLOGY (p. 3)

1. Anatomy is the study of internal and external structure and the physical relationships among body parts. **Physiology** is the study of body function.

LEVELS OF ORGANIZATION IN THE BODY (pp. 3–5)

1. Anatomical structures and physiological mechanisms are arranged in a series of interacting levels of organization. (*Figure 1-1*)

HOMEOSTASIS (pp. 5–6)

1. Homeostasis is the maintenance of a relatively stable internal environment.

2. Symptoms of **disease** appear when failure of homeostatic regulation causes organ systems to malfunction.

NEGATIVE FEEDBACK (p. 5)

3. Negative feedback is a corrective mechanism involving an action that directly opposes a variation from normal limits. (*Figure 1-2*)

POSITIVE FEEDBACK (p. 6)

4. In **positive feedback**, the initial stimulus produces a response that enhances the stimulus.

LEARNING ANATOMY (pp. 7–14)

ANATOMICAL POSITION AND DIRECTIONAL TERMS (p. 7)

1. Standard anatomical illustrations show the body in the upright **anatomical position**. If the figure is shown lying down, it can be either **supine** (faceup) or **prone** (facedown).

2. The use of special directional terms provides clarity when describing relative locations of anatomical structures. (*Figure 1-3; Table 1-1*)

REGIONS OF THE HUMAN BODY (p. 8)

3. In addition to common names, different regions of the body are also known by specific Latin or Greek names. (*Figure 1-4; Table 1-2*)

4. Abdominopelvic quadrants and **abdominopelvic regions** are two different terms used in describing anatomical regions of the body. (*Figure 1-5*)

BODY SECTIONS (p. 10)

5. The three **sectional planes** (**frontal**, or **coronal**, **plane**, **sagittal plane**, and **transverse plane**) describe relationships between the parts of the three-dimensional human body. (*Figures 1-6, 1-7; Table 1-3*)

BODY CAVITIES AND BODY LININGS (p. 12)

6. Body cavities protect delicate organs and permit changes in the size and shape of visceral organs. The **dorsal body cavity** contains the *cranial cavity* (enclosing the brain) and *spinal cavity* (surrounding the spinal cord). The **ventral body cavity** surrounds respiratory, cardiovascular, digestive, urinary, and reproductive organs. (*Figure 1-8*)

7. The **diaphragm** divides the ventral body cavity into the superior **thoracic** and inferior **abdominopelvic cavities**. The thoracic cavity contains two *pleural cavities* (each containing a lung) and a *pericardial cavity* (which surrounds the heart). The **abdominopelvic cavity** consists of the **abdominal cavity** and the **pelvic cavity**. (*Figure 1-8*)

8. *Serous membranes* line the surfaces of the thoracic and abdominopelvic cavities and the organs they contain.

THE PHYSICAL EXAMINATION (pp. 14–15)

1. The physical examination is a vital part of the diagnosis of illness or disease. The common techniques used in a physical examination include **inspection** (vision), **palpation** (touch), **percussion** (tapping), and **auscultation** (listening).

2. The **vital signs** include body temperature, blood pressure, respiratory rate, and heart (pulse) rate. (*Table 1-4*)

Review Questions

MATCHING

Match each item in Column A with the most closely related item in Column B. Use letters for answers in the spaces provided.

	Column A		**Column B**
____	1. physiology	a.	constant internal environment
____	2. homeostasis	b.	lying faceup
____	3. supine	c.	study of structure
____	4. prone	d.	positive feedback
____	5. temperature regulation	e.	back of the body
____	6. anatomy	f.	tissue
____	7. dorsal body cavity	g.	lying facedown
____	8. heart	h.	vital sign
____	9. muscle	i.	serous membrane
____	10. birth	j.	study of body functions
____	11. ventral body cavity	k.	front of the body
____	12. pericardium	l.	brain and spinal cord
____	13. anterior	m.	tapping the body
____	14. posterior	n.	organ
____	15. blood pressure	o.	negative feedback
____	16. percussion	p.	thoracic and abdominopelvic

MULTIPLE CHOICE

17. _____ are terms that apply to the front of the body when in the anatomical position.
(a) anterior, ventral   (b) medial, lateral
(c) posterior, dorsal   (d) back, front

18. A _____ section separates superior and inferior portions of the body.
(a) sagittal   (b) midsagittal
(c) frontal   (d) transverse

19. The relative stability of an organism's internal environment is called _____.
(a) uniformity   (b) homeostasis
(c) equilibrium   (d) constancy

20. The umbilicus is _____ to the chest.
(a) anterior   (b) superior
(c) posterior   (d) inferior

21. The _____ region refers to the thigh.
(a) brachial   (b) femoral
(c) thoracic   (d) cervical

22. The diaphragm is a flat, muscular sheet that divides the ventral body cavity into a superior _____ cavity and an inferior _____ cavity.
(a) pleural, pericardial   (b) abdominal, pelvic
(c) thoracic, abdominopelvic   (d) cranial, thoracic

23. The mediastinum is the region between the _____.
(a) lungs and heart
(b) two pleural cavities
(c) thorax and abdomen
(d) heart and pericardial cavity

24. The membrane that lines the abdominopelvic cavity is the _____.
(a) mesenteries   (b) pleura
(c) pericardium   (d) peritoneum

25. _____ is not a common technique of the physical examination.
(a) inspection   (b) palpation
(c) incision   (d) auscultation

26. _____ is useful in diagnosing heart conditions.
(a) inspection   (b) palpation
(c) percussion   (d) auscultation

TRUE/FALSE

____ 27. Organs consist of more than one type of tissue.

____ 28. The study of anatomy and physiology together is extremely useful in biology because structure and function are closely related.

____ 29. The simplest living structural and functional units of the human body are tissues.

____ 30. When the body is in the anatomical position, the palms of the hands touch the lateral surface of the body.

____ 31. Failure of homeostatic regulation in the body results in disease.

____ 32. The spinal cavity is a subdivision of the ventral body cavity.

____ 33. Vital signs include measurements of body temperature and vision.

SHORT ESSAY

34. Beginning with the molecular level, list in correct sequence the levels of organization from the simplest level to the most complex level.

35. How does negative feedback differ from positive feedback?

36. Describe the position of the body when it is in the anatomical position.

37. You are awakened in the middle of the night by a mosquito in your otic region. Where is the mosquito?

38. In which body cavity would each of the following organs or systems be found?
(a) brain and spinal cord
(b) cardiovascular, digestive, and urinary systems
(c) heart, lungs
(d) stomach, intestines

39. If you begin at the elbow and move proximally 25 cm, medially 20 cm, and inferiorly 30 cm, what surface anatomical landmark will you be near?

40. What cavity is located within the mediastinum?

41. In a physical examination, which technique would be most useful in identifying the following conditions?
(a) skin disorders
(b) tender spots on the body
(c) pneumonia, a lung disorder

APPLICATIONS

42. A hormone called calcitonin from the thyroid gland is released in response to increased levels of calcium dissolved in the blood. If the control of this hormone is by negative feedback, what effect would it have on blood calcium levels?

43. As a surgeon you perform an invasive procedure that requires cutting through the peritoneum. Are you more likely to be operating on the heart or on the stomach?

✔ Answers to Concept Check Questions

(p. 5) **1.** A *cell* is the smallest structural and functional unit of life in all living organisms, including humans. **2.** An *organ* is composed of different tissues that work together to perform specific functions; a tissue is composed of a group of similar cells that perform a specific function.

(p. 6) **1.** The cells making up the human body can function normally only under a narrow range of environmental conditions. *Homeostasis* is the process that prevents potentially harmful changes in the body's internal environment. **2.** *Negative feedback* is the major process that maintains homeostasis in the human body.

Negative feedback acts by reversing the effect of a stimulus to maintain constant internal conditions in the body. **3.** *Positive feedback* works in a manner opposite to negative feedback. It produces a response that magnifies the initial stimulus.

(p. 14) **1.** The elbow is *proximal* to the wrist. The wrist is *distal* to the elbow. **2.** A sagittal section would divide the body into right and left portions. **3.** The surgeon is making an opening into the *abdominopelvic* (or *peritoneal*) cavity.

(p. 14) **1.** Visual inspection. **2.** Body temperature, blood pressure, respiratory rate, and heart (pulse) rate.

CHAPTER
2

Chemistry and the Human Body

CHAPTER OUTLINE

CHAPTER OBJECTIVES

1 Describe the basic structure of an atom.
2 Describe the different ways in which atoms combine to form molecules and compounds.
3 Distinguish between decomposition and synthesis chemical reactions.
4 Distinguish between organic and inorganic compounds.
5 Explain how the chemical properties of water are important to the functioning of the human body.
6 Describe the pH scale and the role of buffers.
7 Describe the various functions of inorganic compounds.
8 Discuss the structure and functions of carbohydrates, lipids, proteins, nucleic acids, and high-energy compounds.
9 Describe the role of enzymes in metabolism.
10 Describe the use of radioisotopes in visualizing organs and the treatment of diseases.

The human body and all objects in its surroundings, alive or not, share a common underlying structure: They are all composed of incredibly small particles called *atoms*. Differences in the types of atoms and how they interact with one another account for the characteristics of all nonliving and living things. The basic functions of the human body involve interactions among atoms. Atoms interact or recombine through *chemical reactions*. Chemical reactions occur throughout our lives in every cell, tissue, and organ in our bodies. The color of our skin, the movements of our legs, the digestion of our food, or reading the words on this page—all are the results of chemical reactions.

Chemistry, the branch of science that deals with atoms and their interactions, has greatly increased our understanding of the human body. Not only does chemistry explain the intricate structures and functions of the body, but knowledge of chemical principles has made possible such advances as genetic engineering, high-tech medical imaging, and the development of new drugs. Understanding the basic principles of chemistry will help you better understand the physical structure of the body as well as the physiological processes that keep us alive.

WHAT IS AN ATOM?

Scientists define **matter** as anything that occupies space and has mass. Matter may occur in any of three familiar states: as a solid, liquid, or gas. All matter is composed of one or more simple substances called **elements**. An element cannot be broken down into any simpler substances by heating or other ordinary physical means. There are over a hundred different types of elements, each with its own individual characteristics. Table 2-1 lists many of the elements important in the human body. The smallest piece of an element, and the simplest unit of matter, is an **atom**. Atoms are so small that one million of them placed end to end would be no longer than a period on this page!

Table 2-1	Principal Elements in the Human Body	
Element	**Symbol**	**Significance**
Major Elements (make up 99% of total body weight)		
Carbon	C	Found in all organic molecules
Hydrogen	H	A component of water and most other compounds in the body
Oxygen	O	A component of water and other compounds; oxygen gas essential for respiration
Nitrogen	N	Found in proteins, nucleic acids, and other organic compounds
Minor Elements (make up part of remaining 1% of total body weight)		
Calcium	Ca	Found in bones and teeth; important for membrane function, nerve impulses, muscle contraction, and blood clotting
Phosphorus	P	Found in bones and teeth, nucleic acids, and high-energy compounds
Potassium	K	Important for proper membrane function, nerve impulses, and muscle contraction
Sodium	Na	Important for membrane function, nerve impulses, and muscle contraction
Chlorine	Cl	Important for membrane function and water absorption
Magnesium	Mg	Required for activation of several enzymes
Sulfur	S	Found in many proteins
Iron	Fe	Essential for oxygen transport and energy capture
Iodine	I	A component of hormones of the thyroid gland

Atomic Structure.
An atom of helium contains two of each subatomic particle: two protons, two neutrons, and two electrons.

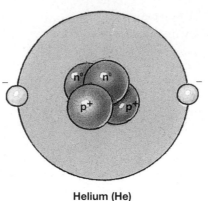

Helium (He)
($2p^+$, $2n^0$, $2e^-$)

Each of the 112 different elements is known worldwide by a unique scientific name. In writing about elements, we often use one- or two-letter abbreviations, known as *chemical symbols*. Most chemical symbols are obviously derived from the names of the elements, but a few (such as *Na* for sodium) come from original Latin names (in this case, *natrium*).

STRUCTURE OF THE ATOM

The structure of an atom determines how it will behave in chemical reactions. Atoms contain three major types of *subatomic particles*: positively charged **protons** (p^+); electrically neutral **neutrons** (n^0); and negatively charged **electrons** (e^-). Protons and neutrons are similar in size and mass, and both are found in the **nucleus**, a dense structure at the center of the atom. Electrons, which are much lighter than other subatomic particles, whirl around the nucleus at high speed, within regions called *electron shells*. All atoms contain protons and electrons, normally in equal numbers. Because each positively charged proton in an atom's nucleus is balanced by a negatively charged electron, atoms themselves are electrically neutral.

Figure 2-1• is a diagram of an atom of helium. This atom contains two protons and two neutrons in the nucleus, and two electrons in its electron shell. The number of protons in an atom defines the element and is known as that element's **atomic number**. (The atomic number of helium, for example, is 2.) As you will see, the number of protons in an atom—and its equal number of electrons—contributes largely to that atom's characteristics.

Hydrogen Atoms.
(a) A typical hydrogen nucleus contains a single proton and no neutrons. **(b)** The nucleus of the hydrogen isotope deuterium (^2H) contains a proton and a neutron. **(c)** The nucleus of the hydrogen isotope tritium (^3H) nucleus contains a pair of neutrons in addition to the proton.

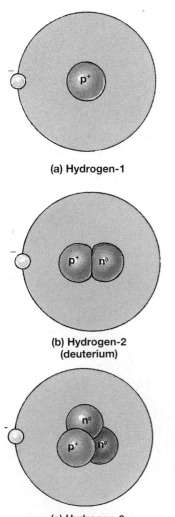

(a) Hydrogen-1

(b) Hydrogen-2
(deuterium)

(c) Hydrogen-3
(tritium)

ISOTOPES

All atoms of a particular element have the same number of protons. However, the number of neutrons in the nucleus can vary. Atoms of an element whose nuclei contain different numbers of neutrons are called **isotopes**. Because the presence or absence of neutrons has very little effect on the chemical properties of an atom, isotopes are basically distinguished from one another by their mass (essentially, their weight). The *mass number*—the total number of protons and neutrons in the nucleus—is used to designate a particular isotope. Figure 2-2• shows the atomic structure of the three isotopes of hydrogen: ^1H (the most abundant form), ^2H (called deuterium), and ^3H (called tritium).

The nuclei of some isotopes, such as ^3H, are unstable, or *radioactive*. Radioactive isotopes, called **radioisotopes**, spontaneously release *radiation* (subatomic particles and energetic rays) from their nuclei. Many recent advances in medicine have involved the use of either naturally occurring or man-made radioisotopes. For example, small amounts of weak radioactive isotopes of various elements such as carbon, nitrogen, and iodine are used as *tracers* to assess the health and status of internal organs. Large doses of radioisotopes are used in nuclear medicine to destroy abnormal or cancerous tissues.

ELECTRON SHELLS

As mentioned earlier, an atom's electrons swirl around its nucleus in certain regions of space called **electron shells**. These regions are like a series of concentric, hollow spheres that might be likened to the layers of an onion. Each electron shell can hold only a limited number of electrons. The more electrons an atom has, the more shells it needs. Some atoms have only one electron shell; others have five, six, or more. Most atoms that make up the human body, however, contain no more than two or three electron shells.

The chemical properties of an atom are determined by the number and arrangement of electrons in its *outermost* electron shell. There are two important things to remember about electron shells: (1) Atoms react with each other in ways that will fill up their outermost shells. The first electron shell is filled when it contains two electrons; the second and third electron shells are filled when they contain eight each. (2) Only the electrons in an atom's outermost shell can be used to form *chemical bonds* (connections) with other atoms. Let's consider a few examples.

Since atoms react to fill their outermost electron shell, it seems reasonable that an atom with a filled outermost electron shell will be stable; that is, it will *not* react with other atoms. For example, a helium atom has two electrons in its single electron shell (see Figure 2-1•). Helium's outer electron shell is full, and therefore helium atoms will neither react with one another nor combine with atoms of other elements. This is why chemists call helium gas *inert* or describe it as an *inert gas*. A hydrogen atom, on the other hand, has only one electron in its single electron shell (see Figure 2-2•). Because its outermost shell is not filled, a hydrogen atom reacts easily with many other types of atoms.

The second electron shell holds up to eight electrons. Carbon (atomic number = 6) has only six electrons. In each carbon atom, the first shell is filled with two electrons while the second shell contains the remaining four electrons (Figure 2-3a•). As you will discover later in this chapter, the fact that carbon's outer electron shell is only half full explains a great deal about why it plays a central role in living systems.

✔ How is it possible for two samples of hydrogen to contain the same number of atoms but have different masses?

✔ Oxygen has an atomic number of eight. How many electrons would you expect in its outermost shell? How many electrons does it need to fill its outermost shell?

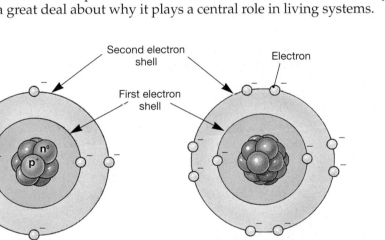

Second electron shell

First electron shell

Electron

Carbon atom
($6p^+$, $6n^0$, $6e^-$)

(a)

Neon atom
($10p^+$, $10n^0$, $10e^-$)

(b)

•**FIGURE 2-3**
Atoms and Electron Shells.
The first (innermost) electron shell can hold only two electrons. The second shell can hold up to eight electrons. **(a)** In a carbon atom, with six protons and six electrons, the third through sixth electrons occupy the second electron shell. **(b)** A neon atom has 10 protons and 10 electrons; thus both the first and second electron shells are filled. Neon, like helium (Figure 2-1) is an inert gas.

CHEMICAL BONDS

An atom with a full outer electron shell is very stable. Chemical reactions—that is, interactions between atoms—occur so that the atoms involved become as stable as possible. Atoms with an unfilled outer electron shell can become stable by sharing, gaining, or losing electrons through the formation of *chemical bonds* with other atoms. There are two primary types of chemical bonds: *ionic bonds* and *covalent bonds*.

When two or more atoms bond together, they form a structure called a **molecule**. Some molecules consist of just two atoms; others may contain hundreds or thousands of atoms. The specific combination of atoms that bond together to make up the molecules of a particular substance is summarized in a *molecular formula*. For example, water has a molecular formula of H_2O. Each water molecule contains two hydrogen atoms and one oxygen atom. Any substance whose molecules contain atoms of more than one element, such as water, is called a **compound**.

IONIC BONDS

As you'll recall, atoms are electrically neutral because the number of protons (each with a +1 charge) is equal to the number of electrons (each with a −1 charge). However, if an atom *loses* an electron, it will have a charge of +1 because there will be one proton without a corresponding electron; losing a second electron would leave the atom with a charge of +2. Similarly, adding one or two extra electrons to the atom will give it a charge of −1 or −2, respectively. Atoms or molecules that have a positive or a negative charge are called **ions**.

If an atom has only one or two electrons in its outer shell, it can "donate" its electron(s) to an atom that needs just one or two electrons. This process has two results: (1) The first atom gets a full outer electron shell (having emptied what was formerly its outermost shell), and the second atom also achieves a full outer shell. (2) Because the first atom *loses* an electron, it becomes a positively charged ion, or *cation*. At the same time, the other atom, which *gains* that electron, becomes a negatively charged ion, or *anion*. Ions formed in this way are bound together by the attraction between their opposite charges. Such a bond is called an **ionic (ī-ON-ik) bond**.

Figure 2-4● shows the steps in the formation of an ionic bond between sodium and chlorine. In this example, a sodium atom donates an electron to a chlorine atom. This creates a *sodium ion* with a +1 charge, and a *chloride ion* with a −1 charge. The two ions do not move apart after the electron transfer, because the positively charged sodium ion is attracted to the negatively charged chloride ion. The combination of these oppositely charged ions forms the ionic compound *sodium chloride*, better known as table salt.

COVALENT BONDS

Rather than donating or receiving electrons, another way an atom can fill its outer electron shell is by *sharing* electrons with other atoms. The result is a molecule held together by **covalent (kō-VĀ-lent) bonds**.

Let's consider hydrogen. Hydrogen atoms have only one electron. Since the first shell holds two electrons, each hydrogen atom needs one more electron to fill its outer shell. However, if two hydrogen atoms *share* their electrons, then the electrons will whirl around *both* nuclei, and each atom will have a complete outer shell. (Indeed, individual hydrogen atoms, as diagrammed in Figure 2-2●, are not found in nature. Instead, we find hydrogen molecules.) The sharing of one pair of electrons creates a *single covalent bond* (see Figure 2-5●).

Many molecules are composed of two or more atoms bonded together by single covalent bonds. For example, a water molecule contains two single covalent

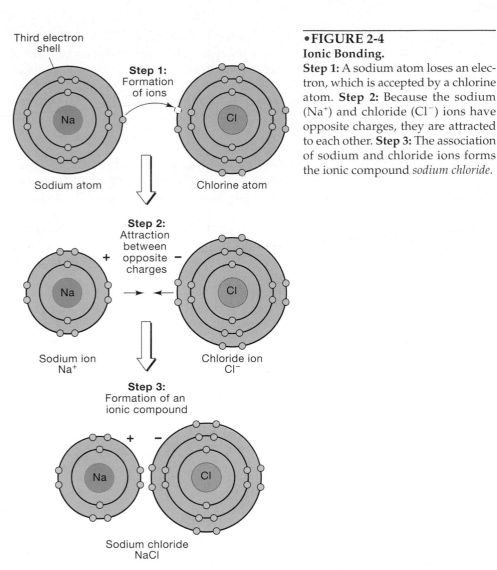

Third electron shell

Step 1:
Formation of ions

Sodium atom

Chlorine atom

Step 2:
Attraction between opposite charges

Sodium ion
Na⁺

Chloride ion
Cl⁻

Step 3:
Formation of an ionic compound

Sodium chloride
NaCl

•FIGURE 2-4
Ionic Bonding.
Step 1: A sodium atom loses an electron, which is accepted by a chlorine atom. **Step 2:** Because the sodium (Na^+) and chloride (Cl^-) ions have opposite charges, they are attracted to each other. **Step 3:** The association of sodium and chloride ions forms the ionic compound *sodium chloride*.

•FIGURE 2-5
Covalent Bonds.
In a molecule of water, two hydrogen atoms share their single electrons with an oxygen atom. This arrangement fills the outermost electron shells of all three atoms. This sharing creates two single covalent bonds.

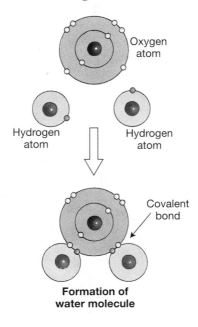

Oxygen atom

Hydrogen atom

Hydrogen atom

Covalent bond

Formation of water molecule

bonds. As Figure 2-5• shows, during the formation of a water molecule each hydrogen atom forms a single covalent bond with the oxygen atom. Oxygen, with an atomic number of 8, has two electrons in its first electron shell and six in the second. The oxygen atom reaches stability by sharing electrons with a pair of hydrogen atoms. This arrangement fills the outer electron shells of both the oxygen atom and the hydrogen atoms.

The number and kind of covalent bonds that hold atoms together in a molecule can be represented in diagram as lines between atoms. For example, the single covalent bond in H_2 is represented as H–H. Like hydrogen, oxygen atoms form oxygen molecules. However, oxygen atoms each require two electrons to reach stability, and they form a *double covalent bond* that is represented as O=O.

Compared with ionic bonds, covalent bonds are very strong because the electrons "tie" the electron shells of the atoms together. In most covalent bonds the individual atoms remain electrically neutral because the electrons are shared equally. Such covalent bonds between carbon atoms create the stable framework of the large molecules that make up most of the structural components of the human body.

✔ Oxygen and neon are both gases at room temperature. Oxygen combines readily with other elements but neon does not. Why?

✔ What kind of bond holds atoms of oxygen together?

✔ Nitrogen atoms need three electrons to fill their outer electron shell. What kind of covalent bond would join two nitrogen atoms together to form a nitrogen molecule, N_2?

CHEMICAL REACTIONS

Living cells remain alive through internal chemical reactions. In every chemical reaction, existing bonds between atoms are broken, and new bonds are formed. In this way, substances we consume as food or breathe in the air can be turned into the structures needed to build or maintain cells, tissues, and organs. However, chemical bonds do not merely hold atoms together. They also represent a form of stored energy. This energy is released when chemical bonds are broken, and stored when bonds are formed. Imagine two blocks of wood held together by a rubber band stretched around them. The tension in the stretched rubber band represents the stored energy of the chemical bond. The energy in the rubber band may be released by slipping one block out (the rubber band returns to its resting size) or by cutting the rubber band (the band flies off!).

metabole, a change
metabolism: a general term that refers to all the chemical reactions underway in the body.

The term **metabolism** (meh-TAB-o-lizm) refers to all the chemical reactions in the body. These reactions release, store, and use the energy of chemical bonds to maintain *homeostasis* and perform essential functions such as growth, tissue repair, and movement.

In a chemical reaction, the interacting participants are called **reactants**, and the new chemicals formed by the reaction are **products**. An arrow indicates the direction of the reaction, from reactants (usually on the left) to products (usually on the right). Most of the metabolic reactions of the human body are one of two types: (1) *decomposition*, which breaks a molecule into smaller fragments; and (2) *synthesis*, which assembles larger molecules from smaller components. These relatively simple reactions can be diagrammed as follows:

$$AB \rightarrow A + B \qquad \text{decomposition reaction}$$

$$A + B \rightarrow AB \qquad \text{synthesis reaction}$$

In both types of reactions, A and B can be individual atoms or individual molecules.

Some chemical reactions are *reversible*, so that if $A + B \rightarrow AB$, then $AB \rightarrow A + B$. Many important biological reactions are freely reversible. Such reactions can be diagrammed as:

$$A + B \leftrightarrow AB$$

✔ In living cells, glucose, a six-carbon molecule, is converted into two three-carbon molecules. What type of chemical reaction does this conversion represent?

✔ Which process—decomposition or synthesis—results in an increase in stored energy?

✔ If the product of a reversible reaction is continuously removed, what do you think the effect will be on the equilibrium?

This equation reminds us that there are really two reactions occurring simultaneously, one a synthesis ($A + B \rightarrow AB$) and the other a decomposition ($A + B \leftarrow AB$). At **equilibrium** (ē-kwi-LIB-rē-um) the two reactions are in balance. As fast as a molecule of AB forms, another degrades into $A + B$. As a result, the numbers of A, B, and AB molecules present at any given moment does not change. Changing the concentrations of one or more of these molecules will temporarily upset the equilibrium. For example, adding additional molecules of A and B will accelerate the synthesis reaction ($A + B \rightarrow AB$). However, as the concentration of AB rises, so does the rate of the decomposition reaction ($AB \rightarrow A + B$), until a new equilibrium is established.

INORGANIC COMPOUNDS

All chemical substances can be broadly categorized as **inorganic** or **organic**. Generally speaking, inorganic compounds are small molecules that do *not* contain carbon atoms. (One notable exception is carbon dioxide, which has long been considered an inorganic compound.) Organic compounds, on the other hand, do contain carbon atoms, and they can be much larger and more complex than inorganic compounds. The human body contains both inorganic and organic compounds. Important inorganic substances in the human body include carbon dioxide, oxygen, water, inorganic acids and bases, and salts.

WATER

As far as we know, water (H_2O) is essential for life. Not only does water account for almost two-thirds of the total weight of the human body, most chemical reac-

tions in the body take place in water. Three general properties of water are particularly important to the functioning of the human body:

1. *Water will dissolve a remarkable variety of inorganic and organic molecules.* A mixture of water and dissolved substances is an example of a *solution*. A **solution** consists of a fluid *solvent* and dissolved *solutes*. In the biological solutions of the human body, the solvent is water and the solutes may be inorganic or organic. As they dissolve, molecules break apart, releasing ions or molecules that become evenly spread out within the solution.

2. *Water can gain or lose a lot of heat without major changes in its temperature.* Because of this property, water in our cells helps maintain our body within an acceptable range. Furthermore, the fact that water is the major component of blood makes our circulatory system ideal for redistributing heat throughout the body. For example, heat absorbed as the blood flows through active muscles will be released when the blood reaches vessels in the relatively cool body surface.

3. *Water is an essential ingredient in the chemical reactions of living systems.* Such reactions commonly involve the use of water in the synthesis and decomposition of various groups of compounds. Many of the chemical compounds that make up the human body are relatively large, complex molecules. Such molecules are built up from many smaller and similar molecules, much as the length of a train is determined by the number of its boxcars.

As Figure 2-6• shows, water molecules are released during the synthesis of large molecules and absorbed during their decomposition. Water molecules form as the small subunit molecules join together. In a sense, the larger molecule forms as water is "wrung out" of the subunits. Such a reaction is also known as a *dehydration* synthesis. The breaking apart, or decomposition, of a large molecule into its subunits requires the replacement of water lost during its formation. This reaction is called *hydrolysis* (hī-DROL-i-sis).

hydro-, water + *lysis*, dissolution
hydrolysis: the breakdown of large molecules into smaller components through the absorption of water

ACIDS AND BASES

The concentration of hydrogen ions in biological solutions, such as blood or other body fluids, is important because hydrogen ions are extremely reactive. In excessive numbers they can disrupt cell and tissue function by breaking chemical bonds and changing the shapes of complex molecules. An **acid** is any substance that breaks apart, or dissociates, in solution to *release* hydrogen ions. A *strong acid* dissociates completely, and the reaction is essentially one-way. Hydrochloric acid

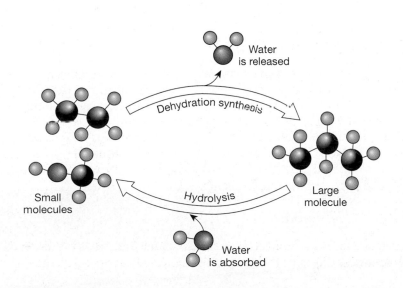

Water is released

Dehydration synthesis

Small molecules

Hydrolysis

Large molecule

Water is absorbed

•FIGURE 2-6
The Role of Water in Body Chemistry.
During dehydration synthesis, larger, more complex molecules are formed by the removal of water molecules. During hydrolysis, complex molecules are broken down by the addition of water molecules.

(HCl), which our stomachs produce to assist in the breakdown of food, is an excellent example of a strong acid:

$$HCl \rightarrow H^+ + Cl^-$$

A **base**, on the other hand, is a substance that *removes* hydrogen ions from a solution. Many common bases are compounds that dissociate in solution to release a hydroxide ion (OH^-). Hydroxide ions react quickly with hydrogen ions to form water molecules and so remove hydrogen ions from solution. Sodium hydroxide (NaOH) is an example of a strong base because in solution it breaks apart completely to form sodium ions and hydroxide ions:

$$NaOH \rightarrow Na^+ + OH^-$$

Strong bases have a variety of industrial and household uses; drain openers and lye are two familiar examples.

Weak acids and *weak bases* do not break down completely, and many reactant molecules remain whole. For the same number of molecules in solution, weak acids and bases have less of an impact on the concentration of hydrogen and hydroxide ions than do strong acids and bases.

pH AND BUFFERS

Because of the potentially damaging effects of large numbers of hydrogen ions in living systems, our bodies must maintain the concentration of hydrogen ions within relatively narrow limits. The concentration of hydrogen ions in a solution is usually reported in terms of the **pH** of that solution. The pH value is a number between 0 (very acidic) and 14 (very basic). An increase or decrease of one unit is equal to a tenfold change in the number of hydrogen (H^+) ions. The lower the pH value, the greater the concentration of hydrogen ions. Pure water has a pH of 7. It is considered a *neutral* solution because it contains equal numbers of hydrogen (H^+) and hydroxide (OH^-) ions. A solution with a pH below 7 is called *acidic* (a-SI-dik), because there are more hydrogen ions than ions that will react with them and remove them from the solution. A pH above 7 is called *basic*, or *alkaline* (AL-kah-lin), because H^+-removing ions outumber the hydrogen ions.

Figure 2-7• compares the pH of some common liquids and body fluids. Blood is slightly alkaline with a pH that normally ranges from 7.35 to 7.45. Variations in pH outside this range can damage cells and disrupt normal cell functions. For

---

•**FIGURE 2-7**
pH and Hydrogen Ion Concentration.
An increase or decrease of one pH unit corresponds to a tenfold change in H^+ concentration.

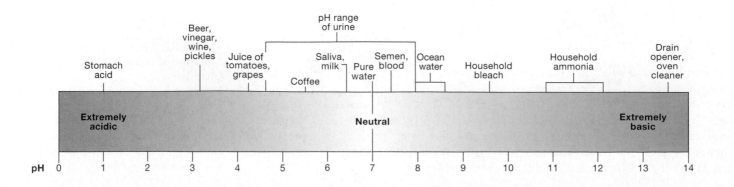

Table 2-2	The Most Common Ions in Body Fluids	
Positive	**Negative**	
Na^+ (sodium)	Cl^- (chloride)	
K^+ (potassium)	HCO_3^- (bicarbonate)	
Ca^{2+} (calcium)	HPO_4^{2-} (biphosphate)	
Mg^{2+} (magnesium)	SO_4^{2-} (sulfate)	

example, a blood pH below 7 can produce coma (a state of unconsciousness), and a blood pH higher than 7.8 usually causes uncontrollable, sustained muscular contractions that can lead to death.

With so many chemical reactions constantly occurring, how does the body control pH levels? The control of pH levels occurs primarily through a variety of chemical *buffers* in body fluids. A **buffer** is a compound that can stabilize pH by either removing *or* releasing hydrogen ions into the solution, as the need arises. An important buffer system in the body is the *carbonic acid–bicarbonate buffer system*. It will be discussed in Chapter 20. Antacids such as Alka-Seltzer® or Rolaids® are bicarbonate-based buffers that tie up excess hydrogen ions in the stomach. Excess hydrogen ions from buffer activity are also removed and excreted by the kidneys.

SALTS

A **salt** is an inorganic compound that dissolves in water but does not release a hydrogen ion or a hydroxide ion. Salts are held together by ionic bonds, and in water they break apart, releasing positively charged and negatively charged ions. For example, table salt (NaCl) in solution breaks down into Na^+ and Cl^- ions, two of the most abundant ions in body fluids.

Salts are examples of compounds known as **electrolytes** (e-LEK-tro-līts) because they dissolve in water, releasing ions that can conduct electricity. Ions of sodium (Na^+), potassium (K^+), calcium (Ca^{2+}), and chloride (Cl^-) are released as electrolytes dissolve in blood and other body fluids. Changes in the body fluid concentrations of these ions will disturb almost every vital function. For example, declining potassium levels will lead to a general muscular paralysis, and rising concentrations will cause weak and irregular heartbeats. The most common ions in the body are listed in Table 2-2.

✔ What is the difference between an acid and a base?

✔ How does an antacid decrease stomach discomfort?

✔ How is a salt different from an acid or base?

ORGANIC COMPOUNDS

Organic chemistry is the study of compounds containing carbon. Organic compounds are larger than inorganic molecules and most are built from characteristic subunit molecules. This section focuses on the four major classes of large organic molecules: *carbohydrates*, *lipids*, *proteins*, and *nucleic acids*. We will also discuss the high-energy compounds that drive many of the chemical reactions under way within our cells.

CARBOHYDRATES

A **carbohydrate** (kar-bō-HĪ-drāt) molecule consists of carbon, hydrogen, and oxygen atoms in a 1:2:1 ratio. Familiar carbohydrates include sugars and starches. Despite their important function as sources of energy, carbohydrates account for less than three percent of the total body weight. The smallest carbohydrates are called *monosaccharides*, and the largest are called *polysaccharides*. The term *sakcharon* means "sugar" in Greek, and the prefixes *mono-* (one), and *poly-* (many) indicate the number of subunits involved.

✚ Clinical Note

What Are Artificial Sweeteners?

Some people cannot tolerate sugar for medical reasons; others avoid it because they do not want to gain weight (excess sugars are stored as fat). Many of these people use artificial sweeteners in their foods and beverages. Familiar examples include NutraSweet™, Saccharin™, and Equal™. These compounds have a very sweet taste, but they either cannot be broken down in the body, or they are used in such small amounts that their breakdown does not contribute to the overall energy balance of the body.

mono-, one + *sakcharon*, sugar
monosaccharide: a simple sugar, such as glucose

poly, many + *sakcharon*, sugar
polysaccharide: a complex sugar consisting of a chain of simple sugars

Simple sugars, or **monosaccharides** (mon-ō-SAK-ah-rīds) are the basic building blocks of large carbohydrate molecules. One example is **glucose** (GLOO-kōs), $C_6H_{12}O_6$, the most important cellular "fuel" in the body.

Complex molecules constructed from many simple carbohydrates are called **polysaccharides** (pol-ē-SAK-ah-rīds). *Starches*, which are synthesized by plants, are important polysaccharides in our diets. **Glycogen** (GLĪ-ko-jen), or *animal starch*, is a polysaccharide composed of glucose molecules. Glycogen is an important energy reserve formed in liver and muscle tissues.

LIPIDS

Lipids (from the Greek *lipos*, meaning "fat"), like carbohydrates, contain carbon, hydrogen, and oxygen but relatively less oxygen than carbohydrates. Familiar lipids include fats, oils, and waxes. Most lipids are *insoluble* (do not dissolve) in water, but special transport mechanisms carry them in the circulating blood. Lipids serve as energy sources and reserves as well as forming essential structural components of all cells. On average, lipids provide roughly twice as much energy as carbohydrates. For example, if you eat a gram of butter, you'll get twice as much energy as you would if you ate a gram of sugar.

The major types of lipids in the body are *triglycerides*, *steroids*, and *phospholipids*. A **triglyceride** (trī-GLI-se-rīd) is made up of a **glycerol** (GLI-se-rol) molecule attached to three fatty acids. **Fatty acids** are long carbon chains with hydrogen atoms attached. The type of fatty acid determines whether the triglyceride is solid or a liquid oil (at room temperature).

Triglycerides are the most common fats in the body. In addition to serving as an energy reserve, fat deposits under the skin serve as insulation, and a mass of fat around a delicate organ, such as a kidney or eye, provides a protective cushion.

Steroids are large lipid molecules composed of connected rings of carbon atoms. *Cholesterol* is probably the best-known steroid. All our cells are surrounded by *cell membranes* that contain cholesterol, and a number of chemical messengers, or *hormones*, are derived from cholesterol.

A **phospholipid** (FOS-fō-lip-id) consists of a triglyceride that has one of its fatty acids replaced by a phosphate molecule. The phosphate portion dissolves in water, whereas the fatty acid portion does not dissolve. As you will learn in Chapter 3, phospholipids are important building blocks of cell membranes.

PROTEINS

Proteins are the most abundant and diverse organic components in the human body. There are roughly 100,000 different kinds performing an amazing variety of essential structural and functional roles. For example, they provide internal support (for cells, tissues, and organs), protective coverings for the body and are responsible for muscular contraction, communication and control between organ systems, and protection from disease.

Proteins are made up of chains of small organic molecules called **amino acids**, all of which contain carbon, hydrogen, oxygen, and nitrogen. There are twenty different kinds of amino acids in the human body. A typical protein contains 1,000 amino acids, but the largest may have 100,000 or more. The shape of a protein determines its function, and the primary factor determining protein shape is the order in which amino acids link together. This *sequence* of amino acids is vital, as changing even one amino acid in a protein may make that protein incapable of performing its normal function.

Enzymes are special proteins that function as *catalysts* (KAT-ah-lists). A **catalyst** is a compound that speeds up a chemical reaction without being changed itself. Many chemical reactions in living cells occur too slowly to be useful without the help of catalysts. Therefore, cells create enzymes to promote various reactions.

Figure 2-8• shows a simple model of enzyme function. The reactants in an enzyme-controlled reaction, called **substrates**, bind to the enzyme at a particular location and interact to form a specific **product**. Successful binding depends on the shapes of the enzyme and substrate molecules, much like fitting a key into a lock.

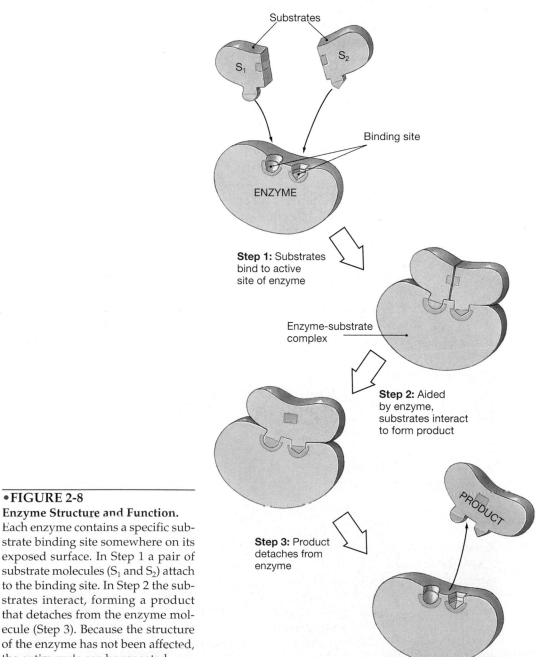

Substrates

S₁ S₂

Binding site

ENZYME

Step 1: Substrates bind to active site of enzyme

Enzyme-substrate complex

Step 2: Aided by enzyme, substrates interact to form product

PRODUCT

Step 3: Product detaches from enzyme

•**FIGURE 2-8**
Enzyme Structure and Function.
Each enzyme contains a specific substrate binding site somewhere on its exposed surface. In Step 1 a pair of substrate molecules (S₁ and S₂) attach to the binding site. In Step 2 the substrates interact, forming a product that detaches from the enzyme molecule (Step 3). Because the structure of the enzyme has not been affected, the entire cycle can be repeated.

The shape of the binding site is determined by the three-dimensional shape of the enzyme molecule. Once the reaction is complete and the products are released, the enzyme is free to catalyze another reaction.

The shape of a protein—and thus its function—can be altered by small changes in its surroundings. For example, very high body temperatures (over 43°C, or 110°F) or drastic changes in the pH of body fluids can cause death, because under these conditions, the cell's proteins lose their proper three-dimensional shape, so that they cannot function properly.

NUCLEIC ACIDS

Nucleic (noo-KLĀ-ik) **acids** are large organic molecules composed of carbon, hydrogen, oxygen, nitrogen, and phosphorus. The two types are **deoxyribonucleic** (dē-ok-si-rī-bō-noo-KLĀ-ik) **acid** and **ribonucleic** (rī-bō-noo-KLĀ-ik) **acid**. You may be more familiar with their commonly used abbreviatons: **DNA** and **RNA**. These molecules store and process information inside living cells. They affect all aspects of body structure and function.

You may have heard about the role of DNA in controlling inherited characteristics such as eye color, hair color, or blood type. In addition, DNA molecules encode the information needed to build *all* the body's proteins. DNA thus regulates not only protein synthesis but all aspects of cellular metabolism. For example, DNA controls the creation and destruction of lipids, carbohydrates, and other vital molecules by controlling the synthesis of the enzymes involved in these reactions.

RNA exists in several different forms, all of which work together to manufacture specific proteins using the information provided by DNA. The relationships among DNA, RNA, and protein synthesis will be discussed in Chapter 3.

A nucleic acid consists of a series of smaller molecules called **nucleotides**. A single nucleotide has three basic components: a sugar, a phosphate group, and a *nitrogen base*. The sugar is always a five-carbon sugar, either *ribose* (in RNA) or *deoxyribose* (in DNA). There are five different nitrogen bases: **adenine** (A), **guanine** (G), **cytosine** (C), **thymine** (T), and **uracil** (U). Both RNA and DNA contain adenine, guanine, and cytosine. Uracil is found only in RNA, and thymine only in DNA.

The structural differences between RNA and DNA are listed in Table 2-3. An important point to note is that an RNA molecule consists of a single chain of nucleotides but a DNA molecule consists of a *pair* of nucleotide chains (Figure 2-9•). The two chains in a DNA molecule are held together by weak bonds between their nitrogen bases. Because of their shapes, adenine can bond only with thymine, and cytosine only with guanine. With the paired nitrogen base acting as "steps," the two strands of DNA twist around one another in a *double helix* that resembles a spiral staircase.

Table 2-3	Comparison of RNA and DNA	
Characteristic	**RNA**	**DNA**
Sugar	Ribose	Deoxyribose
Nitrogen bases	Adenine	Adenine
	Guanine	Guanine
	Cytosine	Cytosine
	Uracil	Thymine
Function	Performs protein synthesis as directed by DNA	Stores genetic information that controls protein synthesis by RNA

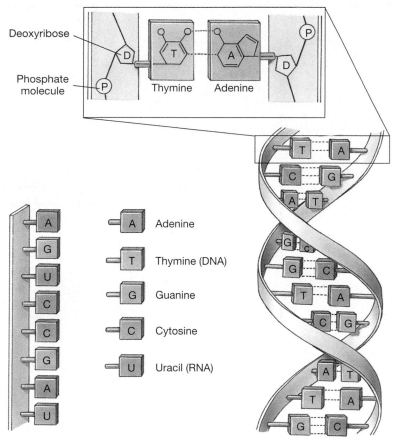

Deoxyribose

Phosphate
molecule

Thymine   Adenine

(a) **RNA molecule**

A   Adenine

T   Thymine (DNA)

G   Guanine

C   Cytosine

U   Uracil (RNA)

(b) **DNA molecule**

•**FIGURE 2-9**
Nucleic Acids: RNA and DNA.
Nucleic acids are long chains of nucleotides. Each molecule starts at the sugar-nitrogen base of the first nucleotide and ends at the phosphate molecule of the last member of the chain. **(a)** An RNA molecule consists of a single nucleotide chain. Its shape is determined by the sequence of nucleotides and the interactions among them. **(b)** A DNA molecule consists of a pair of nucleotide chains linked by weak bonds between paired nitrogen bases.

HIGH-ENERGY COMPOUNDS

Metabolic reactions continuously consume and release energy. Living cells must be able to capture, transfer, and store that energy. Although not all the energy released can be saved (a significant amount is lost as heat), some of the energy released by metabolic reactions is captured in the creation of **high-energy bonds**. A high-energy bond is a covalent bond that stores an unusually large amount of energy. In our cells a high-energy bond usually connects a phosphate molecule to an organic molecule, resulting in a **high-energy compound**.

The most common high-energy compound in cells is **adenosine triphosphate**, better known as **ATP**. This molecule is commonly described as the "energy currency" of cells. ATP contains an organic molecule (adenosine, or A), three phosphate molecules (P) and a high energy bond(~). Within our cells the formation of ATP (A-P-P~P) from **ADP** molecules (*adenosine diphosphate*, A-P-P) and phosphates (P) represents the primary method of energy storage. The reverse reaction provides a mechanism for controlled energy release. The arrangement can be summarized as follows:

$$ATP \leftrightarrow ADP + P + energy$$

When energy sources are available, our cells make ATP from ADP; when energy is required, the reverse reaction occurs.

Table 2-4 reviews the major organic compounds discussed in this chapter. The next chapter considers the combination of these compounds within a living cell.

✔ A food contains organic molecules with the elements C, H, and O in a ratio of 1:2:1. What type of compound is this and what is its major function in the body?

✔ Why does boiling a protein affect its structure and functional properties?

✔ How are DNA and RNA similar?

Table 2-4	Structure and Function of Organic Compounds				
Class	**Elements**	**Basic Subunit(s)**	**Complex Molecule**	**Function(s)**	**Examples**
Carbohydrates	C,H, and O	Monosaccharides (example: glucose)	Polysaccharides (examples: starch, glycogen)	Energy source	 Glucose molecules
Lipids	C,H, and O	Glycerol and fatty acids	Triglycerides (examples: fats and oils)	Energy storage	
Proteins	C,H,O,N, and often S	Amino acids	Proteins	Form structures within the cell and enzymes that catalyze chemical reactions	
Nucleic acids	C,H,O,N, and P	Nucleotides	DNA, RNA	Store and process information for building proteins	
High-Energy Compounds	C,H,O,N, and P	Nucleotides	None	Energy transfer	

THE MEDICAL IMPORTANCE OF RADIOISOTOPES

Many recent advances in medicine have involved the use of radioisotopes (p. 20). The radiation these isotopes release can be used to visualize internal structures as well as to treat diseases. **Nuclear medicine** is the branch of medicine concerned with the use of radioisotopes. Different radioisotopes release radiation in different amounts and at different intensities. The **half-life** of any radioisotope is the time required for a 50 percent reduction in the amount of radiation it emits. The half-lives of radioisotopes range from fractions of a second to thousands of years. Because intense radiation can kill cells and destroy tissues, nuclear medicine typically relies on weakly radioactive isotopes with short half-lives.

RADIOISOTOPES AND ORGAN IMAGING

Radioisotopes are useful because they can be inserted into specific chemical compounds normally found within the body. The resulting radioactive compounds, called **tracers**, can then be introduced into the body and tracked by the radiation they release. In some cases, the labeled compound is swallowed and its uptake, distribution, and excretion is determined by monitoring the radioactivity of samples taken from the digestive tract, body fluids, and waste products. For example, compounds labeled with radioisotopes of cobalt are used to monitor the intestinal absorption of vitamin B_{12}. Usually cobalt-58, a radioisotope with a half-life of 71 days, is used.

In other cases, radioisotopes are injected into the blood or other body fluids to provide information on circulatory anatomy, and the anatomy and function of specific target organs. In **nuclear imaging** the radiation emitted by injected radioisotopes creates an image on a computer monitor or special photographic plate. Such a procedure may be used to identify regions where particular radioactive materials are concentrated or to check the circulation of blood through vital organs. Radioisotopes can also produce pictures of specific organs, such as the liver, spleen, or thyroid, where labeled compounds are removed from the bloodstream and deposited or excreted.

The thyroid gland, for example, secretes chemical "messengers" called *hormones* (Chapter 13) that contain iodine atoms. As a result, the thyroid gland will actively absorb and concentrate radioactive iodine. This gland sits below the larynx (voicebox) on the anterior portion of the neck (Figure 2-10a•). The *thyroid scan* in Figure 2-10b• was taken following the injection of a low dose of iodine-131, a radioisotope with an 8-day half-life. This procedure can provide information on (1) the size and shape of the gland and (2) the amount of absorptive activity under way. Comparing the rate of iodine uptake with the level of injected iodine-131 makes it possible to evaluate how well the gland is functioning.

Radioactive iodine is an obvious choice for imaging the thyroid gland. For most other tissues and organs a radioactive label must be attached to another compound. *Technetium* is the primary radioisotope used in nuclear imaging today. The isotope is artificially produced and has a half-life of 6 hours. This brief half-life significantly reduces the radiation exposure of the patient. Technetium is used in more than 80 percent of all scanning procedures. Technetium scans are performed to examine the spleen, liver, kidneys, digestive tract, bone marrow, and a variety of other organs.

PET (**P**ositron **E**mission **T**omography) scans are based on the same principles as standard radioisotope scans, but the analyses are performed by computer. The scans are much more sensitive, and the computers can reconstruct detailed sections through the body. Among other things, this procedure can analyze blood flow through organs and assess the metabolic activity within specific portions of an

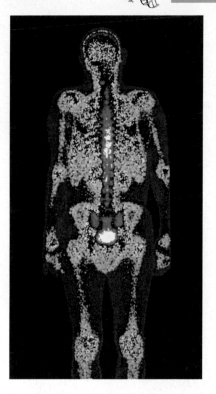

A nuclear image of the human body produced by the radiation released by administered radioisotopes.

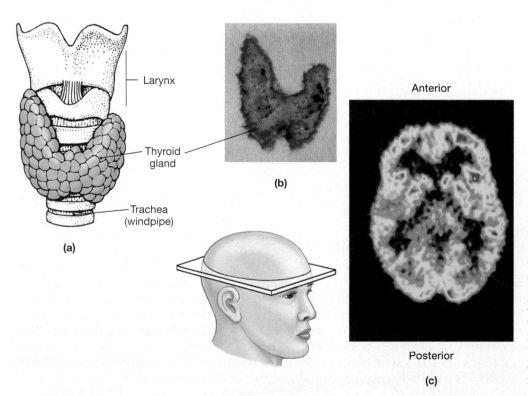

(b)

(a)

Larynx

Thyroid gland

Trachea (windpipe)

Anterior

Posterior

(c)

•**FIGURE 2-10**
Imaging Techniques. (a) The position and contours of the normal thyroid gland as seen in dissection. **(b)** After it has been labeled with radioactive iodine (I-131), the thyroid can be examined by special imaging techniques. In this computer-enhanced image, different intensities indicate differing concentrations of the radioactive tracer. **(c)** A PET scan of the brain in superior view. The light areas indicate regions of increased metabolic activity.

organ as complex as the human brain. Figure 2-10c• is a PET scan of the brain showing activity at a single moment in time. The scan is dynamic, however, and changing patterns of activity can be followed in real time. PET scans can be used to analyze normal brain function as well as to diagnose brain disorders.

RADIOISOTOPES AND DISEASE TREATMENT

Nuclear medicine involving injected radioisotopes has been far more successful in producing useful images than in treating specific disorders. The problem is that relatively large doses of radiation must be used to destroy abnormal or cancerous tissues, and it is very difficult to control circulation of radioisotopes within the body. As a result, both normal and abnormal tissues may be damaged by radiation exposure. For the same reason, it is difficult to control the radiation dosage administered to the diseased tissues. This is a problem because underexposure can have very little effect, whereas overexposure can cause the destruction of adjacent normal tissues.

Radioactive drugs, or *radiopharmaceuticals*, can be effective only if they are delivered precisely and selectively. One success story has been the treatment of *hyperthyroidism*, or thyroid oversecretion. As noted earlier, the thyroid gland selectively concentrates iodine. To treat hyperthyroidism, very large doses of radioactive iodine (Iodine-131) can be administered. The released radiation destroys the abnormal thyroid tissue and stops the excessive production of thyroid hormones. (The individual usually becomes hypothyroid—deficient in thyroid hormone—but this condition can be treated by taking thyroid hormones in tablet form.) This is now the preferred treatment method for most hyperthyroid patients, as it avoids surgery.

A relatively new application of nuclear medicine involves attaching a radioactive isotope to a *monoclonal antibody (MoAb)*. Antibodies are protein molecules produced in the body to provide a defense against disease. A substance that triggers antibody production is called an *antigen*. Monoclonal antibodies are special antibodies that are created in the laboratory to attach to a particular antigen. They are labeled with a radioactive isotope and injected into the body. When the MoAbs bind to their target antigen, the surrounding tissue is exposed to radiation.

Some specific antigens are found on certain types of cancerous tumors, and radiolabeled MoAbs can be created to target these antigens. When injected into the body, the MoAbs travel to the tumor site and attach to the surfaces of cancer cells. Upon attachment, radiation is emitted from the target area. The amount of radiation emitted is low, however, and the procedure is used to produce images rather than to treat the disease. This technique is very sensitive and can detect small tumors for early diagnosis and treatment. Experiments continue, with the eventual goal of using radiolabeled MoAbs to destroy tumor cells.

✔ Cobalt-58 has a half-life of 71 days. How long will it take for its level of radiation to be reduced to one-half of its original amount? to one-fourth of its original amount?

✔ What radioisotope is used in a thyroid scan and the treatment of hyperthyroidism?

CHAPTER REVIEW

Key Words

acid: A compound that breaks apart in solution and releases hydrogen (H^+) ions.

atom: The smallest stable unit of matter.

base: A compound that removes hydrogen ions from solution. Most bases break apart in solution and release hydroxide ions (OH^-).

buffer: A compound that stabilizes the pH of a solution by removing or releasing hydrogen ions.

carbohydrate: Organic compound containing atoms of carbon, hydrogen, and oxygen in a 1:2:1 ratio; examples include sugars and starches.

compound: A molecule containing atoms of two or more elements in combination.

covalent bond (kō-VĀ-lent): A chemical bond between atoms that involves the sharing of electrons.

electrolytes (ē-LEK-trō-līts): Soluble inorganic compounds whose ions will conduct an electric current in solution.

electron: One of the three basic particles of matter; a subatomic particle that carries a negative charge and normally orbits around the positively charged nucleus.

enzyme: A protein that catalyzes (speeds up) a specific biochemical reaction.

ion: An atom or molecule bearing a positive or negative charge due to the loss or gain of an electron.

lipid: An organic compound containing atoms of carbon, hydrogen, and oxygen not in a 1:2:1 ratio; examples include fats, oils, and waxes.

metabolism (me-TAB-ō-lizm): The sum of all of the chemical reactions under way within the human body at a given moment; includes decomposition and synthesis.

molecule: A compound containing two or more atoms that are held together by chemical bonds.

neutron: A subatomic particle that does not carry a positive or negative charge.

pH: A measure of the concentration of hydrogen ions in a solution.

protein: A complex substance made up of amino acids.

proton: A subatomic particle bearing a positive charge.

solute: Materials dissolved in a solution.

solution: A fluid containing dissolved materials.

solvent: The fluid component of a solution.

Selected Clinical Terms

cholesterol: A steroid important in the structure of cellular membranes; high concentrations in the blood increase the risk of heart disease.

nuclear imaging: A procedure in which an image is created on a photographic plate or video screen by the radiation emitted by injected radioisotopes.

nuclear medicine: The branch of medicine concerned with the use of radioisotopes.

PET (positron emission tomography) scan: A nuclear imaging technique in which the emitted radiation is analyzed and the image is created by a computer.

radioisotopes: Isotopes with unstable nuclei, which spontaneously release subatomic particles or radiation.

tracer: A compound labeled with a radioisotope that can be tracked within the body by the radiation it releases.

Study Outline

WHAT IS AN ATOM? (pp. 19–21)

1. Atoms are the smallest units of matter. Elements are simple substances made up of only one type of atom. (*Table 2-1*)

STRUCTURE OF THE ATOM (p. 20)

2. Atoms consist of protons, neutrons, and electrons. (Figure 2-1)

3. The **nucleus** contains protons and neutrons. The number of protons in an atom is its **atomic number**.

ISOTOPES (p. 20)

4. Isotopes are atoms of the same element whose nuclei contain different numbers of neutrons. The *mass number* of an atom is equal to the total number of protons and neutrons in its nucleus. (*Figure 2-2*)

ELECTRON SHELLS (p. 21)

5. Electrons occupy a series of **electron shells** around the nucleus. An atom with an unfilled outermost electron shell is reactive, while one with a filled outermost electron shell is stable, or *inert*. (*Figure 2-3*)

CHEMICAL BONDS (pp. 22–23)

1. Atoms can combine to form a **molecule**; molecules made up of atoms of different elements form a **compound**.

IONIC BONDS (p. 22)

2. An **ionic bond** results from the attraction between **ions**, atoms that have gained or lost electrons. (*Figure 2-4; Table 2-2*)

COVALENT BONDS (p. 22)

3. Some atoms share electrons to form a molecule held together by **covalent bonds**. (*Figure 2-5*)

CHEMICAL REACTIONS (pp. 23–24)

1. Metabolism refers to all the chemical reactions in the body. Our cells capture, store, and use energy to maintain homeostasis and support essential functions.

2. Cells gain energy to power their functions by breaking the chemical bonds of organic molecules in *decomposition* reactions.

3. *Synthesis* reactions result in the formation of new organic molecules and the storage of energy.

4. Reversible reactions consist of simultaneous synthesis and decomposition reactions. At **equilibrium** these two opposing reactions are in balance.

INORGANIC COMPOUNDS (pp. 24–27)

1. Chemical compounds of the body can be broadly classified as **organic** (carbon-based) or **inorganic** (not carbon-based).

WATER (p. 24)

2. Water is the most important inorganic component of the body.

3. Many inorganic and organic compounds dissolve in water.

4. Water has a high heat capacity, and it retains heat for much longer periods of time than other substances.

5. Water is essential for many of the body's chemical reactions. (*Figure 2-6*)

ACIDS AND BASES (p. 25)

6. An **acid** releases hydrogen ions, and a **base** removes hydrogen ions from a solution.

pH AND BUFFERS (p. 26)

7. The **pH** of a solution indicates the concentration of hydrogen ions it contains. Solutions can be classified as **neutral** (pH = 7), **acidic** (pH<7), or **basic (alkaline)** (pH>7) on the basis of pH. (*Figure 2-7*)

8. Buffers maintain pH within normal limits (7.35–7.45) in body fluids by releasing or absorbing hydrogen ions.

SALTS (p. 27)

9. A **salt** is an inorganic compound that contains neither H^+ nor OH^-.

10. Many inorganic compounds, called **electrolytes**, break apart in water to form ions.

ORGANIC COMPOUNDS (pp. 27–32)

1. Organic compounds contain carbon and hydrogen, and usually oxygen as well. Large and complex organic molecules include *carbohydrates, lipids, proteins,* and *nucleic acids.* Organic compounds are the building blocks of cells. (*Table 2-4*)

CARBOHYDRATES (*p. 27*)

2. Carbohydrates are most important as an energy source for metabolic processes. Two major types are **monosaccharides (simple sugars)**, and **polysaccharides**.

LIPIDS (*p. 28*)

3. Lipids are water-insoluble molecules that include fats, oils, and waxes. There are three important classes of lipids: **triglycerides**, **steroids**, and **phospholipids**.

PROTEINS (*p. 28*)

4. Proteins perform a great variety of roles in the body. Structural proteins provide internal support and external protection for the body. Functional proteins are involved with functions such as muscle contraction, regulating chemical reactions, and defending the body.

5. Proteins are chains of **amino acids**.

6. The sequence of amino acids affects the shape of a protein molecule. The shape of a protein determines its function.

7. Enzymes control many chemical reactions within our bodies. **Enzymes** are organic **catalysts**—substances that speed up chemical reactions without themselves being permanently changed.

8. The reactants in an enzymatic reaction, called **substrates**, interact to form a **product** by binding to the enzyme. (*Figure 2-8*)

NUCLEIC ACIDS (*p. 30*)

9. Nucleic acids store and process information at the molecular level. There are two kinds of nucleic acids: **deoxyribonucleic acid (DNA)** and **ribonucleic acid (RNA)**. (*Figure 2-9; Table 2-3*)

10. Nucleic acids are chains of **nucleotides**. Each nucleotide contains a simple sugar, a **phosphate group**, and a **nitrogen base**. The sugar is always **ribose** or **deoxyribose**. The nitrogen bases found in DNA are **adenine**, **guanine**, **cytosine**, and **thymine**. In RNA, **uracil** replaces thymine.

HIGH-ENERGY COMPOUNDS (*p. 31*)

11. Cells store energy in **high-energy compounds** for later use. The most important high-energy compound is **ATP (adenosine triphosphate)**. When energy is available, cells make ATP by adding a phosphate molecule to ADP. When energy is needed, ATP is broken down to ADP and phosphate.

THE MEDICAL IMPORTANCE OF RADIOISOTOPES
(*pp. 32–34*)

1. In **nuclear medicine**, radioisotopes are used to visualize organs and to treat diseases without surgery.

RADIOISOTOPES AND ORGAN IMAGING (*p. 32*)

2. Tracers, or radioisotope-labeled compounds, introduced into the body can be tracked by the radiation they release. Tracers can be measured directly in body fluids to provide information about organ function.

3. In **nuclear imaging** the radiation emitted by injected radioisotopes within an organ creates an image on a computer monitor or special photographic plate. **PET** (**P**ositron **E**mission **T**omography) scans use computers to reconstruct three-dimensional sections through the body.

RADIOISOTOPES AND DISEASE TREATMENT (*p. 34*)

4. *Radiopharmaceuticals,* or radioactive drugs, need to be precisely delivered to their targets to avoid damaging normal tissue.

Review Questions

MATCHING

Match each item in Column A with the most closely related item in Column B. Use letters for answers in the spaces provided.

	Column A	Column B
___	1. atomic number	a. assembly of large molecules
___	2. mass number	b. organic catalyst
___	3. neutron	c. attraction between opposite charges
___	4. covalent bond	d. sharing of electrons
___	5. acid	e. stabilizes pH
___	6. reversible reaction	f. $A + B \leftrightarrow AB$
___	7. element	g. number of protons and neutrons
___	8. ionic bond	h. time for radiation of a radioisotope to decrease by half
___	9. enzyme	i. number of protons
___	10. decomposition	j. carbohydrates, lipids, proteins
___	11. synthesis	k. labeled compound
___	12. organic compounds	l. substance made up of identical atoms
___	13. buffer	m. releases H^+ in solution
___	14. inorganic compounds	n. breakdown of large molecules
___	15. tracer	o. subatomic particle without an electric charge
___	16. half-life	p. water and salts

MULTIPLE CHOICE

17. In atoms, protons and neutrons are found _____.
(a) only in the nucleus
(b) outside the nucleus
(c) within the electron shell
(d) inside and outside the nucleus

18. Isotopes differ from one another in the number of _____.
(a) protons in the nucleus
(b) neutrons in the nucleus
(c) electrons in the outer shells
(d) protons and neutrons in the nucleus

19. The number and arrangement of electrons in an atom's outermost electron shell determines its _____.
(a) atomic weight   (b) atomic number
(c) electrical properties   (d) chemical properties

20. Compounds that break apart in water form _____.
(a) amino acids   (b) ions
(c) sugars   (d) nucleotides

21. A carbon atom can form _____ covalent bonds.
(a) 1   (b) 2
(c) 3   (d) 4

22. The oxygen atoms in a molecule of oxygen are held together by _____.
(a) a single covalent bond   (b) a double covalent bond
(c) a triple covalent bond   (d) an ionic bond

23. All the chemical reactions that occur in the human body are collectively referred to as _____.
(a) anabolism (b) catabolism
(c) metabolism (d) homeostasis

24. Of the following selections, the pH of the most acidic solution is _____.
(a) 6.0 (b) 2.3
(c) 4.5 (d) 1.0

25. Isotopes with _____ radioactivity and _____ half-lives are used to check the structural and functional condition of an organ.
(a) weak; long (b) high; short
(c) weak; short (d) high; long

26. The most frequently used radioisotope in nuclear imaging is *Technetium*. It has a half-life of about
(a) 6 hours (b) 24 hours
(c) 6 days (d) 1 week

TRUE/FALSE

_____ 27. The basic building blocks of carbohydrates are amino acids.

_____ 28. Organic compounds always contain carbon and hydrogen atoms.

_____ 29. Glycogen is an example of a polysaccharide found in our bodies.

_____ 30. Steroid hormones are derived from the compound cholesterol.

_____ 31. The principal high-energy compound produced in the cell is ADP.

_____ 32. Radioactive drugs are called radiopharmaceuticals.

_____ 33. PET scans provide three-dimensional images of living organs.

SHORT ESSAY

34. What are the three kinds of subatomic particles that make up atoms?

35. What are the four most abundant elements by weight in the human body?

36. What is the role of enzymes in chemical reactions?

37. What four major classes of organic compounds are found in the body?

38. What are the three basic components of a DNA nucleotide? What are the three basic components that make up an RNA nucleotide?

39. Nuclear medicine has been more successful in using radioisotopes to form images rather than in the treatment of diseases. Why?

40. Monoclonal antibodies are beginning to be used to destroy tumor cells. What substances are combined with the antibodies that would kill the tumor cells?

APPLICATIONS

41. Concerned about cholesterol, a friend wants to eliminate all dietary fats. On the basis of what you read in this chapter, would you advise against this? Why, or why not?

42. Mr. Smith is admitted to the hospital with a diagnosis of an enlarged thyroid gland. How could the use of radioisotopes help to investigate the underlying cause(s) of this condition?

✔ Answers to Concept Check Questions

(p. 21) **1.** There are three *isotopes* of hydrogen: hydrogen-1 with a mass of 1; hydrogen-2 with a mass of 2; hydrogen-3 with a mass of 3. The heavier sample must contain a higher proportion of the heavier isotopes. **2.** Oxygen has two electrons in its first electron shell and six electrons in its second electron shell. Because an atom's second electron shell is not filled until it contains eight electrons, oxygen needs 20 electrons to fill its outermost electron shell.

(p. 23) **1.** Atoms combine with one another so as to gain a complete set of eight electrons in their outermost electron shell. Oxygen atoms do not have a full outermost electron shell, so they will react with many other elements to reach this stable arrangement. The outermost electron shell of neon is already full and so has little tendency to combine with other elements. **2.** Two oxygen atoms share two electrons each to fill their outermost electron shells and form a *double covalent bond*. **3.** Two nitrogen atoms form a molecule of nitrogen by forming a *triple covalent bond*.

(p. 24) **1.** The breaking of the glucose molecule into two smaller molecules is a *decomposition reaction*. **2.** Because more chemical bonds are being formed, a *synthesis reaction* results in an increase in stored energy. **3.** Removing the product of a reversible reaction would keep its level low compared with the level of the reactants.

The formation of the product molecule would continue, but the reverse reaction would slow down, resulting in a shift in the equilbrium toward the product.

(p. 27) **1.** An *acid* releases hydrogen ions in solution, and a *base* is a substance that removes hydrogen ions from a solution. **2.** Stomach discomfort is often the result of excess stomach acidity ("acid indigestion"). Antacids contain a weak base that neutralizes the excess acid. **3.** Unlike acids and bases, salts do not release hydrogen ions or form hydroxide ions in water. *Salts* dissolve in solution by breaking apart into charged particles called ions. Because salts in solution are capable of conducting an electrical current, they are also called *electrolytes*.

(p. 31) **1.** A C:H:O ratio of 1:2:1 indicates that the molecule is a *carbohydrate*. The body uses carbohydrates chiefly as an energy source. **2.** High temperatures break chemical bonds that maintain the shape of a protein molecule. Because the shape of a protein molecule determines its function, the altered protein molecule cannot function properly. **3.** DNA and RNA are both nucleic acids composed of chains of smaller molecules called *nucleotides*.

(p. 34) **1.** Its radiation will be reduced to one-half of its original amount after 71 days and one-fourth of its original amount after 142 days. **2.** Iodine-131.

CHAPTER

3

The Structure and Function of Cells

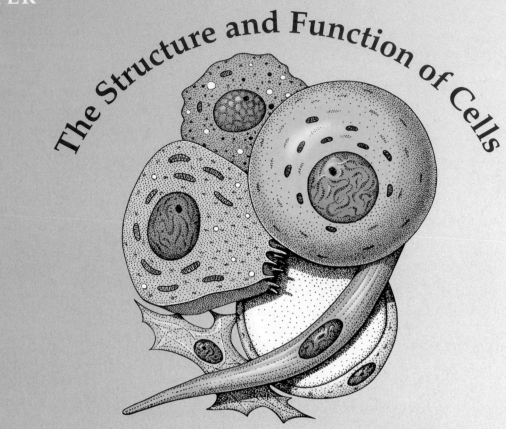

CHAPTER OUTLINE

CHAPTER OBJECTIVES

1 List the functions of the cell membrane and the structures that perform those functions.

2 Describe the ways cells move materials across the cell membrane.

3 Describe the organelles of a typical cell, and indicate their specific functions.

4 Explain the functions of the cell nucleus.

5 Summarize the process of protein synthesis.

6 Describe the process of mitosis, and explain its significance.

7 Define differentiation and explain its importance.

8 Describe disorders caused by abnormal mitochondria or lysosomes.

9 Describe the development of cancer and its possible causes.

The human body is made up of several trillion building blocks called *cells*. However, unlike the identical building blocks of a nonliving structure such as a brick house, cells vary widely in size and appearance. In addition, a single cell can perform many of the same functions as the body. In fact, a cell is the smallest functioning unit of life. Figure 3-1• shows some examples of cells found in the human body. Notice the incredible variety of shapes and sizes! Each of these cell types will be introduced as you study the body systems in later chapters.

Cells combine in many different ways to form all the structures of the body. Each cell maintains its own homeostasis, but homeostasis of the body depends on the combined and coordinated actions of all the body's cells. As you will discover, the specialized structures of each body system are designed to perform unique functions—and this close relationship between structure and function begins at the cellular level.

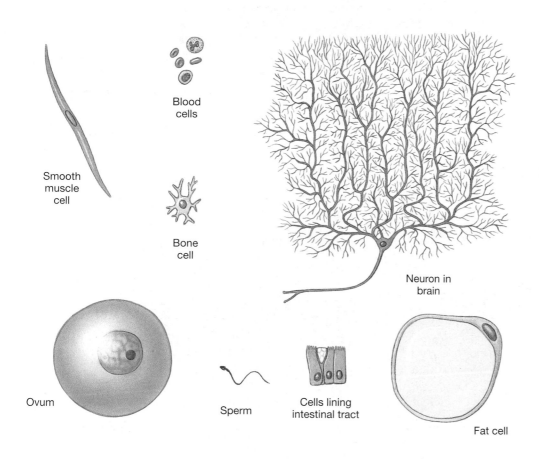

Smooth muscle cell

Blood cells

Bone cell

Neuron in brain

Ovum

Sperm

Cells lining intestinal tract

Fat cell

•**FIGURE 3-1**
The Diversity of Cells in the Human Body.
The cells of the body have many different shapes and a variety of special functions. These examples give an indication of the range of forms; all the cells are shown with the dimensions they would have if magnified approximately 500 times.

AN OVERVIEW OF THE CELL

Although the body is made up of different types of cells, most of those cells have many features in common. Figure 3-2• shows the parts of a "typical" body cell, and Table 3-1 lists their general functions.

Shared features of all human body cells include (1) a *cell membrane*, which forms the boundary of the cell and separates the cell's interior from the watery *extracellular fluid* that surrounds it; (2) *cytoplasm*, which consists of a watery *intracellular fluid* that contains suspended nutrients and other molecules, as well as specialized structures known as *organelles* (or-gan-ELS; little organs); and (3) a *nucleus*, the prominent organelle that serves as "command center" for the cell, containing its hereditary information and controlling all its functions.

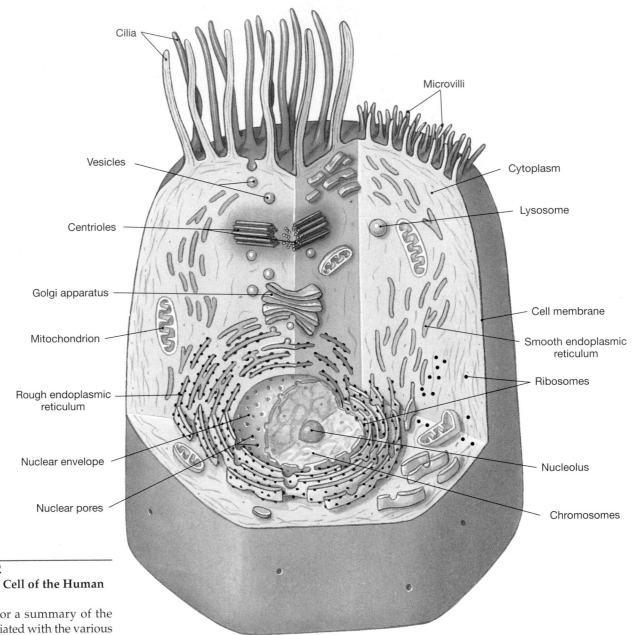

Cilia

Microvilli

Vesicles

Cytoplasm

Centrioles

Lysosome

Golgi apparatus

Cell membrane

Mitochondrion

Smooth endoplasmic reticulum

Rough endoplasmic reticulum

Ribosomes

Nuclear envelope

Nucleolus

Nuclear pores

Chromosomes

•**FIGURE 3-2**
A Generalized Cell of the Human Body.
See Table 3-1 for a summary of the functions associated with the various cell structures.

THE CELL MEMBRANE

The outer boundary of the cell is formed by an extremely thin and delicate **cell membrane**, also called the *plasma membrane*. The cell membrane is a complex and highly ordered structure that is essential for many cell functions. Two of its most important functions are:

1. *Regulation of exchange with the environment*. The cell membrane controls the movement of substances into and out of the cell. For example, nutrients and other needed materials are allowed in, and cell products and wastes are exported out.

2. *Sensitivity*. The cell membrane is the first part of the cell to be affected by changes in the extracellular fluid. It contains a variety of *receptors*, proteins that allow the cell to recognize and respond to specific molecules in its en-

Table 3-1	Functions of a Generalized Cell	

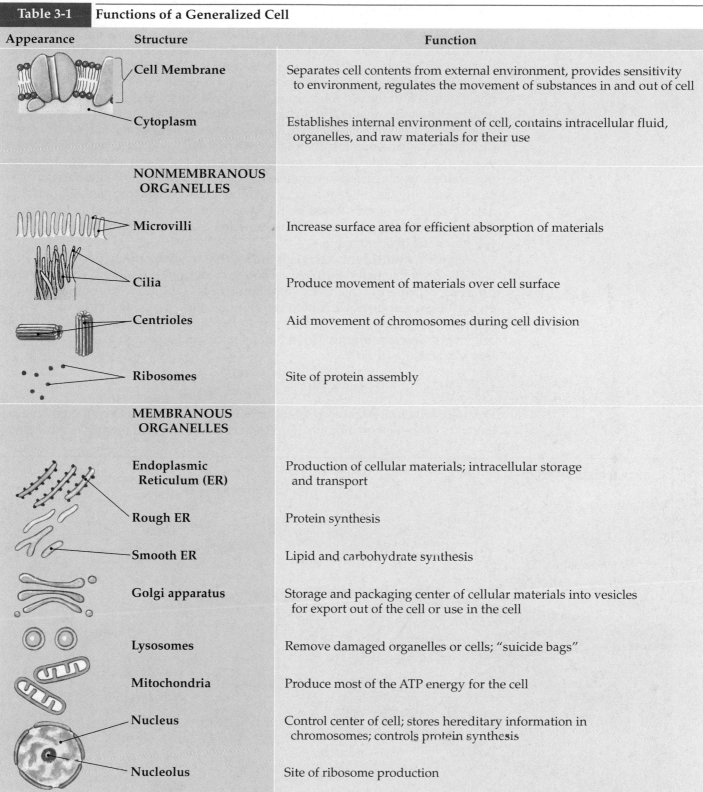

Appearance	Structure	Function
	Cell Membrane	Separates cell contents from external environment, provides sensitivity to environment, regulates the movement of substances in and out of cell
	Cytoplasm	Establishes internal environment of cell, contains intracellular fluid, organelles, and raw materials for their use
	NONMEMBRANOUS ORGANELLES	
	Microvilli	Increase surface area for efficient absorption of materials
	Cilia	Produce movement of materials over cell surface
	Centrioles	Aid movement of chromosomes during cell division
	Ribosomes	Site of protein assembly
	MEMBRANOUS ORGANELLES	
	Endoplasmic Reticulum (ER)	Production of cellular materials; intracellular storage and transport
	Rough ER	Protein synthesis
	Smooth ER	Lipid and carbohydrate synthesis
	Golgi apparatus	Storage and packaging center of cellular materials into vesicles for export out of the cell or use in the cell
	Lysosomes	Remove damaged organelles or cells; "suicide bags"
	Mitochondria	Produce most of the ATP energy for the cell
	Nucleus	Control center of cell; stores hereditary information in chromosomes; controls protein synthesis
	Nucleolus	Site of ribosome production

vironment. The sensation of taste, for example, results when dissolved molecules reach receptors in the taste buds on your tongue. Even your movements depend on a chemically sensitive membrane, for muscle cells begin to contract only after their membrane receptors receive special molecules from nerve cells.

MEMBRANE STRUCTURE

The vital functions of a cell membrane are directly related to its unique structure. The major components of cell membranes are phospholipid molecules. ∞ *[p. 28]* Each phospholipid consists of two parts, sometimes called the "head" and "tail." One end (the "head") dissolves in water, and the other (the "tail") does not. The phospholipids in the cell membrane form two distinct layers, with the water-soluble heads on the outside and the water-insoluble tails facing each other on the inside. As Figure 3-3• shows, there are also other types of molecules (cholesterol, carbohydrates, and several different type of proteins) associated with the cell membrane. The membrane proteins are responsible for many cell membrane functions. For example, some membrane proteins act as receptors for molecules carried by the blood, allowing the cell to interact with other cells. Others, containing open channels or acting as *carriers*, assist in the passage of molecules into or out of the cell. You will encounter these membrane proteins again as you learn about different cell functions in later chapters.

A basic feature of the cell membrane is that it is **selectively permeable**—that is, it allows some materials to pass through freely, but not others. This important property is primarily due to the fatty lipid "tails" within the phospholipid bilayer. Because lipids do not mix with water molecules, many ions and water-soluble compounds cannot cross this fatty portion of the cell membrane. This makes the cell membrane very effective in isolating the cytoplasm from the surrounding watery extracellular fluid.

MEMBRANE PERMEABILITY

Precisely which substances enter or leave the cytoplasm is determined by the *permeability* of the cell membrane. If nothing can cross the cell membrane, the mem-

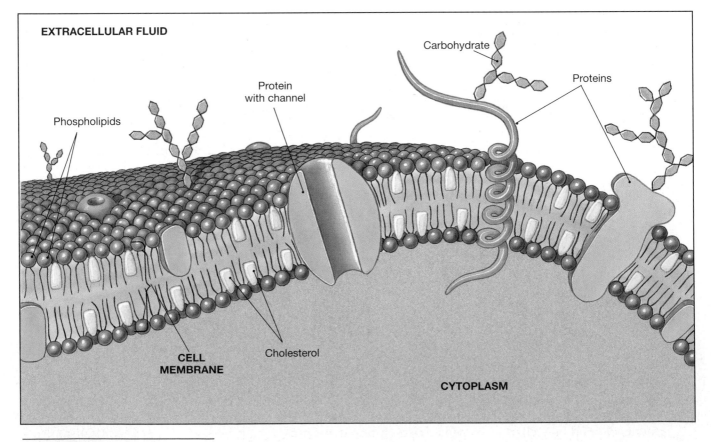

•**FIGURE 3-3**
The Cell Membrane.

brane is said to be *impermeable*. If all substances can cross without difficulty, the membrane is *freely permeable*. As noted previously, cell membranes are *selectively permeable*—they permit the free passage of some materials and restrict the passage of others. Anything that affects a cell's membrane permeability can have serious consequences.

The movement of materials across the cell membrane may be passive or active. *Passive processes* move ions or molecules across the cell membrane without the use of any energy by the cell. *Active transport processes*, discussed later in this chapter, require that the cell expend energy, usually in the form of ATP.

Passive Processes

Passive processes include *diffusion, osmosis, filtration,* and *facilitated diffusion*.

Diffusion. **Diffusion** is the movement of a substance from an area of high concentration to an area of lower concentration. For example, if we place a soluble material, like table salt, in a container of water, it dissolves. Eventually its molecules spread through the solution until they become evenly distributed throughout the container (Figure 3-4●). The difference in concentration of the molecules between the regions where they are close together (high concentration) to where they are absent or far apart (low concentration) represents a *concentration gradient*. Diffusion is often described as proceeding "down a concentration gradient" or "downhill." For any specific *solute* (dissolved substance) in a solution, diffusion continues until all the solute molecules are spread out evenly, a condition known as **equilibrium**.

Diffusion is important in body fluids because it allows cells to pick up or unload various substances without using energy. For example, every cell in the body produces carbon dioxide as a waste product of its normal activity. As the concentration of CO_2 inside the cell increases relative to the extracellular fluid around the cell, diffusion moves CO_2 out of the cell, down its concentration gradient. The CO_2 is then transported away from the cells by the bloodstream. Because of this constant removal of CO_2, a concentration gradient is maintained between the cells and the extracellular fluid, and CO_2 continues to move easily out of each cell.

●**FIGURE 3-4**
Diffusion.
The diffusion of a soluble solid in a beaker of water. The molecules of the solid diffuse from where they are highly concentrated to where they are absent or in a lower concentration. Eventually the molecules of the solid are distributed evenly, and their concentration is the same everywhere.

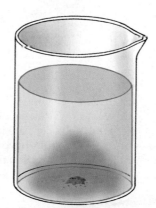

Osmosis. Water is so important to living cells that diffusion of water across the cell membrane has its own name: **osmosis** (oz-MŌ-sis). Osmosis occurs across any selectively permeable membrane that is freely permeable to water but not to solutes. The movement of water molecules occurs in response to differences in the solute concentrations on either side of the membrane. Like all molecules, water molecules will diffuse toward an area of lower water concentration. In the case of solutions, the lower water concentration is where the solute concentration is *higher*. Thus, in osmosis, water molecules diffuse across a membrane toward the area containing the *higher* solute concentration until equilibrium is reached. The greater the initial difference in solute concentrations, the stronger the flow of water molecules. At equilibrium, water molecules continue to move back and forth across the membrane, but there is no net (overall) change in the number of molecules on either side.

Solutions of varying solute concentrations are described as *isotonic, hypotonic,* or *hypertonic* in relation to a cell's internal solute concentration. Each type of solution has a different effect on the overall shape of a cell or the tension of its membrane. Figure 3-5a• shows the appearance of a red blood cell immersed in an isotonic solution. An **isotonic** (*iso-,* equal + *tonos,* tension) solution is one that does not cause a net movement of water into or out of the cell. The solute concentration of the solution is *equal* to the solute concentration within the cell. Therefore, the water is in equilibrium: as one molecule moves out another moves in to replace it. Because of this balanced movement of water there is no overall change in the shape of the cell.

In a **hypotonic** (*hypo-,* below) solution, the concentration of solutes in the solution is *less* than that within a typical living cell. The greater water concentration is outside the cell and therefore water flows *into* the cell by osmosis. If the difference is substantial, the cell swells up like a balloon, as shown in Figure 3-5b•. Ultimately the membrane may burst. In the case of red blood cells, such an event is known as *hemolysis* (the term *hemo,* blood + *lysis,* breaking apart of a cell).

A **hypertonic** (*hyper-,* above) solution is one that contains *more* solutes than a typical living cell. A cell in a hypertonic solution loses water to the sur-

iso-, same + *tonos,* tension
isotonic: an extracellular solution that does not cause the movement of water across a cell membrane

hypo-, below + *tonos,* tension
hypotonic: an extracellular solution that causes the movement of water across a cell membrane and into the cytoplasm

hyper-, below + *tonos,* tension
hypertonic: an extracellular solution that causes the movement of water across a cell membrane and out of the cytoplasm

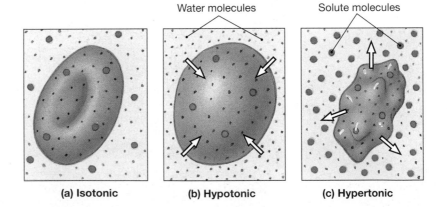

Water molecules   Solute molecules

(a) Isotonic   (b) Hypotonic   (c) Hypertonic

•**FIGURE 3-5**
Osmosis.
White arrows show the direction of water movement due to osmosis. **(a)** Because these red blood cells are immersed in an isotonic saline solution, no osmotic flow occurs and the cells have their normal appearance. **(b)** Immersion in a hypotonic saline solution results in the osmotic flow of water into the cells. The swelling may continue until the cell membrane ruptures. **(c)** Exposure to a hypertonic solution results in the movement of water out of the cells. The red blood cells dehydrate and shrivel.

rounding medium by osmosis, and the cell dehydrates and shrivels (Figure 3-5c•).

Whenever it is necessary to administer large volumes of fluid to people who have had a severe blood loss or who are dehydrated, it is vital that the fluid be isotonic with respect to body cells, especially blood cells. One such fluid, called **normal saline**, is a 0.9 percent solution of sodium chloride (NaCl). It is used because sodium and chloride are the most abundant ions in the extracellular fluid.

Filtration. In **filtration**, water and small solute molecules are forced across a membrane. Molecules of solute are carried along with the water only if they are small enough to fit through the pores, or channels, formed by the membrane proteins. The process is similar to the way a filter in a drip coffeemaker allows the water and small (but important!) molecules from the coffee grounds to pass through. Filtration requires a force to push the fluid. In the coffeemaker, gravity provides this force. In the body, the force comes from *blood pressure*, which is created as the heart pushes blood through the circulatory system. Filtration commonly occurs across the cell membranes of the cells forming the walls of small blood vessels, pushing water and dissolved nutrients into the tissues of the body. Filtration across specialized blood vessels in the kidneys is an essential step in the production of urine.

Fluids administered into the bloodstream are labeled to avoid errors and osmotic problems.

Active Processes

Active processes are involved in both the movement of small molecules and individual ions, as well as the bulk movement of larger amounts of materials.

Active transport. In **active transport**, materials are moved *against* a concentration gradient—that is, from an area of lower concentration to one of higher concentration—using the energy of ATP molecules. All living cells contain ATP-activated carrier proteins that transport molecules, such as glucose and amino acids, and ions across their cell membranes. Especially important are the **ion pumps** that actively transport sodium (Na^+), potassium (K^+), calcium (Ca^{2+}), ions. (As we will see later, the movements of these ions are involved in both muscle contraction and the movement of nerve impulses.) Many of these carrier proteins move a specific ion in one direction only, either in or out of the cell. In a few cases, one carrier protein will move more than one ion at a time. If one ion moves in one direction and the other moves in the opposite direction, the carrier is called an *exchange pump*.

Packaged transport. Bulk materials (large amounts of small molecules or very large particles) that need to be transported into or out of the cell do not pass directly across the cell membrane. Instead, they first must be packaged in small membranous sacs, called *vesicles*.

There are two major processes that bring bulk quantities of extracellular materials into the cell. Both are active, energy-consuming processes. **Pinocytosis** (pi-no-si-TŌ-sis), or "cell drinking," occurs when extracellular fluid and small molecules become trapped in a deep groove or pocket that forms in the cell membrane. The membrane then pinches off, creating a fluid-filled sac inside the cell. **Phagocytosis** (fa-gō-si-TŌ-sis), or "cell eating," produces sacs containing *solid particles*.

Most cells display pinocytosis, but phagocytosis, especially the entrapment of other living or dead cells, is performed only by specialized cells of the immune system, such as white blood cells.

In **exocytosis** (ek-sō-sī-TŌ-sis) a vesicle created inside the cell fuses with the cell membrane and discharges its contents out of the cell. The ejected material may be a secretory product, such as a hormone or mucus, or waste products generated inside the cell.

Many of these transport mechanisms can occur simutaneously within a cell. Table 3-2 reviews and summarizes these mechanisms.

phagein, to eat + *kytos*, cell
phagocytosis: the capture of extracellular objects by the formation of vesicles at the cell surface

exo, out + *kytos*, cell
exocytosis: the ejection of materials from the cytoplasm through the fusion of a vesicle with the cell membrane

✔ What is the difference between active and passive transport processes?

✔ During digestion in the stomach, the concentration of hydrogen ions (H^+) rises to many times the concentration found in the cells of the stomach. What type of transport process could produce this result?

✔ When certain types of white blood cells of the immune system encounter bacteria, they are able to engulf and destroy them. What is this process called?

Table 3-2   Passive and Active Transport Processes

Process	Description	Example
PASSIVE PROCESSES (No cellular energy)		
Diffusion	Movement of substances from areas of higher concentration to lower concentration	Movement of CO_2 out of cell
Osmosis	Movement of water molecules across a selectively permeable membrane from areas of higher water concentration to lower water concentration	High solute concentration outside of cell Movement of water out of cell
Filtration	Movement of water, usually with a solute, caused by internal blood pressure across a membrane	Water and small solutes pushed out of cell
ACTIVE PROCESSES (Cellular energy required)		
Active Transport	Carrier molecule moves specific substances up a concentration gradient	Sodium ions (Na^+) pumped out of cell with energy from ATP
Vesicle Formation		
Pinocytosis	Vesicles form at cell membrane to bring fluids and small molecules into the cell	Water and small molecules enter cell through vesicle
Phagocytosis	Vesicles form at cell membrane to bring solid particles into the cell	Large particles enter cell
Exocytosis	Vesicles fuse with cell membrane to release fluids and/or solids from the cell	Cellular product released from cell

THE CYTOPLASM

Cytoplasm is a general term for the material inside the cell between the cell membrane and the nucleus. It includes the intracellular fluid and tiny structures called organelles.

INTRACELLULAR FLUID

The intracellular fluid has a consistency that varies between that of thin maple syrup and almost-set gelatin. It contains raw materials, finished cellular products, and wastes—essentially all the basic chemical substances associated with life. These include water, ions, proteins, small amounts of carbohydrates, and large reserves of amino acids and lipids. Many of the proteins in intracellular fluid are enzymes that regulate metabolic processes. The carbohydrates are broken down to provide energy, and the amino acids are used to build proteins. Lipids stored in the cell are primarily used as an energy source when carbohydrates are unavailable.

ORGANELLES

Organelles, literally "little organs," are structures that perform specific functions essential to normal cell structure, maintenance, and metabolism (see Figure 3-2• and Table 3-1). Organelles surrounded by a membrane structurally similar to the cell membrane include the *nucleus, mitochondria, endoplasmic reticulum, Golgi apparatus*, and *lysosomes*. A membrane isolates each of these organelles so they can manufacture or store secretions, enzymes, or toxins that may otherwise damage the cell. The *centrioles, ribosomes, microvilli, cilia*, and *flagella*, are organelles not surrounded by a membrane. All of these organelles—except the nucleus—are discussed next. As you will see in the examples provided, the function of any particular cell type is reflected in the organelles it contains.

  Microvilli, cilia, and flagella. All these organelles are associated with extensions of the cell membrane. **Microvilli** are small, brushlike projections of the cell membrane that are supported by a network of protein filaments. Because they increase the surface area of the membrane, microvilli are common features of cells that absorb materials, such as the cells lining the intestinal tract.

  Cilia (singular; *cilium*) are relatively long finger-shaped extensions of the cell membrane that have a complex internal structure. Cilia undergo active movements that require energy. Their coordinated beating moves fluids or secretions across the cell surface. For example, cilia lining the respiratory passageways move sticky mucus and trapped dust particles up toward the throat and away from delicate respiratory surfaces. If the cilia are damaged or immobilized by heavy smoking or some metabolic problem, the cleansing action is lost, and the irritants are no longer removed. As a result, chronic respiratory infections develop.

  Flagella (fla-JEL-ah; singular; *flagellum*, whip) resemble cilia but are much larger. Flagella propel a cell through the surrounding fluid rather than moving the fluid past a stationary cell. The sperm cell is the only human cell that has a flagellum. If the flagella are paralyzed or otherwise abnormal, the man will be sterile because immobile sperm cannot fertilize an egg.

  Centrioles. All animal cells that are capable of dividing contain a pair of short cylindrical structures called **centrioles**. The centrioles create the *spindle fibers* that move DNA strands during cell division. Cells that do not divide, such as mature red blood cells and most nerve cells, do not have centrioles.

  Ribosomes. The **ribosomes** are small organelles that manufacture proteins. These "protein factories" are found scattered throughout the cytoplasm and attached to the *endoplasmic reticulum (ER)*, a membranous organelle discussed next.

Ribosomes are found in almost all cells, but their number varies depending on the type of cell and its activities. We will discuss the ribosomes in more detail when we discuss protein synthesis later in this chapter.

The endoplasmic reticulum. The **endoplasmic reticulum** (en-dō-plaz-mik re-TIK-ū-lum), or **ER**, is a network of intracellular membranes. Its major functions include the manufacture and storage of proteins, carbohydrates, and lipids; and the transport of materials from place to place within the cell.

There are two types of endoplasmic reticulum, **smooth endoplasmic reticulum** (**SER**) and **rough endoplasmic reticulum** (**RER**). The SER, which lacks ribosomes, is the site of lipid and carbohydrate production. The membranes of the RER are studded with ribosomes. The proteins synthesized here are stored within the RER. The amount of endoplasmic reticulum and the proportion of RER to SER vary depending on the type of cell and its ongoing activities. For example, cells of the pancreas that manufacture digestive enzymes (proteins) have an extensive RER, but relatively little SER. The proportion is just the reverse in the cells of the reproductive system that synthesize lipid hormones.

The products of both types of ER enter the inner chambers of their respective endoplasmic reticulum. These molecules are then packaged into small vesicles, that are pinched off from the ER.

The Golgi apparatus. The **Golgi** (GOL-jē) **apparatus** consists of a set of five to six flattened membrane discs. A single cell may contain several sets, each resembling a stack of dinner plates. The Golgi apparatus is involved with the modification, packaging, and distribution of products (proteins, lipids, or carbohydrates) created at the ER. Its products are then moved to the cell surface through the formation, movement, and fusion of vesicles.

Lysosomes. **Lysosomes** (LĪ-so-sōms) perform the cleanup and recycling functions within the cell. They contain digestive enzymes that are activated when the lysosomes fuse with the membranes of damaged organelles, such as mitochondria or fragments of the endoplasmic reticulum.

Lysosomes also function in the defense against disease. Through phagocytosis, cells may bring bacteria, as well as fluids and organic debris, into the cytoplasm within vesicles formed at the cell surface. Lysosomes fuse with vesicles created in this way, and the digestive enzymes then break down the contents and release usable substances such as sugars or amino acids.

Within dead or damaged cells, lysosome membranes disintegrate, releasing active enzymes into the cytoplasm. These enzymes rapidly destroy the proteins and organelles of the cell. Because lysosomes may destroy the cell that houses them, they are sometimes called "suicide organelles."

Mitochondria. **Mitochondria** (mī-tō-KON-drē-ah; singular: mitochondrion) are organelles that provide energy for the cell. Thus, they are often called the "powerhouse" of the cell. The number of mitochondria in a particular cell varies depending on its energy demands. For example, red blood cells have none, but mitochondria may account for one-fifth of the volume of an active liver cell.

lysis, dissolution + *soma*, body
lysosome: cytoplasmic vesicles containing digestive enzymes

✔ Cells lining the small intestine have numerous tiny fingerlike projections on their free surface. What are these structures and what is their function?

✔ Cells in the ovaries and testes contain large amounts of smooth endoplasmic reticulum (SER). Why?

✔ Microscopic examination of a cell reveals that it contains many mitochondria. What does this observation imply about the cell's energy requirements?

THE NUCLEUS

The **nucleus** is the control center for cellular operations, for this is where the genetic material (DNA) is stored. A typical nucleus consists of a double-membrane *nuclear envelope* that separates the nuclear contents from the cytoplasm. Organic compounds within the nucleus include enzymes, RNA and DNA nucleotides, proteins, small amounts of RNA, and DNA.

Most nuclei also contain one to four **nucleoli** (noo-KLĒ-o-lī; singular: *nucleolus*). Nucleoli form where ribosomes are produced. For this reason, nuclei are most apparent in cells that manufacture large amounts of proteins, such as muscle and liver cells.

The transport of materials between the nucleus and the cytoplasm occurs through openings in the nuclear envelope called *nuclear pores*. These pores are large enough to permit the movement of ions, small molecules, and even RNA, but are too small for the passage of proteins and DNA.

The DNA within the cell nucleus is found in structures called **chromosomes**. Each chromosome consists of DNA strands bound to special proteins around which they coil. The degree of coiling determines whether the chromosome is long and thin or short and fat. Chromosomes in a dividing cell are very tightly coiled. In cells that are not dividing, the DNA is loosely coiled, and appears threadlike. A typical cell in the human body contains 23 pairs of chromosomes.

How does the nucleus control cellular operations? It accomplishes this by regulating the production of proteins. Recall that proteins have two general roles. Some proteins form cell structures, and others form enzymes that regulate the chemical reactions of metabolism. The DNA in our chromosomes contain the information needed to make at least 100,000 different proteins.

THE GENETIC CODE

The basic structure of nucleic acids, DNA and RNA, was described in Chapter 2. ∞ *[p. 30]* A single DNA molecule consists of a pair of strands held together by weak bonds between paired nitrogen bases. Information is stored in terms of the number and order of nitrogen bases along the length of the DNA strands. This information storage system is known as the **genetic code**. Although every organism has different DNA, the reading of the genetic code is the same for all organisms.

If we compare the genetic code to written language, the nitrogen bases are like alphabet letters, and they can be interpreted only in groups, like words. In the "language" of the genetic code, however, all words consist of three letters or three nitrogen bases. Each sequence of three nitrogen bases, called a *triplet*, codes for a single amino acid. Different triplets code for each of the twenty different amino acids, all of which are present in the cytoplasm. A **gene** is a specific length of DNA that contains all the triplets needed to produce a specific protein.

PROTEIN SYNTHESIS

Ribosomes, the organelles of protein synthesis, are found in the cytoplasm, whereas the DNA exists in the nucleus. Because of this separation of protein-building sites (ribosomes) and protein-building information (DNA) in a cell, protein synthesis involves two steps, or processes: *transcription* followed by *translation*.

If we continue with our description of DNA as a "language," we can imagine that the genes are like books found in a library reference section. They cannot be "checked out," but they can be copied and the *copy* can be removed. The process of **transcription** copies the information on the gene, so that it can be taken out of the nucleus (the reference section) to ribosomes in the surrounding cytoplasm. This "copy" is created "word by word" (nitrogen base by nitrogen base) with RNA building blocks (nucleotides) and, in its final form, is a complete RNA molecule. This molecule is known as *messenger RNA (mRNA)* because it serves as a messenger, taking the information to the ribosomes.

Translation is the process that uses the information provided by the messenger RNA to make a protein. At the ribosome, where translation occurs, amino acids are brought in and linked together in the exact order specified by the mRNA. The molecule that brings the amino acids to the ribosome is another type of RNA, appropriately called *transfer RNA (tRNA)*. Figure 3-6• shows how the processes of transcription and translation form a protein molecule.

✔ How does the nucleus control the activities of the cell?

✔ In what parts of the cell do the processes of transcription and translation occur?

•FIGURE 3-6
Protein Synthesis.
The production of new proteins requires the completion of two processes, transcription and translation. Transcription takes place in the nucleus as the DNA genetic code is copied into the form of a messenger RNA (mRNA) molecule. Translation occurs in the cytoplasm on ribosomes, the sites of protein synthesis. During translation the genetic code carried by the mRNA is followed to build a chain of specific amino acids into a protein molecule. Transfer RNA (tRNA) molecules ferry amino acids to the ribosome.

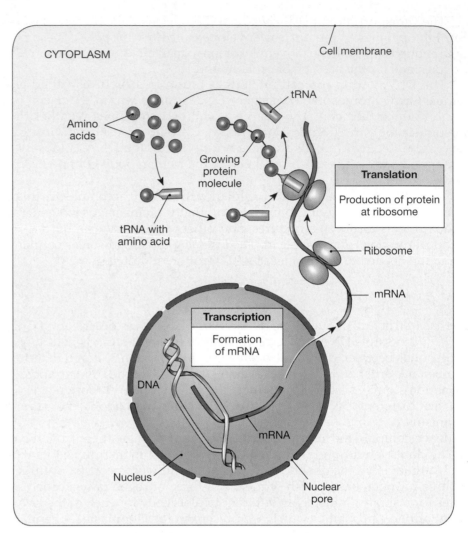

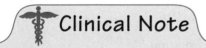 Clinical Note

What Are Mutations?

Mutations are permanent changes in a cell's DNA that affect the nitrogen base sequence of one or more genes. The simplest is a change in a single nitrogen base. With roughly 3 billion pairs of nitrogen bases in the DNA of a human cell, a single mistake might seem relatively unimportant. But a change in the amino acid sequence of a single structural protein or enzyme can prove fatal. For example, several cancers and a potentially lethal blood disorder, *sickle cell anemia*, result from point mutations. Over 100 inherited disorders have been traced to abnormalities in enzyme or protein structure that reflect changes in the nucleotide sequence of specific genes. Mutations can also affect chromosomal structure and even break a chromosome apart.

Mutations are most likely to occur in cells undergoing cell division because a cell must make an exact replica (copy) of its DNA before it divides. (The two copies of the DNA are then distributed to the two daughter cells.) If any mistake exists in the replica, a mutation has occurred. Sometimes a mutation results in a single cell or group of daughter cells. If the mutations occur early in development, however, every cell in the body may be affected. Recent advances in our understanding of the structure and functions of genes have made it possible to diagnose many genetic disorders and even correct a few of them.

CELL GROWTH AND DIVISION

Between the fertilization of an egg and physical maturity a human being goes from a single cell to roughly 75 trillion cells. This amazing increase in number occurs through a form of cellular reproduction called **cell division**. Cell division produces two identical daughter cells, each containing DNA identical to that of the parent cell. Accurate duplication of the cell's genetic material and its distribution to the two new daughter cells is accomplished by a process called **mitosis** (mī-TŌ-sis).

Mitosis occurs during the division of all cells of the body. However, *reproductive cells*, which give rise to sperm or eggs, go through a somewhat different process, called **meiosis** (mī-Ō-sis). Meiosis forms daughter cells with only half the amount of DNA as other body cells. (Remember, these cells may be joined later during fertilization, thereby restoring the proper amount of DNA in the first cell of the new individual.) Meiosis will be described further in Chapter 21.

For most of their lives, cells are in **interphase**, an interval of time between cell divisions when they are growing and performing other normal functions. During interphase, a cell that is going to divide must grow and produce enough organelles and cytoplasm to make two functional cells. Once these preparations have been completed, the cell forms an exact copy of the DNA in the nucleus in a process called *DNA replication*. As a result, two copies of each chromosome exist in the cell. Shortly after DNA replication has been completed, the process of mitosis begins.

THE PHASES OF MITOSIS

Mitosis is divided into four stages: *prophase*, *metaphase*, *anaphase*, and *telophase*. As you read through the descriptions, identify the important features of each phase in Figure 3-7•.

Prophase

As you recall, the DNA within the cell is not always easily visible as distinct chromosomes. During interphase, the DNA exists as long, thin strands extending throughout the nucleus. The hallmark of **prophase** (PRŌ-fāz) is that the chromosomes begin to coil so tightly that they become visible as individual structures, and the nucleoli disappear. As a result of DNA replication during interphase, there are now two copies of each chromosome. Each copy, called a **chromatid** (KRŌ-ma-tid), is connected to its duplicate at a single point.

As the chromosomes appear, the two pairs of centrioles (the original pair was also duplicated) separate and move toward opposite poles of the nucleus. Tubular structures called **spindle fibers** form and extend between the two centriole pairs. Prophase ends with the disappearance of the nuclear envelope.

Metaphase

At the beginning of **metaphase** (MET-a-fāz) the spindle fibers enter the nuclear region, and the chromatids become attached to them. Once attachment is complete, the chromatids move to a narrow central zone of the cell. At the end of metaphase the chromatids are aligned across the middle of the cell.

Anaphase

Anaphase (AN-uh-fāz) begins when each chromatid pair splits and the chromatids separate. The two **daughter chromosomes** are now pulled toward opposite ends of the cell. Anaphase ends when the daughter chromosomes arrive near the centrioles at opposite ends of the cell.

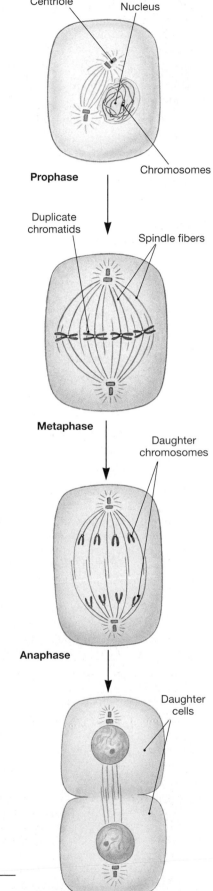

Prophase

Metaphase

Anaphase

Telophase

•**FIGURE 3-7**
Mitosis.

Telophase

During **telophase** (TEL-o-fāz) the new cells prepare to return to the interphase state. Nuclear membranes form around the two nuclei, the nuclei enlarge, and the chromosomes gradually uncoil. Once the chromosomes again become long, thin strands, nucleoli reappear, and the nuclei resemble those of interphase cells.

During telophase, the cytoplasm constricts or pinches in along the central plane of the cell. This is the beginning of the dividing of the cytoplasm that will soon separate the two new cells from each other. This separation marks the end of the process of cell division.

✔ What major event occurs during interphase of cells preparing to undergo mitosis?

✔ What are the four stages of mitosis?

CELL DIVERSITY AND DIFFERENTIATION

Liver cells, fat cells, and nerve cells contain the same chromosomes and genes, but each cell type has a different set of genes available for making proteins. The other genes in the nucleus have been deactivated or "turned off." When a gene is deactivated, the cell loses the ability to create a particular protein and thus to perform any functions involving that protein. Each time another gene switches off and becomes deactivated, the cell's functions become more restricted and specialized. This specialization process is called **differentiation**. Differentation results in the formation of organized collections of cells. These collections are known as *tissues*, and each has different functional roles. The next chapter examines the structure and function of tissues.

✔ How do cells become differentiated, or specialized?

✚ Clinical Note

Drugs and the Cell Membrane

Many medically important drugs affect cell membranes. For example, general anesthetics such as *ether*, *chloroform*, and *nitrous oxide* reduce the sensitivity and responsiveness of nerve cells and muscle cells. Most anesthetics dissolve readily in lipids (fats), and there is a direct correlation between the effectiveness of an anesthetic and its lipid solubility (how easily it dissolves in lipids). Because cell membranes are composed largely of lipids, lipid solubility may speed the drug's passage across the cell membrane and improve its ability to block the movement of ions through protein channels or alter other properties of the cell.

Local anesthetics such as *procaine* and *lidocaine*, as well as *alcohol* and *barbiturate* drugs, are also lipid soluble. These compounds block ion channels and reduce or eliminate the responsiveness of nerve cells to painful (or any other) stimuli. Other drugs interfere with membrane protein receptors for hormones or chemicals that stimulate muscle or nerve cells. *Curare* is a plant extract that interferes with the stimulation of muscle cells. South American Indians use it to coat their hunting arrows so that wounded prey cannot run away. This drug is sometimes given to patients about to undergo surgery to prevent reflexive muscle contractions or twitches while the surgery is being performed.

CELLULAR DISORDERS

ORGANELLES AND DISEASE

Cells perform their varied functions through the coordinated actions of their organelles. All cells have important tasks—some conduct nerve impulses, while others manufacture hormones, build bones, or contract to produce body movements. If important organelles are damaged or otherwise abnormal, cells fail at their particular role, and homeostasis is threatened. Our discussion will highlight disorders that result from abnormal mitochondrial or lysosomal function. Defects in mitochondria and lysosomes are now known to be responsible for a number of diseases.

Mitochondria

There are several inheritable disorders that result from abnormal mitochondrial activity. The mitochondria involved have defective enzymes that reduce their ability to generate energy. Cells throughout the body may be affected, but symptoms involving muscle cells, neurons, and the receptor cells in the eye are most commonly seen because these cells have especially high energy demands.

In some disorders caused by defective mitochondria, the problem appears in only one population of cells. For example, abnormal mitochondria have been found in the nerve cells whose degeneration is responsible for the condition of *Parkinson's disease*, a nervous system disorder characterized by a shuffling gait and uncontrollable tremors.

More often, mitochondria throughout the body are involved. Examples of conditions caused by mitochondrial dysfunction include one class of epilepsies *(myoclonic epilepsy)* and a type of blindness *(Leber's hereditary optic neuropathy)*. These are inherited conditions, but the pattern of inheritance is very unusual. Although men or women may have the disease, only affected women can pass the condition on to their children. All the mitochondria in the body are produced through the replication of mitochondria present in the fertilized egg. Most, if not all, of these mitochondria were provided by the mother, as those carried by the male's sperm disintegrate shortly after fertilization. As a result, children can usually inherit mitochondrial disorders only from their mothers.

Lysosomes

Problems with the production of lysosomes cause more than 30 diseases affecting children. Lysosomes contain a variety of different enzymes, and in each of these conditions one or more of these enzymes are missing. This results in the intracellular buildup of materials normally removed and recycled by lysosomes. Eventually the cell becomes so filled with these waste products that it can no longer function. Three examples of such diseases are Gaucher's disease, Tay-Sachs disease, and glycogen storage disease.

Gaucher's disease is caused by the buildup of a type of *glycolipid* (sugar-lipid) molecule that is found in a variety of cell membranes. This is probably the most common type of lysosomal storage disease. There are two forms of this disease: (1) a form that affects infants, marked by severe nervous system symptoms ending in death, and (2) a form that develops in childhood, with enlargement of the spleen, anemia, pain, and relatively mild nervous system symptoms. Gaucher's disease is most common among the Ashkenazi Jewish population, where it occurs at a frequency of approximately 1 in 1,000 births.

Tay-Sachs disease is another hereditary disorder caused by the inability to break down membrane glycolipids. In this case the glycolipids are a type which are most abundant in neural tissue, especially the brain and spinal cord. Individuals with this condition develop seizures, blindness, dementia, and death, usually by age 3–4. Like Gaucher's disease, Tay-Sachs disease is most common among the Ashkenazi Jewish population, although with less frequency (0.3 cases per 1,000 births).

The **glycogen storage diseases** primarily affect skeletal muscle (muscle attached to bone), heart (cardiac) muscle, and liver cells—the cells that synthesize and store glycogen. ∞ *[p. 28]* In these disorders the cells are unable to use glycogen normally, and large numbers of insoluble glycogen granules build up in the cytoplasm. These granules disrupt the organization of the cell, and also interfere with the transport and synthesis of intracellular materials. In skeletal and heart muscle cells, the buildup leads to a general muscular weakness and potentially fatal heart problems.

CANCER DEVELOPMENT AND GROWTH

Twenty-five percent of all people in the United States develop cancer at some point in their lives. In 1999, an estimated 563,000 people will be killed by some form of cancer in the United States, making this Public Health Enemy Number 2, second only to heart disease.

Tumors and Cancer

A **tumor**, or *neoplasm* ("new growth"), is a tissue mass produced by abnormal cell growth and division. Within normal tissues the rate of mitosis and cell division is balanced by the rate of cell loss. If the rate of cell division becomes greater than that of cell loss, growth occurs; if the cells are abnormal, a tumor forms. Tumors are classified as *benign* or *malignant* based on the characteristics of the tumor cells. In a **benign tumor**, the cells usually remain within a connective tissue capsule. Such a tumor seldom threatens an individual's life, and it can often be removed if its size or position disturbs tissue function. Cells in a **malignant tumor** no longer respond to normal control mechanisms and may spread throughout the body. The term **cancer** refers to an illness characterized by the development of malignant tumors.

Development of Cancer

The steps involved in cancer formation are diagrammed in Figure 3-8•. Initially, the cancer cells are restricted to the initial malignant tumor, known as the **primary tumor**. At first, the growth of the primary tumor distorts the tissue it is developing within, but the surrounding tissue organization remains intact. In most cases, all the cells in the tumor are the daughter cells of a single malignant cell. **Metastasis** (me-TAS-ta-sis), the spread of cancer cells, begins when tumor cells "break out" of the primary tumor and invade the surrounding tissue. They may then enter the lymphatic system and accumulate in nearby lymph nodes. When metastasis involves the penetration of blood vessels, the cancer cells circulate throughout the body.

Responding to cues that are as yet unknown, cancer cells within the circulatory system ultimately escape out of the blood vessels to establish **secondary tumors** at other sites. These tumors enlarge rapidly, and their presence stimulates the growth of blood vessels into the area. The increased circulatory supply provides additional nutrients and further accelerates tumor growth and metastasis.

The growth of blood vessels into the tumor is a vital step in the development and spread of the cancer. Without those vessels, the growth and metastasis of the cancer cells will be limited by the availability of oxygen and nutrients. A peptide called *antiangiogenesis factor* can prevent the growth of blood vessels and can slow the growth of cancers. This peptide, produced in normal human cartilage, can be produced commercially through genetic engineering techniques. It is now being used in experimental cancer therapies.

Organ function begins to deteriorate as the number of cancer cells increases. These cells gradually lose their resemblance to normal tissue cells. They change size and shape, often becoming abnormally large or small. They may not perform their original functions at all, or they may perform normal functions in an unusual way. For example, endocrine cancer cells may produce normal hormones but in excessively large amounts. Cancer cells also use energy very inefficiently. They grow and multiply at the expense of healthy tissues, competing for space and nutrients with normal cells. This competition accounts for the starved appearance of

•**FIGURE 3-8**
The Development of Cancer.

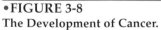

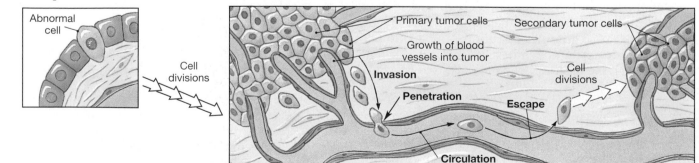

many patients in the late stages of cancer. Death from cancer may occur as a result of the compression of vital organs when nonfunctional cancer cells have killed or replaced the healthy cells in those organs or when the cancer cells have starved normal tissues of essential nutrients.

CAUSES OF CANCER

Relatively few types of cancer are inherited; only 18 types have been identified to date, including two forms of leukemia. Most cancers develop through the interaction of genetic (inherited) and environmental factors, and it is difficult to separate the two completely.

Genetic Factors

Two related genetic factors are involved in the development of cancer: (1) hereditary predisposition and (2) oncogene activation.

An individual born with genes that increase the likelihood of cancer is said to have a hereditary predisposition for the disease. Under these conditions a cancer is not guaranteed, but it becomes a lot more likely. The inherited genes generally affect the ability to neutralize toxins, control mitosis and growth, perform repairs after injury, or identify and destroy abnormal tissue cells. Because these abilities are lost or seriously impaired, body cells become more sensitive to local or environmental factors that would have little effect on normal tissues.

Cancers may also result from mutations of normal somatic cells into cancer cells. (The divisions of *somatic cells* produce all body cells except sperm and eggs; those reproductive cells are produced through the divisions of *sex cells*.) These mutations typically affect genes involved with the regulation of cell growth and division. A mutated gene that causes cancer formation is called an **oncogene** (ON-kō-jēn). In some cases, a viral infection can trigger conversion of a normal gene to an oncogene. For example, one type of human papilloma virus appears to be responsible for many cases of cervical cancer. More than 50 different genes have been identified whose modification can produce a form of cancer.

Our cells also include a variety of anticancer genes—genes that produce proteins that prevent cancer formation. These genes, called *tumor-suppressing genes (TSG)*, inhibit regular cell division and growth. In effect, these genes "put on the brakes" and prevent normal cells from dividing too rapidly or too often. Mutations that inactivate tumor suppressing genes increase the likelihood of developing several blood cell cancers, breast cancer, and ovarian cancer as well as the majority of cancers of the colon and liver.

Environmental Factors

Many cancers can be directly or indirectly attributed to environmental factors called **carcinogens** (kar-SIN-ō-jens). Carcinogens stimulate the conversion of a normal cell to a cancer cell. Common classes of carcinogens include chemicals, radiation, and viruses. Some carcinogens are **mutagens** (MŪ-ta-jens)—that is, they damage DNA strands and may cause chromosomal breakage. Radiation is a mutagen that has carcinogenic effects.

There are many different chemical carcinogens in the environment. Plants manufacture poisons that protect them from insects and other predators, and although their carcinogenic activities are often relatively weak, many common spices, vegetables, and beverages contain compounds that can be carcinogenic if consumed in large quantities. Animal tissues may also store or concentrate toxins, and hazardous compounds of many kinds can be swallowed in contaminated food. A variety of laboratory and industrial chemicals, such as coal tar derivatives and synthetic pesticides, have been shown to be carcinogenic. Cosmic radiation, X-rays, UV radiation, and other radiation sources can also cause cancer. It has been estimated that 70–80 percent of all cancers are the result of chemical and/or environmental factors, and 40 percent are due to a single stimulus: cigarette smoke.

✔ Most mitochondrial diseases are inherited. What is unusual about how mitochondria are passed on to the next generation?

✔ What is the relationship between mitosis and a tumor?

✔ What are the three groups of carcinogens?

Specific carcinogens will affect only those cells capable of responding to that particular physical or chemical stimulus. The responses vary because differentiation produces cell types with specific sensitivities. For example, benzene can produce a cancer of the blood; cigarette smoke, a lung cancer; and vinyl chloride, a liver cancer. Very few stimuli can produce cancers throughout the body; radiation exposure is a notable exception. In general, cells undergoing mitosis are most likely to be vulnerable to chemicals or radiation. As a result, cancer rates are highest where cell divisions occur rapidly, such as the epithelial tissues which cover the body (skin) or line internal surfaces, and lowest where divisions do not normally occur, such as in nervous and muscle tissues.

CHAPTER REVIEW

Key Words

active transport: The energy-requiring absorption or excretion of substances across a cell membrane from an area of low concentration to one of high concentration.

chromosomes: Structures composed of tightly coiled DNA strands that become visible in the nucleus when a cell prepares to undergo mitosis.

cytokinesis (sī-tō-ki-NĒ-sis): The movement of the cytoplasm that separates two daughter cells at the end of mitosis.

cytoplasm: Cellular material that fills the space between the cell membrane and the nucleus.

diffusion: Passive movement of molecules from an area of relatively high concentration to an area of relatively low concentration.

exocytosis (EK-sō-sī-tō-sis): The ejection of cytoplasmic materials from a cell by fusion of a membranous vesicle with the cell membrane.

gene: A portion of a DNA strand that contains the information needed for production of a single protein; found at a particular site on a specific chromosome.

mitochondrion (mī-tō-KON-drē-on): An organelle responsible for generating most of the energy (ATP) required for cellular operations.

mitosis (mī-TŌ-sis): The process by which the DNA of a cell is replicated and divided equally between two daughter cells during cell division

nucleus: Cellular organelle that contains DNA, RNA, and proteins.

osmosis (oz-MŌ-sis): The movement of water across a selectively permeable membrane from a region of lower solute concentration to a region of higher solute concentration.

Selected Clinical Terms

benign tumor: A mass or swelling in which the cells usually remain within a connective tissue capsule.

carcinogen (kar-SIN-ō-jen): An environmental factor that stimulates the conversion of a normal cell to a cancer cell.

malignant tumor: A mass or swelling in which the cells no longer respond to normal control mechanisms but divide rapidly.

metastasis (me-TAS-ta-sis): The spread of cancer cells, from one organ to another, leading to the growth of secondary tumors.

mutagens (MŪ-ta-jenz): Radiation exposure or chemical agents that induce mutations and may be carcinogenic.

oncogene (ON-kō-jen): A cancer-causing gene created by a mutation in a normal gene involved with growth, differentiation, or cell division.

Study Outline

AN OVERVIEW OF THE CELL *(p. 39)*

1. The smallest functional unit of life is a *cell*. The human body is composed of trillions of cells. *(Figure 3-1)*

2. A cell is bathed by *extracellular fluid*. The cell's outer boundary, the **cell membrane**, separates the *cytoplasm* from the extracellular fluid. *(Figure 3-2; Table 3-1)*

THE CELL MEMBRANE *(pp. 40–46)*

1. The cell membrane controls the passage of materials into and out of the cell and is sensitive to the external environment.

MEMBRANE STRUCTURE *(p. 42)*

2. The cell membrane contains lipids, proteins, and carbohydrates. Its major component, phospholipid molecules, form a two-layered structure. *(Figure 3-3)*

MEMBRANE PERMEABILITY *(p. 42)*

3. Cell membranes are **selectively permeable**.

4. Passive processes do not require the expenditure of cellular energy; active processes require energy.

5. Diffusion is the movement of material from an area of higher concentration to an area of lower concentration. Diffusion occurs until **equilibrium** is reached. *(Figure 3-4)*

6. Diffusion of water across a membrane in responses to differences in water concentration is **osmosis**. *(Figure 3-5)*

7. In **filtration**, high pressure forces water across a membrane; if membrane pores are large enough, molecules of solute will be carried along.

8. Active transport processes consume ATP and are independent of concentration gradients. **Ion pumps** are ATP-activated carrier proteins.

9. Movement of materials into the cell can occur by **pinocytosis** ("cell-drinking") and **phagocytosis** ("cell-eating"). Movement out of the cell can occur through **exocytosis**.

THE CYTOPLASM *(pp. 47–48)*

1. The cytoplasm surrounds the nucleus and contains intracellular fluid and organelles.

INTRACELLULAR FLUID *(p. 47)*

2. The intracellular fluid contains raw materials, finished products, and wastes of living cells.

ORGANELLES *(p. 47)*

3. Some **organelles** are surrounded by lipid membranes that isolate them from the intracellular fluid. These organelles include the nucleus, mitochondria, endoplasmic reticulum, the Golgi apparatus, and lysosomes. *(Figure 3-2; Table 3-1)*

4. Other **organelles** are not fully enclosed by membranes. They include the microvilli, centrioles, cilia, flagella, ribosomes, and nucleoli. *(Figure 3-2; Table 3-1)*

5. Microvilli are small projections of the cell membrane that increase the surface area exposed to the extracellular environment.

6. Cilia beat rhythmically to move fluids or secretions across the cell surface.

7. Flagella move a cell through surrounding fluid, rather than moving fluid past a stationary cell.

8. Centrioles direct the movement of chromosomes during cell division.

9. Ribosomes manufacture proteins. They are found in the cytoplasm and attached to the ER.

10. The **endoplasmic reticulum (ER)** is a network of intracellular membranes. There are two types: rough and smooth. **Rough endoplasmic reticulum (RER)** contains ribosomes and is involved in protein synthesis. **Smooth endoplasmic reticulum (SER)** is involved in lipid and carbohydrate synthesis.

11. The **Golgi apparatus** packages cellular products from the ER into membranous sacs called vesicles.

12. Lysosomes are vesicles filled with digestive enzymes. Their functions include ridding the cell of bacteria and debris.

13. Mitochondria, which are surrounded by a double membrane, are the sites of most ATP production within a typical cell.

14. High-energy bonds within adenosine triphosphate, or ATP, provide energy for cellular activities

THE NUCLEUS *(pp. 48–50)*

1. The **nucleus** is the control center for cellular operations. It is surrounded by a *nuclear envelope* containing *nuclear pores*, through which it communicates with the cytoplasm.

2. The nucleus also contains **nucleoli** responsible for ribosome production.

3. The nucleus controls the cell by directing the synthesis of specific proteins using information stored in the DNA of **chromosomes.**

THE GENETIC CODE *(p. 49)*

4. The cell's information storage system, the **genetic code**, is based on a sequence of three nitrogen based, or *triplet* that identifies a single amino acid. Each **gene** consists of all the triplets needed to produce a specific protein.

PROTEIN SYNTHESIS *(p. 49)*

5. Transcription is the process of forming a strand of **messenger RNA (mRNA)**, which carries instructions from the nucleus to the cytoplasm. *(Figure 3-6)*

6. During **translation** a protein is constructed using the information from an mRNA strand. The sequence of nitrogen bases along the mRNA strand determines the sequence of amino acids in the protein. *(Figure 3-6)*

7. Molecules of **transfer RNA (tRNA)** bring amino acids to the ribosomes involved in translation. *(Figure 3-6)*

CELL GROWTH AND DIVISION *(pp. 51–52)*

1. Mitosis is the process by which nonreproductive cells produce new cells. Sex cells (sperm and eggs) are produced by *meiosis*.

2. Most nonreproductive cells spend much of their time in **interphase**. Cells preparing for mitosis undergo *DNA replication* in this phase.

THE PHASES OF MITOSIS *(p. 51)*

3. Mitosis proceeds in four stages: **prophase, metaphase, anaphase**, and **telophase**. *(Figure 3-7)*

4. Mitosis results in the production of two identical daughter cells.

CELL DIVERSITY AND DIFFERENTIATION *(p. 52)*

1. Differentiation produces specialized cells with limited capabilities. These specialized cells form organized collections called *tissues*, each of which has specific functional roles.

CELLULAR DISORDERS *(pp. 52–56)*

ORGANELLES AND DISEASE *(p. 52)*

1. Dysfunctional mitochondria and lysosomes are responsible for some diseases. Mitochondria with defective enzymes have a reduced ability to generate energy. Symptoms are most visible in cells with high energy demands, such as muscle cells, neurons, and the receptor cells in the eye. Defective lysosomes result in the buildup and storage of materials normally removed and recycled from cells.

CANCER DEVELOPMENT AND GROWTH *(p. 53)*

2. Metastasis develops when abnormal cells of a **primary tumor** are distributed by blood vessels of the circulatory system to new locations and they form **secondary tumors**. *(Figure 3-8)*

CAUSES OF CANCER *(p. 55)*

3. The development of cancer is influenced by both genetic and environmental factors. Major environmental factors that act as **carcinogens** include chemicals, radiation, and viruses. Carcinogens that cause mutations are also called **mutagens**.

Review Questions

MATCHING

Match each item in Column A with the most closely related item in Column B. Use letters for answers in the spaces provided.

	Column A		Column B
____	1. nucleus	a.	increase cell surface area
____	2. chromosomes	b.	contains organelles
____	3. mitochondrion	c.	hydrostatic pressure
____	4. filtration	d.	DNA strands
____	5. osmosis	e.	cancer-causing agents
____	6. cytoplasm	f.	contains digestive enzymes
____	7. microvilli	g.	"powerhouse" organelle
____	8. ribosomes	h.	water moves out of cell

___ 9. cell membrane
___ 10. hypertonic solution
___ 11. hypotonic solution
___ 12. isotonic solution
___ 13. lysosomes
___ 14. nucleoli

___ 15. pinocytosis
___ 16. carcinogens
___ 17. tumor

i. abnormal mass of cells
j. cell "drinking"
k. selectively permeable
l. normal saline solution
m. synthesizes ribosomes
n. control center for cellular operations
o. movement of water
p. manufacture proteins
q. water moves into cell

MULTIPLE CHOICE

18. The cell membrane is made up of a double layer of _____.
(a) proteins   (b) phospholipids
(c) DNA   (d) carbohydrates

19. The movement of molecules from an area of greater concentration to an area of lesser concentration best defines _____.
(a) osmosis
(b) diffusion
(c) active transport
(d) filtration

20. Active transport always requires _____.
(a) a concentration gradient
(b) energy
(c) water
(d) electrolytes

21. Structures that perform specific functions within the cell are _____.
(a) organs   (b) organelles
(c) organisms   (d) cytosomes

22. Which of the following cellular structures lacks a membrane? _____.
(a) endoplasmic reticulum
(b) lysosomes
(c) Golgi apparatus
(d) centrioles

23. The synthesis of lipids and carbohydrates occurs chiefly in the _____.
(a) smooth endoplasmic reticulum
(b) rough endoplasmic reticulum
(c) nucleus
(d) nucleolus

24. The _____ is the "recycling" organelle of a cell.
(a) nucleus
(b) rough endoplasmic reticulum
(c) mitochondrion
(d) lysosome

25. The nucleus stores all the information needed to synthesize _____.
(a) carbohydrates   (b) lipids
(c) proteins   (d) phospholipids

26. The number of amino acids coded by a DNA segment made up of 9 nitrogen base pairs is _____.
(a) 9   (b) 6
(c) 3   (d) 1

27. Tumors develop in tissues when the rate of cell division is _____ the rate of cell loss.
(a) equal to   (b) greater than
(c) less than   (d) none of these

28. Tay-Sachs disease is caused by defects in which organelle(s)?
(a) nucleus   (b) mitochondria
(c) endoplasmic reticulum (d) lysosomes

TRUE/FALSE

___ 29. The cytoplasm contains intracellular fluid and organelles.

___ 30. During mitosis the spindle first appears in metaphase.

___ 31. Interphase occurs during the middle of mitosis.

___ 32. Amino acid molecules are carried to ribosomes by tRNA molecules.

___ 33. The ingestion of solid matter by a cell is termed phagocytosis.

___ 34. A mutation is a change in a cell's DNA.

___ 35. Most cancers are inherited.

___ 36. Metastasis is associated with malignant tumors.

SHORT ESSAY

37. What are the four general functions of the cell membrane?

38. How can a cell move materials against a concentration gradient?

39. How does the intracellular fluid differ from the extracellular fluid?

40. Distinguish between transcription and translation.

41. List the stages of mitosis, and briefly describe the events that occur in each.

42. Describe the difference between benign and malignant tumors.

43. Lysosomal diseases are commonly associated with the dangerous buildup of substances in the cell. Why?

APPLICATIONS

44. Defects in mitochondria can lead to disorders that affect the eye and skeletal muscle. Why?

45. Where would a cancer be more likely to start: In the shoulder (deltoid) muscle or in the skin over the same area? Explain your answer.

✔ Answers to Concept Check Questions

(p. 45) **1.** Active transport processes require the use of cellular energy in the form of the high energy bonds of ATP molecules. Passive transport processes (*diffusion*, *osmosis*, and *filtration*) move ions or molecules across the cell membrane without any cellular energy. **2.** In order to transport hydrogen ions (H^+) against their concentration gradient—from a region of lower concentration (the cells lining the stomach) to a region where they are more concentrated (the interior of the stomach)—cellular energy must be expended. An *active transport* process is involved. **3.** This is an example of *phagocytosis*.

(p. 48) **1.** The fingerlike extensions on the surface of intestinal cell are microvilli. They increase the cells' surface area so that they can absorb nutrients more efficiently. **2.** SER functions in the synthesis of lipids such as steroids. Ovaries and testes would be expected to have a great deal of SER because they produce large amounts of steroid hormones. **3.** The function of mitochondria is to produce energy for the cell in the form of ATP molecules. A large number of mitochondria in a cell indicates a high demand for energy.

(p. 49) **1.** The nucleus of a cell contains DNA that codes for production of all the cell's proteins. Some proteins are structural proteins that are responsible for the shape and physical characteristics of the cell. Other proteins are enzymes that govern cellular metabolism, direct the production of cell proteins, and control all of the cell's activities. **2.** Transcription involves copying the sequence of nitrogen bases in a DNA molecule to a mRNA molecule and occurs in the nucleus of a cell. Translation results in the production of a protein molecule from the mRNA molecule. Translation occurs at ribosomes, the sites of protein synthesis within a cell.

(p. 52) **1.** Cells that are preparing for mitosis manufacture additional organelles and duplicate sets of their DNA. **2.** The four stages of mitosis are prophase, metaphase, anaphase, and telophase.

(p. 52) **1.** Cells become differentiated, or specialized for specific functions, as more and more of their genes are deactivated, or "turned-off."

(p. 56) **1.** Mitochondria are capable of their own replication. Most of the mitochondria in the body are produced through the replication of mitochondria present at fertilization. Few if any mitochondria are contributed by the sperm cell. **2.** A tumor is a group or mass of abnormal cells that form as a result of uncontrolled cell division. **3.** Carcinogens include chemicals, radiation, and viruses.

CHAPTER

4

Tissues and Body Membranes

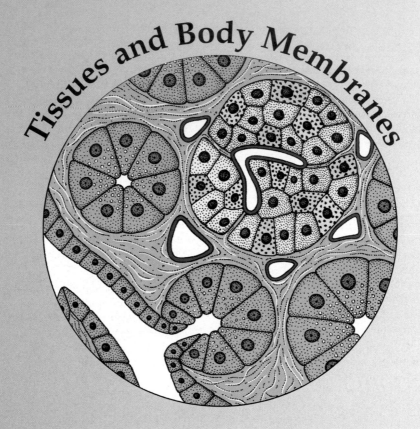

CHAPTER OBJECTIVES

1   Discuss the types and functions of epithelial cells.

2   Describe the relationship between form and function for each epithelial type.

3   Compare the structures and functions of the various types of connective tissues.

4   Explain how epithelial and connective tissues combine to form four different types of membranes, and specify the functions of each.

5   Describe the three types of muscle tissue and the special structural features of each type.

6   Discuss the basic structure and role of nervous tissue.

7   Explain how tissues respond to maintain homeostasis after an injury.

8   Describe how aging affects the tissues of the body.

9   Describe the changes in tissue structure that can lead to a malignant tumor.

10 Describe how cancerous tumors are classified and list the seven warning signs of cancer.

11 Describe the various forms of cancer treatment.

No single body cell contains all the enzymes and the organelles needed to perform the many functions of the human body. Instead, each cell is specialized so it can do a few things very well. Different combinations of specialized cells work together to form **tissues**, which carry out very specific functions within the body. The study, of tissues is called **histology** (derived from the Greek word for tissue, *histos*, his-TOL-ō-jē).

histos, tissue + *logos*, study
histology: the study of tissues

Although there are over 200 different cell types, there are only four basic **tissue types:**

1. *Epithelial* (e-pi-THĒ-lē-al) *tissue* includes *epithelia*, which cover exposed body surfaces and line internal passageways and body cavities, and glands that produce secretions. The surface of the skin is a familiar example of an epithelium.

2. *Connective tissues* fill internal spaces, provide structural support, a means for transport within the body (through blood movement), and energy storage. Bone, fat deposits, and blood are examples of this widely varying tissue.

3. *Muscle tissue* has the ability to contract and produce movement.

4. *Nervous tissue* conducts information from one part of the body to another in the form of electrical impulses.

As you learned in Chapter 1, tissues combine to form the organs of the body. Figure 4-1• shows an example of how the different specializations of the four tissue types enable a particular organ, the heart, to perform its vital function of pumping the blood.

•**FIGURE 4-1**
Tissues of the Heart.

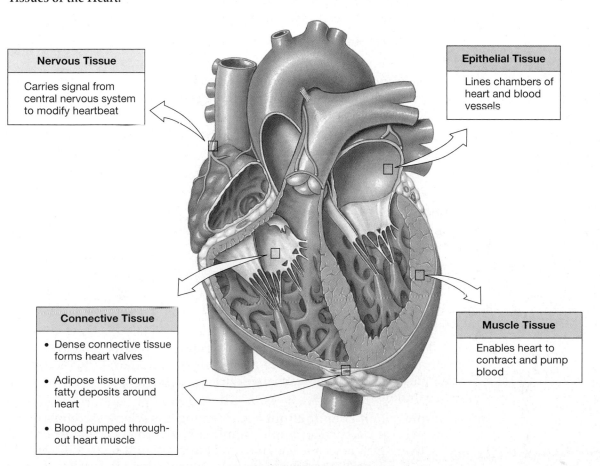

Nervous Tissue

Carries signal from central nervous system to modify heartbeat

Epithelial Tissue

Lines chambers of heart and blood vessels

Connective Tissue

• Dense connective tissue forms heart valves

• Adipose tissue forms fatty deposits around heart

• Blood pumped throughout heart muscle

Muscle Tissue

Enables heart to contract and pump blood

EPITHELIAL TISSUE

An **epithelium** (plural: *epithelia*) is a layer of tightly joined cells that covers an external or internal body surface. All organs within body cavities, as well as all exposed surfaces of the body, are covered by epithelia. The continuous sheetlike structure of an epithelium helps protect these surfaces from abrasion, dehydration, and destruction by chemicals or microorganisms. For example, the epithelium of your skin resists impacts and scrapes, restricts the loss of water, and blocks the invasion of internal tissues by bacteria.

Epithelial tissue also includes **glands**, groups of cells that release, or *secrete*, specialized substances (*secretions*). Glandular secretions include a wide variety of substances such as hormones, digestive enzymes, sweat, and saliva. Epithelial cells that produce secretions are called *gland cells*.

CLASSIFICATION OF EPITHELIA

Epithelia are classified according to the number of cell layers in the sheet and the shape of their outer, or exposed, cells.

There are two types of epithelia based on cell layer; simple and stratified.

1. A **simple epithelium** is composed of a single layer of cells. Because a single layer of cells cannot provide much mechanical protection, simple epithelia are found only in protected areas inside the body. They line internal compartments and passageways, including the body cavities and the interior of the heart and blood vessels.

2. A **stratified epithelium** is made up of several layers of cells and therefore provides a greater degree of protection than simple epithelia. Stratified epithelia are usually found in areas subject to mechanical or chemical stresses, such as the surface of the skin and the linings of the mouth and anus.

The cells at the surface of the epithelium usually have one of the following three basic shapes:

1. In a **squamous** (SKWĀ-mus) **epithelium** the cells are thin and flat. Viewed from the surface, the cells look like fried eggs laid side by side. Viewed from the side, they appear flat.

2. The cells of a **cuboidal epithelium** are cube-shaped or square when viewed from the side.

3. The cells of a **columnar epithelium** are shaped somewhat like rectangular bricks, appearing tall and slender when viewed from the side.

Combinations of the two basic epithelial layers (simple and stratified) and the three possible cell shapes (squamous, cuboidal, and columnar) are used to describe almost every epithelium in the body.

TYPES OF EPITHELIA

The major types of epithelial tissues and examples of where they are found in the body are shown in Figure 4-2•.

- A delicate **simple squamous epithelium** is found in protected regions where absorption takes place or where a slick, slippery surface reduces friction. Examples include the gas exchange surfaces of the lungs and the lining of the ventral body cavities.

- A **simple cuboidal epithelium** occurs in regions where secretion or absorption takes place. For example, simple cuboidal epithelia in the pancreas and salivary glands secrete enzymes and buffers. Simple cuboidal epithelia also line many portions of the kidneys involved in the production of urine.

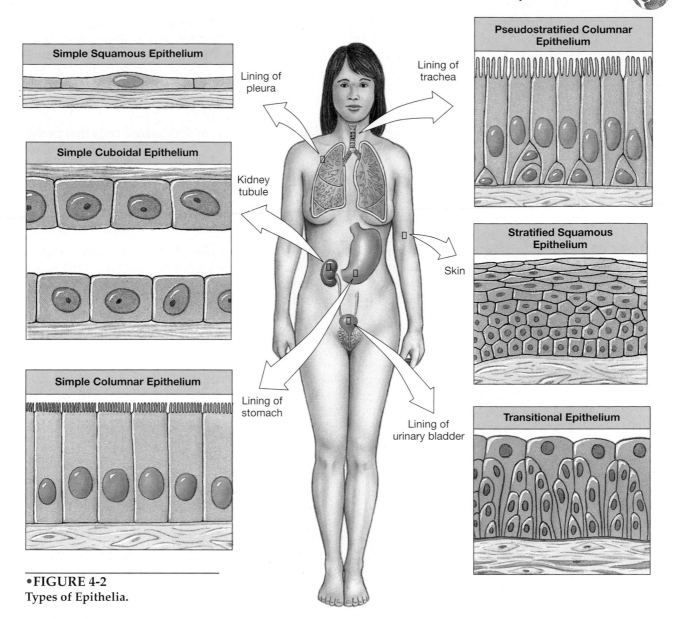

•FIGURE 4-2
Types of Epithelia.

- A **simple columnar epithelium** provides some protection and may also be encountered in areas where absorption or secretion occurs. This type of epithelium lines the stomach, the intestinal tract, and many excretory ducts.

- Portions of the respiratory tract contain a columnar epithelium that includes a mixture of cell types. Because the nuclei are situated at varying distances from the surface, the epithelium has a layered appearance. Because it gives the "false" appearance of being stratified, it is known as a **pseudostratified columnar epithelium**. A ciliated pseudostratified columnar epithelium lines most of the nasal cavity, the trachea (windpipe), and bronchi. This ciliated epithelium moves mucus-trapped irritants away from the lungs and toward the throat for disposal.

- A **stratified squamous epithelium** is found where mechanical stresses are severe. The surface of the skin and the lining of the mouth, esophagus, and anus are good examples.

- A **transitional epithelium** lines the ureters and urinary bladder, where significant changes in volume occur and the lining is frequently stretched. In an empty urinary bladder the epithelium seems to have many layers. The layered appearance results from overcrowding; the actual structure of the epithelium can be seen in the full bladder, when the pressure of the urine has stretched and thinned the lining.

GLANDULAR EPITHELIA

Many epithelia contain *glands*, which are groups of specialized cells that produce secretions. Glands are classified according to where they discharge their secretions.

exo, outside + *krinein*, to secrete
exocrine glands: glands whose secretions are carried by ducts to an epithelial surface

endo, inside + *krinein*, to secrete
endocrine glands: glands whose secretions enter body fluids

- **Exocrine** glands discharge their secretions through a duct (tube) onto the surface of the skin or other epithelial surface. The word *exocrine* means "to secrete outside." Enzymes entering the digestive tract, perspiration on the skin, and milk produced by mammary glands are examples of exocrine secretions.

- **Endocrine** glands release their secretions directly into the surrounding tissues and blood. These secretions, called *hormones*, regulate or coordinate the activities of other tissues, organs, and organ systems. Also known as *ductless glands*, endocrine glands include the pancreas, thyroid, and pituitary gland. (Hormones and endocrine glands will be discussed further in Chapter 13.)

EPITHELIAL REPLACEMENT

✔ You look at a tissue under a microscope and see a simple squamous epithelium. Can it be a sample of the skin surface? Why or why not?

✔ Where would you find exocrine secretions? endocrine secretions?

✔ What is the function of epithelial stem cells?

Life isn't easy for an epithelial cell. Exposed epithelial cells survive for just a day or two, for they are regularly lost or destroyed by disruptive enzymes, toxic chemicals, harmful bacteria, or mechanical abrasion. The only way the epithelium as a whole can survive is by replacing itself over time through the continual division of unspecialized cells known as **stem cells,** or *germinative cells,* found in the deepest layers of the epithelium. Unlike most dividing cells, a stem cell does not produce two identical daughter cells when it divides. Instead, one daughter cell specializes—in this case, to become an exposed epithelial cell—while the other remains a stem cell and prepares for further cell divisions. This ensures a continuous supply of "new" cells to replace those lost at the epithelial surface.

CONNECTIVE TISSUES

Connective tissues are the most diverse of the body's four tissue types. For example, as unrelated as they may seem, bone, blood, and fat are all connective tissues. Despite their widely varying functions and properties, all connective tissues are made up of three basic elements: (1) some kind(s) of specialized cells; (2) extracellular protein fibers produced by the specialized cells; and (3) a *ground substance* that varies in consistency from a fluid to a solid. Together the extracellular protein fibers and ground substance make up the **matrix** that surrounds the cells. Whereas epithelial tissue consists almost entirely of cells, the matrix accounts for most of the volume of connective tissues.

CLASSIFICATION OF CONNECTIVE TISSUES

Three classes of connective tissue are recognized based on the physical properties of their matrix: *connective tissue proper*, which has a syrupy matrix; *fluid connective tissues*, which have a fluid matrix; and *supporting connective tissues*, which have a solid matrix. Figure 4-3• provides a brief overview of this classification.

CONNECTIVE TISSUE PROPER

Connective tissue proper, which includes the most abundant connective tissues in the body, refers to connective tissues with many types of cells and fibers sur-

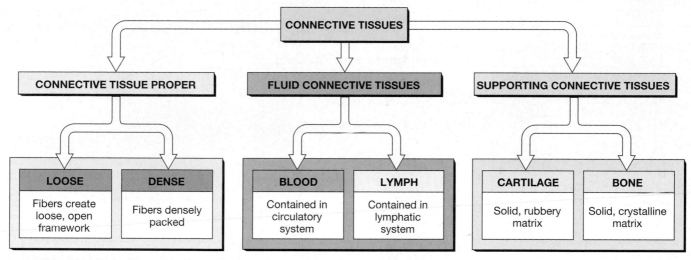

•FIGURE 4-3
Classification of Connective Tissue.

rounded by a syrupy ground substance. The most common fibers in connective tissue proper are long, straight, unbranched **collagen fibers**, which provide strength and flexibility. Connective tissue proper serves to support and protect the body as well as to store energy. Examples include the tissue that underlies the skin (the dermis), fatty tissue, and tendons and ligaments.

Connective tissue proper is divided into loose connective tissues and dense connective tissues.

Loose Connective Tissue

Loose connective tissue, also known as **areolar** (ah-RĒ-o-lar) **tissue**, is the packing material of the body. (*Areolar* means "containing small spaces.") These tissues fill spaces between organs; they provide cushioning and support epithelia. The least specialized connective tissue in the adult body, it contains all the kinds of cells and fibers found in connective tissue proper, in addition to many blood vessels.

Adipose tissue, or fat, is a loose connective tissue containing large numbers of fat cells. Adipose tissue provides another source of padding and shock absorption for the body. It also acts as an insulating blanket that slows heat loss through the skin and functions in energy storage.

Fat cells are metabolically active cells—their lipids are continually being broken down and replaced. When nutrients are scarce, fat cells deflate like collapsing balloons. This is what occurs during a weight-loss program. Because the cells are not killed but merely reduced in size, the lost weight can easily be regained in the same areas of the body.

Dense Connective Tissues

Dense connective tissues are tough, strong, and durable. They resist tension and distortion, and interconnect bones and muscles. They also form *capsules* that enclose joint cavities and surround organs, such as the liver, kidneys, and spleen. Because the dense connective tissues consist mostly of collagen fibers and very little ground substance, they are also referred to as *fibrous tissues*. **Tendons** are cords of dense connective tissue that attach skeletal muscles to bones. The collagen fibers run along the length of the tendon and transfer the pull of the contracting muscle to the bone. **Ligaments** (LIG-a-ments) are bundles of fibers that connect one bone to another. Because ligaments often contain elastic fibers as well as collagen fibers, they can tolerate a modest amount of stretching.

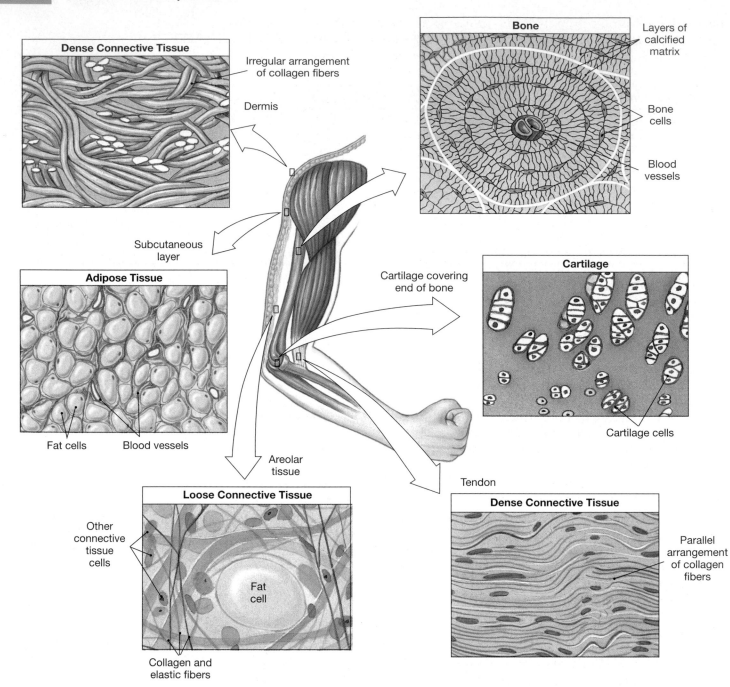

Dense Connective Tissue

Irregular arrangement of collagen fibers

Dermis

Bone

Layers of calcified matrix

Bone cells

Blood vessels

Subcutaneous layer

Adipose Tissue

Fat cells   Blood vessels

Cartilage covering end of bone

Cartilage

Cartilage cells

Areolar tissue

Tendon

Loose Connective Tissue

Other connective tissue cells

Fat cell

Collagen and elastic fibers

Dense Connective Tissue

Parallel arrangement of collagen fibers

•**FIGURE 4-4**

Types of Connective Tissue.
Examples of connective tissue proper and supporting connective tissues are shown here. Fluid connective tissues are not illustrated.

FLUID CONNECTIVE TISSUES

Fluid connective tissues have a distinctive population of cells suspended in a matrix made up of a watery ground substance that contains dissolved proteins. There are two fluid connective tissues, blood and lymph.

Blood, the circulating fluid responsible for the transport of most materials throughout the body, is confined to the vessels of the cardiovascular system. Blood contains *red blood cells*, which carry oxygen; *white blood cells*, which fight infection; and *platelets*, enzyme-filled cell fragments that function in blood clotting. All these are carried in the **plasma** (PLAZ-muh), the watery ground substance of blood, along with various dissolved proteins. Blood and its components will be discussed in Chapter 14.

The lymphatic system works closely with the circulating blood of the cardiovascular system. It picks up fluid that has been squeezed out of the smallest blood vessel and returns it to the blood circulation. When carried by vessels of the lymphatic system, this fluid is known as **lymph**. Between the blood and lymphatic vessels, the leaked fluid fills the spaces around the cells of other tissues and is called *interstitial* (in-ter-STISH-al) *fluid*. Blood, lymph, and interstitial fluids are examples of *extracellular fluid*—fluid that is inside the body but outside of cells. Before the lymph is returned to the blood circulation, it is inspected by different classes of infection-fighting white blood cells within the lymphatic system.

The lymphatic system and its population of infection-fighting cells will be discussed in Chapter 16. The overall balance of fluids in the body will be discussed in Chapter 20.

SUPPORTING CONNECTIVE TISSUES

Cartilage and bone are called **supporting connective tissues** because they provide a strong framework that supports the rest of the body. These tissues have fewer types of cells than other connective tissues, a matrix of dense ground substance, and closely packed fibers. Because it contains mineral deposits, primarily made of calcium, the fibrous matrix of bone is said to be **calcified**. Calcified tissues are typically hard. It is these mineral deposits which give bone its strength and rigidity.

Cartilage

Cartilage is found in the familiar outer portions of the nose and ear, where it provides support and flexibility. It is also located between the bones of the spine and at the ends of limb bones, where it protects the bones by reducing friction and acting as a shock absorber. The matrix of cartilage consists of a firm gel containing embedded fibers. The cartilage-forming cells live in small pockets within the matrix (see Figure 4-4•). Because cartilage lacks blood vessels, these cells must obtain nutrients and eliminate waste products by diffusion through the matrix. However, most structures made of cartilage are covered by a **perichondrium** (pe-re-KON-drē-um). This layer of fibrous connective tissue contains blood vessels that bring nutrients to the surface of the matrix.

peri, around + *chondros*, cartilage
perichondrium: a layer of fibrous connective tissue that surrounds a cartilage

Different types of cartilage are common in movable joints of the body. For example, the knee is an extremely complex joint that contains a considerable amount of cartilage. One type of cartilage covers bony surfaces, and another forms pads within the joint that prevent bone contact when movements are underway. Many sports injuries involve tearing of the cartilage pads. This loss of cushioning places more strain on the cartilages within joints and leads to further joint damage. Because cartilages lack blood vessels, they heal quite slowly.

Bone

The minerals and fibers produced by bone cells consists of hard calcium compounds and flexible collagen fibers. This combination gives bone, or **osseous** (OS-ē-us) **tissue**, truly remarkable properties, making it both strong and resistant to shattering. In its overall properties, bone can compete with the best steel-reinforced concrete.

The general organization of bone can be seen in Figure 4-4•. Bone cells live in small pockets within bone. Like the circular pattern of a target on a shooting range, these pockets are arranged in circles surrounding the blood vessels that branch through the bony matrix. Diffusion cannot occur through the bony matrix, but the bone cells obtain nutrients through cellular extensions that reach blood vessels and other bone cells. Unlike cartilage, bone is constantly being changed, or remodeled. Because of this continual activity, complete repairs can be made even after severe damage has occurred.

✔ Which two types of connective tissue have a liquid matrix?

✔ Lack of vitamin C in the diet interferes with the body's ability to produce collagen. What effect might this have on connective tissue?

✔ The repair capability of cartilage is low compared with that of bone. What accounts for this difference?

BODY MEMBRANES

Epithelia and connective tissues do not function independently in the body. These tissues combine to form membranes that cover and protect other structures and tissues in the body. There are four **body membranes**: the *cutaneous membrane, mucous membranes, serous membranes,* and *synovial membranes* (Figure 4-5•).

CUTANEOUS MEMBRANE

The **cutaneous membrane** of the skin covers the body surface. It consists of a stratified squamous epithelium and the underlying connective tissues. In contrast with serous or mucous membranes, the cutaneous membrane is thick, relatively waterproof, and usually dry. The skin is discussed in Chapter 7.

MUCOUS MEMBRANES

Mucous membranes, also called *mucosae* (singular: *mucosa*), line cavities that open to the exterior, including the digestive, respiratory, reproductive, and urinary

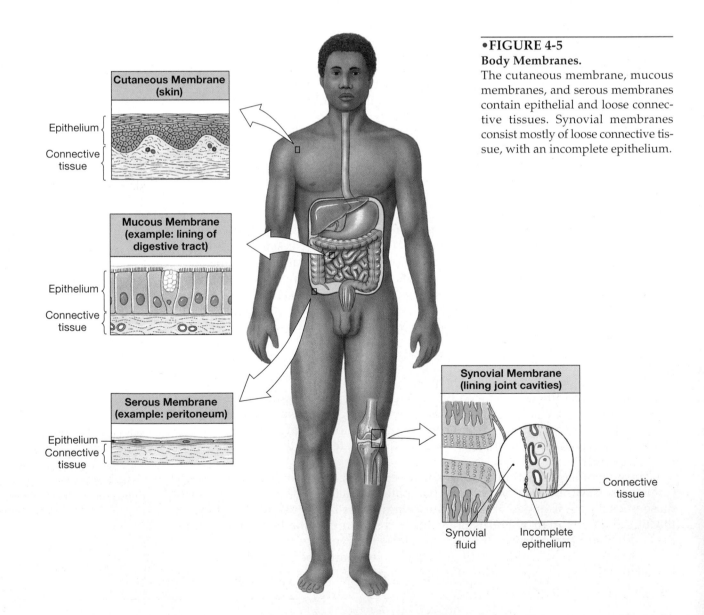

•FIGURE 4-5
Body Membranes.
The cutaneous membrane, mucous membranes, and serous membranes contain epithelial and loose connective tissues. Synovial membranes consist mostly of loose connective tissue, with an incomplete epithelium.

tracts. The epithelial surfaces are kept moist at all times, either by secretions of mucus by mucous glands or by exposure to fluids such as urine or semen.

Many mucous membranes are lined by simple epithelia that perform absorptive or secretory functions, such as the simple columnar epithelium of the digestive tract. However, other types of epithelia may be involved. For example, a stratified squamous epithelium covers the mucous membrane of the mouth, and the mucous membrane along most of the urinary tract contains a transitional epithelium.

SEROUS MEMBRANES

Serous membranes, also called *serosae* (singular: *serosa*), line the sealed, internal cavities of the body. There are three serous membranes, each consisting of a simple epithelium supported by loose connective tissue. The **pleura** (PLOO-ra) lines the pleural cavities and covers the lungs. The **peritoneum** (pe-ri-tō-NĒ-um) lines the peritoneal cavity and covers the surfaces of enclosed organs such as the liver and stomach. The **pericardium** (pe-ri-KAR-dē-um) lines the pericardial cavity and covers the heart.

Every serous membrane has two portions, **parietal** and **visceral**. ∞ *[p. 12]* The parietal portion lines the outer wall of the internal chamber or cavity, and the visceral portion covers organs within the body cavity. For example, the visceral pericardium covers the heart, and the parietal pericardium lines the inner surfaces of the pericardial sac that surrounds the pericardial cavity.

A thin, watery *serous fluid* is formed by fluids diffusing from underlying tissue. The presence of serous fluid on the surfaces of the visceral and parietal membranes minimizes the friction between these opposing surfaces. Excess serous fluid may buildup within the ventral body cavities if serous membranes are injured, irritated, or infected. The increase in fluid volume can compress vital organs and disrupt their normal functioning.

peri, around + *kardia*, heart
pericardium: the serous membrane that lines the pericardial cavity

SYNOVIAL MEMBRANES

Bones of the skeleton contact one another at joints. The connective tissue at a joint may restrict or enhance its movement. At freely movable joints, the bony surfaces do not actually touch. If they did, abrasion and impacts would quickly damage them, and smooth movement would be almost impossible. Instead, the ends of the bones are covered with cartilage and separated by a thick *synovial fluid* produced by the **synovial** (sin-Ō-vē-al) **membrane** that lines the joint cavity. Unlike the other three membranes, the synovial membrane consists primarily of loose connective tissue, and the epithelial layer is incomplete.

✔ Serous membranes produce fluids. What is the function of such fluids?

✔ Why is the same stratified epithelial organization found in the mucous membranes of the pharynx, esophagus, anus, and vagina?

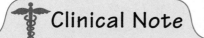 **Clinical Note**

What Can Go Wrong with Serous Membranes?

Infection and chronic irritation can cause the formation of *adhesions* (fibrous connections between opposing membranes that lock them together and prevent motion and friction) and the abnormal buildup of fluid within the lung, heart, or abdominal cavities.

Pleuritis, or *pleurisy,* is an inflammation of the pleural membranes of the lung cavities. At first, the membranes become dry, and there is friction between the opposing membranes. As the inflammation continues, fluid begins to accumulate in the pleural cavities, producing a condition called *pleural effusion.* As fluid builds up, the lungs are compressed and breathing becomes difficult.

Pericarditis is an inflammation of the pericardium. When sudden or severe, the fluid buildup can seriously reduce the efficiency of the heart and restrict blood flow through major blood vessels.

Peritonitis, an inflammation of the peritoneum, can follow an infection of or injury to the peritoneal lining. Peritonitis is a potential complication of any surgical procedure in which the peritoneal cavity is opened. Liver disease, kidney disease, or heart failure can also cause an increase in the rate of fluid movement into the peritoneal cavity. *Ascites* (a-SĪ-tēz), the accumulation of peritoneal fluid, creates a characteristic abdominal swelling. Although often associated with peritonitis, it can have several other causes. The distortion of internal organs by the excess peritoneal fluid often results in shortness of breath, heartburn, and indigestion.

MUSCLE TISSUE

Muscle tissue is specialized in its ability to shorten, or contract. Muscle contraction will be described in more detail in Chapter 9.

There are three different types of muscle tissue: *skeletal muscle, cardiac muscle,* and *smooth muscle* (Figure 4-6●).

SKELETAL MUSCLE TISSUE

Skeletal muscle tissue is made up of very large, long cells, or *muscle fibers*, tied together by loose connective tissue. Each muscle fiber contains several nuclei. The collagen and elastic fibers surrounding each muscle cell and group of muscle cells blend into those of a tendon that conducts the force of contraction, usually to a bone of the skeleton. Contractions of skeletal muscle tissue thus cause bone movement.

Because the contractile proteins within a skeletal muscle fiber are arranged in organized groups, these cells have a series of bands, or *striations*. Skeletal muscles will not usually contract unless stimulated by a signal from a nerve. Because the nervous system provides voluntary control over these contractions, skeletal muscle is known as *striated voluntary muscle*.

CARDIAC MUSCLE TISSUE

Cardiac muscle tissue is found only in the heart. Although they have striations like skeletal muscle fibers, cardiac muscle cells are much smaller. Each cardiac mus-

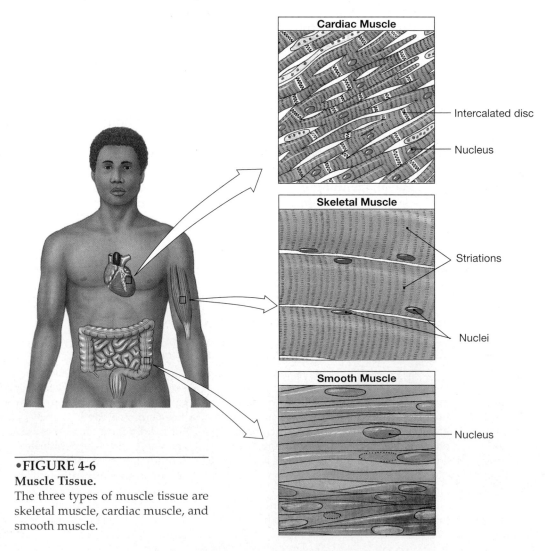

•FIGURE 4-6
Muscle Tissue.
The three types of muscle tissue are skeletal muscle, cardiac muscle, and smooth muscle.

cle cell usually has a single nucleus. Cardiac muscle cells are interconnected at *intercalated discs*, specialized attachment sites that allow free communication between adjacent cells. The muscle cells branch, forming a network that efficiently conducts the force and stimulus for contraction from one area of the heart to another.

Unlike skeletal muscle, cardiac muscle does not rely on nerve activity to start a contraction. Instead, specialized cells, called *pacemaker cells*, establish a regular rate of contraction. Although the nervous system can alter the rate of pacemaker activity, it does not provide voluntary control over individual cardiac muscle cells. Cardiac muscle is known as *striated involuntary muscle*.

SMOOTH MUSCLE TISSUE

Smooth muscle tissue is found in the walls of blood vessels; around hollow organs such as the urinary bladder; and in layers within the walls of the respiratory, circulatory, digestive, and reproductive tracts.

A smooth muscle cell has one nucleus; the cell is small and slender, and tapers to a point at each end. Unlike skeletal and cardiac muscle, the contractile proteins in smooth muscle cells are scattered throughout the cytoplasm so there are no striations.

Smooth muscle cells may contract on their own, or their contractions may be triggered by the nervous system. The nervous system usually does not provide voluntary control over smooth muscle contraction, so smooth muscle is known as *nonstriated involuntary muscle*.

NERVOUS TISSUE

Nervous tissue is specialized for the rapid conduction of electrical impulses that convey information or instructions from one region of the body to another. Most of the nervous tissue (98 percent) is concentrated in the brain and spinal cord, the control centers for the nervous system.

Nervous tissue contains two basic types of cells: **neurons** (NOO-ronz), and **neuroglia** (noo-ROG-lē-a) (Figure 4-7•). The names of these cells come from *neuro-* which means "nerve," and *glia* which means "glue." Neurons, the functional units of neural tissue, transmit information in the form of electrical events at their cell membranes. The several different kinds of neuroglia provide physical support for the neurons, maintain the chemical composition of the tissue fluids, and defend the tissue from infection.

✔ What type of muscle tissue has small, spindle-shaped cells with single nuclei and no obvious banding pattern?

✔ Voluntary control of muscle contraction is restricted to which type of muscle tissue?

✔ Why are skeletal muscle cells and the axons of neurons also called fibers?

✔ In what form is information sent by the neurons from one area of the body to another?

•**FIGURE 4-7**
Nervous Tissue.

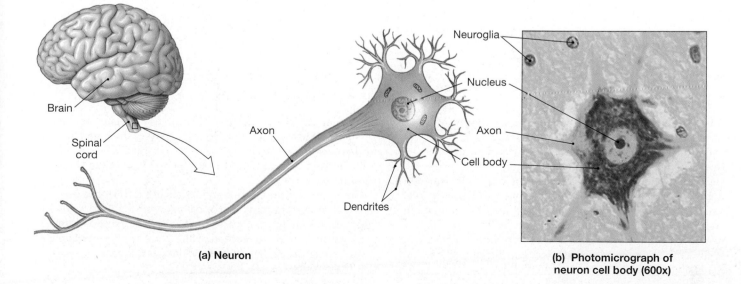

(a) Neuron

(b) Photomicrograph of neuron cell body (600x)

A typical neuron has a **cell body**, which contains the nucleus; a number of branch-like cytoplasmic extensions called **dendrites** (DEN-drīts) (a *dendron* is a tree); and a long single extension called the **axon** (Figure 4-7●). Stimulation alters the properties of the cell membrane, usually near one of the dendrites, eventually producing an electrical impulse that is conducted along the length of the axon. Axons, which may reach a meter in length, are often called *nerve fibers*.

TISSUE INJURY AND REPAIR

Tissues in the body are not independent of each other; they combine to form organs with diverse functions. Any injury to the body affects several different tissue types simultaneously, and these tissues must respond in a coordinated manner to repair themselves and preserve homeostasis.

Restoring homeostasis after a tissue injury involves two related processes. First, the area is isolated from neighboring healthy tissue while damaged cells, tissue components, and any dangerous microorganisms are cleaned up. This phase, which coordinates the activities of several different tissues, is called **inflammation**. Inflammation begins immediately following an injury and produces several familiar sensations, including swelling, warmth, redness, and pain. An *infection* is an inflammation resulting from the presence of pathogens, such as bacteria.

Second, the damaged tissues are replaced or repaired to restore normal function. This repair process is called **regeneration**. Inflammation and regeneration are controlled at the tissue level. The two phases overlap; isolation of the area of damaged tissue establishes a framework that guides the cells responsible for reconstruction, and repairs are underway well before cleanup operations have ended.

TISSUES AND AGING

Tissues change with age, and there is a decrease in the speed and effectiveness of tissue repairs. In general, repair and maintenance activities throughout the body slow down, and a combination of hormonal changes and alterations in lifestyle affect the structure and chemical composition of many tissues. Epithelia get thinner, and connective tissues more fragile. Individuals bruise more easily and bones become relatively brittle; joint pains and broken bones are common complaints. Cardiac muscle cells and neurons are seldom replaced, and cumulative losses from relatively minor damage can contribute to major health problems such as cardiovascular disease or deterioration in mental function. Cancer rates also increase with age because of the individual's cumulative exposure to *carcinogens*, or cancer-causing agents. In future chapters we will consider the effects of aging on specific organs and systems.

✚ Clinical Note

What Is a Biopsy?

A **biopsy** (BĪ-op-sē) is the removal and examination of tissue from the body for the diagnosis of disease. It is commonly used for the diagnosis of cancer because the difference between noncancerous (*benign*) and cancerous (*malignant*) tissue is especially visible. The type of biopsy performed depends on the organ or tissue of concern. For example, small samples of skin or muscle tissue may be removed in skin and muscle biopsies. In an *aspiration biopsy*, a hollow needle is used to suck up tissue from a tumor. The inside of hollow organs, such as the stomach, may be viewed with an instrument called an *endoscope*. Tiny forceps attached to the end of the endoscope are used to snip a part of the organ lining. In other biopsies, a lump in a tissue or organ is removed, or the body cavity may need to be opened to obtain a sample of tissue.

The sample is then processed by slicing it into thin sections. These sections may then be colored with stains to highlight specific structures or features. The processed sections are then viewed with a microscope, which uses a series of glass lenses to magnify images by as much as 1,000 times.

TISSUE STRUCTURE AND DISEASE

Diseases may often affect the structure and organization of tissues, membranes, such as the serous membranes lining body cavities, and organs. **Pathologists** are physicians who specialize in the study of disease processes. Diagnosis, rather than treatment, is usually their main focus. Diagnosis of many diseases relies on the careful examination of samples of tissue, or *biopsies*, for abnormal structural organization. ∞ [p. 72] Such examinations provide information on the nature and severity of diseases. Some changes in tissue structure result from environmental factors or infection; others reflect abnormal gene activity in the tissue cells. It is, however, often difficult to separate these factors since environmental factors can alter gene activity, and a change in genetic activity often alters the local tissue environment.

For example, consider Figure 4-8•, which diagrams the tissue changes induced by exposing the lining of the trachea (windpipe) and its branches to cigarette smoke, a relatively common irritating stimulus. The first abnormality to be observed is **dysplasia** (dis-PLĀ-zē-uh), a change in the normal shape, size, and organization of tissue cells. Dysplasia is usually a response to chronic irritation or inflammation, and the changes are reversible. The normal trachea is lined by pseudostratified, ciliated, columnar epithelium (p. 63). The cilia move a mucous layer that traps foreign particles and moistens incoming air. The drying and chemical effects of smoking first paralyze the cilia, stopping the movement of mucus (Figure 4-8a•). As mucus builds up, the individual coughs to dislodge it (the well-known "smoker's cough").

Epithelia and connective tissues may undergo more radical changes in structure, caused by the differentiation and division of stem cells. **Metaplasia** is a change in structure that dramatically alters the normal functioning of the tissue. In our example, heavy smoking first paralyzes the cilia, and over time the epithelial cells lose their cilia altogether. As metaplasia occurs, a stratified squamous epithelium replaces the normal windpipe epithelium (Figure 4-8b•). Although the protective function remains, the new epithelium completely eliminates the moisturizing and cleaning functions provided by the original ciliated epithelium. The cigarette smoke will now have an even greater impact on more delicate portions of the respiratory tract. Fortunately, metaplasia is reversible, and the epithelium gradually returns to normal once the individual quits smoking.

Dysplasia and metaplasia are reversible responses to tissue stresses. During **anaplasia** (a-nuh-PLĀ-zē-uh) (Figure 4-8c•) there are irreversible structural changes, and tissue organization breaks down. The tissue cells change size and shape, often becoming unusually large or abnormally small. In anaplasia, which occurs in smokers developing one form of lung cancer, the cells divide more frequently, but not all divisions proceed in the normal way. Oncogenes are now activated, and unlike dysplasia and metaplasia, anaplasia is irreversible. ∞ [p. 55]

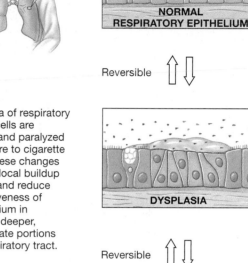

Irritant chemicals and particles in smoke

NORMAL RESPIRATORY EPITHELIUM

Reversible

(a) The cilia of respiratory epithelial cells are damaged and paralyzed by exposure to cigarette smoke. These changes cause the local buildup of mucus and reduce the effectiveness of the epithelium in protecting deeper, more delicate portions of the respiratory tract.

DYSPLASIA

Reversible

(b) In metaplasia, a tissue changes its structure. In this case the stressed respiratory surface converts to a stratified epithelium that protects underlying connective tissues but does nothing for other areas of the respiratory tract.

METAPLASIA

Irreversible

(c) In anaplasia, the tissue cells become tumor cells; anaplasia produces a cancerous tumor.

ANAPLASIA

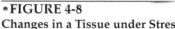

•FIGURE 4-8
Changes in a Tissue under Stress.

CANCER CLASSIFICATION AND TREATMENT

The process of anaplasia leads to the formation of malignant, or cancerous, tumors. As noted earlier, tumors may be benign or malignant. ∞ *[p. 54]* Benign tumors are usually not considered life-threatening because their cells grow in a compact mass within a capsule of dense connective tissue and do not metastasize. If, however, their growth impacts the functions of surrounding organs they are just as harmful as a malignant tumor whose cells spread throughout the body. Examples of benign tumors are included in Table 4-1; most of the names end in *–oma*, which means tumor.

Physicians who specialize in the identification and treatment of cancers are called **oncologists** (on-KOL-o-jists). (The term *onkos* means "mass," referring to the group of cells making up a tumor.) Pathologists and oncologists classify cancers according to their cellular appearance and their sites of origin. Over a hundred kinds have been described, but broad categories are usually used to indicate the location of the primary tumor. For example, **carcinomas** (kar-si-NŌ-mahs) develop in epithelial tissues, **sarcomas** (sar-KŌ-mahs) develop in connective tissue (sarcomas are further distinguished by the location of the connective tissue in which they develop), **leukemias** (loo-KĒ-mē-uhs) develop in blood-forming tis-

Table 4-1	Benign and Malignant (Cancerous) Tumors in the Major Tissue Types
Tissue	**Description**
Epithelia	
Papilloma (pap-i-LŌ-mah)	Benign; a tumor that projects from the surface (a wart)
Nevus (NĒ-vus)	Benign; a small, pigmented tumor of the skin (a mole)
Adenoma (ad-eh-NŌ-mah)	Benign; a tumor of the glandular epithelium
Angioma (an-jē-Ō-mah)	Benign; a tumor of blood vessels or lymph vessels
Carcinoma (kar-si-NŌ-mah)	Any cancer of epithelial tissue
Adenocarcinoma (ad-eh-nō-kar-si-NŌ-mah)	Cancers of glandular epithelium
Angiosarcoma (an-jē-ō-sar-KŌ-mah)	Cancers of blood vessels or lymph vessels
Connective Tissues	
Fibroma (fī-BRŌ-mah)	Benign: a tumor of collagen-producing fibroblast cells
Lipoma (lip-Ō-mah)	Benign; a tumor of adipose (fatty) connective tissue
Liposarcoma (lī-pō-sar-KŌ-mah)	Cancer of adipose tissue
Leukemia (loo-KĒ-mē-ah)	Cancer of blood-forming tissues
Lymphoma (lim-FŌ-mah)	Cancer of lymphoid tissues
Chondroma (kon-DRŌ-mah)	Benign; a tumor of cartilage
Chondrosarcoma (kon-drō-sar-KŌ-mah)	Cancer of cartilage
Osteoma (os-tē-Ō-mah)	Benign; a bone tumor
Osteosarcoma (OS-tē-ō-sar-KŌ-mah)	Cancer of bone
Muscle Tissues	
Myoma (mī-Ō-mah)	Benign; a skeletal muscle tumor
Myosarcoma (mī-ō-sar-KŌ-mah)	Cancer of skeletal muscle tissue
Cardiac sarcoma	Cancer of cardiac muscle tissue
Leiomyoma (lī-ō-mī-ō-mah)	Benign; a smooth muscle tumor
Leiomyosarcoma (lī-ō-mī-ō-sar-KŌ-mah)	Cancer of smooth muscle tissue
Neural Tissues	
Glioma (glī-Ō-mah)	Cancer of neuroglial cells
Neuroma (noo-RŌ-mah)	Cancer of neurons

sues, and **gliomas** (gli-Ō-mahs) form in nervous tissue. Table 4-1 summarizes information concerning benign and malignant tumors (cancers) associated with the major tissues of the body.

Cancer Detection

In general, the odds of survival increase markedly if the cancer is detected early, especially before it undergoes metastasis. Despite the variety of possible cancers, the American Cancer Society has identified seven "warning signs" which indicate a physician should be consulted. These are presented in Table 4-2.

Detection of a cancer often begins during a routine physical examination, when the physician detects an abnormal lump or growth, possibly indicating a tumor. The tumor may be benign or malignant (cancerous) and if malignant, may metastasize rapidly or spread very slowly.

Many laboratory and diagnostic tests are necessary for the correct diagnosis of cancer. Information is usually obtained through examination of a tissue sample, or biopsy, often supplemented by medical imaging, and blood studies. ∞ [p. 12] A biopsy is one of the most significant diagnostic procedures because it permits a direct look at the tumor cells, often making it possible to identify the cell type involved. Not only do malignant cells have an abnormally high rate of cell division, they also are structurally distinct from normal body cells.

Cancer Treatment

It is unfortunate that the media tend to describe cancer as though it were one disease rather than many. This simplistic grouping fosters the belief that some dietary change, air ionizer, or wonder drug will be found that can prevent the affliction. Because cancers have so many separate causes, possible mechanisms, and diverse characteristics, there is no single, universally effective cure.

The goal of cancer treatment is to achieve **remission**. A tumor in remission either ceases to grow or decreases in size. Basically the treatment of malignant tumors must accomplish one of the following to achieve remission:

1. *Surgical removal or destruction of individual tumors.* Tumors containing malignant cells can be surgically removed or destroyed by radiation, heat, or freezing. These techniques are very effective, if the treatment is undertaken before metastasis has occurred. For this reason early detection is important in improving survival rates for all forms of cancer.

2. *Killing metastatic cells throughout the body.* This is much more difficult and potentially dangerous, because healthy tissues are likely to be damaged at the same time. At present the most widely approved treatments are chemotherapy and radiation.

Chemotherapy involves the administration of drugs that will either kill the cancerous tissues or prevent cell divisions. These drugs often affect stem cells in normal tissues, and the side effects are usually unpleasant. For example, because chemotherapy slows the regeneration and maintenance of epithelia of the skin

Table 4-2	Seven Warning Signs of Cancer

Change in bowel or bladder habits

A sore that does not heal

Unusual bleeding or discharge

Thickening or lump in breast or elsewhere

Indigestion or difficulty in swallowing

Obvious change in a wart or mole

Nagging cough or hoarseness

✔ A pathologist examining lung tissue from two patients concludes that one sample shows dysplasia and the other anaplasia. Which patient has cancer?

✔ What is the difference between a carcinoma and a sarcoma?

and digestive tract, patients may lose their hair and experience nausea and vomiting. Several different drugs are often administered simultaneously, or in sequence, because over time cancer cells can develop a resistance to a single drug. Chemotherapy is often used in the treatment of many kinds of metastatic cancer.

Massive doses of radiation are sometimes used to treat advanced cases of *lymphoma*, a cancer of lymphoid tissue that affects the immune system. In this rather drastic procedure enough radiation is administered to kill all the blood-forming cells in the body. After treatment, a transplant of bone marrow will provide a new set of blood stem cells.

CHAPTER REVIEW

Key Words

connective tissue: One of the four primary tissue types; provides a structural framework for the body that stabilizes the relative positions of the other tissue types.

epithelial tissue (e-pi-THĒ-lē-um): One of the four primary tissue types; a layer of cells that forms a superficial covering or an internal lining of a body cavity or vessel.

gland: Group of specialized epithelial cells that produce exocrine or endocrine secretions.

inflammation: A tissue repair mechanism, characterized by swelling, redness, warmth, pain, and some loss of function.

mucous membrane: A sheet of epithelial and connective tissue that lines cavities connected to the exterior of the body; often contains mucous glands.

muscle tissue: One of the four primary tissue types, characterized by cells capable of contraction.

nervous tissue: One of the four primary tissue types; consists of functional units called neurons and supporting neuroglial cells.

neuron (NOO-ron): A cell in nervous tissue, specialized for the conduction of electrical impulses.

serous membrane: A membrane composed of epithelium and underlying loose connective tissue that lines internal cavities of the body such as the pericardial, pleural, and peritoneal cavities.

tissue: A collection of specialized cells and cell products that perform a specific function.

Selected Clinical Terms

anaplasia: An irreversible change to tissue cells that alters their structure and organization.

ascites: An accumulation of peritoneal fluid that creates a characteristic abdominal swelling.

carcinoma: A cancer of epithelial tissues.

chemotherapy: The administration of drugs that will kill cancerous tissues or prevent cell division.

dysplasia: A change in the normal size, shape, and organization of tissue cells; this condition is reversible.

metaplasia: A structural change to tissue cells that significantly alters the character and function of a tissue; this condition is reversible.

oncologists: Physicians who specialize in identifying and treating cancers.

pathologists: Physicians who specialize in diagnosing disease processes.

remission: A stage in which a tumor stops growing or grows smaller; the goal of cancer treatment

sarcoma: A cancer of connective tissues.

Study Outline

INTRODUCTION (p. 61)

1. **Tissues** are collections of specialized cells and cell products that are organized to perform a relatively limited number of functions. The four primary **tissue types** are epithelial tissue, connective tissue, muscle tissue, and nervous tissue. (*Figure 4-1*)

2. Epithelial tissues cover external and internal body surfaces, and form glands.

3. Connective tissues fill internal spaces and provide structural support, a means for communication within the body, and energy storage.

4. Muscle tissue has the ability to contract and produce movement.

5. Nervous tissue analyzes and conducts information from one part of the body to another.

EPITHELIAL TISSUE (pp. 62–64)
CLASSIFICATION OF EPITHELIA (p. 62)

1. Epithelia are classified on the basis of the number of cell layers and the shape of the exposed cells.

2. A **simple epithelium** is made up of a single layer of cells, and a **stratified epithelium** consists of several layers of cells.

3. In a **squamous epithelium** the cells are thin and flat. Cells in a **cuboidal epithelium** resemble little boxes; those in a **columnar epithelium** are taller and more slender.

TYPES OF EPITHELIA (p. 62)

4. **Simple squamous epithelia** line protected internal surfaces; both **simple cuboidal epithelia** and **simple columnar epithelia** line areas of absorption and secretion. (*Figure 4-2*)

5. **Pseudostratified epithelia** covered with cilia line the respiratory tract; **stratified epithelia** cover exposed surfaces; **transitional epithelia** cover surfaces that undergo regular stretching. (*Figure 4-2*)

GLANDULAR EPITHELIA (p. 64)

6. **Exocrine** secretions are released onto body surfaces; **endocrine** secretions, known as *hormones*, are released by gland cells into the surrounding tissues.

EPITHELIAL REPLACEMENT *(p. 64)*

7. Divisions by **stem cells**, or **germinative cells**, continually replace short-lived epithelial cells.

CONNECTIVE TISSUES *(pp. 64–67)*

1. All connective tissues have specialized cells, extracellular protein fibers, and a *ground substance*. The protein fibers and ground substance constitute the **matrix**. *(Figure 4-3)*

CLASSIFICATION OF CONNECTIVE TISSUES *(p. 64)*

2. **Connective tissue proper**, the most abundant connective tissue in the body, has a syrupy matrix. The two types are *loose* and *dense*. *(Figure 4-3)*

3. *Fluid connective tissues* have a watery matrix. The two types are *blood* and *lymph*.

4. *Supporting connective tissues* have a dense or solid matrix. The two types are *cartilage* and *bone*. *(Figure 4-3)*

CONNECTIVE TISSUE PROPER *(p. 64)*

5. **Connective tissue proper** is classified as **loose** or **dense connective tissues**. Loose connective tissue includes **areolar tissue** and **adipose tissue**. *(Figure 4-4)*

6. Most of the volume in dense connective tissue consists of fibers. Dense connective tissues form **tendons** and **ligaments** and also surround internal organs. *(Figure 4-4)*

FLUID CONNECTIVE TISSUES *(p. 66)*

7. **Blood** and **lymph** are connective tissues that contain distinctive collections of cells in a fluid matrix.

8. Blood contains *red blood cells*, *white blood cells*, and *platelets*. Its watery ground substance is called **plasma**.

9. Lymph forms as *interstitial fluid* enters vessels that return lymph to the circulatory system.

SUPPORTING CONNECTIVE TISSUES *(p. 67)*

10. The matrix of **cartilage** consists of a firm gel and cartilage-forming cells. A fibrous **perichondrium** separates cartilage from surrounding tissues. *(Figure 4-4)*

11. Chondrocytes rely on diffusion through the matrix to obtain nutrients.

12. **Bone**, or **osseous tissue**, has a matrix consisting of collagen fibers and calcium salts, which give it unique properties. *(Figure 4-4)*

13. **Osteocytes** obtain nutrients through cellular extensions that reach blood vessels and other osteocytes.

BODY MEMBRANES *(pp. 68–69)*

1. Membranes form a barrier or interface. Epithelia and connective tissues combine to form membranes that cover and protect other structures and tissues. The four types of membranes are *cutaneous*, *mucous*, *serous*, and *synovial*. *(Figure 4-5)*

CUTANEOUS MEMBRANE *(p. 68)*

2. The **cutaneous membrane** covers the body surface. Unlike serous and mucous membranes, it is relatively thick, waterproof, and usually dry.

MUCOUS MEMBRANES *(p. 68)*

3. **Mucous membranes** line cavities that communicate with the exterior. Their surfaces are normally moistened by mucous secretions.

SEROUS MEMBRANES *(p. 69)*

4. **Serous membranes** line internal cavities and are delicate, moist, and very permeable.

SYNOVIAL MEMBRANES *(p. 69)*

5. The **synovial membrane**, located at joints or articulations, produces synovial fluid in joint cavities. Synovial fluid helps lubricate the joint and promotes smooth movement.

MUSCLE TISSUE *(pp. 70–71)*

1. Muscle tissue is specialized for contraction. There are three different types of muscle tissue: **skeletal muscle**, **cardiac muscle**, and **smooth muscle**. *(Figure 4-6)*

SKELETAL MUSCLE TISSUE *(p. 70)*

2. Skeletal muscle tissue contains very large muscle fibers tied together by collagen and elastic fibers. Skeletal muscle fibers have a striped appearance because of the organization of their contractile proteins. The stripes are called *striations*. Skeletal muscle is described as *striated voluntary muscle*.

CARDIAC MUSCLE TISSUE *(p. 70)*

3. Cardiac muscle tissue is found only in the heart. It is characterized as *striated involuntary muscle*.

SMOOTH MUSCLE TISSUE *(p. 71)*

4. Smooth muscle tissue is found in the walls of blood vessels, around hollow organs, and in layers in the walls of various tracts. It is classified as *nonstriated involuntary muscle*.

NERVOUS TISSUE *(pp. 71–72)*

1. **Nervous tissue** is specialized to conduct electrical impulses that convey information from one area of the body to another.

2. Cells in nervous tissue are either neurons or neuroglia. **Neurons** transmit information as electrical impulses in their cell membranes. Several kinds of **neuroglia** serve both supporting and defense functions. *(Figure 4-7)*

3. A typical neuron has a **cell body**, many **dendrites**, and an **axon**.

TISSUE INJURY AND REPAIR *(p. 72)*

1. Any injury affects several tissue types simultaneously, and they respond in a coordinated manner. Homeostasis is restored in two processes: inflammation and regeneration.

2. **Inflammation** isolates the injured area while damaged cells, tissue components, and any dangerous microorganisms are cleaned up.

3. **Regeneration** is the repair process that restores normal function.

TISSUES AND AGING *(p. 72)*

1. Tissues change with age. Repair and maintenance grow less efficient, and the structure and chemical composition of many tissues are altered.

TISSUE STRUCTURE AND DISEASE *(pp. 73–76)*

1. **Pathologists** study disease processes. Their diagnosis of diseases is often based on samples of tissues, or **biopsies**.

2. Tissues undergo a series of visible changes prior to the development of cancer. **Dysplasia** is a change in the normal shape, size, and organization of tissue cells. In **metaplasia**, the change in structure alters the normal functioning of the tissue. Tissue structure completely breaks down in **anaplasia**. Anaplasia is irreversible. *(Figure 4-8)*

CANCER CLASSIFICATION AND TREATMENT *(p. 74)*

3. **Oncologists** specialize in the identification and treatment of cancers.

4. Tumors may be benign or malignant (cancerous). Cancers are classified according to cellular appearance and their sites of origin. **Carcinomas** form in epithelial tissues, **sarcomas** in

connective tissue, **leukemias** in blood, and **gliomas** in nervous tissue. *(Table 4-1)*

5. The detection of cancer commonly relies on biopsies, medical imaging, and blood tests. *(Table 4-2)*

6. Cancer is not a single disease, but many. Different cancers require different treatments. All treatments are aimed at **remission**, stopping the growth of the tumor(s) or decreasing their size. **Chemotherapy** relies on drugs that kill cancerous tissues directly or prevent mitotic divisions.

Review Questions

MATCHING

Match each item in Column A with the most closely related item in Column B. Use letters for answers in the spaces provided.

	Column A		Column B
_____	1. histology	a.	striated, voluntary
_____	2. microvilli	b.	goal of cancer treatment
_____	3. stem cells	c.	endocrine secretion
_____	4. hormones	d.	exocrine secretion
_____	5. adipose cells	e.	study of tissues
_____	6. bone-to-bone attachment	f.	intercalated discs
_____	7. skeletal muscle	g.	nervous system cancer
_____	8. cardiac muscle	h.	contain fat
_____	9. mucus	i.	repair and renewal
_____	10. muscle-to-bone attachment	j.	tendon
_____	11. remission	k.	ligament
_____	12. glioma	l.	absorption

MULTIPLE CHOICE

13. The four basic types of tissue in the body are _____.
(a) epithelial, connective, muscle, nervous
(b) epithelial, blood and lymph, muscle, nervous
(c) neural, connective, muscle, nervous
(d) epithelial, connective, macrophages, nervous

14. The three cell shapes found in epithelial tissue are _____.
(a) simple, stratified, transitional
(b) simple, stratified, pseudostratified
(c) hexagonal, cuboidal, spherical
(d) cuboidal, squamous, columnar

15. The urinary bladder is lined by _____ epithelium.
(a) cuboidal
(b) columnar
(c) simple squamous
(d) stratified squamous
(e) transitional

16. Endocrine glands differ from exocrine glands in that _____.
(a) they have ducts
(b) they are ductless
(c) they consist of epithelia and connective tissue
(d) they produce only mucus

17. The type of tissue that contains a matrix is _____.
(a) epithelial
(b) muscle
(c) nervous
(d) connective

18. Areolar tissue is a _____.
(a) fluid connective tissue
(b) supporting connective tissue
(c) loose connective tissue
(d) dense connective tissue

19. Cartilage is separated from surrounding tissue by a fibrous _____.
(a) perichondrium
(b) periosteum
(c) lacunae
(d) canaliculi

20. The primary function of serous membranes in the body is _____.
(a) to minimize friction between opposing surfaces
(b) to line cavities that communicate with the exterior
(c) to perform absorptive and secretory functions
(d) to cover the surface of the body

21. Large muscle fibers that are multinucleated, striated, and voluntary are found in _____.
(a) cardiac muscle tissue
(b) skeletal muscle tissue
(c) smooth muscle tissue
(d) a, b, and c are correct

22. Axons and dendrites are characteristics of cells found in _____.
(a) epithelial tissue
(b) connective tissue
(c) muscle tissue
(d) nervous tissue

23. A _____ is a cancer of muscle tissue.
(a) papilloma
(b) lipoma
(c) osteoma
(d) myosarcoma

24. In _____, tissues are irreversibly altered.
(a) anaplasia
(b) metaplasia
(c) dysplasia
(d) a, b, and c are correct

TRUE/FALSE

____ 25. Mucous membranes line body cavities that open to the outside.

____ 26. Simple epithelia cover surfaces subjected to mechanical stress.

____ 27. Smooth muscle tissue does not have striations, and its cells have many nuclei.

____ 28. The watery ground substance of blood is called plasma.

____ 29. Contractile proteins are found in muscle tissue.

____ 30. Oncologists are physicians who specialize in diagnosing diseases.

____ 31. Benign tumors are more easily removed by surgery than malignant tumors.

SHORT ESSAY

32. What are the four essential functions of epithelial tissue?

33. What are the three basic components of connective tissues?

34. What fluid connective tissues are found in the human body?

35. Distinguish between the members of each of the following pairs:
(a) serous and mucous secretions
(b) exocrine and endocrine glands
(c) ligaments and tendons

36. What two cell populations make up nervous tissue? What is the function of each?

37. Why is a biopsy so important in diagnosing cancer?

38. A side effect of chemotherapy is the loss of hair. Why?

APPLICATIONS

39. Nancy wanted a flatter abdomen, so she went on a diet and lost ten pounds. Two years later, Nancy regained the ten pounds and her abdomen looks the way it did before the weight was lost. Explain why the extra weight seemingly "returned" to her abdomen.

40. After many years of smoking a pack of cigarettes a day, Mr. Butts suffers from a chronic cough and a biopsy is made of his respiratory tract. The pathology report notes that the tissues of the trachea (windpipe) demonstrate metaplasia. Explain the importance of metaplasia. Also, for what conditions is Mr. Butts at risk?

✔ Answers to Concept Check Questions

(p. 64) **1.** No. A simple squamous epithelium does not provide enough protection against infection, abrasion, and dehydration. **2.** Exocrine secretions would be found on a body surface, either internal or external. They are released by glands with ducts. Endocrine secretions are released into body tissues and the blood from ductless glands. **3.** Epithelial stem cells divide continually to replace the outermost cells lost at an epithelial surface.

(p. 67) **1.** *Blood* and *lymph* are the two types of connective tissue whose cells are contained in a liquid matrix. **2.** Collagen fibers add strength to connective tissue. A vitamin C deficiency would result in a connective tissue that is weak and prone to damage. **3.** Unlike bone, cartilage lacks a direct blood supply, which is necessary for proper healing to occur. Repair materials for damaged cartilage must diffuse from the blood to the cartilage cells, a process that takes a long time and slows the healing process.

(p. 69) **1.** Serous fluid minimizes the friction between the serous membranes covering the surfaces of organs and the surrounding body cavity. **2.** All these regions are subject to mechanical trauma and abrasion—by food (pharynx and esophagus), feces (anus), and intercourse or childbirth (vagina).

(p. 71) **1.** Since cardiac and skeletal muscle are both striated (banded), this must be *smooth muscle* tissue. **2.** Only skeletal muscle tissue is voluntary. **3.** Both skeletal muscle cells and neurons are called *fibers* because they are relatively long and slender. **4.** Information is sent by neurons to different parts of the body as events of electrical activity.

(p. 76) **1.** In dysplasia, cells show some change in normal shape and organization, but the changes are reversible if the stress is removed. In anaplasia, the cells have become quite abnormal in shape and size, and have become cancerous. Anaplasia is irreversible. The second patient has cancer. **2.** A carcinoma is a cancer of the epithelial tissues, and a sarcoma is a cancer of the connective tissues.

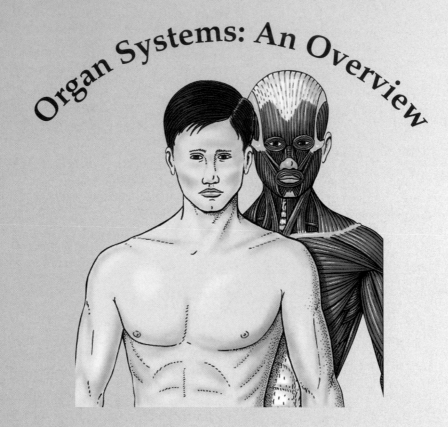

CHAPTER
5

Organ Systems: An Overview

CHAPTER OUTLINE

CHAPTER OBJECTIVES

1   Describe the general features of an organ system.
2   Describe the organs and primary functions of each organ system concerned with support and movement.
3   Describe the organs and primary functions of the nervous and endocrine systems.
4   Describe the organs and primary functions of the cardiovascular and lymphatic systems.
5   Describe the organs and primary functions of each organ system involved in the exchange of materials with the environment.
6   Describe the organs and primary functions of the male and female reproductive systems.

In the past few chapters, we have examined the different levels of organization underlying the structure of the human body. These levels of structure have ranged from the invisible (atoms and molecules) and the barely visible (cells) to the easily visible (tissues). The way in which structures at lower levels combine and interact to form the more complex structures of higher levels is a concept we'll take two steps further in this chapter as we consider organs and organ systems.

AN INTRODUCTION TO ORGAN SYSTEMS

Organs are combinations of tissues that perform complex functions. In *organ systems*, several organs work together to accomplish important functions for the whole body. Structural support, movement, defense, and waste removal are just a few such functions handled by different, specialized organ systems.

Although there is a division of labor among the various organ systems, they do not operate in isolation. As you learn about the individual systems in later chapters, you will see that each organ system interacts with others. For example, the muscular system could not move the body without the structural support of the skeleton. Furthermore, *all* organ systems depend on the cardiovascular (circulatory) system to bring oxygen and nutrients to their tissues and to remove waste products, and all are influenced by the activities of the nervous and endocrine systems. Thus to understand any one organ system, you must know basic information about all of the other organ systems. This chapter provides you with an overview of the 11 organ systems of the human body, focusing on their basic structural and functional characteristics.

SUPPORT AND MOVEMENT

Body support and movement are functions of the integumentary, skeletal, and muscular systems.

THE INTEGUMENTARY SYSTEM

The integumentary (in-teg-ū-MEN-ta-rē; *tegere*, to cover) system, or *integument*, is made up of the **cutaneous membrane**, or *skin*, and associated structures, such as *glands*, *hair*, and *nails*. Few organ systems are as large, as visible, as varied in function, and as unappreciated as the skin. Its surface is continually abused, abraded, and assaulted by microorganisms, sunlight, and a variety of mechanical and chemical hazards.

The basic components of the integumentary system and their functions are shown in Figure 5-1a●. The integument (1) provides protection from the environment; (2) assists in the regulation of body temperature (*thermoregulation*) ∞ *[p. 6]*; (3) contains *sensory receptors* that send information to the nervous system; and (4) excretes water and solutes, including organic wastes and ions. In addition, vitamin D is synthesized in the skin and lipid reserves are stored in adipose tissue under the skin.

THE SKELETAL SYSTEM

The **skeletal system** (Figure 5-1b●) forms an internal supporting framework for the body. It is made up of approximately 206 bones and a large number of cartilages. This system also includes the connective tissues and ligaments that connect the bones at various joints.

The skeletal system can be divided into the **axial skeleton** and the **appendicular skeleton**. The bones of the axial skeleton form the vertical axis of the body and enclose the brain and spinal cord. The appendicular skeleton includes the bones of the limbs as well as those that connect the limbs to the trunk.

•FIGURE 5-1
The Integumentary
and Skeletal Systems.

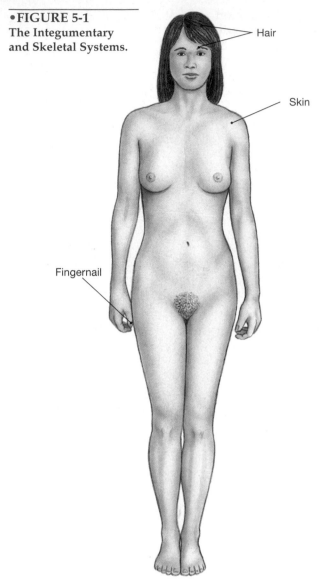

Hair

Skin

Fingernail

(a)   **The Integumentary System**

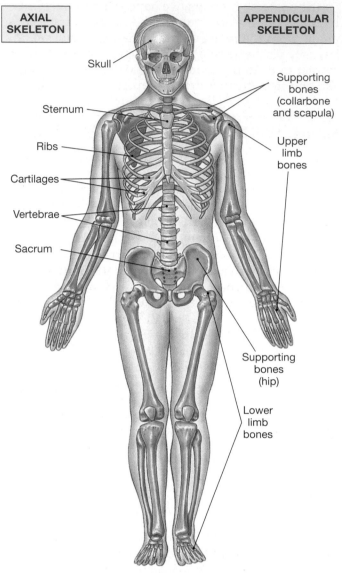

| AXIAL SKELETON | | APPENDICULAR SKELETON |

Skull

Sternum

Ribs

Cartilages

Vertebrae

Sacrum

Supporting bones (collarbone and scapula)

Upper limb bones

Supporting bones (hip)

Lower limb bones

(b)   **The Skeletal System**

Organ	Primary Functions
Skin	
Epidermis	Covers surface, protects underlying tissues
Dermis	Nourishes epidermis, provides strength, contains glands
Hair Follicles	Produce hair
Hairs	Provide sensation, provide some protection for head
Sebaceous glands	Secrete oil that lubricates hair
Sweat Glands	Produce perspiration for evaporative cooling
Nails	Protect and stiffen tips of fingers and toes
Sensory Receptors	Provide sensations of touch, pressure, temperature, pain

Organ	Primary Functions
Bones (206), Cartilages, and Ligaments	Support, protect soft tissues; store minerals
Axial Skeleton (skull, vertebrae, sacrum, ribs, sternum)	Protects brain, spinal cord, sense organs, and soft tissues of chest cavity; supports the body weight over the legs
Appendicular Skeleton (limbs and supporting bones)	Provides internal support and positioning of arms and legs; supports and moves axial skeleton
Bone Marrow	Primary site of blood cell production

The skeletal system supports the body as a whole, as well as individual organs and organ systems. It also directs the forces of skeletal muscle contraction, producing movement. The individual bones store minerals, and some bones, such as those of the skull also protect delicate tissues and organs. In adults, red blood cells are formed in red **bone marrow**, a connective tissue that fills the spaces, or *marrow cavities*, within many bones.

THE MUSCULAR SYSTEM

The **muscular system** includes approximately 700 skeletal muscles in the body that are under voluntary control (Figure 5-2•). Skeletal muscles are organs specialized for contraction. Each skeletal muscle consists of skeletal muscle tissue and connective tissues, such as tendons, that attach the muscle to the skeleton. A skeletal muscle also contains blood vessels that provide circulation and nervous tissue that monitors and controls the muscle's contractions.

The contractions of skeletal muscles can produce movements of body parts (such as waving the hand) or the entire body (walking). Skeletal muscles also maintain posture and balance, support soft tissues, control the openings of the digestive tract, and assist in regulating blood flow through blood vessels. Finally, the heat produced during muscle contraction plays a major role in keeping the body warm.

COMMUNICATION, CONTROL, AND INTEGRATION

THE NERVOUS SYSTEM

The nervous system includes all the nervous tissue of the body. The nervous system performs complex processing and coordinating functions. Although it makes up only 3 percent of total body weight, it contains some 10 billion neurons, and 100 billion supporting, or neuroglial, cells.

The parts of the nervous system, shown in Figure 5-3a•, include the brain, spinal cord, complex sense organs such as the eye and ear, and the nerves that interconnect these organs and link the nervous system with other systems. The brain and spinal cord form the **central nervous system**, or **CNS**. The CNS is the control center for the nervous system, processes information, and provides short-term control over the activities of other organ systems.

The CNS receives information regarding the body's internal and external surroundings through *sensory neurons*. It also directs or adjusts the activities of tissues and organs through *motor neurons*. Both the sensory and motor neurons are part of the **peripheral nervous system**, or **PNS**. The PNS links the CNS

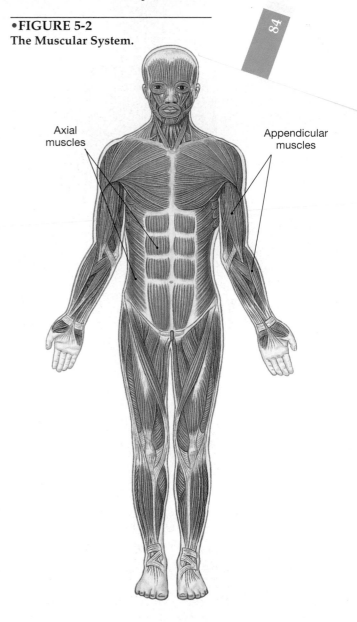

•**FIGURE 5-2**
The Muscular System.

Axial muscles

Appendicular muscles

Organ	Primary Functions
Skeletal Muscles (700)	Provide skeletal movement, control openings of digestive tract, produce heat, support skeletal position, protect soft tissues

✔ The ability to move about and produce heat are major functions of which organ system of the body?

✔ Support, protection of soft tissue, mineral storage, and blood formation are functions of which organ system?

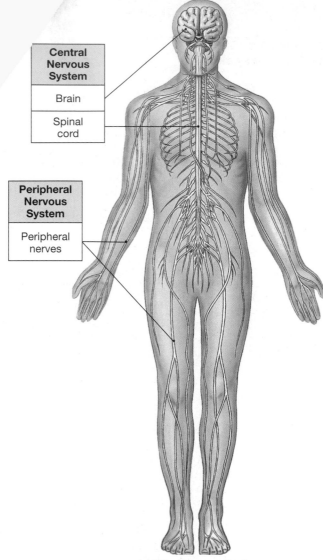

(a) The Nervous System

Organ	Primary Functions
Central Nervous System (CNS)	Control center for nervous system: processes information, provides short-term control over activities of other systems
Brain	Performs complex integrative functions, controls voluntary activities
Spinal cord	Relays information to and from the brain; directs many simple involuntary activities
Peripheral Nervous System (PNS)	Links CNS with other systems and with sense organs

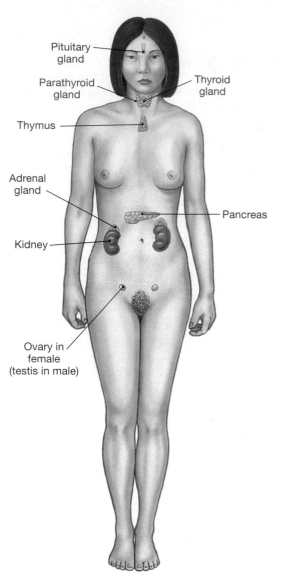

(b) The Endocrine System

Organ	Primary Functions
Pituitary Gland	Controls other glands, regulates growth and fluid balance
Thyroid Gland	Controls tissue metabolic rate and regulates calcium levels
Parathyroid Gland	Regulates calcium levels (with thyroid)
Thymus	Controls white blood cell maturation
Adrenal Glands	Adjust water balance, tissue metabolism, cardiovascular and respiratory activity
Kidneys	Control red blood cell production and elevate blood pressure
Pancreas	Regulates blood glucose levels
Gonads Testes	Support male sexual characteristics and reproductive functions (see Figure 5-7)
Ovaries	Support female sexual characteristics and reproductive functions (see Figure 5-7)

•**FIGURE 5-3**
The Nervous and Endocrine Systems.

with other systems and with sense organs. The *autonomic nervous system*, or *ANS*, includes both CNS and PNS neurons involved in the automatic regulation of physiological functions, such as heart rate and digestive activities.

THE ENDOCRINE SYSTEM

The **endocrine system**, includes ductless glands that produce hormones (Figure 5-3b●). These secretions are distributed by the bloodstream and carry specific instructions that coordinate cellular activities. These chemical messages may be quite general, affecting every cell in the body, or so specific that only a single cluster of cells is affected.

The nervous and endocrine systems represent different approaches to the regulation and coordination of the body's internal operations. The nervous system specializes in analyzing sensory information and providing an immediate, short-term response, often in less than a second after a *stimulus* arrives from a sensory receptor. (A **stimulus** is any change or alteration in the environment that can produce a change in cellular activities.) Regulation by the endocrine system is slower, producing changes in metabolism over seconds, minutes, hours, or even years. Examples of endocrine regulation include the control of blood sugar (glucose) levels and metabolic adjustments to cold and dehydration. Hormones also regulate an individual's growth, sexual characteristics, and the development of sperm and eggs.

✔ Swift responses to stimuli are the function of which organ system?

✔ Which organ system is composed of ductless glands whose secretions coordinate cellular activities of other organ systems?

TRANSPORT AND DEFENSE

THE CARDIOVASCULAR SYSTEM

The **cardiovascular system** consists of the heart, blood, and blood vessels (Figure 5-4a●). This system also includes the bone marrow, where red blood cells are formed. The cardiovascular system transports nutrients, dissolved gases, and hormones to tissues throughout the body. It also carries waste products from the body's tissues to sites of excretion, such as the kidneys. Buffers in the blood help balance the pH of body fluids. Warm blood carries heat from one location to another, aiding in thermoregulation. White blood cells provide defense from disease, and the formation of clots with *fibrin* (a blood protein) restricts the loss of blood through breaks in the skin and slows the spread of disease-causing microorganisms, or *pathogens*, through damaged tissues.

THE LYMPHATIC SYSTEM

The **lymphatic system** consists of a widespread network of lymphatic vessels that collect interstitial fluid and deliver it, as lymph, to the blood vessels of the cardiovascular system. As shown in Figure 5-4b●, the lymphatic system also contains **lymphoid organs** that produce or support large numbers of *lymphocytes* (a type of white blood cell), *plasma cells* (antibody-producing cells), and *phagocytes*. **Lymph nodes** are small lymphoid organs that contain cells sensitive to changes in the composition of the lymph. Larger lymphoid organs include the **thymus gland**, the **tonsils**, and the **spleen**.

The lymphatic system defends the body from disease and toxic substances. Some of its phagocytes roam through the tissues of the body engulfing cellular debris from injuries and attacking pathogens (disease-causing microorganisms). Other phagocytes remain in the lymph nodes to attack damaged cells or pathogens that have evaded the tissue defenses and are transported in the lymph to the nodes. Some lymphocytes circulate through the body and attack abnormal cells or pathogens. Other lymphocytes become plasma cells, producing antibodies that provide a chemical defense.

✔ Which system of the body transports special cells and dissolved materials, such as nutrients, gases, and wastes?

✔ Defending the body from disease and infection is a major function of which organ system of the body?

•FIGURE 5-4
The Cardiovascular
and Lymphatic Systems.

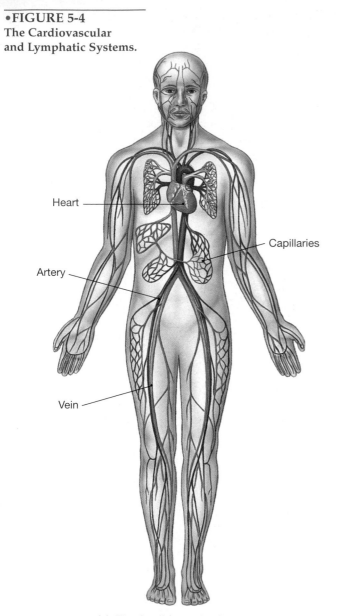

Heart

Capillaries

Artery

Vein

(a) The Cardiovascular System

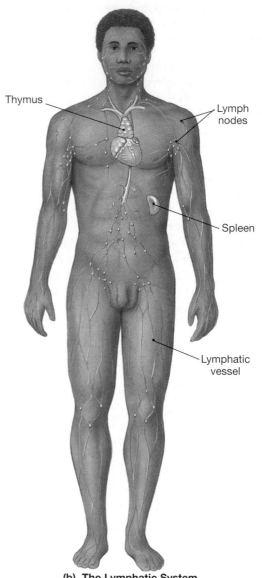

Thymus

Lymph
nodes

Spleen

Lymphatic
vessel

(b) The Lymphatic System

Organ	Primary Functions
Heart	Pumps blood, maintains blood pressure
Blood Vessels	Distribute blood around the body
Arteries	Carry blood from heart to capillaries
Capillaries	Site of exchange between blood and interstitial fluids
Veins	Return blood from capillaries to heart
Blood	Transports oxygen and carbon dioxide, delivers nutrients, removes waste products, assists in defense against disease

Organ	Primary Functions
Lymphatic Vessels	Carry lymph (water and proteins) from body tissues to the veins of the cardiovascular system
Lymph Nodes	Monitor the composition of lymph, stimulate immune response
Spleen	Monitors circulating blood, stimulates immune response
Thymus	Controls development and maintenance of one class of white blood cells (T cells)

EXCHANGE WITH THE ENVIRONMENT

THE RESPIRATORY SYSTEM

The **respiratory system** includes the lungs and the passageways that carry air to them (Figure 5-5•). These passageways begin at the **nasal cavities** and continue through the **pharynx** (throat), **larynx** (voice box), **trachea** (windpipe), and **bronchi** (BRONG-kē; singular, *bronchus*) before entering the lungs. Within the lungs, the bronchi branch over and over, growing ever smaller in diameter. The nasal cavities, trachea, and bronchi are lined with a ciliated pseudostratified epithelium as described in Chapter 4. This epithelium filters, warms, and moistens incoming air, thus protecting the delicate interior gas-exchange surfaces within the lungs. The exchange surfaces, lined by a simple squamous epithelium, form small pockets called **alveoli** (al-VĒ-ō-lī; singular *alveolus*). At these surfaces, gas exchange occurs between the air and the circulating blood.

The primary functions of the respiratory system are the delivery of oxygen to and the removal of carbon dioxide from the blood arriving at the alveoli. It also has a number of secondary functions. For example, by changing the concentration of carbon dioxide in the blood, the respiratory system helps to regulate the pH of the blood. This is important because, as noted in Chapter 2 (p. 26), uncontrolled changes in pH can have disastrous effects on cells and tissues. In addition, the air forced across the vocal cords produces sounds used in communication.

THE DIGESTIVE SYSTEM

The central feature of the **digestive system** is the **digestive tract**, a long tube that begins at the mouth and ends at the anus (Figure 5-6a•). A number of **accessory glands**, including the salivary glands, liver, and pancreas, are associated with the tract along its length. The secretions of the **salivary glands**, **stomach**, **small intestine**, and **pancreas** introduce digestive enzymes that help breakdown food. The **liver**, the largest digestive organ, secretes bile, which is essential to the breakdown of fats. The liver also regulates the nutrient content of the blood and makes important plasma proteins.

The primary function of the digestive system is the breakdown of food for absorption and use by the body. The digestive tract absorbs nutrients, including organic compounds, water, ions, and vitamins. Buffers, ions, and a few metabolic wastes are dumped into the digestive tract by accessory glands, especially the pancreas and liver. The wastes, along with undigested remains of previous meals, are eliminated from the **large intestine** during the process of **defecation**.

•FIGURE 5-5
The Respiratory System.

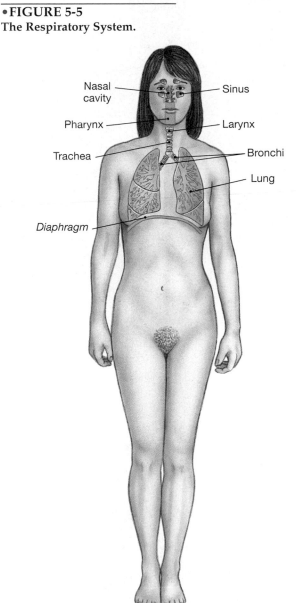

Organ	Primary Functions
Nasal Cavities	Filter, warm, humidify air; detect smells
Pharynx	Chamber shared with digestive tract; conducts air to larynx
Larynx	Protects opening to trachea and contains vocal cords
Trachea	Filters air, traps particles in mucus; cartilages keep airway open
Bronchi	Same as trachea
Lungs	Include airways and alveoli; volume changes responsible for air movement
Alveoli	Sites of gas exchange between air and blood

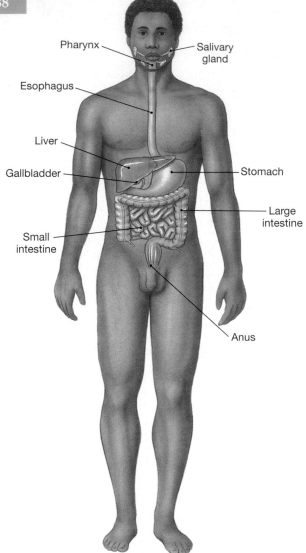

(a) The Digestive System

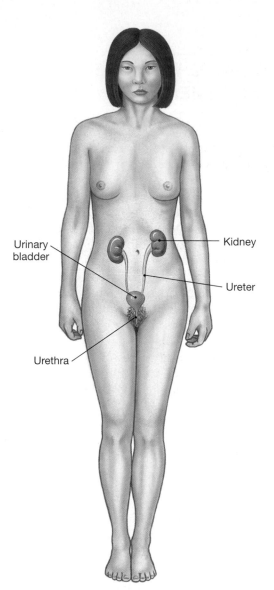

(b) The Urinary System

Organ	Primary Functions
Salivary Glands	Provide lubrication, produce buffers and the enzymes that begin digestion
Pharynx	Passageway connected to esophagus
Esophagus	Delivers food to stomach
Stomach	Secretes acids and enzymes
Small Intestine	Secretes digestive enzymes, absorbs nutrients
Liver	Secretes bile, regulates blood composition of nutrients
Gallbladder	Stores bile for release into small intestine
Pancreas	Secretes digestive enzymes and buffers; contains endocrine cells (see Figure 5-3b)
Large Intestine	Removes water from fecal material, stores wastes

Organ	Primary Functions
Kidneys	Form and concentrate urine, regulate chemical composition of the blood
Ureters	Carry urine from kidneys to urinary bladder
Urinary Bladder	Stores urine for eventual elimination
Urethra	Carries urine to exterior

•**FIGURE 5-6**
The Digestive and Urinary Systems.

THE URINARY SYSTEM

The **urinary system** includes the **kidneys**, the **ureters**, the **urinary bladder**, and the **urethra** (Figure 5-6b•). The major function of the urinary system is the elimination of waste products from the blood through the formation of *urine*. Urine, formed in the kidneys, is transported by the ureters to the urinary bladder for storage. The urethra carries urine from the bladder to the exterior of the body, a process known as **urination**. The kidneys are also important in maintaining the fluid and ion balance of the body, as well as controlling the blood's volume, pressure, and composition.

✔ Delivery of air to sites where gas exchange can occur with the blood is the function of which organ sysytem?

✔ Which organ system functions to eliminate excess water, ions, and waste products from the body?

CONTINUITY OF LIFE

THE REPRODUCTIVE SYSTEM

The **reproductive systems** are responsible for the production of offspring (Figure 5-7•). The two sexes have different, but complementary, roles in reproduction. The male's reproductive system is designed to develop and deliver sperm to the reproductive tract of the female. The female's reproductive system is designed to produce eggs and provide a suitable environment for the developing offspring prior to birth, as well as to nourish the newborn for a short period after birth.

Despite their apparent differences, the reproductive systems of males and females have many similarities. Both produce hormones that affect the development, growth, and maintenance of many other organ systems. Both have paired reproductive organs known as **gonads**—the **testes** of a male and the **ovaries** of a female—in which the reproductive cells, or *gametes,* are produced. Both have a series of ducts (or modified ducts) through which the gametes travel. In males, these ducts include the *epididymis*, *sperm duct*, and *urethra*. In females, eggs travel down the *uterine tube* into the *uterus*, a hollow muscular organ that provides protection and support to developing offspring. The *vagina* serves as a passage both for sperm into the reproductive tract and for birth of the infant. The portions of the reproductive system that are visible at the body surface—the *vulva* in females and the *penis* and *scrotum* in males—are known as the **external genitalia** (jen-i-TĀ-lē-a).

✔ What are the paired reproductive organs of both sexes called?

☤ Clinical Note

What Is an Autopsy?

An **autopsy** is a post-mortem (*mors* means death) examination of a body. An autopsy is usually performed by a pathologist to determine or confirm the cause of death. There are different types of autopsies. In a *complete autopsy*, the tissues and organs of the chest, abdominal, and cranial cavities are removed and examined. In a *biopsy only* autopsy, the organs remain in place and only small fragments or samples of the organs are used for histological (tissue) examination. Other autopsies are limited in their access to the different body cavities. For example, *chest only* autopsies are performed only on the heart and lungs. Autopsies not only clarify the cause(s) of death, they can also reveal major unexpected findings; conditions that were undetected while the individual was alive that may or may not have contributed to their death.

•FIGURE 5-7
The Reproductive System.

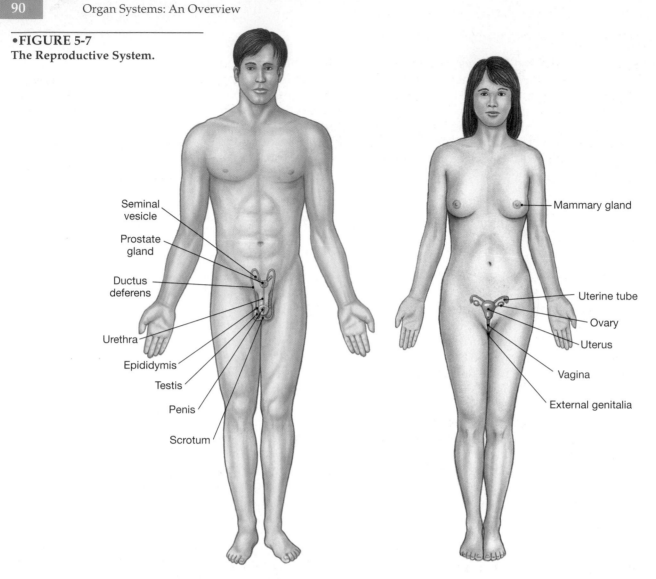

(a) Male

(b) Female

Organ	Primary Functions
Testes	Produce sperm and hormones (see Figure 5-3b)
Accessory Organs	
Epididymis	Site of sperm maturation
Ductus deferens (sperm duct)	Conducts sperm between epididymis and prostate
Seminal vesicles	Secrete fluid that makes up much of the volume of semen
Prostate	Secretes buffers and fluid
Urethra	Conducts semen to exterior
External Genitalia	
Penis	Erectile organ used to deposit sperm in the vagina of a female; produces pleasurable sensations during sexual act
Scrotum	Surrounds and positions the testes

Organ	Primary Functions
Ovaries	Produce ova (eggs) and hormones (see Figure 5-3b)
Uterine Tubes	Deliver ova or embryo to uterus; normal site of fertilization
Uterus	Site of development of offspring
Vagina	Site of sperm deposition; birth canal at delivery; provides passage of fluids during menstruation
External Genitalia	
Clitoris	Erectile organ, produces pleasurable sensations during sexual act
Labia	Contain glands that lubricate entrance to vagina
Mammary Glands	Produce milk that nourishes newborn infant

Clinical Note

What Is an Organ Transplant?

Some types of diseased organs may be replaced with healthy organs through transplant surgery. The major organs that are transplanted include the kidney, liver, pancreas, heart, and lung. The organ is usually removed from a person who has just died. In the U.S., however, some 70 percent of kidneys are taken from living donors.

The earliest major organ transplants were begun with kidneys in the 1950s, but success was limited because of *rejection* of the transplanted kidney. The subsequent development of

drugs that suppress rejection has increased the success of transplant surgery. All organ transplant patients must take such *immunosuppressive drugs* indefinitely. Unfortunately, these drugs also lower a patient's natural immune defenses, leading to an increased risk of infection and certain types of cancer. Other factors leading to increased success in organ transplant surgery include improved methods of matching the tissues of the donor and host (to reduce rejection rates), and better care of the organs between their removal and transplantation.

CHAPTER REVIEW

Key Words

autopsy: The detailed examination of a body after death, normally performed by a pathologist.

cardiovascular: Pertaining to the heart, blood, and blood vessels.

central nervous system (CNS): The brain and spinal cord.

cutaneous membrane: The skin.

defecation (def-e-KĀ-shun): The elimination of wastes from the large intestine.

gametes: Reproductive cells; sperm and egg cells.

gonad: A reproductive organ that produces gametes and hormones.

hormone: A compound secreted by gland cells into the bloodstream that affects the activities of cells in another portion of the body.

marrow: A tissue that fills the internal cavities in a bone; may contain blood-forming cells (red marrow) or fat tissue (yellow marrow).

organ: Structure consisting of several kinds of tissues that together perform a distinct, complex function; examples include brain, kidney, and lung.

organ system: A group of organs working together to accomplish whole-body function; examples include nervous system, urinary system, respiratory system.

pathogen: A disease-causing microorganism.

peripheral nervous system (PNS): All the nervous tissue outside the central nervous system.

stimulus: A change or alteration in the environment that can produce a change in cellular activities.

urination: The discharge of urine.

Study Outline

AN INTRODUCTION TO ORGAN SYSTEMS (p. 81)

1. *Organs* work together in *organ systems* to accomplish important whole body functions.

SUPPORT AND MOVEMENT (pp. 81–83)

THE INTEGUMENTARY SYSTEM (p. 81)

1. The **integumentary system**, or *integument*, includes the **skin**, or *cutaneous membrane*, hair, nails, and glands. (*Figure 5-1a*)

2. Integumentary functions may be summarized as protection, thermoregulation, sensation, excretion, synthesis, and storage.

THE SKELETAL SYSTEM (p. 81)

3. The **skeletal system** includes bones and cartilages connected at joints and stabilized by ligaments and surrounding soft tissues. It may be divided into **axial** and **appendicular** divisions. (*Figure 5-1b*)

4. Skeletal functions include support, locomotion, protection, storage, and blood cell formation.

THE MUSCULAR SYSTEM (p. 83)

5. The **muscular system** includes most of the skeletal muscle tissue in the body. (*Figure 5-2*)

6. The functions of this system include locomotion, maintenance of posture and balance, support of soft tissues, control of entrances and exits of tracts that communicate with the exterior, assistance in blood-flow regulation, stabilization of skeletal elements, and heat production.

COMMUNICATION, CONTROL, AND INTEGRATION (pp. 83–85)

THE NERVOUS SYSTEM (p. 83)

1. The nervous system includes all the nervous tissue in the body. (Figure 5-3a)

2. The primary divisions of the nervous system are the **central nervous system** (CNS), **peripheral nervous system** (PNS), and the *autonomic nervous system* (ANS).

3. The nervous system obtains sensory information from inside and outside the body, integrates that information, and directs or moderates the activity of other organ systems.

THE ENDOCRINE SYSTEM *(p. 85)*

4. The **endocrine system** includes all the glands that secrete their products (hormones) into the bloodstream rather than into ducts. *(Figure 5-3b)*

5. The hormonal secretions may have general or localized effects.

6. Hormonal regulation is slower than nervous regulation but affects many metabolic processes not subject to nervous system control.

TRANSPORT AND DEFENSE *(pp. 85–86)*

THE CARDIOVASCULAR SYSTEM *(p. 85)*

1. The **cardiovascular system** includes the heart, blood, and blood vessels. *(Figure 5-4a)*

2. The functions of this system include distribution of nutrients and heat, transportation of specialized cells, regulation of pH, and prevention of blood loss through the clotting reaction.

THE LYMPHATIC SYSTEM *(p. 85)*

3. The **lymphatic system** consists of lymphatic vessels and **lymphoid organs**, which include the **lymph nodes**, the **thymus**, the **tonsils**, and the **spleen**. *(Figure 5-4b)*

4. The cell population consists of *phagocytes, lymphocytes,* and *plasma cells.*

5. The lymphatic system provides defense against foreign compounds and microorganisms.

EXCHANGE WITH THE ENVIRONMENT *(pp. 87–89)*

THE RESPIRATORY SYSTEM *(p. 87)*

1. The **respiratory system** includes both the conducting passageways and the gas-exchange surfaces of the lungs. *(Figure 5-5)*

2. The primary function of the respiratory system is the exchange of oxygen and carbon dioxide with the environment. Secondary functions include sound production and pH regulation.

THE DIGESTIVE SYSTEM *(p. 87)*

3. The **digestive system** includes the tubular **digestive tract** and the accessory organs (liver and pancreas). *(Figure 5-6a)*

4. With the help of secretions from the accessory glands, the organs of the digestive tract break down ingested food into forms that can be absorbed and used by the cells of the body.

5. Fluids, electrolytes, nutrients, and vitamins are absorbed in the digestive tract; a few wastes are discharged into the tract, and these wastes are eliminated along with the undigested residue.

THE URINARY SYSTEM *(p. 89)*

6. The **urinary system** consists of the **kidneys**, **ureters**, **urinary bladder**, and **urethra**. *(Figure 5-6b)*

7. The urinary system is the primary excretory route for metabolic wastes.

CONTINUITY OF LIFE *(pp. 89–90)*

THE REPRODUCTIVE SYSTEM *(p. 89)*

1. The **reproductive system** includes **gonads**, ducts, accessory glands, and organs. *(Figure 5-7)*

2. The male reproductive system contains **testes**, sperm ducts, accessory glands, and the urethra. The urethra ends at the tip of the penis. The **external genitalia** of the male include the *penis* and *scrotum.*

3. The female reproductive system contains **ovaries**, *uterine tubes,* the *uterus,* and the *vagina.* The **external genitalia** of the female, the *vulva,* surround the vaginal opening.

Review Questions

MATCHING

Match each item in Column A with the most closely related item in Column B. Use letters for answers in the spaces provided.

	Column A		Column B
___	1. ductless glands	a.	integumentary system
___	2. skeletal muscle	b.	skeletal system
___	3. CNS and PNS	c.	muscular system
___	4. spleen	d.	nervous system
___	5. produces gametes	e.	endocrine system
___	6. includes bronchi and alveoli	f.	cardiovascular system
___	7. axial and appendicular divisions	g.	lymphatic system
___	8. liver and pancreas	h.	respiratory system
___	9. skin, hair, and nails	i.	digestive system
___	10. heart, blood, and blood vessels	j.	urinary system
___	11. kidneys	k.	reproductive system

MULTIPLE CHOICE

12. The skin and accessory structures make up the _____.
(a) skeletal system
(b) muscular system
(c) integumentary system
(d) cardiovascular system

13. The skeletal system includes 206 bones and several

_____.
(a) muscles
(b) cartilages
(c) nerves
(d) organs

14. The ability to move about and produce heat are major functions of the _____ of the body.
(a) skeletal system   (c) integumentary system
(b) muscular system   (d) cardiovascular system

15. The secretions of the endocrine system are distributed by way of _____.
(a) ducts from gland to gland
(b) special tubes between organs
(c) the bloodstream
(d) nerve impulses

16. The _____ is not considered a lymphoid organ.
(a) pituitary gland
(b) thymus gland
(c) tonsils
(d) spleen

17. The lungs and the passageways that carry air make up the

_____.
(a) cardiovascular system
(b) lymphatic system
(c) urinary system
(d) respiratory system

18. The central component of the digestive system is the
_____.
(a) digestive tract
(b) stomach
(c) small intestine
(d) pancreas

19. The respiratory system brings _____ into the body and
expels _____ into the surrounding atmosphere.
(a) oxygen, water
(b) water, carbon dioxide
(c) oxygen, carbon dioxide
(d) carbon dioxide, oxygen

20. Sperm are to testes as eggs are to_____.
(a) gonads
(b) ovaries
(c) gametes
(d) the uterus

TRUE/FALSE

___ 21. Urine is produced in the kidneys.

___ 22. Thermoregulation is a major function of the cardiovascular system.

___ 23. Neurons line the internal passages of the respiratory
system and filter dust before it reaches the lungs.

___ 24. The axial skeleton includes the bones of the arms and legs.

___ 25. The muscular system primarily includes cartilage and
skeletal muscle.

___ 26. The cardiovascular system aids in the regulation of
blood pH.

SHORT ESSAY

27. List three primary functions of the skeletal system.

28. Thermoregulation is a function of which organ system(s)?

29. Describe the relationship between the CNS and PNS.

30. Which two organ systems are involved in pH regulation of
body fluids?

✔ Answers to Concept Check Questions

(p. 83) **1.** Organs (muscles) of the muscular system move the
body and produce heat through their contractions. **2.** The skeletal system protects the soft tissues of the body, stores minerals,
and aids in the formation of blood.

(p. 85) **1.** The nervous system enables the body to react swiftly
to various types of stimuli. **2.** The endocrine system is made up
of ductless glands whose secretions are carried throughout the
body, where they alter the activities of organ systems.

(p. 85) **1.** The transport of special cells and dissolved materials occurs through the activities of the cardiovascular system. **2.** The
lymphatic system defends the body from disease and infection.

(p. 89) **1.** The respiratory system delivers air to internal gas exchange sites in the lungs. **2.** The urinary system eliminates excess
water and waste products in the blood from the body.

(p. 89) **1.** The paired reproductive organs of both sexes are called
gonads.

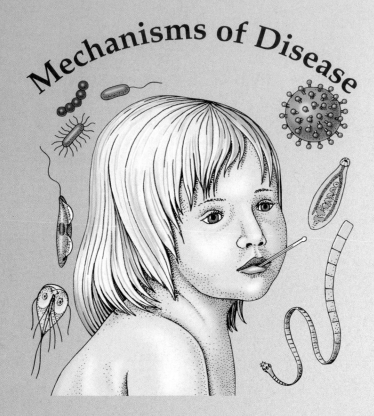

Mechanisms of Disease

CHAPTER OBJECTIVES

1 Describe the relationship between homeostasis and disease.

2 Distinguish among the various types of diseases.

3 Define acute, chronic, and latent diseases.

4 Distinguish between a symptom, sign, and syndrome; and a diagnosis and prognosis.

5 List and describe the major types of disease-causing organisms.

6 List the major pathways that allow pathogens to enter or exit the body.

7 Describe three mechanisms of disease transmission and give examples of transmitted diseases.

8 Distinguish between sterilization and disinfection.

9 Describe the techniques involved in Universal Precautions.

10 List five modes of action that antibiotics have on bacteria.

11 Distinguish between endemic, epidemic, and pandemic diseases.

INTRODUCTION

The 11 organ systems of the human body act both independently and cooperatively to maintain homeostasis. ∞ *[p. 5]* This book highlights the dynamic nature of the interactions among organ systems and the flexibility that makes homeostatic control possible. As you will see, a variety of control mechanisms operate to keep vital characteristics, such as the levels of oxygen, carbon dioxide, glucose, and blood pressure within the relatively narrow limits required for normal function and good health. The ability to maintain homeostatic balance varies with the age, general health, and genetic makeup of the individual.

Disease results from the failure to maintain homeostatic conditions. The disease process initially affects a tissue, an organ, or a system, but it may ultimately lead to changes in the function or structure of cells throughout the body. This chapter will examine the different kinds and causes of diseases, as well as their diagnosis, spread, and control.

DISEASES

Human diseases are caused by infectious organisms, inherited structural or functional genetic defects, environmental factors, or any combination of these factors. The study of the causes or origins of diseases is called **etiology** (e-tē-OL-o-jē).

A CLASSIFICATION OF DISEASES

There are many different types of diseases, and most will be discussed at some point in this text. The major disease categories include:

- **Infectious diseases** are transmitted to a person from the environment or, in the case of a *contagious* infectious disease, from another person. An **infection** is the invasion of the body by another organism, and its subsequent growth and multiplication within body tissues. Infections may be caused by viruses, bacteria, fungi, protozoans, various types of worms, and some arthropods. Infectious diseases are very common—colds and flu are two obvious examples.

- **Inherited diseases** are caused by abnormal genes that are passed on from one generation to the next. They may be due to abnormal chromosomes or errors in the nucleotide sequences of individual genes. Inherited diseases disrupt normal physiology, making normal homeostatic control difficult or impossible. Examples include the *lysosomal storage diseases* (p. 53), and *sickle cell anemia* (discussed in Chapter 14).

- **Neoplastic diseases** are characterized by abnormal cell growth and replication, leading to the formation of benign or malignant tumors. Known causes of neoplastic diseases include genetic factors and environmental factors, such as chemicals, radiation, and viruses. The net result is a loss of normal control mechanisms that operate at the cellular level to control cell growth and the rate of cell division.

- **Immunity-related diseases** develop when immune function either deteriorates, so that the body becomes unable to defend itself, or becomes abnormal, so that immune defenses begin attacking normal tissues. Examples of immunity-related diseases include allergies, such as "hay fever," immune deficiency diseases, such as *AIDS (acquired immune deficiency syndrome)*, and autoimmune disorders such as *myasthenia gravis*. These disorders will be considered in more detail in Chapters 16 and 9, respectively.

- **Degenerative diseases** are disorders associated with aging. Many systems become less adaptable and less efficient as part of the aging process. For example, there are significant reductions in bone mass, respiratory capacity, cardiac efficiency, and kidney filtration as part of the aging process. If older individuals are exposed to stresses that their weakened systems cannot tolerate, symptoms of disease result.

- **Nutritional deficiency diseases** result when an individual's diet is inadequate in terms of the amount or type of proteins, essential amino acids, essential fatty acids, vitamins, minerals, or water. *Kwashiorkor*, a protein deficiency disease, and *scurvy*, a disease caused by vitamin C deficiency, are two examples noted in later chapters. Excessive consumption of high-calorie foods, fats, or fat-soluble vitamins can also cause disease.

- **Endocrine diseases**, or *disorders*, result from excessive or inadequate levels of hormone production. For example, inadequate production of the hormone *insulin* by endocrine cells of the pancreas can lead to one form of *diabetes mellitus*. Chapter 13 will discuss various endocrine disorders.

- **Iatrogenic** (ī-at-rō-JEN-ik) **diseases** result from the activity or treatments of physicians or other health-care providers (*iatros* means "physician" and *gennan* means "to produce.") Iatrogenic diseases include those caused by scar tissue formation after surgery, adverse reactions to drugs, and infections acquired while in a hospital or other medical facility. The latter are also called *nosocomial* (nōs-ō-kō-mē-al) *infections* (*nosos* means "disease" and *komeion* means "take care of.")

- **Environmental diseases** results from trauma or exposure to environmental poisons or toxins.

- **Idiopathic** (id-ē-ō-PATH-ik) **diseases** are diseases whose causes are as yet unknown.

Any specific disease can vary in terms of its intensity and duration. An **acute disease** is one that develops rapidly and is often severe. Examples range from a bad case of the flu to a heart attack. A **chronic disease** persists for an extended period. It may develop gradually, or an acute disease may become chronic. In either case, the effects are initially less severe as those of an acute disease. A chronic disease, can eventually become as life-threatening as any acute disease. Tuberculosis and hypertension (high blood pressure) are examples of chronic diseases. A **latent disease**, or *relapsing disease*, appears, disappears, and reappears over time. Latent diseases can result from exposure to specific environmental factors, or to pathogens. Examples of latent diseases caused by pathogens include *shingles*, caused by the *Herpes varicella* virus, and *malaria*, caused by a protozoan infection.

THE DIAGNOSIS OF DISEASE

pathos, disease + -*logy study of*
pathology: the study of disease

Pathology (pah-THOL-o-jē) is the study of disease, and **pathophysiology** is the study of functional changes caused by disease processes. Different diseases can often produce similar signs and symptoms. For example, a patient with pale skin and complaining of a lack of energy and breathlessness may have (1) respiratory problems that prevent normal oxygen transfer to the blood (as in *emphysema)*, or (2) cardiovascular problems that interfere with normal blood circulation to the rest of the body (as in heart failure) or (3) reduced oxygen-carrying capacity of the blood (as in *anemia*). In such cases, doctors must collect appropriate information to determine the source of the problem. A **diagnosis** is a decision about the nature of an illness. The diagnostic process is often a process of elimination, where a number of potential causes are evaluated and the most likely selected. When uncertainties exist, additional testing may be needed to reach a specific diagnosis.

Returning to the example above, if tests indicate that *anemia* (a low number of red blood cells) is responsible for these symptoms, the specific type of anemia must then be determined so appropriate treatment can begin. After all, the treatment for anemia due to a dietary iron deficiency will be very different from the treatment for anemia due to internal bleeding. Of course, identification of the probable cause of anemia is easier when you are familiar with the physical and chemical structure of red blood cells, and their role in the transport of oxygen.

This brings us to a key concept: *All diagnostic procedures assume an understanding of the normal structure and function of the human body.*

Symptoms, Signs, And Syndromes

When disease processes affect normal functions, the changes are called the *symptoms* and *signs* of the disease. An accurate diagnosis, or identification of the disease, is accomplished through the observation and evaluation of signs and symptoms.

A **symptom** is the patient's subjective perception of a change in normal body function. Examples of symptoms include nausea, malaise (mal-ĀZ) (discomfort), and pain. Symptoms are difficult to measure, and one must rely on asking appropriate questions. Examples of typical questions include:

"When did you first notice this symptom?"
"What does it feel like?"
"Does it come and go, or does it always feel the same?"
"Does anything make it feel better or worse?"

The answers provide information on the duration, sensations, recurrence, and potential triggering mechanisms for the symptoms important to the patient.

Pain, an important symptom of many illnesses, is often an indication of tissue injury. The flow chart in Figure 6-1• demonstrates the types of pain and intro-

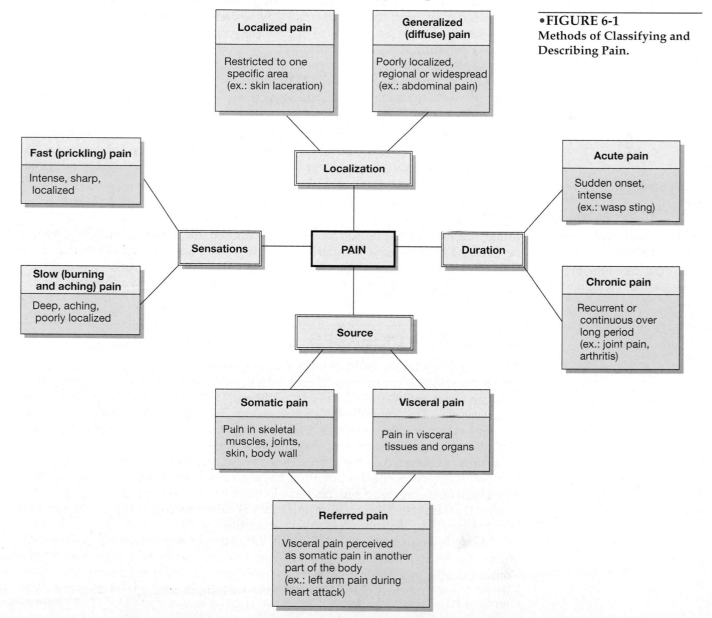

•**FIGURE 6-1**
Methods of Classifying and Describing Pain.

duces some related terms. Pain sensations and pathways are discussed in Chapter 12.

A **sign** is an objective physical characteristic of the disease. Unlike symptoms, signs can be measured and observed through sight, hearing, smell, or touch. The yellow color of the skin (caused by liver dysfunction) or a detectable breast lump are signs of disease.

A **syndrome** (SIN-drōm) is a combination of symptoms and signs that occur together and that are characteristic of a specific disease.

Steps In Diagnosis

A person experiencing serious symptoms usually seeks professional help, and thereby becomes a patient. The consultant, whether a nurse, physician, or emergency medical technician, must determine the need for medical care based on observation and assessment of the patient's symptoms and signs. This is the process of *diagnosis*: the identification of a pathological process by its characteristic symptoms and signs.

Diagnosis is a lot like assembling a jigsaw puzzle. The more pieces (clues) available, the more complete the picture will be. The process of diagnosis is one of deduction and follows an orderly sequence of steps:

1. *The medical history of the patient is obtained:* The past medical history is a concise summary of past illnesses, surgeries, and treatments, general factors that may affect the function of body systems, and the health history of the family. The present medical history is a summary of the current illness—its onset, duration, and symptoms.

2. *A physical examination is performed:* It is a basic part of the diagnostic process. Recall from Chapter 1 (p. 14) that the common techniques involved in a physical examination include *inspection*, *palpation*, *percussion*, and *auscultation*. In addition, a patient's *vital signs* are also recorded.

3. *If necessary, diagnostic procedures are performed:* The physical examination alone may not provide enough information to permit a precise diagnosis. Directed diagnostic procedures can then be used to focus on abnormalities revealed by the history and physical examination.

After The Diagnosis

Once the nature of the illness has been determined, the clinician will make a *prognosis* of the disease. Like a weather forecast, a **prognosis** (prog-NŌ-sis) is a prediction of the probable outcome of the specific disease and the prospect for recovery. The plan of treatment that is then prescribed is called a **therapy**. Therapy means a "service done to the sick."

✔ What is etiology?

✔ What terms are used to distinguish between diseases that progress rapidly and those that develop gradually?

✔ What is the difference between a symptom and a sign of disease?

INFECTIOUS DISEASES

Most people are all too familiar with infectious diseases, thanks to their annual experiences with the common cold and with various forms of the flu. An organism that can cause an infection is called a **pathogen** (PATH-ō-jen). Pathogens come in a variety of sizes—from viruses a few nanometers in diameter to tapeworms that can be a meter or more in length (Figure 6-2●). **Parasites** are relatively large multicellular pathogens. Microscopic pathogens, such as bacteria, viruses, protozoa, and fungi, are often called *microbes*, or *germs*. Microbes are found everywhere in our environment. Some are potential pathogens, whereas others are harmless or even beneficial (for example, bacteria living in our digestive tract that produce essential vitamins). The study of microbes is called **microbiology,** and you will learn more about microbiology in a later section.

Many different terms are used to describe infections or infectious diseases. An infectious disease that can be spread from one individual to another is called a **communicable infectious disease**. Some pathogens can be spread more easily than others. Those that are highly communicable are termed **contagious diseases**. Contagious childhood diseases include *rubella (red measles)* and *chickenpox. Influenza* (flu) is contagious among adults, and especially dangerous for the elderly.

Noncommunicable infectious diseases are not spread from one individual to another; that is, they cannot be "caught" by contact with an infected person. Examples include *food poisoning* (caused by consuming bacterially contaminated food), or *tetanus*, an infection caused by soil bacteria that enter the body through an open wound.

Infectious diseases can also be characterized by their location within the body. A **local infection** is confined to a specific area of the body. The common cold, for example, usually confines itself to the nasal cavities and sinuses. A **systemic infection**, or *generalized infection*, affects most of the body, and the pathogens are widely distributed in many tissues. Typhoid fever, for example, is a systemic infection. *Septicemia*, sometimes called *blood poisoning,* is an example of a systemic infection caused by pathogens multiplying in the bloodstream.

As you proceed through this text, you will encounter many examples of infectious diseases and references to a variety of pathogenic organisms. You will therefore need to become familiar with the major types of pathogens responsible for infectious diseases.

•**FIGURE 6-2**
Representative Pathogens.
(a) A bacterium, with prokaryotic characteristics indicated. Compare with Figure 3-2 (p. 40), which shows a representative eukaryotic cell. **(b)** A typical virus. Each virus has an inner chamber containing nucleic acid, surrounded by a protein capsid or an inner capsid and an outer membranous envelope. **(c)** Protozoan pathogens. Protozoa are eukaryotic, single-celled organisms, common in soil and water. **(d)** Multicellular parasites. Several different groups of organisms are human parasites and many have complex life cycles. Note: These illustrations are not drawn to a common scale.

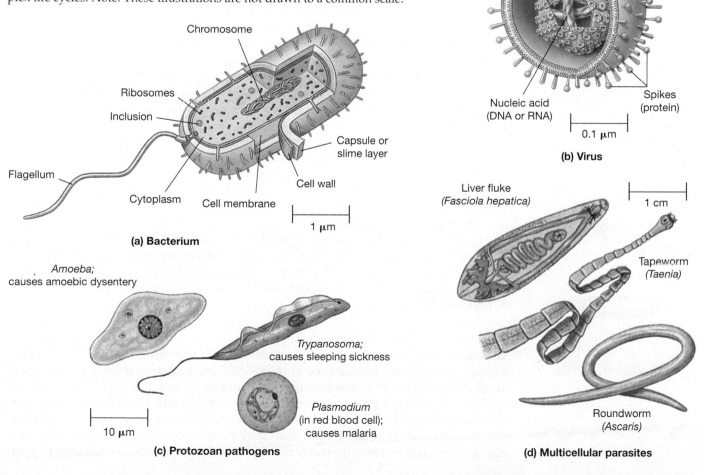

(a) Bacterium

(b) Virus

(c) Protozoan pathogens

(d) Multicellular parasites

AN INTRODUCTION TO PATHOGENS

Chapter 3 of the text presented the structure of a "typical" cell. The cell pictured in Figure 3-2• (p. 40), and described in that chapter, is that of a **eukaryotic** (ū-kar-ē-OT-ik) **cell**. The term, eukaryote, means "true nucleus," and all eukaryotic cells contain a nucleus. All multicellular animals and plants (as well as some single-celled organisms) are composed of eukaryotic cells. Eukaryotic pathogens include *protozoa, fungi, helminths* (worms), and various *arthropods*.

The eukaryotic plan of organization is not the only one found in the living world. **Bacteria** are unicellular organisms, but they lack a nucleus and are not eukaryotic cells. **Viruses** are pathogens that are much simpler in organization than cells.

BACTERIA

bakterion, little rod + *-logy, study of* bacteriology: the study of bacteria

The study of bacteria is called **bacteriology**. All bacteria are *prokaryotic cells*. **Prokaryotic cells** do not have a nucleus or other membranous organelles. Their cell membranes are usually surrounded by a semi-rigid cell wall made of carbohydrate and protein. Figure 6-2a• shows the structure of a representative bacterium.

Bacteria are usually characterized by shape, by their reaction to specific stains, and by their environmental requirements.

Bacterial Classification

Shape.   Bacteria are usually less than 2 µm (micrometer; 1,000 µm = 1 mm) in diameter. Figure 6-3• shows the three basic shapes of bacteria—round, rodlike, and spiral. A round-shaped bacterium is called a **coccus** (KOK-us) and, if more than one, *cocci* (KOK-sī). A rodlike bacterium is a **bacillus** (ba-SIL-us), or *bacilli* (bas-IL-ī) in the plural. Shapes of spiral bacteria vary and so do their names. A **vibrio** (VIB-rē-ō) is comma-shaped, a **spirillum** (spī-RIL-um; plural, *spirilla*) is rigid and wavy, and a **spirochete** (SPĪ-rō-kēt) is shaped like a corkscrew.

Some cocci and bacilli may form various groupings of cells. The Latin names used to describe these arrangements are also used to identify specific bacteria. For example, pairs of cocci bacteria are called *diplococci* (diplo- means double); *streptococci* and *streptobacilli* form twisted chains of cells (strepto- means twisted); and *staphylococci* look like a bunch of grapes (staphylē means a bunch of grapes).

Reaction to Stains.   A **stain** is a dye, and microbiologists use stains to color cellular structures. The *Gram stain* is the most common staining technique used to distinguish among different types of bacteria. The **Gram stain** is a *differential stain* because it relies on the differences in binding strength of two different colored stains on similar cellular structures of various kinds of bacteria. *Gram-positive* cells will stain purple using this technique, whereas *Gram-negative* cells will stain pink.

Environmental requirements.   Some of the environmental factors that affect the growth of bacteria include pH, temperature, and oxygen. Most of the bacteria responsible for human diseases grow well at normal body fluid pH and temperature. Oxygen is not always a requirement for bacterial growth. Bacteria that require oxygen to grow are called **aerobes** and those that do not require oxygen are called **anaerobes**.

Bacterial Replication

Bacteria commonly reproduce by cell division. Generation times, or rates of division, vary among bacteria, from every 20–30 minutes in most types to 12–18 hours in forms such as the tuberculosis-causing *Mycobacterium*. Many bacteria also have the ability to form *spores* under adverse environmental conditions. Spore formation occurs when a portion of the cytoplasm shrinks and becomes surrounded by a thick cell wall. Because they form inside the cell, bacterial spores are called **endospores**. The endospores are highly resistant to extremes of tem-

•**FIGURE 6-3**
Common Bacterial Shapes.

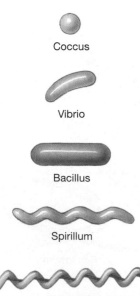

Coccus

Vibrio

Bacillus

Spirillum

Spirochete

perature and moisture. Under the proper conditions, they will split open and cell growth will begin again. *Tetanus* and *anthrax* are diseases that may be spread by bacterial spores.

Many bacteria are quite harmless unless they occur in the wrong place or in extremely high numbers. Many more—including some that live within our bodies—are actually beneficial to us in a variety of ways. Our skin provides a home for many bacteria. These harmless bacteria provide us with indirect benefits because they compete with harmful bacteria, preventing them from attaching and finding sites from which they could invade underlying tissues and produce dangerous infections.

Other bacteria are dangerous pathogens that will destroy body tissues if given the least opportunity. These bacteria are dangerous because as they grow and reproduce they absorb nutrients and release enzymes that damage cells and tissues. A few pathogenic bacteria also release toxic chemicals, or *toxins*.

Two other groups of bacteria, once thought to be viruses because of their extremely small size and living habits, are the *rickettsias* and *chlamydias*. Like viruses, these forms can only live within other living cells. Many are human pathogens. The rickettsias live in both human cells and the cells of lice, fleas, ticks, and mites. These small arthropods, discussed in a later section, spread rickettsial diseases.

Table 6-1 provides some examples of different types of bacteria, the diseases they cause, and the major organ systems that are infected. These and other bacterial infections are discussed in the following chapters of specific organ systems in which they occur.

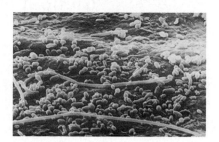

A magnified (40,800×) view of bacteria on human skin.

VIRUSES

Viruses are another microscopic type of pathogen. Their study is called **virology** (vi-ROL-o-jē). Unlike other pathogens, however, viruses are not cells. They are classified as *infectious agents* because they must enter cells (either prokaryotic or

Table 6-1	Examples of Bacterial Diseases and Primary Organ Systems Affected	
Organism	**Disease**	**Affected Organ System**
Bacilli		
Bacillus anthracis	Anthrax	Integumentary System
Mycobacterium tuberculosis	Tuberculosis	Respiratory System
Corynebacterium diphtheriae	Diphtheria	Respiratory System
Cocci		
Staphylococcus aureus	Various skin infections	Integumentary System
Streptococcus pyogenes	Pharyngitis (strep throat)	Respiratory System
Neisseria gonorrheae	Gonorrhea	Reproductive System
Vibrios		
Vibrio cholerae	Cholera	Digestive System
Spirochetes		
Treponema pallidum	Syphilis	Reproductive System, Nervous System
Borrelia burgdorferi	Lyme disease	Skeletal System (joints)
Rickettsias		
Rickettsia prowazekii	Epidemic typhus fever	Cardiovascular System, Integumentary System
Coxiella burnetii	Q fever	Respiratory System
Chlamydias		
Chlamydia trachomatis	Trachoma (eye infections)	Integumentary system
	Lymphogranuloma-venereum (LGV)	Reproductive System

⚕ Clinical Note

Are Any Viruses Helpful?

Viruses are now becoming important as benefactors, as well as adversaries. In genetic engineering procedures, viruses whose nucleic acid structure has been intentionally altered can be used to transfer copies of normal human genes into the cells of individuals with inherited enzymatic disorders. This was the method used to insert the gene for an enzyme missing in patients with *ADA*, or *adenosine deaminase deficiency*. This condition is quite rare and affects only about 20 children worldwide each year. Without the enzyme adenosine deaminase, toxic chemicals build up in cells of the immune system, and as these cells die,

the body's immune defenses break down. The resulting disease is called *severe combined immunodeficiency (SCID)*.

Cystic fibrosis (CF) is a debilitating genetic defect whose most obvious—and potentially deadly—symptoms involve the respiratory system. The underlying problem is an abnormal gene that carries instructions for a chloride ion channel found in cell membranes throughout the body. Researchers have recently treated CF in laboratory animals by inserting the normal gene into a virus that infects cells lining the respiratory passageways. The virus could be given to human patients through an inhalant.

eukaryotic) and use the organelles of those cells to replicate themselves. The only place where viruses can replicate is within living cells.

Viruses come in a variety of sizes and shapes. All are too small to be seen with a light microscope. Figure 6-4• shows the shapes of some viruses and contrasts their sizes with a typical bacteria cell and human liver cell.

Viruses consist of a core of nucleic acid (DNA or RNA) surrounded by a protein coat, or *capsid*. (Some varieties have a membranous outer covering as well.) The structure of a "typical" virus is shown in Figure 6-2b•. Important viral diseases include influenza (flu), yellow fever, some leukemias, AIDS, hepatitis, polio, measles, mumps, rabies, and the common cold. Table 6-2 lists some of these viral diseases, the virus that causes the disease, and the major affected organ system.

PROTOZOA

Protozoa are one-celled, or unicellular, eukaryotic organisms that are abundant in soil and water. Most protozoa are free-living, but some are parasites and cause disease. The protozoa are usually divided into four groups, based on how they move. Figure 6-5• shows an example of each group and their structures used for movement. Protozoan infections occur in red blood cells (*malaria*), and intestinal

•**FIGURE 6-4**

Viruses.

A variety of viruses are shown with a typical bacterial cell, a human liver cell, and a eukaryotic ribosome.

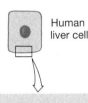

Human liver cell

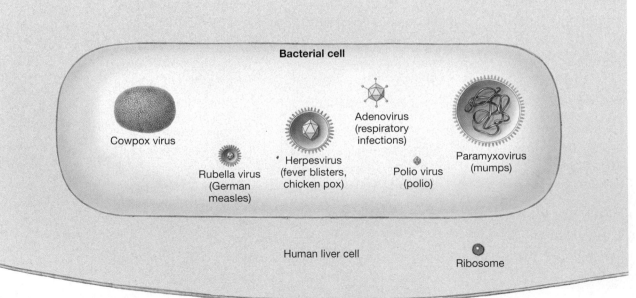

Table 6-2	Examples of Viral Diseases and Primary Organ Systems Affected		
Nucleic Acid Type	**Virus**	**Disease**	**Affected Organ System**
RNA	Influenza A, B, C	Flu	Respiratory System, Reproductive System
	Paromyxovirus	Mumps	Digestive System
	Hepatitis A, C, D, E, G	Infectious hepatitis	Digestive System (liver)
	Rhinovirus	Common cold	Respiratory System
	Human Immunodefiency Virus (HIV)	AIDS	Lymphatic System
DNA	Herpesvirus		
	Herpes simplex 1	Cold sore/fever blister	Integumentary System
	Herpes simplex 2	Genital herpes	Reproductive System
	Varicella-Zoster (VZV)	Chickenpox	Integumentary System
	Varicella-Zoster (VZV)	Shingles	Nervous System
	Hepatitis B	Hepatitis	Digestive System (liver)
	Epstein-Barr	Mononucleosis	Respiratory System

and urinary tracts. Table 6-3 lists examples of important protozoan diseases and the responsible protozoan.

Flagellates

Members of this group have whip-like *flagella* which they use for movement (Figure 6-5a•). Some of the parasitic forms in humans include the *trypanosomes* (which cause African sleeping sickness), *giardias* (diarrhea), and *trichomonads* (vaginal inflammation).

Amoeboids

The **amoeboids** (a-MĒ-boydz) group includes amoeba-like forms that move by extensions of their cytoplasm, or *pseudopodia* (Figure 6-5b•). Most amoeboids are free-living, yet a number of different species can live in the human intestinal tract. One such form, *Entamoeba histolytica*, causes *amoebic dysentery*, a severe diarrhea.

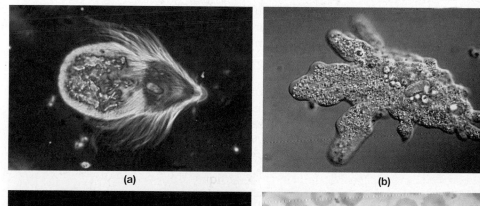

(a)

(b)

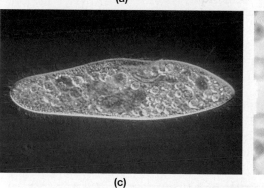

(c)

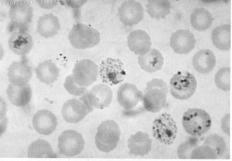

(d)

•**FIGURE 6-5**
Representative Protozoa.
(a) *Trichonympha*, a flagellate from a termite gut. **(b)** *Amoeba proteus* is a free-living form that lives in ponds (350×). **(c)** *Paramecium caudatum*, a free-living ciliate (135×). **(d)** *Plasmodium vivax* (stained cells within red blood cells) causes malaria.

Table 6-3	Examples of Protozoan Diseases and Primary Organ Systems Affected		
Protozoa	**Name (Genus)**	**Disease**	**Affected Organ System**
Flagellates	*Trypanosoma*	African sleeping sickness	Cardiovascular System
	Leishmania	Leishmaniasis	Lymphatic System
	Giardia	Giardiasis	Digestive System
	Trichomonas	Trichomoniasis	Reproductive System
Amoeboids	*Entamoeba*	Amoebic dysentery	Digestive System
Ciliates	*Balantidium*	Dysentery	Digestive System
Sporozoans	*Plasmodium*	Malaria	Cardiovascular System
	Toxoplasma	Toxoplasmosis	Lymphatic System

Ciliates

This is the largest group of protozoans and all have *cilia* over their cell surfaces (Figure 6-5c•). Only one species is known to parasitize humans. It causes diarrhea.

Sporozoans

All of the sporozoans are parasitic and unable to move on their own (Figure 6-5d•). Sporozoans often have complicated life cycles. Malaria is an example of a disease caused by a sporozoan, *Plasmodium vivax*. The *Plasmodium* responsible for malaria requires both mosquito and human hosts for its survival. The life cycle stage within humans infects red blood cells, and this infection causes the alternating chills and fever characteristic of malaria. Infected human red blood cells that are swallowed by biting mosquitoes allow the *Plasmodium* to complete its life cycle in the mosquito. The next generation of *Plasmodium* is then passed on to another unlucky human in the mosquito's next bite.

FUNGI

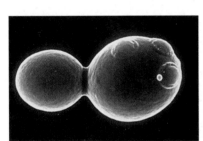

Previous reproductive budding events occurred where the circular scars mark the surface of the larger yeast cell.

Fungi (FUN-jī) (singular *fungus*) are eukaryotic organisms that absorb organic materials from the remains of dead cells. They include the one-celled yeasts and multicellular molds. Mushrooms are familiar examples of very large fungi. Fungi reproduce both asexually and sexually. For example, asexual reproduction in yeasts occurs by the budding of new cells. Fungi also reproduce sexually by forming spores. Unlike bacterial spore formation, in which spores form within cells, fungal spores are formed externally.

In a fungal infection, a microscopic fungus spreads through living tissues, secreting digestive enzymes and toxins, killing cells and absorbing nutrients. Human fungal diseases, called **mycoses** (*myco* means a fungus), vary in their severity. Several relatively common skin conditions (*athlete's foot*) and a few more serious dis-

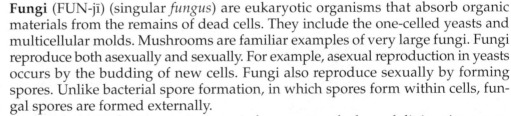

Clinical Note

Hold That Peanut Butter Sandwich!

Some types of molds produce chemical substances called **aflatoxins** (af-lah-TOX-inz). These toxins, considered the most potent cancer-causing compounds yet discovered, may in high concentrations lead to cancer of the liver. Aflatoxins may be found in mold infested foods such as grain and peanuts. One study of peanut butter revealed that 7 percent of the tested samples contained aflatoxins. It has also been shown that similar toxins can be contained in jelly. Preparing a PB&J sandwich from ingredients with high aflatoxin levels using slices of moldy bread could produce a carcinogenic mouthful.

Table 6-4	Examples of Fungal Diseases and Primary Organ Systems Affected	
Organism (Genus)	**Disease**	**Affected Organ System**
Aspergillus	Aspergillosis ("Farmer's lung disease")	Respiratory System
Blastomyces	Blastomycosis	Integumentary System
Histoplasma	Histoplasmosis	Respiratory System
Epidermophyton, Microsporum, and *Trichophyton*	Ringworm tinea capitis (scalp) tinea corporis (body) tinea cruris (groin) tinea unguium (nails)	Integumentary System
Candida	Candidiasis	Integumentary System
Coccidioides	Coccidiomycosis ("San Joaquin valley fever")	Respiratory System

eases (*histoplasmosis*) are the result of fungal infections. Examples of various fungal diseases and the major affected organ systems are listed in Table 6-4.

MULTICELLULAR ANIMALS

Helminths and various *arthropods* make up the larger, multicellular parasitic organisms that can cause diseases. The multiplication of these larger parasites in or on the body is called an **infestation**. **Helminths** are different types of parasitic worms that can live within the body. The major subgroups are **flatworms**, which include the *flukes* and *tapeworms*, and **roundworms**, or *nematodes*. Figure 6-2d● shows some representative helminths. These organisms, which range in size from the microscopic to a meter or more in length, usually cause weakness and discomfort, but do not *by themselves* kill their host. However, complications resulting from the parasitic infection, such as chronic bleeding, malnutrition, or secondary infections by bacterial or viral pathogens, can ultimately prove fatal.

Human fluke infections may be found in the liver (*Clonorchis sinensis*) and lungs, and blood flukes (*Schistosoma japonicum*) spend part of their life cycle in the blood. Tapeworms attach to the intestinal wall and produce long chains of segments filled with eggs. Beef and pork tapeworms, species of *Taenia*, occur in humans.

Roundworms commonly infect the digestive tract. Examples include the pork roundworm, *Trichinella spiralis*, that causes *trichinosis*; the common roundworm, *Ascaris lumbricoides* that causes *ascariasis*; and the pinworm, *Enterobius vermicularis*.

Arthropods make up the largest and most diverse group of living organisms on the Earth. Their shared characteristics include jointed legs and an exoskeleton. They can be found in the soil, on plants, in both fresh and salt water, and as parasites on both plants and animals.

The major arthropods that affect humans are the eight-legged **arachnids**, such as scorpions, spiders, and parasitic *mites* and *ticks*, and the six-legged **insects**, such as *mosquitoes*, some *flies, lice, fleas,* and *bedbugs* (Figure 6-6●). Arthropods may bite humans, producing localized effects, or live as external parasites on humans or other animals. In either of these cases they may also act as *vectors of disease*. A **vector** is an organism that transfers a parasite, or infectious organism, to a new host or another person.

Table 6-5 lists some examples of the diseases caused by multicellular helminth and arthropod parasites.

•FIGURE 6-6
Representative Disease-Carrying Arthropods.
(a) A wood tick, *Dermacentor andersoni*. **(b)** A pubic, or crab, louse (*Phthirus pubis*), holding on to a human pubic hair (55×). Lice suck blood, feeding some five times a day. **(c)** The housefly, *Musca domestica*, can transport microbes on its body (4×). **(d)** The *Aedes* mosquito, a vector for dengue fever. **(e)** A common flea, *Ctenocephalidis canis* (33×).

(a)

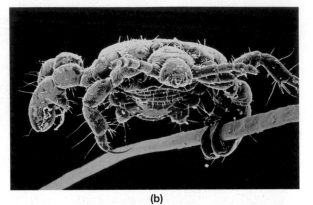

(b)

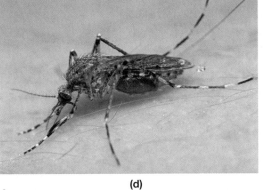

(c)　　　　　(d)　　　　　(e)

Table 6-5	Examples of Diseases Caused by Multicellular Parasites and Primary Organ Systems Affected		
Group	**Organism**	**Disease/Condition**	**Affected Organ System**
Helminths			
Roundworms	*Ascaris*	Intestinal infestation	Digestive System
	Enterobius	Pinworm infestation	Digestive System
	Wucheria	Elephantiasis	Lymphatic System
Flatworms			
Flukes	*Fascioloa, Clonorchis* (liver flukes)	Fascioliasis	Digestive System
	Schistosoma (blood fluke)	Schistosomiasis	Cardiovascular System
Tapeworms	*Taenia*	Tapeworm infestation	Digestive System
Arthropods			
Arachnids (8-legged)	Mites	Vectors of bacterial and rickettsial diseases	Various Systems
	Ticks	Vectors of bacterial and rickettsial diseases	Various Systems
	Spiders, scorpions	Bites may cause inflammation	Integumentary System
Insects (6-legged)	Lice	Vectors of bacterial and rickettsial diseases	Various Systems
	Human louse	Pediculosis	Integumentary System
	Mosquitoes	Vectors of bacterial and rickettsial diseases	Various Systems
	Flies	Passive carriers of bacterial diseases	Various Systems
	Wasps, bees	Bites may cause inflammation	Various Systems

TRANSMISSION OF PATHOGENS

As dangerous or threatening as pathogens may seem, most pathogens that affect humans cannot survive outside the body for very long. For new human infections to spread there must be ways for pathogens to enter and exit the body, thereby spreading their infection from individual to individual.

Portals Of Entry

Pathogens must first enter the body to cause an infection. The sites at which pathogens can enter the body to infect tissues are called **portals of entry**. Common portals of entry are shown in Figure 6-7•. They include breaks in the skin and the

•**FIGURE 6-7**
Portals of Entry for Human Pathogens.

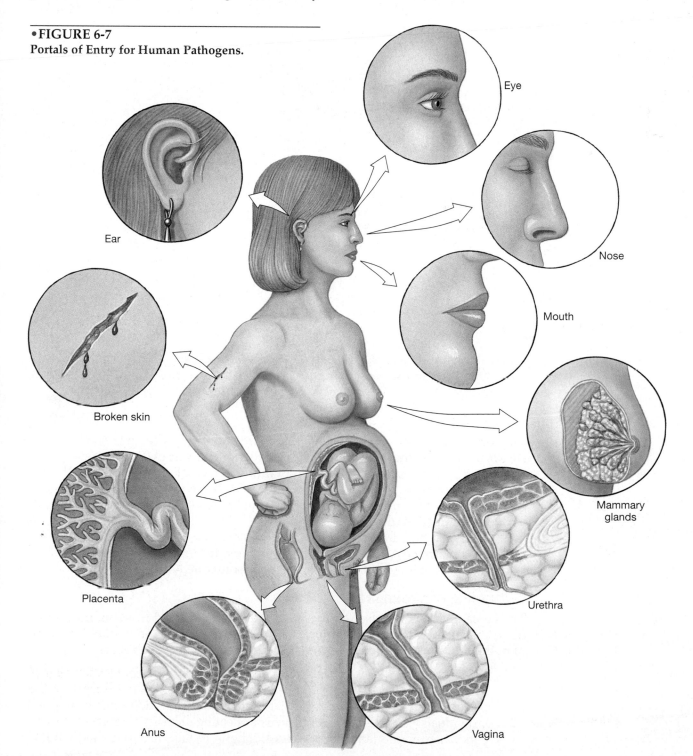

•FIGURE 6-8
Portals of Exit for Human Pathogens.

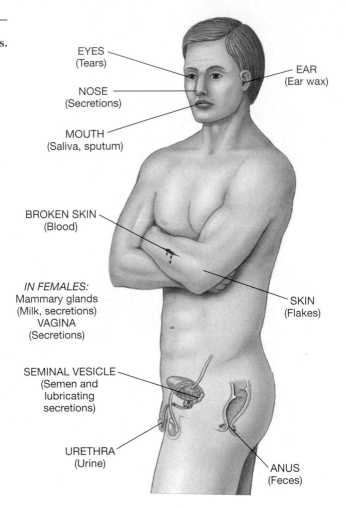

EYES
(Tears)

EAR
(Ear wax)

NOSE
(Secretions)

MOUTH
(Saliva, sputum)

BROKEN SKIN
(Blood)

IN FEMALES:
Mammary glands
(Milk, secretions)
VAGINA
(Secretions)

SKIN
(Flakes)

SEMINAL VESICLE
(Semen and
lubricating
secretions)

URETHRA
(Urine)

ANUS
(Feces)

mucous membranes of the digestive, respiratory, urinary, and reproductive systems. Openings to the outside of the body, such as ears, nose, mouth, eyes, anus, urethra, and vagina, allow microorganisms to enter.

Portals Of Exit

The spread of diseases depends on pathogens leaving an infected individual. The sites where organisms leave the body are called **portals of exit**. Pathogens generally pass through such portals with body fluids or feces. Figure 6-8• notes the different types of body fluids and where they exit. Body fluids may contain many different types of pathogens, and feces are the main portal of exit for the eggs of various helminth worms.

Semen and blood are important means by which viruses can exit the body. Both can contain the AIDS viruses (HIV) and hepatitis viruses. Blood can be an important source of infection to anyone helping someone with an injury.

Spreading of Diseases

The spread of diseases also requires pathogens to be carried from an infected individual to an uninfected individual. Figure 6-9• illustrates three different means of pathogen transmission and examples of such transmitted diseases.

1. *Person-to-person transmission.* **Direct contact** requires body contact between individuals (such as shaking hands, kissing, or sexual contact). **Indirect contact** occurs through nonliving objects that contain pathogens deposited by infected individuals (such as soiled handkerchiefs or eating utensils). **Droplet contact** occurs when an infected person sneezes, coughs, or talks near other individuals.

•FIGURE 6-9
Mechanisms of Disease Transmission.

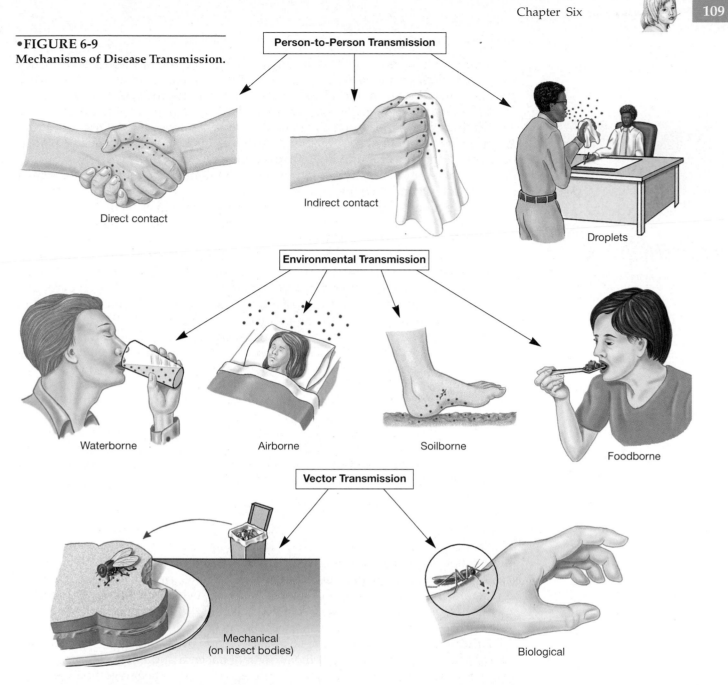

Person-to-Person Transmission

Direct contact

Indirect contact

Droplets

Environmental Transmission

Waterborne

Airborne

Soilborne

Foodborne

Vector Transmission

Mechanical
(on insect bodies)

Biological

2. *Environmental transmission.* Nonliving carriers of infectious pathogens in the environment include water, air, soil, and food.

3. *Vector transmission.* Living organisms that transmit diseases are called vectors. **Mechanical vectors** carry pathogens on their legs or bodies passively. Flies which land on fecal material and then on food are common examples of mechanical vectors. Insects which carry pathogens that must complete part of their life cycle in another organism are called **biological vectors**. The mosquito that carries the *Plasmodium* parasite is an example of a biological vector.

THE CONTROL OF PATHOGENS

Our attempts to control pathogens have often been described as a war between man and microorganism. Sometimes we are ahead, sometimes it's the microorganisms. In this section we will examine some of the methods used in this never-ending conflict. The basic methods of control involve *aseptic techniques*, *chemotherapy*, and *vaccination*.

Aseptic Techniques

Pathogenic organisms may be controlled through the use of **aseptic techniques** such as *sterilization* and *disinfection*. *Sepsis* (SEP-sis) means "decay" and is a toxic condition resulting from the spread of bacteria or their products. In contrast, the condition of **asepsis** (ā-SEP-sis) refers to the lack of pathogenic microorganisms.

Sterilization is the killing or removal of *all* organisms in a material or object. An effective technique used to sterilize surgical instruments is the use of moist heat applied under pressure in an instrument called an **autoclave** (AW-tō-klāv). An autoclave produces both a high pressure and temperatures over 121°C which can kill bacteria and their spores. ∞ *[p. 100]*

Dental instruments are sterilized in a counter-top autoclave.

Disinfection is the *reduction* of pathogenic organisms in or on materials so that they pose no risk of disease. Substances that are applied to nonliving objects are called **disinfectants**. Most disinfectants do not kill bacterial spores. Examples of disinfectants include phenol (carbolic acid) and bleach. Disinfectants that are mild enough to be applied to living tissues are called **antiseptics**. Examples include dilute solutions of some disinfectants, isopropyl or ethyl alcohol, and hydrogen peroxide.

Universal Precautions are a set of aseptic techniques that were developed in 1988 in response to the risk of transmission of AIDS and hepatitis B to personnel in health care facilities. These precautions are called universal because they apply to *all* patients, not just those with AIDS and hepatitis B. They are to be used with blood, semen, and tissue samples, as well as vaginal, cerebrospinal, synovial (joint cavity), pleural (lung), peritoneal (abdomen), pericardial (heart), and amniotic fluids.

The Universal Precautions recommend that health care workers wear latex or vinyl gloves if contact with body fluids is likely, and gowns and masks for protection where splashing of body fluids can occur. Contaminated hypodermic needles and other sharp utensils should be handled as little as possible and discarded immediately into puncture-proof containers. Spills should be cleaned up with gloves and disposable towels, and the areas disinfected with a solution of bleach and water.

Chemotherapy

Chemotherapy is the use of various chemical agents to treat diseases and their symptoms. The term "chemotherapy" is often associated with just cancer treatment, when in fact it includes chemical agents that treat all diseases. Such chemical agents are described as chemotherapeutic agents, or *drugs*. These chemical agents may be naturally produced by organisms, such as *antibiotics*, or artificially created, such as *synthetic drugs*. A wide variety of drugs are used to control bacteria, viruses, protozoans, fungi, and helminths (worms).

Antibiotics ("against life") are substances made by microorganisms that have the capacity to destroy or inhibit the growth of bacteria and other organisms. The first use of antibiotics to control pathogens *and* cure disease began in the 1930s with the discovery of *sulfa drugs*. They were followed by the mass production and use of *penicillin* during the 1940s. The sulfa drugs are all synthetic agents, while penicillin was originally isolated from a fungus but is now produced from synthetic sources.

Antibiotics affect bacteria in a number of different ways. Figure 6-10• shows the major mechanisms of antibiotic activity and their targets within bacteria. These mechanisms include; (1) inhibiting cell wall synthesis, (2) disrupting cell membrane function, (3) inhibiting protein synthesis, (4) inhibiting nucleic acid synthesis, and (5) interfering with cellular metabolism. Because bacteria and human cells are different in terms of cell walls, nuclei, and other traits discussed above, antibiotics directed at bacteria usually have minimal effects on humans.

A drawback to unlimited reliance on antibiotics is that microorganisms may develop **antibiotic resistance**. Such pathogens no longer respond to antibiotics that were once used to control or eliminate them. As a result, there is a continual need to develop new antibiotics to combat resistant strains of bacteria.

Unlike bacteria, viruses do not respond to antibiotics. Chemotherapeutic agents, or drugs, that target viruses, called **antiviral agents**, have only recently been developed. Antiviral agents must work within the host's cells, without severely dam-

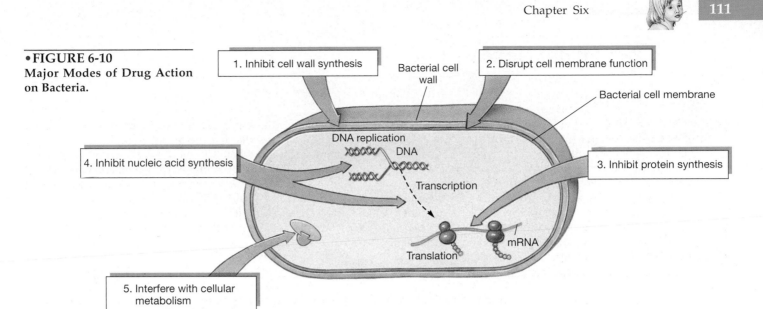

•FIGURE 6-10
Major Modes of Drug Action
on Bacteria.

aging them. As a result, they are designed to interfere with virus replication, rather than killing viruses directly. As with bacteria, viruses may also become resistant to antiviral drugs. This is more of a problem with viruses than bacteria because fewer kinds of antiviral agents have been developed, and they are less widely distributed.

Infections caused by fungi are becoming more common because of increased numbers of individuals with impaired immune systems, as occurs in AIDS patients and in organ transplant recipients treated with immune-suppressing drugs. **Antifungal drugs** can produce quite harmful side effects when given in effective doses. This is because both people and fungi are eukaryotes, and therefore have similar cells. The antifungal drugs currently available affect cell function by interfering with the permeability of the cell membrane and the synthesis of DNA.

Antiprotozoan drugs have been developed to control some diseases (malaria) and infections (vaginal *Trichomonas* infections, and intestinal infections by parasitic amoebas and *Giardia*). Antiprotozoan drugs act by preventing protein synthesis and folic acid synthesis. (Parasitic protozoans require more folic acid than free-living forms.) Some antibiotics have antiprotozoan activity as well.

Some **antihelminth drugs** act as nerve poisons (neurotoxins) that paralyze the muscles of roundworms and pinworms that infect the intestinal tract. Others exert their effects by interfering with the parasitic worm's carbohydrate metabolism.

Vaccination

Vaccination is the administration of a substance, or **vaccine** (vak-SĒN) that stimulates the body's immune system to develop immunity to a specific disease. This process is also called **immunization.** Vaccines contain either pathogens that have been weakened or killed, fragments of pathogens, or toxins released by pathogens.

Epidemiology

The study of the factors involved in the spread of diseases in human populations and communities is called **epidemiology** (ep-i-dē-mē-OL-o-jē). *Epidemiologists* are concerned with such topics as the **incidence,** or frequency of new cases, of infectious diseases in a population; their transmission within and between populations; their causes or origins (etiology); and their control and prevention.

Epidemiologists track a wealth of important data about specific diseases. In addition to recording the *incidence* of a specific disease, they also record its **prevalence,** or number of existing cases within a population. They also monitor the severity of diseases within populations. **Morbidity** (mor-BID-i-tē; from *morbidus* which means sick) is the illness rate, or the ratio of sick people to healthy people in a com-

Clinical Note

What's the CDC?

The **CDC** is a branch of the U.S. Public Health Service. Located in Atlanta, Georgia, the **C**enters for **D**isease **C**ontrol and Prevention is charged with the control and prevention of infectious diseases and other preventable conditions. The CDC's activities are quite varied and include: collecting and publishing statistics on diseases that are a threat to public health, developing guidelines for quarantines and occupational health and safety (see the discussion of *Universal Precautions* on p. 110), cooperating with other national and international health agencies, recommending treatments for antibiotic-resistant organisms, storing rarely used drugs for treating exotic tropical diseases that may be carried to the U.S., and determining when and where vaccines should be used.

The CDC publishes its continually ongoing epidemiologic studies and related statistics in the *Morbidity and Mortality Weekly Report* (MMWR). The MMWR provides statistics and reports on diseases in different parts of the United States and the world. The MMWR and other reports are available online, at *http://www.cdc.gov*

munity. The death rate due to specific causes is called the **mortality** rate. Both are expressed in terms of the numbers of sick (or dead) per 100,000 persons per year.

Epidemiologists use some general terms to distinguish incidences of diseases. A disease is said to be **endemic** if it is continuously present in a particular region but affecting only small numbers of people. Examples of endemic diseases include mumps and chickenpox. An **epidemic** occurs when a disease spreads rapidly, affecting an unusually large number of people within a particular region. Morbidity rates rise as the disease rapidly spreads. A **pandemic** is very widespread and, in fact, may involve the whole world. The greatest pandemic of this century occurred in 1918–19 when swine flu killed some 20–40 million people.

Epidemiologists' control of communicable diseases involves *isolation*, *quarantine*, *immunization*, and *vector control*. In **isolation**, a sick individual is separated from others to prevent any contact. This most often occurs in hospitals. Unlike isolation, **quarantine** involves separating healthy, or not-yet-sick, individuals from the general population. It is intended to prevent the spread of a disease that has not yet produced disease symptoms in infected individuals.

The immunization of large numbers of people with vaccines is an important and effective method to control communicable diseases. Immunizations have drastically reduced the incidences of polio, measles, mumps, diphtheria, and whooping cough in the United States. Immunized people can be exposed to pathogens and not become sick.

Vector control involves eliminating infectious disease vectors, such as insects or rodents. Such control may simply involve covering windows with screens to prevent the entry of biting insects such as mosquitoes and flies. A more complete level of control may require the use of pesticides and insecticides to directly eliminate vectors or their habitats and breeding areas.

✔ What is a pathogen?
✔ What is a systemic infection?
✔ What is a vector?
✔ What is the difference between morbidity and mortality?

CHAPTER REVIEW

Key Words

antibiotic: Chemical agent that selectively kills pathogenic bacteria and other microorganisms.

chemotherapy: The treatment of disease by chemical agents.

disease: The failure to maintain general homeostasis.

disinfectant: An agent that reduces the number of pathogenic organisms in or on materials so that they pose no risk of disease.

epidemiology: The study of the factors which determine the frequencies and distribution of diseases in a population.

host: An organism that harbors or nourishes another organism.

infection: Invasion and colonization of body tissues by pathogenic organisms.

parasite: An organism which lives within or upon another organism and derives benefits at the other's expense.

pathogen: Any disease-producing microorganism.

pathology: The study of the nature of diseases.

prokaryote: An organism consisting of cells that lack membrane-bound nuclei.

sign: A clinical term for objective evidence of the presence of disease.

symptom: Clinical term for subjective abnormality of function due to the presence of disease.

syndrome: A discrete set of symptoms and signs that occur together.

Selected Clinical Terms

acute: Sudden in onset, severe in intensity, and brief in duration.

chronic: Habitual or long-term.

epidemic: A disease which occurs in many people in a region at the same time.

etiology: The study of the causes or origins of diseases.

idiopathic disease: A disease without a known cause.

nosocomial infection: An infection acquired while in a hospital or other medical facility.

pandemic: A widespread epidemic disease.

Study Outline

INTRODUCTION (p. 95)

1. **Disease** is a failure to maintain homeostasis.

DISEASES (pp. 95–98)

1. Human diseases have infectious and noninfectious causes. *Noninfectious diseases* may produce conditions supportive of **pathogens**, organisms that cause *infectious diseases*.

A CLASSIFICATION OF DISEASES (p. 95)

2. **Infectious diseases** are caused by bacteria, viruses, protozoa, fungi, some arthropods, and various types of worms, or *helminths*.

3. **Inherited diseases** are caused by abnormalities in genetic information that disrupt normal homeostatic control mechanisms.

4. **Neoplastic diseases** result from abnormal cell growth, such as tumors, that may be caused by chemicals, radiation, viruses, and other factors.

5. **Immunity-related diseases** occur when the immune system fails to respond normally. Examples include allergies, autoimmune disorders, and AIDS.

6. **Degenerative diseases** occur with aging.

7. **Nutritional deficiency diseases** result from inadequate diets. *Kwashiorkor* is associated with protein deficiency and *scurvy* with vitamin C deficiency.

8. **Endocrine diseases**, or *disorders*, result from the under- or over-abundance of hormone secretions.

9. **Iatrogenic diseases** are caused by the actions or treatments of physicians or other health-care providers. *Nosocomial infections* are infections acquired in a health care facility.

10. **Environmental diseases** are caused by trauma or exposure to poisons or toxins.

11. **Acute infectious diseases** develop and decline rapidly. **Chronic diseases** develop slowly and last a longer time. **Latent diseases** appear, disappear, and reappear again after the initial infection.

THE DIAGNOSIS OF DISEASE (p. 96)

12. **Pathology** is the study of disease; **pathophysiology** is the study of the functional changes caused by disease. A **diagnosis** distinguishes one disease from another.

13. A **symptom** is the perception of a change in normal body function, such as nausea, discomfort, or pain. A **sign** is a visible or otherwise objective evidence of disease. A **syndrome** is a recognizable combination of symptoms and signs. *(Figure 6-1)*

14. An accurate diagnosis includes obtaining a patient's medical history, performing a physical examination, performing diagnostic tests on the individual, and conducting laboratory tests on tissue, blood or other sampled materials.

15. A **prognosis** is a prediction of the effect(s) of a specific disease and the prospect for recovery. The prescribed treatment for recovery is called a **therapy**.

INFECTIOUS DISEASES (p. 98–112)

1. **Communicable infectious diseases** that are spread from person to person are **contagious diseases**. **Noncommunicable infectious diseases** cannot be caught from infected individuals.

AN INTRODUCTION TO PATHOGENS (p. 100)

2. Eukaryotic pathogens include *protozoans, fungi, helminths* (worms), and *arthropods*. Prokaryotic pathogens are *bacteria*. Unlike eukaryotes and prokaryotes, *viruses* do not consist of cells. Most human pathogens are bacteria and viruses.

BACTERIA (p. 100)

3. **Bacteria** are microscopic and consist of prokaryotic cells (cells lacking a membrane-bound nucleus). *(Figure 6-2a)*

4. Bacteria occur in three general shapes; round (**coccus**), rod-shaped (**bacillus**), and spiral (**vibrio**, **spirillum**, and **spirochete**). Some produce **endospores** when conditions become unfavorable. *(Figure 6-3)*

5. Bacteria are distinguished by factors such as size, shape, their reactions to stains or dyes (**Gram stain**), and their physical requirements for growth (factors such as *pH*, *temperature*, and *oxygen*).

6. Many bacteria live on or within our bodies and do us no harm, others are pathogenic. *(Table 6-1)*

VIRUSES (p. 101)

7. **Virology** is the study of viruses. **Viruses** are not cells and require living cells in which to replicate themselves.

8. Viruses consist of an outer protein coat (*capsid*) that contains a core of nucleic acid (DNA or RNA). *(Figures 6-2b, 6-4)*

9. Viruses cause a large number of human diseases, although some non-pathogenic forms are now being used in genetic engineering to transfer genes. *(Table 6-2)*

PROTOZOA (p. 102)

10. Protozoa are one-celled, eukaryotic microorganisms. They are classified into four groups according to their methods of movement; *flagellates, amoeboids, ciliates*, and *sporozoans*. *(Figures 6-2c, 6-5)*

11. Most protozoa are free-living, but some are pathogenic. *(Table 6-3)*

12. The **flagellates** use **flagella** for locomotion and pathogenic forms include the *trypanosomes, giardias*, and *trichomonads*.

13. The **amoeboids** move by **pseudopodia**, or cytoplasmic extensions. One pathogenic form causes amoebic dysentery, a severe diarrhea.

14. Ciliates move by surface **cilia**. Only one species is known to infect humans.

15. Sporozoans cannot move on their own and all are parasitic. *Plasmodium vivax* causes malaria in humans.

FUNGI *(p. 104)*

16. The **fungi** include the molds and one-celled yeasts. Yeasts reproduce by budding and forming spores.

17. Human fungal diseases are called **mycoses**. *(Table 6-4)*

18. Multicellular animals that can cause human diseases include the *helminths* (worms) and *arthropods*.

MULTICELLULAR ANIMALS *(p. 105)*

19. Helminths include parasitic **flatworms** (*flukes* and *tapeworms*) and **roundworms** (*nematodes*). They do not normally kill their human host, but can lead to malnutrition and fatal secondary infections. *(Figure 6-2d)*

20. Arthropods have jointed legs and an exoskeleton. Most cause disease by being a vector for microorganisms. Major arthropod groups that affect humans are the **arachnids** (scorpions, spiders, and *parasitic mites* and *ticks*) and the **insects** (*mosquitoes, flies, lice, fleas,* and *bedbugs*). *(Figure 6-6; Table 6-5)*

TRANSMISSION OF PATHOGENS *(p. 107)*

21. Pathogens enter an individual through **portals of entry**. Portals of entry include the skin, mucous membranes, and openings to the interior of the body. *(Figure 6-7)*

22. Pathogens leave an infected individual through **portals of exit**, often in body fluids or feces. *(Figure 6-8)*

23. Pathogens are transmitted from one individual to another by (1) *person-to-person transmission*, (2) *environmental transmission*, and (3) *vector* (living organism) *transmission*.

24. Person-to-person transmission includes direct, indirect, and droplet contact. **Environmental transmission** occurs through contact with pathogens in water, air, soil, and food. **Vector transmission** includes *mechanical* and *biological vectors*. **Mechanical vectors** carry pathogens passively on their bodies; **biological vectors** carry pathogens that complete part of their life cycle in another organism. *(Figure 6-9)*

THE CONTROL OF PATHOGENS *(p. 109)*

25. *Aseptic techniques, chemotherapy,* and *vaccination* provide the most effective means of pathogen control.

26. Aseptic techniques include *sterilization* and *disinfection*. **Sterilization** kills or removes all pathogenic organisms in a material or object. **Disinfection** reduces the number of pathogenic organisms so that they pose no risk of disease.

27. Universal Precautions are a group of aseptic techniques that are applied to the handling of body fluids of *all* patients. They include the use of gloves, aprons, and masks; cleaning with disinfectants; and the prompt and careful disposal of needles and body fluids.

28. Chemotherapy is the treatment of disease by chemical agents, or *drugs*. Examples of chemical agents include *antibiotics*, produced by organisms, and artificially created *synthetic* drugs.

29. Antibiotics are chemical agents usually made by microorganisms that selectively kill pathogenic organisms. They act on bacteria either by (1) inhibiting cell wall synthesis, (2) disrupting cell membrane function, (3) inhibiting protein synthesis, (4) inhibiting nucleic acid synthesis, or (5) interfering with cellular metabolism. *(Figure 6-10)*

30. Vaccination is the administration of a **vaccine** that stimulates the development of immunity to a specific disease.

31. Epidemiology is the study of the factors involved in the spread of disease in human communities and populations.

32. Morbidity is a measure of illness in a community and **mortality** measures the death rate in a community. General terms that describe the incidence or frequency of disease include: (1) **endemic**, with disease always present in a particular region but affecting small numbers, (2) **epidemic**, where disease occurs in much higher numbers than usual and spreads rapidly within a particular region, and (3) **pandemic**, with a very wide-spread disease.

33. The Centers for Disease Control and Prevention (**CDC**) is a federal government agency that, among its other functions, publishes weekly summaries of the morbidity and mortality rates of infectious diseases that occur in the U.S.

Review Questions

MATCHING

Match each item in Column A with the most closely related item in Column B. Use letters for answers in the spaces provided.

	Column A	Column B
___	1. microbiology	a. helminth
___	2. prokaryote	b. prediction of probable outcome of infection
___	3. pathology	c. *Aedes* mosquito
___	4. chemotherapeutic agent	d. spread of disease to many people at the same time
___	5. flatworm	e. aerobe
___	6. epidemic	f. quickly develops and subsides
___	7. biological vector	g. organism that lacks a membrane-bound nucleus
___	8. acute infectious disease	h. study of microbes
___	9. iatrogenic disease	i. bacillus
___	10. sign	j. drug
___	11. prognosis	k. results from the activity or treatment by physicians
___	12. rod-shaped bacteria	l. study of the nature of disease
___	13. sterilization	m. fungal diseases
___	14. oxygen-requiring	n. visible evidence of disease bacteria
___	15. mycoses	o. removal of all organisms

MULTIPLE CHOICE

16. Parasites that cause infectious diseases are called _____.
(a) microbes   (b) vectors
(c) multicellular pathogens   (d) hosts

17. A(n) disease is characterized by abnormal cell growth, such as tumors.
(a) infectious
(b) neoplastic
(c) nutritional deficiency
(d) iatrogenic

18. Kwashiorkor and scurvy are examples of _____.
(a) nutritional diseases   (b) endocrine diseases
(c) degenerative diseases   (d) infectious diseases

19. A(n) _____ disease appears, disappears, and reappears over time.
(a) acute   (b) chronic
(c) local   (d) latent

20. Visible or other objective physical characteristics of a disease are called _____.
(a) symptoms   (b) signs
(c) syndromes (d) diagnoses

21. Which of the following pathogens do not have eukaryote cells? _____.
(a) bacteria   (b) protozoa
(c) fungi   (d) helminths

22. A spherical bacteria cell is called a _____.
(a) coccus   (b) bacillus
(c) vibrio   (d) spirochete

23. Which of the following groups of pathogens reproduce only in living cells? _____
(a) bacteria   (b) viruses
(c) fungi   (d) helminths

24. Malaria is caused by a protozoan from which group of protozoans?_____
(a) flagellates   (b) amoebas
(c) ciliates   (d) sporozoans

25. An infestation usually refers to the presence of _____.
(a) bacteria   (b) viruses
(c) protozoa   (d) helminths

26. A pandemic is a disease that _____.
(a) is normally present at a low level or incidence
(b) is present in a higher than normal incidence
(c) occurs across a large region.
(d) occurs in an unpredictable manner

TRUE/FALSE

___ 27. Disease is defined as the failure to maintain homeostatic balance.

___ 28. Iatrogenic diseases are caused by the actions or treatments of physicians.

___ 29. All infectious diseases can be transmitted from one person to another.

___ 30. Prokaryote cells contain a nucleus.

___ 31. The Gram stain is a technique used to identify bacteria.

___ 32. Viruses can be controlled by antibiotics.

___ 33. A therapy is the prescribed treatment for recovery from a disease.

SHORT ESSAY

34. List the nine kinds of diseases associated with known causes.

35. What is the difference between a diagnosis and prognosis?

36. List the common portals of entry for pathogens to enter the body.

37. Describe three methods of pathogen transmission.

APPLICATIONS

38. Norma's arms and legs are badly scratched after a fall during a hiking trip. To prevent infections from developing, her physician uses an antiseptic on her wounds before bandaging them. Why didn't the physician use a disinfectant on the wounds?

39. A pharmaceutical company is interested in developing new anti-bacterial drugs from marine organisms. After making extracts from various portions of the collected organisms, they are ready to test the effectiveness of different extracts against a common pathogenic bacterium. Their tests show that an extract from a sponge is quite effective in killing the test bacterium. Wanting to learn how the extract works, the development team lists five ways in which the extract might be affecting the bacteria. What five mechanisms are on their list?

✔ Answers to Concept Check Questions

(p. 98) 1. Etiology is the study of the causes of diseases. 2. Acute diseases develop rapidly and chronic diseases develop gradually. 3. A symptom is the patient's perception of a disease, and a sign is an observable or measurable characteristic of a disease.

(p. 112) 1. A pathogen is an organism that can cause disease. 2. A systemic infection is a generalized infection that affects the body as a whole. 3. A vector is an organism that transfers a parasite, or infectious organism, to a new host or another person. 4. Morbidity refers to the illness rate and mortality to the death rate in a certain area.

CHAPTER
7

The Integumentary System

CHAPTER OUTLINE

CHAPTER OBJECTIVES

1   Describe the general functions of the skin.
2   Compare the structures and functions of the different layers of the skin.
3   Explain what accounts for individual and racial differences in skin, such as skin color.
4   Discuss the effects of ultraviolet radiation on the skin.
5   Discuss the functions of the skin's accessory structures.
6   Describe the mechanisms that produce hair and determine hair texture and color.
7   Summarize the effects of aging on the skin.
8   Describe the major causes, signs and symptoms of several skin disorders.
9   Explain how the skin repairs itself after an injury.
10   What distinguishes first-, second-, and third-degree burns?

The *integumentary system*, or *integument*, consists of the skin, hair, nails, and various glands. It is the most visible organ system of the body, and because of its visibility, most of us devote a lot of time to its upkeep: washing face and hands; brushing, trimming, or shaving hair; clipping nails; taking showers; and applying deodorants, perfumes, and cosmetics.

The integument does much more than help create a first impression, however. Our sense of touch, for example, would be virtually eliminated without it. The integument also performs many essential protective and maintenance functions. The diversity of integumentary structures—from the delicate eyelash to the tough skin of the sole to the milk-secreting glands of the breast—is evidence of the widely varying functions of this surprisingly complex system.

SYSTEM BRIEF

The **integumentary system** (integument means a "covering") is generally considered in two parts: (1) the *skin* and (2) the *accessory structures*. The **accessory structures** include hair, nails, and a variety of exocrine glands.

More than being just a covering for the body, the integumentary system has the following roles:

1. *Protection*. The skin, hair, and nails cover and protect underlying tissues and organs from impacts, chemicals, and infections. The skin also prevents unnecessary loss of body fluids.

2. *Maintenance of normal body temperature*. The skin and exocrine glands regulate heat loss with the surrounding environment.

3. *Storage, excretion, or synthesis*. The skin stores fat and synthesizes vitamin D. Exocrine glands excrete salts, water, and organic wastes in the form of sweat.

4. *Sensory reception*. Specialized nerve endings in the skin detect sensations such as touch, pressure, pain, and temperature.

THE SKIN

The **skin** or *cutaneous membrane* is made up of an outer epithelial covering, or **epidermis** (*epi-*, above), and an underlying **dermis** of connective tissue. Beneath the dermis lies the **subcutaneous layer**, or *hypodermis* made up of loose connective tissue that attaches the skin to deeper structures, such as muscles or bones. (As you read through this chapter, you will notice that anatomists borrowed both the Latin and Greek words for skin: *cutis* and *dermis*, respectively.)

A cross section of the skin, showing all three layers as well as many accessory structures, is shown in Figure 7-1•.

epi, above + *derma*, skin
epidermis: the superfical layer of the skin

sub, below + *cutis*, skin
subcutaneous layer: a layer of loose connective tissue that lies between the skin and deeper structures

THE EPIDERMIS

The epidermis (Figure 7-2•) consists of several different cell layers that form a stratified squamous epithelium. ∞ [p. 63] The deepest layer is the **stratum germinativum** (STRA-tum jer-mi-na-TĒ-vum). This layer marks the boundary between the epidermis and the loose connective tissue of the underlying dermis. The stratum germinativum forms wavy ridges that extend into the dermis. These ridges form the complex rings and loops seen on the thick skin of the hands and feet. Although you may be more familiar with their use in fingerprinting, ridges on the fingers, toes, palms, and soles more importantly serve to increase the surface area of the skin, increasing friction and ensuring a secure grip.

As the outermost skin layer, the epidermis is continually rubbed and scraped by objects in the environment, and its cells are worn away. For this reason, epidermal cells must be constantly replenished. *Stem cells*, which divide to produce some daughter cells that specialize and some that remain as stem cells, are often

stratum, layer + *germinare*, to sprout
stratum germinativum: a layer of stem cells whose divisions produce the daughter cells that form more superficial epidermal layers

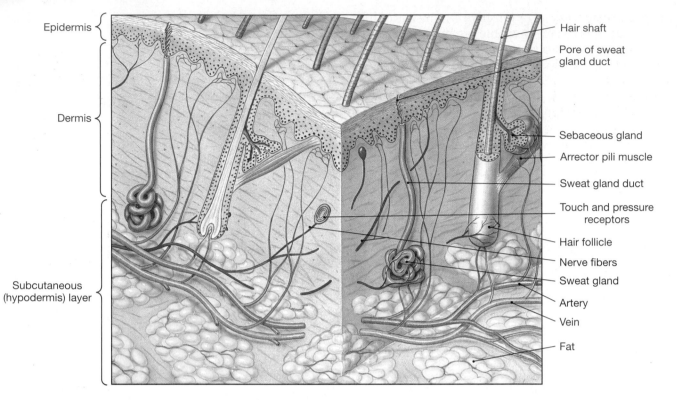

Epidermis

Dermis

Subcutaneous
(hypodermis) layer

Hair shaft

Pore of sweat
gland duct

Sebaceous gland

Arrector pili muscle

Sweat gland duct

Touch and pressure
receptors

Hair follicle

Nerve fibers

Sweat gland

Artery

Vein

Fat

•FIGURE 7-1
Components of the Integumentary System.
Relationships among the main structures of the integumentary system (with the exception of nails, shown in Figure 7-5).

found in tissues that require a constant supply of new cells. (As you learn about other body systems you will see other examples of stem cells.) Not surprisingly, the stratum germinativum is full of stem cells.

Each time an epidermal stem cell divides, one of the daughter cells enters the next, more superficial layer. While these cells become more specialized they are pushed further toward the surface as the stem cells continue to divide. Over this period they increase their production of the fibrous protein **keratin** (KER-a-tin). Keratin is extremely durable and water-resistant. In the human body, keratin not only coats the surface of the skin but forms the basic structure of hair, calluses, and nails. In other animals, keratin forms structures as varied as horns, hooves, beaks, feathers, and reptile scales.

The surface layer of the epidermis is called the **stratum corneum** (KOR-nē-um; *cornu*, horn). This layer is made up of flattened, dead epithelial cells that are tightly joined and filled with large amounts of keratin. It takes about 14 days for a cell to move upward from the stratum germinativum and through intermediate layers of the epidermis before reaching the stratum corneum. The dead cells usually remain in this exposed position for an additional 2 weeks before they are shed or washed away. Thus the deeper portions of the epithelium and underlying tissues are covered by a protective and waterproof barrier. This barrier is composed of tough and expendable cells.

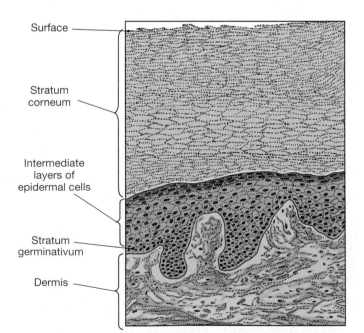

Surface

Stratum
corneum

Intermediate
layers of
epidermal cells

Stratum
germinativum

Dermis

•FIGURE 7-2
The Epidermis in Section.

Skin Color

Two factors influence skin color: (1) the amount of pigment in the epidermis and (2) the amount and composition of the blood supply close to the skin's surface. The primary pigment in the skin is **melanin**. This pigment is brown, yellow-brown, or black, and is produced by specialized epidermal cells called **melanocytes** (Figure 7-3•). Differences in skin color between individuals and even races are not due to different *numbers* of melanocytes but to the different amounts of melanin they produce. *Albino* individuals, who have very white skin, have normal numbers of melanocytes, but their melanocytes are incapable of producing melanin.

The melanin that melanocytes produce is injected into epithelial cells. This colors the entire epidermis and protects the epidermal cells from the potentially damaging rays of the sun. If the melanin is not distributed evenly, *freckles* result. Freckles are discrete accumulations of melanin that appear as flat, pigmented spots on the skin.

Blood supply to the dermis also affects skin color. Blood with abundant oxygen is bright red, and blood vessels in the dermis normally give the skin a reddish tint (most easily seen in lightly pigmented individuals). When those vessels are dilated (expanded), as occurs during inflammation or when one "blushes," the red tones become much more pronounced. When the vessels are temporarily constricted, as when one is frightened, the skin appears relatively pale. If circulation to the skin is reduced for a long period of time, the remaining blood in the skin loses oxygen. Because blood with less oxygen appears darker red, the skin then takes on a bluish coloration called **cyanosis** (sī-a-NŌ-sis; *kyanos*, blue). In individuals of any skin color, cyanosis is most apparent in areas of thin skin, such as the lips, ears, or beneath the nails. It can be a response to extreme cold or a result of circulatory or respiratory disorders, such as heart failure or severe asthma.

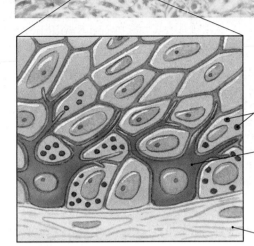

Melanin pigment in epidermal cell

Melanocyte in stratum germinativum

Dermis

•FIGURE 7-3
Melanocytes.
The micrograph and accompanying drawing indicate the position of melanocytes in the deeper layers of the epidermis of a darkly pigmented individual.

The Epidermis and Vitamin D

Although strong sunlight can damage epithelial cells and deeper tissues, limited exposure to sunlight is very beneficial. When exposed to ultraviolet radiation, cells in the lower epidermal layers synthesize a form of **vitamin D**. Vitamin D is essential for bone maintenance and growth.

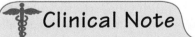 Clinical Note

Melanin and Ultraviolet Radiation

Sunlight contains significant amounts of **ultraviolet (UV) radiation**. A small amount of UV radiation is beneficial; however, too much ultraviolet radiation can cause mild or even serious burns. Melanin helps prevent skin damage by absorbing ultraviolet radiation before it reaches the deep layers of the epidermis and dermis. Within the epidermal cells, melanin concentrates around the nucleus, so it absorbs the UV before it can damage the nuclear DNA. In fact, under conditions of prolonged sun exposure, melanocytes respond by increasing their melanin production, resulting in the temporary darkening of the skin known as a *tan*.

Despite the presence of melanin, damage can still result from repeated long-term exposure to sunlight and its UV radiation. For example, alterations in the underlying connective tissues lead to premature wrinkling, and skin cancers can result from chromosomal damage in stem cells or melanocytes.

THE DERMIS

The second skin layer, the **dermis**, lies beneath the epidermis. Accessory structures derived from the epidermis, such as hair follicles and sweat glands, extend into the dermis (see Figure 7-1●).

The superficial layer of the dermis, to which the stratum germinativum is attached, consists of loose connective tissue that supports the epidermis. It contains the capillaries and nerves that supply the surface of the skin. Dermal blood vessels provide nutrients and oxygen needed by both the epidermis and dermis, and remove wastes. Nerve fibers control blood flow, adjust gland secretion rates, and monitor the sensory receptors of touch, pain, pressure, and temperature that are scattered throughout the dermis.

The deeper layer of the dermis consists of dense connective tissue fibers. Bundles of elastic and collagen fibers blend into those of the upper dermis and also extend into the underlying subcutaneous layer (*hypodermis*). These fibers enable this region to provide support and attachment for the dermis while also allowing considerable flexibility of movement.

THE SUBCUTANEOUS LAYER

Underneath the dermis lies the **subcutaneous layer**, or *hypodermis*. Although the subcutaneous layer is not actually a part of the integument, it is important in stabilizing the position of the skin in relation to underlying tissues and organs.

The subcutaneous layer is made up of loose connective tissue with many fat cells and it contains the major blood vessels that supply the dermis. The subcutaneos layer provides infants and small children with a protective layer of fat over the entire body. This "baby fat" serves to reduce heat loss, provide an energy reserve, and absorb shocks from inevitable tumbles.

The hypodermis is quite elastic and contains no vital organs beneath the blood vessels at its border with the dermis. The lack of vital organs makes *subcutaneous injection* a useful method for administering drugs (thus the familiar term *hypodermic needle*).

✔ Excessive shedding of cells from the outer layer of skin in the scalp causes dandruff. What is the name of this layer of skin?

✔ Why does exposure to sunlight or tanning lamps cause the skin to become darker?

✔ Some criminals sand the tips of their fingers so as not to leave recognizable fingerprints. Would this practice permanently remove fingerprints? Why or why not?

ACCESSORY STRUCTURES

Accessory structures of the integumentary system include hair, glands, and nails.

HAIR

Hairs project above the surface of the skin almost everywhere except over the sides and soles of the feet, the palms of the hands, the sides of the fingers and toes, the lips, and portions of the external genital organs.

Structure of Hair

Hairs originate in tiny complex organs called **hair follicles**. Hair follicles extend deep into the dermis, often projecting into the underlying subcutaneous layer (Figure 7-4●). The epithelium at the base of a follicle surrounds the **hair papilla**, a peglike structure of connective tissue containing capillaries and nerves. Hair is formed as the epithelial stem cells that cover the papilla divide. Keratin, the primary component in these epithelial cells, coats and stiffens the hair.

Ribbons of smooth muscle, called **arrector pili** (a-REK-tor PĪ-lī) muscles (Figures 7-1● and 7-4●), extend from the superficial dermis to the hair follicle. When stimulated, the arrector pili pull on the follicles and elevate the hairs. Contraction of the arrector pili may be caused by emotional states, such as fear or rage, or as a response to cold, producing the characteristic "goose bumps" associated with shivering.

Each individual hair has a **root** that encloses the papilla and an exposed **shaft**. The shaft, which varies in size, shape, and color, is the portion of the hair that is cut or styled. Differences in hair texture result from differences in the shapes of the

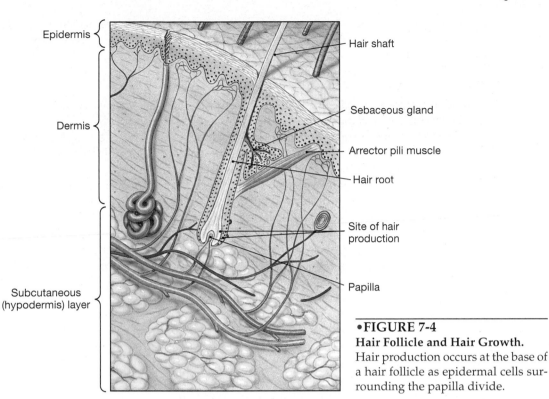

Epidermis

Dermis

Subcutaneous
(hypodermis) layer

Hair shaft

Sebaceous gland

Arrector pili muscle

Hair root

Site of hair
production

Papilla

•FIGURE 7-4
Hair Follicle and Hair Growth.
Hair production occurs at the base of
a hair follicle as epidermal cells surrounding the papilla divide.

hairs (determined by the shape of the hair follicle) and the activity of follicular cells. For example, straight hairs and their follicles are round in cross section, curly hairs are oval or somewhat flattened, and wavy hairs are curved. Differences in hair color among individuals reflect differences in the type and amount of pigment produced by melanocytes at the papilla.

Although hair color and texture are genetically determined, the condition of your hair may also be influenced by hormonal or environmental factors. As pigment production decreases with age the hair color lightens toward gray. White hair results from the presence of air bubbles within the hair shaft. Because the hair itself is quite dead and inert, changes in coloration are gradual. Unless bleach is used, it is not possible for hair to "turn white overnight," as some horror stories would have us believe.

Functions of Hair

The 5 million or so hairs on the human body have important functions. The roughly 100,000 hairs on the head protect the scalp from ultraviolet light, cushion the head from blows, and provide insulation benefits for the skull. The hairs guarding the entrances to the nostrils and external ear canals help prevent the entry of foreign particles and insects, and eyelashes perform a similar function for the surface of the eye.

GLANDS

The skin contains two types of exocrine glands, *sebaceous glands* and *sweat glands* (see Figures 7-1• and 7-4•).

Sebaceous Glands

Sebaceous (se-BĀ-shus) **glands** produce a waxy, oily secretion called **sebum** (SĒ-bum), which lubricates the hair shaft and inhibits bacterial growth within the surrounding area. Sebaceous glands may empty into a hair follicle by means of a short duct or directly onto the surface of the skin. Contraction of the arrector pili muscle that elevates the hair also squeezes the sebaceous gland, forcing out its secretions.

Sebaceous glands are very sensitive to changes in the concentrations of sex hormones, and their secretory activities accelerate at puberty. For this reason an individual with large sebaceous glands may be especially prone to developing

✔ What condition is produced by the contraction of the arrector pili muscles?

✔ A person suffers a burn on the forearm that destroys the epidermis and a portion of the dermis. When the injury heals, would you expect to find hair growing again in the area of the injury?

✔ Describe the functions of the subcutaneous layer.

acne during adolescence. In this condition, sebaceous ducts become blocked and secretions accumulate, causing inflammation and providing a fertile environment for bacterial infection.

Sweat Glands

The skin contains two different types of **sweat glands**, also known as *sudoriferous glands*: *apocrine sweat glands* and *merocrine* (or *eccrine*) *sweat glands*.

Apocrine sweat glands are found in the armpits, around the nipples, and in the groin. Like sebaceous glands, **apocrine sweat glands** empty into hair follicles. At puberty, these coiled tubular glands begin discharging a thick, cloudy secretion that becomes odorous when broken down by bacteria. In other mammals this odor is an important form of communication; for us, whatever function "body odor" may have is usually masked by deodorants and other hygiene products.

Merocrine (eccrine) sweat glands are far more numerous and widely distributed than apocrine glands. Merocrine sweat glands produce the clear secretion called *sweat*, also known as *perspiration*. The adult integument contains around 3 million merocrine sweat glands.

Merocrine sweat glands are coiled tubular glands that open directly onto the surface of the skin. The primary functions of their secretions are to cool the surface of the skin and reduce body temperature. When a person is sweating, all the merocrine glands are working together. At the same time the blood vessels beneath the epidermis are filled with blood, warming the skin and turning it a reddish color. As the moisture evaporates from the skin surface, the skin cools. If body temperature falls below normal, perspiration stops, blood flow to the skin declines, and the cool, dry surfaces release little heat. The role of the skin in the regulation of body temperature was discussed in Chapter 1 (see Figure 1-2•, p. 6).

The perspiration produced by merocrine glands is a clear secretion that is more than 99 percent water, but it does contain a mixture of electrolytes (salts) and waste products such as urea. The electrolytes give sweat its salty taste. When all the merocrine sweat glands are working at maximum, the rate of perspiration may exceed a gallon per hour, and dangerous fluid and electrolyte losses can occur. For this reason marathon runners and other athletes in endurance sports must drink adequate fluids while exercising.

NAILS

Nails form over the tips of the fingers and toes, where they protect the exposed tips. The body of the nail covers the **nail bed**, but nail growth occurs from the **nail root**, an epithelial fold not visible from the surface. The deepest portion of the nail root lies very close to the bone of the fingertip. A portion of the stratum corneum of the fold extends over the exposed nail nearest the root, forming the **cuticle**. Underlying blood vessels give the nail its pink color, but near the root these vessels may be obscured, leaving a pale, moonlike crescent shape known as the **lunula** (LOO-nū-la) (*luna*, moon). The structure of a typical fingernail can be seen in Figure 7-5•.

•FIGURE 7-5
Structure of a Nail.

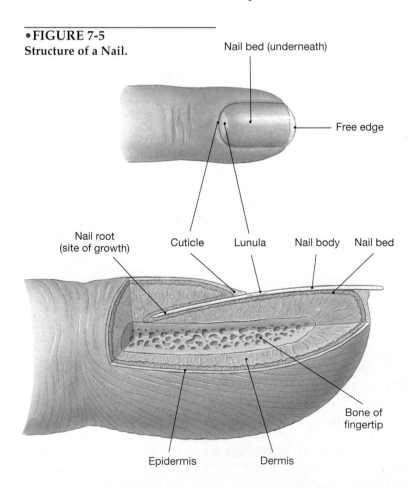

Nail bed (underneath)

Free edge

Nail root (site of growth)   Cuticle   Lunula   Nail body   Nail bed

Bone of fingertip

Epidermis   Dermis

AGING AND THE INTEGUMENTARY SYSTEM

Aging affects all the components of the integumentary system (Figure 7-6•). The major changes include:

1. The epidermis thins as the rate of cell division of the stratum germinativum declines, making the elderly more prone to skin infections or injuries.

2. Melanocyte production declines, and in Caucasians the skin becomes very pale. With less melanin in the skin, the elderly are more sensitive to sun exposure and more likely to experience sunburn.

3. Glandular activity declines, and the skin can become dry and often scaly.

4. Hair follicles stop functioning or produce thinner, finer hairs that are gray or white.

5. The dermis becomes thinner, and the elastic fiber network decreases in size. The skin therefore becomes weaker, and sagging and wrinkling occur. These effects are most pronounced in areas exposed to the sun.

6. The blood supply to the skin is reduced, making the elderly less able to lose body heat. Overexertion or overexposure to warm temperatures can cause dangerously high body temperatures.

7. Skin repairs proceed relatively slowly. For example, it takes 3 to 4 weeks to complete the repairs to a blister site in a person age 18–25. The same repairs at age 65–75 take 6 to 8 weeks. Because repairs are slow, recurrent infections may occur.

✔ What happens when the duct of an infected sebaceous gland becomes blocked?

✔ Deodorants are used to mask the effects of secretions from what type of skin gland?

✔ Older people do not tolerate the summer heat as well as they did when they were young, and are more prone to heat-related illness. What accounts for this change?

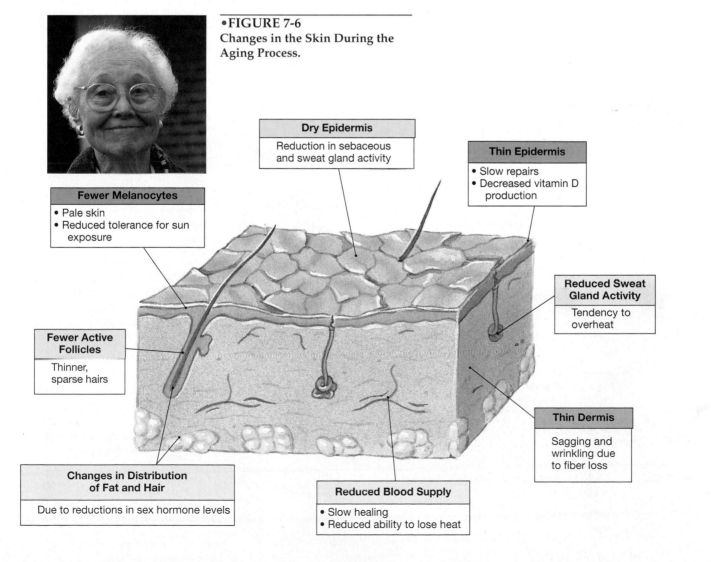

•FIGURE 7-6
Changes in the Skin During the Aging Process.

Dry Epidermis
Reduction in sebaceous and sweat gland activity

Thin Epidermis
• Slow repairs
• Decreased vitamin D production

Fewer Melanocytes
• Pale skin
• Reduced tolerance for sun exposure

Reduced Sweat Gland Activity
Tendency to overheat

Fewer Active Follicles
Thinner, sparse hairs

Thin Dermis
Sagging and wrinkling due to fiber loss

Changes in Distribution of Fat and Hair
Due to reductions in sex hormone levels

Reduced Blood Supply
• Slow healing
• Reduced ability to lose heat

SKIN PROBLEMS

The skin is the most visible organ of the body. As a result, problems with the skin are very noticeable. Trauma that causes a bruise, for example, typically breaks the walls of dermal blood vessels, creating a swollen and discolored area. The general condition of the skin often reflects the health of the individual. Nutritional disorders, stress, allergies, or disorders affecting the cardiovascular, respiratory, or digestive systems will produce changes in skin color, flexibility, elasticity, dryness, or sensitivity. Thus the skin can provide clues to the presence of a more serious illness or disease that might otherwise go unnoticed. For example, extensive bruising without obvious cause may indicate a blood clotting disorder, and a yellow color in the skin and mucous membranes may indicate a serious liver disorder.

EXAMINATION OF THE SKIN

The branch of medicine concerned with the diagnosis and treatment of diseases of the skin is called **dermatology**. Dermatologists use the presence of skin *lesions*, or *skin signs*, to distinguish different disorders and make a diagnosis. **Lesions** (LĒ-zhuns) are changes in skin structure caused by injury or disease. They are also called **skin signs** because they are measurable, visible abnormalities of the skin surface. Figure 7-7• shows the most common skin signs and gives examples of related disorders.

The distribution of lesions may be an important clue to the source of the problem. For example, in shingles (caused by the *herpes zoster* virus) painful blisters develop on the skin. These blisters develop along the path of one or more sensory nerves monitoring the skin. A ring of slightly raised, scaly lesions (called *papules*) on the trunk or scalp is typical of fungal infections. Examples of skin disorders caused by infection or by allergic reactions are included in Table 7-1, with de-

•**FIGURE 7-7**
Skin Signs.

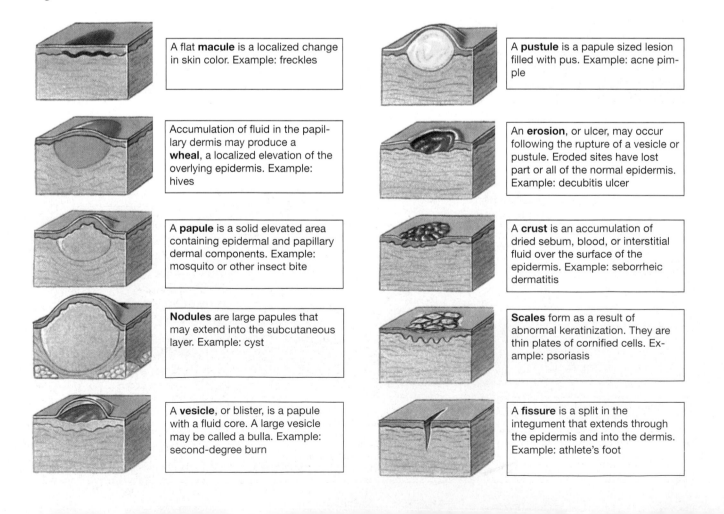

A flat **macule** is a localized change in skin color. Example: freckles

A **pustule** is a papule sized lesion filled with pus. Example: acne pimple

Accumulation of fluid in the papillary dermis may produce a **wheal**, a localized elevation of the overlying epidermis. Example: hives

An **erosion**, or ulcer, may occur following the rupture of a vesicle or pustule. Eroded sites have lost part or all of the normal epidermis. Example: decubitis ulcer

A **papule** is a solid elevated area containing epidermal and papillary dermal components. Example: mosquito or other insect bite

A **crust** is an accumulation of dried sebum, blood, or interstitial fluid over the surface of the epidermis. Example: seborrheic dermatitis

Nodules are large papules that may extend into the subcutaneous layer. Example: cyst

Scales form as a result of abnormal keratinization. They are thin plates of cornified cells. Example: psoriasis

A **vesicle**, or blister, is a papule with a fluid core. A large vesicle may be called a bulla. Example: second-degree burn

A **fissure** is a split in the integument that extends through the epidermis and into the dermis. Example: athlete's foot

Table 7-1	Skin Signs of Various Disorders	
Cause	**Examples**	**Resulting Skin Signs**
Viral infections	Chickenpox	Lesions begin as macules and papules but develop into vesicles
	Measles (rubeola)	A maculopapular rash that begins at the face and neck and spreads to the trunk and extremities
	Erythema infectiosum (Fifth disease)	A maculopapular rash that begins on the cheeks (slapped-appearance) and spreads to the extremities
	Herpes simplex	Raised vesicles that heal with a crust
Bacterial infections	Impetigo	Vesiculopustular lesions with exudate and yellow crusting
Fungal infections	Ringworm	An annulus (ring) of scaly papular lesions with central clearing
Parasitic infections	Scabies	Linear burrows with a small papule at one end
	Lice (pediculosis)	Dermatitis: excoriation (scratches) due to pruritis (itching)
Allergies to medication	Penicillin	Wheals (urticaria or hives)
Food allergies	Eggs, certain fruits	Wheals
Environmental allergies	Poison ivy	Vesicles

scriptions of the related lesions. We consider skin lesions caused by trauma in the section entitled "Classification of Wounds" (p. 131).

Lesions may also involve the accessory organs of the skin. For example, nails can change shape due to an underlying disorder. *Clubbing* of the nails, for example, is a sign of chronic respiratory or cardiovascular disease. In these conditions the fingertips broaden, and the nails become distinctively curved. The condition of the hair can be an indicator of overall health. For example, a person whose diet lacks sufficient protein will over time develop coarse, lightly pigmented hair.

Many different skin disorders produce the same skin signs or uncomfortable sensations. For example, **pruritis** (proo-RĪ-tus), an irritating itching sensation, and **erythema** (er-i-THĒ-ma), or redness (*erythros* means red), are symptoms shared by a variety of skin conditions. Pain is another common symptom of many skin disorders (see Figure 6-1•, p. 97). Although pain is unwelcome, it is important because it lets us know that tissue damage is occurring, and encourages us to do something about it. Without pain sensations, we would be unaware of potentially serious injuries or infections.

Clinical Note

Skin Conditions and Diagnostic Tests

Examples of diagnostic tests used to evaluate skin conditions include the following:

- *Scrapings* of the skin surface, a process often performed to check for fungal infections
- *Culturing* of bacteria collected from a lesion, to help identify the bacteria and decide what drug treatment would be effective
- *Biopsy* of a lesion to look at the tissue structure in detail
- *Skin tests*: Many different disorders can be detected through use of a skin test. In a skin test a small area of skin is exposed to an inactivated pathogen, chemicals released by pathogens, or something that can produce an allergic reaction in sensitive individuals. Exposure may involve either injection into the skin or application to the skin surface. For example, in a *tuberculin test*, a skin test

for *tuberculosis*, non-infectious bacterial antigens are injected into the dermis. This process is called *intradermal* (*intra* means "within") *injection*. If the individual has ever been infected by tuberculosis bacteria, and has developed antibodies to TB, the immune response will attack the injected foreign proteins. This results in erythema (redness) and swelling at the injection site 24–72 hours later. In most states, negative tuberculin tests are required before you can start working as a teacher, cook, or healthcare provider. You can work in these jobs with a positive tuberculin test only if your chest x-ray is normal, with no signs of active TB infection. *Patch testing* is used to check sensitivity to *allergens*, substances that can cause allergic reactions. In a patch test an allergen is applied to the surface of the skin. If erythema, swelling, and/or itching develop, the individual is sensitive to that allergen.

SKIN DISORDERS AND THEIR CAUSES

Figure 7-8• is an overview of skin disorders. Primary skin disorders may be grouped according to their causes into environmental stress and inflammation, infections, tumors, trauma, congenital disorders, nutritional disorders, and degenerative disorders. Secondary skin disorders include effects caused by primary disorders in other organ systems. We will consider representative examples from each category.

•**FIGURE 7-8**
Disorders of the Integumentary System.

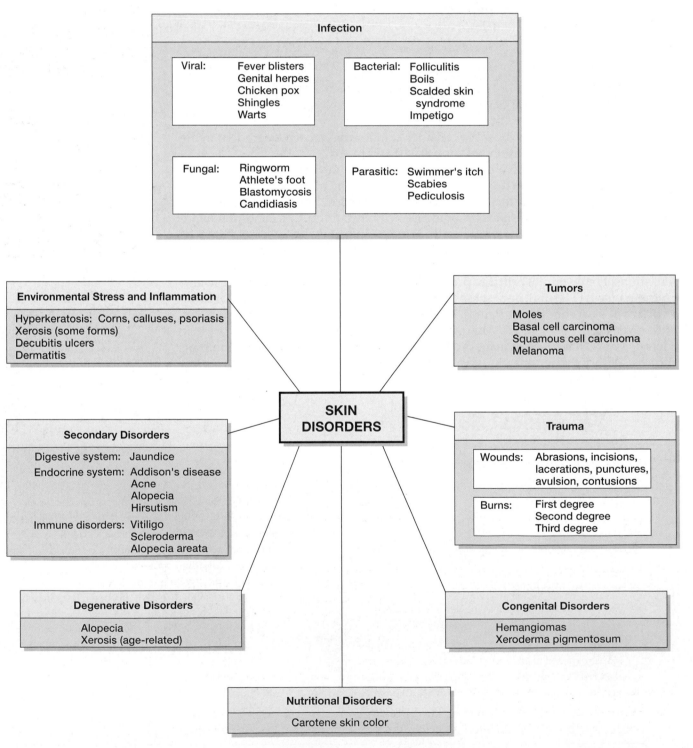

Infection

Viral: Fever blisters
            Genital herpes
            Chicken pox
            Shingles
            Warts

Bacterial: Folliculitis
            Boils
            Scalded skin
             syndrome
            Impetigo

Fungal: Ringworm
            Athlete's foot
            Blastomycosis
            Candidiasis

Parasitic: Swimmer's itch
            Scabies
            Pediculosis

Environmental Stress and Inflammation

Hyperkeratosis: Corns, calluses, psoriasis
Xerosis (some forms)
Decubitis ulcers
Dermatitis

Tumors

Moles
Basal cell carcinoma
Squamous cell carcinoma
Melanoma

SKIN DISORDERS

Secondary Disorders

Digestive system: Jaundice
Endocrine system: Addison's disease
                    Acne
                    Alopecia
                    Hirsutism
Immune disorders: Vitiligo
                    Scleroderma
                    Alopecia areata

Trauma

Wounds: Abrasions, incisions,
         lacerations, punctures,
         avulsion, contusions

Burns: First degree
         Second degree
         Third degree

Degenerative Disorders

Alopecia
Xerosis (age-related)

Congenital Disorders

Hemangiomas
Xeroderma pigmentosum

Nutritional Disorders

Carotene skin color

Environmental Stress and Inflammation

Disorders of Keratin Production. Not all skin signs are the result of infection, trauma, or allergies. Some skin signs are the normal response to environmental stresses. One common response to environmental stress is the excessive production of keratin. This process is called **hyperkeratosis** (hī-per-ker-a-TŌ-sis).The most familiar examples of hyperkeratosis are *calluses* and *corns*. **Calluses** (Figure 7-9a•) are thickened patches that appear on already thick-skinned areas, such as the palms of the hands or the soles or heels of the foot, in response to chronic abrasion and distortion. **Corns** (Figure 7-9b•) are more localized areas of accelerated keratin production that form in areas of thin skin on or between the toes.

Psoriasis (sō-RĪ-a-sis) is a condition characterized by areas of hyperkeratosis. Its cause is unknown. In this condition, areas such as the scalp, elbows, palms, soles, groin, and nails appear to be covered with red patches with small, silvery scales that continuously flake off (Figure 7-9c•). The cells of the stratum germinativum are unusually active, dividing some 10 times faster than normal. Most cases are painless and treatable but can be disfiguring.

Xerosis (ze-RŌ-sis), or dry skin, is a common complaint of the elderly and people who live in arid climates. Under such conditions cell membranes in the outer layers of the skin gradually deteriorate, and the stratum corneum becomes more a collection of scales than a single sheet. The scaly surface is much more permeable than an intact layer of keratin, and the rate of "uncontrolled" water loss increases. In persons afflicted with severe xerosis, the rate of water loss at the skin surface can may reach a liter or more per hour.

Pressure on the skin, another form of environmental stress, can produce *bedsores*, or **decubitis ulcers**. Decubitis means "to lie down" and an **ulcer** is a localized loss of an epithelium. Decubitis ulcers form where dermal blood vessels are compressed against deeper structures such as bones or joints. They may affect bedridden patients or patients with splints and casts. The chronic lack of circulation kills epidermal cells, removing a barrier to bacterial invasion and infection. Eventually, the dermal tissues may break down as well. Bedsores can be prevented or treated by frequently changing body position, thereby relieving the pressure applied to the blood vessels.

Inflammation. Inflammation is a process that helps provide a defense against pathogens and injury. ∞ *[p. 72]* Because of the abundance of sensory receptors in the skin, inflammation can be very painful. **Dermatitis** (der-muh-TĪ-tis) is an inflammation of the skin that primarily involves the upper layer of the dermis. In typical dermatitis, inflammation begins in a portion of the skin exposed to infection or irritated by chemicals, radiation, or mechanical stimuli. Dermatitis may cause no physical discomfort, or it may produce an annoying itch. Other forms of this condition can be quite painful, and the inflammation can spread rapidly across the entire integument.

There are many forms of dermatitis, some of them quite common:

- **Contact dermatitis** generally occurs in response to strong chemical irritants. This may be a direct response or an allergic reaction. It produces an itchy rash that may spread to other areas; poison ivy is an example.

- **Eczema** (EK-se-muh) is a dermatitis that can be triggered by temperature changes, fungus, chemical irritants, greases, detergents, or stress. Both heredity and environmental factors can promote the development of eczema.

- **Diaper rash** is a localized dermatitis caused by a combination of moisture, irritating chemicals from fecal or urinary wastes, and flourishing microorganisms.

- **Urticaria** (ur-ti-KAR-ē-uh), or *hives*, is an extensive acute allergic response to a food, drugs, an insect bite, infection, stress, or other stimulus.

- **Seborrheic** (seb-o-RĒ-ik) **dermatitis** is an inflammation around active sebaceous glands. The affected area becomes red and there is usually some

•FIGURE 7-9
Disorders of Keratin Production.
(a) Calluses. **(b)** Corns. **(c)** Psoriasis.

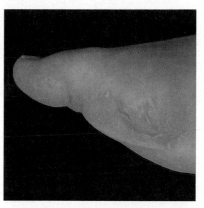

(a) Calluses

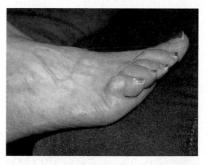

(b) Corns

(c) Psoriasis

epidermal scaling. It is most often associated with the scalp. In babies, mild cases are called "cradle cap." In adults, dandruff results. Anxiety, a superficial fungal infection, and food allergies can contribute to the problem.

Infections

Infections of the skin are caused by various bacteria, viruses, fungi, and parasites. They may also be secondary to arthropod and insect bites and stings.

Bacteria. Most skin bacteria are associated with apocrine sweat glands and sebaceous glands. One of the most common skin bacteria is *Staphylococcus aureus* (staf-il-ō-KOK-us Ō-rē-us). This is the bacteria responsible for "staph" infections that develop in cuts and wounds of the skin. *Staphylococcus* is also responsible for **folliculitis**, or inflammation of a hair follicle. A staph infection may also form a large pustule called a **furuncle** (FU-run-kl), or **boil**, within the skin. An **abscess** (AB-ses) is a pustule that forms in deeper tissues. Many abscesses form as a result of bacterial infection, but they can also result from chronic inflammation or other factors.

In infants, staph bacteria may produce a condition called **scalded skin syndrome**. In this condition, vesicles (blisters) form over large areas of the skin, peel off, and leave wet, red-colored areas. After the lesions dry, the skin returns to normal within 1–2 weeks.

Impetigo is a highly contagious skin infection most commonly found in children. Pustules are formed by staphylococci, streptococci, or both bacteria together. The pustules dry and form crusts. Healing usually occurs without scarring of the skin, although skin pigment may be lost.

Viruses. The *Herpes viruses* group produce a variety of skin disorders. Two types, *Herpes simplex 1* and *Herpes simplex 2*, are common causes of disease. Both primarily affect the skin. Herpes simplex 1 produces fluid-filled blisters usually around the skin and mucous membranes of the mouth. These blisters are commonly called **cold sores**, or **fever blisters**. Herpes simplex 2 is responsible for **genital herpes**, and it produces similar vesicles on the external genitalia. Genital herpes is a common sexually transmitted disease (STD). Another *herpes* virus is the *Varicella zoster* virus that causes two diseases: chickenpox and shingles. **Chickenpox** *(varicella)* is an infectious childhood disease. It usually produces mild cold symptoms, a slight fever, and a rash of small, red macules. The macules form vesicles which dry and become crusts. Early antiviral treatment helps, but symptoms generally disappear within 10 days. The virus, however, has not been eliminated from the body. Instead, it has migrated up the sensory nerves and into the nervous tissue. Later in life, the virus may become activated and produce a pattern of vesicles along the area of skin associated with particular sensory nerves. This painful condition is called **shingles** *(zoster)*. Similar persistence of both types of *Herpes simplex* virus causes multiple recurrences in many people.

Another viral disease of the skin forms **warts**, or **papillomas**. The 30 or so different **human papilloma viruses (HPV)** produce warts not only on the skin, but also in the oral cavity and the respiratory and genital tracts. HPV is transmitted by direct contact. **Dermal warts** form from infected epithelial cells and may appear on the scalp, face, hands, knees, and feet. **Genital warts** form on the penis, and around the vagina and anus. Unlike most dermal warts, genital warts may become malignant and are associated with cervical cancer. Genital warts can be sexually transmitted.

Fungi. Fungal diseases are called *mycoses* (*sing.*, mycosis), and fungal skin diseases are called **dermatomycoses** (DUR-ma-tō-mi-KŌ-sēs). The fungi that cause these diseases grow on skin, hair, and nails in a ring-like lesion (**ringworm**) with a scaly, raised ring around more normal looking skin. Different names are given to ringworm in different areas of the body. For example, *tinea capitis* is on the scalp, *tinea corporis* is on the body, *tinea cruris* is at the groin ("jock itch"), and *tinea unguium* is on a nail. **Athlete's foot** (*tinea pedis*) forms lesions on the foot and between the toes (Figure 7-10a●). These lesions crack and bacteria invade the skin causing the itchy, moist white areas associated with athlete's foot.

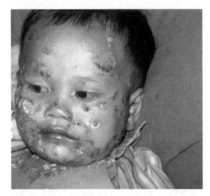

Impetigo lesions

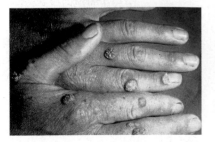

Dermal warts on the hands and fingers.

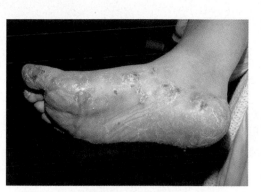

(a) Athlete's foot

(b) Candidiasis infection of the nails

•FIGURE 7-10
Fungal Infections.
(a) Athlete's foot. (b) Candidiasis infection of the nails.

Blastomycosis is produced by *Blastomyces dermatiditis* (blast-Ō-MĪ-sēz derma-TIT-ĭ-dis), a soil fungus that may enter the body through skin wounds or the lungs. In the skin it produces large papules as well as abscesses in the dermis and subcutaneous tissue. It can also spread to the bones and other organs of the body.

Candidiasis (kan-di-DĪ-ah-sis) is produced by a yeast, *Candida albicans* (KAN-did-da AL-bi-kanz), that is a normal inhabitant of the human body. Infections may occur in the skin, mouth (*thrush*), and other organs such as the lungs, heart, and vagina (*vaginitis*). An infection of the skin is characterized by itchy red lesions. A *Candida* infection of the nails is shown in Figure 7-10b•.

Parasites. Some helminth (worm) parasites enter the body through the skin but do not primarily infect the skin. (These parasites are considered in other chapters.) One example of an incomplete helminth invasion of the body through the skin is **swimmer's itch**. This occurs in fresh waters throughout the U.S., and is characterized by redness and itching of the skin. It is caused by the swimming larval stages released by adult schistosome worms that parasitize birds and domestic animals. The immune response kills the larvae and prevents infestations in the body.

Mites, ticks, and lice that live on the skin can produce various disorders though their bites and/or egg-laying activities. **Scabies** (SKĀ-bēz) is produced by the itch mite (*Sarcoptes scabiei*) that burrows into the skin to lay its eggs. Infestations followed by allergic reaction to the itch mite cause redness and intense itching, which can lead to secondary infections.

Lice are insects and **pediculosis** (pe-dik-ū-LŌ-sis) is a lice infestation. The most common is the body louse, *Pediculus humanus*. There are two varieties; one is found living on hair in all climates, and the other lives on the body and clothing in temperate climates. Lice feed on blood and their bites produce redness, dermatitis, and itching—a route for secondary infections by bacteria and fungi. They attach their eggs (called *nits*) to scalp hair. The crab, or pubic, louse (*Phthirus pubis*; thir-us PŪ-bis) lives on pubic hair and its bites produce intense itching (see Figure 6-6b•, p. 106). Scabies and lice are highly contagious and easily transmitted in person-to-person contact.

Table 7-2 summarizes the variety of infections of the integumentary system.

Tumors

Almost everyone has several benign tumors of the skin; *freckles* and *moles* are examples. Skin cancers are the most common form of cancer, and the most common skin cancers are caused by prolonged exposure to the ultraviolet radiation in sunlight. A **basal cell carcinoma** (Figure 7-11a•) is a malignant cancer that originates in the germinativum (basal) layer. ∞ *[p. 117]* This is the most common skin cancer. Roughly two-thirds of these cancers appear in body areas subjected to chronic UV exposure. Genetic factors have been identified that predispose people to this condition. **Squamous cell carcinomas** are less common but almost totally restricted to areas of sun-exposed skin. Metastasis seldom occurs in squamous cell carcinomas and virtually never in basal cell carcinomas, and

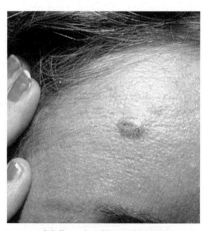

(a) Basal cell carcinoma

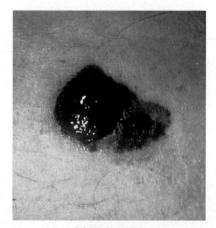

(b) Melanoma

•FIGURE 7-11
Skin Cancers.
(a) Basal cell carcinoma. (b) Melanoma.

Table 7-2	Common Infectious Diseases of the Integumentary System	
Disease	**Organism (Name)**	**Description**
Bacteria		
Folliculitis	*Staphylococcus aureus*	Infections of hair follicles may form pimples, furuncles (boils), and abscesses
Scalded skin syndrome	*Staphylococcus aureus*	In infants; large areas of skin blister, peel off, and leave wet, red areas
Impetigo	*Staphylococci, Streptococci,* or both	Pustules form on skin, dry and become crusts; skin pigment may not reappear after healing
Viruses		
Oral herpes (Fever blisters)	*Herpes simplex 1*	Vesicles (blisters), also called cold sores, form on lips and hands; vesicles disappear but may reappear at various lengths of time
Genital herpes	*Herpes simplex 2*	Lesions as in oral herpes; form on external genitalia; vesicles disappear and reappear
Chickenpox (*varicella*)	*Herpes varicella-zoster*	Childhood disease; small, red macules form vesicles, which dry and become crusts
Shingles (*zoster*)	*Herpes varicella-zoster*	Adults; lesions form a pattern usually on trunk; severe pain often follows attack
Warts	*Human papillomaviruses*	Dermal warts form in the epidermis; Genital warts may become malignant and may be sexually transmitted
Fungi		
Ringworm (*Tinea*)	*Epidermophyton, Microsporum,* and *Trichophyton*	Dry, scaly lesions form on the skin in different parts of the body; scalp (*Tinea capitis*), body (*Tinea corporis*), groin (*Tinea cruris*), foot (*Tinea pedis*), and nails (*Tinea unguium*)
Blastomycosis	*Blastomyces dermatitidis*	Forms pustules and abscesses in the skin; may affect other organs
Candidiasis	*Candida albicans*	Normal inhabitant of the human body; may infect many different organs; red lesions form in skin infections; nails may also become infected
Parasites		
Swimmer's itch	*Schistosoma* worms (flukes)	Itching caused by fresh water larval stages of schistosome worms (flukes) burrowing into skin
Scabies	*Sarcoptes scabiei* (itch mite)	Itch mite burrows and lays eggs in skin in areas between fingers, wrists, armpits, and genitals; entrance marked by tiny, scaly swellings which become red and itchy
Pediculosis	*Pediculus humanus* (human body louse)	Lice infestations on body and scalp; bites produce redness, dermatitis, and itching
	Phthirus pubis (pubic louse)	"Crabs"; lice infestation of the pubic area; their bites produce intense itching

most people survive these cancers. The usual treatment involves surgical removal of the tumor, and at least 95 percent of patients survive five years or more after treatment. (This statistic, the five-year survival rate, is a common method of reporting long-term outlook for a cancer patient.)

Compared with these common and seldom life-threatening cancers, **melanomas** (mel-a-NŌ-mas) are extremely dangerous (Figure 7-11b•). In this condition, cancerous melanocytes grow rapidly and metastasize through the lymphatic system. The outlook for long-term survival changes dramatically, depending on the thickness and cell layer invasion of the tumor when removed. The five-year survival rate is 99 percent if it is small, thin, and superficial; with thick, invasive tumors, the survival rate drops to 14 percent.

To detect melanoma at an early stage, it is essential to know what to look for when you examine your skin. The letters ABCD make it easy to remember the key points of detection:

A is for **A**symmetry: Melanomas tend to be irregular in shape. Typically they are raised; they may ooze or bleed.

B is for **B**order: generally unclear, irregular, and sometimes notched.

C is for **C**olor: generally mottled, with many different colors (tan, brown, black, red, pink, white, and/or blue).

D is for **D**iameter: A growth more than about 5 mm (0.2 in.) in diameter, or roughly the area covered by the eraser on a pencil, is more dangerous.

Fair-skinned individuals who live in the tropics are most susceptible to all forms of skin cancer, because their melanocytes are unable to shield them from the UV radiation. Sun damage can be prevented by avoiding exposure to the sun during the middle of the day and by using a sunblock (not a tanning oil)—a practice that also delays the cosmetic problems of sagging and wrinkling. Everyone who spends any time out in the sun should choose a broad-spectrum sunblock with a sun protection factor (SPF) of at least 15; blondes, redheads, and people with very fair skin are better off with an SPF of 20 to 30. (The risks are the same for those who spend time in a tanning salon or tanning bed.) Hats with brims and panels to shield the neck and face provide added protection.

Trauma

Trauma is a physical injury caused by pressure, impact, distortion, or other mechanical force. Traumatic injuries to the skin are very common, and a number of terms are used to describe them. Such injuries generally affect all components of the skin, and each type of wound presents a different series of problems to clinicians attempting to limit damage and promote healing.

A Classification of Wounds. An *open wound* is an injury producing a break in the epidermis. The major categories of open wounds are illustrated in Figure 7-12•. **Abrasions** are the result of scraping against an object (Figure 7-12a•). Bleeding may be slight, but a considerable area may be open to invasion by microorganisms. **Incisions** are linear cuts produced by sharp objects (Figure 7-12b•). Bleeding may be severe if deep vessels are damaged. The bleeding may help flush the wound, and prompt closure with bandages or stitches can limit the area open to infection while healing is underway. A **laceration** is a jagged, irregular tear in the surface often produced by solid impact or by an irregular object (Figure 7-12c•). Tissue

•**FIGURE 7-12**
Major Types of Open Wounds.

(a) Abrasion

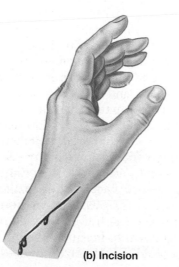

(b) Incision

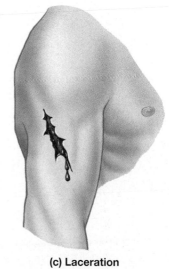

(c) Laceration

•FIGURE 7-12 (continued)
Major Types of Open Wounds.

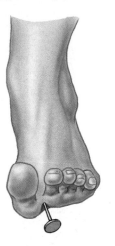

(d) Puncture

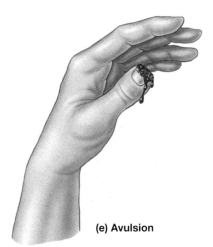

(e) Avulsion

damage is more extensive, and repositioning the opposing sides of the injury may be difficult. Lacerations are prone to infection. **Punctures** result from slender, pointed objects piercing the epithelium (Figure 7-12d•). Little bleeding results, and any microbes delivered under the epithelium are likely to find conditions to their liking. In an **avulsion**, chunks of tissue are torn away by the brute force of an auto accident, explosion, or other incident (Figure 7-12e•). Bleeding may be considerable, and serious internal damage may be present.

Closed wounds may affect any internal tissue. Closed wounds involving the skin are less dangerous than open wounds because the epidermis remains intact, and infection is unlikely. A **contusion** is a "black and blue" area caused by bleeding in the dermis. Most contusions are not dangerous, but contusions of the head, such as "black eyes," may be an external sign of life-threatening bleeding within the skull. Closed wounds affecting internal organs and organ systems are almost always serious threats to life.

Skin Repair. The skin can regenerate effectively even after considerable damage has occurred, because stem cells persist in both the epithelial and connective tissue components. Divisions in the stratum germinativum replace damaged epidermal cells, and stem cells in the loose connective tissue divide to replace lost dermal cells. This replacement process can be slow, and when large surface areas are involved there may be additional problems with infection and fluid loss. The relative speed and effectiveness of skin repair also varies with the type of wound involved. A slender, straight cut, or incision, may heal relatively quickly compared with a deep scrape, or abrasion, which involves a much greater area.

Figure 7-13• shows stages in the regeneration of the skin after an injury. When damage extends through the epidermis and into the dermis, bleeding generally occurs (Step 1). The blood clot, or **scab**, that forms at the surface temporarily restores the integrity of the epidermis and restricts the entry of additional microorganisms while phagocytosis and cleanup operations are underway (Step 2). The bulk of the clot consists of an insoluble network of *fibrin*, a fibrous protein that forms from blood proteins during the clotting response. The color reflects the presence of trapped red blood cells. Cells of the stratum germinativum undergo rapid divisions and begin to migrate along the sides of the wound in an attempt to replace the missing epidermal cells. Meanwhile, phagocytes patrol the damaged area of the dermis and clear away debris and pathogens.

If the wound covers an extensive area, dermal repairs must be under way before epithelial cells can cover the surface. Fiber-producing cells (*fibroblasts*) and connective tissue stem cells divide to produce mobile cells that invade the deeper areas of injury (Step 3). Endothelial cells lining the damaged blood vessels also begin to divide, and capillaries follow the fibroblasts, providing a circulatory supply. The combination of blood clot, fibroblasts, and an extensive capillary network is called *granulation tissue*. Over time, the clot dissolves and the number of capillaries declines (Step 4). Meanwhile, the area is stabilized by the collagen fibers produced by the fibroblasts.

These repairs do not restore the integument to its original condition, however, for the dermis will contain an abnormally large number of collagen fibers and relatively few blood vessels. Severely damaged hair follicles, sebaceous or sweat glands, muscle cells, and nerves are seldom repaired, and they too are replaced by fibrous tissue. The formation of this rather inflexible, fibrous, noncellular **scar tissue** can be considered a practical limit to the healing process.

We do not know what regulates the extent of scar tissue formation, and the process is highly variable. For example, surgical procedures performed on a fetus do not leave scars, perhaps because damaged fetal tissues do not produce the same types of growth factors that adult tissues do. In some adults, most often those with dark skin, scar tissue formation may continue beyond the requirements of tissue repair. The result is a flattened mass of scar tissue that begins at the injury site and grows into the surrounding dermis. This thickened area of scar tissue, called a **keloid** (KĒ-loyd), is covered by a shiny, smooth epidermal surface.

Keloids most commonly develop on the upper back, shoulders, anterior chest, and earlobes. They are harmless, and some aboriginal cultures intentionally produce keloids as a form of body decoration.

Skin repairs proceed most rapidly in young, healthy individuals. For example, barring infection, it might take 3–4 weeks to complete the repairs to a blister site in a young adult. The same repairs at age 65–75 could take 6–8 weeks. This is just

•**FIGURE 7-13**
Skin Repair.

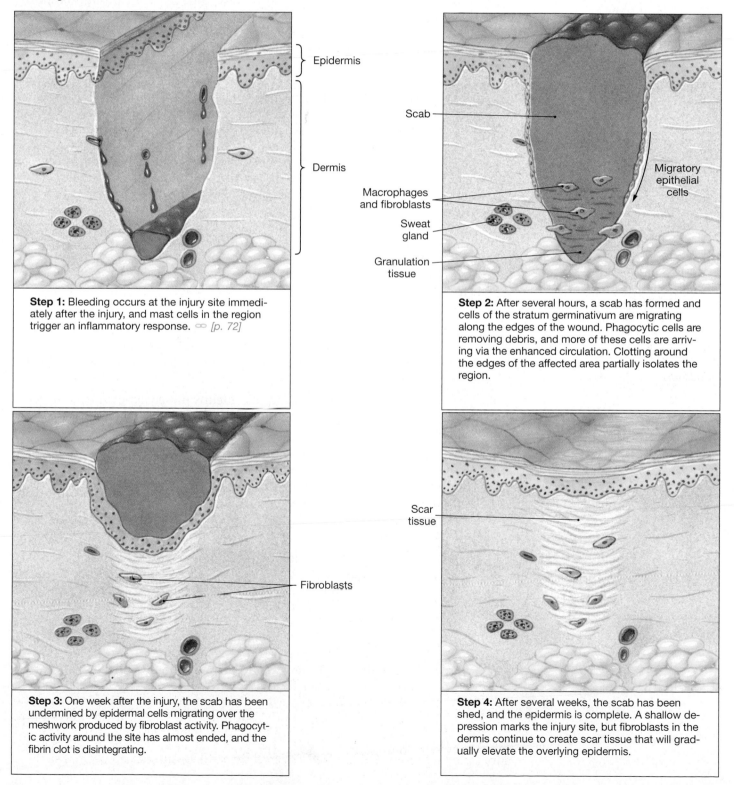

Step 1: Bleeding occurs at the injury site immediately after the injury, and mast cells in the region trigger an inflammatory response. ∞ *[p. 72]*

Step 2: After several hours, a scab has formed and cells of the stratum germinativum are migrating along the edges of the wound. Phagocytic cells are removing debris, and more of these cells are arriving via the enhanced circulation. Clotting around the edges of the affected area partially isolates the region.

Step 3: One week after the injury, the scab has been undermined by epidermal cells migrating over the meshwork produced by fibroblast activity. Phagocytic activity around the site has almost ended, and the fibrin clot is disintegrating.

Step 4: After several weeks, the scab has been shed, and the epidermis is complete. A shallow depression marks the injury site, but fibroblasts in the dermis continue to create scar tissue that will gradually elevate the overlying epidermis.

one example of the changes that occur in the integumentary system as a result of the aging process. Any infection delays healing at all ages.

Burns and Grafts. Burns result from exposure of the skin to heat, radiation, electrical shock, or strong chemical agents. The severity of the burn reflects the depth of penetration and the total area affected.

Table 7-3 is a classification of burns according to affected skin depth. First- and second-degree burns are also called *partial-thickness burns* because damage is restricted to the superficial layers of the skin. Only the surface of the epidermis is affected by a *first-degree burn.* In this type of burn, including most sunburns, the skin reddens and can be painful. The redness results from inflammation of the sun-damaged tissues. In a *second-degree burn,* the entire epidermis and perhaps some of the dermis are damaged. The affected area is red, tender, swollen, and blistered. Accessory structures such as hair follicles and glands are generally not affected. If the blisters rupture at the surface, infection can easily develop. Healing typically takes 1–2 weeks, and some scar tissue may form. *Full-thickness burns,* or *third-degree burns,* destroy the epidermis and dermis, extending into subcutaneous tissues. The affected tissue is generally charred in a thermal (heat) burn or discolored (in a chemical burn). Although the remaining tissues are swollen, such burns are actually less painful than second-degree burns, because sensory nerves are destroyed along with accessory structures, blood vessels, and other dermal components. Extensive third-degree burns cannot repair themselves, because granulation tissue cannot form and epithelial cells are unable to cover the injury site. As a result the burned area remains open to potential infection. The larger the area burned, the more significant the effects on integumentary function. Figure 7-14● presents a standard reference for calculating the percentage of total surface area damaged.

Burns that cover more than 20 percent of the skin surface represent serious threats to life because they affect the following functions:

■ *Fluid and Electrolyte Balance.* Even areas with partial-thickness burns lose their effectiveness as barriers to fluid and electrolyte losses. In full-thickness burns, the rate of fluid loss through the skin may reach five times the normal level.

■ *Thermoregulation.* Increased fluid loss means increased evaporative cooling. More energy must be expended to keep body temperature within acceptable limits.

■ *Protection from Infection.* The epidermal surface, damp from uncontrolled fluid losses, encourages bacterial growth. If the skin is broken at a blister or the site of a third-degree burn, infection is likely. Widespread bacterial infection, or **sepsis** (*septikos* means rotting), is the leading cause of death in burn victims.

With the development of fluid-replacement therapies, infection control methods, and skin grafting techniques, the recovery rate for severely burned patients has improved dramatically. At present, young patients with burns over 80 percent of the body have an approximately 50 percent chance of recovery.

Recent advances in cell culture techniques may improve survival rates further. It is now possible to remove a small section of undamaged epidermis and grow it under controlled laboratory conditions. Over time, the germinative cell divisions produce large sheets of epidermal cells that can then be used to cover the burn area. From initial samples the size of a postage stamp, square yards of epidermis have been grown and transplanted onto body surfaces. Although questions remain concerning the

●FIGURE 7-14
A Quick Method of Estimating the Percentage of Surface Area Affected by Burns.
The method is called the *rule of nines* because the surface area of adults is in multiples of nine. The rule is modified for children because their body proportions are quite different.

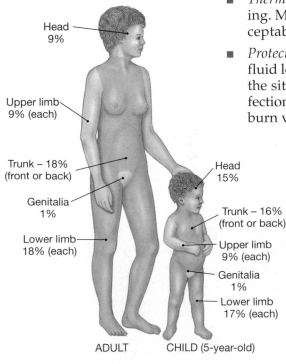

Head
9%

Upper limb
9% (each)

Trunk – 18%
(front or back)

Genitalia
1%

Lower limb
18% (each)

Head
15%

Trunk – 16%
(front or back)

Upper limb
9% (each)

Genitalia
1%

Lower limb
17% (each)

ADULT   CHILD (5-year-old)

Table 7-3	A Classification of Burns	
Classification	**Damage Report**	**Appearance and Sensation**
First-degree burn	*Killed:* superficial cells of epidermis *Injured:* deeper layers of epidermis, papillary dermis	Inflamed, tender
Second-degree burn	*Killed:* superficial and deeper cells of epidermis; dermis may be affected *Injured:* damage may extend into reticular layer of the dermis, but many accessory structures unaffected	Blisters, pain
Third-degree burn	*Killed:* all epidermal and dermal cells *Injured:* hypodermal and deeper tissues and organs	Charred, no sensation at all

strength and flexibility of the repairs, skin cultivation represents a substantial advance in the treatment of serious burns.

Congenital Disorders

One group of congenital skin disorders includes various types of **birthmarks**, or pigmented areas of skin. Such birthmarks are due to tumors that develop in dermal blood vessels, called **hemangiomas** (he-man-jē-Ō-mahs), during development. They may be temporary, such as *capillary hemangiomas*, or permanent and life-long such as **cavernous hemangiomas**, or "port wine stains".

Xeroderma (zē-rō-DER-mah) **pigmentosum** is a rare inherited skin disorder in which the skin is very sensitive to sunlight. Although the skin is normal at birth, exposure to sunlight causes it to become dry, wrinkled, and prematurely aged by age 5. Not only may both benign and malignant tumors develop, but disorders of the eye also appear. Treatment involves avoiding sunlight with clothes and sunblock creams, and surgery or drugs for tumors. A *Baywatch* episode featured a boy with this condition who built sandcastles on the beach at night.

Nutritional Disorders

Skin color is due to the interaction of pigments and dermal blood supply. In addition to the primary pigment, melanin, various amounts of a orange-yellow pigment called **carotene** (KAR-ō-tēn) is present. It is most apparent in the stratum corneum of light-skinned individuals and is also stored in the fatty tissues of the dermis. Carotene is a common pigment in a variety of orange-colored plants, such as carrots and squashes. Vegetarians with light skins with a special liking for carrots can actually have their skin turn orange from consuming an overabundance of carrots. The color change is less striking in people with dark skin.

Degenerative Disorders

Degenerative disorders of the skin are generally associated with aging and were discussed earlier. ∞ *[p. 123]*

Secondary Conditions

Several diseases that have primary impacts on other systems may have secondary effects on the integumentary system. For example, skin color may be affected by disorders in both the digestive and endocrine systems. The endocrine system may also alter skin conditions through its effects on sebaceous glands and acne development, and on hair follicles through hair loss.

Digestive System Disorders. In *jaundice* (JAWN-dis) the liver is unable to excrete bile, and a toxic yellowish pigment accumulates in body fluids. In advanced stages, the skin and whites of the eyes turn yellow.

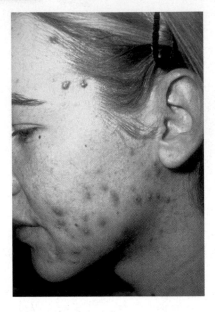

•FIGURE 7-15
Acne.

Endocrine System Disorders. Some tumors affecting the pituitary gland result in the secretion of large amounts of *melanocyte-stimulating hormone* (MSH). This hormone causes a darkening of the skin, as if the individual has an extremely deep bronze tan. In *Addison's disease* the pituitary gland secretes large quantities of ACTH, a hormone that is structurally similar to MSH. The result of ACTH on the skin coloration is also similar to that of MSH.

Levels of sexual hormones change in the course of a lifetime, and these changes can have an impact on the skin. Three examples of conditions associated with sex hormone levels are acne, baldness, and hirsutism. **Acne** is an inflammation of sebaceous glands that commonly affects the face (Figure 7-15•). The condition generally develops at puberty, as sexual hormone production accelerates and stimulates the sebaceous glands. If the ducts become blocked, the secretions accumulate. Inflammation develops, and bacterial infection may occur. Their secretory output may be further encouraged by anxiety, stress, physical exertion, and drugs, but not usually by foods.

Acne generally fades after sex hormone concentrations stabilize. Topical antibiotics, vitamin A derivatives, such as *Retin-A*, or peeling agents may help reduce inflammation and minimize scarring. In cases of severe acne, the most effective treatment usually involves the discouragement of bacteria by the administration of antibiotic drugs. Because oral antibiotic therapy has risks, including the development of antibiotic-resistant bacteria, however, this therapy is not used unless other treatment methods have failed. Truly dramatic improvements in severe cases have been obtained with the prescription drug *Accutane*. This compound is structurally similar to vitamin A, and it reduces oil gland activity on a long-term basis.

Baldness is also known as **alopecia** (al-ō-PĒ-shē-ah). Two factors interact to determine permanent baldness: heredity and hormones. A bald individual inherits a susceptibility for baldness that is triggered by male sex hormones. The inherited condition in men is called **male pattern baldness.** It affects the top of the head and forehead first, only later reducing the hair density along the sides. Many women carry the genetic susceptibility for baldness, but unless major hormonal abnormalities develop, as in certain endocrine tumors, they do not become bald.

If you are a healthy adult, you lose about 50 hairs from your head each day. Several factors affect this rate. Sustained losses of more than 100 hairs per day generally indicate that something is wrong. Temporary increases in hair loss can result from drugs, dietary factors, radiation, vitamin A excess, high fever, stress, or hormonal factors related to pregnancy.

Hirsutism (HER-soot-ism) refers to the growth of hair on women in patterns generally characteristic of men (*hirsutus* means bristly). Because considerable overlap exists between the two genders in terms of normal hair distribution, and because there are significant racial and genetic differences, the definition is more often a matter of personal taste than of objective analysis. Age and sex hormones may play a role, for hairiness increases late in pregnancy and menopause produces a change in body hair patterns. Severe hirsutism is often associated with abnormally high androgen (male sex hormone) production.

Immune System Disorders. Immune disorders can have affects on the skin and its accessory organs. In **vitiligo** (vi-ti-LĪ-gō), patches of skin lose their color due to the loss of melanocytes. The condition develops in about 1 percent of the population, and the incidence increases among individuals with thyroid gland disorders, Addison's disease, and several other immune disorders. It is suspected that vitiligo develops when the immune defenses malfunction and antibodies attack normal melanocytes. The primary problem with vitiligo is cosmetic, especially for individuals with darkly pigmented skin.

Scleroderma is a very rare condition in which there is an overproduction of normal collagen fibers in the face and hands. In some cases, this *fibrosis* (overgrowth of connective tissue) also occurs in internal organs such as blood vessels,

☤ Clinical Note

Drug Patches, An Alternative to Pills

Drugs in oils or other lipid-soluble carriers can penetrate the epidermis. The movement is slow, particularly through the layers of cell membranes in the stratum corneum. But once a drug reaches the underlying tissues it will be absorbed into the circulation. Placing a drug in a solvent that is lipid-soluble can assist its movement through the lipid barriers. Drugs can also be administered by packaging them in *liposomes*, artificially produced lipid droplets. Liposomes containing DNA fragments have been used experimentally to introduce normal genes into abnormal human cells. For example, genes carried by liposomes have been used to alter the cell membranes of skin cancer cells so that the tumor cells will be attacked by the immune system. Another experimental procedure involves the creation of a transient change in skin permeability by administering a brief pulse of electricity. This electrical pulse temporarily changes the positions of the cells in the stratum corneum, creating channels that allow the drug to penetrate.

A useful technique for long-term drug administration involves the placement of a sticky patch containing a drug over an area of thin skin. To overcome the relatively slow rate of diffusion, the patch must contain an extremely high concentration of the drug. This procedure, called **transdermal administration**, has the advantage that a single patch may work for several days, making daily pills unnecessary. Some of the

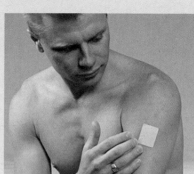

A nicotine skin patch

drugs that are now regularly administered this way include:

Scopolamine, a drug that affects the nervous system, is used to control the nausea associated with motion sickness.

Nitroglycerin, a drug used to improve blood flow within heart muscle and prevent a heart attack.

Estrogen, a female hormone is used to reduce the symptoms of menopause.

Nicotine is used to suppress the urge to smoke cigarettes.

In addition, pain medications and drugs to control high blood pressure may be administered by transdermal patches.

---

lungs, heart, kidney, and digestive system. Scleroderma means "hard skin," and it causes the skin of the face to become hard, like a stiff mask. The disease begins during young to middle age (20s to 40s) and affects women some three times more frequently than men.

Alopecia areata (al-ō-PĒ-shē-ah ar-ē-AH-ta) is a localized hair loss that can affect either gender. The cause is not known, and the severity of hair loss varies from case to case. This condition is associated with several disorders of the immune system. It has also been suggested that periods of stress may promote alopecia areata in individuals who are genetically prone to baldness.

THE TREATMENT OF SKIN DISORDERS

Isolated growths, such as skin tumors or warts, may be surgically removed. Alternatively, abnormal cells may be destroyed by electrical currents (*electrosurgery*) or by freezing (*cryosurgery*). Ultraviolet radiation may help conditions such as acne or psoriasis, whereas use of a sunscreen or sunblock is important in controlling sun-sensitive outbreaks. When large areas of skin are damaged or destroyed, *skin grafts* are usually required.

Drugs that oppose inflammation, such as the steroid *hydrocortisone*, can be applied to the skin to reduce the redness and itching that accompanies many skin conditions. Applying drugs to the skin surface is called *topical application*. Some injected or swallowed drugs (*systemic application*) may also be helpful; aspirin is a familiar systemic drug with **anti-inflammatory** properties. The skin can also be used as a gateway to introduce drugs that target internal systems. This process is called *transdermal medication*.

✔ What do calluses and corns have in common?

✔ What degree burn refers to the destruction of both the epidermis and dermis?

CHAPTER REVIEW

Key Words

accessory structures: Specialized structures of the skin, including hair, nails, and glands.

cutaneous membrane: The skin; includes the epidermis and dermis.

dermis: The deeper of the two primary layers of the skin; the deeper, connective tissue layer beneath the epidermis.

epidermis: The more superficial of the two primary layers of the skin; the epithelium forming the outermost surface of the skin.

hair follicles: The accessory structures responsible for the formation of hairs.

integumentary system: Includes skin, hair, nails, and glands.

keratin (KER-a-tin): Tough, fibrous protein component of nails, hair, calluses, and the general integumentary surface.

melanin (ME-la-nin): Dark pigment produced by the melanocytes of the skin.

nail: Hard, protective structure covering the tips of toes and fingers.

sebaceous glands (se-BĀ-shus): Glands that secrete sebum, usually associated with hair follicles.

sebum (SĒ-bum): A waxy secretion that coats the surfaces of hairs.

stratum (STRĀ-tum): Layer.

subcutaneous layer: The layer of loose connective tissue below the dermis; also called the *hypodermis.*

Selected Clinical Terms

acne: A sebaceous gland inflammation caused by an accumulation of secretions.

cavernous hemangiomas ("port-wine stain"): A birthmark caused by a tumor affecting large blood vessels in the dermis.

dermatitis: An inflammation of the skin that involves the dermis.

furuncle: A boil, resulting from the invasion and inflammation of a hair follicle or sebaceous gland.

lesion: Changes in skin structure caused by injury or disease.

pruritis: An irritating itching sensation, common in skin conditions.

skin signs: Characteristic abnormalities in the skin surface that can assist in diagnosing skin conditions.

ulcer: A localized shedding or loss of an epithelium.

xerosis: "Dry skin," a common complaint of older persons and people who live in arid climates.

Study Outline

SYSTEM BRIEF (p. 117)

1. The **integumentary system** (or the *integument*) is made up of the *skin* and the *accessory structures*.

2. Its functions include (1) protection, (2) maintenance of body temperature, (3) storage, excretion, and synthesis, and (4) sensory reception.

THE SKIN (pp. 117–120)

1. The **cutaneous membrane**, or skin, consists of the **epidermis** and **dermis**. Underneath the skin lies the **subcutaneous layer**. (*Figure 7-1*)

THE EPIDERMIS (p. 117)

2. The epidermis is a stratified squamous epithelial tissue. (*Figure 7-2*)

3. Cell division in the **stratum germinativum** produces new cells to replace the more superficial cells that are continually lost.

4. As epidermal cells age they fill with large amounts of **keratin** to form the outer layer, or **stratum corneum**. Ultimately the cells are shed or lost.

5. The color of the epidermis depends on (1) pigment composition and (2) blood supply. **Melanocytes** protect us from **ultraviolet radiation**. (*Figure 7-3*)

6. Cells within the lower layers of the epidermis produce vitamin D when exposed to sunlight.

THE DERMIS (p. 120)

7. The **dermis** underlies the epidermis and consists of two different types of connective tissues (loose and dense).

8. The dermis supports and nourishes the overlying epidermis through its blood vessels, lymphatics, and sensory nerves. The lower portion of the dermis consists of collagen and elastic fibers that provide structural support and flexibility in the skin.

THE SUBCUTANEOUS LAYER (p. 120)

9. The **subcutaneous layer** or *hypodermis* contains fat cells and stabilizes the skin's position against underlying organs and tissues.

ACCESSORY STRUCTURES (pp. 120–122)

HAIR (p. 120)

1. Hairs originate in complex organs called **hair follicles**. Each hair has a **root** and a **shaft**. (*Figure 7-4*)

2. The **arrector pili** muscles can elevate the hairs.

GLANDS (p. 121)

3. **Sebaceous glands** discharge the waxy **sebum** into hair follicles.

4. **Apocrine sweat glands** produce an odorous secretion; the more numerous **merocrine** (*eccrine*) **sweat glands** produce a watery secretion.

NAILS (p. 122)

5. Nail production occurs at the **nail root**, a fold of epithelial tissue. (*Figure 7-5*)

AGING AND THE INTEGUMENTARY SYSTEM (p. 123)

1. Aging affects all the components of the integumentary system. (*Figure 7-6*)

SKIN PROBLEMS (pp. 124–137)

EXAMINATION OF THE SKIN (p. 124)

1. Dermatologists use the **skin lesions**, or **skin signs**, to diagnosis skin diseases. (*Figure 7-7; Table 7-1*)

SKIN DISORDERS AND THEIR CAUSES (p. 126)

2. Causes of skin disorders range from environmental stress and inflammation, infections, tumors, trauma, congenital conditions, nutritional problems, degenerative disorders, to secondary conditions that result from disorders in other organ systems (digestive, endocrine, and immune). *(Figure 7-8)*

3. Environmental stress is associated with **hyperkeratosis, xerosis**, and **decubitis ulcers**. *(Figure 7-8)*

4. Dermatitis is an inflammation of the skin that primarily involves the dermis.

5. Bacterial skin infections are commonly caused by *Staphylococcus* bacteria. Examples include **folliculitis, furuncles**, or **boils**; and *scalded skin syndrome*. *(Table 7-2)*

6. Examples of viral skin infections include those caused by the *Herpes* viruses and the **human papilloma virus (HPV)**. *(Table 7-2)*

7. Dermatomycoses are fungal infections of the skin. Examples include **ringworm, athlete's foot, blastomycosis**, and **candidiasis** (nail infections). *(Figure 7-10)*

8. Multicellular parasites may live on or burrow into the skin. Examples of skin conditions are **swimmer's itch** caused by helminths (worms); **scabies** caused by the itch mite; and, **pediculosis** caused by the human body louse. *(Table 7-2)*

9. *Freckles* and *moles* are benign tumors of the skin. Most people survive **squamous cell carcinomas** and **basal cell carcinomas** because they rarely metastasize. However, **malignant melanomas**, composed of melanocytes, are dangerous because they can spread rapidly through the lymphatic system. *(Figure 7-11)*

10. Trauma (physical injury) may be caused by pressure, impact, distortion, or other mechanical forces. **Abrasions, incisions, lacerations, punctures**, and **avulsions** are types of open wounds. *(Figure 7-12)*

11. The skin can regenerate effectively even after considerable damage. The process includes the formation of a **scab**. *(Figure 7-13)*

12. Burns are relatively common injuries that damage the epidermis and varying amounts of the dermis. *(Figure 7-14, Table 7-3)*

13. Examples of congenital conditions include birthmarks and inherited disorders. **Birthmarks** are caused by **hemangiomas**, tumors in the dermal blood vessels.

14. Nutritional imbalances such as an excess of carrots can increase the amount of **carotene** pigment in the skin, coloring it orange in light-skinned individuals.

15. Degenerative disorders are usually associated with aging. *(Figure 7-6)*

16. *Jaundice* is a disorder of the liver that causes the skin to appear yellow.

17. Disorders of the endocrine system can affect the skin and hair. Rapid increases in sex hormone production at puberty can cause **acne**, a skin disorder associated with inflammation and blockage of sebaceous gland ducts. *(Figure 7-15)*

18. Male sex hormones (androgens) can affect both men and women. Baldness, or **alopecia**, is more common in men.

19. Skin disorders affected by problems with the immune response include **vitiligo, scleroderma**, and **alopecia areata**.

THE TREATMENT OF SKIN DISORDERS *(p. 137)*

20. Skin tumors, warts and other abnormal cells may be removed surgically. Topical and systemic **anti-inflammatory drugs** can reduce the redness and itching of a variety of skin conditions.

Review Questions

MATCHING

Match each item in Column A with the most closely related item in Column B. Use letters for answers in the spaces provided.

	Column A	Column B
___	1. cutaneous membrane	a. arrector pili
___	2. layer of hard keratin that coats the hair	b. stratum corneum
___	3. carotene	c. closed wound
___	4. melanocytes	d. perspiration
___	5. the layer that contains stem cells	e. lunula
___	6. contractions of these smooth muscles produce "goose bumps"	f. aids bone growth
___	7. epidermal layer of flattened and dead cells	g. hair follicles
___	8. bluish coloration of the skin	h. orange-yellow pigment that accumulates inside skin cells
___	9. apocrine sweat glands	i. cuticle
___	10. merocrine (eccrine) sweat glands	j. pigment cells
___	11. vitamin D	k. inflammation of the skin
___	12. white, crescent-shaped area at the base of the nail	l. skin
___	13. produces hairs	m. produce a cloudy, viscous secretion
___	14. dermatitis	n. stratum germinativum
___	15. contusion	o. cyanosis

MULTIPLE CHOICE

16. The major components of the integumentary system are the accessory structures and the _____.
(a) cutaneous membrane
(b) subcutaneous layer
(c) epidermis
(d) dermis

17. The following are all accessory structures of the integumentary system, *except* _____.
(a) nails
(b) epidermis
(c) hair
(d) sweat glands

18. Each individual hair has a root and a _____.
(a) sweat gland
(b) dermis
(c) shaft
(d) sebaceous follicle

19. Nail production occurs _____.
(a) at the free edge of the nail
(b) in the cuticle
(c) throughout the entire nail
(d) at the nail root

20. The cells making up the stratum corneum contain large amounts of _____.
(a) enzymes
(b) keratin
(c) cellulose
(d) electrolytes

21. Sebaceous glands discharge a secretion called _____.
(a) lumen
(b) matrix
(c) sebum
(d) keratin

22. It takes approximately _____ days for a cell to move from the stratum germinativum to the stratum corneum.
(a) 7
(b) 14
(c) 21
(d) 28

23. The layer that contains the capillaries that supply the surface of the skin is the _____.
(a) epidermis
(b) dermis
(c) subcutaneous layer
(d) hypodermis

24. All of the following are due to excess keratin production except _____.
(a) impetigo
(b) psoriasis
(c) calluses
(d) corns

25. _____ is a condition in which there is a loss of pigment in patches of skin.
(a) alopecia areata
(b) keloid
(c) hemangioma
(d) vitiligo

TRUE/FALSE

___ 26. White hair results from the presence of air bubbles within the hair shaft.

___ 27. Blood with abundant oxygen takes on a bluish color.

___ 28. Sebum is a breeding ground for bacteria.

___ 29. The integumentary system requires exposure to ultraviolet radiation in order to synthesize vitamin D.

___ 30. The differences in skin color between individuals and races is primarily due to different numbers of melanocytes.

___ 31. An avulsion is a type of closed wound.

___ 32. Corns develop on thin-skinned areas of the body.

___ 33. A sunburn is an example of a first-degree burn.

SHORT ESSAY

34. What are the six principal functions of the integumentary system?

35. Why does the epidermis continually shed its cells?

36. How does the ultraviolet radiation in sunlight affect the epidermis? Discuss both positive and negative effects.

37. What is keratin? List three human anatomical structures that are made of keratin.

38. What functions does hair serve in humans?

39. What are the characteristics of the hypodermis? Name four of its functions. Why is it a useful area for administering drugs?

40. Compare the functions of the secretions produced by apocrine sweat glands and merocrine sweat glands.

41. How does aging alter the integumentary system. How do these changes affect the physical condition of elderly people?

42. Describe and give an example of each of the following skin lesions or signs:
(a) macule
(b) wheal
(c) papule
(d) vesicle
(e) pustule
(f) crust

43. Which cells are affected in squamous and basal cell carcinoma, and malignant melanoma?

APPLICATIONS

44. Chanda, a vegetarian, tells her physician about a yellow discoloration of the palms of her hands and the soles of her feet. Further inspection reveals that Chanda's forehead and the area around her nose are also yellow. The physician determines that Chanda has not been exposed to hepatitis (an inflammation of the liver) and that she has normal liver function tests. What might be the problem?

45. Charlie is badly burned in a fireworks accident on the Fourth of July. When he reaches the emergency room, the examining physician determines the severity of the incident as a third-degree burn. The physician orders IV (intravenous) fluids and electrolytes (ions), antibiotics, and a high-nutrient diet for Charlie. Why?

✔ Answers to Concept Check Questions

(p. 120)   **1.** Cells are constantly shed from the outer layers of the *stratum corneum*. **2.** When exposed to the ultraviolet radiation in sunlight or tanning lamps, melanocytes in the epidermis synthesize the pigment *melanin*, darkening the color of the skin. **3.** Sanding the tips of one's fingers will not permanently remove fingerprints. Because the ridges of the fingerprints are formed in layers of the skin that are constantly regenerated, these ridges will eventually reappear.

(p. 121)   **1.** Contraction of the arrector pili muscles pulls the hair follicles erect, depressing the area at the base of the hair and making the surrounding skin appear higher. The result is known as goose bumps or goose pimples. **2.** Hair is derived from the epidermis, so if the epidermis is destroyed by the injury, there will be no hair follicles to produce new hair. **3.** The subcutaneous layer stabilizes the position of the skin in relation to underlying tissues and organs, stores fat, and because its lower region contains few capillaries and no vital organs, provides a useful site for the injection of drugs.

(p. 123)   **1.** If the duct of a sebaceous gland is blocked its secretions accumulate, causing swelling and providing a fertile environment for bacterial infection. **2.** Apocrine sweat glands produce a secretion containing several kinds of organic compounds. Some of these have an odor or produce an odor when broken down by skin bacteria. Deodorants are used to mask the odor of these secretions. **3.** As a person ages, the blood supply to the dermis decreases and merocrine sweat glands become less active. These changes make it more difficult for the elderly to cool themselves in hot weather.

(p. 137)   **1.** Calluses and corns are both skin signs that are the results of excessive production of keratin, or hyperkeratosis. **2.** Third-degree burns, or full-thickness burns, destroy both the epidermis and dermis.

CHAPTER

8

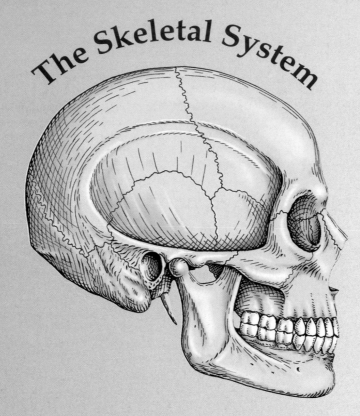

The Skeletal System

CHAPTER OUTLINE

CHAPTER OBJECTIVES

1 Describe the functions of bones and the skeletal system.
2 Compare the structure and function of compact and spongy bones.
3 Discuss the processes by which bones develop and grow.
4 Describe how the activities of the bone cells constantly remodel bones.
5 Discuss the effects of aging on the bones of the skeletal system.
6 Contrast the structure and function of the axial and appendicular skeletons.
7 Identify the bones of the skull.
8 Discuss the differences in structure and function of the various vertebrae.
9 Identify the bones of the limbs.
10 Describe the three basic types of joints.
11 Relate body movements to the action of specific joints.
12 Describe the different causes of bone disorders and give examples of each.
13 Distinguish between degenerative and inflammatory arthritic diseases.

The skeletal system includes the bones of the skeleton and the specialized tissues that interconnect them. Besides supporting the weight of the body and protecting delicate internal organs, bones work together with muscles to produce the movements that let you walk, run, or turn the pages of this book.

This chapter examines the skeletal system from the cellular structure of bone and cartilage to the individual bones themselves. Finally, we show how two or more bones interact at joints, or *articulations*, to permit normal movements.

SYSTEM BRIEF

The skeletal system performs the following functions:

1. **Support**: Bones and cartilages provide structural support for the entire body.

2. **Protection**: Delicate tissues and organs are often surrounded by bones. The ribs protect the heart and lungs; the skull encloses the brain; the vertebrae shield the spinal cord; and the pelvis cradles delicate digestive and reproductive organs.

3. **Leverage**: When a skeletal muscle contracts, it pulls on a bone and produces movement at a specific joint. The location of the muscle attachment site relative to the joint determines the speed and power of the resulting movement.

4. **Storage**: Bones store valuable minerals such as calcium and phosphate. In addition, fat cells within the internal cavities of many bones store lipids as an energy reserve.

5. **Blood cell production**: Red blood cells and other blood elements are produced within the internal cavities of many bones.

BONE STRUCTURE

Bone, or **osseous tissue**, is one of the two supporting connective tissues. ∞ *[p. 67]* (Can you name the other supporting connective tissue?) Like other connective tissues, osseous tissue contains specialized cells in a *matrix* containing extracellular fibers and a ground substance. Remember, it is the matrix that determines the consistency of a connective tissue, and the matrix of bone is dominated by calcium deposits. These compounds account for almost two thirds of a bone's mass and give the tissue its characteristic solid, stony feel. The remaining bone mass is mainly collagen fibers, with living cells providing only around 2 percent of the mass of a bone.

GENERAL FEATURES OF BONE

Human bones have four general shapes: *long, short, flat,* and *irregular*. **Long bones** are longer than they are wide, whereas short bones are of roughly equal dimensions. Examples of long bones include bones of the limbs such as the arm (*humerus*) and thigh (*femur*). **Short bones** include the bones of the wrist (*carpals*) and ankle (*tarsals*). **Flat bones** are thin and relatively broad, such as the *parietal bones* of the skull, the ribs, and the shoulder blades (*scapulae*). **Irregular bones** have shapes that do not fit easily into any other category. An example would be one of the *vertebrae*, the bones of the spinal column.

The typical features of a long bone are shown in Figure 8-1●. A long bone has a central shaft, or **diaphysis** (dī-AF-i-sis), and expanded ends, or **epiphyses** (ē-PIF-i-sēs; singular; *epiphysis*). The diaphysis surrounds a central *marrow cavity*. **Marrow** is the name given to the loose connective tissue that fills such cavities. *Red marrow* is blood-forming tissue, and *yellow marrow* is adipose tissue. The epiphyses of adjacent bones interact at a joint. Within many joints the opposing bony surfaces are covered by smooth and slippery *articular cartilages*.

dia, through + *physis*, growth
diaphysis, growing between; the shaft that separates the two epiphyses of a long bone

epi, upon + *physis*, growth
epiphysis: one expanded end of a long bone

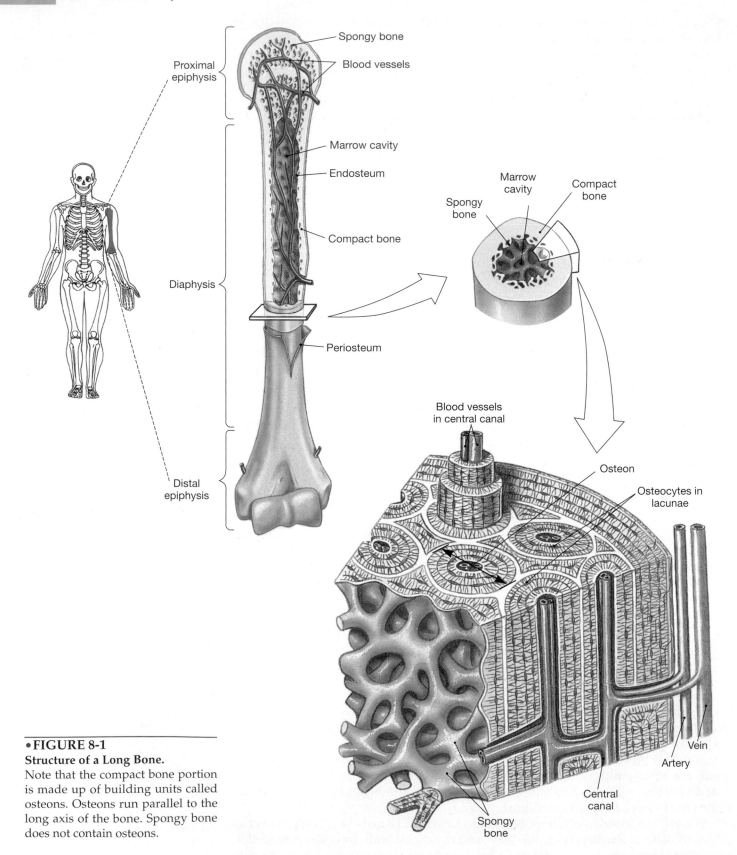

•FIGURE 8-1
Structure of a Long Bone.
Note that the compact bone portion is made up of building units called osteons. Osteons run parallel to the long axis of the bone. Spongy bone does not contain osteons.

Figure 8-1• also shows the two types of bone tissue: *compact bone* and *spongy bone*. **Compact bone** is relatively solid, whereas **spongy bone** is porous and less dense. Both types are present in long bones. Compact bone forms the diaphysis and covers the surfaces of the epiphyses. Each epiphysis contains a core of spongy bone.

The outer surface of a bone is covered by a **periosteum** (per-ē-OS-tē-um), a term that means "around the bone" (Figure 8-1•). The periosteum consists of a fibrous outer layer and an inner cellular layer. A cellular **endosteum** (en-DOS-te-um) (*endo-* means "inside") lines the marrow cavity inside the bone. Blood vessels and nerves penetrate the periosteum to supply the bone and tissues of the bone marrow, and both the periosteum and endosteum are important for bone growth and repair.

peri, around + *osteon*, bone
periosteum: the connective tissue sheath around a bone

endo, inside + *osteon*, bone
endosteum: a layer of cells that lines the internal surfaces of bones

MICROSCOPIC FEATURES OF BONE

Bone Cells

There are three different types of bone cells, each with its own function in the growth or maintenance of bone:

1. **Osteocytes** (OS-tē-ō-sīts) are mature bone cells that are embedded in the matrix. As you will recall, the solid matrix in bone tissue provides strength and rigidity. Osteocytes maintain the density and composition of the bone matrix by removing and replacing the calcium compounds in the surrounding matrix. They also help repair damaged bone. In a mature adult, whose bones are no longer growing, osteocytes are the most numerous bone cells.

   osteon, bone + *cyte*, cell
   osteocyte: a mature bone cell

2. **Osteoblasts** (OS-tē-ō-blasts) are immature bone cells that can develop into osteocytes (the term *blast* means "precursor"). Osteoblasts are bone builders that are responsible for the production of new bone. This process is called *osteogenesis* (os-tē-ō-JEN-e-sis). Osteoblasts produce the matrix of bone. When they become surrounded by this matrix, they become osteocytes.

   osteon, bone + *blastos*, a germ (like a seed)
   osteoblast: immature bone cells responsible for bone growth

3. **Osteoclasts** (OS-tē-ō-klasts) are the demolition team of bone tissue. The term *clast* means "to break," an appropriate name for their function. Osteoclasts within the endosteum and periosteum break down and remove bony matrix. This process is essential for bone remodeling and growth. Osteoclast activity also releases the minerals stored in bone. This process adds calcium and phosphate ions to body fluids. The release of stored minerals from bone is called *resorption*.

   osteon, bone + *klastos*, broken
   osteoclast: a cell that dissolves bone matrix and releases the stored minerals

Bone is a very dynamic tissue, and at any given moment osteoclasts are removing matrix while osteoblasts are adding to it. If osteoblasts add matrix faster than osteoclasts remove it, the bones grow thicker and stronger. If osteoclasts remove matrix faster than osteoblasts deposit it, the bones grow thinner and weaker.

Bone Tissue

Within both compact and spongy bone, osteocytes exist within microscopic pockets called **lacunae** (la-KOO-nē), which means "little lakes." The lacunae are found between thin sheets of calcified matrix. Adjacent lacunae are interconnected by small channels that also link them to nearby blood vessels. Nutrients from the blood and waste products from the osteocytes diffuse through the fluid within these channels, which also contain slender extensions of the osteocytes.

Within compact bone, the basic functional unit is the **osteon** (OS-tē-on), or *Haversian system* (Figure 8-1•). In each osteon, the calcified matrix is arranged in concentric layers around a **central canal** (also known as a *Haversian canal*). The central canal always contains one or more blood vessels, and it may also contain nerves that supply the bone. The central canals are usually oriented along the long axis of the bone. Smaller, transverse canals link the blood vessels of the central canals with those of the periosteum or the marrow cavity.

✔ A sample of bone shows concentric layers surrounding a central canal. Is it from the shaft (diaphysis) or the end (epiphysis) of a long bone?

✔ If the activity of osteoclasts exceeds the activity of osteoblasts in a bone, how will the mass of the bone be affected?

Unlike compact bone, spongy bone has no osteons. It consists of an open network of interconnecting calcified rods or plates. These open areas of spongy bone provide space for red marrow or yellow marrow.

BONE DEVELOPMENT AND MAINTENANCE

A person's height and general body proportions are determined by his or her skeleton. Bone formation is called **ossification**. Ossification begins about 6 weeks after fertilization, when the embryo is approximately 12 mm (1/2 in.) long. Bone formation and growth continue through adolescence, and parts of the skeleton usually do not stop growing until ages 18 to 25.

BONE GROWTH

Before bone formation begins, an embryo has a cartilaginous skeleton. As development proceeds, many of these cartilages are replaced by bones. This process is called **endochondral ossification** (en-dō-KON-dral). Endochondral ossification begins as cartilage cells within a cartilaginous model break down and are then replaced by osteoblasts that begin producing bone matrix. As more cartilage breaks down, the number of osteoblasts increases, and more and more of the cartilage is replaced by bone. Steps in the growth and ossification of a limb bone are diagrammed in Figure 8-2•. Bone formation starts at the shaft surface and then begins within the shaft before it occurs in the ends of the bone.

This process continues until the cartilaginous model has been almost completely replaced. However, until growth is completed, zones of cartilage remain

endo, inside + *chondros*, cartilage
endochondral ossification: bone formation within an existing cartilage

•**FIGURE 8-2**
Endochondral Ossification.

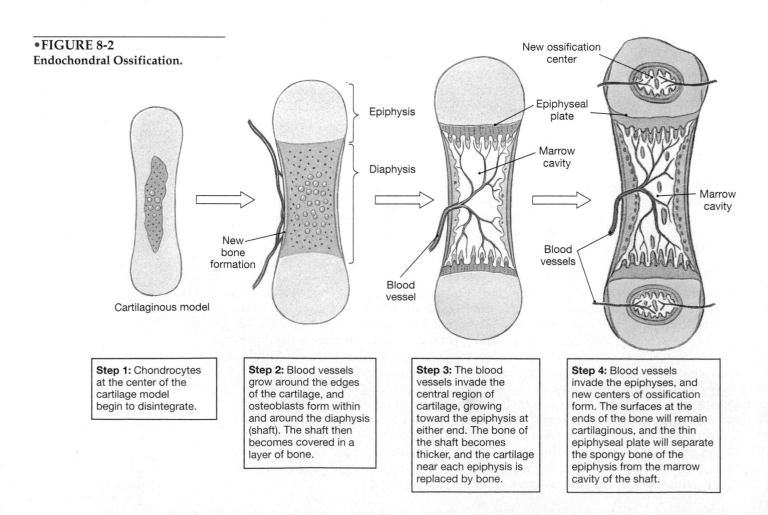

Cartilaginous model

Epiphysis

Diaphysis

New bone formation

Blood vessel

New ossification center

Epiphyseal plate

Marrow cavity

Blood vessels

Marrow cavity

Step 1: Chondrocytes at the center of the cartilage model begin to disintegrate.

Step 2: Blood vessels grow around the edges of the cartilage, and osteoblasts form within and around the diaphysis (shaft). The shaft then becomes covered in a layer of bone.

Step 3: The blood vessels invade the central region of cartilage, growing toward the epiphysis at either end. The bone of the shaft becomes thicker, and the cartilage near each epiphysis is replaced by bone.

Step 4: Blood vessels invade the epiphyses, and new centers of ossification form. The surfaces at the ends of the bone will remain cartilaginous, and the thin epiphyseal plate will separate the spongy bone of the epiphysis from the marrow cavity of the shaft.

between the bone of the shaft and the bone of each epiphysis. These cartilaginous connections are known as the **epiphyseal plates** (see Figure 8-2•). The cartilage at an epiphyseal plate is continually being remodeled. On the shaft side of the epiphyseal plate, osteoblasts are invading the cartilage and converting it to bone. But on the epiphyseal side of the plate, new cartilage is being produced at the same rate. As a result, the epiphyseal plate persists, but the shaft grows longer. This process can be compared to a jogger chasing someone on a bicycle. As long as the jogger and the bicycle are moving at the same speed, the jogger will keep advancing but never catch the cyclist.

When sex hormone production increases at puberty, bone growth accelerates, and the rate of bone production by osteoblasts exceeds the rate of new epiphyseal cartilage formation. In effect, the jogger speeds up and begins to overtake the bicycle. Soon the osteoblasts have completely replaced the cartilage of the epiphyseal plate. This marks the end of the period of sudden growth that accompanies adolescence, and the person becomes physically as well as sexually mature. The location of what was once an epiphyseal plate can still be detected in an X-ray of an adult bone as a distinct *epiphyseal line* that remains after it has stopped growing.

While the bone lengthens, it also grows larger in diameter. The increase in width occurs as osteoblasts form in the cellular layer of the periosteum. These cells begin producing additional bone, and as they become surrounded with calcified matrix, the osteoblasts become osteocytes. As new bone is deposited on the outer surface of the shaft, the inner surface is eroded by osteoclasts in the endosteum. This combination of bone formation on the outside and resorption in the inside increases the diameter of the bone and enlarges the marrow cavity.

BONE MAINTENANCE

The support and storage functions of the skeleton depend on the dynamic nature of bone tissue. Even after the epiphyseal plates have closed, each bone is constantly being modified by the activities of its various cells. Osteocytes, for example, maintain the bony matrix by continually removing and replacing calcium. Osteoclasts and osteoblasts also remain active, continually forming and destroying bony tissue. Normally these activities are balanced, and in adults, about one-fifth of the protein and mineral components of the skeleton are removed and then replaced each year.

Constant mineral turnover gives each bone the ability to adapt to new stresses. When stresses are applied, osteoblasts are stimulated. As a result, heavily stressed bones become thicker, stronger, and develop more pronounced surface ridges. When bones are not stressed, they get thin and brittle because osteoblast activity decreases. One of the benefits of regular exercise is that it stresses the skeleton and helps keep bones strong.

Despite its mineral strength, bone can crack or even break if subjected to extreme loads, sudden impacts, or stresses from unusual directions. Any crack or break in a bone is called a **fracture**.

Fractures are classified according to their external appearance, the site of the fracture, and the type of crack or break in the bone. In a *closed fracture*, or *simple fracture*, the broken bones do not penetrate the skin. In an *open fracture*, or *compound fracture*, the skin is broken. Open fractures are more dangerous than closed fractures because the opening in the skin increases the chances for severe bleeding and bacterial infection.

Fractured bones will usually heal as long as the circulatory supply remains and the cells of the endosteum and periosteum survive the injury. The repair process may take from 4 months to well over a year. In some cases, the repair may be "good as new," with no sign that a fracture ever occurred. Often, however, the bone will be slightly thicker than normal at the fracture site.

🏥 Clinical Note

What Is Osteoporosis?

Osteoporosis (os-tē-ō-por-Ō-sis) is a disease condition caused by an abnormal loss of bone mass that results in extremely fragile bones. Because bones are more fragile, they break easily and do not repair well. For example, a hip fracture may occur when an individual simply stands up, or vertebrae may collapse, distorting the joints between them and putting pressure on spinal nerves.

Sex hormones are important in maintaining normal rates of bone deposition. A significant percentage of women over age 45 suffer from osteoporosis. The condition becomes more common after menopause due to a decline in the levels of estrogen. Because men continue to produce sex hormones until relatively late in life, severe osteoporosis is less common in males below age 60.

Therapies that boost estrogen levels, dietary changes that elevate calcium levels in the blood, and exercise that stresses bones and stimulates osteoblast activity appear to slow, but not completely prevent, the development of osteoporosis.

✔ How could X-rays of the femur be used to determine whether a person had reached full height?

✔ In the Middle Ages, choirboys were sometimes castrated (had their testes removed) before puberty, to prevent their voices from changing. How would this have affected their height?

✔ Why would you expect the arm bones of a weight lifter to be thicker and heavier than those of a jogger?

AGING AND THE SKELETAL SYSTEM

The bones of the skeleton become thinner and relatively weaker as a normal part of the aging process. A reduction in bone mass begins to occur between the ages of 30 and 40. It begins when the rate of bone formation by osteoblasts begins to fall below the rate of bone breakdown by osteoclasts. Once the reduction in bone mass begins, a woman loses roughly 8 percent of her skeletal mass every decade. The rate of bone loss in men is somewhat lower, averaging around 3 percent per decade. Not all parts of the skeleton are equally affected. Epiphyses, vertebrae, and the jaws lose more than their fair share, resulting in fragile limbs, a reduction in height, and the loss of teeth. If the bones become so thin that they can no longer withstand normal stresses, the condition of *osteoporosis* exists.

THE SKELETON

The skeletal system consists of 206 separate bones and a number of associated cartilages. There are two skeletal divisions: the **axial skeleton**, whose 80 bones form the vertical axis of the body, and the **appendicular skeleton**, which includes the bones of the limbs as well as those bones that attach the limbs to the trunk (Figure 8-3●). All together, there are 126 appendicular bones: 32 are associated with each upper limb, and 31 with each lower limb.

SKELETAL TERMS

In addition to its shape, each bone in the skeleton has a variety of other identifying external features. These features, called *bone markings*, are the attachments of tendons and ligaments, or passageways for nerves and blood vessels. A group of specialized terms are commonly used in describing bone markings. For example, a *process* is a general term for a projection or bump on a bone. Table 8-1 contains a short list of these terms, and many specific examples will be seen in the bone illustrations that follow.

THE AXIAL SKELETON

The axial skeleton creates a framework that supports and protects organ systems in the dorsal and ventral body cavities. ∞ *[p. 12]* In addition, it provides an extensive surface area for the attachment of muscles that adjust the positions of the

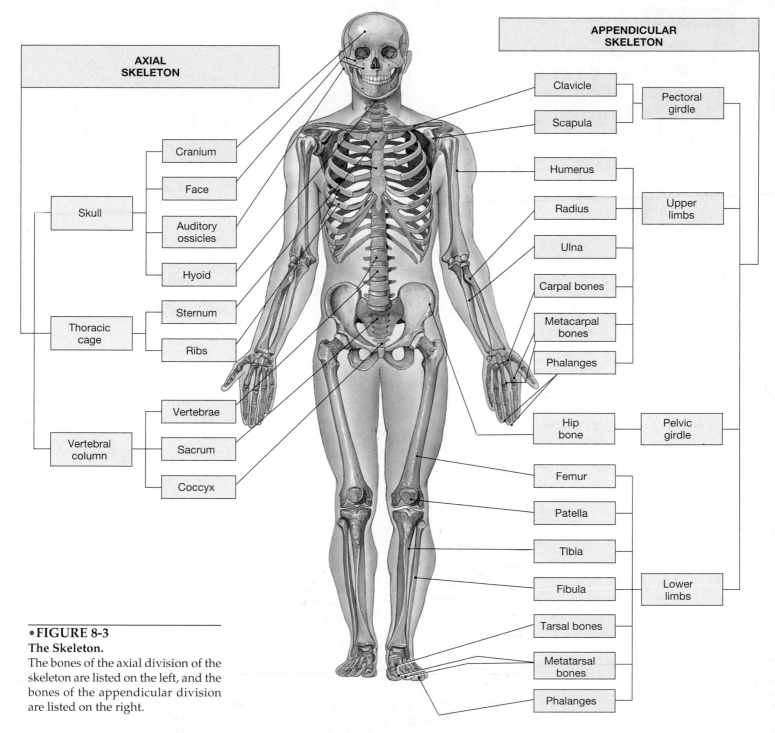

•FIGURE 8-3
The Skeleton.
The bones of the axial division of the skeleton are listed on the left, and the bones of the appendicular division are listed on the right.

head, neck, and trunk. Its 80 bones include the **skull** and other bones associated with the skull, the **vertebral column**, and the **thoracic** (*rib*) **cage**.

The Skull

The bones of the skull protect the brain and support delicate sense organs involved with vision, hearing, balance, smell, and taste. The skull is made up of 22 bones: 8 form the **cranium** (which encloses the brain), and 14 are associated with the *face*. Seven additional bones are associated with the skull: 6 *auditory ossicles*, tiny bones involved in sound detection, and the *hyoid bone*, which is connected by ligaments to the inferior surface of the skull.

Table 8-1	External Features of Bones	
Description	**Term**	**Definition**
Processes formed where tendons or ligaments attach	Trochanter	A large, rough projection
	Tuberosity	A smaller, rough projection
	Tubercle	A small, rounded projection
	Crest	A prominent ridge
	Spine	A pointed process
Processes formed at articulations	Head	Distinct epiphysis, separated from the shaft of a long bone by a narrow neck
	Condyle	A smooth, rounded bump
	Facet	A small, flat surface
Depressions and openings	Fossa	A shallow depression
	Sulcus	A narrow groove
	Foramen	A rounded passageway for blood vessels and/or nerves
	Sinus	A chamber within a bone, normally filled with air

Figure 8-4• presents the cranial and facial portions of the skull. External views of the skull are shown in Figure 8-5•. Figure 8-6• reveals internal features through horizontal (transverse) and sagittal sections. Table 8-2 provides a summary of the names and descriptions of the bones of the skull.

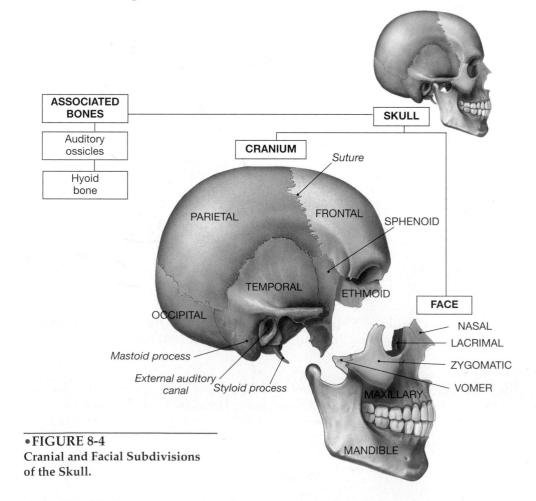

•**FIGURE 8-4**
Cranial and Facial Subdivisions of the Skull.

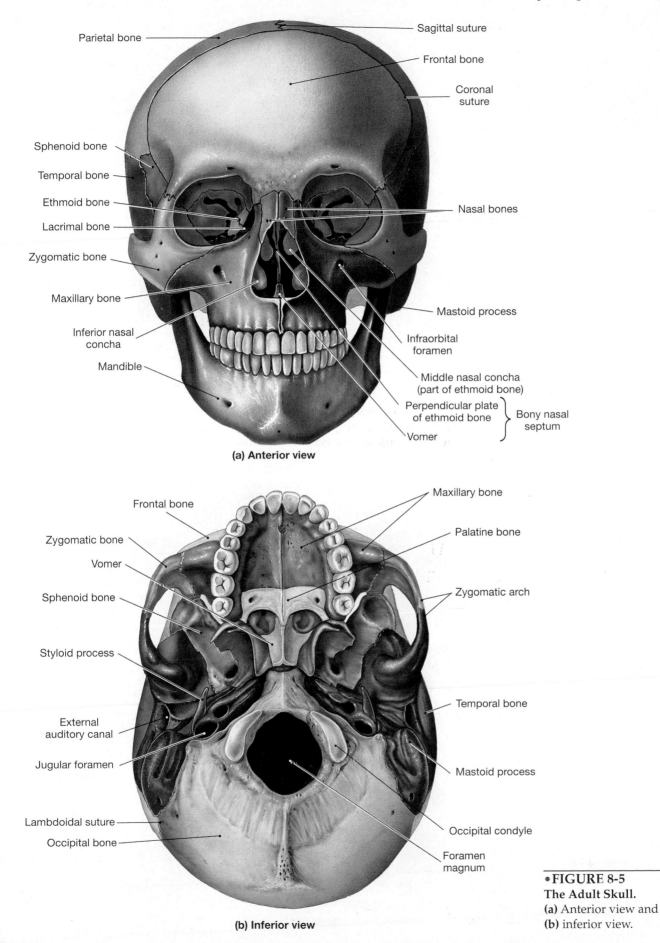

Parietal bone

Sagittal suture

Frontal bone

Coronal suture

Sphenoid bone

Temporal bone

Ethmoid bone

Lacrimal bone

Nasal bones

Zygomatic bone

Maxillary bone

Inferior nasal concha

Mastoid process

Mandible

Infraorbital foramen

Middle nasal concha (part of ethmoid bone)

Perpendicular plate of ethmoid bone

Vomer

Bony nasal septum

(a) Anterior view

Frontal bone

Maxillary bone

Zygomatic bone

Palatine bone

Vomer

Sphenoid bone

Zygomatic arch

Styloid process

External auditory canal

Jugular foramen

Temporal bone

Lambdoidal suture

Occipital bone

Mastoid process

Occipital condyle

Foramen magnum

(b) Inferior view

•**FIGURE 8-5**
The Adult Skull.
(a) Anterior view and
(b) inferior view.

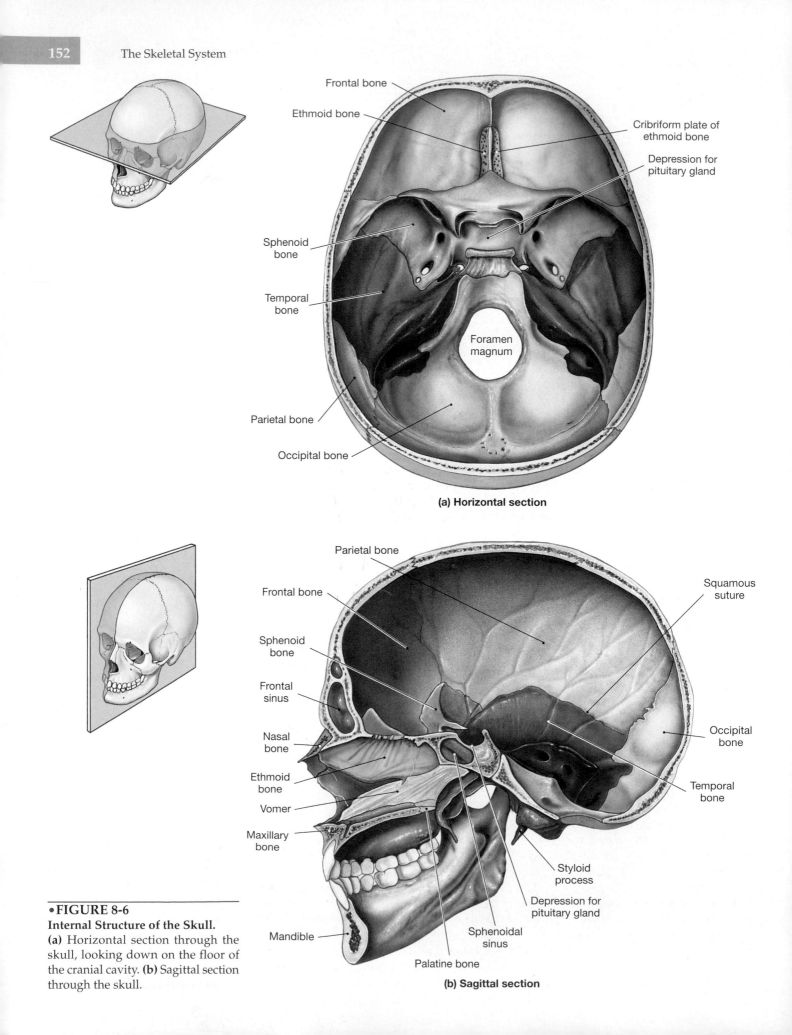

Frontal bone

Ethmoid bone

Cribriform plate of
ethmoid bone

Depression for
pituitary gland

Sphenoid
bone

Temporal
bone

Foramen
magnum

Parietal bone

Occipital bone

(a) Horizontal section

Parietal bone

Frontal bone

Squamous
suture

Sphenoid
bone

Frontal
sinus

Nasal
bone

Occipital
bone

Ethmoid
bone

Vomer

Temporal
bone

Maxillary
bone

Styloid
process

Depression for
pituitary gland

Mandible

Sphenoidal
sinus

Palatine bone

(b) Sagittal section

•FIGURE 8-6
Internal Structure of the Skull.
(a) Horizontal section through the
skull, looking down on the floor of
the cranial cavity. **(b)** Sagittal section
through the skull.

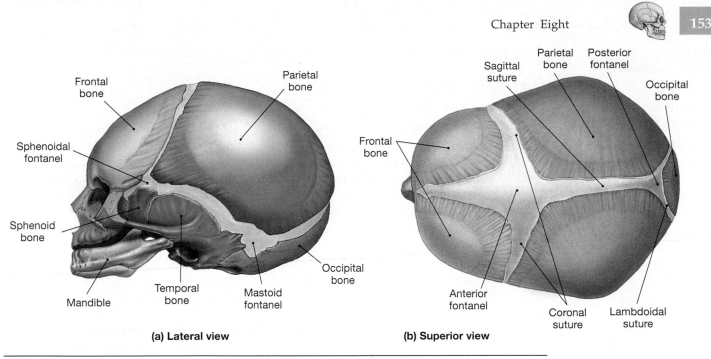

FIGURE 8-7
The Skull of an Infant.
(a) Lateral view. The bones of the infant skull are separated by areas of fibrous connective tissue called fontanels. (b) Superior view. The large anterior fontanel closes when a child is about 18 months old.

The bones of the cranium include the **frontal**, **parietal** (pa-RĪ-e-tal), **occipital**, **temporal**, **sphenoid** (SFĒ-noid), and **ethmoid bones** (see Table 8-2). These bones form the **cranial cavity** that encloses and protects the brain. The outer surface of the cranium provides attachment points for muscles that move the eyes, jaws, and head.

The 14 facial bones protect and support the entrances to the digestive and respiratory tracts. The superficial facial bones (the **lacrimal** (LAK-ri-mal), **nasal**, **maxillary**, and **zygomatic bones**, and the **mandible**) also provide areas for the attachment of muscles that control facial expressions and manipulate food. Of all the facial bones, only the mandible, or lower jaw, is movable. The deeper bones of the face include the **palatines bones**, the **vomer**, and the **inferior nasal conchae** (see Table 8-2).

Sutures are immovable joints between bones. Except for the mandible and its joint with the cranium, all the joints between the skull bones of adults are sutures. At a suture the bones are joined together by dense fibrous connective tissue. The illustrations of the different views of the skull show its four major cranial sutures; the **lambdoidal** (lam-DOYD-al), **coronal**, **sagittal**, and **squamous sutures**.

The sutures so characteristic of the adult skull were not always immovable. At birth, infant skulls lack sutures. Instead, their cranial bones are connected by areas of fibrous connective tissue known as **fontanels** (fon-tah-NELS) (Figure 8-7•). These fontanels, or "soft spots," are quite flexible and permit distortion of the skull without damage. Such distortion normally occurs during delivery and eases the passage of the infant along the birth canal. The fontanels disappear as the cranial bones fuse and interlock during the first 4 years of life.

The frontal, temporal, sphenoid, ethmoid, and maxillary bones contain air-filled internal chambers called **sinuses** (Figure 8-6b•). The *paranasal sinuses*—all the sinuses except for those in the temporal bone—empty into the nasal cavity.

Sinuses make the bones of the face lighter. In addition, the paranasal sinuses are lined by an extensive area of mucous epithelium. Here, mucous secretions are released into the nasal cavities, and the ciliated epithelium passes the mucus back toward the throat, where it is eventually swallowed. Incoming air is moisturized and warmed as it flows across this carpet of mucus, and foreign particles, such as dust or bacteria, become trapped in the sticky mucus and swallowed. This mechanism helps protect more delicate portions of the respiratory tract.

✔ What is the name of the cranial bone of the forehead?

✔ The occipital bone contains a large opening. What is the name of this cranial feature, and what is its function?

✔ What are the functions of the paranasal sinuses?

Table 8-2	Bones of the Skull	
Bone	**Number**	**Comments**
Cranium		
Frontal bone	1	Forms the forehead and the upper surface of each eye socket; contains *frontal sinuses*
Parietal bones	2	Form the roof and the upper walls of the cranium
Occipital bone	1	Forms the posterior and inferior portions of the cranium; contains the *foramen magnum*, the opening through which the spinal cord connects to the brain
Temporal bones	2	Make up the sides and base of the cranium; each contains an *external auditory canal* that leads to the eardrum; contains the auditory ossicles, or ear bones
Sphenoid bone	1	Forms part of the cranial floor and also acts as a bridge uniting the cranial and facial bones; contains a pair of *sphenoidal sinuses*
Ethmoid bone	1	Forms narrow part of the anterior floor of the cranium and part of nasal cavity roof; projections called the *superior* and *middle conchae* extend into the nasal cavity; forms upper portion of the nasal septum; contains the *ethmoidal sinuses*
Face		
Maxillary bones	2	Form (1) the floor and medial portion of the rim of the orbit, (2) the walls of the nasal cavity, and (3) the anterior roof of the mouth, or *hard palate*; contain large *maxillary sinuses*
Palatine bones	2	Form the posterior surface of the *hard palate*, or roof of the mouth
Vomer	1	Forms the inferior portion of the nasal septum
Zygomatic bones	2	Slender bony extension curves laterally and posteriorly to meet a process from the temporal bone to form the *zygomatic arch*, or cheekbone
Nasal bones	2	Form the bridge of the nose
Lacrimal bones	2	Contain the opening for the passageway of the tear duct to the nasal cavity
Inferior nasal conchae	2	Project from the lateral walls of the nasal cavity; with superior and middle conchae of the ethmoid slows and deflects arriving air toward the olfactory (smell) receptors located near the upper portions of the nasal cavity
Mandible	1	Bone of the lower jaw; articulates with the temporal bones
Associated Bones		
Auditory ossicles	6	Three tiny bones enclosed within each temporal bone; transfer sound vibrations from eardrum to hearing receptors in the inner ear
Hyoid bone	1	Suspended under skull by ligaments from the styloid processes of the temporal bones; base for muscles associated with the tongue and *larynx* (voicebox); supports the larynx

The Vertebral (Spinal) Column

The axial skeleton consists of the vertebral (spinal) column and the thoracic (rib) cage. The spinal column protects the spinal cord and provides attachment sites for the ribs as well as the muscles that move the *trunk* (the part of the body to which the limbs are attached) and maintain body posture. The vertebral column is divided into regions based on the structure of the vertebrae, as indicated in Figure 8-8•. Just below the skull are the 7 **cervical vertebrae** of the neck (abbreviated C_1 to C_7). Next, are the 12 **thoracic vertebrae** (T_1 to T_{12}), each of which articulates with one or more pairs of **ribs**. The 5 **lumbar vertebrae** (L_1 to L_5) continue toward the base of the spine, the fifth lumbar vertebra articulating with the **sacrum**, a single bone formed by the fusion of 5 vertebrae. The small **coccyx** (KOK-siks) also consists of fused vertebrae.

Although it may seem like a straight and rigid structure, the spinal column is surprisingly flexible. A side view of the adult spinal column reveals that the vertebrae align to form four **spinal curves** (see Figure 8-8•). These curves add strength to the spine, increasing the skeleton's balance and ability to hold the body upright. The *thoracic* and *sacral curves*, known as the **primary curves**, are present at birth. The *cervical* and *lumbar curves* are known as **secondary curves** because they do not appear until months later. The cervical curve develops as an infant learns to balance

Spinal curves | **Vertebral regions**

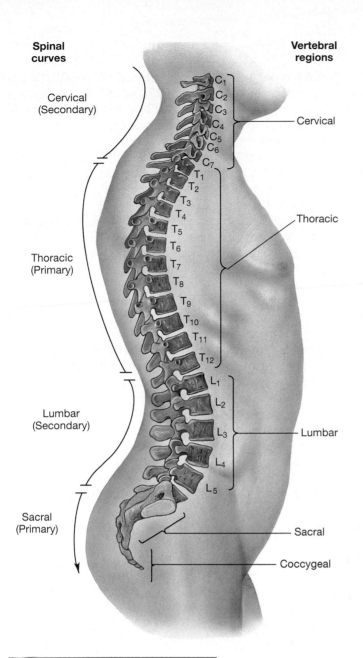

Cervical (Secondary)

Thoracic (Primary)

Lumbar (Secondary)

Sacral (Primary)

C₁ C₂ C₃ C₄ C₅ C₆ C₇ — Cervical

T₁ T₂ T₃ T₄ T₅ T₆ T₇ T₈ T₉ T₁₀ T₁₁ T₁₂ — Thoracic

L₁ L₂ L₃ L₄ L₅ — Lumbar

— Sacral

— Coccygeal

•**FIGURE 8-8**
The Vertebral Column.
The major divisions of the vertebral column, showing the four spinal curves.

its head upright, and the lumbar curve develops with the ability to stand. All four spinal curves are fully developed by the time a child is 10 years old.

All vertebrae share some common features: a relatively massive, weight-bearing **body; transverse processes** that project laterally and serve as sites for muscle attachment; and a **spinous** (or *spinal*) **process**. Each vertebra also has an opening called the **vertebral foramen**, through which the spinal cord passes. Viewed collectively, the vertebral foramina of adjacent vertebrae form the *vertebral canal* that encloses the spinal cord. Despite basic structural similarities throughout the vertebrae of the spine, some regional differences reflect important differences in function. Figure 8-9• shows typical vertebrae from each region.

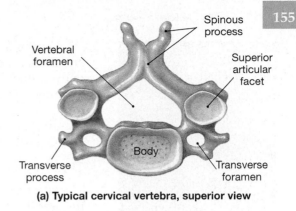

(a) **Typical cervical vertebra, superior view**

All cervical vertebrae contain transverse foramina that protect important blood vessels supplying the brain.

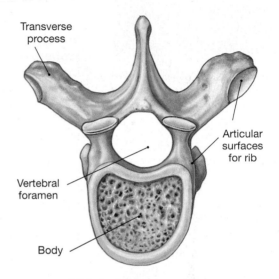

(b) **Typical thoracic vertebra**

A distinctive feature of thoracic vertebrae are articular surfaces on the body and on most of the transverse processes for articulation with one or more pairs of ribs.

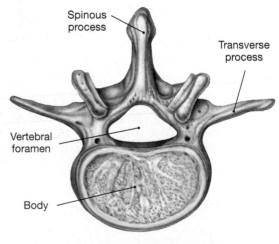

(c) **Typical lumbar vertebra**

The lumbar vertebrae are the most massive and least mobile, for they support most of the body weight. All have a wide and thick vertebral body.

•**FIGURE 8-9**
Typical Cervical, Thoracic, and Lumbar Vertebrae.
All vertebrae are shown in superior view.

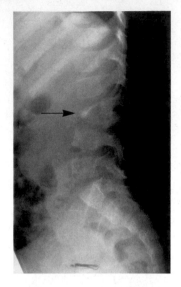

Infection of the spinal column by tuberculosis bacteria can result in the erosion of the intervertebral discs and the collapse of vertebrae.

✔ Intervertebral discs are not found between the first and second cervical vertebrae. How does this fact relate to the function of these vertebrae?

✔ In adults, five large vertebrae fuse to form what single structure?

✔ What bones make up the thoracic cage?

Vertebrae form joints with each other at specialized flattened areas called *facets* and at their vertebral bodies. The vertebral bodies, however, usually do not contact one another directly because an **intervertebral disc** of cartilage lies between them. An intervertebral disc consists of a thick outer layer of fibrous cartilage surrounding a soft, gelatinous core. The intervertebral discs act as shock absorbers, compressing and distorting when stressed. Part of the loss of height that comes with aging results from the decreasing size and resiliency of the intervertebral discs.

The Thoracic Cage

The skeleton of the chest, or *thorax*, consists of the thoracic vertebrae, the ribs, and the sternum. The ribs and the sternum form the **thoracic cage**, or *rib cage*, and establish the shape of the thoracic cavity. The thoracic cage protects the heart, lungs, and other internal organs and serves as an attachment base for muscles involved with breathing (Figure 8-10•).

The adult sternum has three parts: the broad, triangular **manubrium** (ma-NOO-brē-um); an elongate **body**; and the slender, inferior tip of the body, called the **xiphoid** (ZĪ-foid) **process**. Impact or strong pressure can drive the xiphoid process into the liver, causing severe damage. That is why proper hand position is so important when someone is performing cardiopulmonary resuscitation (CPR).

There are 12 pairs of ribs, or *costal bones*. Each pair of ribs is connected posteriorly to one of the 12 thoracic vertebrae. The first seven pairs, called **true ribs**, extend around the entire thoracic cavity and are connected to the sternum by cartilaginous extensions, the **costal cartilages**. Ribs 8 to 12 are called the **false ribs** because rather than attaching directly to the sternum, they attach to the costal cartilage of the seventh rib. The last two pairs of ribs are called **floating ribs** because they have no connection with the sternum at all.

Table 8-3 lists and describes the bones making up the vertebral column and thoracic cage.

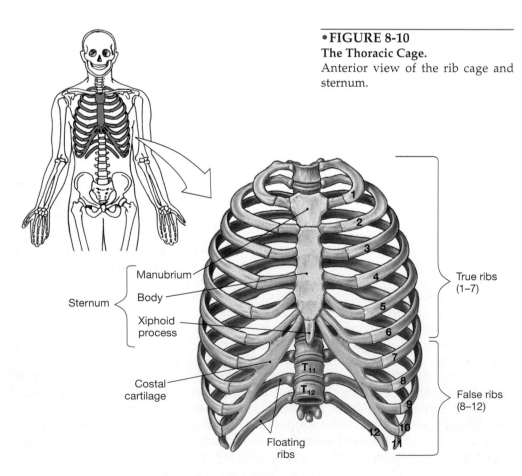

•**FIGURE 8-10**
The Thoracic Cage.
Anterior view of the rib cage and sternum.

Table 8-3	Bones of the Vertebral Column and Thoracic Cage	
Bone (Region)	**Number**	**Comments**
Vertebral Column		
Cervical vertebrae	7	The bones of the neck; contain *transverse foramina* for protection of blood vessels that supply the brain
Thoracic vertebrae	12	Characteristics include large, slender spinous process that points inferiorly; articular surfaces on the vertebral body and, in most cases, on transverse processes for forming joints with one or more pairs of ribs
Lumbar vertebrae	5	Thicker and more oval vertebral body than thoracic vertebrae; stumpy spinous process points posteriorly
Sacrum	1	Made up of five fused vertebrae; protects internal organs; articulates with appendicular skeleton; site of muscle attachment for movement of lower limbs
Coccyx	1	Made up of three to five fused vertebrae; fusion not complete until adulthood
Thoracic Cage		
Sternum	1	Made up of three fused bones: the manubrium, body, and xiphoid process; xiphoid process is usually the last to ossify and fuse
Ribs	24	Same number in each sex
True ribs	14	Each forms portion of anterior body wall and connects to sternum by separate *costal cartilages*
False ribs	6	Do not attach directly to sternum; costal cartilages fuse together and then merge with that of rib 7
Floating ribs	4	Do not attach directly to sternum; no costal cartilages

THE APPENDICULAR SKELETON

The appendicular skeleton includes the bones of the upper and lower limbs as well as the pectoral and pelvic girdles that connect the limbs to the trunk. These bones are all listed and described in Table 8-4.

The Pectoral Girdle

Each upper limb articulates with the trunk at the *shoulder girdle,* or **pectoral girdle**. The pectoral girdle consists of a broad, flat **scapula** (*shoulder blade*) and the short **clavicle** (*collarbone*). The bones of the pectoral girdle are illustrated in Figure 8-11●.

The S-shaped clavicle bone is small, light, and relatively fragile. Furthermore, the joint between the clavicle and sternum forms the *only* direct connection between the axial skeleton and the shoulder girdle. Pressure stress from a simple fall on the hand of an outstretched arm or a sudden jolt to the shoulder is often focused on the clavicle. As a result, broken or fractured clavicles are very common injuries.

The anterior face of the scapula forms a broad triangle. Muscles attach along the edges. Its posterior surface is divided by a long, and easily felt, spine. The shallow, cup-shaped *glenoid cavity* of the scapula articulates with the proximal end of the humerus to form the *shoulder joint*. Two processes of the scapula extend over the glenoid cavity. The **acromion** (a-KRŌ-mē-on) forms the tip of the shoulder and a point of attachment for ligaments that interconnect the scapula and the clavicle. The underlying **coracoid process** provides another attachment site for muscles and ligaments.

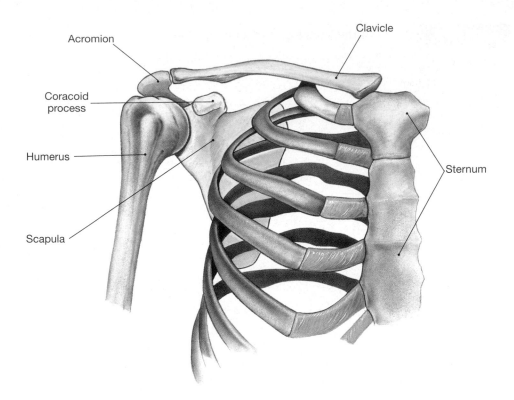

• **FIGURE 8-11**
The Pectoral Girdle.
The clavicle and scapula are the bones
of the pectoral girdle.

Table 8-4	Bones of the Pectoral Girdle and Upper Limb	
Bone (Region)	**Number**	**Comments**
Pectoral Girdle		
Clavicle	2	Articulates with the sternum and the *acromion process* of the scapula; forms the only connection between the pectoral girdle and the axial skeleton
Scapula	2	Provides attachment sites for muscles of the shoulder and arm; the humerus articulates with the scapula at its glenoid cavity, forming the shoulder joint
Upper Limb		
Humerus	2	Various projections mark the attachments of different muscles; depressions at its distal end accept processes of the ulna as the elbow reaches its limits of motion
Radius	2	The lateral bone of the forearm; the proximal joint with the ulna allows the distal radius to roll across the ulna and rotate the hand.
Ulna	2	The medial bone of the forearm; the olecranon process forms the elbow
Wrist and Hand Carpal bones	16	The 8 carpal bones of each wrist form two rows. The proximal row contains (1) *scaphoid*, (2) *lunate*, (3) *triangular* (or *triquetral*), and (4) *pisiform* (PI-si-form). There are also four distal carpal bones: (1) *trapezium*, (2) *trapezoid*, (3) *capitate*, and (4) *hamate*
Metacarpal bones	10	Five metacarpal bones form the palm of each hand
Phalanges	28	Each finger contains three phalanges, and the thumb (*pollex*) contains two

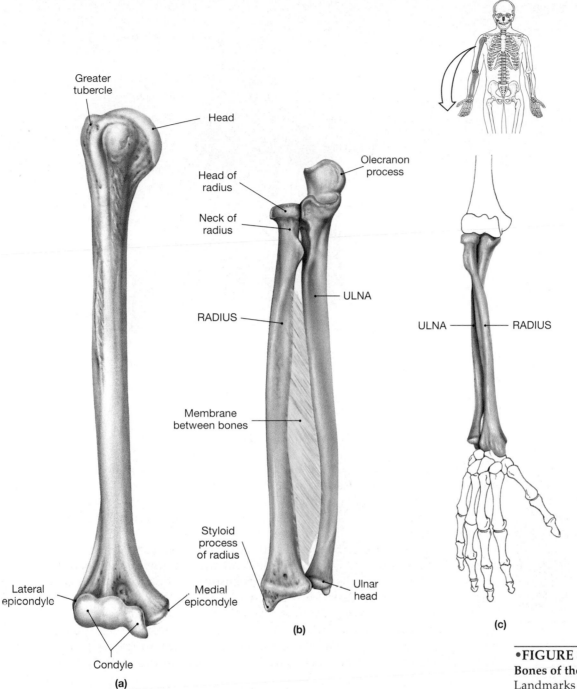

Greater
tubercle

Head

Olecranon
process

Head of
radius

Neck of
radius

ULNA

RADIUS

ULNA — RADIUS

Membrane
between bones

Styloid
process
of radius

Ulnar
head

Lateral
epicondyle

Medial
epicondyle

Condyle

(a)

(b)

(c)

•**FIGURE 8-12**
Bones of the Upper Limb.
Landmarks of the anterior surfaces
of the **(a)** right humerus, and **(b)** right
radius and ulna. **(c)** Note how the radius crosses over the ulna in a movement called *pronation*.

The Upper Limb

The upper limb consists of the arm, forearm, wrist, and hand. You will notice that
the anatomical term *arm* refers only to the proximal segment, rather than to the entire upper limb. The arm contains a single bone, the **humerus**, which extends from
the shoulder joint to the elbow joint. At the elbow joint, the humerus articulates
with the two bones of the forearm, the **radius** (lateral) and **ulna** (medial). These
bones are shown in Figure 8-12•.

In the anatomical position, with the palm facing forward, the radius lies along
the lateral (thumb) side of the forearm. When turning the forearm so that the palm
faces backward, the distal portion of the radius rolls across the ulna (see Figure 8-
12•). In life, a fibrous membrane connects the lateral margin of the ulna to the radius along its entire length. This thin sheet provides added stability to the forearm
as this movement occurs. It also provides a large surface area for the attachment
of muscles that control movements of the wrist and hand.

•FIGURE 8-13
Bones of the Wrist and Hand.
Anterior view of the right wrist and
hand.

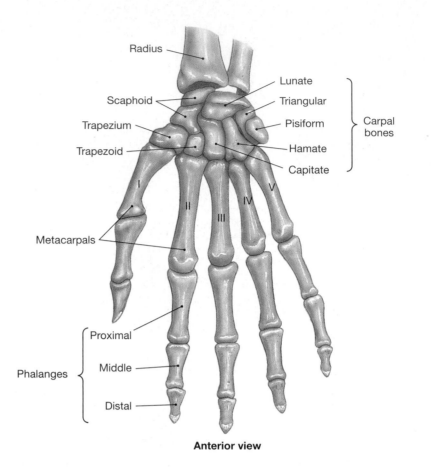

Anterior view

The darkened areas in this bone scan of the hand are areas of infection.

The point of the elbow is formed by a superior projection of the ulna, the **olecra-non** (ō-LEK-ra-non) **process**. The distal end of the ulna is separated from the wrist joint by a pad of cartilage, and only the wide distal portion of the radius articulates with the bones of the wrist. The *styloid process* of the radius helps in stabilizing the wrist joint by keeping the small wrist bones from sliding laterally.

There are 27 bones in the wrist and hand (Figure 8-13•). The 8 bones known as **carpal bones** form the wrist. A fibrous capsule, reinforced by broad ligaments, surrounds the wrist and stabilizes the positions of the individual carpals. Five **metacarpal** (met-a-KAR-pal) **bones** articulate with the distal carpal bones and form the palm of the hand. The metacarpal bones in turn articulate with the finger bones, or **phalanges** (fa-LAN-jēs). There are 14 phalangeal bones in each hand. The thumb has only two phalanges (proximal and distal), whereas the other fingers contain three each (proximal, middle, and distal).

The Pelvic Girdle

The **pelvic girdle** articulates with the thigh bones of the lower limbs. Because of the stresses involved in weight bearing and locomotion, the bones of the pelvic girdle and lower limbs are more massive than those of the pectoral girdle and upper limbs. The pelvic girdle is also much more firmly attached to the axial skeleton. Dorsally, the two halves of the pelvic girdle attach to the sacrum. Ventrally, the pelvic bones are interconnected by a fibrocartilage pad.

The pelvic girdle consists of two large hip bones, or **os coxae** (see Figures 8-3•, 8-14a•). Each hip bone forms through the fusion of three bones: an **ilium** (IL-ē-um), an **ischium** (IS-kē-um), and a **pubis** (PŪ-bis). At the hip joint on either side, the head of the femur (thighbone) articulates with the curved surface of the **ac-etabulum** (as-e-TAB-ū-lum).

The **pelvis** consists of the two hip bones, the sacrum, and the coccyx (see Figure 8-14•). The pelvis is thus a composite structure that includes portions of both the appendicular and axial skeletons. An extensive network of ligaments interconnects these structures and increases the strength of the pelvis.

✔ Where does the only direct connection occur between the pectoral girdle and the axial skeleton?

✔ Which forearm bone is directly involved in the wrist joint?

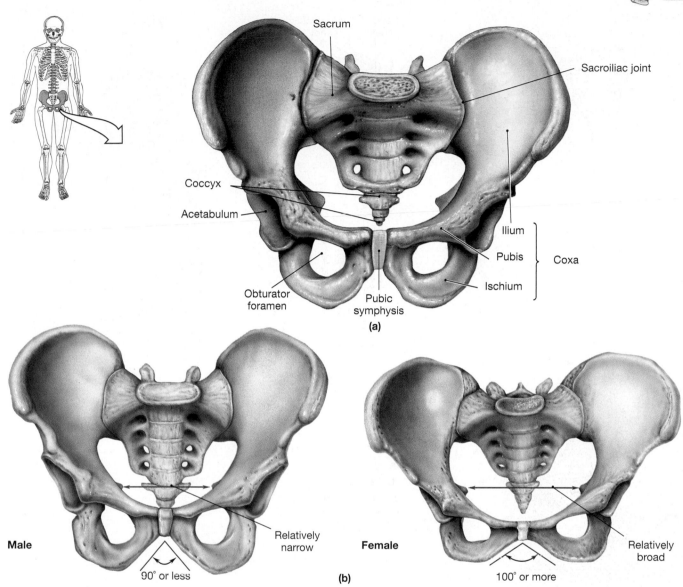

•**FIGURE 8-14**
The Pelvis.
(a) Anterior view of the pelvis. (b) Anatomical differences in the pelvis of a male and female.

The shapes of the male pelvis and female pelvis are somewhat different (Figure 8-14b•). Although some of the differences can be explained by variations between men and women in terms of body size and muscle mass, some features of the female pelvis are necessary adaptations for childbearing. A broader, flatter pelvis helps support the weight of the developing fetus, and a wider pelvic inlet helps ease passage of the newborn through the pelvis during delivery.

The Lower Limb

The lower limb consists of the thigh, the leg, the ankle, and the foot. Notice that the arrangement of bones in the lower limb is similar to that in the upper limb: one bone attaches to the girdle, two bones form the more distal portion, and a number of small bones combine to form the ankle and foot. Also note that the term *leg* refers only to the distal portion of the lower limb, rather than to the limb as a whole.

The **femur**, or *thighbone*, is the longest, heaviest, and strongest bone in the body (Figure 8-15a•). Distally, the femur articulates with the *tibia*, the larger bone of the leg, at the knee joint. The rounded head of the femur joins the pelvis at the acetabulum.

Figure 8-15b• shows the structure of the *tibia* and *fibula*. The **tibia** is the large medial bone of the leg. It is also known as the shinbone. The slender **fibula** par-

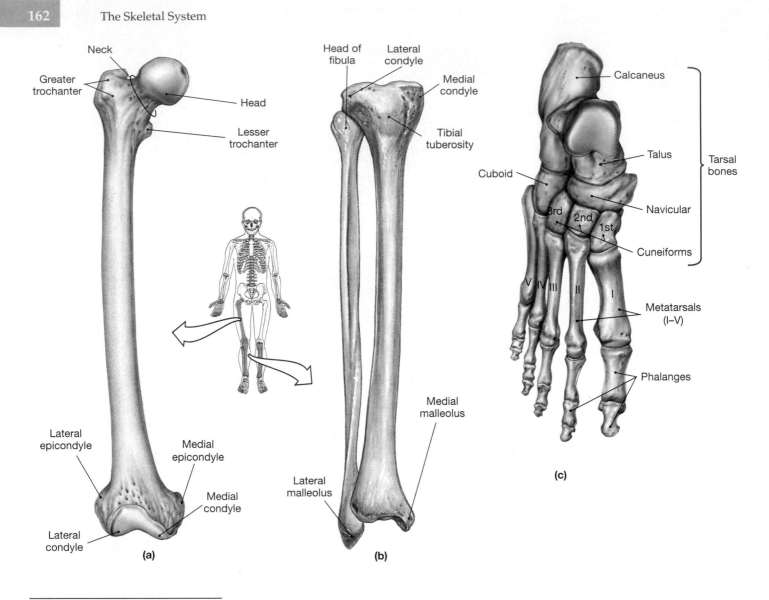

•FIGURE 8-15

Bones of the Lower Limb.

Some features of the anterior surfaces of the right **(a)** femur and **(b)** tibia and fibula. **(c)** Bones of the right foot as viewed from above.

✔ What three bones make up the coxa?

✔ The fibula does not participate in the knee joint nor does it bear weight, but when the fibula is fractured, it is difficult to walk. Why?

✔ While jumping off the back steps at his house, 10-year-old Joey lands on his right heel and breaks his foot. What foot bone is most likely broken?

allels the lateral border of the tibia. The fibula does not participate in the knee joint, and it does not distribute weight to the ankle and foot. However, it is an important surface for muscle attachment, and it helps keep the ankle bones from sliding laterally. A fibrous membrane between the tibia and fibula helps lock the fibula in position and provides additional area for the attachment of muscles that move the ankle and foot.

The ankle includes seven separate **tarsal bones** (Figure 8-15c•). Only the proximal tarsal bone, the **talus**, articulates with the tibia and fibula. The talus then passes the weight to the other tarsal bones, and ultimately to the ground. When standing normally, most of your weight is supported by the large **calcaneus** (kal-KĀ-nē-us), or *heel bone*. The rest of the body weight is passed through other tarsal bones to the **metatarsal bones** that support the sole of the foot. Powerful muscles on the back of the leg are attached to the posterior portion of the calcaneus by the *Achilles tendon*. Contraction of these calf muscles raises the heel and depresses the sole of the foot, as when you stand on tiptoes.

The basic organization of the metatarsal bones and phalanges of the foot resembles that of the metacarpal bones and phalanges of the hand. The metatarsal bones are numbered 1 to 5 (from medial to lateral). The big toe contains two phalanges, whereas each of the other toes contains three phalanges.

Table 8-5 provides a summary of the bones of the pelvic girdle and lower limb.

Table 8-5 **Bones of the Pelvic Girdle and Lower Limb**

Bone (Region)	Number	Comments
Pelvic Girdle		
Hip bone (Os coxa)	2	Each coxa forms through the fusion of three bones; the ilium, ischium, and pubis
Ilium	2	The most superior and largest coxal bone; provides an extensive area for the attachment of muscles, tendons, and ligaments
Ischium	2	The *ilium* fuses with the *ischium* near the superior and posterior margin of the acetabulum; the inferior surface of the ischium supports the body's weight when sitting
Pubis	2	An opening formed by the pubis and ischium, the *obturator foramen*, is covered by connective tissue fibers to which muscles and internal organs attach; the *pubic symphysis* marks the articulation with the pubis of the opposite side; it limits movement between the two pubic bones
Lower Limb		
Femur	2	The longest, heaviest, and strongest bone in the body; the *greater trochanter* and *lesser trochanter* provide attachment sites for large tendons; the distal shaft ends in two large *articular condyles* (*lateral* and *medial*)
Patella	2	The patella, or kneecap, glides over the anterior surface of the articular condyles of the femur
Tibia	2	The large medial bone of the leg; the inferior surface forms a joint with the proximal bone of the ankle; a large process, the *medial malleolus* (ma-LĒ-o-lus; malleolus, "hammer"), provides medial support for the ankle
Fibula	2	Lies lateral to the tibia; the distal *lateral malleolus* provides lateral stability to the ankle
Ankle and Foot		
Tarsal bones	14	Each ankle is made up of seven tarsal bones: the (1) *talus*, (2) *calcaneus*, (3) *navicular*, (4) *cuboid*, (5) *medial cuneiform*, (6) *intermediate cuneiform*, and (7) *lateral cuneiform*
Metatarsal bones	10	Form the sole and ball of the foot
Phalanges		Each toe contains three phalanges, and the great toe contains two

JOINTS

Joints, or **articulations**, exist wherever two bones meet. The function of each joint is closely related to its structure. Each joint reflects a workable compromise between the need for strength and the need for mobility. When movement is not required, joints can be very strong. For example, joints such as the sutures of the skull lock its separate elements together as if they were a single bone. At other joints, movement is more important than strength. At highly mobile joints, the interconnections between bones are looser and the joints weaker. For example, the shoulder joint permits a great variety of arm movements. The range of motion is limited more by the surrounding muscles than by joint structure. The joint itself is relatively weak, however, and shoulder injuries are rather common.

TYPES OF JOINTS

Three types of joints are recognized, based on the range of motion they permit: (1) An immovable joint is a **synarthrosis** (sin-ar-THRŌ-sis; *syn*, together + *arthros*, joint); (2) a slightly movable joint is an **amphiarthrosis** (am-fē-ar-THRŌ-sis; *amphi*, on both sides); and (3) a freely movable joint is a **diarthrosis** (dī-ar-THRŌ-sis; *dia*, through). Freely movable joints are capable of a wide range of movements. As a result, freely movable joints are further subdivided according to the more specific types of movement they permit.

Immovable Joints (Synarthroses)

At a synarthrosis the bony edges of the articulating bones are quite close together, and they are bound to each other by dense fibrous connective tissue. An example is a suture, an immovable joint between the bones of the skull. Another example of a synarthrosis is the joint between a tooth and its bony socket.

Slightly Movable Joints (Amphiarthroses)

An amphiarthrosis permits very limited movement. The bones at such joints are usually farther apart than they are at a synarthrosis, and they are separated by a cartilage. Examples include (1) the articulations between the bodies of adjacent spinal vertebrae, which are separated by intervertebral discs, and (2) the *pubic symphysis*, where a fibrocartilage pad forms the articulation between the right and left pelvic bones (coxae).

Freely Movable Joints (Diarthroses)

Diarthroses permit a wide range of motion. Diarthroses, or **synovial joints**, take advantage of the great friction-reducing properties of synovial fluid (see p. 61). Figure 8-16a● shows the general structure of a synovial joint.

Synovial joints are typically found at the ends of long bones, such as those of the limbs. Under normal conditions the bony surfaces do not contact one another, for they are covered with special **articular cartilages**. The joint is surrounded by a fibrous **joint capsule**, and the inner surfaces of the joint cavity are lined with a synovial membrane. **Synovial fluid** diffuses across the synovial membrane and provides additional lubrication.

In complex joints such as the knee, additional padding in the form of cartilage lies between the opposing articular surfaces. Such shock-absorbing cartilage pads are called **menisci** (men-IS-kē), shown in Figure 8-16b●. (*Meniscus* means "crescent," and refers to the shape of these pads.) Also present in such joints are **fat pads** which protect the articular cartilages and act as packing material. When the bones move, the fat pads fill in the spaces created as the joint cavity changes shape.

•FIGURE 8-16

The Structure of a Synovial Joint.
(a) Diagrammatic view of a simple articulation. **(b)** A simplified sectional view of the knee joint.

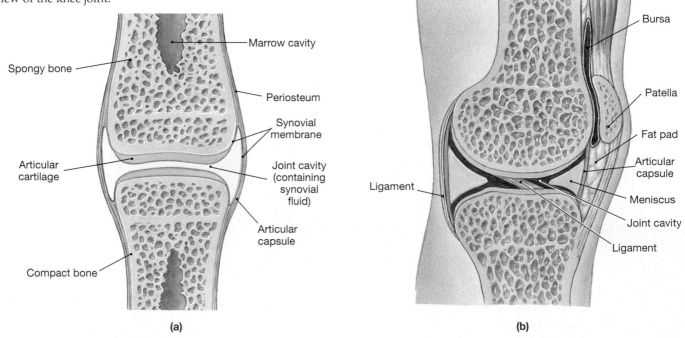

(a)

(b)

The joint capsule that surrounds the entire joint is continuous with the periosteum of each articulating bone. Ligaments that join one bone to another may be found on the outside or inside of the joint capsule. Where a tendon or ligament rubs against other tissues, small pockets of synovial fluid called **bursae** form to reduce friction and act as shock absorbers. Bursae are characteristic of many synovial joints and may also appear around tendons, beneath the skin covering a bone, or within other connective tissues exposed to friction or pressure.

ACTIONS OF MOVABLE JOINTS

Synovial joints are involved in all our day-to-day body movements. Although we commonly describe our movements with general phrases such as "bend the leg" or "raise the arm," it is sometimes helpful to use more specific descriptive terms.

Types of Movement

In **gliding**, two opposing surfaces slide past each other. Gliding occurs between the surfaces of articulating carpal bones (and tarsal bones), and between each clavicle and the sternum. Gliding can involve movement in almost any direction, but the amount of movement is slight.

Most **angular movements** are best remembered in opposing pairs: for example, *flexion* and *extension*, or *adduction* and *abduction*. The descriptions of each movement assume that the individual begins in the anatomical position.

Flexion (FLEK-shun) is a movement that *reduces* the angle between the articulating bones. **Extension** *increases* the angle between the articulating bones (Figure 8-17a•). When you bring your head toward your chest, you flex the neck.

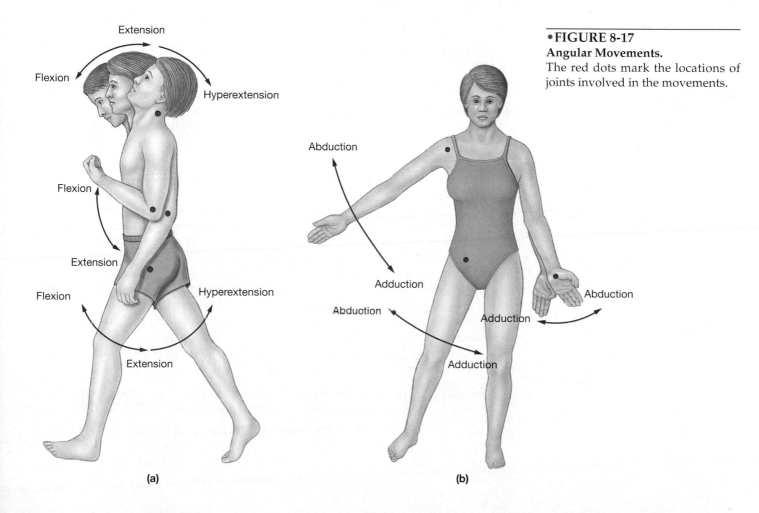

•FIGURE 8-17
Angular Movements.
The red dots mark the locations of joints involved in the movements.

(a)

(b)

•**FIGURE 8-17 (*continued*)**
Angular Movements.

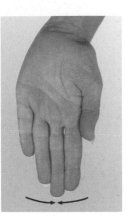

Adduction

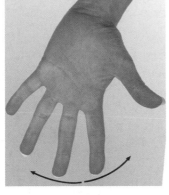

Abduction

(c)

Circumduction

(d)

When you bend down to touch your toes, you flex the spine. Extension reverses these movements. Flexion at the shoulder or hip moves a limb forward (anteriorly), whereas extension moves it backward (posteriorly). Flexion of the wrist moves the palm forward, and extension moves it back. In each of these examples, extension can be continued past the anatomical position, in which case **hyperextension** occurs. You can also hyperextend the head, a movement that allows you to gaze at the ceiling. Hyperextension of other joints is usually prevented or severely limited by ligaments, bony processes, or soft tissues.

Abduction (*ab*, from) is movement *away* from the longitudinal axis of the body. For example, swinging the upper limb away from the trunk is abduction of the limb (Figure 8-17b•); moving it toward the trunk is **adduction** (*ad*, to). Adduction of the wrist moves the heel of the hand toward the body, whereas abduction moves it farther away. Spreading the fingers or toes apart abducts them, because they move *away* from a central digit (finger or toe), as in Figure 8-17c•. Adduction brings the digits together. Abduction and adduction always refer to movements of the appendicular skeleton.

In **circumduction**, the distal portion of a limb moves in a circle while the proximal portion is fixed, as when one is drawing a large circle on a chalkboard (Figure 8-17d•). The fingers may circumduct as well, as when you trace a small circle with the tip of your index finger.

Rotational movements are also described with reference to a figure in the anatomical position. Rotation involves turning around the longitudinal axis of the body or limb, as when you rotate your head to look to one side or rotate your arm and forearm to screw in a lightbulb (Figure 8-18a•). Moving the wrist and hand from palm-facing-front to palm-facing-back is called **pronation** (prō-NĀ-shun) (Figure 8-18b•). The opposing movement, in which the palm is turned forward, is **supination** (sū-pi-NĀ-shun).

There are a number of special terms that apply to specific articulations or unusual types of movement (Figure 8-19•).

■ **Inversion** (*in*, into + *vertere*, to turn), is a twisting motion of the foot that turns the sole inward. The opposite movement is called **eversion** (ē-VER-shun; *e*, out).

■ **Dorsiflexion** is flexion of the ankle and elevation of the sole, as when "digging in the heels." The opposite movement, **plantar flexion** (from *planta*, the sole), extends the ankle and elevates the heel, as when standing on tiptoes.

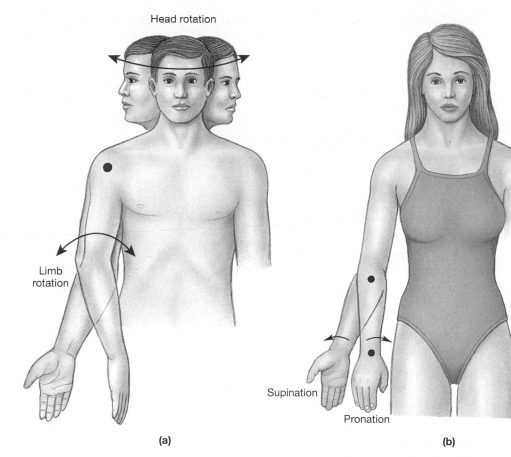

Head rotation

Limb rotation

Supination

Pronation

(a)

(b)

•FIGURE 8-18
Rotational Movements.

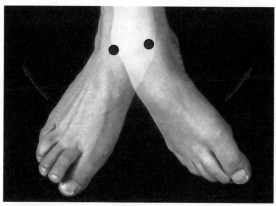

Eversion   Inversion

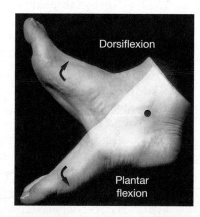

Dorsiflexion

Plantar flexion

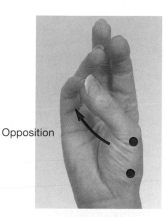

Opposition

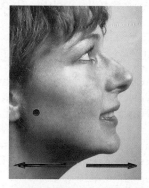

Retraction   Protraction

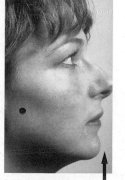

Elevation

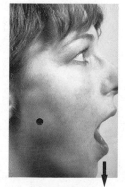

Depression

•FIGURE 8-19
Special Movements.

■ **Elevation** and **depression** occur when a structure moves in a superior or inferior direction. You depress your mandible when you open your mouth, and elevate it as you close it. Another familiar elevation occurs when you shrug your shoulders.

■ **Protraction** involves moving a part of the body anteriorly in the horizontal plane. **Retraction** is the reverse movement. You protract your jaw when you grasp your upper lip with your lower teeth, and you protract your clavicles when you reach for something in front of you..

The Shapes of Movable Joints

Synovial joints can be described as *gliding, hinge, pivot, ellipsoidal, saddle,* or *ball-and-socket* joints based on the shapes of the articulating surfaces. Each type of joint permits a different type and range of motion.

■ **Gliding joints** (Figure 8-20a●) have flattened or slightly curved faces. The relatively flat articular surfaces slide across one another, but the amount of movement is very slight. Although rotation is theoretically possible at such a joint, ligaments usually prevent or restrict such movement. Gliding joints are found at the ends of the clavicles, between the carpal bones, between the tarsal bones, and between the articular facets of adjacent spinal vertebrae.

■ **Hinge joints** (Figure 8-20b●) permit angular movement in a single plane, like the opening and closing of a door. Examples include the elbow, knee, ankle, and interphalangeal joints.

■ **Pivot joints** (Figure 8-20c●) permit rotation only. A pivot joint between the atlas (C_1) and axis (C_2) vertebrae allows you to rotate your head to either side, and another between the radius and the proximal shaft of the ulna permits turning of the hand (pronation and supination).

■ In an **ellipsoidal joint** (Figure 8-20d●), an oval articular face nestles within a depression on the opposing surface. With such an arrangement, angular motion occurs in two planes, along or across the length of the oval. Ellipsoidal joints connect the fingers and toes with the metacarpal and metatarsal bones, respectively.

■ **Saddle joints** (Figure 8-20e●) have articular faces that resemble saddles. Each face is concave on one axis and convex on the other, and the opposing faces nest together. This arrangement permits angular motion, including circumduction, but prevents rotation. The carpometacarpal joint at the base of the thumb is the best example of a saddle joint, and "twiddling your thumbs" demonstrates the possible movements.

■ In a **ball-and-socket joint** (Figure 8-20f●), the round head of one bone rests within a cup-shaped depression in another. All combinations of movements, including circumduction and rotation, can be performed at ball-and-socket joints.Examples include the shoulder and hip joints.

✔ In a newborn infant the large bones of the skull are joined by fibrous connective tissue. What type of joint is this? These bones later grow, interlock, and form immovable joints. What type of joints are these?

✔ Give the proper term for each of the following types of motion: (a) moving your arm away from the midline of the body, (b) turning your palms so that they face forward, (c) bending your elbow.

BONE AND JOINT DISORDERS

The skeletal system has a variety of important functions, including the support of soft tissues, blood cell production, mineral and lipid storage, and, through its relationships with the muscular system, the support and movement of the body as a whole. ∞ *[p. 143]* Skeletal system disorders can thus affect many other systems. The skeletal system is in turn influenced by the activities of other systems. For example, weakness or paralysis of skeletal muscles will lead to a weakening of the associated bones.

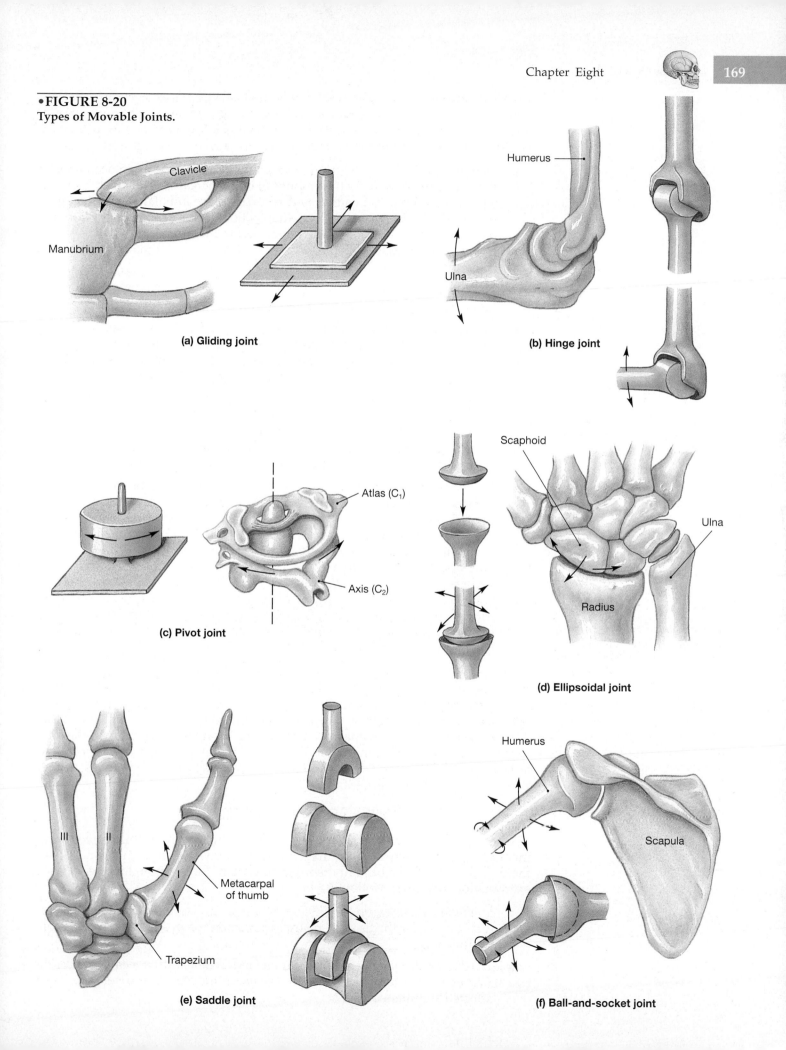

•FIGURE 8-20
Types of Movable Joints.

(a) Gliding joint

(b) Hinge joint

(c) Pivot joint

(d) Ellipsoidal joint

(e) Saddle joint

(f) Ball-and-socket joint

Although bones may seem rigid and permanent structures, the living skeleton is dynamic and undergoes continuous remodeling. The remodeling process involves bone deposition by osteoblasts and the breakdown of bone matrix by osteoclasts. The net result of the remodeling varies with the following five factors:

- *The age of the individual*: During development, bone deposition occurs faster than bone resorption. As the amount of bone increases, the skeleton grows. At maturity, bone deposition and resorption are in balance. As the aging process continues, the rate of bone deposition declines. Bone resorption outpaces bone deposition, and the bones become weaker. This gradual weakening, called *osteopenia*, begins at age 30–40 and may ultimately progress to osteoporosis. ∞ *[p. 148]*

- *The applied physical stresses*: Heavily stressed bones become thicker and stronger, and lightly stressed bones become thinner and weaker.

- *Circulating hormone levels*: Changing levels of growth hormone, androgens and estrogens, thyroid hormones, parathyroid hormone, and calcitonin increase or decrease the rate of mineral deposition in bone. As a result, many disorders of the endocrine system affect the skeletal system.

- *Rates of calcium and phosphate absorption and excretion*: For bone mass to remain constant, the rate of calcium and phosphate loss at the kidneys must be balanced by the rate of calcium and phosphate absorption along the digestive tract. Dietary calcium deficiencies or problems that reduce calcium and phosphate absorption will thus weaken the skeletal system.

- *Genetic or environmental factors*: Genetic or environmental factors may affect the structure of bone or the remodeling process. There are a number of inherited abnormalities of skeletal development. The environment can affect the shape and contours of growing bones as well, producing skeletal abnormalities. For example, some cultures use lashed boards to form an infant's or child's skull to a shape considered fashionable. Environmental stresses can also result in the formation of bone in unusual locations. These bones may develop in a variety of connective tissues exposed to chronic friction, pressure, or impact. For example, cowboys in the nineteenth century sometimes developed small bony plates in the dermis of the thigh, from friction with the saddle.

Figure 8-21• diagrams the major classes of skeletal disorders that affect the structure and function of bones (Figure 8-21a•) and joints (Figure 8-21b•). Traumatic injuries, such as fractures or dislocations, and infections also damage cartilages, tendons, and ligaments. A somewhat different array of conditions affect the soft tissues of the bone marrow. Areas of red bone marrow contain the stem cells for red blood cells, white blood cells, and platelets. Blood diseases characterized by blood cell overproduction or underproduction are usually associated with bone marrow abnormalities.

EXAMINATION OF THE SKELETAL SYSTEM

The bones of the skeleton cannot be seen without relatively sophisticated equipment. ∞ *[p. 12]* However, there are a number of physical signs that can assist in the diagnosis of a bone or joint disorder. Important factors noted in the physical examination include the following:

- *Limitation of movement or stiffness*: Many joint disorders, such as the various forms of arthritis, will restrict movement or produce stiffness at one or more joints.

- *The distribution of joint involvement and inflammation*: In a *monoarthritic condition*, only one joint is affected. In a *polyarthritic condition*, several joints are affected simultaneously.

•FIGURE 8-21
An Overview of Disorders of the Skeletal System.
(a) Bone Disorders. (b) Joint Disorders.

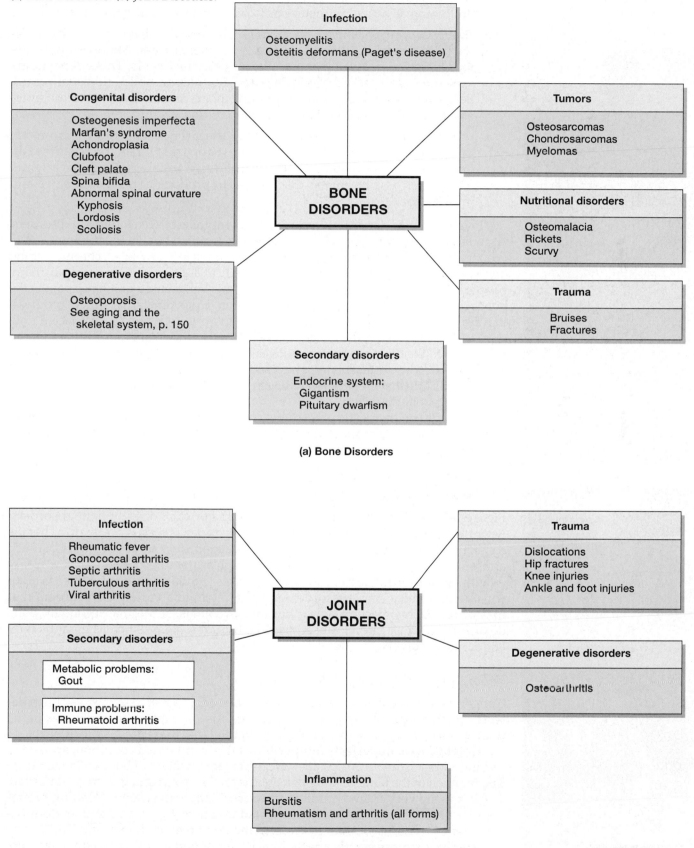

(a) Bone Disorders

(b) Joint Disorders

•FIGURE 8-22
Disorders of Bone Formation.
(a) Marfan's syndrome.
(b) Achondroplasia.

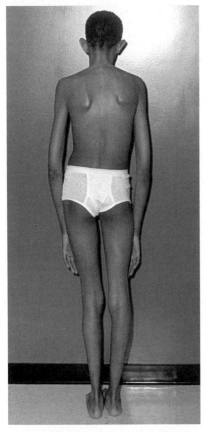

(a)

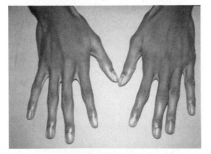

(b)

■ *Sounds associated with joint movement*: *Bony crepitus* (KREP-i-tus) is a crackling or grating sound generated during movement of an abnormal joint. The sound may result from the movement and collision of bone fragments following an articular fracture or from friction at an arthritic joint.

■ *The presence of abnormal bone deposits:* Thickened, raised areas of bone develop around fracture sites during the repair process. Abnormal bone deposits may also develop around the joints in the fingers. These deposits are called *nodules* or *nodes*. When palpated, nodules are solid and usually painless. Nodules, which can restrict movement, commonly form at the interphalangeal joints of the fingers in *osteoarthritis*.

■ *Abnormal posture*: Bone disorders that affect the spinal column can result in abnormal posture. This result is most apparent when the condition alters the normal spinal curvature. A condition involving an intervertebral joint, such as a herniated disc, will also produce abnormal posture and movement.

Pain in the bones and joints are common symptoms of skeletal system disorders. As a result, the presence of pain does not provide much help in identifying a *specific* bone or joint disorder. A person may be able to tolerate low-level, chronic, aching bone or joint pain and therefore not seek medical assistance. (For example, many people tolerate the pain of chronic arthritis.) However, focused or intense pain in the bones or joints usually gets immediate attention. Unfortunately, a slowly progressing and painless skeletal disorder, such as osteoporosis, may remain undetected until a crisis develops. Often the crisis involves a break in the bone; such a fracture develops when the bone has become so weak that it breaks when exposed to stresses easily tolerated by normal bones. The resulting condition, called a *pathologic fracture*, usually indicates that the skeleton has been weakened by severe focal (localized) disease or metabolic disorders.

A number of diagnostic procedures and laboratory tests can be used to obtain additional information about the status of the skeletal system. Table 8-6 summarizes information concerning these procedures.

BONE DISORDERS

Figure 8-21a• provides an overview of the major categories of bone disorders. Representative examples of each of these categories is discussed below. There is considerable overlap between primarily age-related, degenerative conditions, such as osteoporosis, and traumatic conditions, such as fractures; as bones grow weaker with age, the likelihood of fractures increases. Similarly, arthritis, which is often considered to be a single disorder, actually refers to a variety of disorders that may be caused by infection, aging, immune problems, or trauma. Secondary bone disorders are those effects caused by other organ systems, such as gigantism by endocrine problems.

Congenital Disorders

Congenital ("present at birth") diseases, or disorders are also known as "birth defects." Such conditions may be inherited, or caused by infection or damage in the uterus or at birth. In addition, not all congenital disorders are visible at birth.

Osteogenesis imperfecta (im-per-FEK-ta) is an inherited condition, appearing in 1 individual in about 20,000, that affects the organization of bone collagen fibers. The function of the bone-making osteoblasts are impaired and growth is abnormal. The bones are very fragile (think of uncooked spaghetti strands), leading to progressive skeletal deformation and repeated fractures. Ligaments and tendons become very "loose," permitting excessive movement at the joints.

Marfan's syndrome (Figure 8-22a•) is also linked to defective connective tissue structure. Extremely long and slender limbs, the most obvious physical in-

Table 8-6	Examples of Procedures and Tests Used in the Diagnosis of Bone and Joint Disorders	
Diagnostic Procedure	**Method and Result**	**Representative Uses**
X-ray of bone and joint	Standard X-ray; film sheet with radiodense tissues in white on a black background	Detection of fracture, tumor, dislocation, reduction in bone (mass, and bone infection osteomyelitis)
Bone scans	Injected radiolabeled phosphate accumulates in bones, and radiation emitted is converted into an image	Diagnosis of metastatic bone cancer; to detect fractures, infections and degenerative bone diseases
Arthrocentesis	Insertion of a needle into joint for aspiration of synovial fluid	See section on analysis of synovial fluid (below)
Arthroscopy	Insertion of fiber-optic tubing into a joint cavity; attached camera displays joint interior	Detection of abnormalities of menisci and ligaments; useful in differential diagnosis of joint disorders
MRI	Standard MRI produces computer-generated image	Observation of bone and soft tissue abnormalities
Laboratory Test	**Normal Values in Blood Plasma or Serum**	**Significance of Abnormal Values**
Alkaline phosphatase	Adults: 30–85 mIU/ml Children: 60–300+ mIU/ml	Elevated levels in adults may indicate abnormal osteoblast activity, bone cancer, osteitis deformans, multiple myeloma
Calcium	Adults: 8.5–10.5 mg/dl Children 8.5–11.5 mg/dl	Elevated levels occur in bone cancers, multiple fractures, and after prolonged immobility
Phosphorus	Adults: 2.3–4.7 mg/dl Children: 4.0–7.0 mg/dl	Elevated levels occur in acromegaly, parathyroid disorders, and bone tumors
Uric acid	Adult males: 3.5–8.0 mg/dl Adult females: 2.8–6.8 mg/dl	Elevated levels occur with gout
Rheumatoid factor	Adults: A negative result is normal	Roughly 75% of people diagnosed with rheumatoid arthritis have a positive test for this factor
Synovial fluid analysis	WBC: < 200/mm^3 RBC: none Glucose: < 10 mg/dl below serum glucose levels Protein: 1.8 g/dl No uric acid crystals	Elevation of white blood cell count indicates bacterial infection; mild elevation indicates inflammatory process; decreased glucose levels and decreased viscosity of fluid indicates inflammation of joint; uric acid crystals indicate gout

dication of this disorder, result from excessive cartilage formation at the epiphyseal plates. Defective connective tissue is also present in other body organs and weakened blood vessels are often present in individuals with Marfan's syndrome.

Achondroplasia (ā-kon-drō-PLĀ-sē-uh) results from abnormal epiphyseal activity (Figure 8-22b•). In this case the epiphyseal plates of the long bones grow unusually slowly and are replaced by bone early in life. As a result, the individual develops short, stocky limbs. Although other skeletal abnormalities occur, the trunk is normal in size, and sexual and mental development remain unaffected.

Congenital talipes equinovarus (TAL-i-pēz e-kwī-nō-VA-rus) **(clubfoot)** results from an inherited developmental abnormality that affects 2 in 1,000 births. Boys are affected roughly twice as often as girls. One or both feet may be involved, and the condition may be mild, moderate, or severe. The underlying problem is abnormal muscle development that distorts growing bones and joints. In most cases the tibia, ankle, and foot are affected, and the feet are turned medially and inverted. If both feet are involved, the soles face one another. Prompt treatment with casts or other supports in infancy helps alleviate the problem, and fewer than half of the cases require surgery. Kristi Yamaguchi, an Olympic gold medalist in figure skating, was born with this condition.

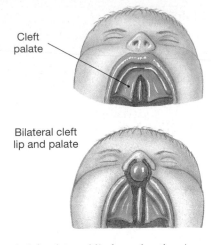

Cleft palate

Bilateral cleft lip and palate

A cleft palate and lip form when there is an incomplete fusion of the maxillary bones and overlying skin.

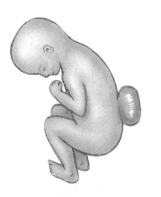

In spina bifida, incomplete development of the vertebrae allow the coverings (meninges) of the spinal cord to bulge outward

A **cleft palate** is a split or opening in the bony palate, or roof of the mouth. It results when the maxillary bones do not fuse properly. A cleft lip may form if the skin overlying the maxillary bones does not fuse. Cleft palates are about half as common as cleft lips. An infant with a cleft lip can nurse normally but because a cleft palate extends into the nasal cavity such infants suck in air and must be fed with bottles. Both conditions can be corrected surgically.

Spina bifida is a congenital defect of the spine in which the bony arches which overlie and form the vertebral foramen do not form normally. The affected vertebrae are usually in the lower back. As a result, part of the spinal cord is exposed. In severe forms, the spinal cord and parts of the brain do not form normally.

Kyphosis, *lordosis*, and *scoliosis* are conditions characterized by abnormal spinal curves. ∞ *[p. 155]* In **kyphosis** (kī-FŌ-sis), the normal thoracic curvature becomes exaggerated posteriorly, producing a "roundback" appearance (Figure 8-23a•). In **lordosis** (lor-dŌ-sis), or "swayback," the abdomen and buttocks protrude due to an anterior exaggeration of the lumbar curvature (Figure 8-23b•). **Scoliosis** (skō-lē-Ō-sis) is an abnormal lateral curvature of the spine (Figure 8-23c•). This lateral deviation may occur in one or more of the movable vertebrae. Scoliosis is the most common distortion of the spinal curvature. This condition may result from developmental problems, such as incomplete vertebral formation, or from muscular paralysis affecting one side of the back. Idiopathic (id-ē-ō-PATH-ik; "of unknown cause") scoliosis generally appears in girls during adolescence, when periods of growth are most rapid. In four out of five cases, the structural or functional cause of the abnormal spinal curvature is impossible to determine. Treatment used to consist of a combination of exercises and braces of doubtful effectiveness. Surgery now corrects the most pronounced cases.

Infection

Infectious diseases that affect the skeletal system become more common as individuals age. This results from the higher incidence of fractures in the elderly, combined with slower healing and the reduction of immune defenses.

•**FIGURE 8-23**
Spinal Deformities.
(a) Kyphosis. **(b)** Lordosis. **(c)** Scoliosis.

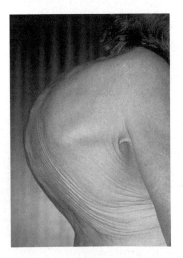

(a)

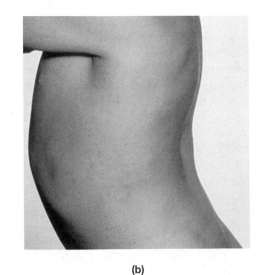

(b)

(c)

Osteomyelitis. Osteomyelitis (os-tē-ō-mī-e-LĪ-tis) is a painful infection of bone and/or marrow (*myelos* means "marrow") generally caused by bacteria, usually *Staphylococcus aureus*. It is more common in children than adults. The long bones of the arms and legs and vertebrae are more often affected in children, while vertebrae and pelvic bones are more affected in adults. Osteomyelitis can lead to dangerous systemic infections.

Paget's Disease. A virus appears to be responsible for **Paget's** (PAJ-ets) **disease**, also known as **osteitis deformans** (os-tē-Ī-tis de-FOR-mans). This condition may affect up to 10 percent of the population over 70. Infected bones weaken and thicken because osteoclast activity accelerates, producing areas of acute osteoporosis, and osteoblasts produce abnormal matrix. The result is a gradual deformation of the skeleton.

Tumors

Any of the cells of the skeletal system can give rise to benign or malignant tumors. Bone cancers are called **osteosarcomas**. **Chondrosarcomas** appear within epiphyseal cartilage or the articular cartilage that overlies the ends of bones within joints. More common than either of these are **myelomas**, cancers of the blood-producing cells within the internal cavity of a bone.

Nutritional Disorders

In **osteomalacia** (os-tē-ō-ma-LĀ-shē-ah), the size of the skeletal elements remains the same, but their mineral content decreases, softening the bones. In this condition the osteoblasts are working hard, but the matrix isn't accumulating enough calcium salts. This can occur in adults or children whose diet contains inadequate levels of calcium, phosphorus, or vitamin D. ∞ *[p. 119]* Vitamin D is necessary for the proper absorption of calcium and phosphorus from the digestive tract. A lack of this vitamin in children (whose bones are growing rapidly) also causes **rickets**, a form of osteomalacia. In this condition, bone formation at the epiphyses is abnormal, and the bending of the soft leg bones gives the individual a "bowlegged" appearance.

Scurvy is a condition in which the amount of bone formation decreases and the bones become more fragile. It is caused by a deficiency of vitamin C. This deficiency interferes with the production of collagen fibers which give bone its flexibility.

Trauma

Bone Injuries. The most common bone injuries are *bone bruises* and *fractures*. A **bone bruise** is an injury to the periosteum that damages connective tissues and blood vessels. Bone bruises usually result in pain and swelling. Fractures, introduced on p. 147, are much more complex injuries involving cracks or breaks in the bone itself. Fractures are classified according to their external appearance, the site of the fracture, and the nature of the crack or break in the bone, as seen in Figure 8-24•. For example, *closed (simple) fractures* are completely enclosed and do not involve a break in the skin. *Open (compound) fractures* project through the skin. Other important fracture types include:

- **Comminuted fractures** shatter the affected area into a multitude of bony fragments.
- **Transverse fractures** break a shaft bone across its long axis.
- **Spiral fractures**, produced by twisting stresses, spread along the length of the bone.
- A **Colles' fracture**, a common break in the distal portion of the radius (the slender bone of the forearm), is typically the result of reaching out to cushion a fall.

•**FIGURE 8-24**
Types of Fractures.

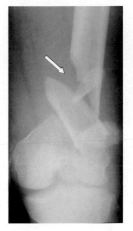

(a) Comminuted fracture of distal femur.

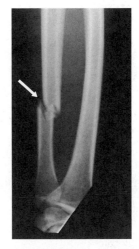

(b) A transverse fracture of the ulna.

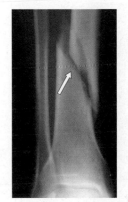

(c) Spiral fracture of tibia.

•FIGURE 8-24 (*continued*)
Types of Fractures.

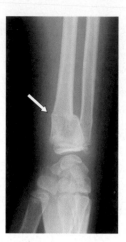

(d) A Colles' fracture.

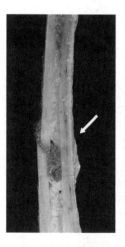

(e) A greenstick fracture.

- In a **greenstick fracture**, only one side of the shaft is broken, and the other is bent. This type generally occurs in children, whose long bones have not ossified fully.

- **Epiphyseal fractures** tend to occur where the bone matrix is undergoing calcification, replacing epiphyseal cartilage (see Figure 8-2, p. 146). A clean transverse fracture along this line generally heals well. Unless carefully treated, fractures between the epiphysis and the epiphyseal plate can permanently halt further longitudinal growth; surgery is often required.

- A **Pott's fracture** occurs at the ankle and affects both bones of the lower leg.

- **Displaced fractures** produce new and abnormal arrangements of bony elements.

- **Nondisplaced fractures** retain the normal alignment of the bony elements or fragments.

- **Compression fractures** occur in vertebrae subjected to extreme stresses, as when you land on your seat in a fall.

Bone Repair.   Bones will usually heal even after they have been severely damaged. Steps in the repair process, which may take from 3 weeks (fingers) to well over a year (femur), are diagrammed in Figure 8-25•:

Step 1:   In even a small fracture, many blood vessels are broken and extensive bleeding occurs. Pooling and clotting of the blood forms a swollen area called a **fracture hematoma** (*hemato* means blood, and *tumere* means to swell) that closes off the injured blood vessels.

Step 2:   When a fracture occurs, cells of the periosteum and endosteum activate and migrate into the fracture zone. There they form an **external callus** (*callum*, hard skin) and **internal callus**, respectively. At the center of the external callus, cells differentiate into chondrocytes and build blocks of cartilage.

Step 3:   As the repair continues, osteoblasts replace the central cartilage of the external callus with spongy bone. When this process is complete, the external and internal calluses form a continuous brace of spongy bone at the fracture site.

Step 4:   Remodeling of the spongy bone at the fracture site may continue from a period of 3 weeks (fingers) to well over a year (femur). When the re-

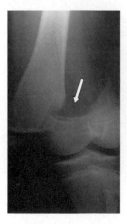

(f) An epiphyseal fracture of the femur.

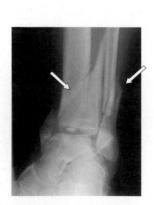

(g) A Pott's fracture.

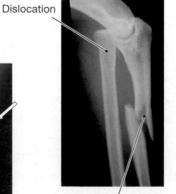

Dislocation

Fracture

(h) A displaced ulnar fracture; the radius has been dislocated.

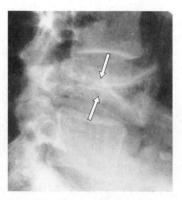

(i) A compression fracture of a vertebra.

modeling is complete, the fragments of dead bone and the spongy bone of the calluses will be gone, and only living compact bone will remain. The repair may be "good as new," with no sign that a fracture ever occurred, but often the bone will be slightly thicker and stronger than normal at the fracture site.

Degenerative Disorders

Osteoporosis is a condition that produces a reduction in bone mass resulting in extremely fragile bones (also see Clinical Note, p. 148). The excessive fragility of the bones frequently leads to breakage, and subsequent healing is impaired. Vertebrae may collapse, distorting the vertebral articulations and putting pressure on spinal nerves. Osteoporosis can also develop as a secondary effect of many cancers. Cancers of the bone marrow, breast, or other tissues release a chemical known as *osteoclast-activating factor*. This compound increases both the number and activity of bone-consuming osteoclasts and produces a severe osteoporosis.

Secondary Disorders

Endocrine System. A variety of endocrine or metabolic problems can also result in characteristic skeletal changes. **Gigantism** results from an overproduction of growth hormone before puberty. (The world record for height is 272 cm, or 8 ft. 11 in., with a weight of 216 kg, or 475 lb.) The opposite extreme is **pituitary growth**

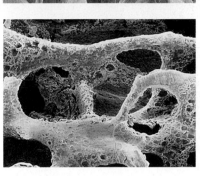

Normal spongy bone (top); spongy bone weakened by osteoporosis (bottom).

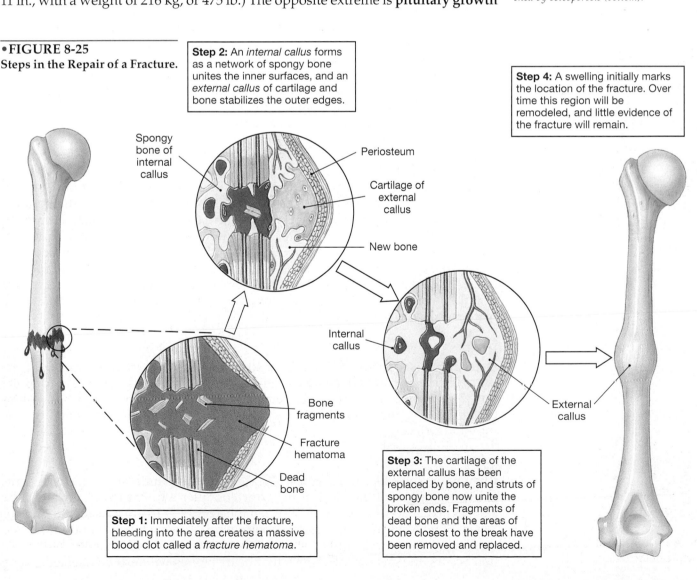

•FIGURE 8-25
Steps in the Repair of a Fracture.

Step 1: Immediately after the fracture, bleeding into the area creates a massive blood clot called a *fracture hematoma*.

Step 2: An *internal callus* forms as a network of spongy bone unites the inner surfaces, and an *external callus* of cartilage and bone stabilizes the outer edges.

Step 3: The cartilage of the external callus has been replaced by bone, and struts of spongy bone now unite the broken ends. Fragments of dead bone and the areas of bone closest to the break have been removed and replaced.

Step 4: A swelling initially marks the location of the fracture. Over time this region will be remodeled, and little evidence of the fracture will remain.

Spongy bone of internal callus

Periosteum

Cartilage of external callus

New bone

Internal callus

Bone fragments

Fracture hematoma

Dead bone

External callus

failure (*pituitary dwarfism*), in which inadequate growth hormone production leads to reduced epiphyseal activity and abnormally short bones. This form of growth failure is becoming increasingly rare in the United States because children can be treated with human growth hormone.

JOINT DISORDERS

The various causes of joint disorders are shown in Figure 8-21b•. Joints may be affected by inflammation, degeneration, infection, secondary disorders such as immune and metabolic problems, and trauma.

Inflammation

Bursitis. *Bursae* are small pockets of synovial fluid that reduce friction during joint or tendon movement. ∞ *[p. 165]* When bursae become inflamed, causing pain in the affected area whenever the tendon or ligament moves, a condition of **bursitis** exists. Inflammation can result from the friction associated with repetitive motion, pressure over the joint, irritation by chemical stimuli, infection, or trauma. Bursitis associated with repetitive motion typically occurs at the shoulder; for example, musicians, golfers, pitchers, painters, and tennis players may develop bursitis at this location. The most common pressure-related bursitis is a *bunion*. Bunions form over the base of the great toe as a result of the friction and distortion there, often by tight shoes with pointed toes.

Rheumatism and Arthritis. *Rheumatism* is a general term that indicates pain and stiffness affecting the skeletal system, the muscular system, or both. There are several major forms of rheumatism. One of them, **arthritis** (ar-THRĪ-tis) ("joint inflammation"), includes all the rheumatic diseases that affect synovial joints.

Proper working of synovial joints requires healthy articular cartilages. When an articular cartilage has been damaged, the matrix breaks down and the exposed cartilage changes from a slick, smooth gliding surface to a rough feltwork of bristly collagen fibers. The change in surface texture increases the amount of contact and friction between adjacent bones, damaging the cartilage further and producing pain. Eventually the central area of the articular cartilage may completely disappear, exposing the underlying bone.

The various diseases of arthritis are usually considered as either inflammatory or degenerative in nature. Inflammatory diseases start with the inflammation of synovial tissues, and damage later spreads to the articular surfaces. Inflammation can be caused by many factors, such as infections by bacteria or viruses, immune problems, or metabolic problems that are discussed in other sections. Degenerative diseases begin at the articular cartilages, and modification of the underlying bone and inflammation of the joint occur secondarily. Advanced stages of inflammatory and degenerative forms of arthritis produce an inflammation that spreads into the surrounding area.

Degenerative Disorders

Osteoarthritis (os-tē-ō-ar-THRĪ-tis), also known as *degenerative arthritis* or *degenerative joint disease (DJD)*, generally affects older individuals. In the U.S. population, 25 percent of women and 15 percent of men over 60 years of age show signs of this disease. The condition seems to result from cumulative wear and tear on the joint surfaces.

Infection

Rheumatic (roo-MA-tik) **fever** is a disease that causes inflammation of the joints and other tissues of the body. It may develop after a throat infection caused by streptococcal bacteria. Children between the ages of 5–15 are most commonly affected. Symptoms include high fever, joint pain, a distinctive full-body rash, and swelling of one or more of the larger joints of the body. The longer the inflammation occurs, the greater the chance that heart tissues will also be affected. Many times this may not be recognized and heart problems may show up 10–20 years after the infection.

Arthritis may result secondarily from infections by bacteria and viruses. For example, *gonococcal arthritis* may result from the bacterial disease *gonorrhea* (a sexually transmitted disease), *septic arthritis* is commonly due to infections by *Staphylococcus aureus*, and *tuberculous arthritis* may result from cases of *tuberculosis*. Viral arthritis has been associated with *German measles (rubella)* and infections by the *hepatitis B* virus.

Secondary Disorders

Metabolic Problems. Gout, or *gouty arthritis*, is a condition that results from too much uric acid in the blood and other body fluids. The formation of crystals of uric acid in the space within the synovial fluid, synovial membrane, or joint capsule causes pain and inflammation. The most commonly affected joint is the metatarsal/phalangeal joint of the big toe. The condition is about 10 times more common in men than women. Treatment involves anti-inflammatory medication and reducing uric acid levels by drugs or reducing the amount of meats and fats in the diet.

Immune Problems. Rheumatoid arthritis is an inflammatory condition that affects roughly 2.5 percent of the adult U.S. population. It is an *autoimmune* disorder in which the body's immune response is directed at its own tissues. Although rheumatoid arthritis can affect any joint of the body, joints of the fingers, wrists, and toes are most commonly affected. The synovial membrane becomes swollen and inflamed, a condition known as **synovitis** (sī-nō-VĪ-tis), and the matrix of the articular cartilage begins to break down. The joints swell, stiffen, become painful, and can become deformed.

Complete rest was routinely prescribed for arthritis patients in the past. Unfortunately, this lack of movement could result in **ankylosis** (ang-ki-LŌ-sis), an immobility due to fusion between the articulating surfaces of bones within a joint. Today, arthritis treatment includes regular exercise, physical therapy, drugs that reduce inflammation, such as aspirin, and even immune suppressants, such as *methotrexate*, which may slow the progression of the disease. Surgical procedures can realign or redesign the affected joint. In extreme cases involving the hip, knee, elbow, or shoulder, the defective joint can be replaced by an artificial one. Joint replacement has the advantage of eliminating the pain and restoring full range of motion. **Prosthetic (artificial) joints,** such as those shown in Figure 8-26•, are weaker than natural ones, but elderly people seldom stress them to their limits.

Degenerative changes comparable to those seen in arthritis may result from joint immobilization. When motion ceases, so does the circulation of synovial fluid, and the cartilages begin to suffer. Continuous passive motion (CPM) of any injured joint appears to encourage the repair process by improving the circulation of synovial fluid. The movement is often performed by a physical therapist or a machine during the recovery process.

Trauma

Dislocations. A **joint dislocation**, or **luxation** (luks-Ā-shun), occurs when articulating bone surfaces are forced out of position. This displacement can damage the articular cartilages, tear ligaments, or distort the joint capsule. The damage accompanying a **partial dislocation**, or **subluxation**, is less severe.

Hip Fractures. Hip fractures are most often suffered by individuals over age 60, when osteoporosis has weakened the thigh bones. In recent decades, there has also been an increase in hip fractures among young, healthy professional athletes, who subject their hips to extreme forces. These are complex injuries, often involving fracture-dislocation of the hip and sometimes accompanied by pelvic fractures. For individuals with osteoporosis, healing proceeds very slowly. In addition,

•**FIGURE 8-26**
Artificial Joints.

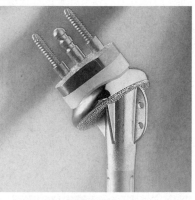

(a) Shoulder

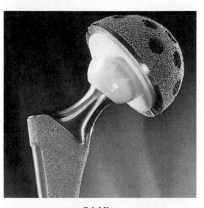

(b) Hip

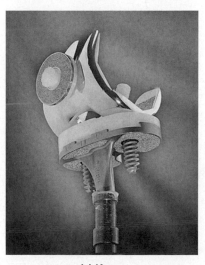

(c) Knee

•**FIGURE 8-27**
Arthroscopy and MRI Scans of the Knee.
(a) Arthroscopic view of a damaged knee, showing the torn edge of an injured meniscus. **(b)** A transverse MRI scan of the knee joint. **(c)** A frontal MRI scan of the knee joint.

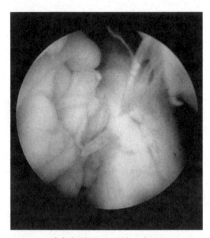

(a) Arthroscopic view

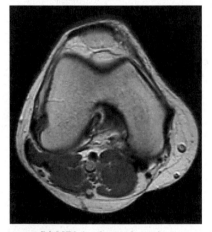

(b) MRI, horizontal section

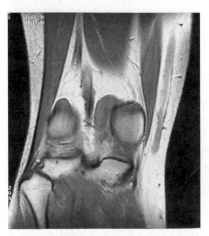

(c) MRI, frontal section

the large muscles that surround the joint can easily prevent proper alignment of the bone fragments. Hip fractures generally heal well if the joint can be stabilized; steel frames, pins, screws, or some combination of those devices may be needed to preserve alignment and permit healing to proceed normally.

Knee Injuries. Knee injuries are most common among relatively young and active individuals. Athletes place tremendous stresses on their knees. Ordinarily, the medial and lateral menisci move as the position of the femur changes. Placing a lot of weight on the knee while it is partially flexed can trap a meniscus between the tibia and femur, resulting in a break or tear in the cartilage. In the most common injury, the lateral surface of the lower leg is driven medially, tearing the medial meniscus. In addition to being quite painful, the torn cartilage may restrict movement at the joint. It can also lead to chronic problems and the development of a "trick knee," a knee that feels unstable. Sometimes the meniscus can be heard and felt popping in and out of position when the knee is extended. Less common knee injuries involve tearing one or more stabilizing ligaments or damaging the patella. Torn ligaments tend to be difficult to correct surgically, and healing is slow.

Total knee replacements are rarely performed on young people, but they are becoming increasingly common among elderly patients with severe arthritis.

Injuries to the Ankle and Foot. The ankle and foot are subjected to a variety of stresses during normal daily activities. In a **sprain**, a ligament is stretched to the point at which some of the collagen fibers are torn. The ligament remains functional, and the structure of the joint is not affected. The most common cause of a *sprained ankle* is a forceful inversion of the foot that stretches the lateral ligament. An ice pack is generally required to reduce swelling. With rest and support, the ankle should heal in about three weeks.

In more serious incidents, the entire ligament may be torn apart, or the connection between the ligament and the lateral malleolus may be so strong that the bone breaks instead of the ligament. In general, a broken bone heals more quickly and effectively than does a completely torn ligament. A dislocation may accompany such injuries.

Inspection and Treatment of Joint Injuries

An **arthroscope** uses fiber optics within a narrow tube to permit exploration of a joint without major surgery. Optical fibers are thin threads of glass or plastic that conduct light. The fibers can be bent around corners, so they can be introduced into a knee or other joint and moved around, enabling the physician to see what is going on inside the joint. If necessary, the apparatus can be modified to perform surgical modification of the joint at the same time. This procedure, called **arthroscopic surgery**, has greatly improved the treatment of knee and other joint injuries. Figure 8-27a• is an arthroscopic view of the interior of an injured knee, showing a damaged meniscus. New tissue-culturing techniques may someday permit the replacement of the meniscus or even the articular cartilage.

An arthroscope cannot show the physician soft tissue details outside the joint cavity, and repeated arthroscopy eventually leads to the formation of scar tissue and other joint problems. Magnetic resonance imaging (MRI) is a cost-effective and noninvasive method of viewing, without injury, and examining soft tissues around the joint. Figures 8-27b• and 8-27c• are MRI views of the knee joint. Note the image clarity and the soft tissue details visible in these scans.

✔ How does a lack of vitamin D in the diet affect bone development?

✔ What is the difference between rheumatism and arthritis?

CHAPTER REVIEW

Key Words

amphiarthrosis: An articulation that permits a small degree of movement.

appendicular skeleton: Pertaining to the bones of the upper or lower limbs.

articulation (ar-tik-ū-LĀ-shun): A joint between bones.

axial skeleton: Pertaining to the bones of the skull, vertebral column, and rib cage.

diarthrosis: freely movable joint; a synovial joint.

ligament (LI-ga-ment): Dense band of connective tissue fibers that attach one bone to another.

marrow: A tissue that fills the internal cavities in a bone; may be dominated by blood element–forming cells (red marrow) or adipose tissue (yellow marrow).

ossification: The replacement of other tissue with bone.

osteoblast: A cell that produces the fibers and matrix of bone.

osteoclast: A cell that dissolves the fibers and matrix of bone.

osteocyte (OS-tē-ō-sīt): A bone cell responsible for the maintenance and turnover of the mineral content of the surrounding bone.

osteon (OS-tē-on): The basic structural unit of compact bone, consisting of osteocytes organized around a central canal and separated by rings of calcified matrix.

periosteum (pe-rē-OS-tē-um): Layer of fibers and cells that surrounds a bone.

synarthrosis: A joint that does not permit movement.

synovial (sī-NŌV-ē-ul) **fluid:** Substance secreted by synovial membranes that lubricates most freely movable joints.

Selected Clinical Terms

arthritis: Rheumatic diseases that affect the synovial joints, especially the articular cartilages on the ends of bones.

bony crepitus: An abnormal crackling or popping sound produced during movement of a joint.

bursitis: Inflammation of a bursa, causing pain whenever the associated tendon or ligament moves.

fracture: A crack or break in a bone.

gout: A metabolic problem that results from too much uric acid in the blood and other body fluids. Causes pain and inflammation within joints.

luxation: Dislocation of a joint.

osteomalacia: A condition where the size of the bones remains the same, but their mineral content decreases, softening the bones.

osteomyelitis: A painful infection of bone and/or marrow generally caused by bacteria.

osteoporosis: A disease condition characterized by abnormal loss of bone mass, resulting in fragile bones.

rickets: A form of osteomalacia that occurs in children, causing bones to weaken and resulting in a "bowlegged" appearance.

scurvy: A condition in which the amount of bone formation decreases and the bones become more fragile. Caused by a deficiency in Vitamin C.

sprain: A condition caused when a ligament is stretched to the point where some collagen fibers are torn. The ligament remains functional, and the structure of the joint is not affected.

subluxation: A partial dislocation of a joint.

Study Outline

SYSTEM BRIEF *(p. 143)*

1. The skeletal system includes the bones of the skeleton and the cartilages, ligaments, and other connective tissues that stabilize or interconnect bones. Its functions include structural support, protection, leverage, storage, and blood cell production.

BONE STRUCTURE *(pp. 143–146)*

1. Osseous tissue is a supporting connective tissue with a solid **matrix**.

GENERAL FEATURES OF BONE *(p. 143)*

2. General shapes of bones include long bones, short bones, flat bones, and irregular bones.

3. The features of a long bone include a **diaphysis**, **epiphyses**, and a central marrow cavity. (*Figure 8-1*)

4. There are two types of bone: **compact bone** and **spongy bone**.

5. A bone is covered by a **periosteum** and lined with an **endosteum**.

MICROSCOPIC FEATURES OF BONE *(p. 145)*

6. There are three types of bone cells. **Osteocytes** maintain the structure of existing bones. **Osteoblasts** produce the matrix of new bone through the process of *osteogenesis*. **Osteoclasts** dissolve the bony matrix and release the stored minerals.

7. Osteocytes occupy **lacunae**, spaces in bone. The lacunae are surrounded by layers of calcified matrix.

8. The basic functional unit of compact bone is the **osteon**, containing osteocytes arranged around a **central canal**. (*Figure 8-1*)

BONE DEVELOPMENT AND MAINTENANCE *(pp. 146–148)*

1. Ossification is the process of converting other tissues to bone.

BONE GROWTH *(p. 146)*

2. In the process of **endochondral ossification**, a cartilaginous model is gradually replaced by bone. (*Figure 8-2*)

3. There are differences among bones and among individuals regarding the timing of epiphyseal closure.

4. Normal osteogenesis requires a reliable source of minerals, vitamins, and hormones.

BONE MAINTENANCE *(p. 147)*

5. The organic and mineral components of bone are continually recycled and renewed through the process of remodeling.

6. The shapes and thicknesses of bones reflect the stresses applied to them. Mineral turnover allows bone to adapt to new stresses.

7. Calcium is the most common mineral in the human body, with roughly 99 percent of it located in the skeleton.

8. A **fracture** is a crack or break in a bone.

AGING AND THE SKELETAL SYSTEM *(p. 148)*

9. The bones of the skeleton become thinner and relatively weaker as aging occurs.

THE SKELETON *(pp. 148–163)*

1. The skeletal system consists of the axial skeleton and the appendicular skeleton. The **axial skeleton** can be subdivided into the **skull**, the **thoracic** (rib) **cage**, and the **vertebral column**.

2. The **appendicular skeleton** includes the limbs and the pectoral and pelvic girdles that attach the limbs to the trunk. *(Figure 8-3)*

SKELETAL TERMS *(p. 148)*

3. *Bone markings* are used to describe and identify specific bones. *(Table 8-1)*

THE AXIAL SKELETON *(p. 148)*

4. The skull can be divided into *cranial* and *facial* subdivisions. The **cranium** encloses the **cranial cavity**, a division of the dorsal body cavity that encloses the brain. *(Figure 8-4)*

5. The cranial bones include the **frontal bone, parietal bones, occipital bone, temporal bones, sphenoid bone,** and **ethmoid bone.** *(Figures 8-4, 8-5, 8-6; Table 8-2)*

6. The facial bones include the left and right **maxillary bones** (*maxillae*), **palatine bones, vomer, zygomatic bones, nasal bones, lacrimal bones, inferior nasal conchae,** and the **mandible.** *(Figures 8-4, 8-5, 8-6; Table 8-2)*

7. The **auditory ossicles** lie within the temporal bone. The **hyoid bone** is suspended below the skull by ligaments from the styloid processes of the temporal bones. *(Figure 8-4; Table 8-2)*

8. The nasal septum divides the nasal cavities. Together the **frontal, sphenoid, ethmoid,** and **maxillary sinuses** make up the *paranasal sinuses,* which drain into the nasal cavities. *(Figures 8-5, 8-6)*

9. Fibrous connections at **fontanels** permit the skulls of infants and children to continue growing. *(Figure 8-7)*

10. There are 7 **cervical vertebrae**, 12 **thoracic vertebrae** (which articulate with ribs), and 5 **lumbar vertebrae** (which articulate with the sacrum). The **sacrum** and **coccyx** consist of fused vertebrae. *(Figure 8-8; Table 8-3)*

11. The spinal column has four **spinal curves** that accommodate the unequal distribution of body weight and keep it in line with the body axis. *(Figure 8-8)*

12. A typical vertebra has a **body** and a **vertebral foramen**. Adjacent vertebral bodies are separated by an **intervertebral disc.**

13. Cervical vertebrae are distinguished by the *transverse foramina* on either side. Thoracic vertebrae articulate with the ribs. The lumbar vertebrae are the most massive and least mobile, and they are subjected to the greatest strains. *(Figure 8-9)*

14. The **sacrum** protects reproductive, digestive, and excretory organs. It articulates with the **coccyx**. *(Figures 8-3, 8-8; Table 8-3)*

15. The skeleton of the thorax consists of the thoracic vertebrae, the ribs, and the sternum. The ribs and sternum form the **rib cage**. *(Figure 8-10; Table 8-3)*

16. Ribs 1 to 7 are **true ribs**. Ribs 8 to 12 lack direct connections to the sternum and are called **false ribs**; they include two pairs of **floating ribs**. *(Figure 8-10; Table 8-3)*

17. The sternum consists of a **manubrium**, a **body**, and a **xiphoid process**. *(Figure 8-10; Table 8-3)*

THE APPENDICULAR SKELETON *(p. 157)*

18. Each upper limb articulates with the trunk at the *shoulder,* or **pectoral**, *girdle,* which consists of the **scapula** and **clavicle**. *(Figures 8-3, 8-11; Table 8-4)*

19. The clavicle and scapula position the shoulder joint, help move the arm, and provide a base for arm movement and muscle attachment.

20. The scapula articulates with the **humerus** at the shoulder joint. *(Figure 8-12a; Table 8-4)*

21. Distally the humerus articulates with the radius and ulna at the elbow joint.

22. The radius and ulna are the bones of the forearm. The **olecranon** process forms the point of the elbow. *(Figure 8-12b,c; Table 8-4)*

23. The bones of the wrist form two rows of **carpal bones**. The distal carpal bones articulate with the **metacarpal bones** of the hand. Four of the fingers contain three **phalanges**; the thumb has only two. *(Figure 8-13; Table 8-4)*

24. The **pelvic girdle** consists of two **coxae**; each coxa forms through the fusion of an **ilium**, an **ischium**, and a **pubis**. *(Figure 8-14; Table 8-5)*

25. The **pelvis** consists of the coxae, the sacrum, and the coccyx.

26. The **femur**, or *thighbone*, is the longest bone in the body. It articulates with the **tibia** at the knee joint. *(Figure 8-15a,b)*

27. The *medial malleolus* of the tibia and the *lateral malleolus* of the fibula articulate with the **talus** at the ankle joint. *(Figure 8-15b)*

28. The ankle includes seven **tarsal bones**; only the **talus** articulates with the tibia and fibula. When we are standing normally, most of our weight is transferred to the ground through the **calcaneus**; the rest is passed through the **metatarsal bones**. *(Figure 8-15c)*

JOINTS *(pp. 163–168)*

1. Articulations (joints) exist wherever two bones interact.

TYPES OF JOINTS *(p. 163)*

2. Immovable joints are **synarthroses**, slightly movable joints are **amphiarthroses**, and those that are freely movable are called **diarthroses**.

3. An example of a synarthrosis is a *suture.*

4. An example of an amphiarthrosis is the *pubic symphysis.* *(Figure 8-14)*

5. The bony surfaces at diarthroses, or **synovial joints**, are covered by **articular cartilages**, lubricated by **synovial fluid**, and enclosed within a **joint capsule**. Other synovial structures can include **menisci, fat pads,** and various ligaments. *(Figure 8-16)*

ACTIONS OF MOVABLE JOINTS *(p. 165)*

6. Important terms that describe dynamic motion are **flexion, extension, hyperextension, rotation, circumduction, abduction,** and **adduction**. *(Figure 8-17)*

7. The bones in the forearm permit **pronation** and **supination**. *(Figures 8-12c, 8-18)*

8. The ankle undergoes **dorsiflexion** and **plantar flexion**. Movements of the foot include **inversion** and **eversion**. **Opposition** is the thumb movement that enables us to grasp objects. *(Figure 8-19)*

9. Protraction involves moving something forward; **retraction** involves moving it back. **Depression** and **elevation** occur when we move a structure down and up. *(Figure 8-19)*

10. Synovial joints include **gliding joints, hinge joints, pivot joints, ellipsoidal joints, saddle joints,** and **ball-and-socket joints**. *(Figure 8-20)*

BONE AND JOINT DISORDERS *(pp. 168–180)*

1. Bone is continuously recycled and renewed through the process of **remodeling**. Factors affecting this process include age, physical stress, hormone levels, rates of calcium and phosphate exchange, and genetic or environmental factors.

EXAMINATION OF THE SKELETAL SYSTEM *(p. 170)*

2. Signs of bone or joint disorders include limited joint movements, joint inflammation, sounds of joint movements (**bony crepitus**), abnormal bone deposits, and abnormal posture.

3. A common symptom of bone and joint disorders is pain. *(Table 8-6)*

BONE DISORDERS *(p. 172)*

4. Bone disorders may be caused by **congenital factors** (genetic abnormalities or anatomical problems), **infection**, **endocrine problems**, **nutritional problems**, **tumors**, **trauma**, **degenerative problems**, or as **secondary problems** of the *endocrine system* or *metabolism*. *(Figure 8-21a)*

5. Examples of inherited disorders of the skeletal system include **osteogenesis imperfecta**, **Marfan's syndrome**, and **achondroplasia**. *(Figure 8-22)*

6. Examples of congenital problems with skeletal development include **congenital talipes equinovarus (clubfoot)**, **cleft palate**, **spina bifida**, and abnormal spinal curvatures (**kyphosis, lordosis,** and **scoliosis**). *(Figure 8-23)*

7. **Osteomyelitis** is an infection of the bone or bone marrow caused by bacteria. **Paget's disease** is an infection caused by a virus.

8. Vitamin deficiencies may lead to a loss of minerals in the skeleton, a condition called **osteomalacia. Rickets**, one form of osteomalacia, is due to a lack of vitamin D. In **scurvy**, a lack of vitamin C, decreases the production of collagen and bones become brittle.

9. Malignant tumors are associated with bone (**osteosarcomas**), epiphyseal cartilage or articular cartilages (**chondrosarcomas**), or bone marrow (**myelomas**).

10. A trauma is a physical injury. **Bone bruises** damage the periosteum of a bone and *fractures* are a crack or break in a bone. *(Figure 8-24)*

11. Repair of a fracture involves the formation of a **fracture hematoma**, an **external callus**, and an **internal callus**. *(Figure 8-25)*

12. Osteoporosis is usually a degenerative, age-related condition that produces a reduction in bone mass resulting in extremely fragile bones. It can also develop as a secondary effect of some cancers.

13. Selected endocrine problems are **gigantism** and **pituitary growth failure**.

JOINT DISORDERS *(p. 178)*

14. Joint disorders involve both inflammatory and degenerative problems. Major categories include inflammation, degenerative problems, infection, secondary disorders, and trauma. *(Figure 8-21b)*

15. Bursitis is an inflammation of bursae, pockets of synovial fluid that are associated with movable joints.

16. *Rheumatism* is a general term that indicates pain and stiffness affecting the skeletal system, the muscular system, or both. **Arthritis** is a form of rheumatism that affects synovial joints.

17. Osteoarthritis, or *degenerative arthritis* or *degenerative joint disease (DJD)*, is associated with the wear and tear of joints that accompany aging.

18. Rheumatic fever is an inflammatory condition that affects joints and other tissues, especially those of the heart. It develops after an infection by streptococcal bacteria.

19. *Gonococcal, septic,* and *tuberculous arthritis* may develop secondarily after infection by various types of bacteria.

20. A metabolic disorder, **gout**, or *gouty arthritis*, is caused by the formation of crystals of uric acid within joints.

21. Rheumatoid arthritis is an autoimmune disease. The synovial membrane becomes swollen and inflamed, a condition known as **synovitis**.

22. In extreme cases of the breakdown of articular cartilages involving the hip, knee, elbow, or shoulder, the defective joint is replaced by a **prosthetic (artificial) joint.** *(Figure 8-26)*

23. In a **joint dislocation**, or **luxation**, the articulating bone surfaces are forced out of position within their joint capsule.

24. Fractures of the hip generally involve fracture-dislocation of the hip, which may be accompanied by pelvic fractures.

25. Knee injuries commonly involve damage to the cartilaginous menisci and/or structural ligaments.

26. Injuries to the ankle and foot often involve ligaments and dislocations.

27. Arthroscopic surgery permits the treatment of knee and other joint injuries without opening up the entire joint capsule. *(Figure 8-27)*

Review Questions

MATCHING

Match each item in Column A with the most closely related item in Column B. Use letters for answers in the spaces provided.

	Column A		Column B
___	1. elbow and knee	a.	abduction
___	2. diaphysis	b.	heel bone
___	3. auditory ossicles	c.	ball-and-socket joints
___	4. moving the hand into a palm-back position	d.	joint inflammation
___	5. osteoblasts	e.	hinge joints
___	6. C_1	f.	bone covering that increases the thickness of bone
___	7. periosteum	g.	immovable joints
___	8. hip and shoulder	h.	bone shaft
___	9. patella	i.	functional unit of compact bone
___	10. calcaneus	j.	bone-producing cells
___	11. sutures	k.	a cervical vertebra
___	12. moving the hand into a palm-front position	l.	pronation
___	13. osteoclasts	m.	short stature
___	14. raising the arm laterally	n.	supination
___	15. osteon	o.	kneecap
___	16. arthritis	p.	bone-dissolving cells
___	17. achondroplasia	q.	ear bones

MULTIPLE CHOICE

18. The bones of the skeleton store energy reserves as lipids in areas of _____.
(a) red marrow   (b) yellow marrow
(c) the matrix of bone tissue   (d) the ground substance

19. The two types of osseous tissue are _____.
(a) compact bone and spongy bone
(b) dense bone and compact bone
(c) spongy bone and cartilage
(d) a, b, and c are correct

20. The cells that maintain mature compact bone are _____.
(a) lacunae   (b) osteocytes
(c) osteoblasts   (d) osteoclasts

21. The axial skeleton consists of the bones of the _____.
(a) pectoral and pelvic girdles
(b) skull, thorax, and vertebral column
(c) arm and hand, legs and feet
(d) limbs, pectoral girdle, and pelvic girdle

22. The appendicular skeleton consists of the bones of the _____.
(a) pectoral and pelvic girdles
(b) skull, thorax, and vertebral column
(c) arm, legs, hands, and feet
(d) limbs, pectoral girdle, and pelvic girdle

23. Of the following sets of bones, _____ lists *only* bones of the cranium.
(a) mandible, parietal, occipital, sphenoid
(b) frontal, occipital, zygomatic, parietal
(c) occipital, sphenoid, temporal, parietal
(d) mandible, maxillary, nasal, zygomatic

24. Of the following bones, the _____ is unpaired.
(a) vomer   (b) maxillary bone
(c) palatine bone (d) nasal bone

25. At the glenoid cavity, the scapula articulates with the proximal end of the _____.
(a) humerus   (b) radius
(c) ulna   (d) femur

26. Each coxa of the pelvic girdle consists of three fused bones, the _____.
(a) ulna, radius, humerus
(b) hamate, capitate, trapezium
(c) femur, tibia, fibula
(d) ilium, ischium, pubis

27. The function of the synovial fluid is _____.
(a) to nourish chondrocytes
(b) to provide lubrication
(c) to absorb shock
(d) a, b, and c are correct

28. Standing on tiptoe is an example of a movement called _____.
(a) elevation   (b) dorsiflexion
(c) plantar flexion   (d) retraction

29. A _____ is a complete joint dislocation.
(a) subluxation   (b) fracture hematoma
(c) luxation   (d) displaced fracture

30. _____ is a type of arthritis that commonly affects the big toe.
(a) gout   (b) osteoarthritis
(c) rheumatoid arthritis   (d) bursitis

TRUE/FALSE

____ 31. In anatomical position, the ulna lies medial to the radius.

____ 32. The red marrow is the site of production of red blood cells and other blood elements.

____ 33. Joints typically found at the end of long bones are amphiarthroses.

____ 34. Abduction and adduction always refer to movements of the appendicular skeleton.

____ 35. Both the scapula and clavicle form joints with the axial skeleton.

____ 36. Rickets is a disease in which the bones weaken because of a lack of vitamin C in the diet.

____ 37. Osteosarcomas are cancers of the bone.

SHORT ESSAY

38. What are the five primary functions of the skeletal system?

39. During the growth of a long bone, how is the epiphysis forced farther from the shaft?

40. What two primary functions are performed by the thoracic cage?

41. What is the difference between the *pelvic girdle* and the *pelvis*?

42. While working at an excavation, an archaeologist finds several small skull bones. She examines the frontal, parietal, and occipital bones and concludes that the skulls are those of children not yet 1 year old. How can she tell their ages from examining the bones?

43. Describe four congenital disorders of the spine.

44. Place the steps involved in repairing a fracture into the correct sequence:
(a) Osteoblasts replace central cartilages with spongy bone
(b) A fracture hematoma forms at the site of the injury
(c) An external callus encircles the bone
(d) Osteoclasts and osteoblasts remodel the injured area

45. What are bunions? Would you expect them to be more common in men or women? Explain your answer.

APPLICATIONS

46. Joyce fractures a bone. Although no bones protrude through her skin, an X-ray image reveals multiple bone fragments in the injured area. What type of fracture did she sustain? Is it open or closed?

47. During a basketball game, Lisa injured her right knee when she jumped to retrieve the ball and then landed off-balance on her right leg. Since then she has pain and limited mobility of her right knee joint. What did Lisa likely injure?

48. Diane, a 65-year-old woman, is brought into the emergency room by her daughter, Mary. Diane has had severe back pain since she fell yesterday. Her daughter insists it was a minor fall and is confused about the severity of the pain, and tenderness over the thoracic spine. X-ray studies of the thoracic vertebrae reveal compression fractures of T_{10}, T_{11}, and T_{12}. The X-rays also reveal a decreased bone density in all other vertebrae. Laboratory tests are within normal limits. The physician diagnoses the fracture as a vertebral fracture most likely due to:
(a) osteomalacia
(b) osteomyelitis
(c) osteoporosis
(d) Paget's disease

✔ Answers to Concept Check Questions

(pp. 145–146)   **1.** Concentric layers of bone around a central canal indicate a *Haversian system* that makes up compact bone. Since the ends (epiphyses) of long bones are mostly spongy bone, this sample most likely came from the shaft (diaphysis). **2.** Because osteoclasts break down, or demineralize, bone, the bone would have less mineral content and be weaker.

(p. 148)   **1.** Long bones of the body, like the femur, have a plate of cartilage (the *epiphyseal plate*) that separates the epiphysis from the diaphysis as long as the bone is growing in length. An X-ray would reveal whether the epiphyseal plate was still present. If it was, then growth was still occurring; if not, the bone had reached its adult length. **2.** The increase in the male sex hormone, testosterone, that occurs at puberty contributes to an increased rate of bone growth and the closure of the epiphyseal plates. Since the source of testosterone, the testes, is removed in castration, the boys would be expected to have a longer, though slower, growth period and be taller than if they had not been castrated. **3.** The larger arm muscles of the weight lifter apply more mechanical stress to the bones of the arm. In response to the stress, the bones will grow thicker. We would expect the jogger to have heavier thigh bones for similar reasons.

(p. 153)   **1.** The *frontal bone* of the cranium lies under the forehead. **2.** The opening in the occipital bone is called the *occipital foramen*. It provides a passage for the spinal cord that connects to the brain. **3.** The paranasal sinuses function to make some of the heavier skull bones lighter, and their epithelial linings produce mucus.

(p. 156)   **1.** An intervertebral disc between the first and second vertebrae would not allow the head to rotate, or turn, to the right and left. **2.** In adults, the five sacral vertebrae fuse to form a single sacrum. **3.** The *thoracic cage* is made up of the thoracic vertebrae, the ribs, and the sternum.

(p. 160)   **1.** The only direct connection between the pectoral girdle and the axial skeleton occurs between the clavicle and the *manubrium* of the sternum. **2.** Only the *radius* of the forearm is involved in forming the wrist joint.

(p. 162)   **1.** The three bones that make up the coxa are the ilium, ischium, and pubis. **2.** Although the fibula is not part of the knee joint nor does it bear weight, it is an important point of attachment for many leg muscles. When the fibula is fractured, these muscles cannot function properly to move the leg and walking is difficult and painful. **3.** Joey has most likely fractured the *calcaneus* (heel bone).

(p. 168)   **1.** At first, the joint is a type of amphiarthrotic joint. Later, when the bones interlock, they form sutural (synarthrotic) joints. **2.** (a) abduction; (b) supination; (c) flexion.

(p. 180)   **1.** Calcium (and phosphorus) minerals are part of the matrix that makes bones solid and stiff. Vitamin D is required for the absorption of these minerals from the digestive tract. A lack of vitamin D results in softer, more flexible bones. **2.** Rheumatism is a more general term than arthritis. Rheumatism refers to pain and stiffness associated with the skeletal and/or muscular systems. Arthritis is a type of rheumatism that affects synovial, or freely-movable, joints.

CHAPTER
9

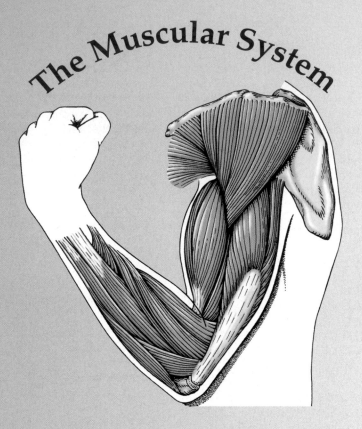

The Muscular System

CHAPTER OUTLINE

CHAPTER OBJECTIVES

1 Describe the properties and functions of muscle tissue.
2 Describe the organization of skeletal muscle.
3 Describe the structure of a sarcomere.
4 Describe the sliding-filament model of muscle contraction.
5 Describe the relationship between muscle contractions and motor units.
6 Contrast isotonic and isometric muscle contractions.
7 Distinguish between aerobic and anaerobic exercise.
8 Identify the major axial muscles of the body.
9 Identify the major appendicular muscles of the body.
10 Describe the effects of exercise and aging on muscle tissue.
11 Describe the two common symptoms of muscle disorders.
12 Describe the major disorders of the muscular system.

It is hard to imagine what our lives would be like without muscle tissue. We would be unable to sit, stand, walk, speak, or grasp objects. Blood would not circulate, because there would be no heartbeat to propel it through blood vessels. The lungs could not rhythmically empty and fill, nor could food move through the digestive tract.

Three different types of muscle tissue move the body and the materials within it. *Skeletal muscle tissue* moves the body by pulling on the bones of the skeleton. *Cardiac muscle tissue* contracts the chambers of the heart, pushing blood through the vessels of the circulatory system. *Smooth muscle tissue* pushes and squeezes fluids and solids along the digestive tract and performs a variety of other functions. All three muscle tissues share the following properties: **excitability**, the ability to respond to stimulation; **contractility**, the ability to shorten; and **elasticity**, the ability of a muscle to rebound toward its original length after a contraction.

This chapter focuses on skeletal muscle tissue. It begins by examining the microscopic structure of skeletal muscle tissue and ends with an overview of the major muscle groups of the body.

SYSTEM BRIEF

The muscular system is made up of some 700 skeletal muscles. Together, they make up about 45 percent of the body weight of a male and 35 percent of the weight of a female. These muscles perform the following important functions:

1. **Produce movement**: Contracting muscles attached to the bones of the skeleton produce movement.

2. **Maintain posture and body position**: Constant muscular contractions permit us to sit upright without collapsing and stand without toppling over.

3. **Maintain body temperature**: Contracting muscles generate heat that keeps our body temperature in the normal range.

THE STRUCTURE OF SKELETAL MUSCLE

Skeletal muscles are the organs of the muscular system and, like any organ, contain a variety of tissues. Each skeletal muscle contains connective and neural tissues as well as skeletal muscle tissue. A typical skeletal muscle appears in Figure 9-1•.

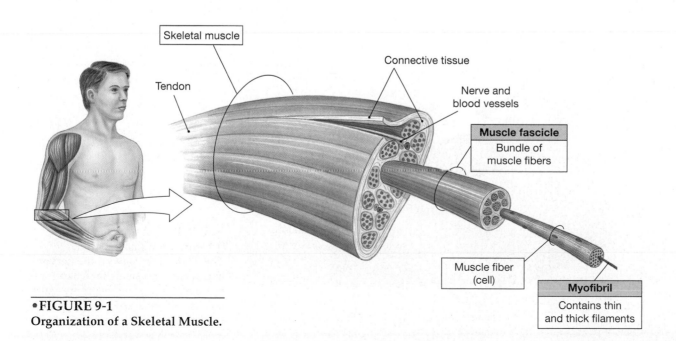

•**FIGURE 9-1**
Organization of a Skeletal Muscle.

Connective tissue is an important part of each muscle. Three layers (outer, middle, and inner) are found in all skeletal muscles. The outer layer surrounds the entire muscle, separating it from surrounding tissues and organs. The connective tissue of the middle layer divides the skeletal muscle into a series of compartments, each containing a group of muscle cells. Each group is called a **fascicle** (FA-si-kul), from the Latin word *fasciculus*, meaning "a bundle." The middle connective tissue layer also contains blood vessels and nerves that supply the surrounding fascicles. Within a fascicle, an inner layer of connective tissue surrounds each skeletal muscle cell and ties adjacent muscle cells together.

At each end of the muscle, the connective tissue layers usually converge to form cordlike **tendons.** Tendons attach skeletal muscles to bones. ∞ *[p. 65]* Any contraction of the muscle will pull on the tendon and, in turn, on the attached bone. Sometimes muscles attach to bones or other muscles through broad sheets of connective tissue rather than through cordlike tendons. Such a sheet is called an **aponeurosis** (ap-ō-noo-RŌ-sis).

STRUCTURE OF A MUSCLE FIBER

Skeletal muscle tissue consists of elongated cells called *muscle fibers*. Muscle fibers, shown in Figure 9-1●, are quite different from the "typical" cells described in Chapter 3. ∞ *[p. 40]* One obvious difference is their enormous size. For example, a skeletal muscle fiber from a thigh muscle could have a diameter of 100 μm (0.1 mm) and a length equal to that of the entire muscle (30 to 40 cm). In addition, each skeletal muscle fiber is *multinucleate*, containing hundreds of nuclei just beneath the cell membrane.

Like a package of spaghetti noodles, each muscle fiber contains hundreds of long, cylindrical structures called **myofibrils.** (*Myo* is Latin for "muscle.") Each my-

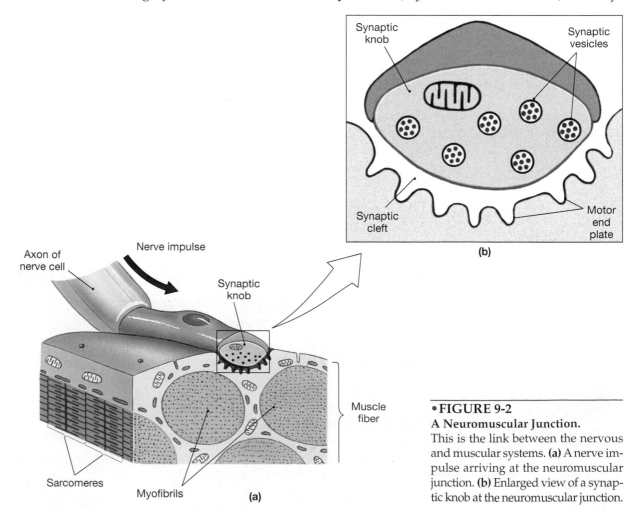

●**FIGURE 9-2**
A Neuromuscular Junction.
This is the link between the nervous and muscular systems. **(a)** A nerve impulse arriving at the neuromuscular junction. **(b)** Enlarged view of a synaptic knob at the neuromuscular junction.

ofibril (one of the "noodles") contains bundles of threadlike proteins, or *filaments*, responsible for the contraction of the muscle fiber. There are two types of filaments: **thin filaments,** consisting primarily of the protein *actin*, and **thick filaments,** composed primarily of the protein *myosin*. The filaments in each myofibril are organized in repeating units called *sarcomeres*.

THE NERVE CELL–MUSCLE CELL LINK

Skeletal muscles normally contract in response to a signal from the nervous system. Skeletal muscles are controlled by nerve cells called *motor neurons*. The link between a motor neuron and a skeletal muscle fiber occurs at a specialized connection called a **neuromuscular junction**. As Figure 9-2• shows, the neuromuscular junction contains a *synaptic knob*, the expanded tip of a process from the motor neuron. The synaptic knob contains membranous sacs (*synaptic vesicles*), each filled with **neurotransmitter** molecules. Binding of a neurotransmitter to receptors at the muscle fiber will trigger an electrical impulse in the cell membrane of the muscle fiber. A thin space, the *synaptic cleft*, separates the synaptic knob from the *motor end plate* of the muscle fiber. Folds in the muscle cell membrane in this region increase the surface area of the muscle fiber exposed to neurotransmitter released by the synaptic knob. When neurotransmitter molecules trigger the formation of an electrical impulse in the muscle fiber membrane, that impulse stimulates the release of calcium ions stored within the endoplasmic reticulum of the muscle fiber. These calcium ions then trigger the contraction of the muscle fiber.

A false-color SEM image of a neuromuscular junction.

HOW A MUSCLE FIBER CONTRACTS

Sarcomeres (SAR-kō-mērs) are the working units of a skeletal muscle fiber. Each myofibril contains thousands of sarcomeres. The regular arrangement of thick and thin filaments within each sarcomere produces the striped appearance of a myofibril. All the myofibrils are arranged parallel to the long axis of the cell, with their sarcomeres lying side by side. As a result, the entire muscle fiber has a banded, or *striated*, appearance that corresponds to the bands of the individual sarcomeres (see Figure 4-6•, p. 70).

sarco-, muscle + *meros*, part
sarcomere: the smallest contracting unit in a muscle fiber; each myofibril consists of a chain of sarcomeres

SARCOMERE STRUCTURE

The structure of the sarcomere provides a key to understanding how a muscle fiber contracts. Figure 9-3a• shows a diagrammatic view of an individual sarcomere at

•**FIGURE 9-3**
Filament Sliding and Contraction of a Sarcomere.
(a) The sarcomere at rest, when the thick and thin filaments are not interacting. **(b)** When a contraction occurs, the free ends of the thin filaments move closer together as cross-bridges form and pivot toward the center of the sarcomere.

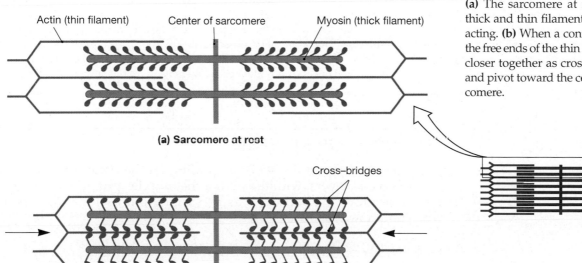

Actin (thin filament)   Center of sarcomere   Myosin (thick filament)

(a) Sarcomere at rest

Cross–bridges

(b) Sarcomere contracting

rest. The thick filaments lie in the center of the sarcomere. Thin filaments are attached to either end of the sarcomere and extend toward the center of the sarcomere, passing among the thick filaments. Sliding of the thin filaments toward the center of the sarcomere causes the unit to shorten, or contract (Figure 9-3•). This explanation of muscle contraction is called, not surprisingly, the *sliding filament model*.

SARCOMERE FUNCTION

What causes filament sliding? Sliding occurs after the **cross-bridges**, or "heads," of the thick filaments attach to the thin filaments. Each cross-bridge then pivots at its base, pulling the thin filament toward the center of the sarcomere. This pivoting, called a *power stroke*, shortens the sarcomere (Figure 9-3b•). The sliding of thin filaments continues as each cross-bridge detaches, pivots, and reattaches, much like an individual pulling in a rope grabs, pulls, releases and then grabs again, hand over hand. A molecule of ATP must bind to the myosin head before it can detach from the thin filament; the energy provided by the ATP will be used in the next power stroke.

What triggers a contraction? Cross-bridge formation can occur only in the presence of calcium ions. Calcium ions are normally stored within the endoplasmic reticulum of the muscle fiber. The calcium ions are released into the cytoplasm around the myofibrils when one or more electrical impulses travel across the surface of the muscle fiber, in response to the arrival of neurotransmitter at the motor end plate. As soon as the impulse, or impulses, have passed, the calcium ions are pumped back into the endoplasmic reticulum. This is an active transport process that requires ATP. ∞ *[p. 45]* The muscle contraction ends when calcium ion concentrations in the cytoplasm return to their normal low levels. Under these circumstances, cross-bridge formation can no longer occur.

Each time ATP is broken down, some of the energy is released as heat. Muscle contractions release large amounts of heat, and skeletal muscle activity is largely responsible for maintaining body temperature. That is why we shiver when we get cold—the muscle contractions generate heat that can elevate (or at least stabilize) body temperature.

✔ Why does skeletal muscle appear striated when viewed with a microscope?

✔ Where would you expect to find the greatest concentration of calcium ions in resting skeletal muscle?

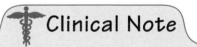

Clinical Note

What Is Rigor Mortis?

When death occurs, circulation stops, and the skeletal muscles are deprived of nutrients and oxygen. Within a few hours, the skeletal muscle fibers have run out of ATP, and the endoplasmic reticulum becomes unable to remove calcium ions from the cytoplasm. Calcium ions diffusing into the cell from the extracellular fluid or leaking out of their storage area (the endoplasmic reticulum) then trigger a sustained contraction. Without ATP, the myosin cannot detach from the cross-bridges, and the muscle locks in the contracted position. All the body's skeletal muscles are involved, and the individual becomes "stiff as a board." This physical state, called **rigor mortis**, lasts until the muscle fibers begin to decompose 15–25 hours later.

THE WORK OF MUSCLE

Skeletal muscles perform work by pulling on the bones of the skeleton. The bones serve as levers, with the pull of the muscles providing the force to lift a weight or move the body. The nerve supply to muscles allows us to control when and how we move.

HOW NERVES AND MUSCLES WORK TOGETHER

During normal movements, whenever a muscle fiber contracts, it contracts fully, and it always produces the same amount of pull, or *tension*. There is no mechanism to control the amount of tension produced by the contraction of a single muscle fiber: The muscle fiber is either ON (contracting completely) or OFF (relaxing). This is known as the **all-or-none principle**.

Our skeletal muscles would not be very useful to us, however, if the entire muscle could contract only completely or not at all. Fortunately, the all-or-none principle applies only to individual muscle fibers—not to an entire muscle. The amount of tension produced in a skeletal muscle *as a whole* is determined by how many of its muscle fibers are stimulated at one time.

A typical skeletal muscle contains tens of thousands of muscle fibers. Each muscle fiber receives instructions from a motor neuron at a single neuromuscular junction. Although some motor neurons control only a few muscle fibers, most motor neurons control hundreds of muscle fibers. All the muscle fibers controlled by a single motor neuron form a **motor unit**.

When we decide to perform a particular movement, specific groups of motor neurons are stimulated. The amount of tension in a muscle depends on how many motor units are then "recruited"; the greater the number called into action, the stronger the muscle's contraction. The size of the motor units within a muscle determines the degree of control available over the tension produced. In the muscles of the eye, where precise control is quite important, each motor neuron may control as few as two or three muscle fibers. On the other hand, we have less precise control over the force exerted by our leg muscles, where one motor neuron may control as many as 2,000 muscle fibers.

Some of the motor units within any particular muscle are always active, even when the entire muscle is not contracting. The contractions of these motor units do not produce enough tension to cause movement, but they do tense and firm the muscle. This background tension in a resting skeletal muscle is called **muscle tone**. A muscle with little muscle tone appears limp and soft, whereas one with moderate muscle tone is quite firm and solid.

In space, muscles do not need to overcome gravity. Unless special exercises are performed, muscles gradually weaken and there is a loss of muscle mass.

HOW MUSCLES AND BONES WORK TOGETHER

Bones and muscles are closely interconnected. The physical relationships among bones and muscles, especially the location of muscle attachments relative to the joints involved, determine the power, speed, and direction of body movements. These principles can best be understood in mechanical terms. A **lever** is a rigid structure—such as a crowbar—that moves on a fixed point, called the **fulcrum,** in response to an applied force. In the body each bone is a lever, each joint is a fulcrum, and our muscles apply the forces that produce movement. A force that resists or opposes movement is called a **resistance**. Body weight or an external weight (like a brick in the hand) are examples of resistances that can oppose movement.

Three classes of levers are found in the body. They differ in the placement of the fulcrum (the joint) in relation to the applied force (the muscle attachment) and the position of the resistance (the weight to be moved). An everyday example of the use of a lever is shown in Figure 9-4a•. The most common levers in the body are *third-class levers* (Figure 9-4b•). In this type of lever, the applied force lies between the fulcrum and the resistance. The point of force application and the resistance move in the same direction, but the resistance moves farther. Thus third-class levers increase the speed and range (distance traveled) of body movements. Figure 9-4c• shows how this type of lever action is used to raise the forearm and hand.

Origins, Insertions, and Actions

Consider a muscle that extends between two articulating bones. Typically, when that muscle contracts, one bone will move while the other remains fixed in position. The end of the muscle attached to the stationary bone is called the **origin**, and the end attached to the bone that moves is called the **insertion.** The

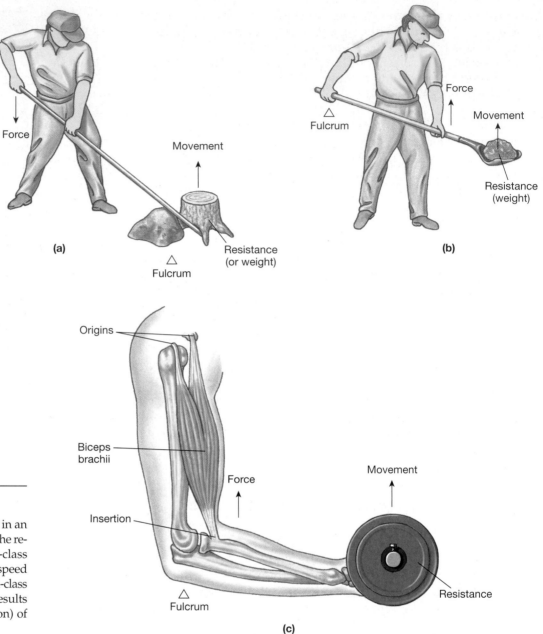

•FIGURE 9-4
Levers and Muscle Action.
(a) First-class lever action results in an increase in force (or strength) on the resistance (the stump). (b) Third-class lever action results in increased speed and range of movement. (c) Third-class lever action at the elbow joint results in the upward movement (flexion) of the forearm and the barbell.

action is the specific movement produced when the muscle contracts. We usually indicate the action in terms of the joint involved. Figure 9-4c● shows that for the *biceps brachii* muscle, there are two points of origin on the scapula and a single insertion on the radius of the forearm. We thus say that the biceps brachii originates on the scapula and inserts on the radius. The primary movement produced when the biceps brachii contracts is flexion of the elbow. (The various types of body movements were discussed in Chapter 8. See Figures 8-17 to 8-19● on pp. 165–167.)

Skeletal muscles do not work in isolation. Generally, individual muscles or groups of muscles work in pairs; as one contracts, the other relaxes. Muscles can be grouped and described according to their primary actions. A muscle whose contraction is chiefly responsible for producing a particular movement is a **prime mover**. The *biceps brachii,* which produces flexion of the elbow, is an example of a prime mover. **Antagonists** are prime movers whose actions oppose that of the muscle under consideration. For example, the *triceps brachii* is a prime mover that extends the elbow and is an antagonist of the biceps brachii.

Isotonic and Isometric Contractions

Muscle contractions may be classified as *isotonic* or *isometric* based on the pattern of tension produced in the muscle and its overall change in length. In an **isotonic** contraction, enough tension is generated to produce movement, and tension remains at that same level until the movement stops, and relaxation occurs. During isotonic contractions, the muscle shortens and the bone at the insertion is pulled toward the origin. Lifting an object off a desk, walking, running, and so forth involve isotonic contractions.

Contracting muscles do not always produce movement, however. Have you ever strained to lift something so heavy you were unable to move it, no matter how hard you tried? Although your muscles were contracting, no movement was produced. Contractions in which the muscle tension increases but the muscle length stays the same are called **isometric**, from the Greek for "same measure." Familiar examples of isometric contractions include pushing against a wall or trying to pick up a car. Although these are rather unusual movements, less obvious isometric contractions are actually quite common. For example, the muscle contractions that keep our bodies upright when standing or sitting involve the isometric contractions of muscles that oppose gravity.

Normal daily activities involve a combination of isotonic and isometric muscular contractions. As you sit reading this text, isometric contractions stabilize your vertebrae and maintain your upright position. When you next turn a page, the movements of your arm, forearm, hand, and fingers are produced by isotonic contractions.

MUSCLES AND EXERCISE

Muscle performance improves with exercise. Athletes use different training methods depending upon whether their sport is primarily supported by **aerobic** (oxygen-requiring) or **anaerobic** (without oxygen) energy production.

Aerobic sports involve sustained activities such as jogging and distance swimming. Aerobic exercises are designed to extend the length of time over which a muscle can continue to contract. Aerobic training exercises stimulate an increase in the number of mitochondria and the amount of *myoglobin,* an oxygen-binding pigment, within muscle fibers. The presence of additional mitochondria improves the ability of the muscle fibers to generate the ATP needed to support contractions. **Myoglobin** (MĪ-ō-glō-bin) binds oxygen molecules inside a muscle fiber, just as a related pigment, *hemoglobin,* binds oxygen in the blood. Myoglobin provides an oxygen reserve that can be used during a contraction.

Anaerobic sports include such activities as a 50-yard dash or swim, a pole vault, or a weight-lifting competition. The frequent, brief, and intensive training workouts used by athletes engaged in such sports stimulate an increase in the number of myofibrils and sarcomeres within their muscle fibers. The result is an enlargement, or **hypertrophy**, of the stimulated muscles. The effects are dramatic—for example, champion weight lifters and bodybuilders have hypertrophied muscles. We are born with almost all the muscle fibers we will ever have, and hypertrophy results from the enlargement of existing muscle fibers, rather than from the formation of new muscle fibers.

Muscle Fatigue

Skeletal muscle is considered *fatigued* when it can no longer contract, despite continued stimulation. Muscle fatigue may be caused either by a lack of ATP or by the buildup of *lactic acid.*

What causes a lack of ATP? Skeletal muscle fibers, like most cells in the body, generate ATP through oxygen-requiring chemical reactions in their mitochondria. Because these reactions use oxygen, this process of energy production is called **aerobic metabolism**. If a muscle's contractions require no more ATP than the amount generated by its mitochondria, and oxygen supplies remain adequate, the muscle fiber can function aerobically for long periods. Under these conditions, sufficient ATP is avail-

iso, same + *tonos,* tension
isotonic contraction: a muscle contraction where tension rises until movement begins, and then remains at that level throughout the contraction

✔ A motor unit from a skeletal muscle contains 1500 muscle fibers. Would this muscle be involved in fine, delicate movements or powerful, gross movements? Explain.

✔ Is it possible for a muscle to contract without shortening? Explain.

✔ How would you distinguish between the origin and insertion of a particular muscle?

hyper, over + *trophē,* nourishment
hypertrophy: an enlargement of skeletal muscles in response to repeated stimulation

aero, air + *bios,* life
aerobic metabolism: chemical reactions that require oxygen for the generation of ATP; the oxygen is obtained from the air at the lungs and distributed by the cardiovascular system

able, and fatigue occurs only after relatively long periods. For example, the muscles of long-distance athletes may become fatigued only after several hours of moderate exertion, when the muscle fibers run out of organic compounds suitable for mitochondria.

An athlete sprinting the 100-yard dash, however, may suffer from the muscle fatigue associated with lactic acid buildup. A muscle working at *peak* levels of exertion consumes ATP much faster than its mitochondria can produce it. The additional ATP is produced through **anaerobic** (non-oxygen–requiring) **metabolism**, in a process called *glycolysis* (glī-KOL-i-sis). In glycolysis, glucose molecules are split in half to form pyruvic acid molecules and a small amount of ATP. This reaction occurs in the cytoplasm, not in the mitochondria. Although glycolysis can supply straining muscles with ATP, problems develop as the excess pyruvic acid is converted to **lactic acid**. Lactic acid releases hydrogen ions that cause a dangerous decline in pH in and around the muscle fiber. After a relatively short time (seconds to minutes), the pH change inactivates key enzymes, and the muscle can no longer function normally.

Muscle Recovery

After exertion, conditions inside the muscle gradually return to normal levels during the *recovery period*. During this period, lactic acid is removed and the muscle fibers rebuild their energy reserves (especially glycogen). In addition, the body as a whole must lose the heat generated during intense muscular contraction; that's why you continue to perspire after an exercise period ends.

During the recovery period, the body's oxygen demand goes up considerably. The extra oxygen is used by mitochondria (1) in liver cells, where ATP is required for the conversion of lactic acid to glucose, and (2) in muscle fibers, where ATP is needed to restore the cell's reserves of ATP and glycogen to pre-exertion levels. The additional oxygen required during the recovery period is often called the **oxygen debt**. While repaying an oxygen debt the body's breathing rate and depth of breathing must increase. This is why you continue to breathe heavily for some time after you stop exercising.

Muscle Atrophy

A common fitness slogan is "use it or lose it." Skeletal muscles that are not "used," or stimulated, on a regular basis will become smaller and weaker. This process is called **atrophy.** Individuals whose muscles are paralyzed by spinal injuries first lose muscle tone and then undergo gradual muscle atrophy. Even a temporary reduction in muscle use can lead to muscular atrophy. This can easily be seen by comparing limb muscles before and after a cast has been worn. Muscle atrophy is reversible in its early stages, but once muscle fibers die, they are replaced by fibrous tissues, and the atrophy is permanent. That is why physical therapy is crucial to individuals temporarily unable to move normally.

a-, without + *aeros*, air + *bios*, life
anaerobic metabolism: chemical reactions that can generate ATP in the absence of oxygen

a-, without + *trophē*, nourishment
atrophy: the degeneration of unstimulated muscle fibers

✔ Why would a sprinter experience muscle fatigue before a marathon runner would?

✔ Which activity would be more likely to create an oxygen debt, swimming laps or lifting weights?

MAJOR MUSCLES OF THE BODY

The muscular system includes all the nearly 700 skeletal muscles that can be controlled voluntarily. Figure 9-5• shows some of the major superficial muscles of the human body.

In the last chapter we described the skeletal system by dividing it into axial and appendicular divisions. We will use the same divisions in our discussion of the muscular system.

■ The **axial muscles** are associated with the axial skeleton only. They position the head and spinal column and also move the rib cage, assisting in the movements that make breathing possible. Roughly 60 percent of the skeletal muscles in the body are axial muscles.

■ The **appendicular muscles** stabilize or move the bones of the appendicular skeleton.

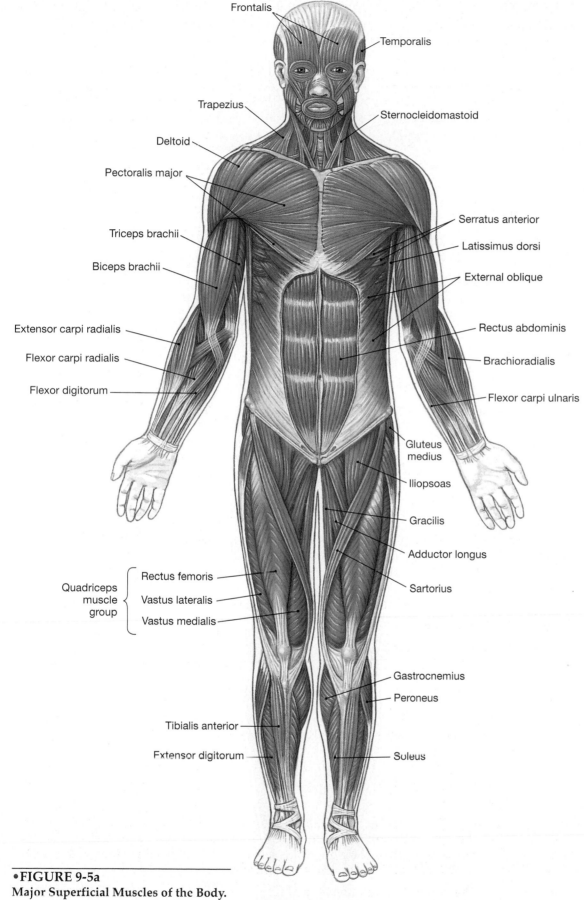

Frontalis

Temporalis

Trapezius

Sternocleidomastoid

Deltoid

Pectoralis major

Serratus anterior

Triceps brachii

Latissimus dorsi

Biceps brachii

External oblique

Extensor carpi radialis

Rectus abdominis

Flexor carpi radialis

Brachioradialis

Flexor digitorum

Flexor carpi ulnaris

Gluteus
medius

Iliopsoas

Gracilis

Adductor longus

Quadriceps
muscle
group
{
  Rectus femoris
  Vastus lateralis
  Vastus medialis
}

Sartorius

Gastrocnemius

Peroneus

Tibialis anterior

Extensor digitorum

Soleus

•FIGURE 9-5a
Major Superficial Muscles of the Body.
Anterior view.

•FIGURE 9-5b
Major Superficial Muscles of the Body.
Posterior view.

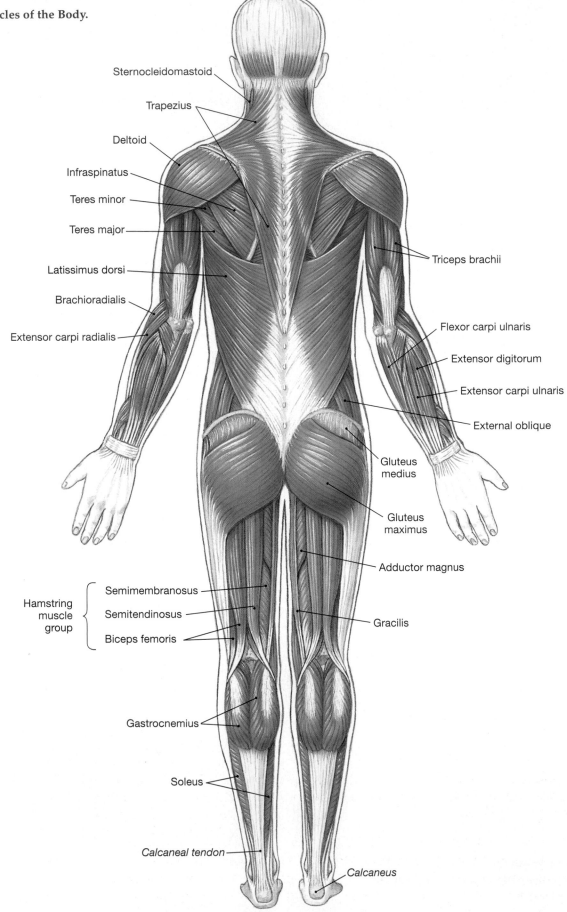

Sternocleidomastoid

Trapezius

Deltoid

Infraspinatus

Teres minor

Teres major

Latissimus dorsi

Brachioradialis

Extensor carpi radialis

Triceps brachii

Flexor carpi ulnaris

Extensor digitorum

Extensor carpi ulnaris

External oblique

Gluteus medius

Gluteus maximus

Adductor magnus

Hamstring muscle group

Semimembranosus

Semitendinosus

Biceps femoris

Gracilis

Gastrocnemius

Soleus

Calcaneal tendon

Calcaneus

Table 9-1	Muscle Terminology		
Terms Indicating Direction Relative to Axes of the Body	**Terms Indicating Specific Regions of the Body***	**Terms Indicating Structural Characteristics of the Muscle**	**Terms Indicating Actions**
Anterior (front)	Abdominis (abdomen)	**Origin**	**General**
Externus (superficial)	Carpi (wrist)	Biceps (two heads)	Abductor
Inferioris (inferior)	Cleido/clavius (clavicle)	Triceps (three heads)	Adductor
Lateralis (lateral)	Costalis (ribs)	Quadriceps (four heads)	Depressor
Medialis/medius (medial, middle)	Femoris (femur)	**Shape**	Extensor
	Ilio- (ilium)	Deltoid (triangle)	Flexor
Obliquus (oblique)	Lumborum (lumbar region)	Orbicularis (circle)	Levator
Posterior (back)	Oculo- (eye)	Serratus (serrated)	Pronator
Rectus (straight, parallel)	Oris (mouth)	Teres (long and round)	Rotator
Superficialis (superficial)	Psoas (loin)	Trapezius (trapezoid)	Supinator
Superioris (superior)	Radialis (radius)	**Other Striking Features**	**Specific**
Transversus (transverse)	Temporalis (temples)	Gracilis (slender)	Buccinator (trumpeter)
	Thoracis (thoracic region)	Latissimus (widest)	Sartorius (like a tailor)
	Tibialis (tibia)	Longissimus (longest)	
	Ulnaris (ulna)	Longus (long)	
		Magnus (large)	
		Major (larger)	
		Maximus (largest)	
		Minor (smaller)	
		-tendinosus (tendinous)	
		Vastus (great)	

For other regional terms, refer to Figure 1-4, p. 9, which deals with anatomical landmarks.

As you learn about the individual muscles and muscle groups keep in mind what you already know about the bones of the skeleton and the types of movements discussed in Chapter 8. Muscle names are descriptive. Understanding the meaning of the muscle names will thus help you remember and identify them. The name of a muscle may refer to (1) its location (for example, the *temporalis* attaches to the temporal bone); (2) its action (for example, the *flexor digitorum* flexes the fingers); or (3) its shape (for example, the *deltoid* muscle has the shape of the Greek letter *delta* [Δ]. Examples are given in Table 9-1.

THE AXIAL MUSCLES

The major axial muscles can be placed into three groups. The first group, the *muscles of the head and neck*, includes muscles responsible for facial expressions, chewing, and swallowing. The second group, the *muscles of the spine*, includes flexors and extensors of the spinal column. The third group, the *muscles of the trunk*, form the muscular walls of the thoracic and abdominopelvic cavities, and the floor of the pelvic cavity. Table 9-2 summarizes the location and actions, or movements, of the major axial muscles.

Muscles of the Head and Neck

The muscles of the head and neck are shown in Figure 9-6• and listed in Table 9-2. The muscles of the face originate on the surface of the skull and insert into the dermis of the skin. When they contract, the skin moves, changing our facial ex-

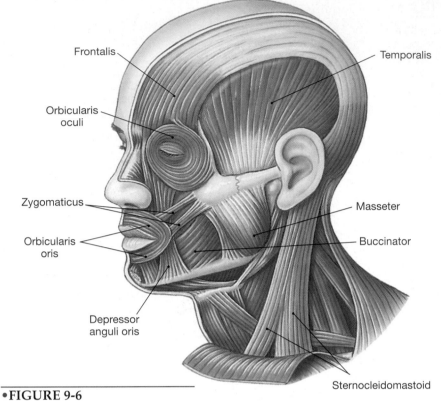

Frontalis

Temporalis

Orbicularis
oculi

Zygomaticus

Masseter

Buccinator

Orbicularis
oris

Depressor
anguli oris

Sternocleidomastoid

•FIGURE 9-6
Muscles of the Head and Neck.

pression. The largest group of facial muscles is associated with the mouth. The **orbicularis oris** (or-bik-ū-LA-ris ŌR-is) constricts the opening, and other muscles move the lips or the corners of the mouth. The **buccinator** (BUK-si-nā-tor), another muscle associated with the mouth, compresses the cheeks, as when pursing the lips and blowing forcefully. During chewing, contraction and relaxation of the buccinator moves food back across the teeth from the space inside the cheeks. The chewing motions are primarily produced by contractions of the **masseter** (MAS-se-tur), with assistance from the **temporalis** (tem-pō-RA-lis).

Smaller groups of muscles control movements of the eyebrows and eyelids, the scalp, the nose, and the external ear. Other muscles of the neck control the position of the larynx, lower the mandible, tense the floor of the mouth, and provide a stable foundation for muscles of the tongue and pharynx. One of these, the **sternocleidomastoid** (ster-nō-klī-dō-MAS-toyd) extends from the clavicles and the sternum to the mastoid region of the skull. It can rotate the head or flex the neck.

Muscles of the Spine

The muscles of the spine are covered by more superficial back muscles, such as the trapezius and latissimus dorsi (see Figure 9-5b•). Spinal extensor muscles include the **erector spinae** (the *sacrospinalis, iliocostalis, longissimus,* and *spinalis*), which keep the spine and head erect (Figure 9-7• and Table 9-2). When contracting together, these muscles extend the spinal column. When only the muscles on one side contract, the spine is bent laterally. The erector spinae are opposed by the *quadratus lumborum* and the various abdominal muscles that flex the spinal column.

Muscles of the Trunk

The muscles of the trunk include (1) the **external** and **internal intercostals** and **obliques,** (2) the **transversus abdominis,** (3) the **rectus abdominis,** (4) the muscular **diaphragm** that separates the thoracic and abdominopelvic cavities, and (5) the muscles that form the floor of the pelvic cavity.

✔ If you were contracting and relaxing your masseter muscle, what would you probably be doing?

✔ Which muscle would best be described as the "kissing" muscle?

•FIGURE 9-7
Muscles of the Spine.

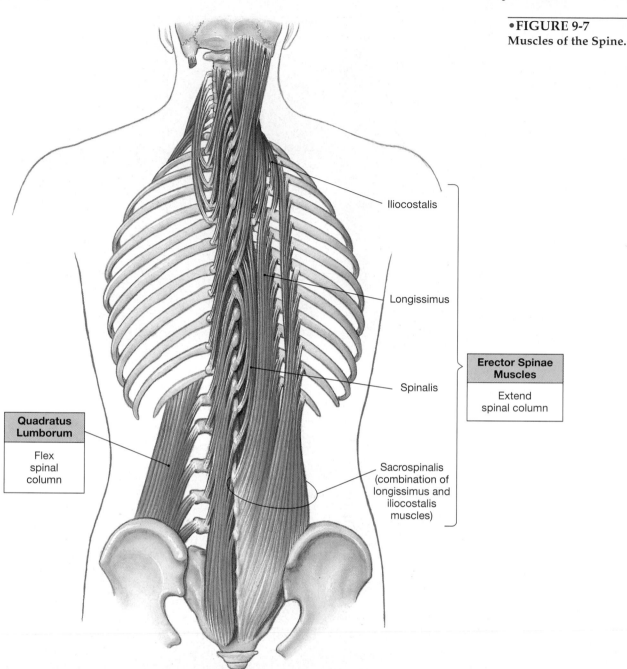

Iliocostalis

Longissimus

Erector Spinae Muscles

Extend spinal column

Spinalis

Quadratus Lumborum

Flex spinal column

Sacrospinalis (combination of longissimus and iliocostalis muscles)

The dome-shaped diaphragm is important in breathing. Because of its shape and its position inferior to the lungs, contraction of the diaphragm increases the size of the pleural cavities and expands the lungs. This action pulls air into the respiratory system. When it relaxes, the diaphragm moves upward, compressing the lungs and pushing air out of the respiratory system. The intercostal muscles, located between the ribs, are also involved in breathing. The external intercostals raise the rib cage and expand the lungs, and the internal intercostals lower the rib cage and compress the lungs.

The floor of the pelvic cavity is called the **perineum** (pe-ri-NĒ-um). It is formed by a broad sheet of muscles that extend from the sacrum and coccyx to the ischium and pubis. Prominent muscles in this region include the **levator ani**, which supports most of the organs within the pelvic cavity, and the **urethral** and **anal sphincter muscles**, which provide voluntary control of urination and defecation. Both superficial and deep muscles of the perineum are shown in Figure 9-8•. Note that there are no differences between the deep muscles in males and females.

•**FIGURE 9-8**
Muscles of the Male and Female Pelvic Floor.
There are no differences in the deep muscles between male and female.

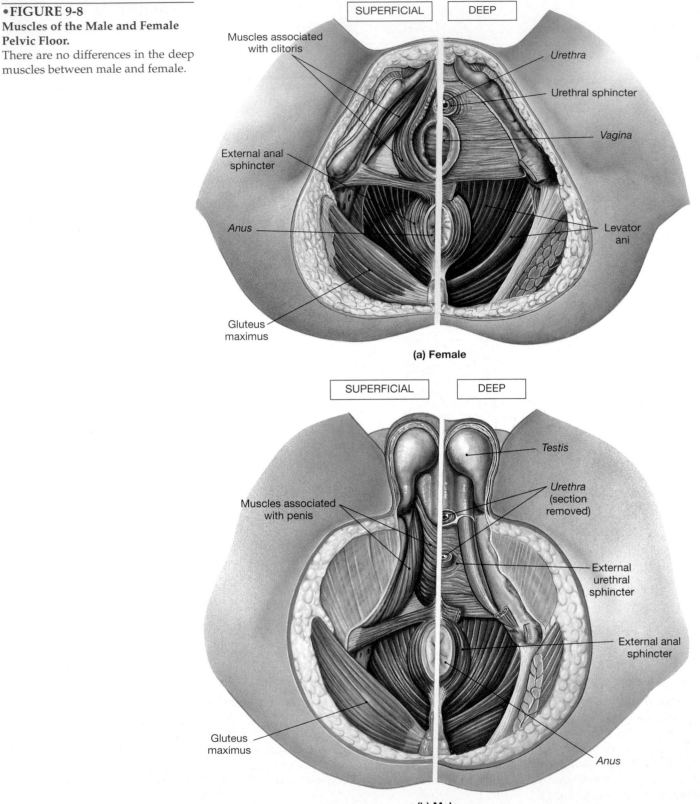

SUPERFICIAL   DEEP

Muscles associated with clitoris

Urethra

Urethral sphincter

Vagina

External anal sphincter

Anus

Levator ani

Gluteus maximus

(a) Female

SUPERFICIAL   DEEP

Muscles associated with penis

Testis

Urethra (section removed)

External urethral sphincter

External anal sphincter

Gluteus maximus

Anus

(b) Male

✔ Which muscles of the spine and trunk would you use while doing sit-ups?

✔ The perineum forms the floor of what body cavity?

The oblique muscles are broad muscular sheets that form the walls of the abdomen (Figure 9-9• and Table 9-2). Contraction of the oblique muscles can compress organs in the abdominopelvic cavity or rotate the spinal column, depending on whether one side or both sides are contracting. The *rectus abdominis* is an important flexor of the spinal column.

•FIGURE 9-9
Muscles of the Trunk.

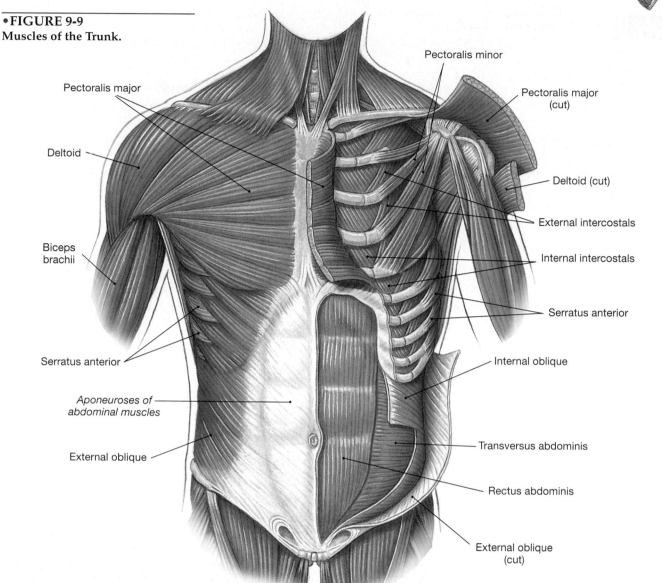

Table 9-2	Axial Muscles	
Region/Muscle	**Location**	**Action**
Muscles of the Head and Neck		
Frontalis	Forehead	Raises eyebrows, wrinkles forehead
Orbicularis oculi	Encircles eyelids	Closes eye
Buccinator	Lines the cheek	Pushes food into teeth, blowing, sucking
Orbicularis oris	Encircles lips	Purses lips; "kissing" muscle
Zygomaticus	Upper cheek	Raises corner of mouth; "smiling" muscle
Depressor anguli oris	Side of chin	Pulls corner of mouth down
Temporalis	Side of head above ear	Raises jaw
Masseter	Rear of cheek	Raises jaw
Sternocleidomastoid	Side of neck	Together, both flex the neck; alone, turns head to one side
Muscles of the Spine		
Erector spinae	Along back from head to sacrum	Extend spinal column
Quadratus lumborum	Hip to lower vertebrae and adjacent surface of ribs	Flexes spinal column

Table 9-2	Axial Muscles (*continued*)	
Region/Muscle	**Location**	**Action**
Muscles of the Trunk		
Intercostals	Between each rib and adjacent ribs	Externals raise ribs; internals lower ribs
Diaphragm	Separates thoracic and abdominal cavities	Expands and reduces size of lung cavities
Obliques (external and internal) and transversus abdominis	Make up three layers of the abdominal wall	Compress abdomen
Rectus abdominis	Anterior surface of abdominal wall, from ribs to pelvis	Flexes spinal column; the "sit-up" muscle
Muscles of the Pelvic Floor		
Urethral sphincter	Encircles urethra	Closes urethra; provides voluntary control of urination
Anal sphincter	Encircles anus	Closes anal opening; provides voluntary control of defecation
Levator ani	Forms much of the pelvic floor	Supports internal organs

THE APPENDICULAR MUSCLES

The appendicular muscles include (1) the muscles of the shoulders and upper limbs and (2) the muscles of the pelvic girdle and lower limbs. There are few similarities between the two groups because the functions and required ranges of motion are very different.

Muscles of the Shoulder and Upper Limb

The large, superficial **trapezius** (tra-PĒ-zē-us) muscle covers the back and portions of the neck, reaching to the base of the skull (see Figures 9-5a• and b•). This muscle forms a broad diamond. Its actions are quite varied because specific regions can be made to contract independently.

On the chest, the **serratus** (ser-RA-tus) **anterior** originates along the anterior surfaces of several ribs and inserts along the vertebral border of the scapula. When the serratus anterior contracts, it pulls the shoulder forward.

The primary muscles of the arm are shown in Figure 9-5•. They include (1) the superficial muscle of the shoulder, the **deltoid** (DEL-toyd) (Figure 9-5a•), which is the major abductor of the arm; (2) the **infraspinatus** (in-fra-spī-NĀ-tus), **teres** (TER-ēz) **major**, and **teres minor** (Figure 9-5b•), which rotate the arm and are often called the muscles of the *rotator cuff*; and (3) the **pectoralis** (pek-tō-RA-lis) **major** (Figure 9-5a•) and the **latissimus dorsi** (la-TIS-i-mus DOR-sē) (Figure 9-5b•). The pectoralis major flexes the shoulder, and the latissimus dorsi extends it. These two muscles also work together to adduct and rotate the arm. All these muscles provide substantial support for the loosely built shoulder joint. The muscles of the rotator cuff are a frequent site of sports injuries. Powerful, repetitive arm movements, such as pitching a fastball at 95 mph for nine innings, can place intolerable stresses on the muscles of the rotator cuff, leading to muscle strains (a tear or break in the muscle), bursitis (inflammation of bursae), and other painful injuries.

Although most of the muscles that insert on the forearm and wrist originate on the humerus, there are two noteworthy exceptions. The **biceps brachii** and the **triceps brachii** insert on the forearm and originate on the scapula. Their primary actions are on the elbow. For example, the triceps brachii extends the elbow during push-ups, while the biceps brachii makes a prominent bulge as it both flexes the elbow and supinates the forearm.

Other important muscles of the upper limb include the following:

- The **brachioradialis** flexes the elbow, opposed by the triceps brachii.

- The **flexor carpi ulnaris** and the **flexor carpi radialis** work together to produce flexion of the wrist. Note that the flexor muscles are on the anterior and medial surfaces of the forearm.

- The **extensor carpi radialis** and the **extensor carpi ulnaris** produce extension of the wrist. Note that the extensor muscles are on the posterior and lateral surfaces of the forearm.

- The **flexor digitorum** and **extensor digitorum** muscles perform flexion and extension of the fingers.

The muscles of the forearm provide strength and crude control of the palm and fingers. Fine control of the hand involves small muscles that originate on the carpal and metacarpal bones. No muscles originate on the phalanges, and only tendons extend across the distal joints of the fingers.

Table 9-3 summarizes the major muscles of the shoulder and upper limb.

✔ What muscles are you using when you pull your shoulders forward?

✔ Baseball pitchers sometimes suffer from rotator cuff injuries. What muscles are involved in this type of injury?

✔ Injury to the flexor carpi ulnaris would impair what two movements?

Muscles of the Hip and Lower Limb

The muscles of the lower limb can be divided into three groups: (1) muscles that move the thigh, (2) muscles that move the leg, and (3) muscles that affect the ankles and feet.

Table 9-3	Muscles of the Shoulder and Upper Limb	
Region/Muscle	**Location**	**Action**
Shoulder		
Trapezius	Upper back, back of head and neck, to clavicle and scapula	Varied movements of scapula; elevates clavicle; extends head and neck
Serratus anterior	Between ribs and scapula	Pulls shoulder forward (protraction)
Muscles That Move the Arm		
Deltoid	Tip of shoulder to humerus	Raises (abducts) arm
Latissimus dorsi	Middle of back to humerus	Lowers (adducts) arm and extends shoulder
Pectoralis major	Upper chest to humerus	Flexes shoulder and adducts arm
Infraspinatus	Shoulder blade of scapula to humerus	Lateral rotation of humerus
Teres minor	Posterior, inferior edge of scapula to humerus	Lateral rotation of humerus
Teres major	Posterior, inferior edge of scapula to humerus	Medial rotation of humerus
Muscles That Move the Forearm		
Biceps brachii	Anterior surface of arm	Flexes the elbow
Brachioradialis	Extends from lower humerus and adjacent forearm to lateral tip of radius	Flexes the elbow
Triceps brachii	Posterior surface of arm	Extends elbow; opposes biceps brachii
Muscles That Move the Hand and Fingers		
Flexor carpi muscles	Anterior side of forearm	Flex wrist
Extensor carpi muscles	Posterior side of forearm	Extend wrist
Flexor and extensor digitorum muscles	Anterior and posterior sides of the forearm tendons extending to bones of the fingers	Flex and extend the fingers at the interphalangeal joints

Table 9-4	Muscles of the Lower Limb	
Group/Muscle	**Location**	**Action**
Muscles That Move the Thigh		
Gluteus maximus	Large buttock muscle	Extends hip
Gluteus medius	Under the gluteus maximus between hip and upper femur	Abducts thigh
Adductors	Medial (inner) region of thigh	Adduct thigh and flex hip
Gracilis	Medial (inner) region of thigh	Adducts thigh and flexes knee
Iliopsoas	Anterior, superior thigh	Flexes hip
Muscles That Move the Leg		
Hamstring Muscles 　Biceps femoris 　Semimembranosus 　Semitendinosus	Posterior surface of thigh; extends from lower pelvis to tibia and fibula	Flexors of the knee
Sartorius	Crosses anterior thigh; extends from pelvis to tibia	Flexes hip and knee

The following muscles that move the thigh may be located in Figure 9-5•, pp. 195–196:

■ **Gluteal muscles** cover the lateral surface of the ilia. The **gluteus maximus** is the largest and most posterior of the gluteal muscles, which extend the hip and rotate and abduct the thigh.

■ The adductors of the thigh include the **adductor magnus**, the **adductor longus**, and the **gracilis** (GRAS-i-lis). When an athlete suffers a *pulled groin*, the problem is a strain in one of these adductor muscles.

■ The largest flexor of the hip is the **iliopsoas** (il-ē-ō-SŌ-us) muscle. It inserts at the greater trochanter of the femur.

The general pattern of muscle distribution in the lower limb is that extensors are found along the anterior and lateral surfaces of the limb, and flexors lie along the posterior and medial surfaces.

■ The flexors of the knee include three muscles collectively known as the *hamstrings* (the **biceps femoris**, the **semimembranosus** (sem-ē-mem-bra-NŌ-sus), and the **semitendinosus** (sem-ē-ten-di-NŌ-sus) and the **sartorius** (sar-TŌR-ē-us).

■ Collectively the extensors of the knee are known as the **quadriceps femoris**. The three **vastus** muscles (only two are shown in Figure 9-5•) and the **rectus femoris** insert on the patella, which is attached to the tibia by the *patellar ligament*.

Most of the muscles that move the ankle produce the plantar flexion involved with walking and running.

■ The large **gastrocnemius** (gas-trok-NĒ-mē-us) of the calf is assisted by the underlying **soleus** (SŌ-lē-us) muscle. These muscles share a common tendon, the **calcaneal tendon**, or *Achilles tendon*.

■ The **peroneus** (per-Ō-nē-us) muscle produces eversion as well as plantar flexion, as when you twist the sole of your foot outward or point your toes.

■ The large **tibialis** (tib-ē-A-lis) **anterior** opposes the gastrocnemius and soleus, and dorsiflexes the foot, as when you "dig in your heels" and pull on a rope while playing tug-of-war.

Table 9-4	Muscles of the Lower Limb (*continued*)	
Group/Muscle	**Location**	**Action**
Quadriceps Muscles Rectus femoris Vastus lateralis Vastus medialis	Anterior surface of thigh; extends from lower pelvis to tibia and fibula	Extensors of the knee
Muscles That Move the Foot		
Tibialis anterior	Anterior and lateral side of tibia; extends to foot	Dorsiflexion of foot ("digging in the heels")
Gastrocnemius	Posterior leg ("calf") between femur and heel	Flexes knee, plantar flexes foot ("on tiptoe")
Peroneus	Posterior leg; extends from fibula and tibia to foot	Plantar flexes and everts foot (turns sole outward)
Soleus	Posterior leg; extends from fibula and tibia to heel	Plantar flexes and inverts foot (turns sole inward)
Flexor and extensor digitorum muscles (several)	Anterior and posterior leg; extends to bones of toes	Flex and extend toes at interphalangeal joints

Important muscles that move the toes originate on the surface of the tibia, the fibula, or both. Table 9-4 notes four muscles that move the toes, although only the relatively superficial **extensor digitorum** is shown in Figure 9-5a•. The other digital muscles lie deep within the leg. Several smaller muscles involved with fine movement of the toes originate on the tarsal and metatarsal bones.

✔ You often hear of athletes suffering a "pulled hamstring." To what does this phrase refer?

✔ How would you expect a torn calcaneal tendon to affect movement of the foot?

AGING AND THE MUSCULAR SYSTEM

As the body ages, there is a general reduction in the size and power of all muscle tissues. The effects on the muscular system can be summarized as follows:

1. *Skeletal muscle fibers become smaller in diameter.* The overall effect is a reduction in muscle strength and endurance, and a tendency to fatigue more rapidly.

2. *Skeletal muscles become smaller and less elastic.* Aging skeletal muscles develop increasing amounts of fibrous connective tissue, a process called *fibrosis.* Fibrosis makes the muscle less flexible, and the collagen fibers can restrict movement and circulation.

3. *The tolerance for exercise decreases.* A lower tolerance for exercise results in part from the tendency for rapid fatigue and in part from a decreased ability to control body temperature, which leads to overheating.

4. *The ability to recover from muscular injuries decreases.* When an injury occurs, repair capabilities are limited, and scar tissue formation is the usual result.

The rate of decline in muscular performance is the same in all individuals, regardless of their exercise patterns or lifestyle. Therefore to be in good shape late in life, one must be in *very* good shape early in life. Regular exercise helps control body weight, strengthens bones, and generally improves the quality of life at all ages. Extremely demanding exercise is not as important as regular exercise. In fact, extreme exercise in the elderly may lead to problems with tendons, bones, and joints. Although it has obvious effects on the quality of life, there is no clear evidence that exercise prolongs life expectancy.

MUSCULAR SYSTEM DISORDERS

Skeletal muscles contract only under the command of the nervous system. For this reason, observation of muscular activity not only provides direct information about the muscular system, but indirect information about the nervous system as well. Two common symptoms of muscular disorders are weakness and pain in the affected skeletal muscles. Because the muscular and nervous systems work together so closely, these symptoms can have a variety of causes. Muscular system disorders can be classified as *primary* (resulting from problems within the muscular system) or *secondary* (resulting from problems with the nervous system or, less commonly, with other systems). Figure 9-10• shows the major sources of primary and secondary muscular disorders that can produce either or both of these symptoms.

SIGNS AND SYMPTOMS OF MUSCLE DISORDERS

Signs

When a physical exam is underway, *signs* (aspects of a disorder that can be observed) from facial expressions, posture, speech, and the manner of walking or moving, can provide important information. Muscle weakness, a common sign of muscle disease, can be checked by applying an opposite force against a specific action. For example, an examiner might exert a gentle extending force on a patient's forearm while asking the patient to flex the elbow. Important signs of muscle disorders include the following:

- A **muscle mass**, or an abnormal dense region, may sometimes be seen or felt within a skeletal muscle. A muscle mass may result from torn muscle tissue, a parasitic infection, or bone deposited within the muscle.

•**FIGURE 9-10**
Some Causes of Muscular System Disorders.

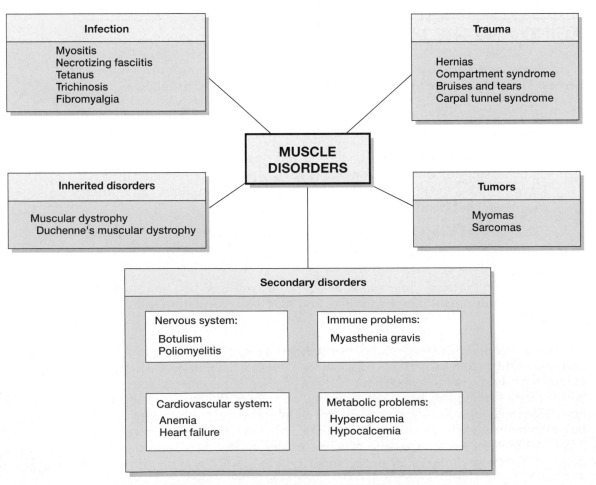

- Abnormal contractions may indicate problems with the muscle tissue or its connection with the nervous system. A **muscle spasm** is a sudden, strong, involuntary contraction. An example is *hiccups*, which occurs when the diaphragm goes into spasm.

- **Muscle flaccidity** exists when the relaxed skeletal muscle appears soft and loose, and its contractions are very weak or absent.

- **Ptosis** (TŌ-sis) is a drooping of the upper eyelid. It may be seen in some muscle disorders, such as *myasthenia gravis* (discussed below), or following damage to the cranial nerve that stimulates a muscle of the eyelid.

- **Muscle atrophy** is skeletal muscle deterioration, or *wasting*, due to disuse, immobility, or interference with the normal links between nerve cells and muscle cells. For example, it is the loss of motor neurons that causes muscle atrophy following infection by the polio virus.

- Abnormal patterns of muscle movement, such as *tics* or *tremors*, and **muscular paralysis** are generally caused by nervous system disorders.

Symptoms

A common symptom of muscle disorders is pain. (Remember, symptoms are an individual's subjective perception of a change in normal body function.) The potential causes of muscle pain include the following:

- *Muscle trauma*: A **trauma** is a wound or injury. Examples of injuries to a skeletal muscle include a laceration, a deep bruise or crushing injury, a muscle tear, and a damaged tendon.

- *Muscle infection*: Skeletal muscles may be infected by viruses, bacteria, or colonized by parasitic worms. These infections generally produce pain that is restricted to the involved muscles. Diffuse muscle pain may develop in the course of other infectious diseases, such as influenza or measles.

- *Related problems with the skeletal system*: Muscle pain may result from skeletal problems, such as arthritis or a sprained ligament.

- *Problems with the nervous system*: Muscle pain may be experienced due to inflammation of sensory neurons or stimulation of pain pathways in the central nervous system.

Table 9-5 provides information about some of the important diagnostic procedures and tests for disorders of the muscular system.

PRIMARY DISORDERS

Trauma

There are many different types of muscle trauma, or injury. Some, like *hernias, compartment syndromes,* and activity-related *sports injuries*, are the immediate result of severe stresses. Other traumatic injuries, such as *carpal tunnel syndrome*, are the result of long-term or repetitive exposure to relatively minor stresses.

Hernias. Forceful contractions of the abdominal muscles can produce high pressures within the abdominopelvic cavity. Those pressures can produce a *hernia*. A **hernia** develops when a visceral organ protrudes in an abnormal way through an opening in a muscular wall. Two types of hernias are *inguinal (groin) hernias* and *diaphragmatic hernias*. Late in the fetal development of males, the testes descend into the scrotum by passing through the abdominal wall at the inguinal canals. In adult males, the inguinal canals serve as passageways for the spermatic cords to reach the testes. The spermatic cords contain the sperm ducts and associated blood vessels. In an inguinal hernia (Figure 9-11•) the

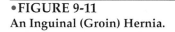

•**FIGURE 9-11**
An Inguinal (Groin) Hernia.

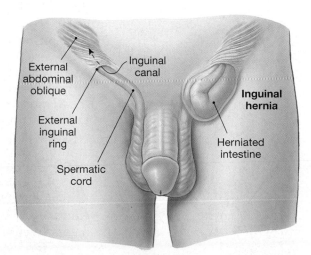

External
abdominal
oblique

Inguinal
canal

External
inguinal
ring

**Inguinal
hernia**

Spermatic
cord

Herniated
intestine

Table 9-5	Examples of Procedures and Tests Used in the Diagnosis of Muscle Disorders	
Diagnostic Procedure	**Method and Result**	**Representative Uses**
Muscle biopsy	Removal of a small amount of affected muscle tissue	Determination of muscle disease; also used to detect cyst formation or larvae in trichinosis
Electromyography	Insertion of a probe that transmits information to an instrument that measures the electrical activity in contracting muscles	Abnormal EMG readings occur in disorders such as myasthenia gravis, amyotrophic lateral sclerosis (ALS), and muscular dystrophy
MRI	Standard MRI	Useful in the detection of muscle diseases and associated soft tissue damage
Laboratory Test	**Normal Values in Blood Plasma or Serum**	**Significance of Abnormal Values**
Aldolase	Adults: less than 8 U/l 22–59 mU/l (SI units)	Elevated levels occur in muscular dystrophy but not in myasthenia gravis or multiple sclerosis
Aspartate aminotransferase	Adults: 7–40 U/l Children: 15–55 U/l	Elevated levels occur in some muscle diseases
Creatine phosphokinase (CPK, or CK)	Adults: 30–180 IU/l	Elevated levels occur in muscular dystrophy and myositis
Lactate dehydrogenase (LDH)	Adults: 100–190 U/l	Elevated levels occur in some muscle diseases and in lactic acidosis
Electrolytes Potassium	Adults: 3.5–5.0 mEq/l	Decreased levels of potassium can cause muscle weakness
Calcium	Adults: 8.5–10.5 mg/dl	Decreased calcium levels can cause muscle tremors and tetany; increased levels can cause muscle flaccidity

inguinal canal enlarges, and the abdominal contents such as a mesentery, small intestine, or (more rarely) the urinary bladder are forced into the inguinal canal. If the herniated structures become trapped or twisted, surgery may be required to prevent serious complications. Inguinal hernias are not always caused by unusually high abdominal pressures. Injuries to the abdomen or inherited weakness of the canal may have the same effect.

The esophagus and major blood vessels pass through an opening (or *hiatus*) in the diaphragm, the voluntary muscle that separates the thoracic and abdominopelvic cavities. In a diaphragmatic hernia, also called a *hiatal hernia* (hī-Ā-tal), abdominal organs slide into the thoracic cavity, typically through the esophageal hiatus, the opening used by the esophagus. The severity of the condition will depend on the location and size of the herniated organ(s). When clinical complications develop, they generally occur because abdominal organs that have pushed into the thoracic cavity are exerting pressure on structures there. As is the case with inguinal hernias, a diaphragmatic hernia may result from elevated abdominal pressure, congenital factors (those existing at birth) or from an injury that weakens or tears the diaphragmatic muscle.

Compartment Syndromes. In the limbs, groups of skeletal muscles are isolated in compartments formed by sheets of dense connective tissue. Blood vessels and nerves traveling to specific muscles within the limb enter and branch within the appropriate compartments. When a crushing injury, severe contusion, or muscle strain occurs, the blood vessels within one or more compartments may be damaged. These compartments then become swollen with blood and fluid leaked from damaged vessels. Because the connective tissue partitions are very strong, the accumulated fluid cannot escape, and pressure rises within the affected compartments. Eventually compartment pressures may become so high that they compress nearby blood vessels and eliminate the circulatory supply to the muscles and nerves of the compartment. This compression produces a condition of **ischemia** (is-KĒ-mē-uh), or "blood starvation," known as **compartment syndrome**.

Slicing into the compartment along its longitudinal axis or implanting a drain are emergency measures used to relieve the pressure. If such steps are not taken, the contents of the compartment will suffer severe damage. Nerves in the affected compartment will be destroyed after 2–4 hours of ischemia, although they can regenerate to some degree if the circulation is restored. After 6 hours or more, the muscle tissue will also be destroyed, and no regeneration will occur. The muscles will be replaced by scar tissue, and shortening of the connective tissue fibers may result in *contracture,* a permanent contraction of an entire muscle following the atrophy of individual muscle fibers.

Sports Injuries. **Sports injuries** affect people involved in contact sports, such as football, and other, less aggressive activities such as jogging. Muscles and bones respond to increased use by enlarging and strengthening. Poorly conditioned individuals are therefore more likely than people in good condition to subject their bones and muscles to intolerable stresses.

A partial list of activity-related conditions includes the following:

- **Bone bruise:** Bleeding within the periosteum that covers a bone
- **Bursitis:** Inflammation of the bursae (fluid-filled sacs) around one or more joints
- **Muscle cramps**: Prolonged, involuntary, and painful muscular contractions
- **Myositis:** Inflammation of a skeletal muscle
- **Sprains**: Tears or breaks in ligaments or tendons
- **Strains**: Tears in muscles
- **Stress fractures**: Cracks or breaks in bones subjected to repeated stresses or trauma
- **Tendinitis:** Inflammation of the connective tissue within or around a tendon

Carpal Tunnel Syndrome. **Carpal tunnel syndrome** is a common example of a group of disorders caused by repetitive movements of the arms, hands, and fingers. Carpal tunnel syndrome results from inflammation and swelling of the tendon sheath surrounding the flexor tendons of the palm. Such an inflammation is called **tenosynovitis**. The swelling compresses the *median nerve* where it passes across the anterior surface of the wrist between the tendons and the carpal bones. Symptoms include pain, especially when flexing the wrist, and a tingling sensation or numbness on the palm and the first three fingers. This condition often strikes those engaged in typing, working at a computer keyboard, or playing the piano. Treatment involves administration of anti-inflammatory drugs such as aspirin, injection of anti-inflammatory agents, such as glucocorticoids (steroid hormones produced by the adrenal cortex), and use of splints to prevent wrist flexion and to stabilize the region.

Infection

Muscle infection results in inflammation (myositis) and eventual breakdown in the structure of skeletal muscles. We will consider only a few representative examples: *necrotizing fasciitis,* a bacterial infection characterized by a breakdown in the connective tissues of skeletal muscles; *tetanus,* caused by an infection from soil bacteria; *trichinosis,* characterized by the colonization of muscles by parasites; and *fibromyalgia,* a muscle disorder that may be linked to viral infections.

Several bacteria, such as the *streptococci,* produce enzymes that break down connective tissues. **Streptococcus A** bacteria are involved in many human diseases, most notably "strep throat," an infection of the pharynx. In most cases the immune response is sufficient to contain and ultimately defeat these bacteria before extensive tissue damage has occurred.

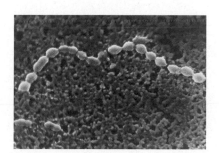

Streptococcal bacteria.

However, in 1994 tabloid newspapers had a field day recounting stories of "killer bugs" and "flesh-eating bacteria" that terrorized residents of Gloucester, England. The details were horrific—minor cuts become major open wounds, with interior connective tissues dissolving. Although in fact there were only seven reported cases, five of the victims died. The pathogen responsible was a strain of Streptococcus A that overpowered immune defenses and swiftly invaded and destroyed soft tissues. Moreover, the pathogens eroded their way along the connective tissue wrapping, or *fasciae* (FASH-ē-ē), that covers skeletal muscles and other organs. The term for this condition is **necrotizing fasciitis**. (*Necrosis* refers to the death of cells due to disease or injury.)

Tetanus is an infectious disease caused by a common soil bacteria, *Clostridium tetani* (klō-STRID-ē-um TE-tan-ē). When children are told to be careful of rusty nails, parents are not worrying about the rust or the nail, but infection by this bacteria. *C. tetani* thrives in environments with low oxygen, and a deep puncture wound (as from a nail) carries a much greater risk of tetanus than a shallow, open cut that bleeds freely. When introduced into the body by a break in the skin, the bacteria release a toxin that spreads and interferes with motor neuron activity. The result is sustained, powerful involuntary contractions of skeletal muscles throughout the body.

After exposure to the bacteria, symptoms generally develop in less than 2 weeks. The most common complaints are headache, muscle stiffness, and difficulty swallowing. Because it soon becomes difficult to open the mouth due to jaw muscle spasms, this disease is also called *lockjaw*. Severe tetanus has a 40–60 percent mortality rate. Immunization is effective in this disease and only about 100 cases occur each year in the U.S. Childhood immunization is given routinely in the U.S. in the *DPT vaccination* (*D*iphtheria, *P*ertussis (whooping cough), and *T*etanus) and tetanus booster shots are recommended every 10 years.

Trichinosis (trik-i-NŌ-sis) results from infection by the parasitic nematode (roundworm) *Trichinella spiralis*, whose life history is diagrammed in Figure 9-12●. Symptoms include diarrhea, weakness, and muscle pain. These symptoms are caused by the invasion of larval worms into skeletal muscle. The parasites create small pockets within its connective tissue layers. Muscles of the tongue, eyes, diaphragm, chest, and lower limbs are most often affected. Larvae are common in the flesh of pigs, horses, dogs, and other mammals. The larvae are killed by cooking, and people are most often exposed by eating undercooked pork. Once eaten, the larvae mature within the human intestinal tract, where they mate and produce eggs. The new generation of larvae then migrates through the body tissues

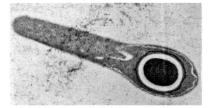

The tetanus bacteria and its spores are common in soil.

●FIGURE 9-12
Life Cycle of the Pork Tapeworm (*Trichinella spiralis*).

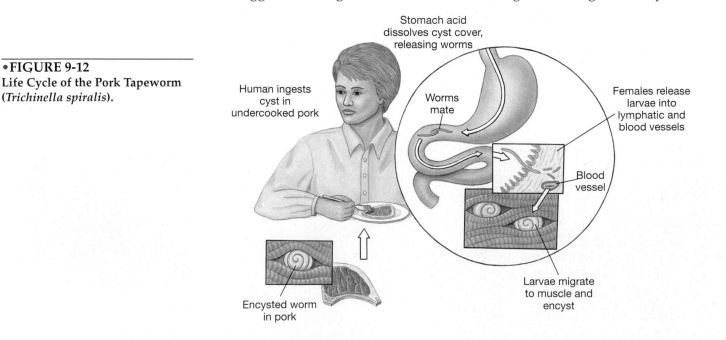

Stomach acid dissolves cyst cover, releasing worms

Human ingests cyst in undercooked pork

Worms mate

Females release larvae into lymphatic and blood vessels

Blood vessel

Larvae migrate to muscle and encyst

Encysted worm in pork

to reach the muscles, where they complete their early development. The migration and subsequent settling produce a generalized achiness, muscle and joint pain, and swelling in infected tissue.

Myalgia (mī-AL-jē-uh) refers to muscular pain (*-algia* means "pain") and is a common symptom of a wide variety of infections. **Fibromyalgia** is an inflammatory musculoskeletal (muscles, joints, and ligaments) disorder characterized by a distinctive pattern of symptoms, but not muscle weakness. Fibromyalgia may be caused by a virus, because pain and an absence of muscle weakness are symptoms of other viral diseases.

Symptoms of fibromyalgia include chronic aches, pain, and stiffness, and several tender points at specific, characteristic locations. The four most common tender points are (1) the inside of the knee, (2) the lateral or outer side of the elbow, (3) the lower back of the head, and (4) the upper chest at the junction between the second rib and its costal cartilage. An additional clinical criterion is that the pains and stiffness cannot be explained by other mechanisms. Patients suffering from this condition invariably report chronic fatigue; they sleep poorly and feel tired on awakening.

Treatment at present is limited to relieving symptoms. For example, anti-inflammatory medications may help relieve pain, other drugs can be used to promote sleep, and exercise programs help maintain normal range of motion.

Tumors

Primary tumors of muscle tissue may be benign or malignant. **Myomas** are benign tumors of muscle tissue; more specifically, *leiomyomas* form in smooth muscle and *rhabdomyomas* form in skeletal muscle. Malignant, cancerous tumors are called **sarcomas**; *leiomyosarcomas* (smooth muscle) and rhabdomyosarcomas (skeletal muscle). Secondary tumors rarely develop in muscle tissue.

Inherited Disorders

There are five different **muscular dystrophies** (DIS-trō-fēz) and all are rare, inherited diseases that produce progressive muscle weakness and deterioration. The most common and best understood is **Duchenne's muscular dystrophy (DMD)**. This form appears in childhood, commonly between the ages of 3 and 7. The condition generally affects only males; the incidence is roughly 30 per 100,000 male births. A progressive muscular weakness develops, and most individuals die before age 20 due to respiratory paralysis. Skeletal muscles are primarily affected, although for some reason the facial muscles continue to function normally. In later stages of the disease, the facial muscles and cardiac muscle tissue may also become involved.

The inheritance of DMD is sex-linked: Women carrying the defective genes are unaffected, but each of their male children will have a 50 percent chance of developing DMD. Now that the specific location of the responsible gene has been identified, it is possible to determine whether or not a woman is carrying the defective gene. It is also possible to use a prenatal test to determine if a fetus has this condition.

✔ High abdominal pressures may lead to what kinds of injuries?

✔ Myalgia is a common symptom of a wide variety of infections. What is myalgia?

SECONDARY DISORDERS

Problems in other body systems may disrupt the normal functioning of the muscular system. For example, damage to nerve cells that innervate skeletal muscles will result in paralysis of muscle fibers, even though those fibers may be otherwise normal. Muscle contraction may also become abnormal if circulatory or respiratory problems deprive skeletal muscles of oxygen. We will consider three examples of secondary muscle disorders: *botulism*, caused by a bacterial toxin which interferes with neuromuscular junctions; *polio*, which results from a viral infection of nerve cells; and *myasthenia gravis*, a muscle disorder that results from a malfunction of the immune system.

Nervous System Problems

Botulism is caused by a toxin produced by *Clostridium botulinum* (bot-ū-LĪ-num), an anaerobic, soil bacteria closely related to the tetanus bacteria. ∞ *[p. 210]* The *botulinus toxin* prevents the release of neurotransmitter at neuromuscular junctions, and, as a result, produces a severe and potentially fatal paralysis of the skeletal muscles. The botulism bacteria does not grow in the body but can grow (and produce toxin) in the low oxygen environment of honey or a sealed jar or can. *C. botulinum* can grow in improperly-preserved foods such as home-preserved vegetables, canned tuna or beets, smoked fish, and cold soups.

Symptoms usually begin 12–36 hours after eating a contaminated meal that has not been heated enough to break down the toxin. Symptoms are often vision-related, such as seeing double or a painful sensitivity to bright lights. These symptoms are followed by slurred speech and an inability to stand or walk. The major risk of botulinus poisoning is respiratory paralysis and death by suffocation. Mortality rates up to 70 percent occur in untreated cases, but treatment with respiratory support and an antitoxin reduces that figure. The overall mortality rate in the United States is about 10 percent. Small amounts of botulinus toxin have been injected into spasming muscles for the relief of otherwise untreatable facial tics.

Poliomyelitis (pō-lē-ō-mī-e-LĪ-tis), or simply **polio**, is an infectious disease caused by *polioviruses*. Although there are no clinical symptoms in roughly 95 percent of infected individuals, others develop a flu-like illness, or a brief *meningitis* (an inflammation of the protective membranes that surround the central nervous system). In other cases, the virus attacks motor neurons in the central nervous system.

In this third form of the disease, the individual develops a fever 7–14 days after infection. The fever subsides but recurs roughly a week later, accompanied by muscle pain, cramping, and muscle paralysis of one or more limbs. If the neurons that control respiration are infected respiratory paralysis and death may also occur. The survivors of paralytic polio may develop progressive muscular weakness 20–30 years after the initial infection. This *postpolio syndrome* is characterized by fatigue, muscle pain and weakness, and, in some cases, muscular atrophy. There is no cure for this condition, although rest seems to help.

Polio has been almost completely eliminated from the U.S. population due to a successful immunization program that began in the 1950s. However, it is still a serious risk for non-vaccinated individuals in other parts of the world.

The vaccine used for years has been the oral *Sabin vaccine*, preferred for ease of administration and longer-lasting immune stimulation than the injectable vaccine. As polio becomes rarer, the injectable vaccine may be preferred due to fewer side effects.

Immune Problems

Myasthenia gravis (mī-as-THĒ-ne-uh GRA-vis) is characterized by a general muscular weakness that tends to be most pronounced in the muscles of the arms, head, and chest. The first symptom is generally a weakness of the eye muscles and drooping eyelids (*ptosis*). Facial muscles are commonly weak as well, and the individual develops a peculiar smile that resembles a snarl. As the disease progresses, weakness in the jaw and throat muscles causes problems with chewing and swallowing, and as the neck muscles weaken it becomes difficult to hold the head upright.

The muscles of the upper chest and arms are affected next. All the voluntary muscles of the body may ultimately be involved. Severe myasthenia gravis produces respiratory paralysis, with a mortality rate of 5–10 percent. However, the disease does not always progress to such a life-threatening stage. For example, roughly 20 percent of people with the disease experience no other symptoms other than eye problems.

The primary cause of myasthenia gravis appears to be a malfunction of the immune system. The problem develops when the immune response attacks the neuromuscular junctions (specifically, the neurotransmitter receptors of the motor end plate) as if they were foreign substances. For unknown reasons, 1.5 times as many women than men are affected. The typical age at onset is 20–30 for women, versus over 60 for men. Estimates of the incidence of this disease in the United States range from 2 to 10 cases per 100,000 population.

Cardiovascular Problems

The cardiovascular system delivers oxygen and nutrients to the skeletal muscles and removes carbon dioxide, lactic acid, and heat produced during normal metabolism and periods of active contractions. Interference with this delivery system by blocked blood vessels or low numbers of oxygen-carrying red blood cells (a condition called *anemia*) can result in muscle weakness even though the muscles are structurally sound. *Heart failure* results in reduced circulation and compromised function of muscle tissue throughout the body.

Metabolic Problems

Metabolic problems affecting muscle tissue are often related to disorders of the urinary, digestive, or endocrine systems. Some of the most serious are triggered by changes in the ion concentration of the extracellular fluid that surrounds muscle cells. For example, if calcium levels are elevated (*hypercalcemia*), muscles become excitable, contractions are more powerful and prolonged, and the muscles undergo spasms. In extreme cases, the cardiac muscle of the heart goes into an extended state of contraction that is usually fatal. When calcium levels are low (*hypocalcemia*), muscle contractions weaken.

CHAPTER REVIEW

Key Words

aerobic: Requiring the presence of oxygen.

anaerobic: Without oxygen.

aponeurosis: A broad sheet of connective tissue that may serve as the origin or insertion of a skeletal muscle.

atrophy: Wasting away of muscle tissue from lack of use.

fascicle: A small bundle of muscle fibers

glycolysis (glī-KOL-i-sis): An anaerobic process that breaks down glucose to form pyruvic acid and small amounts of ATP; occurs in the cytoplasm, not in the mitochondria.

insertion: In a muscle, the point of attachment to the bone that is more movable.

isometric contraction: A muscular contraction characterized by increasing tension but no change in length.

isotonic contraction: A muscular contraction during which tension increases and then remains stable as the muscle shortens.

lactic acid: Acid produced during anaerobic glycolysis; the amount of lactic acid can build up in muscle tissue and cause fatigue.

motor unit: All the muscle fibers controlled by a single motor neuron.

myofibrils: Collections of thin and thick filaments that lie within skeletal muscle fibers and cardiac muscle cells.

neuromuscular junction: A functional connection between a motor neuron and a skeletal muscle fiber.

origin: In a muscle, the point of attachment to the bone that is less movable.

prime mover: In a muscle group, the muscle primarily responsible for performing a specific action.

sarcomere: The smallest contractile unit of a skeletal muscle cell; lie end to end within myofibrils.

sphincter: Muscular ring that contracts to close the entrance or exit of an internal passageway.

Selected Clinical Terms

carpal tunnel syndrome: A disorder resulting in inflammation and swelling of the tendon sheath surrounding the flexor tendons of the wrist; caused by repetitive movements of the hand, and fingers.

compartment syndrome: A condition of ischemia, or "blood starvation," to one of the muscle compartments in the limbs.

hernia: Abnormal protrusion of a body part or organ through a body cavity wall.

ischemia: An inadequate blood supply to a region of the body.

muscle cramps: Prolonged, involuntary, and painful muscular contractions or spasms.

muscular dystrophies: Inherited muscular disorders that produce progressive muscle weakness and deterioration.

myalgia (mī-AL-jē-uh): Muscular pain.

myasthenia gravis: An immune disorder which results in a general muscular weakness due to the loss of functional neuromuscular junctions.

polio: A viral disease that causes the destruction of motor neurons and is characterized by muscle paralysis and atrophy.

rigor mortis: A state following death during which muscles are locked in the contracted position, making the body extremely stiff.

sprains: Tears or breaks in ligaments or tendons.

strains: Tears in muscles.

tendinitis: Inflammation of the connective tissue within and around a tendon.

Study Outline

INTRODUCTION (p. 187)

1. There are three types of muscle tissue: skeletal muscle, cardiac muscle, and smooth muscle. All muscle tissues share the properties of **excitability**, **contractility**, and **elasticity**. The muscular system includes all the skeletal muscle tissue that can be controlled voluntarily.

SYSTEM BRIEF (p. 187)

1. Skeletal muscles attach to bones directly or indirectly and perform these functions: (1) produce skeletal movement, (2) maintain posture and body position, and (3) maintain body temperature.

THE STRUCTURE OF SKELETAL MUSCLE (pp. 187–189)

1. Layers of connective tissue surround and interconnect each muscle fiber, each bundle of muscle fibers, and the entire muscle. **Tendons** at the ends of muscles attach them to bones. *(Figure 9-1)*

STRUCTURE OF A MUSCLE FIBER (p. 188)

2. A muscle cell contains numerous **myofibrils** that run the length of the cell. The myofibrils contain **thin filaments** (*actin*) and **thick filaments** (*myosin*) that aid in contraction. *(Figure 9-1)*

3. The filaments in each myofibril are organized into short, repeating units called **sarcomeres.**

THE NERVE CELL–MUSCLE CELL LINK (p. 189)

4. Muscles contract in response to signals from the nervous system.

5. Each skeletal muscle fiber is controlled by a neuron at a **neuromuscular junction**. *(Figure 9-2)*

6. Neurotransmitter molecules released at the neuromuscular junction cause an electrical impulse to develop in the muscle fiber that leads to the release of calcium ions.

HOW A MUSCLE FIBER CONTRACTS (pp. 189–190)

1. Sarcomeres are the functional, contracting units of muscle fibers.

SARCOMERE STRUCTURE (p. 189)

2. The relationship between the thick and thin filaments changes as the muscle contracts and shortens. The ends of the sarcomeres move closer together as the thin filaments slide past the thick filaments. *(Figure 9-3)*

SARCOMERE FUNCTION (p. 190)

3. The contraction process involves temporary binding between myosin on the thick filament and actin on the thin filaments. The presence of calcium ions permits the myosin **cross-bridges** to attach to the thin filaments.

4. Muscle contraction requires calcium ions and large amounts of ATP.

THE WORK OF MUSCLE (pp. 190–194)

1. During normal muscular activity, the number of activated muscle fibers determines the tension developed by an entire skeletal muscle.

HOW NERVES AND MUSCLES WORK TOGETHER (p. 190)

2. There is no mechanism to regulate the amount of tension produced in the contraction of an individual muscle fiber: it is either ON (producing tension) or OFF (relaxed). This is known as the **all-or-none principle**.

3. The number and size of a muscle's **motor units** indicate how precisely controlled its movements are.

4. An increase in muscle tension is produced by increasing the number of motor units.

HOW MUSCLES AND BONES WORK TOGETHER (p. 191)

5. Bones and muscles produce body movements through lever action. Leverage in the body usually results in an increase in speed and distance moved, not an increase in strength. *(Figure 9-4)*

6. Each muscle may be identified by its **origin**, **insertion**, and primary action. A muscle that works in opposition to a **prime mover** is called an **antagonist**.

7. Normal activities usually include both **isotonic** and **isometric** contractions.

MUSCLES AND EXERCISE (p. 193)

8. Physical training that is based on **anaerobic exercises** can lead to **hypertrophy** (enlargement) of the stimulated muscles.

9. Aerobic exercises seek to extend the time over which a muscle can continue to contract while supported by mitochondrial activities. Muscle fibers respond by increasing the number of mitochondria and concentrations of **myoglobin**.

10. A **fatigued** muscle can no longer contract, because of changes in pH, a lack of energy, or other problems.

11. Muscle **recovery** begins immediately after a period of muscle activity and continues until conditions inside the muscle have returned to pre-exertion levels. The **oxygen debt** created during exercise is the amount of oxygen used in recovery to restore normal conditions.

12. Inadequate stimulation causes muscles to undergo **atrophy**.

MAJOR MUSCLES OF THE BODY (pp. 194–205)

1. The muscular system includes approximately 700 skeletal muscles that can be voluntarily controlled. *(Figure 9-5; Table 9-1)*

2. The **axial muscles** position the head and spinal column, and move the ribcage. The **appendicular muscles** stabilize or move components of the appendicular skeleton.

THE AXIAL MUSCLES (p. 197)

3. The axial muscles fall into three groups based on location and/or function. These groups include muscles of the (1) head and neck, (2) spine, and (3) trunk and pelvic floor.

4. The muscles of the head and neck include the **orbicularis oris**, **buccinator**, **frontalis** and **occipitalis**, **masseter**, **temporalis**, and **sternocleidomastoid**. *(Figure 9-6; Table 9-2)*

5. The superficial muscles of the spine, or **erector spinae**, extend the spine. The *quadratus lumborum* flexes the spine. *(Figure 9-7; Table 9-2)*

6. The muscular floor of the pelvic cavity is called the **perineum**. The **levator ani** supports the organs of the pelvic cavity, and the **urethral** and **anal sphincters** control the movement of materials through the urethra and anus. *(Figure 9-8; Table 9-2)*

7. The trunk muscles in the thoracic region include the **intercostal** and **transversus abdominis** muscles. The **external intercostals** and the **internal intercostals** are important in breathing movements of the ribs. The **diaphragm** is also important to breathing. *(Figure 9-9; Table 9-2)*

THE APPENDICULAR MUSCLES *(p. 202)*

8. The **trapezius** affects the position of the shoulder, head, and neck. The **serratus anterior** inserts on the scapula and pulls the shoulders anteriorly. *(Figure 9-5; Table 9-3)*

9. The **deltoid** is an important arm abductor. The **teres major**, **infraspinatus**, and **teres minor** rotate the arm. *(Figure 9-5; Table 9-3)*

10. The **pectoralis major** flexes the shoulder, and the **latissimus dorsi** extends it. Both muscles adduct and rotate the arm. *(Figure 9-5; Table 9-3)*

11. The primary actions of the **biceps brachii, brachialis**, and **brachioradialis** flex the elbow, and the **triceps brachii** extends the elbow. The **flexor carpi** muscles cooperate to flex the wrist. They are opposed by the **extensor carpi** muscles. *(Figure 9-5; Table 9-3)*

12. **Gluteal muscles** cover the lateral surface of the ilium. They extend the hip and abduct and rotate the thigh. *(Figure 9-5; Table 9-4)*

13. Three adductors of the thigh are the **adductor magnus, adductor longus**, and **gracilis**. *(Figure 9-5; Table 9-4)*

14. The **iliopsoas** muscle is a powerful flexor of the hip. *(Figure 9-5; Table 9-4)*

15. The flexors of the knee include the **gracilis, sartorius**, and the *hamstrings*, the **biceps femoris, semimembranosus**, and **semitendinosus**. *(Figure 9-5; Table 9-4)*

16. The knee extensors are known as the **quadriceps femoris**. This group includes the **vastus** muscles and the **rectus femoris**. *(Figure 9-5; Table 9-4)*

17. The **gastrocnemius** and **soleus** muscles produce plantar flexion. The **peroneus** muscle produces eversion as well as plantar flexion. The **tibialis anterior** performs dorsiflexion. *(Figure 9-5; Table 9-4)*

18. Control of the toes is provided by muscles originating on the tibia and fibula and at the tarsal and metatarsal bones. *(Figure 9-5; Table 9-4)*

AGING AND THE MUSCULAR SYSTEM *(p. 205)*

1. The aging process reduces the size, elasticity, and power of all muscle tissues. Exercise tolerance and the ability to recover from muscular injuries both decrease.

MUSCULAR SYSTEM DISORDERS *(pp. 206–213)*

1. Muscular system disorders can be classified as primary (resulting from problems within the muscular system) or secondary (resulting from problems with other body systems). *(Figure 9-10)*

SIGNS AND SYMPTOMS OF MUSCLE DISORDERS *(p. 206)*

2. Signs of muscle disorders include **muscle mass, muscle spasm, muscle flaccidity** (weakness), **ptosis, muscular paralysis**, and **muscle atrophy**.

3. Pain is a common *symptom* of muscle disorders. Possible causes of muscle pain include: *muscle trauma, muscle infection,* and *related problems with the skeletal* and *nervous systems. (Table 9-5)*

PRIMARY MUSCLE DISORDERS *(p. 207)*

4. Primary muscle disorders include trauma (injury), infection, tumors, and inherited disorders.

5. Examples of muscle trauma include: **hernias; compartment syndrome; sports injuries** such as **bone bruises, bursitis, muscle cramps, myositis, sprains, strains, stress fractures**, and **tendinitis**; and, **carpal tunnel syndrome**, or **tenosynovitis**. *(Figure 9-11)*

6. Examples of muscle infections include **necrotizing fasciitis** and **tetanus** (bacterial infections), **trichinosis** (roundworm infection), and **fibromyalgia** (possible viral infection). *(Figure 9-12)*

7. **Myomas** are benign tumors of muscle tissue, and sarcomas are malignant tumors. Secondary tumors rarely develop in muscle tissue.

8. Examples of inherited disorders include different forms of **muscular dystrophy**; all lead to a weakening and deterioration of muscle tissue. **Duchenne's muscular dystrophy (DMD)** is a sex-linked disease and more common in males.

SECONDARY DISORDERS *(p. 211)*

9. **Botulism** is caused by a bacterial toxin which affects neuromuscular junctions and causes muscle relaxation. **Poliomyelitis (polio)** is a viral disease of the motor neurons which stimulate skeletal muscles.

10. **Myasthenia gravis** results from an improper immune response to the body's neuromuscular junctions.

11. The cardiovascular system can produce muscle weakness through a lack of oxygen by low numbers of red blood cells (*anemia*) or heart muscle weakness (*heart failure*).

12. Metabolic problems may cause changes in the concentrations of ions in and around muscles. When calcium levels are elevated (*hypercalcemia*), muscle contractions are more powerful and prolonged. Under conditions of low calcium (*hypocalcemia*), muscle contractions are weak.

Review Questions

MATCHING

Match each item in Column A with the most closely related item in Column B. Use letters for answers in the spaces provided.

	Column A		Column B
___	1. myoglobin	a.	resting tension
___	2. fascicle	b.	binds oxygen
___	3. atrophy	c.	thin filaments
___	4. anaerobic	d.	enlargement
___	5. actin	e.	contractile units
___	6. myosin	f.	sudden, involuntary contraction

___ 7. extensor of the knee

g. thick filaments

___ 8. sarcomeres

h. muscle bundle

___ 9. abductor of the arm

i. inflammation of skeletal muscle

___ 10. muscle tone

j. without oxygen

___ 11. flexor of the knee

k. quadriceps femoris

___ 12. hypertrophy

l. muscle degeneration

___ 13. endoplasmic reticulum

m. deltoid

___ 14. spasm

n. stores calcium ions needed for contraction

___ 15. myositis

o. hamstring muscles

MULTIPLE CHOICE

16. A skeletal muscle contains _____ .
(a) connective tissues
(b) blood vessels and nerves
(c) skeletal muscle tissue
(d) a, b, and c are correct

17. Skeletal muscle contains bundles of muscle fibers called _____ .
(a) fascicles
(b) spindle fibers
(c) myofibrils
(d) sarcomeres

18. The contraction of a muscle will exert a pull on a bone because _____ .
(a) muscles are attached to bones by ligaments
(b) muscles are directly attached to bones
(c) muscles are attached to bones by tendons
(d) a, b, and c are correct

19. The smallest functional units of a muscle fiber are _____ .
(a) actin filaments
(b) myofibrils
(c) sarcomeres
(d) myofilaments

20. Which of the following muscles is involved in the process of biting? _____
(a) masseter
(b) orbicularis oculi
(c) zygomaticus
(d) frontalis

21. If muscle contractions use ATP at or below the maximum rate of ATP generation by mitochondria, the muscle fiber _____ .
(a) produces lactic acid
(b) functions aerobically
(c) functions anaerobically
(d) relies on the process of glycolysis

22. The _____ muscle is involved in breathing.
(a) diaphragm
(b) trapezius
(c) deltoid
(d) rectus abdominis

23. The type of contraction in which the tension rises but the resistance does not move is called _____ .
(a) a reflex action
(b) extension
(c) an isotonic contraction
(d) an isometric contraction

24. The muscular partition separating the thoracic and abdominopelvic cavities is the _____ .
(a) perineum
(b) diaphragm
(c) rectus abdominis
(d) external oblique

25. If a prime mover produces flexion, its antagonist will be _____ .
(a) abduction
(b) adduction
(c) pronation
(d) extension

26. Atrophy of a muscle is characterized by _____ .
(a) an increase in the diameter of its muscle fibers
(b) an increase in the number of its muscle fibers
(c) a reduction in the size of muscle
(d) an increase in tension

27. The neuromuscular junction occurs between a _____ and a _____ .
(a) neuron; skeletal muscle fiber
(b) neuron; cardiac muscle cell
(c) smooth muscle cell; skeletal muscle fiber
(d) neuron; neuron

28. _____ is a parasitic infection of skeletal muscle by a roundworm.
(a) trichinosis
(b) fibromyalgia
(c) myasthenia gravis
(d) hypocalcemia

29. _____ is a muscle disorder in which the muscles become progressively weaker and deteriorate, and is much more common in males than in females.
(a) polio
(b) myoma
(c) myasthenia gravis
(d) Duchenne's muscular dystrophy

TRUE/FALSE

___ 30. Actin filaments extend the entire length of a sarcomere.

___ 31. High calcium levels within the muscle cell cytoplasm inhibit muscle fiber contraction.

___ 32. The origin of a skeletal muscle remains stationary while its insertion moves.

___ 33. Normal resting tension in a skeletal muscle is called muscle tone.

___ 34. Anaerobic exercises will result in extreme hypertrophy of muscle tissue.

___ 35. Compartment syndrome is a type of ischemia that occurs in skeletal muscles.

___ 36. A strain is a tear in a ligament or tendon.

___ 37. The tetanus toxin paralyzes muscles in a relaxed state.

SHORT ESSAY

38. List three functions of skeletal muscle.

39. Describe the three characteristic properties of all muscle tissue.

40. What is the functional difference between the axial musculature and the appendicular musculature?

41. The muscles of the spine include many dorsal extensors but few ventral flexors. Why?

42. What are the effects of aging on the muscular system?

43. What is carpal tunnel syndrome?

44. How are muscle contractions affected by conditions of hypercalcemia and hypocalcemia?

APPLICATIONS

45. While lifting his 8-year-old daughter, Jim experiences a sharp pain in his groin. He can feel a small bulge in his inguinal region. a. What is your diagnosis? b. Describe possible causes of this problem. c. Would you expect it to be more common in men or women, or equally common in both? Explain your answer.

46. Mr. West, 57 years old, has noticed that his eyelids sag, sometimes blocking his vision. He also has occasional difficulty with chewing and swallowing his food. Explain a possible cause for these symptoms.

✔ Answers to Concept Check Questions

(p. 190) **1.** Skeletal muscle appears striated, or striped, under a microscope because it is composed of thin (actin) and thick (myosin) protein filaments arranged in such a way that they produce a banded appearance in the muscle. **2.** You would expect to find the greatest concentration of calcium ions within the endoplasmic reticulum of a resting muscle cell.

(p. 193) **1.** A muscle motor unit with 1500 fibers is most likely from a large muscle involved in powerful, gross body movements. Muscles that control fine or precise movements, such as movement of the eye or the fingers, have only a few fibers per motor unit. **2.** There are two types of muscle contractions, *isotonic* and *isometric*. In an isotonic contraction, tension remains constant and the muscle shortens. In isometric contractions, however, the same events of contraction occur, but instead of the muscle shortening, the tension in the muscle increases. **3.** The origin of a muscle is the end that remains stationary during an action. The insertion of a muscle is the end that moves during an action.

(p. 194) **1.** The sprinter requires large amounts of energy for a relatively short burst of activity. To supply this demand for energy, the muscles switch to anaerobic metabolism. Anaerobic metabolism is not as efficient in producing energy as aerobic metabolism, and the process also produces acidic waste products. The combination of less energy and the waste products contributes to fatigue. Marathon runners derive most of their energy from aerobic metabolism, which is more efficient and does not produce the level of waste products that anaerobic metabolism does. **2.** We would expect activities that require short periods of strenuous activity to produce a greater oxygen debt because this type of activity relies heavily on energy production by anaerobic metabolism. Since lifting weights is more strenuous over the short term, we would expect this type of exercise to produce a greater oxygen debt than swimming laps, which is an aerobic activity.

(p. 198) **1.** Contraction of the masseter muscle raises the mandible, while the mandible depresses when the muscle is relaxed. These movements occur in the process of chewing. **2.** The "kissing muscle" is the *orbicularis oris,* which compresses and purses the lips.

(p. 200) **1.** The *quadratus lumborum* muscles of the spine and the *rectus abdominis* muscles of the trunk act to flex the spine during sit-ups. **2.** The perineum forms the floor of the pelvic cavity.

(p. 203) **1.** When you pull your shoulders forward, you are contracting your *serratus anterior* muscles. **2.** The rotator cuff muscles include the *infraspinatus, teres major,* and *teres minor.* **3.** Injury to the *flexor carpi ulnaris* would impair the ability to flex and adduct the hand.

(p. 205) **1.** The hamstring refers to a group of three muscles that collectively function in flexing the leg. The hamstring muscles are the *biceps femoris, semimembranosus,* and *semitendinosus.* The *sartorius* also aids flexion of the leg. **2.** The Achilles (calcaneal) tendon attaches the *soleus* and *gastrocnemius* muscles to the calcaneus (heel bone). A torn Achilles tendon would make extension of the foot difficult, and the opposite action, flexion, would be more pronounced as a result of less action from the soleus and gastrocnemius.

(p. 211) **1.** Elevated abdominal pressures can lead to inguinal and/or diaphragmatic (hiatal) hernias. **2.** Myalgia is muscle pain.

CHAPTER
10

The Nervous System 1:

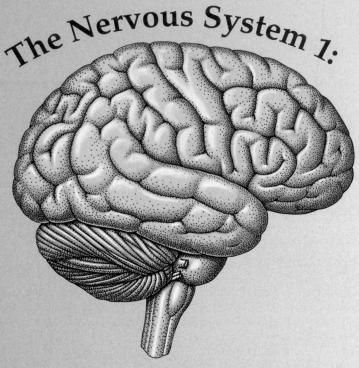

Neurons and the Spinal Cord

CHAPTER OBJECTIVES

1 Describe the overall organization and
   functions of the nervous system.
2 Describe the differences in the structure and
   function of neurons and neuroglia.
3 What is the relationship between neurons and
   nerves?
4 Describe the structure of a synapse and how
   an action potential (nerve impulse) passes
   from one neuron to another.
5 Describe the structure and function of the
   spinal cord.
6 Describe the essential elements of a reflex arc.
7 Describe the structures and functions of the
   sympathetic and parasympathetic divisions of
   the ANS.
8 Describe nervous system disorders associated
   with demyelination.
9 Describe common disorders of the spinal cord
   and spinal nerves.

The nervous system is among the smallest organ systems in terms of body weight, yet it is by far the most complex. Although it is often compared to a computer, it is much more complicated and versatile. For example, the nervous system contains 20 billion information-processing cells, called *neurons*, each of which may receive input from 100,000 different sources! As in a computer, the rapid flow of information and high processing speed depend on electrical activity. Unlike a computer, however, our brains are capable of reworking their own electrical connections as we acquire new information in the process of learning.

Along with the *endocrine system* (discussed in Chapter 13), the nervous system coordinates the body's responses to changing environmental conditions. The nervous system controls relatively swift but brief responses to such changes, whereas the endocrine system usually controls processes that are slower but longer lasting. For example, the nervous system adjusts your body position and moves your eyes across this page, while the endocrine system regulates the daily rate of energy use by your entire body.

This chapter first examines neurons and other cells of the nervous system. It then examines the basic structure of the spinal cord, its nerves, and its vital functions. As you will see, the spinal cord does more than carry information back and forth to the brain. It also acts on its own through preprogrammed actions, or *reflexes*. The brain and its varied functions are considered in the next chapter.

SYSTEM BRIEF

The functions of the nervous system include (1) monitoring the body's internal and external environments, (2) integrating sensory information, and (3) directing or coordinating the responses of other organ systems.

The nervous system is divided into a **central nervous system** (**CNS**) and a **peripheral nervous system** (**PNS**). The CNS consists of two organs, the *brain* and *spinal cord*. The CNS as a whole is responsible for reacting to sensory information and issuing motor commands. Higher functions, such as intelligence, memory, and emotion, are based in the brain. The PNS is made up of the *cranial nerves* and *spinal nerves* that lie outside the CNS.

Figure 10-1 • shows how the different functions of the nervous system are linked together. All communication between the CNS and the rest of the body occurs by way of the PNS. The PNS is divided into two divisions based on the direction of information flow relative to the CNS. Afferent (from the same word root as *ferry*) means "to carry to," and the **afferent division** of the PNS brings sensory information to the CNS. Efferent means "to carry from," and the **efferent division** of the PNS carries motor commands from the CNS to muscles and glands.

Motor commands to skeletal muscles control body movements. The neurons involved form the **somatic nervous**

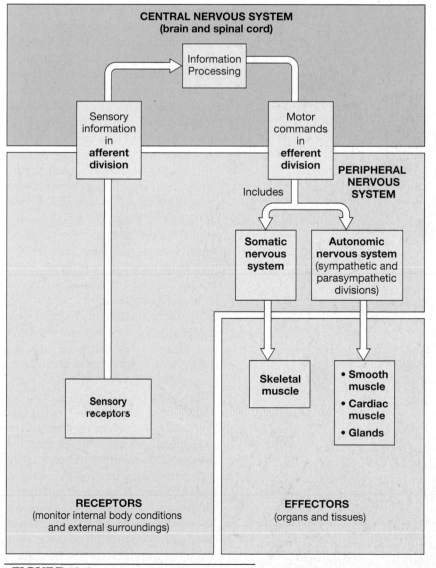

•**FIGURE 10-1**

Organization of the Nervous System.

system. (*Soma* means "body.") Motor commands to smooth muscle, cardiac muscle, and glands control or modify their activities. You do not have conscious, voluntary control over these motor commands, which are issued at the subconscious level. The neurons involved with these motor commands form the **autonomic nervous system** (**ANS**). The ANS is further divided into *sympathetic* and *parasympathetic divisions*. The structure and function of the ANS and its divisions are discussed at the end of this chapter.

THE STRUCTURE OF NEURONS AND NERVES

Neural tissue, introduced in Chapter 4, consists of two kinds of cells, *neurons* and neuron-supporting cells called *neuroglia*. ∞ *[p. 71]* **Neurons** (*neuro*, nerve) are the basic units of the nervous system. All functions of the nervous system involve communications among neurons and with other cells. The **neuroglia** (noo-RŌ-glē-a; literally, "nerve glue") provide a supporting framework for neurons and protect them by (1) regulating the chemical composition of the extracellular fluid and (2) engulfing dead or invading cells by phagocytosis. ∞ *[p. 45]* Neuroglia, also called *glial cells*, far outnumber the neurons.

NEURONS

Sensory neurons convey information from outside and inside the body to neurons within the brain and spinal cord. These neurons are part of the afferent division of the PNS. Sensory neurons extend between a *receptor* and the CNS (spinal cord or brain). A receptor may be part of a sensory neuron, or it may be a specialized epithelial or connective tissue cell that contacts a sensory neuron. Receptors are sensitive to changes in the environment. Some receptors are sensitive to a variety of stimuli. For example, one receptor might provide sensations of touch, pressure, or pain, depending on the strength of the stimulus. Other receptors are very specialized; for example, the visual receptors in the eyes are sensitive only to light. Typically, when a receptor is stimulated, a sensory neuron carries the information to the brain or spinal cord.

Motor neurons, which form the efferent division of the PNS, carry instructions from the CNS to other tissues, organs, or organ systems. The organs and tissues controlled by motor neurons are called *effectors* because their activities have an effect on the body. Effectors include skeletal muscles, which we can consciously control, and cardiac muscle, smooth muscle, and glands, whose activities are regulated at the subconscious level.

Interneurons are located entirely within the brain and spinal cord. As the name implies (*inter* means "between"), these neurons form connections between other neurons. Interneurons are responsible for the distribution and analysis of sensory information and the coordination of motor commands. These complex processes are known as *integrative functions*.

Figure 10-2• shows a typical motor neuron. Its cell body contains a relatively large, round nucleus with a prominent nucleolus. The cell body also contains the organelles that provide energy and synthesize organic compounds. The numerous mitochondria, ribosomes, and membranes of the rough endoplasmic reticulum (RER) give the neuron's cytoplasm a coarse, grainy appearance and a gray color. Typically, the cell body does not contain centrioles, which are organelles that form the spindle fibers that move chromosomes during cell division. ∞ *[p. 52]* Most neurons lose their centrioles as they mature and are thus unable to divide. That is why most neurons lost to injury or disease cannot be replaced.

Extensions projecting from the cell body include a variable number of **dendrites** and a single large **axon.** The cell membrane of the dendrites and cell body is sensitive to chemical, mechanical, or electrical stimulation. This stimulation can lead to the formation of *nerve impulses*, electrical events associated with the movement of ions across the cell membrane. Once formed, nerve impulses are conducted along the cell membrane of the axon. At the tip of an axon, or at the tips

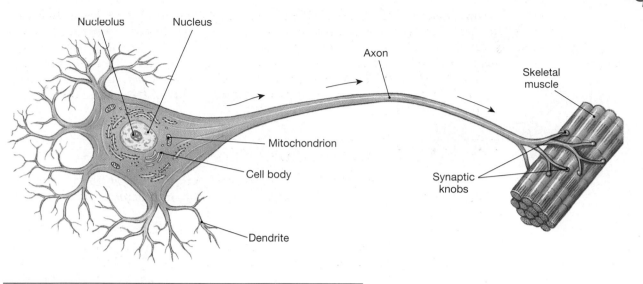

Nucleolus   Nucleus

Axon

Skeletal muscle

Mitochondrion

Cell body

Synaptic knobs

Dendrite

•**FIGURE 10-2**
Structure of a Motor Neuron.
Structure of a motor neuron that innervates a skeletal muscle. The arrows show the direction of movement of nerve impulses.

of its branches, are expanded structures called **synaptic knobs**. A synaptic knob is part of a **synapse**, a site where a neuron communicates with another cell. A neuromuscular junction, described in Chapter 9, is an example of a synapse. ∞ *[p. 188]*

NEUROGLIA

Neuroglia are abundant around and among the neurons. Unlike neurons, neuroglia have no information-processing ability of their own. Their primary role is the protection and support of neurons. Neuroglia are found in both the CNS and PNS. Three basic types of neuroglia in the central nervous system are shown in Figure 10-3•.

1. **Astrocytes** (AS-trō-sīts), from the same Greek root as the word astronomy, are star-shaped cells. They are the largest and most numerous neuroglia. Astrocyte processes form a supporting framework for the CNS. Astrocyte processes also form a blanket around capillaries in the brain. These processes are essential to the maintenance of the *blood–brain barrier*, which keeps chemicals in the general circulation from entering the extracellular fluid of the brain.

   aster, star + *cyte*, cell
   astrocytes: star-shaped cells important for structural support and maintenance of the blood–brain barrier on the CNS

2. **Oligodendrocytes** (o-li-gō-DEN-drō-sītz) wrap pancakelike extensions of their cell membranes around axons. The axon sheath formed by multiple layers of cell membrane is called **myelin** (Figure 10-3•). An axon coated with myelin is said to be *myelinated*. Myelinated axons conduct nerve impulses much faster than do uncoated (*unmyelinated*) ones. Myelin is lipid-rich, and areas of the brain and spinal cord containing myelinated axons are glossy white. These areas make up the **white matter** of the CNS, whereas areas of **gray matter** are dominated by neuron cell bodies.

   oligos, few + *dendron*, tree + *cyte*, cell
   oligodendrocytes: supporting cells within the CNS that coat axons with myelin

3. **Microglia** (mī-KROG-lē-uh) are the smallest of the neuroglia in the CNS. Microglia are phagocytic white blood cells that entered the CNS during its development.

   micro, small + *glia*, glue
   microglia: small phagocytic cells found inside the CNS

   In the PNS (the nervous tissue outside the brain and spinal cord), the most important glial cells are **Schwann cells** (Figure 10-4•). Schwann cells form the myelin that covers the axons of neurons outside the CNS. In creating a myelin sheath, a Schwann cell wraps itself around a segment of a single axon. Areas between adjacent Schwann cells are unmyelinated. These small unmyelinated segments are called *nodes*, and the much larger myelinated areas are called *internodes* because they exist between (*inter-*) the nodes.

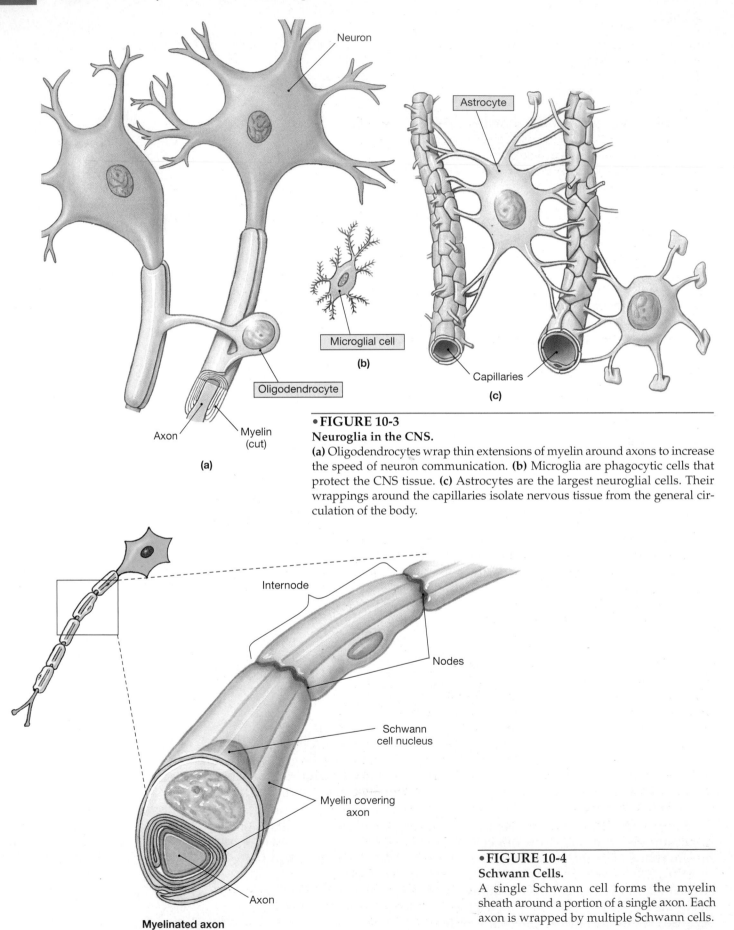

Neuron

Astrocyte

Microglial cell

(b)

Capillaries

(c)

Oligodendrocyte

Axon

Myelin
(cut)

(a)

•FIGURE 10-3
Neuroglia in the CNS.
(a) Oligodendrocytes wrap thin extensions of myelin around axons to increase the speed of neuron communication. **(b)** Microglia are phagocytic cells that protect the CNS tissue. **(c)** Astrocytes are the largest neuroglial cells. Their wrappings around the capillaries isolate nervous tissue from the general circulation of the body.

Internode

Nodes

Schwann
cell nucleus

Myelin covering
axon

Axon

Myelinated axon

•FIGURE 10-4
Schwann Cells.
A single Schwann cell forms the myelin sheath around a portion of a single axon. Each axon is wrapped by multiple Schwann cells.

NERVES

You may already have a basic understanding of a *nerve* as a structure of the nervous system that forms a connection between a part of the body and the brain or spinal cord. For example, if you "pinch a nerve" in your back, you will experience pain there; if you cut a nerve in an accident, you may no longer be able to feel or move some part of your body. In a dissection, nerves are whitish stringlike structures, easily visible to the naked eye. But what *are* nerves exactly?

Nerves are bundles of axons in the PNS. The bundled axons may be those of sensory neurons or motor neurons. Because axons are very long and slender, they are often called *nerve fibers*. Nerves are sheathed in connective tissue, and each nerve contains blood vessels that supply nutrients and oxygen to its axons and Schwann cells. There are a total of 43 pairs of nerves attached to the CNS: 12 pairs of *cranial nerves* connected to the brain, and 31 pairs of *spinal nerves* connected to the spinal cord.

✔ How would damage to the afferent division of the nervous system interfere with communication between the PNS and CNS?

✔ What type of neuroglial cell would you expect to find in large numbers in brain tissue from a person suffering from an infection of the CNS?

NEURON FUNCTION

Neurons are communication specialists. For example, sensory neurons relay information from receptors to the CNS. The distance involved can be considerable; a neuron that monitors pain receptors on the sole of the foot must carry pain sensations to the spinal cord, a distance of over a meter. How does this "information" travel from place to place? In neurons, information travels in the form of nerve impulses that are conducted along axons. To understand how a nerve impulse is triggered and conducted, you must learn a little more about the properties of the cell membrane.

THE MEMBRANE POTENTIAL

The cell membrane acts to regulate the passage of substances into and out of the cell. ∞ *[p. 42]* Because it is not freely permeable to charged particles (such as ions), the cell membrane separates charged particles on the inside of the cell from those on the outside. The extracellular fluid is dominated by sodium ions (Na^+) and chloride ions (Cl^-). The intracellular fluid is dominated by potassium ions (K^+) and negatively charged proteins (Pr^-). In a resting cell, due to differences in relative permeability, there is an excess of positive charges on the outside of the cell membrane and an excess of negative charges on the inside. This uneven distribution of charges on the two sides of the membrane is called a *membrane potential*. Each resting cell has a characteristic membrane potential. When that cell is disturbed in some way, the membrane potential can change—for example, ion channels can open or close, changing the distribution of ions across the membrane.

THE ACTION POTENTIAL

An **action potential** is a sudden change in the membrane potential at one spot on the membrane surface. An action potential develops when special ion channels open, allowing the sudden movement of sodium ions across the cell membrane. So many sodium ions cross the membrane that for a moment the membrane potential reverses—the inner membrane surface contains an excess of positive ions, and the outer surface contains an excess of negative charges. This reversal is short-lived, because the ion channels close almost immediately. Although the influx of sodium ions occurs at a localized spot on the cell membrane, the momentary reversal in membrane potential at that spot opens ion channels in neighboring portions of the membrane. Thus the opening of ion channels at point A leads to opening of channels at point B, and then at point C, and so on, in a chain reaction. In this way, the action potential is conducted across the entire cell membrane.

Any stimulus that opens the special ion channels at one site in the membrane can produce an action potential that will be conducted across the membrane surface. The most common stimulus is the arrival of a chemical that binds to the channel proteins and triggers their opening. As in the contraction of a muscle fiber ∞ *[p. 189]*, an **all-**

✔ Give two examples of cell types capable of conducting an action potential.

✔ How would a chemical that blocks the passage of positively charged sodium ions across the cell membrane affect a neuron?

or-none principle applies to action potentials in neurons: A given stimulus either triggers a full-blown action potential or none at all. All action potentials are of equal strength and duration, regardless of the strength or nature of the stimulus involved.

Action potentials are conducted along the axons of neurons and across the membranes of muscle cells. In a skeletal muscle fiber, an action potential begins at the neuromuscular junction and spreads over the entire membrane surface. In a neuron, an action potential usually begins at the cell body and travels along the length of the axon toward the synaptic knobs. Although the basic mechanism of action potential conduction along unmyelinated and myelinated axons is the same, conduction along myelinated axons is *much* faster (by about five to seven times) because the action potential "jumps" from node to node.

THE SYNAPSE

Every neuron communicates with other neurons or with other types of cells. The communication occurs at synapses. At a synapse, an action potential is relayed from one cell to another. A synapse consists of three basic parts: (1) the **synaptic cleft**, which is the space between the two cells; (2) the synaptic knob of the **presynaptic neuron**, which is the transmitting neuron; and (3) the adjacent portion of the **postsynaptic neuron** (or effector cell), which receives the message. *Pre* means "before" and *post* means "after." Therefore, you might think of the information as existing in the *pre*synaptic neuron *before* it travels across the synaptic cleft, and it reaches the *post*synaptic neuron *after* it has crossed the synaptic cleft.

How does the action potential cross the synapse to activate another cell? Figure 10-5● shows a synapse and the structures involved in transmitting information from one neuron to another. Upon arrival of the action potential at a synaptic knob, the presynaptic neuron releases a chemical called a **neurotransmitter**. Neurotransmitter molecules are stored in vesicles in the synaptic knob, and they are released through the process of exocytosis. ∞ *[p. 45]* The neurotransmitter diffuses across the synaptic cleft until it reaches receptors in the membrane of the receiving (postsynaptic) cell. Neurotransmitter then binds to these receptors. This chemical binding opens ion channels that allow the movement of sodium ions across the postsynaptic cell membrane. This ion movement triggers the formation of an action potential in the postsynaptic cell.

The synaptic cleft contains an enzyme that rapidly inactivates the neurotransmitter. As a result, the ion channels do not remain open for long. Thus the arrival of a single action potential at the synaptic knob results in the generation of a single action potential in the postsynaptic cell membrane.

⚕ Clinical Note

Drugs, Poisons, and Synapses

Many drugs and poisons (toxins) work by interfering with normal synapse functions. These substances may interfere with the synthesis and release of neurotransmitters, prevent the inactivation of the neurotransmitter in the synaptic cleft, or prevent its binding to receptors on the membrane of the target (postsynaptic) cell. For example, the drug Procaine (Novocain™) that is used as a local anesthetic by dentists prevents the stimulation of sensory nerve cells by reducing the movement of sodium ions into the cells. Another common drug is *nicotine*, the active ingredient in cigarette smoke. In low doses, nicotine makes voluntary muscles easier to stimulate, but in high doses it may cause paralysis. Nicotine stimulates postsynaptic cells by binding to receptor sites for the neurotransmitter ACh. Because there are no enzymes to remove the nicotine molecules from the receptor sites, its effects are relatively long lasting.

Toxins released by various organisms cause many types of poisonings. *Botulism* results from eating contaminated canned or smoked food. The toxin, produced by bacteria, prevents the release of ACh at synapses and causes the widespread paralysis of skeletal muscles. The venom of the black widow spider has the opposite effect. It causes a massive release of ACh that produces intense muscular cramps and spasms.

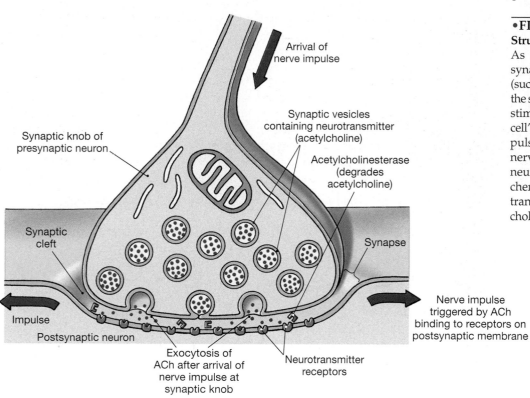

Arrival of
nerve impulse

Synaptic knob of
presynaptic neuron

Synaptic vesicles
containing neurotransmitter
(acetylcholine)

Acetylcholinesterase
(degrades
acetylcholine)

Synaptic
cleft

Synapse

Impulse

Nerve impulse
triggered by ACh
binding to receptors on
postsynaptic membrane

Postsynaptic neuron

Exocytosis of
ACh after arrival of
nerve impulse at
synaptic knob

Neurotransmitter
receptors

•FIGURE 10-5
Structure and Function of a Synapse.
As the nerve impulse reaches the synapse, neurotransmitter molecules (such as acetylcholine) are released into the synaptic cleft. The neurotransmitter stimulates receptors in the receiving cell's membrane to initiate a nerve impulse. The continual production of nerve impulses by the postsynaptic neuron is prevented by the rapid chemical breakdown of the neurotransmitter by an enzyme (acetylcholinesterase, in this example).

✔ What type of chemical is required for an action potential to move from one neuron to another neuron at a synapse?

✔ When a batch of neurotransmitter molecules arrives at the postsynaptic cell membrane, why doesn't it produce an unlimited number of action potentials, rather than just one?

There are many different neurotransmitters. One common neurotransmitter is **acetylcholine**, or **ACh**. ACh is released at neuromuscular junctions. ∞ *[p. 189]* The enzyme responsible for its inactivation is called *acetylcholinesterase*. Another common neurotransmitter is **norepinephrine** (nōr-ep-i-NEF-rin), or **NE**. Also called *noradrenaline*, it is important in the brain and in portions of the autonomic nervous system. Both ACh and NE generally have stimulatory effects on postsynaptic cells. **Dopamine** (DŌ-pah-mēn) and **serotonin** (ser-ō-TŌ-nin) are CNS neurotransmitters whose effects are usually inhibitory.

THE SPINAL CORD AND SPINAL NERVES

The spinal cord serves as the major pathway for impulses to and from the brain. The spinal cord also integrates information on its own, controlling spinal reflexes that occur without any brain involvement.

STRUCTURE OF THE SPINAL CORD

The adult spinal cord extends from the base of the brain to the first or second lumbar vertebrae. It measures approximately 45 cm (18 in.) in length and about 14 mm (1/2 in.) in width. Long, hairlike nerves extend from the lower tip of the spinal cord and form the *cauda equina* (KAW-da ek-WĪ-na), a term that in Latin means "horse's tail" (Figure 10-6a•).

The entire spinal cord consists of 31 segments. Each segment is connected to a pair of **spinal nerves**. One spinal nerve in each pair supplies a portion of the left side of the body, the other a corresponding portion of the right side. The spinal nerves are discussed in the next section.

As Figure 10-6b• shows, the most striking feature of a cross section of the spinal cord is a rough H- or butterfly-shaped area around the narrow central canal. This is the *gray matter* of the spinal cord. The gray matter is composed of the cell bodies of neurons and supporting neuroglial cells. Many of the cell bodies are those of motor neurons whose axons pass out of the spinal cord to join the spinal nerves.

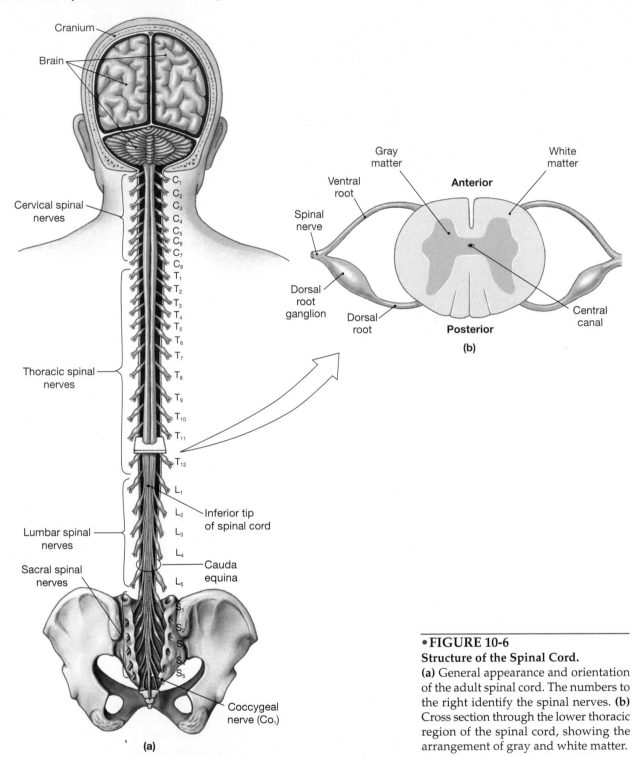

•FIGURE 10-6
Structure of the Spinal Cord.
(a) General appearance and orientation of the adult spinal cord. The numbers to the right identify the spinal nerves. **(b)** Cross section through the lower thoracic region of the spinal cord, showing the arrangement of gray and white matter.

Some are concerned with the motor control of skeletal muscles, and others regulate the activities of smooth muscle, cardiac muscle, and glands. Other cell bodies are those of interneurons, neurons lying entirely within the central nervous system.

The gray matter is surrounded by *white matter*, which contains myelinated and unmyelinated axons. It is the glossy white color of the myelin that gives white matter its name. The axons of the white matter form bundles (called *tracts*) that carry either sensory data or motor commands. **Ascending tracts** carry sensory information toward the brain, and **descending tracts** convey motor commands from the brain into the spinal cord.

✔ What portion of the spinal cord would be affected by a disease that damages myelin sheaths?

THE SPINAL NERVES

The 31 pairs of spinal nerves can be grouped according to where they leave the vertebral canal (see Figure 10-6a•). There are 8 pairs of cervical nerves (C_1–C_8), 12 pairs of thoracic nerves (T_1–T_{12}), 5 pairs of lumbar nerves (L_1–L_5), 5 pairs of sacral nerves (S_1–S_5), and 1 pair of coccygeal nerves (Co_1).

All spinal nerves are classified as **mixed nerves** because they contain the axons of both sensory and motor neurons. However, the motor and sensory fibers are separate inside the vertebral canal, where each spinal nerve divides into two branches. The posterior branch, called the *dorsal root*, contains the axons of sensory neurons that bring information to the spinal cord. The anterior branch, called the *ventral root*, contains the axons of motor neurons that carry commands to muscles or glands. As you can see in Figure 10-6b•, each dorsal root of a spinal nerve is also associated with a *dorsal root ganglion*. A **ganglion** is any collection of neuron cell bodies in the PNS. In a dorsal root ganglion, the cell bodies are those of sensory neurons. The cauda equina consists of spinal nerve roots that extend inferior to the tip of the spinal cord. The spinal nerves they form leave the vertebral canal between vertebrae in the lower lumbar and sacral regions.

Each pair of spinal nerves serves a specific region or area of the body surface, known as a **dermatome**. Dermatomes, shown in Figure 10-7•, are clinically important because damage or infection of a spinal nerve or dorsal root ganglion will produce a characteristic loss of sensation in the skin. For example, in *shingles*, a virus that infects dorsal root ganglia causes a painful rash whose distribution on the skin corresponds to that of the affected sensory nerves.

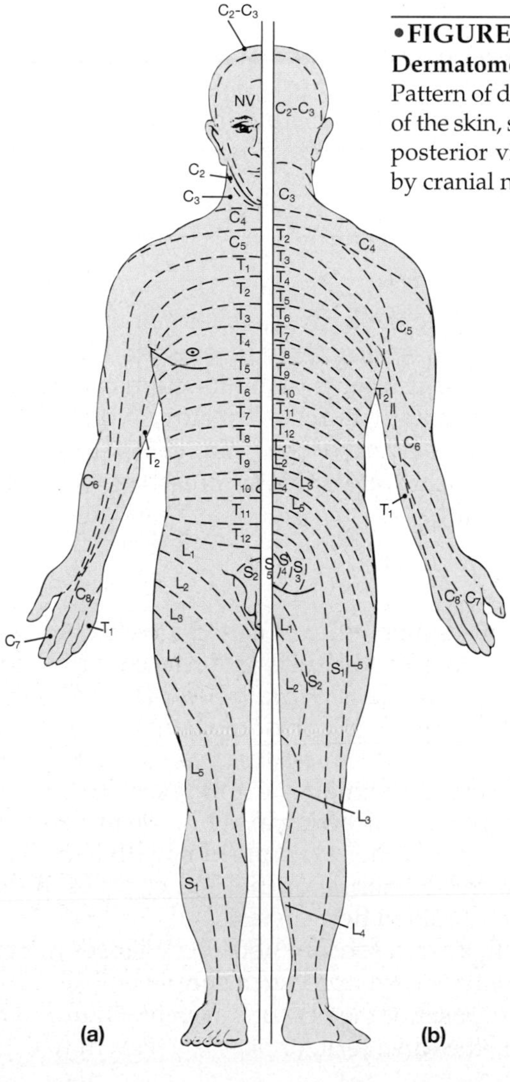

•FIGURE 10-7
Dermatomes.
Pattern of dermatomes on the surface of the skin, seen in **(a)** anterior and **(b)** posterior views. (The face is served by cranial nerves, not spinal nerves.)

Nerve Plexuses and Major Spinal Nerves

Each of the spinal nerves divides to form major and minor branches. The branches and their smaller divisions ensure that the head, neck, trunk, and limbs are innervated by sensory and motor neurons. The *thoracic nerves* and their branches innervate the chest wall. The remaining spinal nerves form networks called **nerve plexuses** (PLEK-sus-ēs; singular: *plexus* [PLEK-sus]) before they divide into smaller branches. (A *plexus* is a Latin term that means "braid.") There are three major plexuses on each side of the body: the *cervical plexus*, the *brachial plexus*, and the *lumbosacral plexus*.

1. The branches of the cervical plexus supply the muscles of the neck, the skin of the neck and back of the head, and the diaphragm muscle, an important muscle in respiration. The major nerve of this plexus is the *phrenic nerve*, which innervates the diaphragm.

2. The brachial plexus innervates the shoulder and upper limb. One major nerve of this plexus is the *median nerve*, which controls flexor muscles and the pronator muscles of the forearm. This nerve travels across the posterior portion of the olecranon process. When you hit your "funny bone," you are really compressing this nerve.

3. The lumbosacral plexus innervates the lower limbs. The largest nerve of this plexus is the *sciatic nerve*. The sciatic nerve passes along the back of the femur and divides into two branches posterior to the knee. Each branch continues on to the foot. The sciatic nerve and its branches control flexion of the knee, plantar flexion of the foot, and flexion of the toes. It also carries sensory information from the skin of the thigh, leg, and foot.

Have you ever experienced a limb that has "fallen asleep"? This is a relatively common example of a peripheral *nerve palsy*. A nerve palsy is characterized by regional losses of sensory and motor function, and it is the result of nerve damage or compression. Mild and temporary palsies can occur after leaning or sitting in an uncomfortable position.

REFLEXES

Conditions inside or outside the body can change rapidly and unexpectedly. **Reflexes** are automatic responses, triggered by specific stimuli, that preserve homeostasis by making rapid adjustments in the functions of organs or organ systems. For example, if you accidentally touch a hot stove, you will pull back your hand before you are consciously aware of the pain. Reflex responses also regulate internal functions such as heart rate or blood pressure and coordinate complex activities such as swallowing.

THE REFLEX ARC

The neural "wiring" of a single reflex is called a **reflex arc**. A reflex arc begins at a receptor and ends at an effector, such as a muscle or gland. Figure 10-8● diagrams the five steps involved in a neural reflex: (1) *arrival of a stimulus and activation of a receptor*, (2) *activation of a sensory neuron*, (3) *processing in the CNS*, (4) *activation of a motor neuron*, and (5) *response by an effector* (muscle or gland).

A reflex response usually removes or opposes the original stimulus. In the case of our stove example, a contracting arm muscle pulled your hand away from a painful stimulus. This is another example of negative feedback. ∞ *[p. 5]* By opposing potentially harmful external or internal changes in the environment, reflexes play an important role in homeostasis.

In the simplest reflex arc, a sensory neuron synapses directly on a motor neuron. Because there is only one synapse, such simple reflexes control the most rapid, stereotyped motor responses of the nervous system. Figure 10-8● shows one of the best known examples, the **knee jerk**, or *patellar reflex*, in which a sharp rap on the

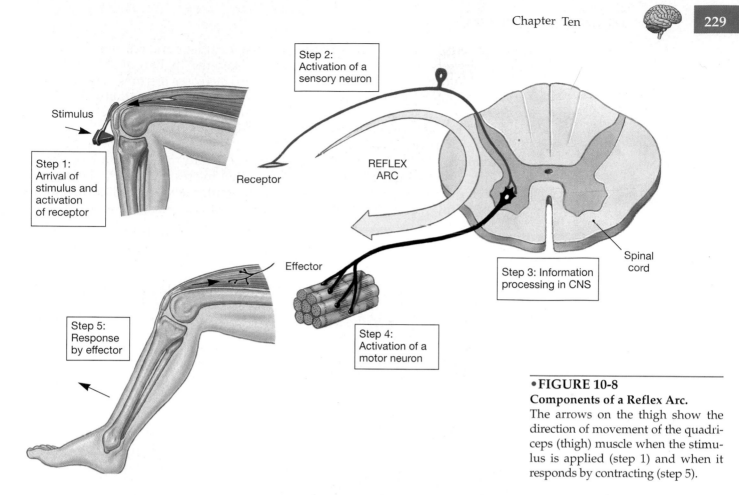

Step 2:
Activation of a
sensory neuron

Stimulus

Step 1:
Arrival of
stimulus and
activation
of receptor

Receptor

REFLEX
ARC

Spinal
cord

Step 3: Information
processing in CNS

Effector

Step 5:
Response
by effector

Step 4:
Activation of a
motor neuron

•FIGURE 10-8
Components of a Reflex Arc.
The arrows on the thigh show the
direction of movement of the quadriceps (thigh) muscle when the stimulus is applied (step 1) and when it
responds by contracting (step 5).

patellar tendon produces a noticeable kick. Many other reflexes may have at least
one interneuron placed between the sensory neuron and the motor neuron. Due
to the larger number of synapses, these reflexes have a longer delay between a
stimulus and response; however, they can produce far more complicated responses. In general, the larger the number of interneurons involved, the more
complicated the response. For example, reflexes involved with maintaining balance depend on interneurons that control and coordinate the activities of many different muscle groups.

✔ What is the minimum number of
neurons needed for a reflex arc?

THE AUTONOMIC NERVOUS SYSTEM

As noted earlier, the efferent division of the PNS carries motor commands issued
at both the conscious and subconscious levels. The autonomic nervous system, or
ANS, regulates the activities of smooth muscle, cardiac muscle, and glands. The
regulation occurs at the subconscious level, outside of our conscious awareness and
control. Nevertheless, the ANS is responsible for many of the activities of the nervous system that ensure our immediate or long-term survival. In fact, without the
ANS, a simple night's sleep would be a life-threatening event.

The ANS functions differ from those of the somatic nervous system which is responsible for the voluntary control of skeletal muscles. The ANS also has a unique
organization. In the somatic nervous system, the axon of a motor neuron extends
directly from the CNS through the PNS to the skeletal muscle. In the ANS, by contrast, there is always a synapse and an additional neuron between the CNS and
the peripheral effector. These additional neurons are called *ganglionic neurons* because they exist in *autonomic ganglia* located outside the CNS. Motor neurons from
the CNS, known as *preganglionic neurons*, send their axons to the autonomic ganglia, where they synapse on ganglionic neurons. The axons of the ganglionic neurons leave the ganglia and innervate cardiac muscle, smooth muscle, and glands.

The ANS consists of two divisions, the sympathetic and parasympathetic divisions. The **sympathetic division** is often called the "fight or flight" system because it usually stimulates tissue metabolism, increases alertness, and generally prepares the body to deal with emergencies. The **parasympathetic division** is often regarded as the "rest and repose" system because it conserves energy and promotes sedentary activities, such as digestion.

The sympathetic and parasympathetic divisions affect target organs through the release of specific neurotransmitters—generally, acetylcholine or norepinephrine—by the ganglionic neurons. Whether the result is a stimulation or inhibition of activity depends on the neurotransmitter released and the membrane receptor involved on the target cell. Three general patterns are worth noting:

1. All preganglionic neurons, whether sympathetic or parasympathetic, are *cholinergic*: their synaptic terminals release acetylcholine (ACh). The effects of ACh are always excitatory.

2. Parasympathetic ganglionic neurons are also cholinergic, but the ACh effects may be excitatory or inhibitory, depending on the nature of the membrane receptor on the target cell.

3. Most sympathetic ganglionic neurons are *adrenergic*: they release norepinephrine (NE) at their synaptic terminals. The effects of NE release are usually excitatory.

THE SYMPATHETIC DIVISION

The sympathetic division is extensive and it innervates many different organs (Figure 10-9•). As a result, activation of a single sympathetic motor neuron inside the CNS can produce a complex and coordinated response. The sympathetic division is made up of the following components:

1. *Preganglionic neurons located between segments T_1 and L_2 of the spinal cord.*

2. *Ganglia near the vertebral column.* There are two different types of sympathetic ganglia: paired and unpaired. Paired *sympathetic chain ganglia* lie on either side of the vertebral column. These ganglia contain neurons whose axons continue to travel with the spinal nerves; they control various effectors in the body wall and inside the thoracic cavity. Three single ganglia lie in front of the vertebral column. These contain neurons that innervate tissues and organs in the abdominopelvic cavity.

3. Specialized neurons in the interior of each adrenal gland, a region known as the *adrenal medulla.* ∞ [p. 84]

As Figure 10-9• shows, the axons of neurons within the sympathetic chain ganglia that are carried by the spinal nerves stimulate sweat gland activity and arrector pili muscles (producing "goose bumps"), accelerate blood flow to skeletal muscles, release stored lipids from adipose tissue, and dilate the pupils. Other axons accelerate the heart rate, increase the force of cardiac contractions, and dilate the respiratory passageways. The sympathetic innervation provided by the unpaired ganglia reduces blood flow and energy use by visceral organs not important to short-term survival (such as the digestive tract).

In addition, sympathetic stimulation of the adrenal medulla region releases the neurotransmitters norepinephrine (NE) and epinephrine (E) into surrounding capillaries, which carry them throughout the body. In general, their effects resemble those described in the preceding paragraph. However, (1) they also affect cells not innervated by the sympathetic division, and (2) their effects last much longer than those produced by direct sympathetic innervation. All these changes prepare the individual for sudden, intensive physical activity.

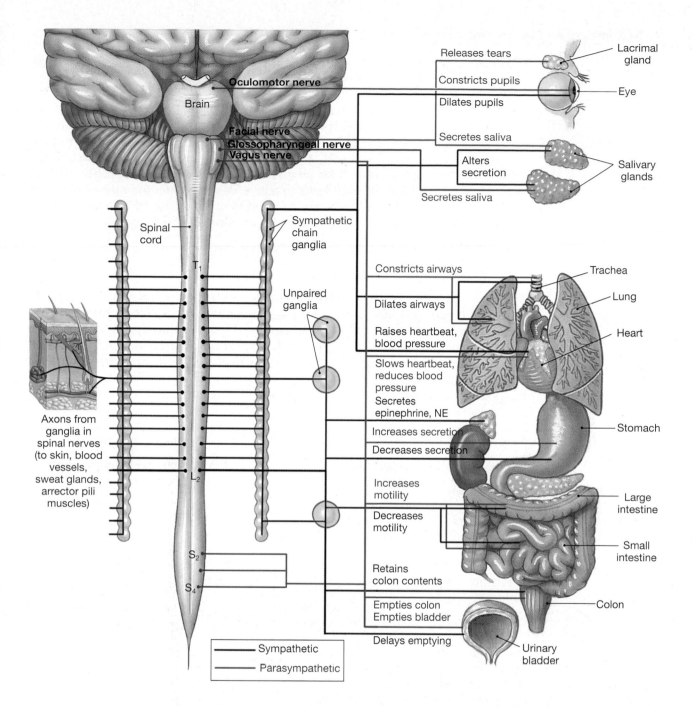

•FIGURE 10-9

Labels in figure: Releases tears — Lacrimal gland; Constricts pupils; Dilates pupils — Eye; Oculomotor nerve; Brain; Secretes saliva; Facial nerve; Glossopharyngeal nerve; Vagus nerve; Alters secretion — Salivary glands; Secretes saliva; Spinal cord; Sympathetic chain ganglia; T₁; Unpaired ganglia; Constricts airways — Trachea; Dilates airways — Lung; Raises heartbeat, blood pressure — Heart; Slows heartbeat, reduces blood pressure; Secretes epinephrine, NE; Increases secretion; Decreases secretion — Stomach; Axons from ganglia in spinal nerves (to skin, blood vessels, sweat glands, arrector pili muscles); L₂; Increases motility; Decreases motility — Large intestine; Small intestine; S₂; Retains colon contents; S₄; Empties colon / Empties bladder — Colon; Delays emptying — Urinary bladder; Sympathetic; Parasympathetic

THE PARASYMPATHETIC DIVISION

Compared with the longer-lasting and more widespread effects of sympathetic stimulation, the effects of parasympathetic stimulation are relatively brief in duration and restricted to specific organs and sites. The parasympathetic division, also shown in Figure 10-9•, includes the following types of neuron groups:

1. *Preganglionic neurons in the brain and in sacral segments S₂ to S₄ of the spinal cord.*

2. *Ganglionic neurons in peripheral ganglia located within or adjacent to the target organs.* As a result of this closeness, the effects of parasympathetic stimulation are more specific and localized than those of the sympathetic division.

The axons of preganglionic parasympathetic neurons leave the brain within four different cranial nerves (nerves connected to the brain, described in Chapter 11) before synapsing within autonomic ganglia. One of these cranial nerves is the *vagus nerve.* The vagus nerves provide parasympathetic innervation to ganglia within organs of the thoracic and abdominopelvic cavity as distant as the last seg-

The Autonomic Nervous System and Its Major Targets.
The distribution of the sympathetic division is shown in black, and that of the parasympathetic division is shown in red. Note that four nerves (cranial nerves) from the brain carry most of the parasympathetic fibers. Each division's distribution is shown on only one side of the body. (The connections of the sympathetic division to somatic structures are shown on the left, and the connections of both divisions to visceral organs are shown on the right.)

ments of the large intestine. *Vagus* means "to wander," and the vagus nerves provide roughly 75 percent of all parasympathetic outflow and innervate most of the organs in the thoracic and abdominopelvic cavities.

Among other things, the parasympathetic division constricts the pupils, increases secretions by the digestive glands, increases smooth muscle activity of the digestive tract, stimulates defecation and urination, constricts respiratory passageways, and reduces heart rate and contraction force (see Figure 10-9●). These functions center on relaxation, food processing, and energy absorption. Stimulation of the parasympathetic division leads to a general increase in the nutrient level of the blood. Cells throughout the body respond to this increase by absorbing nutrients and using them to support growth and the storage of energy reserves.

RELATIONSHIP BETWEEN THE SYMPATHETIC AND PARASYMPATHETIC DIVISIONS

✔ What advantage does dual innervation of the heart provide?

✔ What physiological changes would you expect to observe in a dental patient who is about to have a root canal procedure done and who is quite anxious about the procedure?

The sympathetic division has widespread impact, reaching visceral and somatic structures throughout the body, whereas the parasympathetic division innervates only visceral structures serviced by the cranial nerves or lying within the abdominopelvic cavity. Although some organs are innervated by one division or the other, most vital organs receive **dual innervation**—that is, instructions from both autonomic divisions. Where dual innervation exists, the two divisions often have opposing effects, keeping the body's homeostatic mechanisms working in a dynamic balance. Table 10-1 provides examples of different organs and the effects of either single or dual innervation.

Table 10-1	The Effects of the Sympathetic and Parasympathetic Divisions of the ANS on Various Organs	
Structure	**Sympathetic Innervation Effect**	**Parasympathetic Innervation Effect**
Eye	Dilation of pupil	Constriction of pupil
	Focusing for distance vision	Focusing for near vision
Skin		
Sweat glands	Increases secretion	None (not innervated)
Arrector pili muscles	Contraction, erection of hairs	None (not innervated)
Tear Glands	None (not innervated)	Secretion
Cardiovascular System		
Blood vessels	Vasoconstriction or vasodilation (varies depending on receptor type)	None (not innervated)
Heart	Increases heart rate, force of contraction, and blood pressure	Decreases heart rate, force of contraction, and blood pressure
Adrenal Glands	Secretion of epinephrine and norepinephrine by adrenal medullae	None (not innervated)
Respiratory System		
Airway diameter	Increases	Decreases
Respiratory rate	Increases	Decreases
Digestive System		
General level of activity	Decreases	Increases
Liver	Stimulates glycogen breakdown, glucose synthesis and release	Stimulates glycogen synthesis
Skeletal Muscles	Increases force of contraction, glycogen breakdown	None (not innervated)
Urinary System		
Kidneys	Decreases urine production	Increases urine production
Bladder	Constricts sphincter, relaxes urinary bladder	Tenses urinary bladder, relaxes sphincter to eliminate urine
Reproductive System	Increased glandular secretion; ejaculation in male	Erection of penis (male) or clitoris (female)

AN INTRODUCTION TO DISORDERS OF THE NERVOUS SYSTEM

The disorders described in this chapter are examples of conditions that either (1) result from changes in neuron and glial cell structure, or (2) are primarily restricted to the spinal cord and spinal nerves. The basic features of a neurological examination and the classification of nervous system disorders will be considered in the next chapter, after we have completed our discussion of the nervous system.

NERVOUS TISSUE DISORDERS

Demyelination Disorders

Demyelination is the destruction of the myelin sheath that encloses an axon in the CNS or PNS. Demyelination disorders are linked by this common symptom, although the cause of the demyelination differs in each of these disorders.

Heavy metal poisoning. Chronic exposure to heavy metal ions, such as arsenic, lead, or mercury, can damage glial cells and produce progressive demyelination. As demyelination occurs, nerve impulses slow, affected axons deteriorate, and the condition becomes irreversible. Historians note several interesting examples of heavy metal poisoning with widespread social impact. For example, lead contamination of drinking water has been cited as one factor in the decline of the Roman Empire. Well into the nineteenth century, mercury used in the preparation of felt presented a serious occupational hazard for those employed in the manufacture of stylish felt hats. Over time, mercury absorbed through the skin and across the lungs accumulated in the CNS, producing neurological damage that affected both physical and mental function. (This effect is the source of the expression "mad as a hatter.") More recently, Japanese fishermen working in Minamata Bay, Japan, collected and consumed seafood contaminated with mercury discharged from a nearby chemical plant. Levels of mercury in their systems gradually rose to the point at which clinical symptoms appeared in hundreds of people. Making matters worse, mercury contamination of developing embryos and fetuses caused severe, crippling birth defects.

Diphtheria. Diphtheria (dif-THĒ-rē-a) is a disease that results from a bacterial infection of the nose, throat and skin. The bacteria form a tough, fibrous membrane over the palate and throat (*diphthera* means "leather"). In addition to restricting airflow and sometimes damaging the respiratory surfaces, the bacteria produce a powerful toxin that injures the kidneys and adrenal glands, among other tissues. In the nervous system, diphtheria toxin damages Schwann cells and destroys myelin sheaths in the PNS. This demyelination leads to sensory and motor problems that may ultimately produce a fatal paralysis. The toxin also affects cardiac muscle cells, and heart enlargement and heart failure may occur. The fatality rate for untreated cases ranges from 35 to 90 percent. Because an effective vaccine exists, cases are relatively rare in countries with adequate health care. Russia experienced a diphtheria epidemic in the 1990s, as the economy deteriorated.

Multiple sclerosis. Multiple sclerosis (skler-Ō-sis) or **MS**, is an autoimmune disease that produces muscular paralysis (*sclerosis* means "hardness") and sensory losses through demyelination of axons in the optic nerve, brain, and/or spinal cord. The demyelination is apparently the result of an immune response that targets myelin. Common symptoms include partial loss of vision and problems with speech, balance, and general motor coordination. The time between incidents and the degree of recovery vary from case to case. In about one-third of all cases, the disorder is progressive, and each incident leaves a greater degree of functional impairment. More common in women than in men (3 women for every 2 men), the average age at the first attack is 30–40 years. Treatment with interferon, a chemical messenger of the immune system, has slowed progression in some cases.

Tumors

Neuromas are tumors that originate from neurons. They may affect any nerve in the body. Symptoms include pain in parts of the body served by the nerve, or, as the tumor increases in size, the area may become weak and numb. **Gliomas**, or *brain tumors*, originate in the supporting glial cells of the brain.

A **neuroblastoma** is a tumor of cells of the sympathetic ("fight-or-flight") nervous system including the adrenal medulla (the interior of the adrenal gland). Neuroblastomas occur most often in children younger than 10 years old. Common symptoms include hard, painless masses in the abdomen, chest, or neck, or exaggerations of the normal responses of the sympathetic system, such as elevated heart rate and blood pressure, and weight loss, pain, and irritability. Neuroblastomas range from completely curable to highly malignant metastatic cancers.

SPINAL CORD DISORDERS

Trauma

At the outset, any severe injury to the spinal cord produces a period of sensory and motor paralysis termed **spinal shock**. The skeletal muscles become flaccid; normal reflexes do not occur; and the brain no longer receives sensations of touch, pain, heat, or cold. The location and severity of the injury determine how long these symptoms persist and how completely the individual recovers.

Violent jolts, such as those associated with blows or gunshot wounds near the spinal cord, may cause **spinal concussion** without visibly damaging the spinal cord. Spinal concussion produces a period of spinal shock, but the symptoms are only temporary and recovery may be complete in a matter of hours. More serious injuries, such as vertebral fractures, generally involve physical damage to the spinal cord. In a **spinal contusion**, the white matter of the spinal cord may degenerate at the site of injury. Gradual recovery over a period of weeks may leave some functional losses. Recovery from a **spinal laceration** by vertebral fragments or other foreign bodies tends to be far slower and less complete. **Spinal compression** occurs when the spinal cord becomes squeezed or distorted within the vertebral canal. In a **spinal transection**, the spinal cord is completely severed. At present, surgical procedures cannot repair a severed spinal cord.

Many spinal cord injuries involve some combination of compression, laceration, contusion, and partial transection. Relieving pressure and stabilizing the affected area through surgery (such as *spinal fusion*, the immobilization of adjacent vertebrae) may prevent further damage and allow the injured spinal cord to recover as much as possible.

Extensive damage to the spinal cord at or above the fifth cervical vertebra will eliminate sensation and motor control of the upper and lower limbs. The resulting paralysis is called **quadriplegia**. Damage that is lower, at the thoracic region of the spinal cord, can result in **paraplegia**, the loss of motor control of the lower limbs.

Actor Christopher Reeve was paralyzed after a fall from a horse fractured cervical vertebrae C_1 and C_2.

SPINAL NERVE DISORDERS

Neuritis (nū-RĪ-tis) is the inflammation of a nerve. Formerly restricted in its use to referring to a nerve inflammation caused by an infection, it now also refers to damaged nerves and other disorders of nerves. **Neuropathy** (nū-rop-uh-thē) is a term still used for disorders of peripheral nerves.

Infection

In **shingles**, or **herpes zoster**, the *Herpes varicella-zoster* virus attacks neurons within the dorsal roots of spinal nerves and sensory ganglia of cranial nerves. This disorder produces a painful rash whose distribution corresponds to that of the affected sensory nerves (Figure 10-10•). Shingles develops in adults who were first exposed to the virus as children. The initial infection produces symptoms known as chickenpox. After this encounter, the virus remains dormant within neurons of the ante-

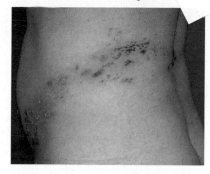

rior gray matter of the spinal cord. It is not known what triggers reactivation of this pathogen. Fortunately for those affected, attacks of shingles generally heal and leave behind only unpleasant memories. Most people suffer only a single episode of shingles in their adult lives. However, the problem may recur in people with weakened immune systems, including those with AIDS or some forms of cancer. Treatment typically involves large doses of the antiviral drug *acyclovir* (*Zovirax*).

The condition traditionally called **leprosy**, now more commonly known as **Hansen's disease**, is an infectious disease caused by a bacterium, *Mycobacterium leprae* (LEP-rē). It is a disease that progresses slowly, and symptoms may not appear for up to 30 years after infection. The bacterium invades peripheral nerves, especially those in the skin, producing initial sensory losses. Over time motor paralysis develops, and the combination of sensory and motor loss can lead to recurring injuries and infections. The eyes, nose, hands, and feet may develop deformities as a result of neglected injuries (Figure 10-11•). There are several forms of this disease; peripheral nerves are always affected, but some forms also involve extensive lesions of the skin and mucous membranes.

Only about 5 percent of those exposed to *Mycobacterium leprae* develop symptoms; people living in the tropics are at greatest risk. There are about 2,000 cases in the United States, and an estimated 12–20 million cases worldwide. The disease can generally be treated successfully with drugs, and early detection and treatment can prevent deformities. Treated individuals are not infectious, and the practice of confining "lepers" in isolated compounds has been discontinued.

The viral disease called **polio** causes paralysis due to the destruction of somatic motor neurons. This disorder, introduced in Chapter 9, has been virtually eliminated in the Western Hemisphere. ∞ *[p. 212]* Immunization continues because polio remains in other parts of the world. The disease could be brought into the United States at any time, leading to an epidemic among unimmunized children. The World Health Organization hopes to eliminate polio worldwide in the next decade through an intensive immunization program.

Peripheral Palsies

Peripheral nerve palsies, or *peripheral neuropathies*, are characterized by regional losses of sensory and motor function as the result of nerve trauma. *Brachial palsies* result from injuries to the brachial plexus or its branches. *Crural (leg) palsies* involve the nerves of the lumbosacral plexus. ∞ *[p. 228]* Palsies may appear for several reasons. The pressure palsies are especially interesting; a familiar but mild example is the experience of having an arm or leg "fall asleep." The limb becomes numb, and afterward an uncomfortable "pins-and-needles" sensation, or *paresthesia* (par-es-THĒ-zē-ah), accompanies the return to normal function. The cause is thought to be compression of the blood supply to the nerve.

These incidents are seldom of clinical significance, but they provide graphic examples of the effects of more serious palsies that can last for days to months. In *radial nerve palsy*, pressure on the back of the arm interrupts the function of the radial nerve, so the extensors of the wrist and fingers are paralyzed. This condition

•**FIGURE 10-10**
Shingles.
The distribution of the rash in this viral condition follows the paths of infected sensory nerves.

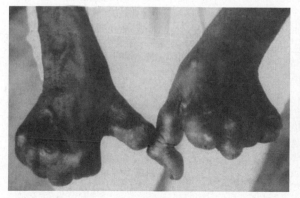

(a)   (b)

•**FIGURE 10-11**
Hansen's Disease.
(a) The distal portions of the limbs are gradually deformed as untreated Hansen's disease progresses. **(b)** Facial features are also affected, often starting with degenerative changes around the eyes and at the nose and ears.

is also known as "Saturday night palsy," for falling asleep on a couch with your arm over the seat back (or beneath someone's head) can produce the right combination of pressures. Students may also be familiar with *ulnar palsy*, which can result from prolonged contact between elbow and desk. The ring and little fingers lose sensation, and the fingers cannot be adducted. *Carpal tunnel syndrome*, a neuropathy resulting from compression of the *median nerve* at the wrist, was considered in Chapter 9 (p. 209).

Men with large wallets in their hip pockets may develop symptoms of *sciatic compression* after they drive or sit in one position for extended periods. As nerve function declines, the individuals notice some lumbar or gluteal pain, a numbness along the back of the leg, and a weakness in the leg muscles. Similar symptoms result from compression of nerve roots that form the sciatic nerve by a distorted lumbar intervertebral disc. This condition is termed **sciatica** (sī-AT-i-kah), and one or both legs may be affected, depending on the site of compression. Finally, sitting with your legs crossed may produce symptoms of a **peroneal palsy**. Sensory losses from the top of the foot and side of the leg are accompanied by a decreased ability to dorsiflex the ankle or evert the foot.

✔ How may prolonged exposure to mercury or other heavy metals affect the nervous system?

✔ What do brachial, crural, and peroneal palsies have in common?

CHAPTER REVIEW

Key Words

action potential: Nerve impulse; caused by a reversal in a cell's membrane potential that is conducted along the membrane surface.

autonomic nervous system (ANS): Efferent division of the nervous system involved in the unconscious regulation of visceral functions; includes components of the CNS and PNS.

central nervous system (CNS): The brain and spinal cord.

dual innervation: Innervation from both autonomic divisions; type of innervation of most vital organs.

ganglion/ganglia: A collection of neuron cell bodies outside the CNS.

nerve: Bundles of axons and associated blood vessels surrounded by connective tissue; *cranial nerves* are connected to the brain; *spinal nerves* are attached to the spinal cord.

neuroglia (noo-RŌ-glē-a): Cells that support and protect the neurons; literally, "nerve glue."

neuron: A cell in the nervous tissue specialized for cell-to-cell communication through its ability to conduct electrical impulses.

neurotransmitter: Chemical compound released by one neuron to affect the membrane potential of another neuron or effector cell.

parasympathetic division: One of the two divisions of the autonomic nervous system; generally responsible for activities that conserve energy and lower the metabolic rate.

peripheral nervous system (PNS): All the nervous tissue outside the CNS.

reflex: A rapid, automatic response to a stimulus.

reflex arc: The receptor, sensory neuron, motor neuron, and effector involved in a particular reflex; interneurons may or may not be present.

somatic nervous system: The portion of the PNS that includes the axons of motor neurons innervating skeletal muscles.

spinal nerve: One of 31 pairs of peripheral nerves that originate on the spinal cord from anterior (ventral) and posterior (dorsal) roots.

sympathetic division: Division of the autonomic nervous system responsible for "fight or flight" reactions; primarily concerned with the elevation of metabolic rate and increased alertness.

synapse (SIN-aps): Site of communication between a neuron and some other cell; if the other cell is not a neuron, the term *neuroeffector junction* is often used.

Selected Clinical Terms

demyelination: Destruction of the myelin sheaths around axons in the CNS and PNS.

diphtheria: A disease of the respiratory tract caused by a bacterial infection. Among other effects, the bacterial toxins damage Schwann cells and cause PNS demyelination.

Hansen's disease *(leprosy):* A bacterial infection that begins in sensory nerves of the skin and gradually progresses to a motor paralysis of the same regions.

multiple sclerosis (MS): A disease marked by recurrent incidents of demyelination affecting axons in the optic nerve, brain, and/or spinal cord.

palsies: Regional losses of sensory and motor functions as a result of nerve trauma or compression; also called peripheral neuropathies.

paraplegia: Paralysis involving loss of motor control of the lower but not the upper limbs.

paresthesia: An abnormal tingling sensation, usually described as "pins and needles," that accompanies sensory return after a temporary palsy.

sciatica (sī-AT-i-kah): The painful result of compression of the roots of the sciatic nerve.

shingles: A condition caused by infections of neurons in dorsal root ganglia by the virus *Herpes varicella-zoster*. The primary symptom is a painful rash along the affected sensory nerves.

spinal shock: A period of depressed sensory and motor function following any severe injury to the spinal cord.

Study Outline

INTRODUCTION (p. 219)

1. The nervous system and the endocrine system coordinate the body's responses to a changing environment. The nervous system provides a rapid response, whereas the endocrine system controls long-term changes and responses.

SYSTEM BRIEF (pp. 219–220)

1. The **nervous system** is made up of the **central nervous system (CNS)** (the *brain* and *spinal cord*) and the **peripheral nervous system (PNS)** (all the neural tissue outside the CNS).

2. The PNS is divided into an *afferent division*, which brings sensory information to the CNS, and an *efferent division*, which carries motor commands to muscles and glands. The efferent division includes the **somatic nervous system** (voluntary control over skeletal muscle contractions) and the **autonomic nervous system (ANS)** (automatic, involuntary regulation of smooth muscle, cardiac muscle, and glandular activity). *(Figure 10-1)*

THE STRUCTURE OF NEURONS AND NERVES (pp. 220–223)

1. There are two types of cells in neural tissue: **neurons**, which are responsible for information transfer and processing, and **neuroglia**, or *glial cells*, which provide a supporting framework and act as phagocytes.

NEURONS (p. 220)

2. **Sensory neurons** deliver information to the CNS. Sensory information is provided by sensory neurons or sensory receptors.

3. **Motor neurons** stimulate or modify the activity of a peripheral tissue, organ, or organ system.

4. **Interneurons** may be located between sensory and motor neurons; they analyze sensory inputs and coordinate motor outputs.

5. A typical neuron has a cell body, an **axon**, and several branching **dendrites**. *(Figure 10-2)*

6. **Synaptic knobs** occur at the ends of axons. *(Figure 10-2)*

NEUROGLIA (p. 221)

7. Three common types of neuroglia in the CNS are (1) **astrocytes** (largest and most numerous); (2) **oligodendrocytes**, which are responsible for the **myelination** of CNS axons; and (3) **microglia**, phagocytic white blood cells. *(Figure 10-3)*

NERVES (p. 223)

8. The axons of neuron cell bodies in the PNS are covered by myelin wrappings of **Schwann cells**. Bundles of axons (nerve fibers) in the PNS form **nerves**. *(Figure 10-4)*

NEURON FUNCTION (pp. 223–225)

THE MEMBRANE POTENTIAL (p. 223)

1. Communication between one neuron and another neuron or other type of cell occurs at their cell membranes through a change in *membrane potential*. The membrane potential exists due to the uneven distribution of positive and negative ions across the cell membrane.

THE ACTION POTENTIAL (p. 223)

2. A reversal in the membrane potential results when large numbers of sodium ions enter a cell. In an **action potential**, this reversal of charges is conducted along the membrane surface.

3. Action potentials move faster along myelinated axons than along unmyelinated axons. In a myelinated axon the action potential "jumps" from one node to the next.

THE SYNAPSE (p. 224)

4. A **synapse** is a site where intercellular communication occurs through the release of chemicals called **neurotransmitters**.

5. Neurotransmitters released from the **presynaptic neuron** cross the **synaptic cleft** to receptor sites on the **postsynaptic neuron**. *(Figure 10-5)*

THE SPINAL CORD AND SPINAL NERVES (pp. 225–228)

1. The spinal cord carries messages to and from the brain and also controls spinal reflexes on its own.

STRUCTURE OF THE SPINAL CORD (p. 225)

2. The spinal cord has 31 segments, each associated with a pair of **spinal nerves.** *(Figure 10-6a)*

3. A section of the spinal cord contains two differently colored neural tissues. The *white matter* contains myelinated and unmyelinated axons, while the *gray matter* contains cell bodies of neurons and glial cells. *(Figure 10-6b)*

4. The white matter contains **ascending tracts** that relay information from the spinal cord to the brain, and **descending tracts** that carry information from the brain to the spinal cord.

THE SPINAL NERVES (p. 227)

5. There are 31 pairs of spinal nerves: 8 cervical, 12 thoracic, 5 lumbar, 5 sacral, and 1 coccygeal.

6. Spinal nerves are **mixed nerves**; they contain both sensory and motor axons (fibers).

7. Each pair of spinal nerves monitors a region of the body surface called a **dermatome**. *(Figure 10-7)*

8. A complex, interwoven network of nerves is called a **nerve plexus**. The three large plexuses are the *cervical plexus*, the *brachial plexus*, and the *lumbosacral plexus*.

REFLEXES (pp. 228–229)

1. A nerve **reflex** is an automatic, involuntary motor response that helps preserve homeostasis by rapidly adjusting the activities of organs or organ systems.

THE REFLEX ARC (p. 228)

2. A **reflex arc** is the neural "wiring" of a single reflex.

3. There are five steps involved in a neural reflex: (1) arrival of a stimulus and activation of a receptor, (2) activation of a sensory neuron, (3) information processing, (4) activation of a motor neuron, and (5) response by an effector. *(Figure 10-8)*

THE AUTONOMIC NERVOUS SYSTEM (pp. 229–232)

1. The autonomic nervous system (ANS) coordinates cardiovascular, respiratory, digestive, excretory, and reproductive functions.

2. **Preganglionic neurons** in the CNS send axons to synapse on *ganglionic neurons* in *autonomic ganglia* outside the CNS. The axons of the ganglionic neurons innervate cardiac muscle, smooth muscle, glands, and adipose tissue.

3. Sympathetic motor neurons from the thoracic and lumbar segments form the **sympathetic division** ("fight or flight" system) of the ANS. Parasympathetic motor neurons in the brain and sacral segments of the spinal cord form the **parasympathetic division** ("rest and repose" system). *(Figure 10-9)*

THE SYMPATHETIC DIVISION (p. 230)

4. The sympathetic division consists of preganglionic neurons between spinal cord segments T_1 and L_2, ganglionic neurons in paired (*sympathetic chain ganglia*) and unpaired ganglia near the vertebral column, and specialized ganglionic neurons in the *adrenal medullae*. (*Figure 10-9*)

5. During sympathetic activation, neurons within the adrenal medullae of the adrenal glands secrete epinephrine and norepinephrine into the bloodstream.

6. In a crisis, the entire sympathetic division responds, producing increased alertness, a feeling of energy and euphoria, increased cardiovascular and respiratory activity, and general elevation in muscle tone.

THE PARASYMPATHETIC DIVISION (p. 231)

7. The parasympathetic division includes preganglionic neurons in the brain and sacral segments of the spinal cord, and ganglionic neurons in ganglia located within or next to target organs. (*Figure 10-9*)

8. The effects produced by the parasympathetic division center on relaxation, food processing, and energy absorption.

9. The effects of parasympathetic stimulation are usually brief and restricted to specific sites.

RELATIONSHIP BETWEEN THE SYMPATHETIC AND PARASYMPATHETIC DIVISIONS (p. 232)

10. The sympathetic division has widespread impact, reaching visceral and somatic structures throughout the body.

11. The parasympathetic division innervates only visceral structures serviced by cranial nerves or lying within the abdominopelvic cavity. Organs with **dual innervation** receive instructions from both divisions. (*Table 10-1*)

AN INTRODUCTION TO DISORDERS OF THE NERVOUS SYSTEM (pp. 233–236)

NERVOUS TISSUE DISORDERS (p. 233)

1. **Demyelination disorders** are characterized by the demyelination of axons in the CNS and PNS. Demyelination may have different causes. **Heavy metal poisoning** results from demyelination by ions of lead, arsenic, and mercury.

2. **Diphtheria,** a bacterial disease of the respiratory tract, also affects the nervous system. A toxin released by the diphtheria bacteria destroys myelin sheaths of PNS axons.

3. In **multiple sclerosis,** axons in the optic nerve, brain, and spinal cord become demyelinated.

4. Tumors of the nervous system include **neuromas** of neurons and **gliomas** of glial cells. A **neuroblastoma** is a tumor of the adrenal glands and/or the sympathetic nervous system.

SPINAL CORD DISORDERS (p. 234)

5. Any severe injury to the spinal cord produces **spinal shock,** a period of sensory and motor paralysis. A **spinal concussion** does not show visible damage to the spinal cord. More serious injuries include **spinal contusions, spinal lacerations, spinal compressions,** and **spinal transections.**

6. Paralysis of the upper and lower limbs is **quadriplegia; paraplegia** is a paralysis of the lower limbs.

SPINAL NERVE DISORDERS (p. 234)

7. **Neuritis** and **neuropathy** refer to nerve inflammation and nerve disorders.

8. **Shingles,** or **herpes zoster,** is caused by a virus which attacks neurons within the dorsal roots of spinal nerves and sensory ganglia of cranial nerves. A painful rash is distributed over the body in a pattern whose distribution corresponds to that of the affected sensory nerves. (*Figure 10–10*)

9. **Leprosy,** or **Hansen's disease,** is an infectious disease caused by a bacterium. The bacterium invades peripheral nerves, especially those in the skin, producing sensory losses initially and motor paralysis later. The disease progresses slowly, and symptoms may not appear for up to 30 years after infection. (*Figure 10–11*)

10. **Polio** is a viral disease that infects and destroys somatic motor neurons, leading to paralysis.

11. **Peripheral nerve palsies,** or **peripheral neuropathies,** are regional losses of sensory and motor function as the result of nerve injury or compression.

12. **Sciatica** is a condition resulting from compression of nerve roots that form the sciatic nerve by a distorted lumbar intervertebral disc. One or both lower limbs may be affected.

Review Questions

MATCHING

Match each item in Column A with the most closely related item in Column B. Use letters for answers in the spaces provided.

	Column A		Column B
___	1. neuroglia	a.	masses of neuron cell bodies
___	2. microglia	b.	carry sensory information to the brain
___	3. sensory neurons	c.	bacterial infection
___	4. parasympathetic division	d.	controls contractions of skeletal muscles
___	5. ganglia	e.	controls smooth and cardiac muscle, and glands
___	6. sympathetic division	f.	cells of the efferent division
___	7. ascending tracts	g.	multiple sclerosis
___	8. descending tracts	h.	coat CNS axons with myelin
___	9. oligodendrocytes	i.	"rest and repose"
___	10. somatic nervous system	j.	carry motor commands to spinal cord
___	11. motor neurons	k.	coat PNS axons with myelin
___	12. autonomic nervous system	l.	supporting cells of the nervous system
___	13. Schwann cells	m.	cells of the afferent division
___	14. Hansen's disease	n.	"fight or flight"
___	15. demyelination	o.	phagocytic cells

MULTIPLE CHOICE

16. Regulation by the nervous system provides _____ .
(a) relatively slow but long-lasting responses to stimuli
(b) swift, long-lasting responses to stimuli
(c) swift but brief responses to stimuli
(d) relatively slow, short-lived responses to stimuli

17. The peripheral nervous system (PNS) is made up of _____ .
(a) all of the neural tissue outside of the CNS
(b) the brain and spinal cord
(c) the brain and spinal nerves
(d) a, b, and c are correct

18. The efferent division of the PNS _____ .
(a) brings sensory information to the CNS
(b) carries motor commands to the muscles and glands
(c) integrates, processes, and coordinates sensory information
(d) is the source of sensory information for the body

19. Spinal nerves are called mixed nerves because _____ .
(a) they contain sensory and motor fibers
(b) they exit at intervertebral foramina
(c) they are associated with a pair of dorsal root ganglia
(d) they are associated with a pair of dorsal and ventral roots

20. Which of these is *not* a response of the sympathetic division of the ANS? _____
(a) dilation of the pupils (b) dilation of the airways
(c) increased heart rate (d) increased digestive activity

21. All preganglionic neurons release _____ at their synaptic terminals, and the effects are always _____ .
(a) norepinephrine; inhibitory (b) norepinephrine; excitatory
(c) acetylcholine; excitatory (d) acetylcholine; inhibitory

22. Approximately 75 percent of parasympathetic nerve impulses occur over the _____ .
(a) vagus nerves (b) glossopharyngeal nerves
(c) facial nerves (d) oculomotor nerves

23. Axon demyelination is caused by all of the following except _____ .
(a) bacterial toxins (b) lead
(c) arsenic (d) a neuroma

24. In adults, the chickenpox virus may produce a disease called _____ .
(a) spinal shock (b) polio
(c) shingles (d) Hansen's disease

TRUE/FALSE

____ 25. The three major parts of a neuron are the cell body, dendrites, and axon.

____ 26. In the CNS, the myelin coverings are made by Schwann cells.

____ 27. A substance called acetylcholinesterase stimulates the release of the neurotransmitter acetylcholine.

____ 28. The cell bodies of motor neurons and interneurons are located in the gray matter of the spinal cord.

____ 29. Ganglia of the parasympathetic division are located near or in visceral effectors.

____ 30. Paraplegia is a loss of motor control of the upper and lower limbs.

____ 31. Sciatica is a condition that affects the leg.

SHORT ESSAY

32. What three functional groups of neurons are found in the nervous system? What role does each fill?

33. State the all-or-none principle regarding action potentials (nerve impulses).

34. What steps are involved in transferring a nerve impulse from one neuron to another at a synapse?

35. Myelination of peripheral neurons occurs rapidly through the first year of life. How can this process explain the increased abilities of infants during their first year of life?

36. Compare the general effects of the sympathetic and parasympathetic divisions of the ANS.

37. A neuroblastoma may form in what division of the autonomic nervous system?

38. How might a low fat diet affect the development of the nervous system in a young child?

APPLICATIONS

39. Four-year-old Nicole lives with her parents in an old apartment building where the paint is peeling, and Nicole has gotten into the habit of eating paint chips. Lately she has been showing signs of increased tiredness and fatigue and is having trouble walking. You know that some older buildings were painted with a lead-based paint. Discuss a possible cause of Nicole's condition and why it could produce these symptoms.

40. A patient is brought to the emergency room after a work-related accident. The physician who examines him mentions the terms "spinal concussion" and "spinal contusion." Distinguish between these terms. Which is more serious?

✔ Answers to Concept Check Questions

(*p. 223*) **1.** The afferent division of the nervous system is composed of nerves that carry sensory information to the brain and spinal cord. Damage to this division would interfere with a person's ability to experience a variety of sensory stimuli. **2.** Microglial cells are small phagocytic cells that are found in increased number in damaged and diseased areas of the CNS.

(*p. 224*) **1.** Only neurons and skeletal muscle fibers have membranes that conduct action potentials. **2.** A reversal of a neuron's membrane potential involves the opening of the sodium channels and the rapid movement of sodium ions into the cell. If the sodium channels were blocked, the cell membrane of a neuron would not be able to reverse and conduct an action potential, or nerve impulse.

(*p. 225*) **1.** A chemical called a *neurotransmitter* is required for the passage of a nerve impulse across a synapse. **2.** The number of nerve impulses generated at a synapse is limited by enzymes within the synapse that break down the neurotransmitter that is released by the presynaptic neuron.

(*p. 226*) **1.** The white matter of the spinal cord consists of axons covered by myelin sheaths. Their destruction would interfere with the flow of nerve impulses carrying sensory information up the spinal cord and nerve impulses carrying motor commands down the spinal cord to effectors such as skeletal muscles.

(*p. 229*) **1.** The minimum number of neurons in a reflex arc are two: a sensory neuron to bring impulses to the CNS and a motor neuron to bring about a response to the sensory input.

(*p. 232*) **1.** Dual innervation of the heart by both divisions of the autonomic nervous system produces opposite effects on heart function, which help maintain relatively rapid homeostatic control of this important organ. **2.** A patient who is anxious about impending root canal surgery would probably exhibit some or all of the following changes: a dry mouth, increased heart rate, increased blood pressure, increased rate of breathing, cold sweats, an urge to urinate or defecate, change in motility of the stomach (i.e., "butterflies in the stomach"), and dilated pupils. These changes would be the result of anxiety or stress, causing an increase in sympathetic stimulation.

(*p. 236*) **1.** Heavy metals, such as arsenic, lead, and mercury, cause the destruction of the myelin sheath of axons. **2.** All of these are pressure palsies and result from the compression of nerves. All produce the temporary loss of sensory and motor responses of the affected areas.

The Nervous System 2:

The Brain and Cranial Nerves

CHAPTER OUTLINE

THE BRAIN   *p. 241*

THE CEREBRUM   *p. 245*

THE LIMBIC SYSTEM   *p. 247*

THE DIENCEPHALON   *p. 248*

THE BRAIN STEM   *p. 248*

THE CEREBELLUM   *p. 249*

THE CRANIAL NERVES   *p. 249*

AGING AND THE NERVOUS SYSTEM   *p. 252*

A SUMMARY OF THE DISORDERS OF THE NERVOUS SYSTEM   *p. 253*
Clinical Note: Brain Waves

CHAPTER OBJECTIVES

1   Name the major regions of the brain.
2   Describe the three membrane layers that cover the brain.
3   Describe the relationship of cerebrospinal fluid with the ventricles of the brain.
4   Locate the motor, sensory, and association areas of the cerebral cortex, and discuss their functions.
5   Describe the location and functions of the limbic system.
6   Describe the functions of the diencephalon.
7   Describe the functions of the brain stem components.
8   Describe the functions of the cerebellum.
9   Identify the cranial nerves, and relate each pair to its major functions.
10   Summarize the effects of aging on the nervous system.
11   Describe the three symptoms characteristic of many nervous system disorders.
12   Describe common disorders of the brain and cranial nerves.

THE BRAIN

The brain is probably the most fascinating and mysterious organ of the body. We know relatively little about its structural and functional complexities. What we *do* know is that the brain's activity is responsible for our dreams, passions, plans, and memories—in short, for what distinguishes us as human.

The adult human brain is far larger and more complex than the spinal cord. It contains almost 98 percent of the neural tissue in the body. An average adult brain weighs 1.4 kg (3 lbs.) and has a volume of 1,200 cubic centimeters (cc) (71 cubic in.). Brain size varies considerably among individuals, and the brains of males are generally about 10 percent larger than those of females, because of greater average body size. There is no correlation between brain size and intelligence; people with the smallest (750 cc) and largest (2,100 cc) brains are functionally normal.

The brain consists of gray and white matter. The gray matter, which contains neuroglia and neuron cell bodies, forms separate *nuclei* inside the brain and also forms the surface of the brain in some areas. This superficial layer of gray matter is called *neural cortex*. White matter, made up of bundles of axons, form tracts that link nuclei with one another and with neurons of the neural cortex and spinal cord. Just as the spinal cord communicates with the periphery through spinal nerves, the brain communicates with other body tissues and organs by means of cranial nerves.

BASIC STRUCTURE AND FUNCTION

There are six major regions in the adult brain. The six regions are the (1) *cerebrum*, (2) *diencephalon*, (3) *midbrain*, (4) *pons*, (5) *cerebellum*, and (6) *medulla oblongata*. Figure 11-1• shows the locations of these brain regions and their general functions.

The **cerebrum** (SER-e-brum or se-RĒ-brum) is divided into large left and right **cerebral hemispheres**. A thick layer of neural cortex covers the surfaces of the cerebral hemispheres. Conscious thought processes, sensations, intellectual functions, memory storage and retrieval, and complex motor patterns originate in this layer of neural cortex. Figure 11-1a• also shows the major regions, or *lobes*, of the cerebral hemispheres.

The hollow **diencephalon** (dī-en-SEF-a-lon) connects the cerebral hemispheres with the rest of the brain (Figure 11-1b•). Its sides form the **thalamus**, which contains relay and processing centers for incoming sensory information. The **hypothalamus** (*hypo-*, below) forms the floor of the diencephalon. The hypothalamus contains centers involved with emotions, the coordination of ANS activities, and hormone production. A narrow stalk connects the hypothalamus to the *pituitary gland*, which is part of the endocrine system. The hypothalamus is the primary link between the nervous and endocrine systems. (The endocrine system is discussed in Chapter 13.)

The midbrain, pons, and medulla oblongata form the **brain stem.** The brain stem contains important processing centers and relay stations for information passing to or from the cerebrum or cerebellum and also controls vital functions such as breathing, and heart and digestive activities.

The **midbrain** receives sensory information from the eyes and ears and controls reflexive responses that adjust muscle tone and posture. It also contains centers whose activity keeps us conscious and alert.

The term **pons** refers to a bridge, and the pons of the brain connects the cerebellum to the brain stem. In addition to tracts and relay centers, this region of the brain also contains nuclei involved with somatic and visceral motor control.

The pons is also connected to the **medulla oblongata**, the segment of the brain that is attached to the spinal cord. The medulla oblongata relays sensory information to the thalamus and other brain stem centers. It also contains centers concerned with the regulation of respiration and the control of heart rate, blood pressure, and digestive activities.

•FIGURE 11-1

The Brain.
(a) Lateral view of the left surface of the brain showing the cerebrum and its specialized regions, and the cerebellum. (b) The diencephalon and portions of the brain stem are hidden by the overlying cerebrum. The medulla oblongata connects the brain with the spinal cord.

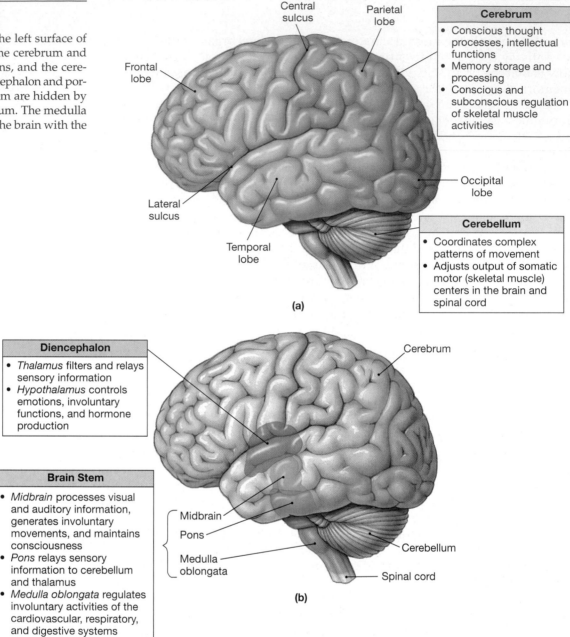

Central sulcus

Parietal lobe

Frontal lobe

Cerebrum
- Conscious thought processes, intellectual functions
- Memory storage and processing
- Conscious and subconscious regulation of skeletal muscle activities

Lateral sulcus

Occipital lobe

Temporal lobe

Cerebellum
- Coordinates complex patterns of movement
- Adjusts output of somatic motor (skeletal muscle) centers in the brain and spinal cord

(a)

Diencephalon
- *Thalamus* filters and relays sensory information
- *Hypothalamus* controls emotions, involuntary functions, and hormone production

Cerebrum

Brain Stem
- *Midbrain* processes visual and auditory information, generates involuntary movements, and maintains consciousness
- *Pons* relays sensory information to cerebellum and thalamus
- *Medulla oblongata* regulates involuntary activities of the cardiovascular, respiratory, and digestive systems

Midbrain

Pons

Medulla oblongata

Cerebellum

Spinal cord

(b)

✔ List the names of the six regions of the brain. Which is the largest?

✔ Which region of the brain is the connecting link to the spinal cord?

The large cerebral hemispheres and the **cerebellum** (se-re-BEL-um) almost completely cover the brain stem. The cerebellum has two **cerebellar hemispheres** covered by neural cortex. The cortical neurons adjust ongoing movements on the basis of sensory information and stored memories of previous movements. Cerebellar activity occurs at the subconscious level. It adjusts postural muscles, maintains balance and equilibrium, and directs learned movements such as swinging a golf club or playing the piano.

THE MENINGES

The spinal cord and brain are extremely delicate and must be protected against shocks, infection, and other dangers. In addition to neuroglial cells within the neural tissue, the CNS is protected by membranes called the *meninges*. The **meninges** (men-IN-jēz) cover and cushion the brain and spinal cord (Figure 11-2•). There are three cranial meningeal layers: the *dura mater*, the *arachnoid*, and the *pia mater*.

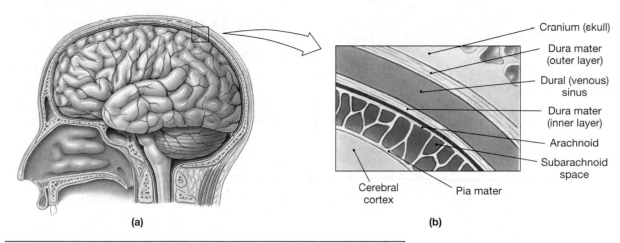

•FIGURE 11-2
Meninges of the Brain.
(a) The meninges surrounding the brain are continuous with those covering the spinal cord. **(b)** An enlarged view of the three layers of the meninges covering the surface of the brain.

The tough, fibrous **dura mater** (DŪ-ra MA-ter; Latin, "hard mother") forms the outermost covering of the central nervous system. The dura mater surrounding the brain consists of two layers. The inner and outer layers are separated by an area of loose connective tissue that contains tissue fluids and blood vessels. The large veins known as *dural sinuses* are found within this area.

The second meningeal layer is called the **arachnoid** (a-RAK-noyd). Its name refers to the "spidery" web of collagen and elastic fibers that fills the underlying *subarachnoid space*. The subarachnoid space is filled with cerebrospinal fluid. Cerebrospinal fluid acts as a shock absorber, and it also transports dissolved gases, nutrients, chemical messengers, and waste products.

arachne, spider + *eidos*, resemblance *arachnoid membrane*: the meningeal layer that separates the dura mater from the subarachnoid space

Below the subarachnoid space is the innermost meningeal layer, the **pia mater** (PĒ-ah MA-ter; Latin, "delicate mother"). This layer, firmly attached to the neural tissue of the CNS, supports the blood vessels serving the brain and spinal cord. These blood vessels supply the neural cortex as well as deeper structures. The extensive circulatory supply of the brain is extremely important, because the brain has a very high rate of metabolism. When you are resting quietly, your brain uses 20 times more oxygen than an equal mass of skeletal muscle.

THE BLOOD–BRAIN BARRIER

Neural tissue in the CNS is protected from the general circulation by the *blood–brain barrier*. This barrier is maintained by the neuroglial cells called astrocytes. ∞ *[p. 221]* Where they contact the processes of astrocytes, the CNS capillaries are impermeable to most ions and compounds circulating in the bloodstream. In general, only lipid-soluble compounds can diffuse freely into the interstitial fluid of the brain and spinal cord. Water-soluble compounds cannot cross the capillary lining without the assistance of specific carrier molecules. For example, there are separate transport systems that pump circulating glucose across the blood–brain barrier and into the interstitial fluid of the brain.

Neurons have a constant need for glucose that must be met regardless of its relative concentration in blood. Even when the blood concentration is low, glucose continues to be transported from the blood into the neural tissue of the CNS. In contrast, only trace amounts of circulating epinephrine, norepinephrine, or other stimulatory chemicals enter these tissues, because the blood–brain barrier rejects them. This prevents massive, uncontrolled stimulation of neurons in the brain and spinal cord whenever the sympathetic nervous system is activated.

THE VENTRICLES OF THE BRAIN

The brain and spinal cord are hollow, and their internal cavities are filled with cerebrospinal fluid. The passageway within the spinal cord is the narrow central canal. The brain has a central passageway that expands to form four chambers called **ventricles** (VEN-tri-kuls). As Figure 11-3• shows, there are two **lateral ventricles**, one in each cerebral hemisphere. There is no direct connection between them, but both open into the **third ventricle** of the diencephalon. Instead of a ventricle, the midbrain has a slender canal known as the *cerebral aqueduct*, which connects the third ventricle with the **fourth ventricle** of the pons and superior portion of the medulla oblongata. Within the medulla oblongata the fourth ventricle narrows and joins the central canal of the spinal cord.

Cerebrospinal fluid, or **CSF**, provides cushioning for delicate neural structures. It also provides support, because the brain essentially floats in the cerebrospinal fluid. A human brain weighs about 1,400 g in air, but only about 50 g when supported by the cerebrospinal fluid. The CSF also transports nutrients, chemical messengers, and waste products as it circulates from the ventricles of the brain to the dural sinuses, where it is reabsorbed.

CSF is produced in each of the four ventricles by a capillary network called the **choroid plexus** (see Figure 11-4•). The capillaries of the choroid plexus, covered by neuroglial cells, secrete about 500 ml of CSF each day. Figure 11-4• diagrams the circulation of cerebrospinal fluid. After forming at the choroid plexus, CSF circulates among the different ventricles and diffuses along the central canal. There is also a continual flow of CSF from the fourth ventricle into the subarachnoid space. Once inside the subarachnoid space, the CSF circulates around the spinal cord and across the surfaces of the brain. Between the cerebral hemispheres, slender extensions of the arachnoid penetrate the inner layer of the dura mater and project into the dural sinus. Diffusion across these arachnoid extensions returns cerebrospinal fluid to the circulatory system.

Because free exchange occurs between the interstitial fluid and CSF, changes in CNS function may produce changes in the composition of the CSF. Samples of the CSF can be obtained through a *lumbar puncture*, or *spinal tap*, providing useful clinical information concerning CNS injury, infection, or disease.

✔ The blood–brain barrier is maintained by what kinds of cells?

✔ How would decreased diffusion of CSF into the venous sinus affect its volume in the ventricles?

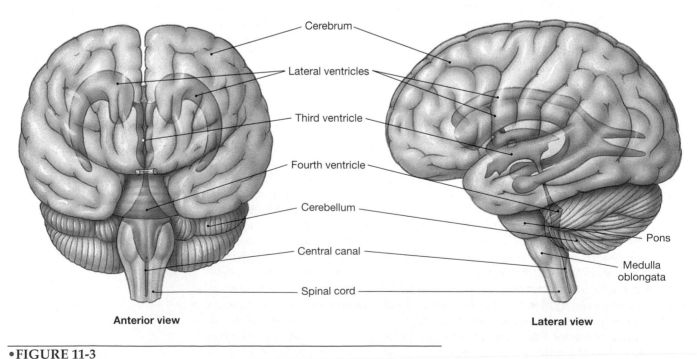

Anterior view

Lateral view

Cerebrum

Lateral ventricles

Third ventricle

Fourth ventricle

Cerebellum

Central canal

Spinal cord

Pons

Medulla oblongata

•**FIGURE 11-3**

Ventricles of the Brain.

Anterior and lateral views of the ventricles as seen through a transparent brain.

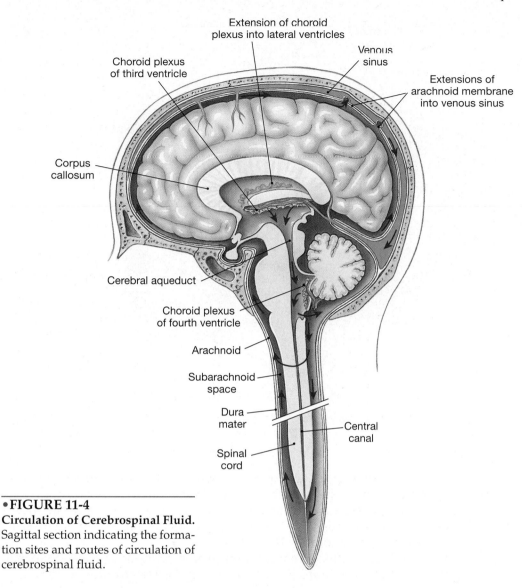

Extension of choroid
plexus into lateral ventricles

Venous
sinus

Choroid plexus
of third ventricle

Extensions of
arachnoid membrane
into venous sinus

Corpus
callosum

Cerebral aqueduct

Choroid plexus
of fourth ventricle

Arachnoid

Subarachnoid
space

Dura
mater

Central
canal

Spinal
cord

•FIGURE 11-4
Circulation of Cerebrospinal Fluid.
Sagittal section indicating the formation sites and routes of circulation of
cerebrospinal fluid.

THE CEREBRUM

The **cerebrum**, the largest region of the brain, is where conscious thought and intellectual functions originate. Much of the cerebrum is involved in receiving somatic
sensory information and then exerting voluntary or involuntary control through
the commands of somatic motor neurons. In general, we are aware of these events.
However, most sensory processing and all autonomic activities involve other areas
of the brain, and these activities occur outside of our conscious awareness.

STRUCTURE OF THE CEREBRAL HEMISPHERES

Figure 11-5• is a diagrammatic view of the left cerebral hemisphere. A thick blanket of neural cortex known as the **cerebral cortex** covers the cerebrum. This outer
surface forms a series of elevated ridges, or **gyri** (JĪ-rī), separated by shallow depressions, called **sulci** (SUL-sī), or deeper grooves, called **fissures**. Gyri increase the
surface area of the cerebral hemispheres and the number of neurons in the cortex.

The cerebrum is divided into left and right cerebral hemispheres by a *longitudinal fissure.* Each cerebral hemisphere is divided into lobes named after the
overlying bones of the skull. The locations of the **frontal**, **temporal**, **parietal**, and
occipital lobes are shown in Figures 11-1a, p. 242, and 11-5•.

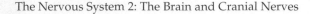

•FIGURE 11-5
The Cerebral Hemispheres.
Major anatomical landmarks and functional regions of the left cerebral hemisphere.

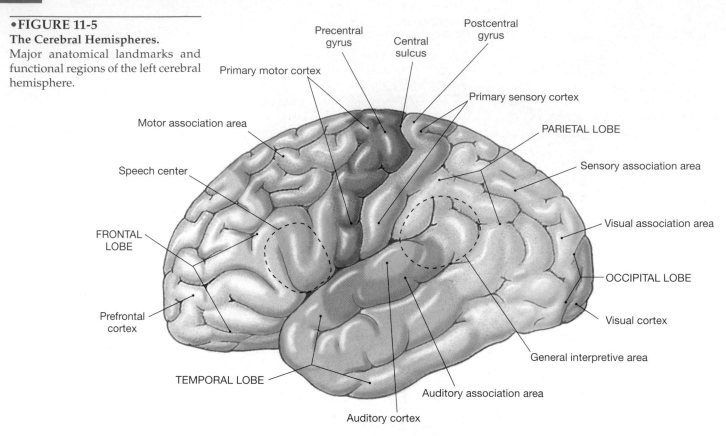

corpus, body + *callosus*, hard
corpus callosum: a thick band of white matter that interconnects the two cerebral hemispheres

In each lobe, some regions are concerned with sensory information and others with motor commands. Each cerebral hemisphere receives sensory information and generates motor commands that affect the opposite side of the body. For example, the left cerebral hemisphere controls the right side of the body, whereas the right cerebral hemisphere controls the left side.

The various regions of the cerebral cortex are interconnected by the white matter that lies beneath the cerebral cortex. This white matter interconnects areas within a single cerebral hemisphere and links the two hemispheres across the **corpus callosum** (see Figure 11-4•). The corpus callosum carries enormous amounts of information. For example, its 200 million axons carry an estimated 4 billion nerve impulses per second! Other tracts (bundles of axons) link the cerebral cortex with the diencephalon, brain stem, cerebellum, and spinal cord.

FUNCTIONS OF THE CEREBRAL HEMISPHERES

Figure 11-5• also details the major motor and sensory regions of the cerebral cortex. The **central sulcus** separates the motor and sensory portions of the cortex. The *precentral gyrus* of the frontal lobe forms the anterior margin of the central sulcus, and its surface is the **primary motor cortex**. Neurons of the primary motor cortex direct voluntary movements by controlling somatic motor neurons in the brain stem and spinal cord.

The *postcentral gyrus* of the parietal lobe forms the posterior margin of the central sulcus, and its surface contains the **primary sensory cortex**. Neurons in this region receive somatic sensory information from touch, pressure, pain, taste, and temperature receptors. We are consciously aware of these sensations only when the diencephalon relays the sensory information to the primary sensory cortex.

The **visual cortex** of the occipital lobe receives visual information, and the **auditory cortex** and **olfactory cortex** of the temporal lobe receive information concerned with hearing and smell respectively.

The sensory and motor regions of the cortex are connected to nearby **association areas** that interpret sensory data or coordinate a motor response. For example, the *premotor cortex,* or *motor association area*, is responsible for coordinating learned movements, such as the eye movements involved in reading. The differences between motor and sensory association areas are most apparent after localized brain damage. For example, someone with damage to the premotor cortex might understand written letters and words but be unable to read due to an inability to track along the lines on a printed page. In contrast, someone with a damaged **visual association area** can scan the lines of a printed page but cannot figure out what the letters mean.

Other "higher-order" centers receive information through axons from many different association areas. Three of these regions, also shown in Figure 11-5•, control extremely complex motor activities and perform complicated analytical functions.

1. The **general interpretive area** receives information from all the sensory association areas. This center is present in only one hemisphere, usually the left. Damage to this area affects the ability to interpret what is read or heard, even though the words are understood as individual entities. For example, an individual might understand the meaning of the words "sit" and "here" but be totally bewildered by the instruction "Sit here."

2. The **speech center** (*Broca's area*) lies in the same hemisphere as the general interpretive area. The speech center regulates the patterns of breathing and vocalization needed for normal speech. The corresponding regions on the opposite hemisphere are not "inactive," but their functions are less well defined. Damage to the speech center can manifest itself in various ways. Some people have difficulty speaking even when they know exactly what words to use; others talk constantly but use all the wrong words.

3. The **prefrontal cortex** of the frontal lobe coordinates information from the association areas of the entire cortex. In doing so it performs such abstract intellectual functions as predicting the future consequences of events or actions. The prefrontal cortex also has connections with other portions of the brain, such as the limbic system, discussed later.

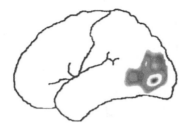

Seeing Words

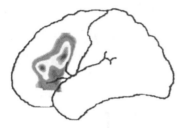

Generating Verbs

PET scans of the left side of the brain showing the different areas used for different language-related tasks.

Lying deep within the central white matter of each cerebral hemisphere are **cerebral nuclei**. They play an important role in the control of skeletal muscle tone and the coordination of learned movement patterns. These nuclei do not initiate a movement—that decision is a voluntary one—but once a movement is underway, the cerebral nuclei provide pattern and rhythm. For example, during a simple walk the cerebral nuclei control the cycles of arm and thigh movements that occur between the time the decision is made to "start walking" and the time the "stop" order is given.

THE LIMBIC SYSTEM

The **limbic** (LIM-bik) **system** includes several groups of neurons and tracts of white matter along the border (*limbus* means "border") between the cerebrum and diencephalon. This system is concerned with (1) the sense of smell and (2) long-term memory storage. One part of the limbic system, the **hippocampus**, plays a vital role in learning and the storage of long-term memories. Damage to the hippocampus in Alzheimer's disease interferes with memory storage and retrieval. The limbic system also includes centers within the hypothalamus responsible for (1) emotional states, such as rage, fear, and sexual arousal, and (2) the control of reflexes that can be consciously activated, such as the chewing, licking, and swallowing movements associated with eating.

✔ Mary suffers a head injury that damages her primary motor cortex. Where is this area located?

✔ After suffering a head injury in an automobile accident, David has difficulty comprehending what he hears or reads. This might indicate damage to what portion of the brain?

THE DIENCEPHALON

The diencephalon (see Figure 11-1•, p. 242) provides switching and relay centers that integrate the conscious and unconscious sensory and motor pathways. It contains a central chamber, the *third ventricle*, that is filled with cerebrospinal fluid. The top of the diencephalon contains a choroid plexus and the *pineal gland*. The pineal gland will be discussed in Chapter 13. The sides of the diencephalon make up the thalamus; below the thalamus is the hypothalamus.

THE THALAMUS

The **thalamus** is the final relay point for all the sensory information, other than smell, that reaches our conscious awareness. It acts as a filter, passing on only a small portion of the arriving sensory information. Only about 1 percent of arriving sensory information reaches our conscious awareness. The rest is relayed to other areas of the brain, where it affects centers that operate at the subconscious level. The thalamus also plays a role in the coordination of motor commands issued at the conscious or subconscious level.

THE HYPOTHALAMUS

The **hypothalamus** (1) contains centers associated with the emotions of rage, pleasure, pain, thirst, hunger, and sexual arousal; (2) adjusts and coordinates the activities of autonomic centers in the pons and medulla oblongata; (3) coordinates neural and endocrine activities; (4) produces a variety of hormones, including *antidiuretic hormone (ADH)* and *oxytocin*; (5) maintains normal body temperature; and (6) coordinates voluntary and autonomic functions.

THE BRAIN STEM

The brain stem is made up of the midbrain, the pons, and the medulla oblongata (see Figure 11-1b•, p. 242).

THE MIDBRAIN

The midbrain contains various nuclei (groups of neuron cell bodies) and ascending and descending tracts. Different functions are associated with the different nuclei. For example, some of these groups of neurons direct the reflex responses to sudden visual and auditory stimuli (such as a flash of light or a loud noise). The midbrain also contains motor nuclei for two cranial nerves concerned with eye movements. Other nuclei assist the cerebrum and cerebellum by regulating muscle tone and adjusting posture.

The midbrain also contains the headquarters of the **reticular formation**, one of the most important brain stem components. The reticular formation extends the length of the brain stem. Within the midbrain, the reticular formation contains the *reticular activating system*, or *RAS*. This system directly affects the activity of the cerebral cortex. When the RAS is inactive, so are we; when the RAS is stimulated, so is our state of attention or wakefulness.

reticulum, a network
reticular formation: an extensive interconnected network of nuclei that extends throughout the brain stem; the portion in the midbrain includes the reticular activating system

THE PONS

The pons links the cerebellum with the midbrain, diencephalon, cerebrum, and spinal cord. One group of nuclei within the pons includes the sensory and motor nuclei for four of the cranial nerves. Other nuclei are concerned with the subconscious control of the pace and depth of respiration. Tracts passing through the pons link the cerebellum with the brain stem, cerebrum, and spinal cord.

THE MEDULLA OBLONGATA

The medulla oblongata physically connects the brain with the spinal cord, and many of its functions are directly related to this fact. For example, all communication be-

Clinical Note

Brain Waves

The electrical activity of the brain is often monitored to assess brain activity. The brain contains billions of nerve cells, and their activity generates an electrical field that can be measured by placing electrodes on the brain or outer surface of the head. The electrical activity changes constantly as different nuclei and regions of the cerebral cortex are stimulated or quiet down. A printed record of the electrical activity of the brain is called an *electroencephalogram (EEG)*. The electrical patterns observed on the EEG are called **brain waves.**

There are four major types of brain waves. *Alpha waves* are found in normal adults under resting conditions but are re-placed by *beta waves* during times of concentration. *Theta waves* appear in the brains of children and in stressed adults and may indicate a brain disorder. *Delta waves* appear in the brains of infants and in the brains of adults during deep sleep or in cases of brain damage.

Electrical activity in each of the two cerebral hemispheres is usually synchronized, or in step. A lack of synchrony can be used to detect localized areas of damage. For example, a tumor or injury affecting one hemisphere often changes the pattern of activity in that hemisphere. An EEG would then show that the two hemispheres are no longer "in step."

tween the brain and spinal cord involves tracts that ascend or descend through the medulla oblongata. These tracts often synapse in the medulla oblongata, in sensory or motor nuclei that act as relay stations. In addition to these nuclei, the medulla ob-longata contains sensory and motor nuclei associated with five of the cranial nerves.

Many vital autonomic functions are controlled by the medulla oblongata. Reflex centers in this region receive inputs from cranial nerves, the cerebral cortex, and other portions of the brain stem. The output of these centers controls or adjusts the activi-ties of one or more peripheral systems. For example, the *cardiovascular centers* adjust heart rate, the strength of cardiac contractions, and the flow of blood through pe-ripheral tissues. The *respiratory rhythmicity centers* set the basic pace for respiratory movements, and their activities are adjusted by the respiratory centers of the pons.

THE CEREBELLUM

The cerebellum (see Figure 11-1•) performs two important functions: (1) it makes rapid adjustments in muscle tone and position to maintain balance and equilibri-um by modifying the activity within the brain stem, and (2) it programs and fine-tunes all our movements. These functions are performed indirectly, by regulating motor activity in the cerebrum and brain stem. The cerebellum may be perma-nently damaged by trauma or stroke, or temporarily affected by drugs such as al-cohol. These alterations can produce *ataxia* (a-TAK-sē-a), a disturbance in balance. Ataxia means "a lack of order."

THE CRANIAL NERVES

The cranial nerves are nerves of the peripheral nervous system that connect to the brain rather than the spinal cord. The 12 pairs of cranial nerves, shown in Figure 11-6•, are numbered according to their position along the axis of the brain. The prefix N designates a cranial nerve, and Roman numerals are used to distinguish the in-dividual nerves. For example, N I refers to the first cranial nerve, the *olfactory nerve.*

DISTRIBUTION AND FUNCTION

Functionally, each cranial nerve can be classified as primarily sensory, primarily motor, or mixed (sensory and motor). In addition, cranial nerves also distribute au-tonomic fibers to PNS ganglia, just as spinal nerves deliver them to ganglia along the spinal cord. Examples include N III, VII, IX, and X. The distribution and func-tions of the cranial nerves are described next and summarized in Figure 11-6•.

✔ Damage to nuclei of the thala-mus would interfere with the functions of which of the senses?

✔ What area of the diencephalon would be stimulated by changes in body temperature?

✔ The medulla oblongata is one of the smallest sections of the brain, yet damage there can cause death, whereas similar damage in the cerebrum might go unnoticed. Why?

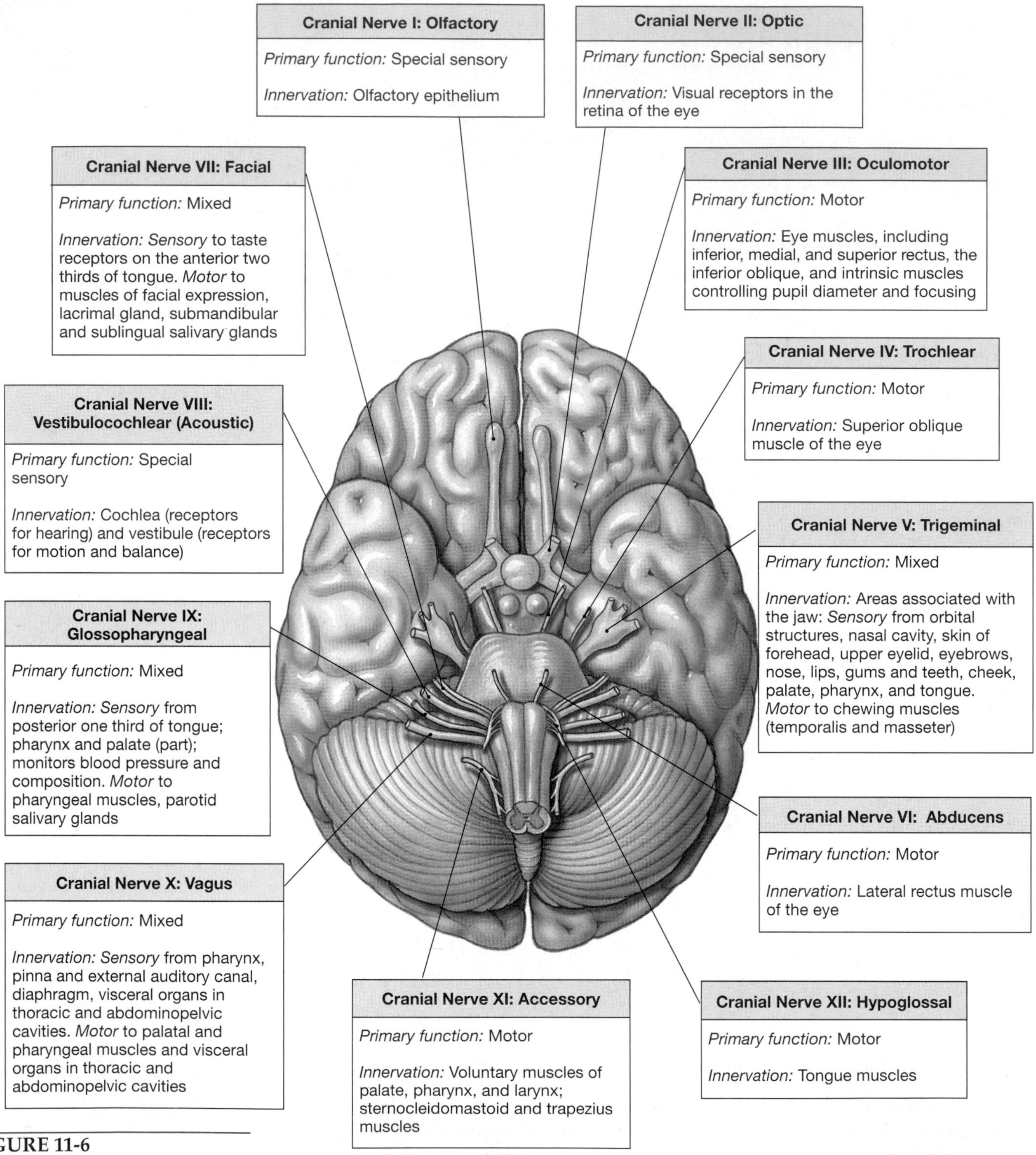

Cranial Nerve I: Olfactory

Primary function: Special sensory

Innervation: Olfactory epithelium

Cranial Nerve II: Optic

Primary function: Special sensory

Innervation: Visual receptors in the retina of the eye

Cranial Nerve VII: Facial

Primary function: Mixed

Innervation: Sensory to taste receptors on the anterior two thirds of tongue. *Motor* to muscles of facial expression, lacrimal gland, submandibular and sublingual salivary glands

Cranial Nerve III: Oculomotor

Primary function: Motor

Innervation: Eye muscles, including inferior, medial, and superior rectus, the inferior oblique, and intrinsic muscles controlling pupil diameter and focusing

Cranial Nerve IV: Trochlear

Primary function: Motor

Innervation: Superior oblique muscle of the eye

Cranial Nerve VIII: Vestibulocochlear (Acoustic)

Primary function: Special sensory

Innervation: Cochlea (receptors for hearing) and vestibule (receptors for motion and balance)

Cranial Nerve V: Trigeminal

Primary function: Mixed

Innervation: Areas associated with the jaw: *Sensory* from orbital structures, nasal cavity, skin of forehead, upper eyelid, eyebrows, nose, lips, gums and teeth, cheek, palate, pharynx, and tongue. *Motor* to chewing muscles (temporalis and masseter)

Cranial Nerve IX: Glossopharyngeal

Primary function: Mixed

Innervation: Sensory from posterior one third of tongue; pharynx and palate (part); monitors blood pressure and composition. *Motor* to pharyngeal muscles, parotid salivary glands

Cranial Nerve VI: Abducens

Primary function: Motor

Innervation: Lateral rectus muscle of the eye

Cranial Nerve X: Vagus

Primary function: Mixed

Innervation: Sensory from pharynx, pinna and external auditory canal, diaphragm, visceral organs in thoracic and abdominopelvic cavities. *Motor* to palatal and pharyngeal muscles and visceral organs in thoracic and abdominopelvic cavities

Cranial Nerve XI: Accessory

Primary function: Motor

Innervation: Voluntary muscles of palate, pharynx, and larynx; sternocleidomastoid and trapezius muscles

Cranial Nerve XII: Hypoglossal

Primary function: Motor

Innervation: Tongue muscles

•**FIGURE 11-6**
The Cranial Nerves.
Diagram of the attachments of the 12 pairs of cranial nerves on the underside of the brain, their function(s), and their connections with specific body structures.

Olfactory Nerves (N I)

The first pair of cranial nerves, the olfactory nerves, are the only cranial nerves attached to the cerebrum. The rest start or end within nuclei of the diencephalon or brain stem. These nerves carry special sensory information responsible for the sense of smell.

Optic Nerves (N II)

The **optic nerves** carry visual information from the eyes to nuclei in the thalamus. From there, the information is relayed to the visual cortex of the cerebral hemispheres and to various centers in the brain stem.

Oculomotor Nerves (N III)

Each **oculomotor nerve** innervates four of the six muscles that move an eye. (These muscles are detailed in Chapter 12.) These nerves also carry autonomic (*parasympathetic*) fibers to intrinsic eye muscles (muscles inside the eyeball) that control the shape of the lens and the amount of light entering the eye.

oculo, eye + *motorius*, moving
oculomotor nerve: the primary nerve controlling the muscles that move the eye

Trochlear Nerves (N IV)

The **trochlear** (TRŌK-lē-ar) **nerves** are the smallest cranial nerves. Each nerve innervates one of the six muscles that move an eye.

Trigeminal Nerves (N V)

The **trigeminal** (trī-JEM-i-nal) **nerves** are very large cranial nerves. These nerves provide sensory information from the head and face and motor control over the chewing muscles. As the name implies, the *trigeminal* has three major branches. The *ophthalmic branch* provides sensory information from the orbit around the eye, the nasal cavity and sinuses, and the skin of the forehead, eyebrows, eyelids, and nose. The *maxillary branch* provides sensory information from the lower eyelid, upper lip, cheek, and nose. It also monitors the upper gums and teeth, the palate (roof of the mouth), and portions of the pharynx (throat). The *mandibular branch*, the largest of the three, provides sensory information from the skin of the temples, the lower gums and teeth, the salivary glands, and the anterior portions of the tongue. It also provides motor control over the chewing muscles.

tri, three + *geminus*, twin
trigeminal nerve: the largest cranial nerve, with three major branches

Abducens Nerves (N VI)

Each **abducens** (ab-DOO-senz) **nerve** innervates the last of the six oculomotor muscles. The nerve reaches the bony orbit around the eye together with the oculomotor and trochlear nerves (N III and N IV).

abducens, to abduct or draw away
abducens nerve: the cranial nerve that controls the muscle that turns an eye away from the nose, so that it looks to the side

Facial Nerves (N VII)

The **facial nerves** are mixed nerves of the face. The motor fibers produce facial expressions by controlling the superficial muscles of the scalp and face and deep muscles near the ear. The sensory fibers monitor the state of contraction in the facial muscles, provide deep pressure sensations over the face, and carry taste information from receptors on the anterior two thirds of the tongue.

Vestibulocochlear Nerves (N VIII)

The **vestibulocochlear nerves** monitor the sensory receptors of the inner ear. Each vestibulocochlear nerve has two branches. A **vestibular nerve** originates in the part of the inner ear concerned with balance sensations, and it conveys information on position, movement, and balance. The second branch is called the **cochlear** (KOK-lē-ar) **nerve**, and it monitors the portion of the inner ear responsible for hearing.

Glossopharyngeal Nerves (N IX)

The **glossopharyngeal** (glos-ō-fah-RIN-je-al) **nerves** are mixed nerves. The sensory portion of each nerve provides taste sensations from the posterior third of the tongue and monitors the blood pressure and dissolved gas concentrations within a major blood vessel of the neck (the *carotid artery*). The motor portion of N IX controls the pharyngeal muscles involved in swallowing.

glossum, tongue + *pharyngeus*, pharynx
glossopharyngeal nerve: a mixed cranial nerve innervating structures of the tongue and neck

Vagus Nerves (N X)

The **vagus** (VĀ-gus) **nerves** provide sensory information from the external ear canals, the diaphragm, taste receptors in the pharynx, and from visceral receptors along the respiratory tract and digestive tract from the esophagus to the last portions of the large intestine. This sensory information is vital to the autonomic control of visceral function, but the sensations usually fail to reach the cerebral cortex. As a result, we are not consciously aware of the sensations that trigger various autonomic responses. The motor components of the vagus include parasympathetic fibers that control skeletal muscles of the soft palate and pharynx and affect cardiac muscle, smooth muscle, and glands throughout the areas monitored the vagus. Its far-flung activities reflect its Latin root word, *vagus*, which means "to wander."

Accessory Nerves (N XI)

The **accessory nerves**, sometimes called the *spinal accessory nerves*, differ from other cranial nerves in that some of their axons originate in the gray matter of the first five cervical vertebrae. The *medullary branch* of each accessory nerve innervates the voluntary swallowing muscles of the soft palate and pharynx, and the laryngeal muscles that control the vocal cords and produce speech. The *spinal branch* controls the *sternocleidomastoid* and *trapezius* muscles. ∞ *[pp. 198, 202]*

Hypoglossal Nerves (N XII)

The **hypoglossal** (hī-pō-GLOS-al) **nerves** provide voluntary control over the skeletal muscles of the tongue.

Few people are able to remember the names, numbers, and functions of the cranial nerves without a struggle. Many use phrases in which the first letter of each word represents the cranial nerves, such as *Oh, Once One Takes The Anatomy Final, Very Good Vacations Are Heavenly.*

✔ Damage to which of the cranial nerves do you think could result in death?

✔ John is experiencing problems in moving his tongue. His doctor tells him it is due to pressure on a cranial nerve. Which cranial nerve is involved?

✔ What symptoms would you associate with damage to the abducens nerve (N VI)?

AGING AND THE NERVOUS SYSTEM

The aging process affects all the body systems, and the nervous system is no exception. Structural and functional changes begin shortly after maturity (probably by age 30) and accumulate over time. Although an estimated 85 percent of the elderly (above age 65) lead relatively normal lives, there are noticeable changes in mental performance and CNS functioning.

Common age-related anatomical changes in the nervous system include the following:

1. *A reduction in brain size and weight, primarily from a decrease in the volume of the cerebral cortex.* The brains of elderly individuals have narrower gyri and wider sulci than those of young persons.

2. *A reduction in the number of neurons.* Brain shrinkage has been linked to a loss of neurons, primarily in the cerebral cortex.

3. *A decrease in blood flow to the brain.* With age, fatty deposits gradually accumulate in the walls of blood vessels. Like a kink in a garden hose, these deposits reduce the rate of blood flow through arteries. (This process, called *arteriosclerosis*, affects arteries throughout the body; it is discussed further in Chapter 15.)

4. *Changes in the structure of synapses in the brain.* The degree of branching between neurons and the number of synapses decreases, and the rate of neurotransmitter production declines.

5. *Changes in and around CNS neurons.* Many neurons in the brain begin accumulating abnormal intracellular deposits, such as pigments or abnormal proteins that have no known function. There is evidence that these changes occur in all aging brains, but when present in excess they seem to be associated with clinical abnormalities.

As a result of these anatomical changes, neural function is impaired. For example, memory consolidation, the conversion of short-term memory to long-term memory (such as repeating a telephone number), often becomes more difficult. It often becomes harder to recall memories, especially those of the recent past. The sensory systems, notably hearing, balance, vision, smell, and taste, become less acute. Light must be brighter, sounds louder, and smells stronger before they are perceived. Reaction times are slowed, and many reflexes become weaker or even disappear. There is a decrease in the precision of motor control, and it takes longer to perform a given motor pattern than it did 20 years earlier.

For roughly 85 percent of the elderly, these changes do not interfere with their abilities to function in society. But for as yet unknown reasons, many become incapacitated by progressive CNS changes.

A SUMMARY OF THE DISORDERS OF THE NERVOUS SYSTEM

As you should now realize, the nervous system is the most complex of all the body's systems. Neural tissue is extremely delicate, and when homeostatic mechanisms break down under the stress of genetic or environmental factors, infection, or trauma, symptoms of neurological disorders appear.

DIAGNOSIS OF NEUROLOGICAL DISORDERS

There are hundreds of different disorders of the nervous system. Figure 11-7• introduces several major categories of nervous system disorders. Examples of disorders that affect the spinal cord and spinal nerves were discussed in Chapter 10.

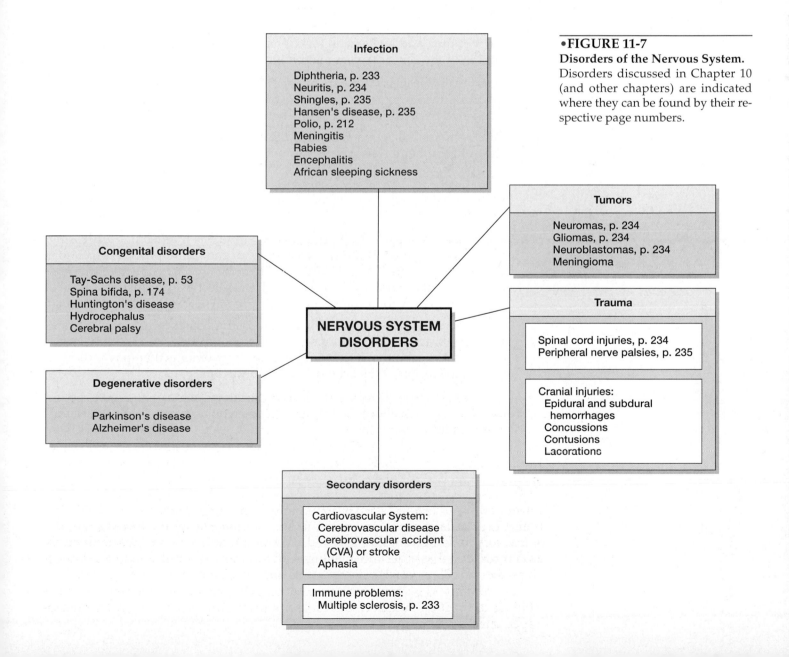

•FIGURE 11-7
Disorders of the Nervous System. Disorders discussed in Chapter 10 (and other chapters) are indicated where they can be found by their respective page numbers.

Infection
Diphtheria, p. 233
Neuritis, p. 234
Shingles, p. 235
Hansen's disease, p. 235
Polio, p. 212
Meningitis
Rabies
Encephalitis
African sleeping sickness

Tumors
Neuromas, p. 234
Gliomas, p. 234
Neuroblastomas, p. 234
Meningioma

Congenital disorders
Tay-Sachs disease, p. 53
Spina bifida, p. 174
Huntington's disease
Hydrocephalus
Cerebral palsy

NERVOUS SYSTEM DISORDERS

Trauma
Spinal cord injuries, p. 234
Peripheral nerve palsies, p. 235

Cranial injuries:
Epidural and subdural hemorrhages
Concussions
Contusions
Lacerations

Degenerative disorders
Parkinson's disease
Alzheimer's disease

Secondary disorders
Cardiovascular System:
Cerebrovascular disease
Cerebrovascular accident (CVA) or stroke
Aphasia

Immune problems:
Multiple sclerosis, p. 233

The Symptoms of Neurological Disorders

Despite the number and complexity of neurological disorders, there are a few common symptoms that accompany many of them:

- *Headache:* Roughly 90 percent of headaches are either *tension headaches*, usually due to muscle tension, or *migraine headaches*, which can be due to both neurological and circulatory factors. Neither of these conditions is life threatening.

- *Muscle weakness:* Muscle weakness can have an underlying neurologic basis. ∞ *[p. 206]* The examiner must determine the origin of the symptom. Myopathies (muscle disease) must be differentiated from neurologic diseases such as demyelinating disorders, neuromuscular junction dysfunction, and peripheral nerve damage.

- *Paresthesias:* Loss of feeling, numbness, or tingling sensations may develop after damage to (1) a sensory nerve or (2) sensory pathways inside the CNS. The effects may be temporary or permanent. For example, a pressure palsy may last a few minutes, whereas the paresthesia that develops distal to an area of severe spinal cord damage will probably be permanent.

The Neurological Examination

A *neurological examination* attempts to trace the source of the problem, by assessing sensory, motor, and intellectual functions of the nervous system. Examples of factors noted in the physical examination include:

- *State of consciousness:* There are many different levels of consciousness, ranging from unconscious and incapable of being aroused, to fully alert and attentive, to hyperexcitable.

- *Reflex activity:* The general state of the nervous system, and especially the state of peripheral sensory and motor innervation, can be checked by testing specific reflexes. For example, the knee-jerk reflex will not be normal if there has been damage to associated segments of the lumbar spinal cord, their spinal nerve roots, or the peripheral nerves involved in the reflex. ∞ *[p. 228]*

- *Abnormal speech patterns:* Normal speech involves intellectual processing, motor coordination at the speech centers of the brain, precise respiratory control, regulation of tension in the vocal cords, and adjustment of the muscles of the palate and face. Problems with the selection, production, or use of words commonly follows damage to the cerebral hemispheres, as in a stroke.

- *Abnormal motor patterns:* An individual's posture, balance, and way of walking, or *gait*, are useful indicators of the level of motor coordination. Physicians also ask about abnormal involuntary movements that may indicate a *seizure*, a temporary disorder of cerebral function.

A number of diagnostic procedures and laboratory tests can be used to obtain additional information about the status of the nervous system. Table 11-1• summarizes information about these procedures.

DISORDERS OF THE BRAIN

Infection

Encephalitis (en-sef-ah-LĪ-tis) is an inflammation of the brain (*enkephalos* means brain). It is usually associated with viruses. In many cases, the meninges of the spinal cord and brain are also involved, a condition called **meningitis**. Both viruses and bacteria may infect the meninges. In addition to bacterial and viral infections, the central nervous system may also be infected by protozoan parasites.

Table 11-1   Examples of Procedures and Tests Used in the Diagnosis of Nervous System Disorders

Diagnostic Procedure	Method and Result	Representative Uses
Lumbar puncture (spinal tap)	Needle aspiration of CSF from the subarachnoid space in the lumbar area of the spinal cord	See CSF analysis for diagnostic uses
Skull X-ray	Standard X-ray	Detection of fracture and possible sinus involvement
Electroencephalography (EEG)	Electrodes placed on the scalp detect electrical activity of the brain, and EEG produces graphic record	Detection of abnormalities in frequency and amplitude of brain waves, due to cranial trauma or neurological disorders such as seizures
Computerized tomography (CT) scan of the brain	Standard CT; contrast media are commonly used	Detection of tumors, cerebrovascular abnormalities, such as aneurysms (weakened areas in vessel walls), scars, strokes, or areas of edema
Cerebral angiography and digital subtraction angiography	Dye is injected into an artery of the neck, and the movement of the dye is observed in a series of X-rays; digital subtraction angiography transfers information to a computer for image enhancement	Detection of abnormalities in the cerebral vessels, such as aneurysms or blockages
Positron emission tomography (PET) scan	Radiolabeled compounds injected into the circulation accumulate at specific areas of the brain; the radiation emitted is monitored by a computer that generates a reconstructed image	Determination of blood flow to the brain; detection of focal points of brain activity; also useful in the diagnosis of Parkinson's disease and Alzheimer's disease
Magnetic resonance imaging (MRI)	Standard MRI; contrast media are commonly used to enhance visualization	Detection of brain tumors, hemorrhaging, edema, spinal cord injury, and other structural abnormalities

Laboratory Test	Normal Values	Values
Analysis of CSF		
Pressure of CSF	<200 cm H_2O	Pressure higher than 200 cm H_2O is considered abnormal possibly indicating hemorrhaging, tumor formation, or infection
Color of CSF	Clear and colorless	Increased turbidity suggests hemorrhage or faulty puncture technique
Glucose in CSF	50–75 mg/dl	Decreased levels are found when CSF tumors or infections are present
Protein in CSF	15–45 mg/dl	Elevated levels occur in some infectious processes, such as meningitis and encephalitis; may be elevated during inflammation or following tumor formation
Cells present in CSF	No RBCs present; WBC count should be less than 5 per mm^3	RBCs appear with subarachnoid hemorrhage; neutrophil count increases in bacterial infections, such as bacterial meningitis; lymphocyte count increases in viral meningitis
Culture of CSF	Microorganism infecting brain or spinal cord can be cultured for identification and determination of antibiotic sensitivity	Determination of causative agent in meningitis or brain abscess

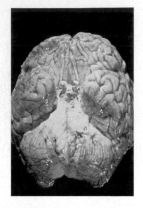

•FIGURE 11-8
Meningitis.
The brain of a patient who died of *Streptococcus pneumoniae* meningitis. Note the accumulation of pus from the infection.

Bacteria. Examples of infections of the brain and central nervous system caused by bacteria include meningitis and *brain abscesses*. Different forms of meningitis strike different age groups and are caused by different bacteria. For example, meningitis caused by *Haemophilus influenzae* (hē-MOF-il-us in-flū-EN-zī) occurs most often during the first year of life and rarely affects children over 5 years old. Untreated children who recover from this infection often have some degree of mental retardation. Vaccination has dramatically reduced the incidence of *Haemophilus* meningitis. *Neisseria meningitidis* (nī-SĒ-rē-a me-NIN-ji-tī-dis) infects individuals between 5–40 years, and is called *meningococcal* meningitis. These bacteria infect the upper throat, pass into the blood, and are carried to the meninges. If they also spread to all parts of the body, death can occur within hours. During World War II, this was the leading cause of death by infectious disease of U.S. armed forces. Treatment involves antibiotics. Adults over the age of 40 (and occasionally children) may contract *Streptococcus* meningitis, which is caused by *Streptococcus pneumoniae* (noo-MŌ-nē-ī). Bacteria that initially infect the lungs, sinuses, and ears can pass through the blood to the meninges (Figure 11-8•).

The most common clinical assessment involves checking for a "stiff neck" by asking the patient to touch the chin to the chest. Meningitis affecting the cervical portion of the spinal cord results in a marked increase in the muscle tone, or tension, of the extensor muscles of the neck. So many motor units become activated that voluntary or involuntary flexion of the neck becomes painfully difficult, if not impossible.

An **abscess** is a localized collection of pus lying within an enclosed tissue space. A **brain abscess** may be caused by different types of bacteria. They reach the brain through wounds to the head or by blood from other infected areas. As they grow, they compress the brain, affecting the functions of the compressed regions.

Viruses. Examples of viral infections of the brain and central nervous system include viral meningitis, *rabies,* and different varieties of encephalitis. **Viral meningitis** is caused by different types of viruses, including the *mumps* virus. In contrast to untreated bacterial meningitis, which is usually fatal, viral meningitis is usually not fatal.

Rabies is an acute disease of the central nervous system. The rabies virus infects and kills mammals worldwide. Not all mammals are susceptible to rabies, and bats, for example, can harbor the virus and not become ill. Because rabies can also infect salivary glands, it is usually transmitted by animal bites. The virus multiplies first in the injured area of a bite, then enters nerves and is carried to the spinal cord (Figure 11-9•).

During the first few days after exposure, the individual may experience headache, fever, muscle pain, nausea, and vomiting. The victim then enters a phase marked by extreme excitability, hallucinations, muscle spasms, and disorientation. There is difficulty in swallowing, and the accumulation of saliva makes the individual appear to be "foaming at the mouth." Within 10–14 days of the first symptoms, coma and death occur.

Treatment, which must begin almost immediately after exposure, consists of injections that contain antibodies against the rabies virus followed by a series of vaccinations against rabies. This postexposure treatment may not be sufficient after a massive infection, which can lead to death in as little as four days. Individuals such as veterinarians or field biologists who are at high risk of exposure commonly take a pre-exposure series of vaccinations. These injections bolster the immune defenses and improve the effectiveness of the postexposure treatment. Without preventive treatment, rabies infection in humans is always fatal.

There are four major types of encephalitis in the U.S., each caused by a different virus. The names given the various forms of encephalitis indicate their geographic distributions. For example, *eastern equine encephalitis* occurs most often in the eastern U.S., *western equine encephalitis* is in the western U.S., *Venezuelan*

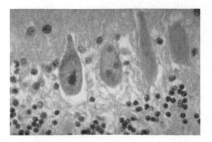

•FIGURE 11-9
The Rabies Virus.
Clusters of the rabies viruses in nervous tissue indicate infection.

equine encephalitis is in Florida, Texas, Mexico, and South America. *St. Louis encephalitis* occurs in the central U.S. All are introduced into the body by the bites of infected mosquitoes. Horses are intermediate hosts for the first three forms, and English sparrows are associated with St. Louis encephalitis. Initially, the virus lives in the skin, then spreads to lymph nodes, the blood, and, finally, the CNS where the neurons are destroyed. Eastern equine encephalitis is the most serious and is fatal in 50–80 percent of all cases. Treatment of encephalitis is aimed at treating the symptoms. Much attention is given to mosquito control in the areas of infection.

Protozoa. **African sleeping sickness** is a disease caused by flagellated protozoans that are carried to humans by flies. Two varieties of the protozoan *Trypanosoma bruzei* (trī-pan-ō-SŌ-ma BREW-sē) are responsible for this condition, also called *trypanosomiasis*. When humans are bitten by an infected *tsetse* (tset-sē) *fly*, hundreds of the protozoans enter the blood (Figure 11-10•). After growing in the blood, they take up residence in the lymph nodes, and then finally invade the CNS. Initially, symptoms include shortness of breath, anemia, and weakness, which steadily worsen. Symptoms of CNS infection include headache, tremors, an unsteady gait, and then, paralysis, coma, and death. Treatment involves the use of different drugs depending on the stage of the illness at the time of diagnosis. There is no vaccine to prevent infection.

A summary of diseases of the nervous system are listed in Table 11-2. Also included in the table are infectious diseases discussed in Chapter 10.

•FIGURE 11-10
The Pathogen Causing African Sleeping Sickness.
African sleeping sickness is caused by *Trypanosoma brucei*, the protozoan shown in this blood smear.

Trauma

Cranial trauma is a head injury resulting from harsh contact with another object. Head injuries account for over half the deaths attributed to all of the different

---

Table 11-2	Examples of Infectious Diseases of the Nervous System	
Disease	**Organism(s)**	**Description**
Bacterial Diseases		
Hansen's disease (leprosy)	*Mycobacterium leprosae*	Progresses slowly; invades nerves and produces sensory loss and motor paralysis; cartilage and bone may degenerate
Bacterial meningitis		Inflammation of the spinal or cranial meninges
	Haemophilus influenzae	*Haemophilus meningitis*; usually infects children (age 2 months–5 years); vaccine available
	Neisseria meningitidis	*Meningococcal meningitis*; usually infects children and adults (age 5–40 years); treatment with antibiotics
	Streptococcus pneumoniae	*Streptoccocal meningitis*; usually infects adults over 40; high mortality (40%)
Brain abscesses	Various bacteria	Infection increases in size and compresses the brain
Viral Diseases		
Poliomyelitis	Polioviruses	Polio has different forms, only one attacks motor neurons leading to paralysis of limbs and muscle atrophy; vaccine available
Rabies	Rabies virus	Virus invades the central nervous system through peripheral nerves; untreated cases fatal; treatment involves rabies antitoxin
Encephalitis	Different encephalitis viruses	Inflammation of the brain; fever and headache; no vaccine available; transmission occurs by mosquitoes
		Eastern equine encephalitis is most lethal (50–75%)
Parasitic Diseases		
African sleeping sickness	*Trypanosoma brucei*	Caused by a flagellated protozoan; infection occurs through bite of tsetse fly; infects blood, lymph nodes, and then nervous system; symptoms include headache, tiredness, weakness, and paralysis, before coma and death; no vaccine available

types of trauma. There are roughly 8 million cases of cranial trauma each year in the United States, and over a million of these involve intracranial hemorrhaging (such as an *epidural hemorrhage*), concussion, contusion, or laceration of the brain. The characteristics of spinal concussion, contusion, and laceration were given on p. 234; comparable descriptions are applied to injuries of the brain.

Concussions typically accompany even minor head injuries. A concussion involves a temporary loss of consciousness and some degree of amnesia. Physicians examine any concussed individual quite closely and may X-ray the skull to check for skull fractures or cranial bleeding. Mild concussions produce a brief interruption of consciousness and little memory loss. Severe concussions produce extended periods of unconsciousness and abnormal neurological functions. Severe concussions are typically associated with contusions (bruises) or lacerations (tears) of the brain tissue; the possibilities for recovery vary with the areas affected. Extensive damage to the reticular formation of the brain stem may produce a permanent state of unconsciousness, whereas damage to the lower brain stem generally proves fatal. ∞ *[p. 248]*

A severe head injury may damage blood vessels within the meninges and cause bleeding into the cranial cavity. The most serious is a break in an artery, because the high blood pressure results in extensive bleeding and tissue distortion. If blood is forced between the dura mater and the cranium, then the condition is called an **epidural hemorrhage**. As it pushes against the dura mater, the blood distorts the underlying tissues of the brain. The individual loses consciousness from minutes to an hour after injury, and death quickly follows in untreated cases. Damaged veins do not produce serious symptoms so quickly because blood pressure is lower in veins and there is less blood loss and tissue distortion. In such cases, unconsciousness may occur from hours to days or even weeks after the original injury. In **subdural hemorrhages**, blood leaking from veins usually accumulates in the lower layer of the dura mater. Subdural hemorrhages are roughly twice as common as epidural hemorrhages.

Tumors

A benign tumor of the meninges is called a **meningioma**. It forms in cells of the arachnoid layer and may become attached to the dura mater. Because they are separated from underlying nervous tissue, they can often be completely removed surgically. Tumors of the central nervous system and those associated with the autonomic nervous system were discussed in Chapter 10 (p. 234).

Congenital Disorders

Examples of inherited nervous system disorders discussed previously are Tay-Sachs disease (see p. 53) and spina bifida (p. 174). Another example of an inherited disorder is Huntington's disease. Two congenital problems include hydrocephalus and cerebral palsy.

Huntington's disease is an inherited disease marked by a progressive deterioration of mental abilities. There are approximately 25,000 Americans with this condition. In Huntington's disease, the cerebral nuclei and frontal lobes of the cerebral cortex degenerate. The cause of this deterioration is not known, but the genetic abnormality has been identified. The first signs of the disease generally appear in early to middle adulthood, between the ages of 35–50. As you would expect in view of the areas affected, the symptoms involve difficulties in performing voluntary and involuntary patterns of movement and a gradual decline in intellectual abilities that eventually lead to *dementia* and death. Screening tests can now detect the presence of the abnormal gene responsible for Huntington's disease. Unfortunately, there is no effective treatment. The children of a person with Huntington's disease have a 50% risk of receiving the gene and developing the disease.

Hydrocephalus, or "water on the brain," refers to an abnormal buildup of cerebrospinal fluid. Infants have flexible skulls that expand as a result (Figure 11-11●). In adults with hydrocephalus, rising fluid pressure inside the cranium distort the brain and produce a variety of symptoms. The problem is generally due to some interference in the normal circulation of cerebrospinal fluid (CSF), such that CSF builds up in the ventricles of the brain. Treatment generally involves the installation of a *shunt*, a bypass that either avoids the blockage site or drains the excess CSF. In children, the shunt is removed if further growth of the brain eliminates the blockage or when further development of the meninges permits normal recycling of CSF. Meningeal development is usually sufficiently advanced by the time the child reaches 3 years of age.

Cerebral palsy refers to a number of disorders that affect voluntary motor movements. They appear during infancy or childhood, and persist throughout the life of the individual. The cause may be trauma associated with premature or stressful birth, maternal infection or exposure to drugs, including alcohol, or a genetic defect in the development of the motor pathways. Problems with labor and delivery may cause an interruption in blood flow and oxygen supply to the fetus. CNS functions are affected if oxygen concentrations in the fetal blood decline significantly for as little as 5–10 minutes. Abnormalities in motor skills, posture and balance, memory, speech, and learning abilities may characterize cerebral palsy.

Degenerative Disorders

Parkinson's disease is a brain disorder that causes a general increase in muscle tone. Voluntary movements become hesitant and jerky, a condition called **spasticity**. The characteristics of spasticity arise because movement cannot occur until one muscle group manages to overpower its antagonistic muscle groups. Individuals with Parkinson's disease show spasticity during voluntary movement and a continual tremor when at rest. A **tremor** represents a tug of war between antagonistic muscle groups that produces a background shaking of the limbs, in this case at a frequency of 4–6 cycles per second. Individuals with Parkinson's disease also have difficulty starting voluntary movements. Even changing one's facial expression requires intense concentration, and the individual acquires a blank, static expression. Other symptoms include a rigid posture and a slow, shuffling walk. In the late stages of this condition, other CNS effects, such as depression, hallucinations, and dementia (a loss of mental abilities) commonly appear. Most individuals with Parkinson's disease are elderly.

Parkinson's disease is caused by the degeneration of neurons in the cerebral nuclei. Normally, these neurons control the degree of activation at the motor cortex when voluntary movements are underway. As a result, muscle movements are smooth and continuous, not jerky. The degeneration of the cerebral nuclei results in reduced ability to control the precise amount of muscle activity.

In a normal individual, synapses within the cerebral nuclei are continuously releasing the neurotransmitter *dopamine*. Symptoms of Parkinson's disease develop when levels of dopamine decline. Although there is no cure, providing

•FIGURE 11-11
Hydrocephalus.
This infant has untreated hydrocephalus, a condition generally caused by impaired circulation or impaired removal of cerebrospinal fluid. The buildup of CSF leads to distortion of the brain and enlargement of the cranium. Such conditions may be treated by inserting a shunt that bypasses the blockage and drains the excess CSF.

the cerebral nuclei with dopamine can significantly reduce the symptoms for two-thirds of Parkinson's patients. The most common procedure involves the oral administration of the drug *L-DOPA* (*levodopa*), a compound that can cross the blood-brain capillaries and is then converted to dopamine. A more prolonged and effective reduction in the symptoms of Parkinson's has resulted from the experimental transplantation of dopamine-producing cells and fetal brain tissues.

Alzheimer's disease is the most common age-related, degenerative condition. It is a progressive disorder characterized by the loss of higher cerebral functions. This is the most common cause of **senile dementia**, or "senility." The first symptoms usually appear after 60 years of age, although the disease occasionally affects younger individuals. Alzheimer's disease has widespread impact on the elderly; an estimated 2 million people in the United States, including 5–8 percent of those over 65, have some form of the condition. It causes approximately 100,000 deaths each year.

In its characteristic form, Alzheimer's disease produces a gradual deterioration of mental organization. The individual loses memories, verbal and reading skills, and emotional control. As memory losses continue to accumulate, problems become more severe. The affected person may forget relatives, a home address, or how to use the telephone. The loss of memory affects both intellectual and motor abilities, and a patient with severe Alzheimer's disease has difficulty performing even the simplest motor tasks. Although by this time individuals with the disease are relatively unconcerned about their mental state or motor abilities, the condition can have devastating and emotional and economic effects on the immediate family, as the patient needs daily, round-the-clock care.

Secondary Disorders

Cardiovascular Problems. **Cerebrovascular diseases** are circulatory disorders that interfere with the normal blood supply to the brain. The severity of the condition depends on the involved vessel and the degree to which oxygen or nutrients are blocked. A *stroke*, or **cerebrovascular accident (CVA)**, occurs when the blood supply to a portion of the brain is shut off. Affected neurons begin to die in a matter of minutes. The loss of neurons is accompanied by a loss of brain function. Muscle weakness or paralysis is often the most dramatic sign of a stroke.

Aphasia is a disorder affecting language use. It affects the abilities to speak, read, write, or understand speech. In *global aphasia*, speech and comprehension are severely impaired. Global aphasia often results from damage to large areas of the cortex including the speech and language areas. Lesser degrees of aphasia commonly follow a smaller stroke. There is no initial period of global aphasia, and the individual can understand spoken and written words. The problems encountered with speaking or writing gradually fade. Many individuals with minor aphasia recover completely.

Immune Problems. Multiple sclerosis, or MS, is a disease characterized by recurrent incidents of demyelination that affects axons in the optic nerve, brain, and/or spinal cord. MS was discussed in Chapter 10 (p. 233).

DISORDERS OF THE CRANIAL NERVES

Testing Cranial Nerves

A variety of different tests are used to monitor the condition of specific cranial nerves. For example,

■ The olfactory nerve (I) is assessed by asking the subject to distinguish among different aromatic odors.

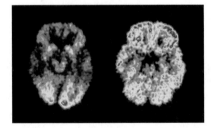

PET scan of the brain of an Alzheimer's patient (left) compared with a normal brain (right). Metabolic activity is high in red areas and low in blue areas.

✔ What disease is transmitted by the tsetse fly?

✔ What inherited disorder does not produce symptoms until adulthood?

- Cranial nerves II, III, IV, and VI are assessed while the vision and movement of the eyes are checked. First, the person is asked to hold the head still and track the movement of the examiner's finger with the eyes. For the eyes to track the finger through the visual field, the oculomotor muscles and their associated cranial nerves must be functioning normally. For example, if the person cannot track with the right eye a finger that is moving from left to right, there may be damage to the right *lateral rectus* muscle or to N VI on the right side.

- Cranial nerve V, which provides motor control over the muscles of chewing, can be checked by asking the person to clench the teeth. The jaw muscles are then palpated; if motor components of N V on one side are damaged, the muscles on that side will be weak or flaccid. Sensory components of N V can be tested by lightly touching areas of the forehead and side of the face.

- The facial nerve (N VII) is checked by watching the muscles of facial expression or asking the person to perform particular facial movements. Wrinkling the forehead, raising the eyebrows, pursing the lips, and smiling are controlled by the facial nerve. If a branch of N VII has been damaged, there will be muscle weakness or drooping on the affected side. For example, the corner of the mouth may sag and fail to curve upward when the person smiles. Special sensory components of N VII can be checked by placing solutions known to stimulate taste receptors on the anterior two-thirds of the tongue.

- The glossopharyngeal and vagus nerves (N IX and X) can be evaluated by watching the person swallow something. Examination of the soft palate arches and *uvula* for normal movement is also important.

- The accessory nerve (N XI) can be checked by asking the person to shrug the shoulders. Atrophy of the sternocleidomastoid or trapezius muscles may also indicate problems with the accessory nerve.

- The hypoglossal nerve (XII) can be checked by having the person extend the tongue and move it from side to side.

Cranial Nerve Problems

The Trigeminal Nerve (N V). **Tic douloureux** (doo-loo-ROO), or **trigeminal neuralgia**, is a painful condition affecting the face that causes the individual to wince or twitch (*douloureux* means painful). Sufferers complain of severe, almost totally debilitating pain that arrives with a sudden, shocking intensity and then disappears. It is stimulated by contact with the lip, tongue, or gums. In most cases, only one side of the face is involved, and the pain is along the sensory path of the maxillary and mandibular branches of the trigeminal nerve. This condition generally affects adults over 40 years of age; the cause is unknown. The pain can often controlled by drug therapy, but surgery is sometimes required.

The Facial Nerve (N VII). **Bell's palsy** is a cranial nerve disorder that results from a loss of function in the facial nerve. This condition is probably an immune-related inflammation following a viral infection. Symptoms include paralysis of facial muscles (including the eye and mouth) on the affected side and loss of taste sensations from the anterior two-thirds of the tongue. The condition may be painful. In most cases, normal nerve function returns after a few weeks or months.

CHAPTER REVIEW

Key Words

cerebral cortex: An extensive area of neural cortex covering the surface of the cerebral hemispheres.

cerebrospinal fluid (CSF): Fluid bathing the internal and external surfaces of the CNS; secreted by the choroid plexus.

cerebrum (SER-e-brum, se-RĒ-brum): The largest portion of the brain; consists of two cerebral hemispheres and includes the cerebral cortex (gray matter), white matter, and the cerebral nuclei.

cranial nerves: Peripheral nerves originating at the brain.

diencephalon (dī-en-SEF-a-lon): A division of the brain that includes the thalamus and hypothalamus.

limbic (LIM-bik) system: Group of nuclei and centers in the cerebrum and diencephalon that are involved with emotional states, memories, and behavioral drives.

meninges (men-IN-jēz): Three membranes that surround the surfaces of the CNS; the dura mater, pia mater, and arachnoid.

neural cortex: The gray matter found at the surface of the cerebral and cerebellar hemispheres.

nucleus: A mass of gray matter in the central nervous system.

tract: A bundle of axons inside the central nervous system.

Selected Clinical Terms

Alzheimer's disease: A progressive disorder marked by the loss of higher brain functions.

aphasia: A disorder affecting language comprehension and expression.

cerebral palsy: A disorder that affects voluntary motor performance and arises in infancy or early childhood as a result of prenatal trauma, drug exposure, or a congenital defect.

encephalitis: Inflammation of the brain.

hydrocephalus: A condition characterized by the abnormal accumulation of fluid in the cranium and, in infants, an enlarged skull.

meningitis: Inflammation of the spinal or cranial meninges.

Parkinson's disease: A condition characterized by a pronounced increase in muscle tone.

spasticity: A condition characterized by hesitant, jerky voluntary movements and increased muscle tone.

tremor: Repetitive shaking of the limbs resulting from a "tug of war" between antagonistic muscle groups.

Study Outline

THE BRAIN *(pp. 241–245)*
BASIC STRUCTURE AND FUNCTION *(p. 241)*

1. There are six regions in the adult brain: cerebrum, diencephalon, midbrain, pons, cerebellum, and medulla oblongata. *(Figure 11-1)*

2. Conscious thought, intellectual functions, memory, and complex involuntary motor patterns originate in the **cerebrum**. The **cerebellum** adjusts voluntary and involuntary motor activities based on sensory data and stored memories. *(Figure 11-1a)*

3. The walls of the **diencephalon** form the **thalamus**, which contains relay and processing centers for sensory data. The **hypo-**

thalamus contains centers involved with emotions, autonomic functions, and hormone production. *(Figure 11-1b)*

4. Three regions make up the **brain stem**: the **midbrain** processes visual and auditory information and generates involuntary somatic motor responses; the **pons** connects the cerebellum to the brain stem and is involved with somatic and visceral motor control; the **medulla oblongata,** which connects the brain and the spinal cord, relays sensory information and regulates autonomic functions. *(Figure 11-1b)*

THE MENINGES *(p. 242)*

5. Special covering membranes, the **meninges**, protect and support the spinal cord and delicate brain. The *cranial meninges* (the *dura mater, arachnoid,* and *pia mater*) are continuous with those of the spinal cord, the *spinal meninges. (Figure 11-2)*

6. The **dura mater** is the outermost layer covering the brain and spinal cord.

7. Beneath the inner surface of the dura mater lies the **arachnoid** (the second meningeal layer) and the *subarachnoid space,* which contains cerebrospinal fluid. The fluid acts as a shock absorber and a medium for absorbing and transporting dissolved gases, nutrients, chemical messengers, and waste products.

8. The **pia mater**, the innermost meningeal layer, is attached directly to the underlying neural tissue.

THE BLOOD–BRAIN BARRIER *(p. 243)*

9. The *blood–brain barrier* isolates nervous tissue from the general circulation.

10. The blood–brain barrier is generally permeable to lipid-soluble compounds. Water-soluble compounds require specific carriers to enter the interstitial fluid of the brain and spinal cord.

THE VENTRICLES OF THE BRAIN *(p. 244)*

11. The brain contains four chambers called **ventricles**. Cerebrospinal fluid continually circulates from the ventricles and central canal of the spinal cord into the subarachnoid space of the meninges that surround the CNS. *(Figure 11-3)*

12. **Cerebrospinal fluid (CSF)** (1) cushions delicate neural structures; (2) supports the brain; and (3) transports nutrients, chemical messengers, and waste products. CSF returns to the venous circulation by diffusion through extensions of the arachnoid layer into the dural sinuses. *(Figure 11-4)*

13. The **choroid plexus** is the site of cerebrospinal fluid production.

THE CEREBRUM *(pp. 245–247)*
STRUCTURE OF THE CEREBRAL HEMISPHERES *(p. 245)*

1. The cortical surface contains **gyri** (elevated ridges) separated by **sulci** (shallow depressions) or deeper grooves (**fissures**). The **longitudinal fissure** separates the two **cerebral hemispheres**. The **central sulcus** marks the boundary between the **frontal lobe** and the **parietal lobe**. Other sulci form the boundaries of the **temporal lobe** and the **occipital lobe**. *(Figures 11-1a, 11-5)*

2. Each cerebral hemisphere receives sensory information and generates motor commands that concern the opposite side of the body. However, there are significant functional differences between the two hemispheres.

3. The axons of the central white matter interconnect areas of neural cortex within each cerebral hemisphere. The left and right cerebral hemispheres are connected by axons making up the **corpus callosum**. *(Figure 11-4)*

FUNCTIONS OF THE CEREBRAL HEMISPHERES *(p. 246)*

4. The **primary motor cortex** of the *precentral gyrus* directs voluntary movements. The **primary sensory cortex** of the *postcentral gyrus* receives somatic sensory information from touch, pressure, pain, taste, and temperature receptors.

5. **Association areas**, such as the **visual association area** and **somatic motor association area** *(premotor cortex)*, control our ability to understand sensory information and coordinate a motor response.

6. "Higher-order" integrative centers receive information from many different association areas and direct complex motor activities and analytical functions. *(Figure 11-5)*

7. The **general interpretive area** receives information from all the sensory association areas. It is present in only one hemisphere, usually the left. *(Figure 11-5)*

8. The **speech center** regulates the patterns of breathing and vocalization needed for normal speech. *(Figure 11-5)*

9. The **prefrontal cortex** coordinates information from the secondary and special association areas of the entire cortex and performs abstract intellectual functions. *(Figure 11-5)*

10. The **cerebral nuclei** lie within the central white matter and aid in the coordination of learned movement patterns and other somatic motor activities.

THE LIMBIC SYSTEM *(p. 247)*

1. The **limbic system** includes the **hippocampus**, which is involved in memory and learning, and other nuclei that control reflex movements associated with eating. The functions of the limbic system involve emotional states and related behavioral drives.

THE DIENCEPHALON *(p. 248)*

1. The diencephalon provides the switching and relay centers necessary to integrate the conscious and unconscious sensory and motor pathways. Its upper portion contains the **pineal gland** and a vascular network that produces cerebrospinal fluid.

THE THALAMUS *(p. 248)*

2. The **thalamus** is the final relay point for ascending sensory information. It acts as a filter, passing on only a small portion of the arriving sensory information to the cerebrum, relaying the rest to the cerebral nuclei and centers in the brain stem.

THE HYPOTHALAMUS *(p. 248)*

3. The **hypothalamus** contains centers that control emotions and behavioral drives, coordinate voluntary and autonomic functions, regulate body temperature, and secrete hormones.

THE BRAIN STEM *(pp. 248–249)*

1. The brain stem is made up of the midbrain, pons, and medulla oblongata. *(Figure 11-1b)*

THE MIDBRAIN *(p. 248)*

2. The midbrain contains sensory nuclei that receive visual and auditory information. Other nuclei integrate information from the cerebrum and issue involuntary motor commands related to muscle tone and posture. Still others determine whether we are conscious and alert, or unconscious.

THE PONS *(p. 248)*

3. The pons contains nuclei for several cranial nerves, nuclei concerned with involuntary control of respiration, and ascending and descending tracts.

THE MEDULLA OBLONGATA *(p. 248)*

4. The medulla oblongata connects the brain to the spinal cord. It contains reflex centers that adjust heart rate and blood flow (the *cardiovascular centers*) and breathing (the *respiratory rhythmicity center*).

THE CEREBELLUM *(p. 249)*

1. The cerebellum oversees the body's postural muscles and adjusts voluntary and involuntary movements. The cerebellar hemispheres consist of *cerebellar cortex* (gray matter) formed into folds that enclose white matter.

THE CRANIAL NERVES *(pp. 249–252)*

1. There are 12 pairs of cranial nerves (N I–N XII) attached to the underside of the brain. *(Figure 11-6)*

DISTRIBUTION AND FUNCTION *(p. 249)*

2. In terms of function, each nerve may be classified as sensory, motor, or mixed (both).

3. Cranial nerves concerned primarily with sensory functions include the **olfactory** (N I), **optic** (N II), and **vestibulocochlear** (N VIII) **nerves**.

4. Cranial nerves concerned with motor functions include the **oculomotor** (N III), **trochlear** (N IV), **abducens** (N VI), **accessory** (N XI), and **hypoglossal** (N XII) **nerves**.

5. Mixed nerves include the **trigeminal** (N V), **facial** (N VII), **glossopharyngeal** (N IX), and **vagus** (N X) **nerves**.

AGING AND THE NERVOUS SYSTEM *(p. 252)*

1. Age-related changes in the nervous system include (1) reduction in brain size and weight, (2) reduction in the number of neurons, (3) decreased blood flow to the brain, (4) reduced number of synapses in the brain, and (5) chemical changes within and outside neurons.

A SUMMARY OF THE DISORDERS OF THE NERVOUS SYSTEM *(pp. 253–262)*

DIAGNOSIS OF NEUROLOGICAL DISORDERS *(p. 253)*

1. Common symptoms of neurological disorders, include *headaches, muscle weakness,* and *paresthesia*. *(Figure 11-7)*

2. A *neurological examination* attempts to determine the sources of nervous system disorders. Factors noted in the physical examination include the individual's *state of consciousness; reflex activity; abnormal speech patterns;* and *abnormal muscle strength and motor patterns*. *(Table 11-1)*

DISORDERS OF THE BRAIN *(p. 256)*

3. **Encephalitis** is an inflammation of the brain; **meningitis** is an inflammation of the meninges that surround the spinal cord and brain.

4. Several different bacteria cause meningitis: including *Hemophilus, Meningococcus,* and *Streptococcus*. *(Figure 11-8; Table 11-2)*

5. A **brain abscess** may result from infections by various bacteria. *(Table 11-2)*

6. Viral infections of the central nervous system include viral meningitis, *rabies,* and different varieties of encephalitis. **Viral meningitis** results in much fewer fatalities than bacterial meningitis. *(Table 11-2)*

7. Rabies is caused by a virus which infects neurons of the central nervous system. *(Figure 11-9; Table 11-2)*

8. Viral encephalitis is transmitted through mosquito bites. *(Table 11-2)*

9. African sleeping sickness, or *trypanosomiasis*, is caused by flagellated protozoans that are carried to humans by flies.*(Figure 11-10; Table 11-2)*

10. Cranial trauma, or head injuries, may involve intracranial hemorrhaging, concussion, contusion, or laceration of the brain.

11. Meningiomas are benign tumors that develop within the meninges.

12. Huntington's disease is an inherited disease characterized by a progressive deterioration of mental abilities.

13. Hydrocephalus, "water on the brain," refers to the buildup of cerebrospinal fluid (CSF) in the ventricles of the brain. An enormously expanded skull may appear in infants. *(Figure 11-11)*

14. Cerebral palsy is a congenital disorder characterized by disorders in voluntary motor movements.

15. Parkinson's disease is a brain disorder of the elderly that causes a general increase in muscle tone.

16. Alzheimer's disease is the most common age-related, degenerative condition of the nervous system.

17. Problems of the cardiovascular system can affect the nervous system. **Cerebrovascular diseases** interfere with the normal blood supply to the brain. A *stroke*, or **cerebrovascular accident (CVA)**, cuts off the blood supply to a portion of the brain. Strokes may produce muscle weakness, paralysis, or **aphasia,** an inability to understand or express language.

DISORDERS OF THE CRANIAL NERVES *(p. 267)*

18. Various types of tests are used to assess the functioning of cranial nerves.

19. Tic douloureux, or **trigeminal neuralgia**, is a painful condition affecting the face that is due to a disorder in cranial nerve V, the trigeminal nerve.

20. Bell's palsy is a cranial nerve disorder that results from an inflammation of cranial nerve VII, the facial nerve.

Review Questions

MATCHING

Match each item in Column A with the most closely related item in Column B. Use letters for answers in the spaces provided.

	Column A		Column B
___	1. dura mater	a.	innermost meningeal layer
___	2. cerebellum	b.	relay center for sensory information to cerebrum
___	3. hypothalamus	c.	connects the brain to the spinal cord
___	4. olfactory nerve	d.	outermost covering of brain and spinal cord
___	5. nuclei	e.	bacterial infection
___	6. pia mater	f.	masses of neuron cell bodies in the CNS
___	7. thalamus	g.	adjusts ongoing movements and postural muscles
___	8. optic nerve	h.	contains centers for maintaining consciousness
___	9. medulla oblongata	i.	link between nervous and endocrine systems
___	10. vestibulocochlear	j.	sensory, vision nerve
___	11. hypoglossal nerve	k.	sensory, smell
___	12. choroid plexus	l.	motor, tongue move movements
___	13. midbrain	m.	cranial nerve disorder
___	14. meninigitis	n.	produces CSF
___	15. Bell's palsy	o.	equilibrium, hearing

MULTIPLE CHOICE

16. Conscious thought processes and all intellectual functions originate in the _____ .
(a) cerebellum   (b) cerebrum
(c) corpus callosum   (d) medulla oblongata

17. The structural and functional link between the cerebral hemispheres and the brain stem is the _____ .
(a) diencephalon   (b) midbrain
(c) neural cortex   (d) medulla oblongata

18. Regulation of autonomic function, such as heart rate and blood pressure, occurs in the _____ .
(a) cerebrum   (b) thalamus
(c) cerebellum   (d) medulla oblongata

19. The part of the brain that coordinates rapid, automatic adjustments that maintain balance and equilibrium is the

_____ .
(a) cerebrum   (b) pons
(c) cerebellum   (d) medulla oblongata

20. The ventricles in the brain are filled with _____ .
(a) blood   (b) cerebrospinal fluid
(c) air   (d) nervous tissue

21. Reading, writing, and speaking depend on the activities in the _____ .
(a) left cerebral hemisphere
(b) right cerebral hemisphere
(c) prefrontal cortex
(d) central sulcus

22. Establishment of emotional states and related behavioral drives are roles of the _____ .
(a) pineal gland   (b) cerebellum
(c) limbic system   (d) thalamus

23. _____ is not an infectious disease of the nervous system.
(a) rabies
(b) cerebral palsy
(c) Western equine encephalitis
(d) African sleeping sickness

24.The _____ nerve is involved in tic douloureux.
(a) optic nerve   (b) trigeminal
(c) facial   (d) vagus

TRUE/FALSE

___   25. The two cerebral hemispheres are connected by the corpus callosum.

___   26. Cerebrospinal fluid is reabsorbed into the blood in the cerebral arteries.

____ 27. The part of the brain that regulates body temperature and autonomic functions is the hypothalamus.

____ 28. Three of the 12 cranial nerves control muscles that move the eyes.

____ 29. Gyri increase the surface area of the cerebral hemispheres.

____ 30. A brain abscess is the result of a viral infection.

____ 31. Encephalitis is transmitted by mosquitoes.

SHORT ESSAY

32. Stimulation of what part of the brain would produce sensations of hunger and thirst?

33. What major part of the brain is associated with respiratory and cardiac activity?

34. List three important functions of the cerebrospinal fluid.

35. What three layers make up the cranial and spinal meninges?

36. A police officer has just stopped Bill on suspicion of driving while intoxicated. Bill is first asked to walk the yellow line on the road and then asked to place the tip of his index finger on the tip of his nose. How would these activities indicate Bill's level of sobriety? What part of the brain is being tested by these activities?

37. Describe the three symptoms common to many nervous disorders.

38. What is aphasia?

APPLICATIONS

39. Mrs. Fujimoto, 73 years old, has been finding it increasingly difficult to control her movements, and even while sitting still she experiences a slight tremor. Identify her condition and discuss the parts of the brain involved.

40. Julie is mountain climbing with a group of friends when she slips, falls, and bumps the left side of her head on a rock. She gets up slowly, and is dazed but otherwise appears unhurt. She feels able to proceed, and the climb continues. An hour later, Julie gets a severe headache and experiences a ringing in her ears. She starts having trouble speaking, and soon loses consciousness. Before medical personnel can reach the scene, Julie dies. What was the likely cause of death?

✔ Answers to Concept Check Questions

(p.242) **1.** The six regions of the brain are the cerebrum, the diencephalon, the midbrain, the pons, the cerebellum, and the medulla oblongata. The cerebrum is the largest region of the human brain. **2.** The medulla oblongata is the connecting link between the brain and spinal cord.

(p.244) **1.** The astrocytes, a type of neuroglia, help maintain the blood–brain barrier. The blood–brain barrier restricts the movement of water-soluble molecules across the capillaries into the extracellular fluid of the brain. **2.** Diffusion into the venous sinus is the means by which CSF reenters the bloodstream. If this process decreased, excess fluid would accumulate in the ventricles, and the volume of fluid in the ventricles would increase.

(p.247) **1.** The primary motor cortex is located in the precentral gyrus of the frontal lobe of the cerebrum. Neurons of the primary motor cortex direct voluntary movements of the body. **2.** Damage to the left parietal lobe would interfere with processing by the general interpretive area of the cerebrum.

(p.249) **1.** Thalamic nuclei are involved with all ascending sensory information except olfactory (sense of smell). Damage to thalamic nuclei would interfere with such senses as sight, taste, hearing, balance, pain, and hot and cold temperatures. **2.** Changes in body temperature would stimulate nuclei in the hypothalamus, a division of the diencephalon. **3.** Even though the medulla oblongata is small, it contains many vital reflex centers including those that control breathing and regulate the heart and blood pressure. Damage to the medulla oblongata can result in cessation of breathing or changes in heart rate and blood pressure that are incompatible with life.

(p.252) **1.** Damage to the vagus nerve (cranial nerve X) can result in death, since it has motor fibers to regulate breathing, heart rate, and blood pressure. **2.** The glossopharyngeal nerve (cranial nerve IX) controls swallowing muscles and provides sensory information from the tongue. **3.** Since the abducens nerve (cranial nerve VI) controls lateral movement of the eyes via the lateral rectus muscles, we would expect an individual with damage to this nerve to be unable to move his eyes laterally.

(p. 261) **1.** The tsetse fly transmits African sleeping sickness, or trypanosomiasis. It is caused by a protozoan that eventually invades and infects the central nervous system. **2.** The symptoms of Huntington's disease are not expressed until adulthood.

CHAPTER

12

The Senses

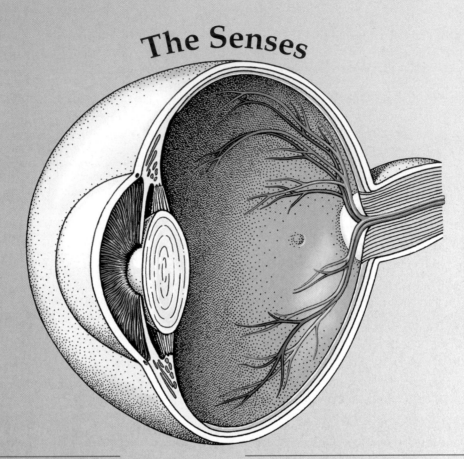

CHAPTER OBJECTIVES

CHAPTER OBJECTIVES

1 Distinguish between the general and special senses.

2 Identify the receptors for the general senses and describe how they function.

3 Describe the receptors and processes involved in the sense of smell.

4 Discuss the receptors and processes involved in the sense of taste.

5 Identify the parts of the eye and their functions.

6 Explain how we are able to see objects and distinguish colors.

7 Discuss the receptors and processes involved in the sense of equilibrium.

8 Describe the parts of the ear and their roles in the process of hearing.

9 Describe the assessment of general senses and pain relief by non-narcotic and narcotic drugs.

10 Describe examples of disorders in the senses of smell, taste, vision, equilibrium, and hearing.

Our picture of the world around us is based on information provided by our different senses, and although we may not realize it, this picture is incomplete. For example, colors invisible to us guide insects to flowers, and sounds and smells we cannot detect are regular information to dogs, cats, and dolphins.

The **general senses** are senses of temperature, pain, touch, pressure, and body position. The sensory receptors for the general senses are scattered throughout the body. (As you may recall from Chapter 10, sensory receptors are either specialized neurons or other cells that communicate with sensory neurons.) The **special senses** are smell (*olfaction*), taste (*gustation*), balance (*equilibrium*), hearing, and seeing (vision). The sensory receptors for the five special senses are found in complex *sense organs*.

SENSORY RECEPTORS

Sensory receptors detect changes in conditions inside or outside the body. Many sensory receptors are the processes of sensory neurons. When these receptors are stimulated, nerve impulses are generated. These impulses are carried to the CNS along the axons of the sensory neurons. A sensory receptor may also be another type of cell. In that case, the specialized receptor cell communicates with a sensory neuron, and stimulation of the receptor usually stimulates the neuron, leading to nerve impulse generation just as if the neuron itself were the receptor. All types of sensory information are carried by sensory neurons in the form of nerve impulses. We perceive a specific sensation as touch, light, or sound only when the impulses arrive at specific regions of the brain. For example, touch sensations are carried to the primary sensory cortex, visual sensations are delivered to the visual cortex, and auditory sensations arrive at the auditory cortex. The conscious awareness of a sensation is called *perception*.

If our sensory receptors are continually keeping track of internal and external conditions, why aren't we overwhelmed by all the arriving information? There are two major reasons:

1. Most arriving sensory information gets relayed to centers in the brain stem and diencephalon. It never reaches the cerebral cortex, and thus we are never consciously aware of these sensations. Only about 1 percent of the information provided by sensory neurons reaches the cerebral cortex and our conscious awareness.

2. Over time, a process called **adaptation** occurs. Adaptation is a reduction in the sensory response to a constant stimulus. In other words, adaptation reduces our awareness of stimuli that do not change. Think about the last time you sat in a hot tub or stood on a noisy city street. When you first enter a hot tub, it seems very hot, but your discomfort quickly fades as adaptation occurs. Similarly, the city sounds at first seem much too loud, but soon adaptation "tunes out" the background noise to the point that you can carry on a conversation with a neighbor.

TYPES OF SENSORY RECEPTORS

Sensory receptors are sometimes grouped according to the nature of the stimulus that excites them. *Pain receptors* are sensitive to tissue damage and injury; *thermoreceptors* are sensitive to temperature; *mechanoreceptors* are sensitive to movement, whether distortion, light touch, or deep pressure; *chemoreceptors* are sensitive to chemicals; and *photoreceptors* are sensitive to light.

Mechanoreceptors provide our perceptions of touch, pressure, vibration, body position, hearing, and equilibrium (balance). Chemoreceptors provide us with our special senses of smell and taste. Various types of photoreceptors provide us with the ability to see in black and white as well as in color.

THE GENERAL SENSES

The **general senses**—pain, heat, cold, touch, pressure, and body position—are provided by receptors that are scattered throughout the body and are relatively simple in structure. All the receptors for the general senses are found in the skin. Examples are shown in Figure 12-1●.

PAIN

Pain receptors are especially common in the skin, in joints, in the outer covering of bones, and around the walls of blood vessels. There are no pain receptors in the neural tissue, and there are few pain receptors in deep tissues or in most visceral organs.

Pain receptors are the simplest type of sensory receptors. They consist of **free nerve endings,** the terminal branches of a sensory neuron. Figure 12-1● shows free nerve endings in the superficial layer of the skin (the *epidermis*).

There are two types of pain sensations, although there is no way to distinguish one pain receptor from another. **Fast pain**, or *prickling pain*, reaches our conscious awareness very quickly. We become aware of the specific area affected. Examples of injuries producing fast pain sensations would be a deep cut or injection. **Slow pain**, or *burning and aching pain*, reaches our awareness more slowly, and we become aware only of the general area affected. A deep muscle bruise or ache produces slow pain sensations. Many injuries produce both sorts of pain—a sharp, intense pain as well as a deep aching pain—due to stimulation of both classes of pain receptors.

TEMPERATURE

The temperature receptors for heat and cold are free nerve endings that are scattered immediately beneath the surface of the skin. Separate **thermoreceptors** exist for hot and cold, although cold receptors are more common.

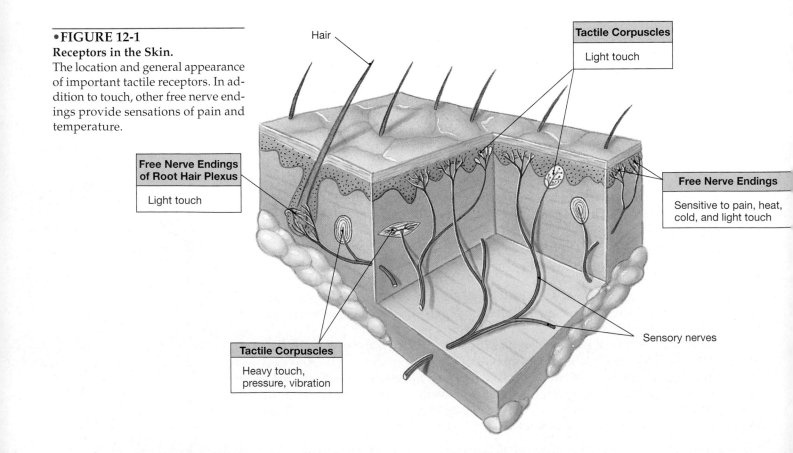

●FIGURE 12-1
Receptors in the Skin.
The location and general appearance of important tactile receptors. In addition to touch, other free nerve endings provide sensations of pain and temperature.

Hair

Tactile Corpuscles
Light touch

Free Nerve Endings of Root Hair Plexus
Light touch

Free Nerve Endings
Sensitive to pain, heat, cold, and light touch

Tactile Corpuscles
Heavy touch, pressure, vibration

Sensory nerves

Temperature receptors are very active when the temperature is changing, but they quickly adapt to a stable temperature. When we enter an air-conditioned classroom on a hot summer day or a warm lecture hall on a brisk fall evening, the temperature seems unpleasant at first, but the discomfort fades as adaptation occurs.

TOUCH AND PRESSURE

There are several different types of touch and pressure receptors. They provide sensations that the brain perceives as fine (light) touch, crude (heavy) touch and pressure, and vibration. Fine touch and pressure receptors provide detailed information about a source of stimulation, including its exact location, shape, size, texture, and movement. Crude touch and pressure receptors provide poor localization and little additional information about the stimulus.

The receptors responsible for providing touch and pressure sensations are called **tactile receptors**. Tactile receptors range in structure from free nerve endings to specialized **tactile corpuscles** with accessory cells and supporting structures. Figure 12-1• shows different types of tactile receptors in the skin.

BODY POSITION

Proprioceptors (prō-prē-ō-SEP-tors) monitor the position of joints, the tension in tendons and ligaments, and the state of muscular contraction. Tendon proprioceptors monitor the strain on a tendon and inhibit muscular contractions that might tear the tendon. Sensory receptors within skeletal muscle monitor the length of a skeletal muscle.

Information from these "position sensors" is relayed to centers in the brain, where it is integrated with balance information from the inner ear. The combination gives us precise information about our position in space and the status of our joints and muscles. This information is essential to the maintenance of posture and the coordination of muscular activity.

proprius, one's own + *capio*, to take
proprioceptors: receptors that monitor the position and stress levels at joints, muscles, and tendons

✔ List the general senses of the body.

✔ What is the role of proprioceptors in the body?

THE SPECIAL SENSES

The sensory receptors for the five special senses are housed within *sense organs*, where they are protected from environmental hazards. The tissues of a sense organ also effectively limit the number and type of stimuli that can contact the receptors. For example, the placement and structure of the eyes make it very difficult for any stimulus other than light to reach the visual receptors. There are five **special senses:** smell, taste, balance (equilibrium), hearing, and seeing (vision).

SMELL

The sense of smell, or *olfaction*, is provided by paired **olfactory organs**. These organs, shown in Figure 12-2•, are located in the roof of the nasal cavity on either side of the nasal septum. Each olfactory organ consists of an **olfactory epithelium**, which contains the **olfactory receptors** and supporting cells. Beneath the olfactory epithelium, large *olfactory glands* produce a mucus that covers the epithelium. The mucus prevents the buildup of potentially dangerous or overpowering stimuli and keeps the area moist and clean. Once the stimulating compounds reach the olfactory organs, they must be dissolved in the mucus before they can stimulate the olfactory receptors.

The olfactory receptors are highly modified, ciliated neurons. Altogether there are somewhere between 10 million and 20 million olfactory receptors packed into an area of roughly 4.8 cm^2 (0.75 in.^{2}). Nevertheless, our olfactory sensitivities cannot compare with those of dogs, cats, or fishes. A German shepherd sniffing for smuggled drugs or explosives has an olfactory receptor surface 72 times greater than that of the nearby customs inspector.

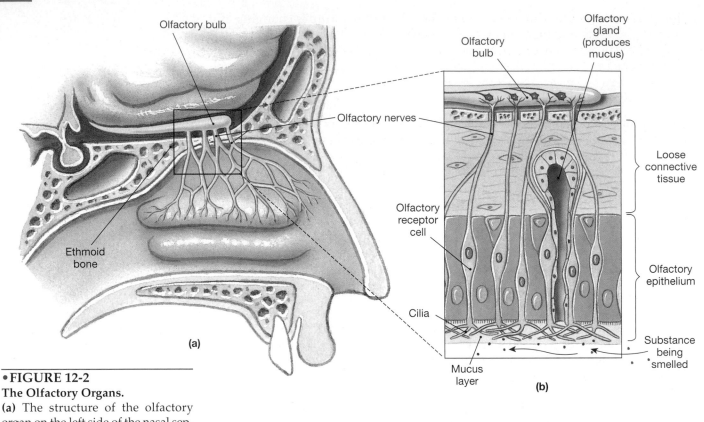

•FIGURE 12-2
The Olfactory Organs.
(a) The structure of the olfactory organ on the left side of the nasal septum. (b) The olfactory epithelium.

Olfactory receptors are stimulated when dissolved chemicals bind to receptor molecules on the surfaces of the cilia. This interaction produces nerve impulses that are relayed toward the central nervous system along the olfactory nerve (N I). Olfactory sensations are the only type of sensory information that reaches the cerebral cortex directly, without being relayed by the diencephalon. Olfactory information is also distributed to the hypothalamus, especially to centers involved with emotions and behavior. This explains the intensity of the responses produced by certain smells. The perfume industry understands this quite well, spending billions of dollars to develop odors that trigger sexual attraction.

TASTE

Taste receptors, or **gustatory** (GUS-ta-tōr-ē) **receptors,** are distributed mainly over the superior surface of the tongue. They are clustered together in individual **taste buds**. Taste buds lie along the sides of epithelial projections called **papillae** (pa-PIL-lē). This location protects them from abrasion by food and from the mechanical stress associated with chewing and swallowing. Figure 12-3• shows the taste buds within one of the large papillae located at the base of the tongue.

Each taste bud contains slender sensory receptors, known as **gustatory cells**, and supporting cells. Each gustatory cell extends slender *taste hairs* into the surrounding fluids through a narrow **taste pore**. The mechanism of taste reception is similar to that of smell. Dissolved chemicals contacting the taste hairs stimulate a change in the taste cell that leads to the release of chemical transmitters by the gustatory cell. These chemicals stimulate a sensory neuron, leading to the generation of nerve impulses. Three different cranial nerves monitor the taste buds—the seventh (facial), ninth (glossopharyngeal), and tenth (vagus).

There are four **primary taste sensations**: sweet, salt, sour, and bitter. Each taste bud shows a particular sensitivity to one of these tastes, and the sensory map of the tongue in Figure 12-3a• shows that each is concentrated in a different area.

✔ What would account for the greater sensitivity of the sense of smell by dogs as compared with people?

✔ If you completely dry the surface of the tongue, then place salt or sugar crystals on it, they cannot be tasted. Why not?

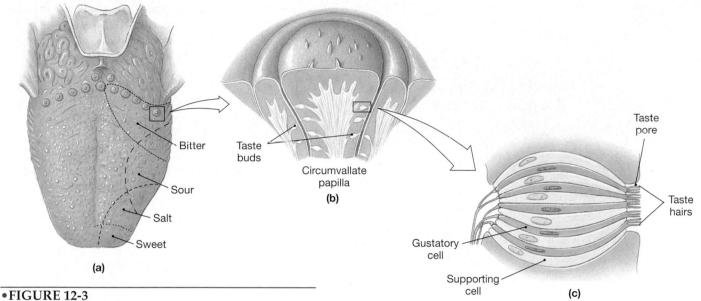

•FIGURE 12-3
Gustatory Reception.
(a) Gustatory receptors are located within the papillae of the
tongue. (b) Taste buds within the epithelium of the tongue.
(c) Diagram of a taste bud.

VISION

Our eyes are elaborate sense organs that contain the sensory receptors that permit
sight. The various accessory structures of the eye provide protection, lubrication,
and support for our sensory receptors.

Accessory Structures of the Eye

The **accessory structures of the eye** include the (1) eyelids, (2) a mucous membrane
(*conjunctiva*) covering the eye, (3) various exocrine glands, and (4) the six extrinsic eye muscles.

The eyelids can close firmly to protect the delicate surface of the eye. By their
continual blinking movements, the eyelids keep the anterior surface of the eye lubricated and free from dust and debris. The hairs of the eyelashes help prevent foreign particles and insects from reaching the surface of the eye.

The anterior surface of the eye is covered by the **conjunctiva** (kon-junk-TĪ-
va), a mucous membrane that is continuous with the inner lining of the eyelids.
The conjunctiva contains many free nerve endings, and it is very sensitive. The
painful condition of *conjunctivitis*, or "pinkeye," results from irritation of the conjunctiva. The reddish color results from dilation of the blood vessels beneath the
conjunctival surface.

Several different types of glands protect the eye and its accessory structures.
The eyelids contain sweat glands and sebaceous glands that lubricate the eyelashes and the edge of the eyelid. These glands sometimes become infected by
bacteria, producing a painful localized swelling known as a *sty*.

Above the eyeball, a dozen or more ducts from the **lacrimal gland**, or tear
gland, empty into the pocket between the eyelid and the eye. Lacrimal secretions
are watery, slightly alkaline, and contain an enzyme that attacks bacteria. Tears reduce friction, remove debris, prevent bacterial infection, and provide nutrients
and oxygen to the conjunctiva covering the eye. The blinking movement of the eyelids sweeps the tears across the surface of the eye. Two small pores at the inner connection of the lower and upper eyelid are connected to the **nasolacrimal duct**,
which carries tears to the nasal cavity. Figure 12-4• shows the lacrimal glands and
ducts that make up the **lacrimal apparatus** of the eye.

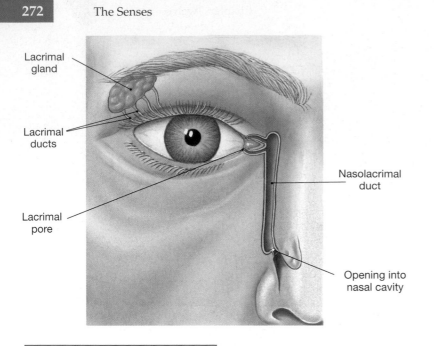

Six **extrinsic** (external) **eye muscles** originate on the bony surfaces of the orbit and insert on the outside of the eyeball. These are the *inferior rectus, lateral rectus, medial rectus, superior rectus, inferior oblique*, and *superior oblique*. Figure 12-5• shows these muscles and how their contractions move the eyeball.

Structure of the Eye

Each eye is roughly spherical, with a diameter of nearly 2.5 cm. The eyeball is hollow, and the lens divides the interior into two cavities (Figure 12-6a•). The large *posterior cavity*, also called the *vitreous chamber*, contains the gelatinous *vitreous body*. The vitreous body helps maintain the shape of the eye. The smaller *anterior cavity* is filled with a clear, liquid *aqueous humor* ("watery liquid"). The aqueous humor circulates within the anterior cavity and around the edges of the vitreous body in the posterior cavity.

•**FIGURE 12-4**
The Lacrimal Apparatus.

The wall of the eye is made up of three distinct layers. The outer layer, the **sclera** (SKLER-a), is made up mostly of tough, fibrous connective tissue. This layer provides protection and serves as an attachment site for the extrinsic eye muscles. The sclera is visible as the "white of the eye." A portion of the sclera, the **cornea** (KŌR-nē-a), is transparent. It is the first structure of the eye through which light passes.

The second layer, the **choroid**, lies inside the sclera. It contains numerous blood vessels that supply eye tissues. Toward the front of the eye, it forms the *iris* and muscular *ciliary body*. The iris regulates the amount of light entering the eye, and the ciliary body controls the shape of the lens, an essential part of the focusing process.

•**FIGURE 12-5**
The Extrinsic Eye Muscles.

Visible through the transparent cornea, the colored **iris** contains blood vessels, pigment cells, and two sets of *intrinsic* (internal) *eye muscles*. When these muscles contract, they change the diameter of the central opening, or **pupil**, of the iris. Pupillary dilation and constriction are controlled by the autonomic nervous system in response to sudden changes

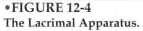

(a) Lateral view, left eye

(b) Anterior view, left eye

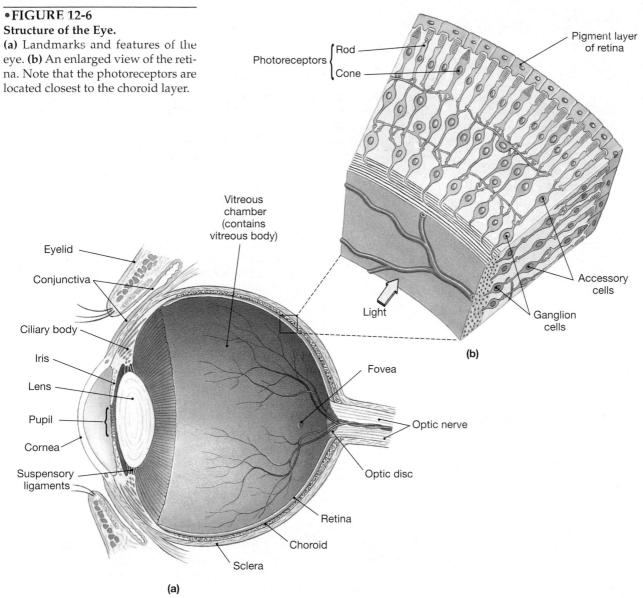

•FIGURE 12-6
Structure of the Eye.
(a) Landmarks and features of the eye. **(b)** An enlarged view of the retina. Note that the photoreceptors are located closest to the choroid layer.

in light intensity. Exposure to bright light produces a rapid decrease in the pupil's diameter. A sudden reduction in light levels slowly opens or slowly dilates the pupil.

Along its outer edge, the iris attaches to the **ciliary body**, most of which consists of the *ciliary muscle*, a muscular ring that projects into the interior of the eye. The ciliary muscle is attached to collagen fibers (the *suspensory ligaments*) extending from the ciliary body to the lens. These fibers position the lens so that light passing through the center of the pupil passes through the center of the lens. Contraction and relaxation of the ciliary muscle changes the shape of the flexible lens.

The inner layer of the eye, the **retina**, contains an outer layer of pigment cells and the photoreceptors that respond to light. Figure 12-6b• shows an enlarged view of the retina and the two types of photoreceptors, **rods** and **cones**. Rods do not discriminate among different colors of light. They are very light sensitive and enable us to see in dim light. Cones provide us with color vision. There are three types of cones (blue, green, and red), and their stimulation in various combinations provides the perception of different colors. Cones also give us sharper, clearer images than rods, but they require more intense light than rods. If you sit outside at sunset (or sunrise) you will probably be able to tell when your visual system shifts from cone-based vision (clear images in full color) to rod-based vision (relatively grainy images in black and white).

☤ Clinical Note

What Do Glasses and Contact Lenses Really Do?

In the normal eye, when the ciliary muscles are relaxed and the lens is flattened, a distant image will be focused on the retina (Figure 12-7a •). This condition is called *emmetropia*, which means "proper eye." However, irregularities in the shape of the lens or cornea can affect the clarity of the visual image. This condition, called *astigmatism*, can usually be corrected by glasses or special contact lenses.

It is estimated that 47 percent of the people in the United States wear glasses or contact lenses. Figure 12-7 • diagrams two other common problems that are corrected by glasses or contact lenses. If the eyeball is too deep, the image of a distant object will form in front of the retina and be blurry and out of focus (Figure 12-7b •). Vision at close range will be normal, because the lens will be able to round up as needed to focus

the image on the retina. As a result, such individuals are said to be "nearsighted." Their condition is more formally termed **myopia** ("to shut the eye"). Figure 12-7c • shows how a corrective lens can help focus the image on the retina.

If the eyeball is too shallow, **hyperopia** ("far eye") results (Figure 12-7d •). The ciliary muscles must contract to focus even a distant object on the retina, and at close range the lens cannot provide enough refraction. These individuals are said to be "farsighted" because they can see distant objects most clearly. Older individuals become farsighted as their lenses become less elastic; this form of hyperopia is called **presbyopia** ("old man eye"). Hyperopia can be treated with corrective lenses that place the image in its proper position on the retina (Figure 12-7e •).

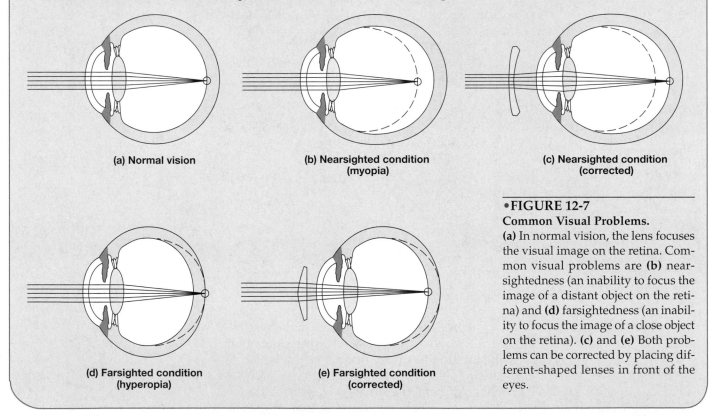

(a) Normal vision

(b) Nearsighted condition
(myopia)

(c) Nearsighted condition
(corrected)

(d) Farsighted condition
(hyperopia)

(e) Farsighted condition
(corrected)

•**FIGURE 12-7**
Common Visual Problems.
(a) In normal vision, the lens focuses the visual image on the retina. Common visual problems are **(b)** nearsightedness (an inability to focus the image of a distant object on the retina) and **(d)** farsightedness (an inability to focus the image of a close object on the retina). **(c)** and **(e)** Both problems can be corrected by placing different-shaped lenses in front of the eyes.

Persons unable to distinguish certain colors have a form of *color blindness*. Color blindness occurs because one or more classes of cones are absent or nonfunctional. In the most common condition, the red cones are missing, and the individual cannot distinguish red light from green light. Color blindness is about 15 times more common in men than women. Total color blindness is extremely rare.

Rods and cones are not evenly distributed across the retina. There are no rods in the region where the visual image arrives after passing through the cornea and lens. In this region lies an area called the **fovea** (FŌ-vē-a), a central, shallow depression where cones are highly concentrated (Figure 12-6a •). The fovea is the center of color vision and the site of sharpest vision; when you look directly at an object its image falls on this portion of the retina. It is not a large area—if you extend your arm and look at your hand, the apparent diameter of the fovea is roughly equal to the width of your three middle fingers.

Clinical Note

What Is 20/20 Vision?

In rating the clarity of vision, or visual acuity, a person whose vision is rated 20/20 is seeing details at a distance of 20 feet as clearly as a "normal" individual would. Vision noted as 20/15 is better than average, for at 20 feet the person is able to see details that would be clear to a normal eye only at a distance of 15 feet. Conversely, a person with 20/30 vision must be 20 feet from an object to discern details that a person with normal vision could make out at a distance of 30 feet.

When visual acuity falls below 20/200, even with the help of glasses or contact lenses, the individual is considered to be legally blind. There are probably fewer than 400,000 legally blind people in the United States; more than half are over 65 years of age.

Rod and cone cells communicate with accessory neurons. Stimulation of the accessory neurons in turn causes stimulation of *ganglion cells* (Figure 12-6b•). The nerve fibers of the ganglion cells deliver visual sensations to the brain.

The axons of the ganglion cells converge at the **optic disc**, a circular region medial to the fovea. The optic disc is the start of the optic nerve (N II) (Figure 12-6a•). Blood vessels that supply the retina pass through the center of the optic nerve and emerge on the surface of the optic disc. There are no photoreceptors at the optic disc, and because light striking this area goes unnoticed, it is commonly called the **blind spot**. You do not "notice" a blank spot in the visual field because involuntary eye movements keep the visual image moving, and the brain fills in the missing information.

Structure and Function of the Lens

The lens lies behind the cornea. Light entering the eye passes through the superficial conjunctiva and then through the cornea, aqueous humor, the lens, and the vitreous body before it reaches the retina. The primary function of the lens is to focus the image on the sensory receptors in the retina. Focusing depends on a bending, or **refraction**, of the light from the object. The lens provides the needed refraction by changing its shape. The transparent lens normally has a spherical shape. Because it is somewhat elastic, however, an outside force can pull the lens into the shape of a flattened oval.

The lens is held in place by the suspensory ligaments that originate at the ciliary body (see Figure 12-6a•). Smooth muscle fibers in the ciliary body encircle the lens. When we are viewing a nearby object, the ciliary muscles contract, and the inner edge of the ciliary body moves toward the lens. This movement reduces the tension in the suspensory ligaments, and the lens becomes more spherical. When we are viewing a distant object, the ciliary muscle relaxes, the suspensory ligaments pull at the edges of the lens, and the lens becomes relatively flat. This process of changing the shape of the lens to keep the image focused on the retina is called **accommodation**.

Visual Perception

The pathway to vision begins at the photoreceptors and ends at the visual cortex of the cerebral hemispheres. Along the way, visual information is also relayed to reflex centers in the brain stem. For example, visual information at the *pineal gland* helps establish a daily pattern of activity that is tied to the day/night cycle. This *circadian* ("day/night") *rhythm* affects metabolic rate, endocrine function, blood pressure, digestive activities, the awake/sleep cycle, and other processes.

✔ What layer of the eye would be the first to be affected by inadequate tear production?

✔ When the lens is very round, are you looking at an object that is close to you or distant from you?

✔ If a person was born without cones in her eyes, would she still be able to see? Explain.

EQUILIBRIUM AND HEARING

The senses of equilibrium and hearing are provided by the *inner ear*, a sensory structure located in the temporal bone of the skull. ∞ *[p. 153]* The basic receptor for both of these senses are ciliated cells called *hair cells*. Movement of the cilia causes the hair cells to produce nerve impulses. The different arrangements of surrounding structures explain how hair cells can respond to different stimuli and account for two different senses:

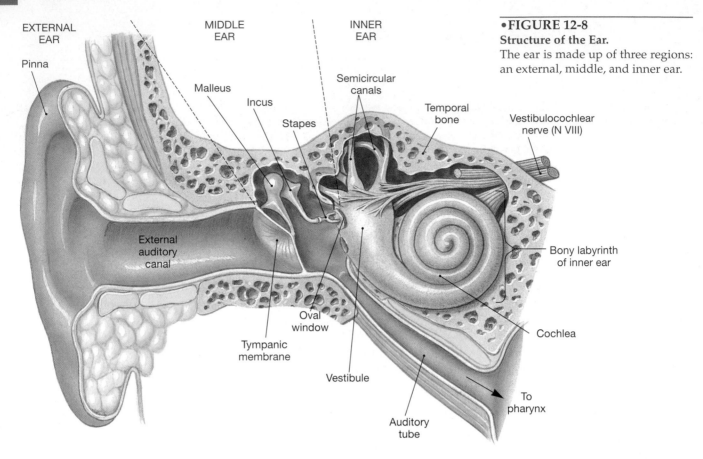

EXTERNAL
EAR

Pinna

MIDDLE
EAR

Malleus

Incus

Stapes

INNER
EAR

Semicircular
canals

Temporal
bone

•FIGURE 12-8
Structure of the Ear.
The ear is made up of three regions:
an external, middle, and inner ear.

Vestibulocochlear
nerve (N VIII)

Bony labyrinth
of inner ear

External
auditory
canal

Cochlea

Tympanic
membrane

Oval
window

Vestibule

Auditory
tube

To
pharynx

- Equilibrium, which informs us of the position of the body in space by monitoring gravity, acceleration, and rotation.

- Hearing, which enables us to detect and interpret sound waves.

Structure of the Ear

The ear is composed of three regions (Figure 12-8•). The *external ear* is the visible portion of the ear, and it collects and directs sound waves to the eardrum. The *middle ear* is a chamber located within a thickened portion of the temporal bone. Structures within the middle ear collect and amplify sound waves and transmit them to a portion of the inner ear concerned with hearing. The *inner ear* also contains the sensory organs responsible for equilibrium sensations.

The **external ear** includes the fleshy **pinna**, or earlobe, which surrounds the entrance to the **external auditory canal**. The earlobe, which is supported by elastic cartilage, protects the opening of the canal and funnels sounds into the canal. **Ceruminous** (se-ROO-mi-nus) **glands**, modified sweat glands, along the external auditory canal secrete a waxy material, called *cerumen* ("earwax"), that slows the growth of microorganisms and reduces the chances of infection. In addition, small, outwardly projecting hairs help prevent the entry of foreign objects or insects. The external auditory canal ends at the eardrum, also called the **tympanic membrane**. The tympanic membrane is a thin sheet that separates the external ear from the middle ear.

The **middle ear** is filled with air. It is open to the upper portion of the pharynx through a duct called the **auditory tube**, or *Eustachian tube*. The auditory tube serves to equalize the pressure on either side of the eardrum. Unfortunately, it also provides a path for microorganisms to enter and cause middle ear infections.

The middle ear contains three tiny ear bones, the *auditory ossicles*. These bones connect the tympanic membrane with the receptors of the inner ear. Named for their shape, the **malleus** ("hammer") is attached to the inside surface of the tympanum. The middle bone, the **incus** ("anvil"), attaches the malleus to the inner **stapes** ("stirrup"). The base of the stapes is attached to the *oval window*, a mem-

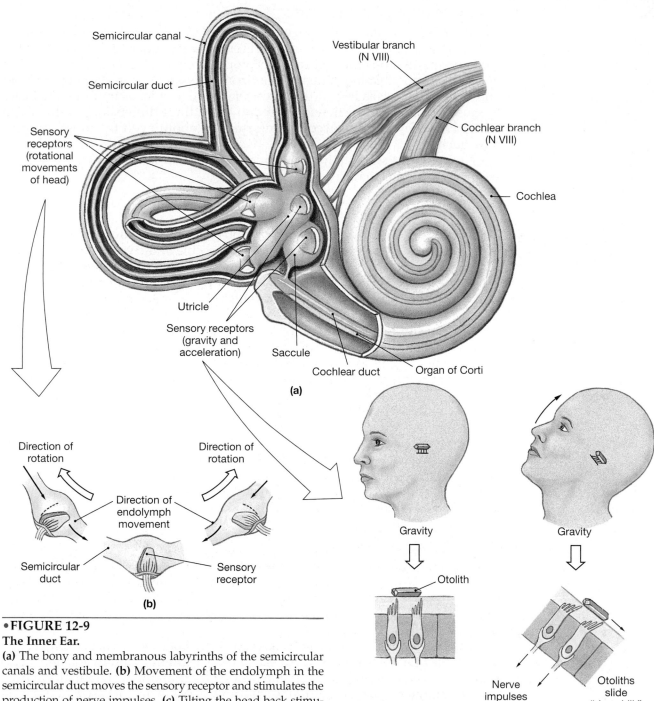

•FIGURE 12-9
The Inner Ear.
(a) The bony and membranous labyrinths of the semicircular canals and vestibule. **(b)** Movement of the endolymph in the semicircular duct moves the sensory receptor and stimulates the production of nerve impulses. **(c)** Tilting the head back stimulates the hair cells of the sensory receptors in the utricle and saccule of the vestibule.

brane of the inner ear. Vibration of the eardrum causes the three ear bones to move, and rocks the stapes against the oval window.

As Figure 12-8• shows, the **inner ear** consists of the **vestibule** (VES-ti-būl), **semicircular canals**, and **cochlea** (KO-klē-a). The senses of equilibrium and hearing are provided by the receptors within the inner ear.

The receptors for both of these senses lie within a network of fluid-filled tubes and chambers known as the **membranous labyrinth** (Figure 12-9a•). The membranous labyrinth contains a fluid called **endolymph** (EN-dō-limf). The **bony labyrinth** is a shell of dense bone (see Figure 12-8•) that surrounds and protects the membranous labyrinth. Between the bony and membranous labyrinths flows another fluid, the **perilymph** (PER-i-limf).

Function of the Semicircular Canals

The three semicircular canals provide information concerning rotational movements of the head. Each semicircular canal encloses a slender *semicircular duct* that contains endolymph and a sensory receptor. Each semicircular duct responds to one of three possible movements: a horizontal rotation, as in shaking the head "no"; front to back, as in nodding "yes"; and side to side, as in tilting the head. Because these three directions correspond to the three dimensions in the world around us, they provide accurate information about even the most complex movements.

Figure 12-9b • shows how a movement of the head can stimulate the sensory receptors in a semicircular duct.

Function of the Vestibule

The vestibule includes a pair of membranous sacs, the **saccule** (SAK-ūl) and the **utricle** (Ū-tre-kl. Receptors in the utricle and saccule of the vestibule provide the sensations of gravity and acceleration. If you stand with your head tilted these receptors will report the angle involved and whether your head tilts forward, backward, or to the side. These receptors also give you the impression of increasing speed, as when your car accelerates.

The sensory receptors of the utricle and saccule lie under a layer of densely packed mineral crystals called *otoliths* (literally, "ear stones"). When the head is in the normal, upright position the weight of the otoliths pushes the sensory hairs downward rather than to one side or another. When the head is tilted, the pull of gravity on the otoliths shifts the mass to the side. This shift stimulates the receptor and causes a change in nerve impulse activity that signals the CNS that the head is no longer level (see Figure 12-9c •).

Nerve impulses from the sensory receptors of the vestibule and semicircular canals are passed on to neurons whose axons form the **vestibular branch** of the vestibulocochlear nerve, N VIII. The sensory information arriving from each side of the head is relayed to the cerebellum and the cerebral cortex, where it provides a conscious awareness of position and movement.

Function of the Cochlea

The bony cochlea contains the **cochlear duct** of the membranous labyrinth. Receptors within the cochlear duct provide the sense of hearing. The cochlear duct sits sandwiched between a pair of perilymph-filled chambers, and the entire complex is coiled around a central bony hub. The spiral arrangement resembles that of a snail shell, a *cochlea* in Latin.

The walls of the bony labyrinth consist of dense bone everywhere except at two small areas near the base of the cochlear spiral. The **round window** is a thin membrane that separates perilymph within the cochlea from the air within the middle ear. The margins of the **oval window** are firmly attached to the base of the stapes. When sound waves cause the eardrum to vibrate back and forth, the movements are conducted over the malleus and incus to the stapes. Movement of the stapes ultimately leads to the stimulation of receptors within the cochlear duct, and we hear the sound. The receptors for hearing lie within a structure called the **organ of Corti**, which extends along the entire length of the coiled cochlear duct.

Figure 12-10 • summarizes the process of hearing in six basic steps.

- *Step 1:* Sound waves enter the external auditory canal and travel toward the tympanic membrane.
- *Step 2:* Vibration of the tympanic membrane causes movement of the three auditory ossicles.
- *Step 3:* Movement of the stapes at the oval window creates pressure waves in the perilymph of the upper duct.

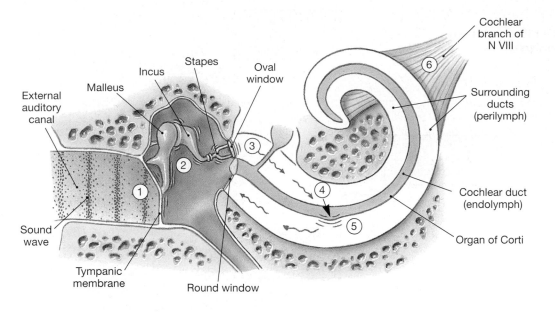

•**FIGURE 12-10**
The Process of Hearing.
Steps in the reception of sound.

- *Step 4:* The pressure waves cause vibrations in the cochlear duct on their way to the round window. High-frequency, or high-pitched, sounds cause vibrations in the cochlear duct near the oval window. Low-frequency, or low-pitched, sounds cause vibrations farther away from the oval window.

- *Step 5:* Vibration of the cochlear duct causes movement of the sensory receptor cells within the organ of Corti. The louder the sound, the stronger the vibration and the greater the number of stimulated receptor cells.

- *Step 6:* Information concerning the region and intensity of stimulation is relayed to the CNS over the cochlear branch of N VIII.

Nerve impulses then pass to regions within the brain stem that coordinate automatic reflexes to different sounds. For example, reflexes can rapidly change the position of the head in response to a sudden loud noise. Sounds are perceived when the thalamus relays the sensations to the auditory cortex of the cerebral hemispheres.

✔ What structure permits the pressure of the middle ear to be equalized with the external pressure?

✔ What inner ear structures sense and inform the brain of rotational movements of the head?

SENSORY DISORDERS

The best way to understand sensory disorders is to group the special senses together and examine their individual components. Every sensory system includes sensory receptors, sensory neurons, ascending tracts in the spinal cord, sensory nuclei in the brain, and specific areas of the cerebral cortex. Figure 12-11• organizes various sensory disorders into a "troubleshooting" format, similar to that used to diagnose problems with cars or other mechanical devices.

THE GENERAL SENSES

Assessing Tactile Sensitivities

The area monitored by a single receptor cell is called its **receptive field** (Figure 12-12•). Sensitivity to light touch in different regions of the body can be checked by gentle contact with a fingertip or a slender wisp of cotton. The **two-point discrimination test** provides a more detailed sensory map of tactile receptors. Two fine points of a drawing compass, a bent paper clip, or other object are applied to the skin surface simultaneously. The subject then describes the contact. When the points fall within a single receptive field, the individual will report only one point of contact. For example, a normal individual loses two-point discrimination at

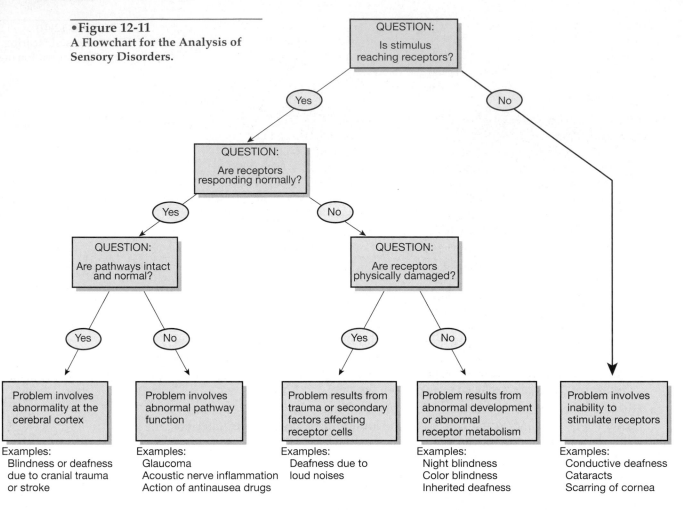

•Figure 12-11
A Flowchart for the Analysis of
Sensory Disorders.

QUESTION:
Is stimulus
reaching receptors?

Yes

No

QUESTION:
Are receptors
responding normally?

Yes

No

QUESTION:
Are pathways intact
and normal?

QUESTION:
Are receptors
physically damaged?

Yes

No

Yes

No

Problem involves
abnormality at the
cerebral cortex

Problem involves
abnormal pathway
function

Problem results from
trauma or secondary
factors affecting
receptor cells

Problem results from
abnormal development
or abnormal
receptor metabolism

Problem involves
inability to
stimulate receptors

Examples:
  Blindness or deafness
  due to cranial trauma
  or stroke

Examples:
  Glaucoma
  Acoustic nerve inflammation
  Action of antinausea drugs

Examples:
  Deafness due to
  loud noises

Examples:
  Night blindness
  Color blindness
  Inherited deafness

Examples:
  Conductive deafness
  Cataracts
  Scarring of cornea

1 mm (0.04 in.) on the surface of the tongue, at 2–3 mm (0.08–0.12 in.) on the lips, at 3–5 mm (0.12–0.20 in.) on the backs of the hands and feet, and at 4–7 cm (1.6–2.75 in.) over the trunk.

Vibration receptors are tested by applying the base of a tuning fork to the skin. Damage to an individual spinal nerve produces insensitivity to vibration along the paths of the related sensory nerves. If the sensory loss results from spinal cord damage, the injury site can typically be located by walking the tuning fork down the spinal column, resting its base on the vertebral spines.

Descriptive terms are used to indicate the degree of sensitivity in the area considered. **Anesthesia** implies a total loss of sensation; the individual cannot perceive

•Figure 12-12
Sensory Receptors and Receptive Fields.
The receptive field is the specific area monitored by a sensory receptor cell.

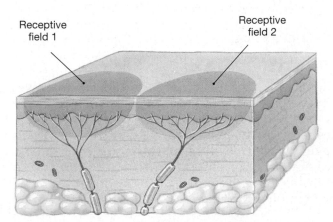

Receptive
field 1

Receptive
field 2

touch, pressure, pain, or temperature sensations from that area. **Hypesthesia** is a reduction in sensitivity, and *paresthesia* is the presence of abnormal sensations, such as the pins-and-needles sensation when a limb "falls asleep" due to pressure on a peripheral nerve (see p. 235).

Pain Management

Pain management poses a number of problems for physicians. Painful sensations can result from tissue damage or sensory nerve irritation; it may originate where it is perceived, represent a false signal generated along the sensory pathway, or be referred from another location. For example, heart pain is often perceived as originating in the upper chest and left arm (Figure 12-13•).

Acute pain is often the result of tissue injury; the cause is apparent, and treatment is usually effective in relieving the pain. **Chronic pain** is more difficult to categorize and treat. It includes pain from an injury that persists after tissue structure has been repaired, pain from a chronic disease, such as cancer, and pain without an apparent cause.

Anesthetics and *analgesics* are two terms used when discussing pain suppression or pain relief. Although anesthesia refers to the total loss of sensation, it is generally used in referring to the absence of pain during painful procedures such as surgery. An **anesthetic** is an agent that suppresses pain in such conditions. Two types of anesthesia are used in medical procedures. In **local anesthesia**, pain is suppressed in only part of the body and the patient is conscious. In **general anesthesia**, the patient is made to lose consciousness. **Analgesia** is the relief of pain, and an **analgesic** is a drug that relieves pain without an individual losing their sensitivity to other stimuli, such as touch or pressure.

There are two main categories of analgesic drugs: **non-narcotic** (such as aspirin and acetaminophen) and **narcotic** (such as morphine and codeine). Non-narcotic drugs are used in treating mild to moderate pain, such as headaches. Narcotic drugs are used when pain is much more severe. They act in a variety of ways; for example, some analgesics reduce inflammation and the swelling that causes pain, whereas others reduce stimulation of neurons or interneurons along pain pathways.

Tissue injury, inflammation, or infection results in damage to cell membranes and the release of various molecules. One type of fatty acid that is released is converted into *prostaglandin* molecules. These prostaglandin molecules attach to receptor sites on sensory neurons in the area, which then pass on a nerve impulse to the brain where it is interpreted as pain. Aspirin and related analgesics reduce inflammation and suppress pain by blocking the production of prostaglandins. In contrast, acetaminophen and narcotic drugs block the perception of painful stimuli by suppressing activity along the pain pathways.

SMELL

Disorders of the sense of smell may occur by head injury or through normal, age-related changes. If an injury to the head damages the olfactory nerves (cranial nerve I), then the sense of smell may be impaired. Unlike other populations of neurons, the olfactory receptor cells are regularly replaced by the division of stem cells. Despite this process, the total number of receptors declines with age, and the remaining receptors become less sensitive. As a result, the elderly have difficulty detecting odors in low concentrations. This explains why your grandmother may overdo her perfume, and your grandfather's aftershave may seem so strong—they must use more of the odorous solution to be able to smell it themselves.

TASTE

Disorders of the sense of taste can be caused by problems with olfactory receptors, damage to taste buds, damage to cranial nerves, and age-related changes. The sense of smell also makes a large contribution to our sense of taste. As a result, conditions that affect the olfactory receptors, such as the common cold, also cause a

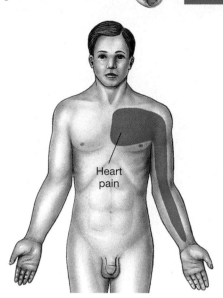

•**Figure 12-13**
Referred Pain.
In *referred pain*, pain sensations originating in visceral organs are often perceived as involving specific regions of the body surface.

reduced sense of taste. A reduced sense of taste may also occur through damage to taste buds by inflammation or infections of the mouth.

Three cranial nerves provide taste sensations from the taste buds to the brain: the facial (N VII), the glossopharyngeal (N IX), and vagus (N X) nerves. Possible external and internal sources of damage to these cranial nerves may occur through trauma or compression by a tumor.

The sense of taste becomes dulled with age because of a reduction in the number and sensitivity of taste buds. We begin life with more than 10,000 taste buds, but the number begins to decline dramatically by age 50. For this reason the elderly often complain that their food tastes bland and unappetizing.

VISION

We rely more on our vision than any other sense, and disorders of the eye can have a large impact on our day-to-day lives. Major disorders of the eye can be grouped into trauma, infection, tumors, congenital disorders, degenerative disorders, and secondary disorders due to problems in other systems. Physicians who specialize in the eye and its diseases are called **ophthalmologists** (of-thal-MOL-o-jists) (*ophthalmos* means "eye").

Trauma

Injury to the cornea is common and generally minor, but extremely painful. Scratches of the cornea are called **corneal abrasions** and may be due to particles rubbing the cornea or to the over-long use of contact lenses. Abrasions can lead to a sloughing or opening in the cornea, called a **corneal ulcer**. Chemical spills of acids or alkalis can "burn" the eye and damage the cornea. Flushing with large amounts of water are necessary after such accidents.

Trauma to the head or eyes may also affect the retina, causing a **retinal tear**. **Retinal detachment** often follows a retinal tear. Retinal detachment occurs when the neural portion of the retina loses contact with the choroid. Surgery is required to repair a detached retina.

Infection

Bacteria. *Conjunctivitis* is an inflammation of the conjunctiva. ∞ *[p. 271]* There are three basic kinds of conjunctivitis caused by bacteria: *Ophthalmia neonatorum, bacterial conjunctivitis* ("pinkeye"), and *trachoma*. Each is caused by different types of bacteria. *Ophthalmia neonatorum*, or **conjunctivitis of the newborn,** may be found in both babies and adults (Figure 12-14●). Babies can become infected as they pass through a birth canal in which *Neisseria gonorrhoeae* or *Chlamydia trachomatis* are present. The bacteria produce a pus-forming infection, which can cause *keratitis* (inflammation of the cornea), corneal ulcers, and blindness. Adults can become infected by transferring the bacteria from the genitals to the eyes. Topical antibiotics are used in the prevention and treatment of this disease.

Bacterial conjunctivitis, also called "pinkeye," is a very common eye infection. It may be caused by a number of common, resident bacteria of the body, such as *Staphylococcus aureus* and *Streptococcus pneumoniae*. Symptoms of pinkeye include redness, itchy eyes, and a pus-filled discharge. It is highly contagious and can spread quickly through schools, or other institutions where large numbers of children are present. Treatment is with antibiotics.

Trachoma is a serious conjunctivitis caused by the bacteria *Chlamydia trachomatis* (kla-MID-ē-a tra-KŌ-ma-tis) (Figure 12-15●). Trachoma means "roughness," and the infected conjunctiva swell and become covered by thickened, round swellings giving it a rough texture. Untreated, destruction of the cornea occurs, and blindness follows. This disease is uncommon in the U.S., but it is the leading cause of preventable blindness in other parts of the world.

Viruses. In addition to causing various skin infections, the *Herpes simplex* virus can also infect the eye. It produces an inflammation of the cornea and the con-

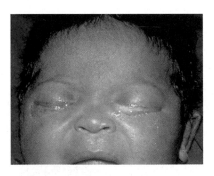

•Figure 12-14
Conjunctivitis.
Conjunctivitis of the newborn. Adults can also become infected.

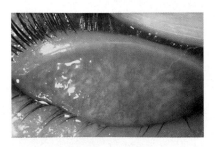

•Figure 12-15
Trachoma.
Trachoma is a particularly severe form of bacterial conjunctivitis.

junctiva, a condition called **keratoconjunctivitis** (ker-ah-tō-kon-junk-ti-VĪ-tis). In this condition, blisters appear on the cornea and inner eyelids.

Parasites. Onchocerciasis (ong-ko-ser-SĪ-a-sis), or **river blindness**, occurs in Central America and Africa. It is caused by a roundworm *Onchocerca volvulus* (ong-kō-SER-ka vol-VYOO-lus). The parasite enters the body as a larval stage through the bite of a blackfly. The larval stage invades different tissues of the body, including the eyes. The worms mature and reproduce, and when another blackfly bites it will ingest more larval worms, which then reproduce in the fly. The worms cause skin abscesses and damage to the blood vessels in the eye that result in blindness. A drug, *ivermectin*, can be used to kill the worms, but the dying worms release toxins that can cause a severe reaction, called *anaphylactic shock*, which in turn can result in death.

Tumors

Malignant tumors can form in different layers of the eye. The most common primary malignant tumor of the eye is **melanoma of the choroid** (the middle, or second, layer of the eye). Depending on its location, it may or may not cause a decrease in vision. As a result, it may not be detected early in its development if regular eye examinations are not done. This tumor can spread and form secondary tumors in other parts of the body.

Retinoblastoma is a malignant tumor of the retina that usually appears during the first three years of life. It may be an inherited condition. A sign of retinoblastoma is the presence of a whiteness in the pupil. The affected eye may be blind and the eyes may become misaligned from each other (a condition called *strabismus*). Retinoblastoma can spread from the eye to the optic nerve and the brain. Treatment may involve removal of the eye and/or radiation therapy.

Congenital Disorders

Congenital disorders affect many aspects of the eye and vision. Three conditions already discussed include variations in the shape of the cornea (**astigmatism**), and shape of the eyes (**myopia**, or "nearsighted" and **hyperopia** "farsighted" conditions), and **color blindness**. ∞ *[p. 274]* Normal vision also requires that the eyes view the same object simultaneously. Deviations in the alignment of the eyes to each other is called **strabismus**. In this condition, one or both eyes turn inward or outward. In convergent strabismus (*cross-eye*), the eye(s) turn inward, and in divergent strabismus (*walleye*), the eye(s) turn outward. Because the eyes are not aligned, double vision occurs. Vision from the affected eye is suppressed by the brain.

A **retinopathy** (ret-i-NOP-ah-thē) is a disease or disorder of the retina. **Retinitis pigmentosa** refers to a collection of inherited retinopathies, that usually do not appear until after adolescence. In this disorder, the rods and cones of both eyes degenerate. One of the first symptoms is poor vision in dim light. A ring-shaped area of blindness also develops on the periphery of the field of vision that slowly extends inward. The rate of the degeneration is quite variable in different individuals.

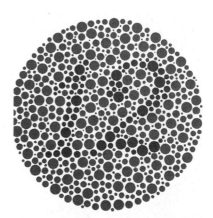

Part of a standard test for color vision. Persons with one form of color-blindness cannot see the number within this circle.

Degenerative Disorders

Age-related, degenerative disorders often include problems with the lens. *Presbyopia* and *cataracts* are examples. As we age, the lens becomes less elastic and we become more "farsighted," a condition called **presbyopia**, an age-related form of hyperopia. ∞ *[p. 274]*

The lens of the eye is normally transparent. A **cataract** is a loss of lens transparency. Cataracts can result from drug reactions, injuries, or radiation, but *senile cataracts* are the most common form. During aging, the lens becomes less elastic, takes on a yellowish hue, and loses its transparency. Both eyes usually develop cataracts. As the lens becomes opaque, or cloudy, the person needs brighter and brighter lights, and visual clarity fades. The only way to restore clear vision is through cataract surgery. The opaque lens is removed and replaced with a plastic implant. Such surgery is quite successful.

Glaucoma is an increase in the fluid pressure within the eye. The pressure may become so excessive that distortion of the retina and optic disc lead to partial or complete blindness. The pressure builds because problems develop in the removal of aqueous humor from the eye. ∞ *[p. 272]* Symptoms of glaucoma are not striking, and usually include a gradual loss of peripheral vision that is often not noticed by the individual. Glaucoma is one of the major eye disorders in those over 60, and a leading cause of blindness in the U.S. Treatment is possible early in the development of glaucoma before there is a loss of vision.

Secondary Disorders

Cardiovascular Disorders. **Retinal occlusion** is the blockage of the blood supply of the retina. A retinal artery occlusion may affect the main artery or its branches. Such blockages may result in permanent blindness or partial loss of vision. Retinal vein occlusion affects the main vein carrying blood away from the retina. It may lead to blindness.

The fovea lies within the **macula**, a region of the retina where there are no rods, only cones (p. 274). This is the sharpest area of vision. **Macular degeneration** is an age-related disorder whose cause is unknown. This condition is associated with the uncontrolled growth and proliferation of blood vessels in the choroid. Leakage of blood from these abnormal vessels leads to retinal scarring and a gradual loss of vision. Approximately 200,000 new cases occur in the U.S. each year; this number has been increasing steadily as the proportion of elderly persons in the population increases. There is no cure for macular degeneration, but lasers can sometimes be used to cauterize leaky blood vessels and slow the progress of the disease.

Immune Problems. **Uveitis** (ū-vē-Ī-tis) is an autoimmune disorder that impairs vision. It is an inflammation of the middle layer of the eye (the choroid layer), the iris, and the ciliary body.

One type of retinopathy develops as a complication of *diabetes mellitus*, an autoimmune disorder of the pancreas. In **diabetic retinopathy,** the capillaries within the eye leak fluid and hemorrhage into the retina. Further growth and hemorrhaging of retinal capillaries impairs vision. Diabetic retinopathy is the leading cause of blindness in the U.S.

Nutritional Problems. Vitamin A deficiencies may lead to **xerophthalmia** (ze-rof-THAL-me-ah), a dryness of the cornea and conjunctiva that can severely damage the cornea and result in corneal abrasions. **Night blindness**, an inability to see well in dim light, can also develop because of vitamin A deficiency.

EQUILIBRIUM AND HEARING

Equilibrium Problems

Vertigo is an illusion of movement (either one's self or in one's surroundings). This meaning distinguishes it from "dizziness," a sensation of light-headedness and disorientation that typically precedes a fainting spell. Vertigo can have a number of causes that alter the function of the inner ear receptor complex, the vestibular branch of the vestibulocochlear (acoustic) nerve, or sensory nuclei and pathways in the central nervous system.

Any event that sets endolymph into motion can stimulate the equilibrium receptors and produce vertigo. Placing an ice pack in contact with the temporal bone or flushing the external auditory canal with cold water may chill the endolymph in the outermost portions of the labyrinth and establish a temperature-related circulation of fluid. A mild and temporary vertigo is the result. The consumption of excessive quantities of alcohol or exposure to certain drugs can also produce vertigo by changing the composition of the endolymph or disturbing the hair cells. Many people experience vertigo when they have high fever, or take a whirling carnival ride.

Probably the most common cause of vertigo is *motion sickness*. Its unpleasant symptoms include headache, sweating, flushing of the face, nausea, and vomiting. It is thought that the condition results when processing centers in the brain receive conflicting sensory information from the eyes and inner ear. The drugs commonly administered to prevent motion sickness appear to depress activity in the brain stem.

Hearing Problems

Infection. Otitis media is an infection of the middle ear, most commonly of bacterial origin. Acute otitis media typically affects infants and children and is occasionally seen in adults. The pathogens tend to gain access via the auditory tube, generally during an upper respiratory infection. As the pathogen population rises in the tympanic cavity, white blood cells rush to the site, and the middle ear becomes filled with pus. If untreated, the tympanic membrane may rupture, producing a characteristic oozing from the external auditory canal. The bacteria can generally be controlled by antibiotics, the pain reduced by analgesics, and the swelling reduced by decongestants. In the United States it is rare for otitis media to progress to the stage at which rupture of the tympanic membrane occurs.

The mastoid process of the temporal bone (p. 151) contains a number of small air-filled chambers known as the *mastoid air cells*. These chambers open into the middle ear. If the pathogens leave the middle ear and invade the mastoid air cells, symptoms of **mastoiditis** appear. The connecting passageways are very narrow, and as the infection progresses, the subject experiences severe earaches, fever, and swelling behind the ear in addition to symptoms of otitis media. The same antibiotic treatment is used to deal with both conditions; the particular antibiotic selected depends on the identity of the bacterium involved. The major risk of mastoiditis is the spread of the infection to the brain along the connective tissue covering the facial nerve (N VII). Should this occur, prompt antibiotic therapy is needed. If the problem remains, the person may have to undergo surgical procedures that open and drain the mastoid air cells or drain the middle ear through a surgical opening in the tympanic membrane.

Hearing Deficits. It is estimated that over 6 million people in the U.S. have at least a partial hearing deficit, or deafness. **Conductive deafness** results from interference with the normal transfer of vibrations from the tympanum to the oval window. Examples of interference include excess wax or trapped water in the external auditory canal, scarring or perforation of the tympanum, or immobilization of one or more auditory ossicles by fluid or a tumor.

In **nerve deafness**, the problem lies within the cochlea or somewhere along the auditory pathway. The vibrations are reaching the oval window, but the receptors either cannot respond or their response cannot reach the central nervous system. For example, very loud noises can cause nerve deafness by damaging the sensory cilia on the receptor cells. Bacterial or viral infections may also kill receptor cells and damage sensory nerves.

In this hearing test, the child's responses to sounds of varying frequency and intensity are recorded.

Bone conduction tests are used to discriminate between conductive and nerve deafness. If you put your fingers in your ears and talk quietly, you can still hear yourself, because the bones of the skull conduct the sound waves to the cochlea, bypassing the middle ear. In one bone conduction test, the physician places a vibrating tuning fork against the skull. If the subject hears the sound of the tuning fork when it is in contact with the skull but not as well when it is held next to the opening of the external auditory canal, the problem must lie within the external or middle ear. If the subject remains unresponsive to either stimulus, the problem must be at the receptors or along the auditory pathway.

Several effective treatments exist for conductive deafness. A hearing aid overcomes the loss in sensitivity by increasing the intensity of stimulation. Surgery may repair the tympanic membrane, or free damaged or immobilized ossicles. Artificial ossicles may also be implanted if the original ones are damaged beyond repair.

There are few possible treatments for nerve deafness. Mild conditions may be overcome by the use of a hearing aid if some functional receptor cells remain. In a *cochlear implant*, a small, battery-powered device is inserted beneath the skin behind the mastoid process. Small wires run through the round window to reach the cochlear nerve; the implant "hears" a sound and stimulates the nerve

✔ What are drugs called that relieve pain without affecting the sense of touch?

✔ What is the loss of transparency of the lens called?

✔ Why is mastoiditis so dangerous?

directly. At present over 25,000 people are using cochlear implants. A new approach involves inducing the regeneration of hair cells of the organ of Corti. Researchers working with mammals other than humans have been able to induce hair cell regeneration both in cultured hair cells and in live animals. This is a very exciting area of research, and there is hope that it may ultimately lead to an effective treatment for human nerve deafness.

Age-Related Disorders. Hearing is not as uniformly affected by aging as other senses. However, the tympanum loses some of its elasticity and it becomes more difficult to hear high-pitched sounds. The progressive loss of hearing associated with aging is called **presbycusis** (prez-bē-KŪ-sis).

CHAPTER REVIEW

Key Words

accommodation: Alteration in the curvature of the lens of the eye to focus an image on the retina.

auditory: Pertaining to the sense of hearing.

choroid: Middle layer in the wall of the eye.

cochlea (KŌ-klē-a): Spiral portion of the bony labyrinth of the inner ear that surrounds the organ of hearing.

fovea (FŌ-vē-a): Portion of the retina providing the sharpest vision, with the highest concentration of cones.

gustation (gus-TĀ-shun): The sense of taste.

iris: A contractile structure made up of smooth muscle that forms the colored portion of the eye.

lacrimal gland: Tear gland; located above and to the lateral side of the eye.

lens: The transparent body lying behind the iris and pupil and in front of the vitreous body.

olfaction: The sense of smell.

proprioception (prō-prē-ō-SEP-shun): Awareness of the positions of bones, joints, and muscles.

pupil: The opening in the center of the iris through which light enters the eye.

retina: The innermost layer of the eye, lining the vitreous chamber; contains the photoreceptors.

sclera (SKLER-a): The fibrous, outer layer of the eye forming the white area and the transparent cornea.

semicircular canals: Tubular portions of the inner ear connected to the vestibule and responsible for sensing rotational movements of the head.

Selected Clinical Terms

analgesia: Relief from pain.

analgesic: An agent that relieves pain without causing a loss of sensitivity to other stimuli, such as touch, temperature, or pressure.

anesthesia: Loss of feeling or sensation. Usually applied to the loss of pain, as when it is induced in the performance of surgery.

anesthetic: An agent that is used to abolish the sensation of pain.

cataract: A reduction in lens transparency.

conductive deafness: Deafness resulting from conditions in the outer or middle ear that block the transfer of vibrations from the tympanic membrane to the oval window.

nerve deafness: Deafness resulting from problems with the cochlea or along the auditory pathway.

presbycusis: The loss of hearing associated with aging.

retinopathy (ret-i-NOP-ah-thē): A disease or disorder of the retina.

vertigo: An illusion of movement.

Study Outline

INTRODUCTION (p. 267)

1. The **general senses** are temperature, pain, touch, pressure, vibration, and body position. The **special senses** include smell, taste, sight, balance, and hearing.

SENSORY RECEPTORS (p. 267)

1. A **sensory receptor** is a specialized cell that, when stimulated, sends a sensation to the CNS.

2. The CNS interprets the nature of the arriving sensory information based on the area of the brain stimulated.

3. **Adaptation** is a reduction in sensitivity to a constant stimulus.

TYPES OF SENSORY RECEPTORS (p. 267)

4. Sensory receptors may be grouped according to the type of the stimulus that excites them: *pain receptors, thermoreceptors* (hot or cold temperatures), *mechanoreceptors* (various movements), *chemoreceptors* (chemicals), and *photoreceptors* (light).

THE GENERAL SENSES (pp. 268–269)

1. Receptors for the **general senses** are distributed throughout the body.

PAIN (p. 268)

2. The simplest pain receptors are **free nerve endings**. Most are associated with the skin. *(Figure 12-1)*

3. There are two types of these painful sensations: **fast pain**, or *prickling pain*, and **slow pain**, or *burning and aching pain*.

TEMPERATURE (p. 268)

4. Separate **thermoreceptors** respond to cold and hot temperatures. Adaptation to a constant temperature occurs fairly rapidly.

TOUCH AND PRESSURE *(p. 269)*

5. Tactile receptors are mechanoreceptors sensitive to fine and crude touch, pressure, and vibration. Structurally, they may simply be free nerve endings or more complex **tactile corpuscles**. *(Figure 12-1)*

BODY POSITION *(p. 269)*

6. Proprioceptors monitor the position of joints, tension in tendons and ligaments, and the state of muscular contraction.

THE SPECIAL SENSES *(pp. 269–279)*

1. The receptors for the **special senses** are located in specialized areas or in *sense organs*.

SMELL *(p. 269)*

2. The **olfactory organs** contain the **olfactory epithelium** with **olfactory receptors** (modified nerve cells sensitive to chemicals dissolved in the overlying mucus) and supporting cells. Their surfaces are coated with the mucus secretions of the *olfactory glands*. *(Figure 12-2)*

TASTE *(p. 270)*

3. Gustatory (taste) receptors are clustered in **taste buds** that are associated with epithelial projections (**papillae**) on the dorsal surface of the tongue. *(Figure 12-3)*

4. Each taste bud contains **gustatory cells**, which extend taste hairs through a narrow **taste pore**. *(Figure 12-3)*

5. The **primary taste sensations** are sweet, salt, sour, and bitter.

6. The taste buds are monitored by three different cranial nerves.

VISION *(p. 271)*

7. The **accessory structures** of the eye include the eyelids, a mucous membrane, various exocrine glands, and the extrinsic eye muscles.

8. An epithelium called the **conjunctiva** covers the exposed surface of the eye except over the transparent **cornea**.

9. The secretions of the **lacrimal gland** bathe the conjunctiva; these secretions contain an enzyme that attacks bacteria. Tears reach the nasal cavity after passing through the **nasolacrimal duct**. *(Figure 12-4)*

10. Six **extrinsic eye muscles** control external eye movements: the *inferior* and *superior rectus*, *lateral* and *medial rectus*, and *superior* and *inferior obliques*. *(Figure 12-5)*

11. The interior of the eye is divided into a small *anterior cavity* (anterior to the lens) and a large *posterior cavity* (posterior to the lens). A liquid *aqueous humor* bathes the lens and the posterior cavity contains the *vitreous body*, a gelatinous mass that helps stabilize the shape of the eye and supports the retina. *(Figure 12-6a)*

12. The eye has three layers. Most of the outer surface is covered by the **sclera** (a dense fibrous connective tissue), which is continuous with the cornea. *(Figure 12-6a)*

13. The second layer, called the **choroid**, includes blood vessels, the *iris*, and the *ciliary body*. The ciliary body contains the *ciliary muscle* and processes that attach to the *suspensory ligaments* of the lens. *(Figure 12-6a)*

14. The inner **retina** consists of an outer pigment layer and an inner layer containing the photoreceptors. *(Figure 12-6b)*

15. There are two types of photoreceptors: **rods** and **cones**. Rods are very light-sensitive but cannot discriminate among different colors; cones give us color vision. Many cones are densely packed within a region of the retina called the **fovea**, the site of sharpest vision. *(Figure 12-6a)*

16. From the photoreceptors, the information is relayed to accessory cells, then to *ganglion cells*, and to the brain over the optic nerve, which forms at the **optic disc**. *(Figure 12-6b)*

17. The lens, held in place by the suspensory ligaments, focuses a visual image on the retinal receptors. Light is **refracted** (bent) when it passes through the cornea and lens. During **accommodation** the shape of the lens changes to focus an image on the retina. *(Figure 12-7)*

18. Visual information is relayed from photoreceptors to the visual cortex of the occipital lobe, which contains a sensory map of the field of vision.

EQUILIBRIUM AND HEARING *(p. 275)*

19. The senses of equilibrium and hearing are provided by mechanoreceptors in the *inner ear* called *hair cells*.

20. The ear consists of three regions with different functions. The **external ear** collects sound waves, the **middle ear** amplifies them, and the **inner ear** provides the senses of hearing and balance, or equilibrium.

21. The external ear includes the **pinna**, which surrounds the entrance to the **external auditory canal**, which ends at the **tympanic membrane** ("eardrum"). *(Figure 12-8)*

22. The middle ear communicates with the nasopharynx through the **auditory tube** *(Eustachian tube)*. The middle ear encloses and protects the *auditory ossicles* that connect the tympanic membrane with the receptor complex of the inner ear. *(Figure 12-8)*

23. The inner ear can be subdivided into the **vestibule,** the **semicircular canals** (receptors in the vestibule and semicircular canals provide the sense of equilibrium), and the **cochlea** (these receptors provide the sense of hearing). *(Figures 12-8, 12-9)*

24. The sensory receptors of the inner ear lie within a network of fluid-filled tubes and chambers known as the **membranous labyrinth.** The membranous labyrinth contains a fluid called **endolymph.**

25. Each of the semicircular canals contains a *semicircular duct*, whose sensory receptors provide sensations of rotation. The hair cells become stimulated by endolymph movement. *(Figure 12-9b)*

26. The vestibule includes a pair of membranous sacs, the **saccule** and **utricle**, whose receptors provide sensations of gravity and linear acceleration. Hair cells are situated so that they are in contact with *otoliths* (mineral crystals). When the head tilts, the mass of otoliths shifts, and the resulting distortion in the sensory hairs stimulates the receptors. *(Figure 12-9c)*

27. The vestibular receptors activate sensory neurons whose axons form the **vestibular branch** of the vestibulocochlear nerve (N VIII).

28. The cochlea of the inner ear contains the **cochlear duct**, an elongated portion of the membranous labyrinth. *(Figures 12-9, 12-10)*

29. Sound waves travel toward the tympanic membrane, which vibrates; the auditory ossicles conduct the vibrations to the inner ear. Movement at the **oval window** applies pressure to the perilymph surrounding the cochlear duct. The resulting pressure waves distort the hair cells of the **organ of Corti** that lies within the cochlear duct. *(Figure 12-10)*

30. The stimulated receptors send nerve impulses to the CNS over the **cochlear branch** of the vestibulocochlear nerve (N VIII). *(Figure 12-10)*

SENSORY DISORDERS *(pp. 279–286)*

1. Sensory disorders can be diagnosed by an analysis of a sense organ's components. *(Figure 12-11)*

THE GENERAL SENSES (p. 279)

2. The sensitivity of tactile receptors (touch, pressure, and vibration) can be assessed through a **two-point discrimination test** and a tuning fork test. *(Figure 12-12)*

3. Anesthesia, hypesthesia, and paresthesia are terms used in the description of sensitivity. **Anesthesia** is a total loss of sensation, **hypesthesia** is a reduction in sensitivity, and *paresthesia* is the presence of abnormal sensations ("pins and needles.")

4. Pain is described as **acute** or **chronic**; chronic pain includes referred pain, pain from visceral organs that appears to be from another location. *(Figure 12-13)*

5. An **anesthetic** suppresses pain. In a **local anesthesia** pain is suppressed in only part of the body and the patient is conscious. In **general anesthesia**, the patient is made unconscious. *(Figure 12-13)*

6. Analgesia is the relief of pain, and an **analgesic** is a drug that relieves pain without an individual losing their sense of touch.

7. The sensation of pain often begins with the production of *prostaglandins* in damaged tissues. Non-narcotic drugs suppress pain by blocking the production of prostaglandins (aspirin) or by blocking the pain impulses directly in the brain (acetaminophen).

SMELL (p. 281)

8. Trauma and age-related disorders affect the sense of smell.

TASTE (p. 281)

9. Damage and loss of taste buds, damage to cranial nerves, and age-related changes can impair the sense of taste.

VISION (p. 282)

10. An **ophthalmologist** specializes in the eye and its diseases.

11. Examples of traumatic injuries to the cornea include **corneal abrasions** and **corneal ulcers**, and, to the retina, a **retinal tear** and a **detached retina**.

12. *Conjunctivitis* is an inflammation of the conjunctiva, the mucous membrane covering the white of the eye and the inner eyelids. *(Figures 12-14, 12-15)*

13. Keratoconjunctivitis is an inflammation of the cornea and the conjunctiva caused by the *Herpes simplex* virus.

14. A roundworm parasite causes **onchocerciasis**, or **river blindness**, in Central America and Africa.

15. Two malignant tumors of the eye are **melanoma of the choroid** and **retinoblastoma**.

16. Congenital disorders include **astigmatism, myopia, hyperopia, color blindness, strabismus**, and **retinitis pigmentosa**.

17. Degenerative disorders of the eye include *presbyopia, cataracts,* and *glaucoma*.

18. Secondary disorders include problems caused by the cardiovascular system, immune responses, and nutrition.

EQUILIBRIUM AND HEARING (p. 284)

19. Vertigo is an illusion of movement. It can be caused by disorders that affect the inner ear, the vestibule and semicircular canals, the vestibular branch of the vestibulocochlear (acoustic) nerve, or nuclei and pathways in the central nervous system. *Motion sickness* is an example of vertigo.

20. Otitis media, an infection of the middle ear, is usually caused by bacteria, and generally affects infants and children.

21. Two alternate sets of problems can lead to deafness. **Conductive deafness** is caused by interference with the normal transfer of vibrations from the tympanum to the oval window. **Nerve deafness** is caused by problems within the cochlea or somewhere along the auditory pathway.

22. The age-related loss of hearing is called **presbycusis.**

Review Questions

MATCHING

Match each item in Column A with the most closely related item in Column B. Use letters for answers in the spaces provided.

	Column A		Column B
___	1. myopia	a.	color vision
___	2. outer eye layer	b.	free nerve endings
___	3. chemoreceptors	c.	sclera and cornea
___	4. proprioceptors	d.	rotational movements
___	5. cones	e.	provide information on joint position
___	6. accommodation	f.	provide sense of smell and taste
___	7. tympanum	g.	site of sharpest vision
___	8. thermoreceptors	h.	active in dim light
___	9. lacrimal glands	i.	eardrum
___	10. olfaction	j.	change in lens shape to focus image on retina
___	11. fovea	k.	nearsightedness
___	12. tactile receptors	l.	eye specialist
___	13. semicircular ducts	m.	cross-eye
___	14. rods	n.	secretes tears
___	15. strabismus	o.	provide sense of touch
___	16. ophthalmologist	p.	smell

MULTIPLE CHOICE

17. The anterior, transparent part of the outer eye layer is called the _____ .
(a) iris   (b) cornea
(c) retina   (d) sclera

18. The malleus, incus, and stapes are tiny bones located in the _____ .
(a) outer ear   (b) middle ear
(c) inner ear   (d) cochlea

19. The organ of Corti contains the sensory receptors responsible for the sense of _____ .
(a) vision   (b) smell
(c) hearing   (d) taste

20. When chemicals dissolve in the nasal cavity, they stimulate _____ .
(a) gustatory cells
(b) rod cells
(c) olfactory hair cells
(d) tactile receptors

21. A reduction in sensitivity in the presence of a constant stimulus is called _____ .
(a) adaptation   (b) sensory coding
(c) responsiveness (d) perception

22. Each eye is moved by _____ extrinsic muscles.
(a) 2   (b) 4
(c) 6   (d) 8

23. The _____ transmits vibrations directly to the oval window.
(a) utricle   (b) incus
(c) malleus   (d) stapes

24. _____ is a reduction in sensitivity.
(a) paresthesia   (b) hypesthesia
(c) anesthesia   (d) analgesia

25. _____ does not lead to partial or full blindness.
(a) retinitis pigmentosa   (b) glaucoma
(c) presbyopia   (d) trachoma

TRUE/FALSE

____ 26. The conjunctiva is a thin membrane that covers the lens.

____ 27. Each inner ear contains three semicircular canals.

____ 28. The inner ear is located in the sphenoid bone of the skull.

____ 29. The ciliary muscle changes the shape of the lens.

____ 30. The fovea is the part of the retina that contains only rods.

____ 31. Retinopathy is a disorder or disease of the retina.

____ 32. Onchocerciasis is transmitted from person to person by mosquitoes.

SHORT ESSAY

33. Distinguish between the general and special senses.

34. Jane makes an appointment with the optometrist for a vision test. Her test results are reported as 20/15. What does this test result mean? Is a rating of 20/20 better or worse?

35. After attending a Fourth of July celebration with fireworks, Joyce finds it difficult to hear normal conversation, and her ears keep "ringing." What is causing her hearing problems?

36. After riding the express elevator from the twentieth floor to the ground floor, for a few seconds you still feel as if you are descending, even though you have obviously come to a stop. Why?

37. Describe two problems with the eye that are associated with vitamin A deficiency.

38. Given identically prepared tacos, the grandmother describes her taco as not spicy enough, yet her granddaughter complains hers is too spicy. Explain a possible cause for their different reactions.

APPLICATIONS

39. Mr. Young visits his doctor because he is having trouble with his vision. The physician performs a series of tests and determines the following: Stimuli are reaching Mr. Young's sensory receptors; the receptors respond normally; and the nerve pathways are intact and normal. Classify Mr. Young's sensory problem, and give an example of a condition that could cause the problem.

40. Mr. Romero, 62, has trouble hearing people during conversations, and his family persuades him to have his hearing tested. How can the physician determine whether Mr. Romero's problem results from conductive deafness or nerve deafness?

✔ Answers to Concept Check Questions

(p. 269) **1.** The general senses of the body include temperature, pain, touch, pressure, and proprioception (body position). **2.** Proprioceptors are the sense receptors of body position. They monitor the position of joints, the tension in tendons and ligaments, and the state of muscular contraction.

(p. 270) **1.** The greater the number of olfactory receptors, the greater the sensitivity of the sense of smell. Dogs have some 72 times more olfactory receptors than people do. **2.** The taste receptors (taste buds) are sensitive only to molecules and ions that are in solution. If you dry the surface of the tongue, there is no moisture for the sugar molecules or salt ions to dissolve in, so they will not stimulate the taste receptors.

(p. 275) **1.** The first layer of the eye to be affected by inadequate tear production would be the conjunctiva. Drying of this layer would produce an irritated, scratchy feeling. **2.** When the lens is round, you are looking at something closer to you. **3.** A person born without cones would still be able to see as long as they had functioning rod cells. Since cone cells provide color vision, a person with only rods would see only black and white.

(p. 279) **1.** The auditory tube, or Eustachian tube, provides an opening to the middle ear chamber and permits the equalization of pressure on the sides of the eardrums. **2.** Receptors in the three semicircular ducts are stimulated by rotational movements of the head.

(p. 286) **1.** Analgesics. **2.** A cataract. **3.** Mastoiditis is an inflammation or infection of the mastoid air cells within the mastoid process of the temporal bone. An infection in the mastoid region could reach the cranial meninges and brain by means of the facial nerve (N VII).

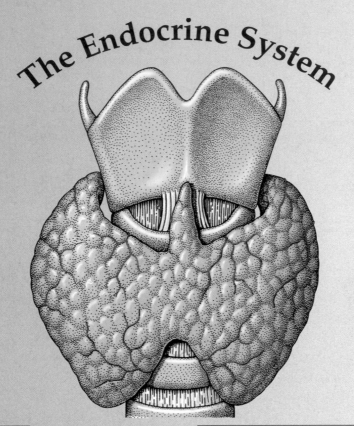

CHAPTER 13

The Endocrine System

CHAPTER OUTLINE

CHAPTER OBJECTIVES

1   Contrast the response times to changing conditions by the nervous and endocrine systems.

2   List the types of molecules that form the two main groups of hormones.

3   Explain the general actions of hormones.

4   Describe how endocrine organs are controlled by negative feedback.

5   Discuss the location, hormones, and functions of the following endocrine glands and tissues: pituitary, thyroid, parathyroids, thymus, adrenals, pancreas, testes, ovaries, and pineal gland.

6   Briefly describe the functions of the hormones secreted by the kidneys, heart, digestive system, and adipose tissue.

7   Explain how the endocrine system responds to stress.

8   Describe the causes of symptoms of most endocrine disorders.

9   Describe examples of disorders of the major endocrine glands.

The body adapts to changing conditions through a combination of rapid, short-term and slower, long-term responses. As we learned in the last two chapters, the nervous system allows us to respond quickly to changing internal and external stimuli. In contrast, the responses of the **endocrine system** are slower and longer lasting. Some of the endocrine responses aid in maintaining homeostasis, such as the control of blood sugar (glucose) and calcium ion levels, while others promote permanent structural changes, such as those associated with growth and development.

This chapter introduces the wide range of structures and functions of the endocrine system. It will also show the close working relationship between the nervous and endocrine systems.

SYSTEM BRIEF

The endocrine system includes all the endocrine cells and tissues of the body. As noted in Chapter 4, *endocrine cells* are secretory cells that release their secretions into the bloodstream, where they can reach all parts of the body. In contrast, the secretions of *exocrine cells* travel through ducts to specific locations. ∞ *[p. 64] Endocrine glands* are ductless glands made up of groups of endocrine cells and other tissues. They form the organs of the endocrine system.

The cells of endocrine glands secrete chemicals called **hormones**, a word that means "to set in motion." Hormones are chemical compounds that travel in the circulatory system and affect the activities of other cells.

The major glands of the endocrine system are shown in Figure 13-1•. Some of these organs, such as the pituitary gland, only secrete hormones; others, such as the pancreas, have many other functions in addition to hormone secretion.

Table 13-1 lists the major glands of the endocrine system and the hormones they secrete, and contains brief remarks about their general effects on body functions.

•FIGURE 13-1
The Endocrine System.

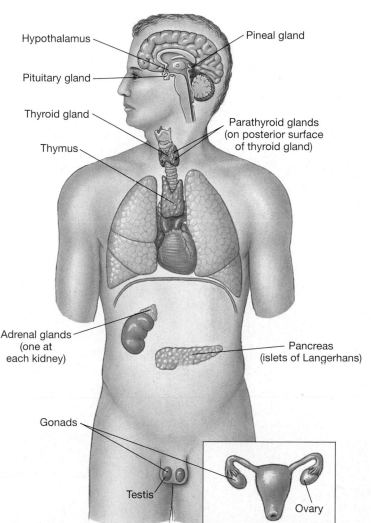

HORMONES

CHEMICAL MAKEUP

Most hormones are composed of amino acids, the building blocks of proteins. ∞ *[p. 28]* These **amino acid–based hormones** range in size from hormones derived from single amino acids to large protein hormones. This is by far the largest category of hormones, and it includes the hormones of most endocrine organs other than the reproductive organs (ovaries, testes, and placenta) and the outer portion of the adrenal gland (the adrenal cortex).

Lipid-based hormones include *steroid hormones* and *prostaglandins*. **Steroid hormones** are built from molecules of cholesterol. Steroid hormones are released by the reproductive organs and the adrenal cortex. **Prostaglandins** are produced from fatty acids by most tissues of the body. Locally, prostaglandins coordinate cellular activities within the

Table 13-1   Endocrine Hormones: Their Sources and Effects

Gland/Hormone	Effects
Hypothalamus	
Releasing hormones	Stimulate hormone production in anterior pituitary
Inhibiting hormones	Prevent hormone production in anterior pituitary
Anterior Pituitary	
Thyroid-stimulating hormone (TSH)	Triggers release of thyroid hormones
Adrenocorticotropic hormone (ACTH)	Stimulates adrenal cortex cells to secrete glucocorticoids
Follicle-stimulating hormone (FSH)	*Female*: promotes egg development; stimulates estrogen production
	Male: promotes sperm production
Luteinizing hormone (LH)	*Female*: produces ovulation (egg release); stimulates ovaries to produce estrogen and progesterone
	Male: stimulates testes to produce androgens (e.g., testosterone)
Prolactin (PRL)	Stimulates mammary gland development and production of milk
Growth hormone (GH)	Stimulates cell growth and division
Posterior Pituitary	
Antidiuretic hormone (ADH)	Reduces water loss at kidneys
Oxytocin	Stimulates contraction of smooth muscles of uterus and release of milk in females; stimulates prostate gland contraction in males
Thyroid	
Thyroxine	Stimulates general rate of body metabolism
Calcitonin	Reduces calcium ion levels in blood
Parathyroid	
Parathyroid hormone (PTH)	Increases calcium ion levels in the blood
Thymus	
Thymosins	Stimulate development of white blood cells (lymphocytes) in early life
Adrenal Cortex	
Glucocorticoids	Stimulate glucose synthesis and storage (as glycogen) by liver
Mineralocorticoids	Cause the kidneys to retain sodium ions and water and excrete potassium ions
Androgens	Produced in both sexes, but functions uncertain
Adrenal Medullae	
Epinephrine (E)	Also known as *adrenaline*; stimulates use of glucose and glycogen and release of lipids by adipose tissue; increases heart rate and blood pressure
Norepinephrine (NE)	Also known as *noradrenaline*; effects similar to epinephrine
Pancreas	
Insulin	Decreases glucose levels in blood
Glucagon	Increases glucose levels in blood
Testes	
Testosterone	Promotes production of sperm and development of male sex characteristics
Ovaries	
Estrogens	Support egg development, growth of uterine lining, and development of female sex characteristics
Progesterones	Prepare uterus for arrival of developing embryo and support of further embryonic development
Pineal Gland	
Melatonin	Delays sexual maturation; establishes day/night cycle
Adipose Tissue	
Leptin	Promotes weight loss; stimulates metabolism

tissue. For this reason, they are sometimes called "local hormones." However, some prostaglandins enter the circulation and have more widespread effects on body processes.

ACTIONS ON CELLS

Each hormone has specific **target cells** that respond to its presence. As Figure 13-2• shows, such cells have the appropriate receptor molecules in their cell membranes. Because the target cells may be anywhere in the body, a single hormone can affect the activities of multiple tissues and organs simultaneously.

All the structures and functions of cells depend on proteins. Structural proteins determine the general shape and internal structure of a cell, and enzymes (which are proteins as well) control its metabolic activities. Hormones alter the workings of a target cell by changing the dominant *types, activities,* or *quantities* of important enzymes and structural proteins. For example, the hormone *testosterone,* the dominant sex hormone in males, stimulates the production of enzymes and proteins in skeletal muscle fibers, increasing muscle size and strength.

CONTROL OF HORMONE SECRETION

The rate of hormone secretion varies from hour to hour and day to day. The control of hormone secretion is based on negative feedback. ∞ *[p. 5]* Recall that in negative feedback processes, actions are taken that reduce or eliminate the effects of

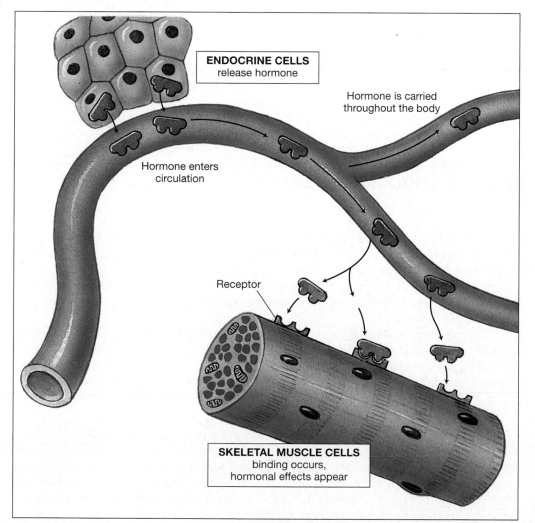

•**FIGURE 13-2**
Hormones and Target Cells.
For a hormone to affect a target cell, that cell must have receptors that can bind the hormone and initiate a change in cellular activity. This hormone affects skeletal muscle cells because the muscle cells have the appropriate receptors.

ENDOCRINE CELLS
release hormone

Hormone is carried
throughout the body

Hormone enters
circulation

Receptor

SKELETAL MUSCLE CELLS
binding occurs,
hormonal effects appear

a stimulus. Hormone secretions occur in response to changes in the composition of blood or surrounding tissue fluid. Once in circulation, the hormone stimulates a target cell to restore normal conditions (homeostasis).

Some endocrine glands respond directly to changes in the composition of body fluids. For example, the parathyroid and thyroid glands control calcium ion levels in this manner. When circulating calcium levels decline, the parathyroid glands release *parathyroid hormone*, and target cells respond by elevating blood calcium ion levels. When calcium levels rise, the thyroid gland releases *calcitonin*, and target cells respond by lowering blood calcium ion levels.

Other endocrine glands respond indirectly to changes in the composition of body fluids. The activities of these endocrine glands are regulated by the hypothalamus. As noted in Chapter 11, the hypothalamus provides the link between the nervous and endocrine systems.

The Hypothalamus and Hormone Secretion

Neurons in the hypothalamus regulate the activities of the nervous and endocrine systems in three ways (see Figure 13-3●).

1. The hypothalamus contains autonomic nervous system centers that control the endocrine cells of the adrenal medullae through sympathetic innervation. When the sympathetic division of the ANS is activated, the adrenal medullae release hormones into the bloodstream. These hormones, epinephrine and norepinephrine, increase heart rate, and blood pressure, and free up energy reserves for immediate use.

2. The hypothalamus itself acts as an endocrine organ, releasing the hormones *ADH* and *oxytocin*. ADH reduces water losses at the kidneys; oxytocin stimulates smooth muscle contractions in the uterus and mammary glands (women) and the prostate gland (men).

3. The hypothalamus also secretes **regulatory hormones**, special hormones that regulate the secretions of endocrine cells in the anterior pituitary gland. There are two classes of regulatory hormones: *Releasing hormones (RH)* stimulate the production of pituitary hormones, and *inhibiting hormones (IH)* prevent the synthesis and secretion of pituitary hormones.

✔ How does a hormone affect a target cell?

✔ What do the regulatory hormones of the hypothalamus regulate?

MAJOR ENDOCRINE GLANDS

THE PITUITARY GLAND

hypo, below + *physis*, growth
hypophysis: the pituitary gland, which lies beneath the diencephalon and is connected to the hypothalamus

The pituitary gland, or **hypophysis** (hī-POF-i-sis), secretes eight different hormones. Many of the pituitary's hormones "turn on" other endocrine glands, and are called *-tropins*. This name comes from the Greek word *tropos*, which means "to turn" or "change." As a result, the pituitary is often called the "master gland" of the body. Figure 13-3● shows the hormones released by the pituitary gland and their target organs.

The pituitary is a small, oval gland that lies under the brain, where it is nestled within a depression in the sphenoid bone of the skull. ∞ *[p. 153]* It is connected to the overlying hypothalamus by a slender stalk called the **infundibulum** (in-fun-DIB-ū-lum). The pituitary gland is divided into distinct anterior and posterior regions.

infundibulum: a funnel-shaped structure, in this case the stalk attaching the pituitary gland to the hypothalamus

Anterior Pituitary

The endocrine cells of the **anterior pituitary** are surrounded by a network of capillaries. Those capillaries are also connected to another capillary network within the hypothalamus. These two networks, along with the connecting blood vessels, make up the *hypophyseal portal system*. The two groups of capillaries provide a direct route that carries hormones from the hypothalamus to the anterior pituitary gland, and from there into the general circulation.

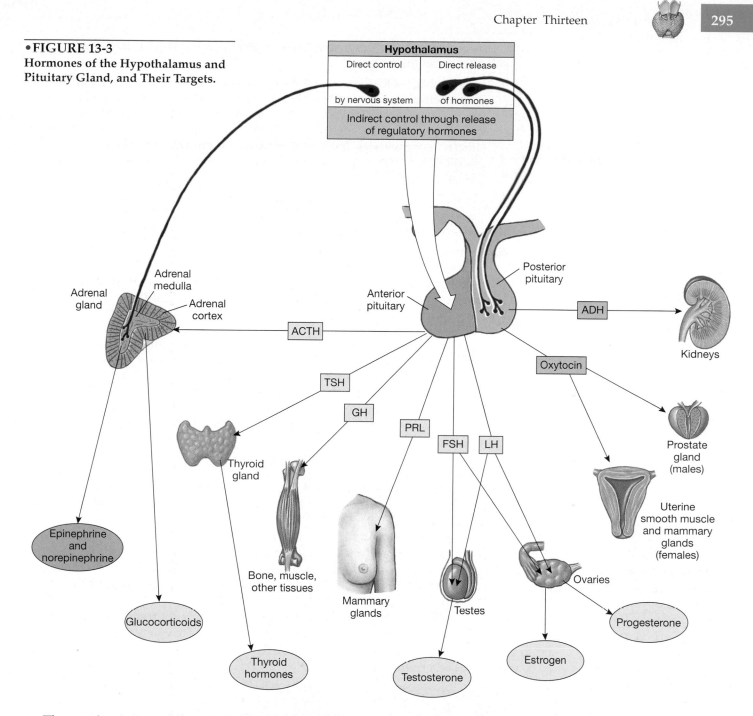

•FIGURE 13-3
Hormones of the Hypothalamus and
Pituitary Gland, and Their Targets.

The regulatory hormones produced by the hypothalamus control the secretions of the anterior pituitary gland. An endocrine cell in the anterior pituitary may be controlled by releasing hormones, inhibiting hormones, or some combination of the two. These regulatory hormones are released by neurons in the hypothalamus and carried directly to the anterior pituitary gland by the hypophyseal portal system.

Six hormones are produced by the anterior pituitary. Of these, four regulate the production of hormones by other endocrine glands.

1. **Thyroid-stimulating hormone (TSH)** targets the thyroid gland and triggers the release of thyroid hormones.

2. **Adrenocorticotropic hormone (ACTH)** stimulates the release of steroid hormones by the adrenal cortex. In particular, it stimulates the cells producing hormones called **glucocorticoids** (gloo-kō-KOR-ti-koyds).

3. **Follicle-stimulating hormone (FSH)** promotes oocyte development in women and stimulates the secretion of **estrogens**, steroid hormones pro-

duced by cells of the ovary. In men, FSH production supports sperm production in the testes.

4. **Luteinizing** (LOO-tē-in-ī-zing) **hormone** (**LH**) induces ovulation (egg release) in women and promotes the ovarian secretion of estrogens and the **progestins** (such as *progesterone*) that prepare the body for possible pregnancy. In men the same hormone stimulates the production of male sex hormones by the *interstitial cells* of the testes. FSH and LH are called **gonadotropins** (gō-nad-ō-TRŌ-pinz) because they regulate the activities of the male and female sex organs (gonads).

5. **Prolactin** (prō-LAK-tin), or **PRL**, which means "before milk," stimulates the development of the mammary glands and their production of milk in the female. It has other stimulatory effects on cell growth and development in both sexes.

6. **Growth hormone** (**GH**), also called *human growth hormone* (hGH) or *somatotropin*, stimulates overall body growth through cell growth and cell division by increasing the rate of protein synthesis. Its greatest effects are on muscular and skeletal development, especially in children.

Posterior Pituitary

The **posterior pituitary** stores hormones produced by two different groups of neurons within the hypothalamus. One group manufactures antidiuretic hormone *(ADH)* and the other, *oxytocin*. Their secretions are transported within axons that extend through the infundibulum to the posterior pituitary.

Antidiuretic hormone (**ADH** or vasopressin) is released when there is a rise in the concentration of electrolytes (ions) in the blood or a fall in blood volume or pressure. ADH acts to decrease the amount of water lost at the kidneys. With losses minimized, any water absorbed from the digestive tract will be retained, reducing the concentration of electrolytes. ADH also causes the constriction of blood vessels, which helps increase blood pressure. Alcohol interferes with the production of ADH, which explains why the excretion of urine increases after the consumption of alcoholic beverages.

In women, **oxytocin**, which means "quick childbirth," stimulates smooth muscle cells in the uterus and special cells surrounding the secretory cells of the mammary glands. The stimulation of uterine muscles by oxytocin helps maintain and complete normal labor and childbirth. After delivery, oxytocin triggers the release of milk from the breasts (see Chapter 22).

In the male, oxytocin stimulates the smooth muscle contraction in the walls of the *prostate gland*. This action may be important prior to ejaculation.

Figure 13-3• and Table 13-1 summarize important information about the hormones of the pituitary gland.

THE THYROID GLAND

The thyroid gland is located just below the **thyroid** ("shield-shaped") **cartilage**, or "Adam's apple" in the neck (Figure 13-4•). Its left and right lobes wrap around the *larynx*, or "voice box." The thyroid gland has a deep red color because of its large number of blood vessels.

The thyroid gland contains large numbers of spherical **thyroid follicles**. Thyroid follicles release two main types of hormones into the circulation. The most important of these is **thyroxine** (thī-ROKS-in). Thyroxine production depends on a regular supply of iodine. In many parts of the world, inadequate dietary iodine leads to an inability to synthesize thyroid hormones. Under these conditions, the thyroid follicles swell up, resulting in an enlarged thyroid gland, or *goiter*. This is seldom a problem in the United States because the typical American diet provides roughly three times the minimum daily requirement of iodine, thanks to the addition of iodine to table salt ("iodized salt").

Thyroxine affects almost every cell in the body. It stimulates energy production in cells, resulting in an increase in cellular metabolism and oxygen consumption. In growing children, thyroid hormones are essential to normal development of the

anti, against + *dia*, through + *ouresis*, urination
antidiuretic hormone: a hormone that reduces the water content of urine

✔ Why is the pituitary gland referred to as the "master gland" of the body?

✔ How would dehydration affect the level of ADH released by the posterior pituitary?

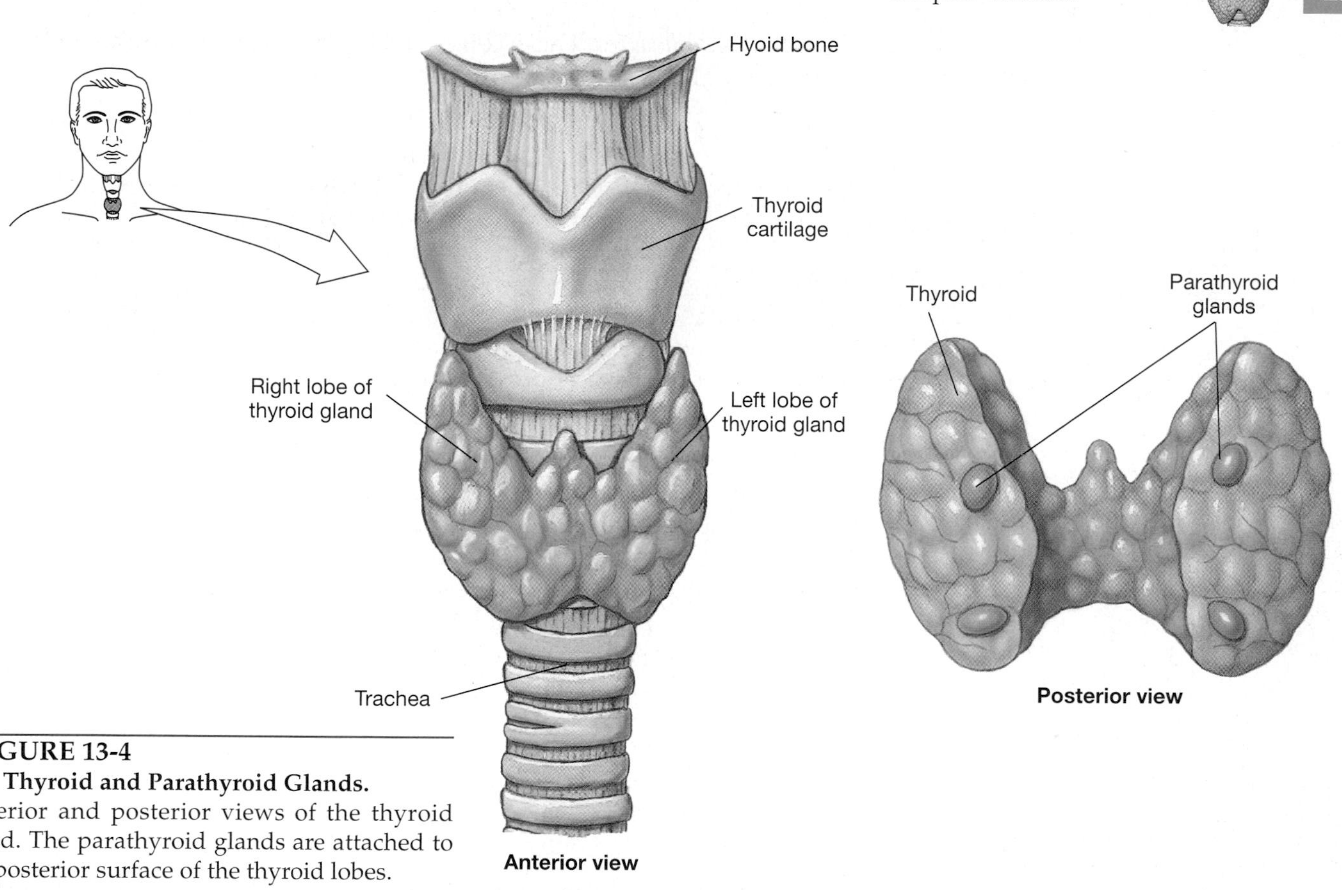

•FIGURE 13-4
The Thyroid and Parathyroid Glands.
Anterior and posterior views of the thyroid
gland. The parathyroid glands are attached to
the posterior surface of the thyroid lobes.

skeletal, muscular, and nervous systems. They also help them adapt to cold temperatures. When the metabolic rate increases, cells consume more energy and more
heat is generated, replacing the heat lost to a chilly environment.

The thyroid gland also produces the hormone **calcitonin** (kal-si-TŌ-nin) (**CT**).
Calcitonin helps regulate calcium ion concentrations in body fluids. This hormone is
released when the calcium ion concentration of the blood rises above normal. The target organs are the bones, the digestive tract, and the kidneys. Calcitonin reduces calcium levels by stimulating bone-building cells (osteoblasts), reducing calcium
absorption by the intestine, and stimulating calcium excretion at the kidneys. The resulting reduction in the calcium ion concentrations eliminates the stimulus and "turns
off" calcitonin secretion.

THE PARATHYROID GLANDS

Two tiny pairs of parathyroid glands are embedded in the posterior surfaces of the
thyroid gland (Figure 13-4•). The parathyroid glands produce **parathyroid hormone (PTH)** when the calcium concentration falls below normal. Although
parathyroid hormone acts on the same target organs as does calcitonin, it produces the opposite effects. PTH stimulates bone-dissolving cells (osteoclasts), promotes intestinal absorption of calcium, and reduces urinary excretion of calcium
ions until blood concentrations return to normal.

THE THYMUS

The thymus is embedded in a mass of connective tissue inside the thoracic cavity, just posterior to the sternum. In a newborn infant the thymus is relatively enormous, often extending from the base of the neck to the upper border of the heart.
As the child grows, the thymus continues to enlarge slowly, reaching its maximum size just before puberty. After puberty it gradually diminishes in size.

The thymus produces several hormones, collectively known as **thymosins**
(thī-MŌ-sins). The thymosins promote the development and maturation of white

✔ What element needs to be supplied in the diet for the thyroid to
manufacture thyroxine hormone?

✔ Which glands regulate the concentration of calcium in the body?

blood cells called *lymphocytes*. These cells, which play a key role in the body's immune defenses, will be discussed further in Chapter 16.

THE ADRENAL GLANDS

Each kidney is topped by an **adrenal gland**. Each adrenal gland can be divided into two parts, an outer **adrenal cortex** and an inner **adrenal medulla**.

Adrenal Cortex

The adrenal cortex has a grayish yellow coloration because of the presence of stored lipids, especially cholesterol and various fatty acids. The adrenal cortex produces three different classes of steroid hormones, collectively called *corticosteroids*.

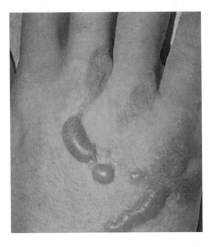

The irritation of poison ivy dermatitis can be controlled by using glucocorticoid creams, such as hydrocortisone.

1. **Glucocorticoids:** The **glucocorticoids** (gloo-kō-KOR-ti-koyds) (**GC**) are steroid hormones that affect glucose metabolism through a *glucose-sparing effect*. These hormones speed up the rates of glucose synthesis and glycogen formation, especially in the liver. In addition, fatty tissue responds by releasing fatty acids into the blood, and other tissues begin to break down fatty acids and proteins for energy instead of glucose. The net result of all this activity is that more glucose is available to the brain. This is important because glucose is the sole source of energy for neurons.

    Glucocorticoids also have *anti-inflammatory activity*; they suppress the activities of white blood cells and other components of the immune system. Glucocorticoid creams are often used to control irritating allergic rashes, such as those produced by poison ivy, and injections of glucocorticoids may be used to control more severe allergic reactions. *Cortisol* (KOR-ti-sol), also called *hydrocortisone*, is one of the most important glucocorticoids.

2. **Mineralocorticoids:** Corticosteroids known as **mineralocorticoids** (min-er-al-ō-KOR-ti-koyds) (**MC**) affect the concentrations of sodium and potassium ions in body fluids. **Aldosterone** (al-DOS-ter-ōn), the main mineralocorticoid, targets kidney cells that regulate the composition of urine. It causes the retention of sodium ions and water, reducing fluid loss in the urine. Aldosterone also reduces sodium and water loss at the sweat glands, salivary glands, and along the digestive tract. The sodium ions recovered are exchanged for potassium ions, so aldosterone also lowers potassium ion concentrations in body fluids.

3. **Androgens:** The adrenal cortex in both sexes produces small quantities of sex hormones called **androgens**. Androgens are produced in large quantities by the testes of males, and the importance of the small adrenal production in both sexes remains uncertain.

Adrenal Medulla

Each adrenal medulla has a reddish brown coloration partly because of the many blood vessels in this area. Its cells are targets of sympathetic nerve fibers that extend from the spinal cord. ∞ *[p. 230]*

The adrenal medullae contain secretory cells that produce **epinephrine** *(adrenaline)* and **norepinephrine** *(noradrenaline)*. These hormones are normally released at a low rate, but nerve impulses from the sympathetic nervous system speed up their rate of secretion dramatically. The sudden release of these hormones rapidly prepares the body for emergency "fight-or-flight" situations.

Epinephrine and norepinephrine speed up the use of cellular energy and free up energy reserves. They accomplish this by targeting receptors on skeletal muscle, fat, and liver cells. Skeletal muscles release glucose from their glycogen reserves and, in turn, produce energy from the glucose. This increases muscular power and endurance. In fatty tissue, stored fats are broken down, releasing fatty acids, and in the liver, glycogen molecules are converted to glucose. The fatty acids and glucose are then released into the bloodstream for use by other body tissues. The heart responds to epinephrine and norepinephrine with an increase in the rate and strength of cardiac contractions. This elevates blood pressure.

THE PANCREAS

The **pancreas** lies between the stomach and small intestine (Figure 13-5a•). It is a slender, usually pink organ with a lumpy consistency that contains both exocrine and endocrine cells. The **exocrine pancreas**, discussed further in Chapter 18, produces large quantities of *pancreatic juice* that is secreted into the digestive tract.

Cells of the **endocrine pancreas** form clusters known as **pancreatic islets**, or the *islets of Langerhans* (LAN-ger-hanz) (Figure 13-5b•). Like small islands, the islets are scattered among the exocrine cells that make up most of the pancreas. Each islet contains different cell types that produce different hormones. The two most important hormones are **glucagon** (GLOO-ka-gon) and **insulin** (IN-su-lin). Glucagon and insulin regulate blood glucose concentrations in the same way that parathyroid hormone and calcitonin control blood calcium levels.

Regulation of Blood Glucose

When glucose levels in the blood rise, certain islet cells release insulin, which stimulates the transport of glucose across cell membranes into most body cells. As cells absorb glucose from the bloodstream, circulating glucose concentrations decline.

In summary, insulin lowers blood glucose by shifting the glucose into cells. While glucose levels remain high most cells use it as an energy source instead of breaking down fatty acids or amino acids. The energy generated by the breakdown of glucose molecules is then used to build proteins and enhance energy reserves in the form of fats or glycogen. Thus insulin stimulates cell growth throughout the body. In adipose tissues, fat cells enlarge as they synthesize additional triglycerides; in the liver and in skeletal muscles, glycogen formation accelerates.

When blood glucose concentrations fall below normal, other islet cells release glucagon, and energy reserves are used. Skeletal muscles and liver cells break down glycogen to release glucose, fatty tissues release fatty acids, and proteins are broken down into their component amino acids. The fatty acids are absorbed by many cells, and used for energy production instead of glucose. The liver absorbs the amino acids and converts them to glucose, which can be released into the circulation. As a result, blood glucose concentrations rise toward normal levels.

Some cells, such as brain and kidney cells, do not have insulin receptors. These cells can absorb and use glucose without the presence of insulin. When blood glucose levels fall below normal, and insulin secretion is minimal, other tissues stop using glucose. The glucose that remains in circulation then remains available for brain and kidney cells.

•FIGURE 13-5
The Pancreas and Its Endocrine Cells.
(a) Orientation of the pancreas. (b) A pancreatic islet surrounded by exocrine-secreting cells.

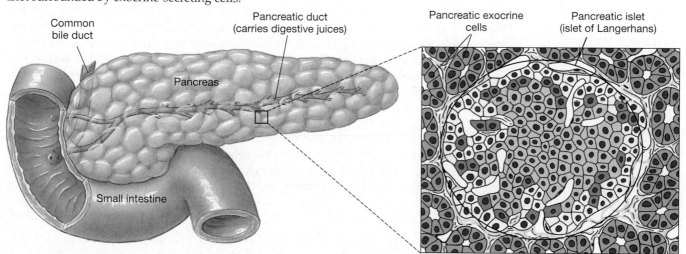

Common bile duct

Pancreatic duct (carries digestive juices)

Pancreatic exocrine cells

Pancreatic islet (islet of Langerhans)

Pancreas

Small intestine

(a)

(b)

⚕ Clinical Note

Diabetes Mellitus

Diabetes mellitus (dī-a-BĒ-tēs mel-Ī-tus) is characterized by high glucose concentrations in the blood and urine. The high levels of glucose in the kidney limits its ability to conserve water, and as a result excessive amounts of urine are produced. Diabetes mellitus occurs when islet cells produce inadequate amounts of insulin, insulin-producing cells die, or when cells have too few insulin receptors (or abnormal receptors). Insulin stimulates cells to absorb glucose from their surroundings. When insulin levels are low, or cells are unable to detect it, the cells start breaking down fatty acids and even proteins to obtain energy. Because most tissues can no longer absorb glucose, circulating glucose levels climb after every meal, and then remain elevated between meals. Over time, the combination of metabolic changes and water losses cause a variety of problems in virtually all body systems. Some examples of these problems include disturbances of vision; kidney failure; nervous system disorders; and an increased risk of heart attacks.

Under normal conditions, glucose is conserved at the kidneys, and glucose is not lost in the urine. A lack of insulin or an inability to respond normally to insulin will prevent most cells from absorbing glucose, even after a meal rich in sugars. As a result, the glucose concentration in the blood can rise to very high levels (several times normal levels). When this happens, the kidneys can no longer prevent the loss of glucose in the urine. This is a characteristic symptom of the condition called *diabetes mellitus*.

THE REPRODUCTIVE ORGANS

The male and female reproductive organs, the testes and ovaries, produce sex cells and hormones. Details of the structure of the reproductive organs and how hormones control reproductive function will be considered in Chapter 21.

The Testes

In the male, the *interstitial cells* of the testes produce the steroid hormones known as **androgens** (from the word *andros*, meaning male). **Testosterone** (tes-TOS-ter-ōn) is the most important androgen. During embryonic development, the production of testosterone causes the development of male reproductive ducts and external genital organs, such as the penis and scrotum. Later in life testosterone stimulates the production of sperm cells, maintains the secretory glands of the male reproductive tract, and determines other male traits, called *secondary sexual characteristics*, such as muscle mass, sexual drive, and the distribution of facial hair and body fat.

The Ovaries

In the ovaries, eggs develop in specialized structures called *follicles*, under stimulation by FSH. Follicle cells surrounding the developing eggs produce **estrogens** (ES-trō-jenz). These steroid hormones support the development of the eggs and stimulate the growth of the uterine lining during the uterine cycle. They are also responsible for determining female *secondary sexual characteristics* such as breast development and body fat distribution. After ovulation has occurred, the cells making up the follicle also begin to secrete **progesterone** (prō-JES-ter-ōn). Progesterone prepares the uterus for the arrival of a developing embryo. Along with other hormones, it also causes an enlargement of the mammary glands.

During pregnancy the placenta itself functions as an endocrine organ, working together with the ovaries and the pituitary gland to promote normal fetal development and delivery. The placenta secretes estrogen, progesterone, and other hormones. The presence of a placental hormone in the urine is the basis of home pregnancy tests. Placental hormones are discussed in Chapter 22.

THE PINEAL GLAND

The pineal gland lies in the roof of the diencephalon superior to the third ventricle. This gland synthesizes the hormone **melatonin** (mel-a-TŌ-nin). Visual information is relayed to the pineal gland, and light/dark cycles affect the rate of melatonin production. Melatonin production is lowest during daylight hours and highest in the dark of night.

Melatonin has several functions. It plays a role in the timing of puberty and human sexual maturation. In addition, because its rate of secretion varies with the day/night cycle, the pineal gland helps set a basic 24-hour rhythm to bodily processes.

OTHER HORMONE SOURCES

THE KIDNEYS

The kidneys release three hormones. One is important in balancing the levels of calcium, and the other two help regulate blood pressure and blood volume.

Calcitriol (kal-si-TRĪ-ol) is a hormone secreted by the kidney in response to the presence of parathyroid hormone (PTH). Calcitriol stimulates the absorption of calcium ions along the digestive tract.

Erythropoietin (e-rith-rō-POY-e-tin), or **EPO**, is released by the kidney in response to low oxygen levels in kidney tissues. Its name, derived from terms that mean "red making," refers to its function: EPO stimulates the production of red blood cells by the bone marrow. The increase in the number of erythrocytes increases the amount of oxygen carried to body tissues.

Renin is released by kidney cells in response to a drop in blood volume and/or blood pressure. Once in the bloodstream, renin acts as an enzyme, and it starts a chemical chain reaction that leads to the formation of the hormone *angiotensin II*. **Angiotensin** (an-jē-ō-TEN-sin) **II** has several functions, including the stimulation of aldosterone production by the adrenal cortex. The renin–angiotensin system will be considered further in Chapters 15 and 20.

THE HEART

The endocrine cells in the heart are cardiac muscle cells in the walls of the *atria*, chambers that receive venous blood. If the blood volume becomes too great, these cardiac muscle cells are excessively stretched. Under these conditions, they release **atrial natriuretic** (nā-trē-ū-RET-ik) **peptide (ANP)**. This hormone lowers blood volume by increasing the loss of sodium and water at the kidneys and suppressing thirst. We will consider the actions of this hormone in greater detail when we discuss the control of blood pressure and volume in Chapter 15.

THE DIGESTIVE SYSTEM

In addition to the pancreatic islets, the lining of the digestive tract produces a variety of endocrine secretions that are essential to the normal breakdown and absorption of food. Although the pace of digestive activities can be affected by the autonomic nervous system, most digestive processes are controlled locally. The various components of the digestive tract communicate with one another by means of hormones. These hormones will be considered in Chapter 18.

THE ADIPOSE TISSUE

Leptin is a recently discovered weight-control hormone produced by adipose tissue. Its name means "slender." Released into the blood by fat cells, leptin binds to the appetite control centers of the hypothalamus. It acts to control weight gain by (1) suppressing appetite and (2) stimulating metabolic processes that burn energy.

HORMONES AND STRESS

Any condition within the body that threatens its steady state, or homeostasis, is a form of **stress**. Causes of stress may be (1) physical, such as illness or injury; (2) emotional, such as depression or anxiety; (3) environmental, such as extreme heat or cold;

✔ What effect would elevated cortisol levels have on the level of glucose in the blood?

✔ How do insulin and glucagon control the levels of glucose in blood?

✔ What hormones are secreted by the testes and ovaries?

✔ Increased amounts of light would reduce the production of which hormone?

✔ Low oxygen concentrations at the kidneys stimulate the release of EPO. What is the function of this kidney hormone?

✔ How would an increase in the amount of atrial natriuretic peptide affect the volume of urine?

☤ Clinical Note

Anabolic Steroids and Athletes

In order to enhance their performances (and future earnings potential), significant numbers of athletes around the world have turned to the use of **anabolic steroids**. These are synthetic molecules of the male hormone testosterone. Normally, testosterone stimulates muscle building and secondary sex effects during puberty.

Although the use of anabolic steroids is banned by many national and international athletic organizations, it is estimated that 10 to 20 percent of male high school athletes, up to 30 percent of college and professional athletes, and up to 80 percent of bodybuilders use anabolic steroids to "bulk up." Known health risks associated with such use include premature stoppage of bone growth, liver problems (including jaundice and tumors), enlargement of the prostate gland, shrinkage of the testes, and sterility. A link to heart attacks and strokes has also been suggested.

In males, high doses of anabolic steroids can also depress the normal production of testosterone. This effect, which can be permanent, can occur by interference with the production of the hypothalamic releasing hormone that stimulates LH secretion.

Anabolic steroids also add muscle mass to female bodies. However, women taking these hormones can develop irregular menstrual periods, reduced breasts, changes in body hair distribution (including baldness), and a lowered voice. In both genders, androgen abuse may cause a depression of the immune system.

or (4) metabolic, such as starvation. The stresses produced may be opposed by specific homeostatic adjustments. For example, as discussed in Chapter 1. ∞ *[p. 6]* a decline in body temperature will result in responses, such as changes in the pattern of circulation or shivering, that attempt to restore normal body temperature.

In addition, the body has a general, or standard, response to different types of stress that involves both the endocrine and nervous systems.

The first, or immediate, response to the stress involves the sympathetic division of the autonomic nervous system. During this phase, energy reserves are mobilized, mainly in the form of glucose, and the body prepares for any physical activities needed to eliminate or escape from the source of the stress with increases in heart and breathing rates. Epinephrine is the dominant hormone of the alarm phase, and its secretion by the adrenal medullae accompanies the sympathetic activation that produces the "fight or flight" response discussed in Chapter 10. ∞ *[p. 230]*

If a stress lasts longer than a few hours, other hormones become involved in maintaining the higher energy demands placed on the body. Although epinephrine, growth hormone, and thyroid hormones are released, the dominant hormones of this period are the glucocorticoids. As discussed earlier, glucocorticoids are important in maintaining adequate levels of glucose in the blood, primarily for use by the nervous tissue of the brain and spinal cord. At the same time, however, these secretions have side effects that decrease the immune response and increase the chances of infection.

After extended periods of time under continual stress, organ systems begin to fail due to a lack of lipids, a lack of glucocorticoids, high blood pressure, a failure to balance electrolytes, and mounting damage to vital organs. Although a single cause, such as heart failure, may be listed as the official cause of death, the underlying problem is the inability to support the endocrine and metabolic responses to stress.

✔ What is a negative side effect of continuing glucocorticoid release during periods of stress?

DISORDERS OF THE ENDOCRINE SYSTEM

Endocrine disorders result from the overproduction or underproduction of hormones or from abnormal target cells, which may have too few or too many hormone receptors. Each hormone has specific targets that may be distributed throughout the body. As a result, the effects of an endocrine disorder can be very diverse, and it is not always easy to see the link to the endocrine system.

DIAGNOSIS OF ENDOCRINE DISORDERS

The effects of endocrine disorders are interesting because they show us the normally "silent" or "invisible" contributions made by the endocrine system to maintain normal homeostasis and good health. Figure 13-6• provides an overview of the major classes of endocrine disorders by their functional effects. Note that some of the disorders affect more than one function.

Symptoms of Endocrine Disorders

Symptoms of endocrine disorders can be caused by either an abnormal level of hormone production (*hyposecretion* or *hypersecretion*) or by abnormal sensitivity of the target tissues. Knowledge of the individual endocrine organs and their functions make it possible to predict the symptoms of specific endocrine disorders. For example, thyroid hormones affect the metabolic rate of the body, along with body heat production, perspiration, activity, and heart rate. An elevated metabolic rate, increased body temperature, weight loss, nervousness, excessive perspiration, and an increased or irregular heartbeat are common symptoms of hyperthyroidism. Conversely, a low metabolic rate, decreased body temperature, weight

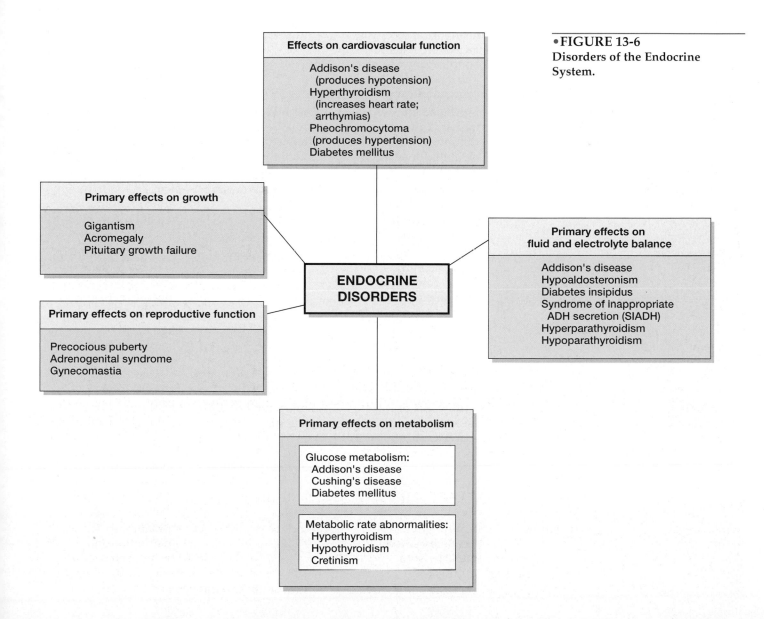

•FIGURE 13-6
Disorders of the Endocrine System.

gain, lethargy (drowsiness), dry skin, and a reduced heart rate typically accompany hypothyroidism. The symptoms associated with over- and underproduction of major hormones are summarized in Table 13-2.

Table 13-2	**Symptoms of Hormone Underproduction and Overproduction**			
Hormone	**Underproduction Syndrome**	**Principal Symptoms**	**Overproduction Syndrome**	**Principal Symptoms**
Growth hormone (GH)	Pituitary dwarfism growth failure	Retarded growth, abnormal fat distribution, low blood glucose hours after a meal	Gigantism, acromegaly	Excessive growth
Antidiuretic hormone (ADH)	Diabetes insipidus	Polyuria, dehydration, thirst	SIADH (syndrome of inappropriate ADH secretion)	Increased body weight and water content
Thyroid hormones	Hypothyroidism, myxedema, cretinism	Low metabolic rate, body temperature; impaired physical and mental development	Hyperthyroidism, Graves' disease	High metabolic rate and body temperature
Parathyroid hormone (PTH)	Hypopara-thyroidism	Muscular weakness, neurological problems, formation of dense bones, tetany due to low blood Ca^{2+} concentrations	Hyperparathyroidism	Neurological, mental, muscular problems due to high blood Ca^{2+} concentrations; weak and brittle bones
Insulin	Diabetes mellitus (Type 1)	High blood glucose, impaired glucose utilization, dependence on lipids for energy; glycosuria	Excess insulin production or administration	Low blood glucose levels, possibly causing coma
Mineralocorticoids (MCs)	Hypoaldosteronism	Polyuria, low blood volume, high blood K^+, low blood Na^+ concentrations	Hyperaldosteronism	Increased body weight due to Na^+ and water retention; low blood K^+ concentration
Glucocorticoids (GCs)	Addison's disease	Inability to tolerate stress, mobilize energy reserves, or maintain normal blood glucose concentrations	Cushing's disease	Excessive breakdown of tissue proteins and lipid reserves; impaired glucose metabolism
Epinephrine (E), norepinephrine (NE)	None identified		Pheochromocytoma	High metabolic rate, body temperature, and heart rate; elevated blood glucose levels
Estrogens (females)	Hypogonadism	Sterility; lack of secondary sexual characteristics	Adrenogenital syndrome	Overproduction of androgens by the adrenal cortex leads to masculinization
			Precocious puberty	Premature sexual maturation and related behavioral changes
Androgens (males)	Hypogonadism	Sterility; lack of secondary sexual characteristics	Precocious puberty	Premature sexual maturation and related behavioral changes

The Physical Examination

The first step in the diagnosis of an endocrine disorder is the physical examination. Several disorders produce characteristic physical signs that reflect abnormal hormone activities. For example, in *Cushing's disease* fat accumulates in the lower posterior cervical area (causing a "buffalo hump"), and in the face (a "moonface"), but the limbs become relatively thin. In *acromegaly*, the facial features become distorted due to excessive growth at the mandible and brow ridges, and the lower jaw protrudes, a sign known as *prognathism* (PROG-nah-thizm). The hands and feet also become enlarged.

These signs are very useful, but many other signs and symptoms related to endocrine disorders are less obvious. For example, the condition of **polyuria**, or increased urine production, may be the result of hyposecretion of ADH (a condition called *diabetes insipidus*), or from diabetes mellitus (p. 300); a symptom such as **hypertension** (high blood pressure) can be caused by a variety of cardiovascular or endocrine problems. In these instances, many diagnostic decisions are based on blood tests, which can confirm the presence of an endocrine disorder by detecting abnormal levels of circulating hormones, followed by tests that determine whether the primary cause of the problem lies within the endocrine gland or the target tissues. Table 13-3 provides some examples of important blood tests and other tests used in the diagnosis of endocrine disorders.

The discussion that follows will cover representative disorders, described in Table 13-2, of the pituitary gland, the thyroid gland, the parathyroid glands, the adrenal glands. Before we begin you may want to refer to Table 13-1 for a review of the normal effects of the hormones released by these glands.

THE PITUITARY GLAND

The pituitary gland secretes a variety of hormones that affect the body in one of two ways. Some hormones, such as ACTH, TSH, FSH, and LH, stimulate other endocrine glands. Disorders associated with this group will be discussed later in this chapter, with the specific gland involved. The other group of hormones, which includes growth hormone, antidiuretic hormone, and prolactin, exert their effects directly on tissues throughout the body.

Growth Hormone Disorders

Growth hormone stimulates protein synthesis and cell division, which, in children, lead to normal muscular and skeletal growth and development. If there is an overproduction, or hypersecretion, of GH before puberty when the epiphyseal plates within the bones are not closed, then there will be an increase in height, weight, and muscle mass in the individual. In extreme cases this is called **gigantism**. In **acromegaly**, an excessive amount of GH is released after puberty when most of the epiphyseal plates are already closed. Cartilages and small bones respond to the hormone, however, resulting in abnormal growth at the hands, feet, lower jaw, skull, and clavicle (Figure 13-7●). Pituitary surgery to reduce the number of GH-producing cells stops progress of the disease.

An undersecretion (hyposecretion) of GH in children leads to a condition called **pituitary growth failure**. These individuals have a short stature and larger-than-normal fat reserves. Normal growth patterns can be restored by the administration of GH. In the past, GH was extracted and purified from corpses at considerable expense. It is now obtained in large quantities from bacteria using genetic engineering techniques.

●**Figure 13-7**
Acromegaly.
Acromegaly results from the hypersecretion of growth hormone in adults.

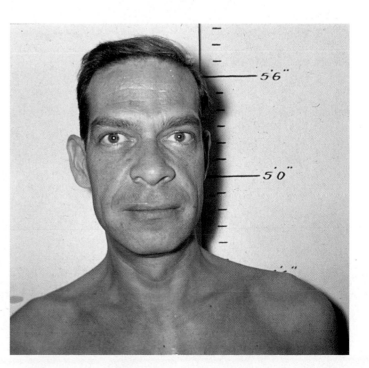

Table 13-3	Representative Diagnostic Procedures and Laboratory Tests for Disorders of the Endocrine System

Diagnostic Procedure	Method and Result	Representative Uses
Thyroid		
Thyroid scanning	Radionuclide given by mouth accumulates in the thyroid, giving off radiation captured to create an image of the thyroid	To determine size, shape, and abnormalities of the thyroid gland
Radioactive iodine uptake test (RAIU)	Radioactive iodine is ingested and trapped by the thyroid; detector determines the amount of radioiodine taken up over a period of time	To determine hyperactivity or hypoactivity of the thyroid gland
Parathyroid		
Ultrasound examination of parathyroid glands	Standard ultrasound	To determine structural abnormalities of the parathyroid gland, such as enlargement
Adrenal		
CT scan of adrenal gland	Standard cross-sectional CT	To determine abnormalities in adrenal size or shape
Adrenal angiography	Injection of radiopaque dye for examination of the vascular supply to the adrenal gland	To detect tumors and adrenal enlargement

Laboratory Test	Normal Values in Blood Plasma or Serum	Significance of Abnormal Values
Pituitary		
Growth hormone	>10 ng/ml	<10 ng/ml of GH suggests deficiency (common in hypopituitarism and pituitary dwarfism)
Plasma ACTH	Morning: 20–80 pg/ml Late afternoon: 10–40 pg/ml	Increased in stress, hypofunction of adrenal and pituitary hyperactivity. Decreased in Cushing's syndrome and carcinoma of adrenal gland
Serum TSH	Adults: <3 ng/ml	Elevated in hyperpituitarism or hypothyroidism. Decreased in hypopituitarism or hyperthyroidism
Serum LH	Premenopausal females: 3–30 mIU/ml Females, midcycle: 30–100 mIU/ml	Elevated by pituitary tumors and hyperpituitarism. Decreased levels occur in hypopituitarism and adrenal tumors
Thyroid		
Free serum thyroxine	Adults: 1.0–2.3 ng/dl	Elevated in hyperthyroidism. Decreased with hypothyroidism
Calcitonin (plasma)	Adult males: <40 pg/ml Adult females: <20 pg/ml	Elevated in carcinoma of thyroid
Parathyroid		
Serum parathyroid hormone	Adults: PTH-N 400–900 pg/ml PTH-C 200–600 pg/ml	Increased in hyperparathyroidism and hypercalcemia. Decreased in hypoparathyroidism (PTH values are used to determine chronic nature)
Plasma cortisol	Adults, morning: 5–23 μg/dl Adults, afternoon: 3–13 μg/dl	Increased in adrenal hyperactivity, Cushing's syndrome, stress, and steroid use. Decreased in Addison's disease and pituitary hypofunction
Serum aldosterone	Adults: 4–30 ng/dl supine: decreased <1 ng/dl elevated >9 ng/dl	Increased in dehydration, hyperactivity of adrenal, and hyponatremia (low blood sodium). Decreased in hypernatremia (high blood sodium) and adrenal hypoactivity

Table 13-3	Representative Diagnostic Procedures and Laboratory Tests for Disorders of the Endocrine System *(continued)*	
Laboratory Test	**Normal Values in Blood Plasma or Serum**	**Significance of Abnormal Values**
Adrenal		
Serum Na$^+$	Adults: 135–145 mEq/l	Increased in dehydration and hypoaldosteronism; decreased in adrenocortical insufficiency and hypoaldosteronism
Serum K$^+$	Adults: 3.5–5.0 mEq/l	Increased in hypoactivity of adrenal glands and hypoaldosteronism. Decreased in hyperactivity of adrenal glands and aldosteronism
Urine hydroxycorticosteroid	Adults: 2–12 mg/24 hr	This breakdown product of cortisol is increased with Cushing's syndrome. Decreased in Addison's disease
Pancreas		
Glucose tolerance test (monitors serum glucose following ingestion by a patient who has fasted for 12 hours)	Adults: Results after ingestion of 75–100 g of glucose: 0 hr 75–115 mg/dl 2 hr <140 mg/dl	Increased in diabetes mellitus, Cushing's syndrome, alcoholism, and infections. Decreased in hyperinsulinism and hypoactivity of adrenal glands
Serum insulin	Adults: 5–25 μU/ml	Increased in early Type 2 diabetes mellitus and obesity. Decreased in Type 1 diabetes mellitus

ADH Disorders

ADH acts on the kidneys to reduce the volume of water in urine. The underproduction of ADH, then, leads to the loss of excessive amounts of water in urine, or **polyuria**. Individuals with this condition, known as **diabetes insipidus**, are constantly thirsty, but the fluids they drink are not retained by the body. Treatment is available as a synthetic form of ADH that is administered in a nasal spray. In some cases of diabetes insipidus, the kidneys do not respond to the ADH even though it is at normal levels in the blood. An overproduction of ADH, called *syndrome of inappropriate ADH secretion (SIADH)*, results in increased water retention, increased body weight, and low sodium ion concentrations in body fluids.

Tumors and the Pituitary Gland

Abnormal secretions of the pituitary gland may be due to different factors, such as inherited conditions, development abnormalities, and tumors. Tumors of the pituitary gland are not common, but when they form they often result in over- or undersecretion of hormones. Because there are many different types of hormone-secreting cells in the pituitary, tumors can have many affects depending on which cells make up the tumor. For example, a *prolactinoma* is a benign tumor of the pituitary gland that overproduces prolactin. In females, it causes *galactorrhea*, milk production at times other than just before childbirth and during breastfeeding. In males, this disorder may cause *impotence* (failure to achieve or maintain an erection) and breast enlargement.

Tumors may also affect systems other than the endocrine system. Because of its position at the base of the brain, the swelling of pituitary tumors can press against the optic nerves and affect vision.

THE THYROID GLAND

Disorders of the thyroid gland include *goiters*, and the underproduction or overproduction of thyroid hormones. A **goiter** (Figure 13-8•) is an enlargement of the

•**Figure 13-8**
Goiter.
A goiter is an enlarged thyroid gland that is usually caused by a deficiency of iodine.

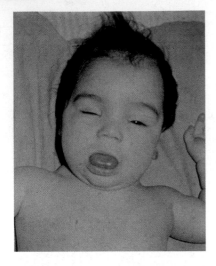

•Figure 13-9
Cretinism.
Cretinism is due to an underproduction of thyroid hormone in infancy.

thyroid gland. ∞ [p. 296] Often caused by an inadequate amount of iodine in the diet, a goiter may also form in connection with hyperthyroidism or hypothyroidism.

Hypothyroidism

Inadequate production of thyroid hormones is called **hypothyroidism**. Hypothyroidism in an infant produces a condition known as **cretinism** (KRĒ-tin-ism). This condition is characterized by inadequate skeletal and nervous system development, and a metabolic rate about 40 percent lower than normal (Figure 13-9•). Cretinism that develops later in childhood will retard growth and mental development and delay puberty.

Adults with hypothyroidism are slow-moving, and dislike cold temperatures. The symptoms of adult hypothyroidism, or **myxedema** (miks-e-DĒ-muh), include subcutaneous swelling, dry skin, hair loss, low body temperature, muscular weakness, and slowed reflexes.

Hyperthyroidism

Hyperthyroidism occurs when there is an overproduction of thyroid hormones. The metabolic rate climbs, and the skin becomes flushed and moist with perspiration. Blood pressure and heart rate increase. The effects on the CNS make the individual restless, excitable, and subject to shifts in mood and emotional states. Despite the drive for increased activity, the person has limited energy reserves and tires easily.

In **Graves' disease**, excessive thyroid activity leads to goiter and the symptoms of hyperthyroidism. Protrusion of the eyes, or **exophthalmos** (eks-ahf-THAL-mus) may also appear, for unknown reasons. Graves' disease has a genetic autoimmune basis and affects many more women than men. Treatment may involve the use of antithyroid drugs, surgical removal of portions of the glandular mass, or destruction of part of the gland by exposure to radioactive iodine.

Hyperthyroidism may also result from inflammation or, rarely, thyroid tumors. In extreme cases the individual's metabolic processes accelerate out of control. During a *thyrotoxic* crisis, or "thyroid storm," the subject experiences an extremely high fever, rapid heart rate, and the malfunctioning of a variety of physiological systems.

Thyroid gland function is controlled by the pituitary gland, so thyroid gland disorders may not only be due to disorders in the thyroid itself, but also to alterations in the pituitary or even the hypothalamus. Abnormal thyroid gland function may also result from other factors, such as inherited conditions, development abnormalities, autoimmune responses, and tumors.

THE PARATHYROID GLANDS

The thyroid gland and the parathyroid glands regulate the amount of calcium in body fluids. Calcitonin released from the thyroid causes a decrease in calcium concentrations when levels are unusually high. However, in the healthy, non-pregnant adult calcitonin plays little role in the regulation of calcium ion concentrations. Parathyroid hormone (PTH), which is released by the four parathyroid glands, increases the calcium concentration in body fluids. Changes in PTH levels cause significant variations in calcium ion levels.

Hypoparathyroidism

Hypoparathyroidism, the underproduction of PTH, results in a decrease in body fluid calcium ion concentrations. The most obvious symptoms involve neural and muscular tissues. The nervous system becomes more excitable, and the affected individual may experience *tetany*, prolonged muscle spasms affecting the limbs and face (tetany is not the same as tetanus, which is caused by a bacteria).

Treatment is difficult because at present PTH can be obtained only by extraction from the blood of individuals with normal PTH levels. Thus PTH is extremely costly, and because supplies are very limited, PTH administration is not used to treat this condition, despite its probable effectiveness. As an alternative, a dietary combination of vitamin D and calcium can be used to elevate body fluid calcium concentrations.

Hypoparathyroidism often occurs after neck surgery, especially the removal of the thyroid gland (*thyroidectomy*). Fortunately there are four parathyroids, and the secretion of even a portion of one gland can maintain normal calcium concentrations.

Hyperparathyroidism

Hyperparathyroidism, the overproduction of PTH, results in an abnormally high concentration of calcium ions in body fluids. Calcium is released from the skeleton, weakening the bones. CNS function is depressed, thinking slows, memory is impaired, and the individual often experiences emotional swings and depression. Nausea and vomiting occur, and in severe cases the patient may go into a coma. Muscle function deteriorates, and skeletal muscles become weak. Other tissues are also affected as calcium begins to crystallize in joints, tendons, and the dermis. In the kidney, calcium deposits may produce masses called *kidney stones*, which block filtration and conduction passages involved in urine formation.

Hyperparathyroidism most commonly results from a tumor of the parathyroid gland. Treatment of hyperparathyroidism involves the surgical removal of the overactive tissue.

THE ADRENAL GLANDS

The cortex of each adrenal gland secretes a group of hormones different from that of the medulla (p. 298). We will first examine the endocrine disorders of the adrenal cortex, those associated with the production of glucocorticoids, mineralocorticoids, and androgens, and then disorders of the adrenal medulla.

Disorders of the Adrenal Cortex

Glucocorticoids. **Addison's disease** is characterized by an inadequate production of glucocorticoid hormones. It may result from inadequate levels of ACTH from the anterior pituitary or from the inability of the adrenal cells to synthesize the necessary hormones. Affected individuals become weak and lose weight due to a combination of appetite loss, low blood pressure (*hypotension*), and low blood volume (*hypovolemia*). They cannot adequately mobilize energy reserves, and their blood glucose concentrations fall sharply within hours after a meal. Stresses cannot be tolerated, and a minor infection or injury may lead to a sharp and fatal decline in blood pressure. If the disorder is caused by defective adrenal cells, then ACTH levels in the blood remain high. The ACTH stimulates melanocytes and increased melanin pigmentation darkens the skin (Figure 13-10•).

Cushing's disease results from the overproduction of glucocorticoids; most cases reflect oversecretion of ACTH by the pituitary gland. The symptoms resemble those of a protracted and exaggerated response to stress. (Hormones and the stress response were discussed on p. 301.) Glucose metabolism is suppressed, and lipids and amino acids are mobilized in excess of the existing demand. The energy reserves are shuffled around, and the distribution of body fat changes. Adipose tissues in the cheeks and around the base of the neck become enlarged, producing a "moon-faced" appearance (Figure 13-11•). The demand for amino acids falls most heavily on the skeletal muscles, which respond by breaking down their contractile proteins. This response reduces muscular power and endurance. The skin becomes thin, and may develop stretch marks or changes in pigmentation. The chronic administration of large doses of steroids can produce symptoms similar to those of Cushing's disease, and physicians avoid such treatment when possible. However, athletes who use steroids to "bulk up" may develop symptoms of this disorder (p. 302).

Mineralocorticoids (Aldosterone). **Hypoaldosteronism** (hī-pō-al-do-STER-ōn-izm), occurs when the adrenal cortex fails to produce enough aldosterone. Low

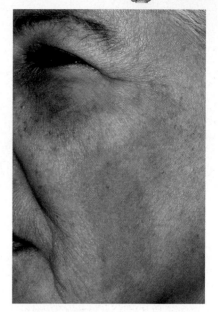

•**Figure 13-10**
Addison's Disease.
Addison's disease is caused by the underproduction of glucocorticoids. The increase in pigment on the cheek results from the stimulation of melanocytes by ACTH.

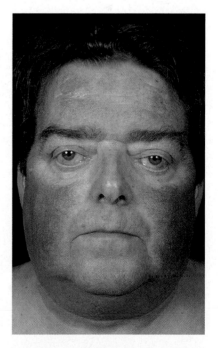

•**Figure 13-11**
Cushing's Disease.
Cushing's disease is caused by the overproduction of glucocorticoids. One sign is the accumulation of adipose tissue in the cheeks and at the base of the neck.

aldosterone levels lead to excessive losses of water and sodium ions at the kidneys (polyuria), and the water loss in turn leads to low blood volume and a fall in blood pressure. Aldosterone also affects the concentration of potassium ions, yet in a way opposite to sodium ions. So, as sodium ions are lost because of low levels of aldosterone, potassium ions are gained. The resulting changes in electrolyte concentrations affect membrane potentials, eventually causing dysfunctions in neural and muscular tissues.

Hypersecretion of aldosterone results in the condition of **hyperaldosteronism**, or *aldosteronism*. Under continued aldosterone stimulation, the kidneys retain water and sodium ions, and potassium ions are lost in the urine. This leads to high blood volume and high blood pressure (*hypertension*) and low levels of potassium ions in body fluids. Again, changes in electrolyte concentrations interfere with membrane potentials, especially cardiac muscle cells and neurons.

Androgens. The development of a tumor in the area of the adrenal cortex that produces androgens (male hormones) can increase their secretion so much that typical male sexual secondary characteristics develop in females. The symptoms of this condition, called **adrenogenital syndrome**, include the development of male body and facial hair patterns, fatty tissue deposits, and muscular development.

Disorders of the Adrenal Medullae

A **pheochromocytoma** (fē-ō-krō-mō-sī-TŌ-mah) is a tumor that produces epinephrine in massive quantities. The tumor generally develops within an adrenal medulla but may also involve other sympathetic ganglia. The most dangerous symptoms are rapid and irregular heartbeat and high blood pressure; other symptoms include uneasiness, sweating, blurred vision, and headaches. This condition is rare, and surgical removal of the tumor is the most effective treatment.

THE PANCREAS

The endocrine cells of the pancreas regulate the levels of glucose in the blood through secretions of insulin and glucagon. The most common endocrine disorder of the pancreas (and the endocrine system) is diabetes mellitus, a condition characterized by **hyperglycemia** (high blood sugar levels) and **glycosuria** (glī-kō-SOO-rē-a), or high levels of glucose in the urine (see p. 300). Other disorders of the pancreas will be discussed in Chapter 18.

Diabetes Mellitus

There are two forms of diabetes mellitus, **insulin-dependent diabetes mellitus (IDDM)**, or **Type 1 diabetes**, and **non-insulin dependent diabetes mellitus (NIDDM)**, or **Type 2 diabetes**.

Insulin-Dependent (Type 1) Diabetes Mellitus. The primary cause of insulin-dependent diabetes is inadequate or absent insulin production. Type 1 diabetes most commonly appears in individuals under 30–40 years of age. Because it typically appears in childhood, it has been called *juvenile-onset diabetes*. This form of diabetes is apparently an autoimmune disorder, in which the insulin-producing cells of the pancreas are destroyed by an improper immune response by the body.

The high blood glucose concentrations characteristic of diabetes mellitus limit the ability of the kidneys to conserve water, due to the amount of glucose in the urine. As a result, the individual urinates frequently (polyuria) and may become dehydrated. The chronic dehydration leads to disturbances of neural function (blurred vision, tingling sensations, disorientation, fatigue) and muscle weakness.

Probably because of unstable glucose levels, problems with the circulatory system are also common. For example, the growth and hemorrhaging of capillaries at the retina can cause partial or complete blindness. This condition, called **diabetic retinopathy**, was discussed in Chapter 12. ∞ *[p. 284]* (A related problem is a reduction in the clarity of the lens, producing *cataracts*.) Circulation to the limbs may also be disrupted. A reduction in blood flow to the feet can lead to a loss of sensation, tissue death, ulcer formation, infection, and the loss of toes.

Despite high blood concentrations, glucose cannot enter endocrine or most other tissues without insulin, and these tissues react as if glucose were in short supply. Peripheral tissues break down lipids and proteins to obtain the energy needed to continue functioning. The breakdown of large numbers of fatty acids promotes the generation of smaller metabolic acids whose accumulation can cause a dangerous reduction in blood pH. This condition, called *acidosis*, commonly triggers vomiting. In severe cases it can progress to a fatal coma.

Long-term treatment involves a combination of dietary and exercise control and the administration of insulin, either by injection, using a small hypodermic needle, or by infusion, using a small tube that is permanently inserted into a blood vessel. The most practical method of infusion is an *insulin pump*, a battery operated portable device roughly the size of a cigarette pack. Insulin therapy is complicated by the fact that tissue glucose demands vary with meals, physical activity, emotional state, stress, and other factors that are hard to assess or predict. Dietary control, including regulation of the type of food, time of meals, and amount consumed, can help reduce oscillations in blood glucose levels. Modern insulin pumps can be programmed by the user to compensate for changes in activity and eating patterns. However, it remains difficult to maintain stable and normal blood glucose levels over long periods of time, even with an insulin pump.

Since 1990, pancreas transplants have been used to treat diabetes in the U.S. The procedure is generally limited to gravely ill patients already undergoing kidney transplantation. The graft success rate over 5 years is roughly 50%.

Non-Insulin Dependent (Type 2) Diabetes Mellitus. Type 2 diabetes typically affects genetically susceptible obese individuals over 40 years of age. Because of the age factor, this condition is also called *maturity-onset diabetes*. Type 2 diabetes is far more common than Type 1 diabetes, occurring in about 7 percent of the U.S. population. There are 500,000 new cases diagnosed each year in the United States alone, and 90 percent of these cases involve obese individuals. Unexpectedly, a significant number of these new cases are being diagnosed in obese, school-age children.

In Type 2 diabetes, insulin levels are normal or elevated, but peripheral tissues do not respond normally to insulin and glucose levels. Treatment consists of weight loss, exercise, and dietary restrictions that may improve insulin production and tissue response. The drug *metformin* (*Glucophage*®), which received FDA approval in 1995, lowers plasma glucose concentrations, primarily by reducing glucose synthesis and release by the liver. The use of metformin in combination with other drugs affecting glucose metabolism and insulin response reduce the complications and early death associated with Type 2 diabetes.

THE REPRODUCTIVE ORGANS

The primary regulators of the testes and ovaries are FSH and LH from the anterior pituitary gland. Abnormally low production of either of these hormones will produce **hypogonadism** (hī-po-GŌ-nad-izm). Children with this condition will not undergo sexual maturation, and adults with hypogonadism will not produce functional sperm or egg cells. In contrast, the overproduction of these hormones can lead to **precocious puberty**, the early onset of sexual maturation. The signs vary depending on whether the child has testes or ovaries.

The Testes

Overproduction of androgens can cause precocious (premature) puberty as early as 5 or 6 years of age. Not only do secondary sex characteristics such as facial hair, increased muscle mass and body size begin to appear, but there may be behavioral changes such as an increase in aggressiveness and assertiveness.

The Ovaries

Similarly, in females, an overproduction of estrogens can cause precocious puberty. Female secondary sex characteristics such as the development of body hair, new regions of fat deposits, development of breasts, and related behavioral changes also occur.

✔ A disorder in which gland produces a low metabolic rate and inability to tolerate cold temperatures?

✔ Why are weak bones characteristic of hyperparathyroidism?

CHAPTER REVIEW

Key Words

adrenal cortex: Outer portion of adrenal gland that produces steroid hormones.

adrenal medulla: Core of the adrenal gland; secretes epinephrine and norepinephrine into the blood following sympathetic activation.

endocrine gland: A gland that secretes hormones into the blood.

glucagon (GLOO-ka-gon): Hormone secreted by cells of the pancreatic islets; increases blood glucose concentrations.

hormone: A compound secreted by one cell that travels through the circulatory system to affect the activities of specific cells in another portion of the body.

hypophyseal (hī-po-FI-sē-al) **portal system:** Network of vessels that carry blood from capillaries in the hypothalamus to capillaries in the anterior pituitary gland.

hypothalamus: The region of the brain involved with the unconscious regulation of organ functions, emotions, drives, and the coordination of nervous and endocrine functions.

insulin: Hormone secreted by the cells of the pancreatic islets; causes a reduction in blood glucose concentrations.

negative feedback: Corrective mechanism that opposes or negates a variation from normal limits.

pancreas: Digestive organ containing exocrine and endocrine tissues; exocrine portion secretes pancreatic juice, endocrine portion secretes the hormones insulin and glucagon.

pituitary gland: The hypophysis, or "master gland," situated in the seat of the sphenoid bone and connected to the hypothalamus; secretes eight different hormones.

steroid: A ring-shaped lipid structurally related to cholesterol.

Selected Clinical Terms

Addison's disease: A condition caused by the hyposecretion of glucocorticoids in the adrenal gland; characterized by inability to mobilize energy reserves and maintain normal blood glucose levels.

cretinism: A condition caused by hypothyroidism in infancy; marked by inadequate skeletal and nervous system development, and a low metabolic rate.

Cushing's disease: A condition caused by hypersecretion of glucocorticoids by the adrenal gland; characterized by excessive breakdown and relocation of fat (lipid) reserves and proteins.

diabetes insipidus: A disorder that develops when the pituitary gland no longer releases adequate amounts of ADH or when the kidneys cannot respond to ADH.

diabetes mellitus: A disorder characterized by chronically elevated blood glucose concentrations.

exophthalmos (eks-ahf-THAL-mus): Protruding of the eyes from their sockets; a common symptom of hypersecretion of thyroid hormones in Graves' disease.

glycosuria (glī-kō-SOO-rē-a): The presence of glucose in the urine, usually as the result of abnormally high blood glucose concentrations.

goiter: An enlargement of the thyroid gland.

myxedema: A condition caused by the hyposecretion of thyroid hormones in an adult; symptoms include subcutaneous swelling, hair loss, dry skin, low body temperature, muscle weakness, and slowed reflexes.

polyuria: The production of excessive amounts of urine; a symptom of diabetes insipidus and both forms of diabetes mellitus.

Study Outline

INTRODUCTION *(p. 291)*

1. In general, the nervous system performs short-term "crisis management," whereas the **endocrine system** regulates longer-term, ongoing body processes.

SYSTEM BRIEF *(p. 291)*

1. *Endocrine cells* release chemicals called **hormones** that alter the metabolic activities of many different tissues and organs simultaneously. *(Figure 13-1; Table 13-1)*

HORMONES *(pp. 291–294)*
CHEMICAL MAKEUP *(p. 291)*

1. Hormones can be divided into two groups based on chemical structure: **amino acid–based** and **lipid-based**.

2. Amino acid–based hormones range in size from single amino acid molecules to large proteins.

3. Lipid-based hormones include **steroid hormones** (built from cholesterol molecules) and **prostaglandins** (derived from fatty acids.)

ACTIONS ON CELLS *(p. 293)*

4. Hormones exert their effects by modifying the activities of **target cells** (cells that are sensitive to that particular hormone). *(Figure 13-2)*

5. Hormones alter a target cell by changing the *types*, *activities*, or *quantities* of its important enzymes and structural proteins.

CONTROL OF HORMONE SECRETION *(p. 293)*

6. The simplest control of endocrine secretions involves the direct negative feedback of changes in the extracellular fluid on the endocrine cells.

7. The most complex control of secretion involves the hypothalamus. The hypothalamus regulates the activities of the nervous and endocrine systems by three mechanisms: (1) Its autonomic centers exert direct neural control over the endocrine cells of the adrenal medullae; (2) it acts as an endocrine organ itself by releasing hormones into the circulation; (3) it secretes **regulatory hormones** that control the activities of endocrine cells in the pituitary gland.

MAJOR ENDOCRINE GLANDS *(pp. 294–301)*
THE PITUITARY GLAND *(p. 294)*

1. The pituitary gland, or **hypophysis**, releases eight important protein hormones. *(Figure 13-3)*

2. The hypothalamus releases regulatory factors into the *hypophyseal portal system*, which carries them to target cells in the anterior pituitary.

3. The rate of regulatory hormone secretion by the hypothalamus is regulated through negative feedback mechanisms. *(Figure 13-3)*

4. The six hormones of the **anterior pituitary** include (1) **thyroid-stimulating hormone (TSH)**, which triggers the release of thyroid hormones; (2) **adrenocorticotropic hormone**

(ACTH), which stimulates the release of **glucocorticoids** by the adrenal gland; (3) **follicle-stimulating hormone (FSH),** which stimulates **estrogen** secretion and egg development in women and sperm production in men; (4) **luteinizing hormone (LH),** which causes ovulation and **progestin** production in women and **androgen** production in men; (5) **prolactin (PRL),** which stimulates the development of the mammary glands and the production of milk; and (6) **growth hormone (GH),** which stimulates cell growth and replication. *(Figure 13-3; Table 13-1)*

5. The **posterior pituitary** secretes **antidiuretic hormone (ADH)** and **oxytocin.** ADH decreases the amount of water lost at the kidneys. In women, oxytocin stimulates smooth muscle cells in the uterus and contractile cells in the mammary glands. In men, it stimulates prostatic smooth muscle contractions. *(Figure 13-3; Table 13-1)*

THE THYROID GLAND *(p. 296)*

6. The thyroid gland lies near the **thyroid cartilage** of the larynx and consists of two lobes. *(Figure 13-4)*

7. The thyroid gland contains numerous **thyroid follicles.** Thyroid follicles primarily release iodine-containing **thyroxine** hormone.

8. Thyroxine increases the rate of cellular metabolism.

9. The thyroid follicles also produce **calcitonin (CT),** which helps regulate calcium ion concentrations in body fluids.

THE PARATHYROID GLANDS *(p. 297)*

10. Four parathyroid glands are embedded in the posterior surface of the thyroid gland. They produce **parathyroid hormone (PTH)** in response to lower than normal concentrations of calcium ions. Together with the thyroid gland, the parathyroid glands maintain calcium ion levels within relatively narrow limits. *(Figure 13-4)*

THE THYMUS *(p. 297)*

11. The thymus produces several hormones called **thymosins,** which play a role in developing and maintaining normal immune defenses.

THE ADRENAL GLANDS *(p. 298)*

12. A single **adrenal gland** lies above each kidney. The adrenal gland is made up of an outer **adrenal cortex** layer and an inner **medulla.**

13. The adrenal cortex manufactures steroid hormones called *corticosteroids.* The cortex produces (1) **glucocorticoids (GCs),** the hormones that affect glucose metabolism; (2) **mineralocorticoids (MCs),** principally **aldosterone,** which regulates sodium ion, potassium ion, and water losses at the kidneys, sweat glands, digestive tract, and salivary glands; and (3) *androgens* of uncertain significance.

14. The adrenal medulla produces **epinephrine** and **norepinephrine.**

THE PANCREAS *(p. 299)*

15. The **pancreas** contains both exocrine and endocrine cells. The exocrine pancreas secretes an enzyme-rich fluid that travels to the digestive tract. Cells of the **endocrine pancreas** group in clusters called **pancreatic islets** *(islets of Langerhans)* and produce the hormones **glucagon** and **insulin.** *(Figure 13-5)*

16. Insulin lowers blood glucose by increasing the rate of glucose uptake and utilization; glucagon raises blood glucose by increasing the rates of glycogen breakdown and glucose manufacture in the liver.

THE REPRODUCTIVE ORGANS *(p. 300)*

17. The *interstitial cells* of the male testis produce **androgens** (steroid hormones). The androgen **testosterone** is the most important sex hormone in the male.

18. In women, ova develop in *follicles;* follicle cells surrounding the eggs produce **estrogens.** After ovulation, the follicle cells reorganize and release a mixture of estrogens and **progesterone.** If pregnancy occurs, the *placenta* functions as an endocrine organ.

THE PINEAL GLAND *(p. 300)*

19. The pineal gland synthesize **melatonin.** Melatonin appears to (1) slow the maturation of sperm, eggs, and reproductive organs and (2) establish daily 24-hour rhythms.

OTHER HORMONE SOURCES *(p. 301)*

THE KIDNEYS *(p. 301)*

1. Endocrine cells in the kidneys produce three hormones that regulate calcium metabolism, blood volume, and blood pressure.

2. Calcitriol stimulates calcium and phosphate ion absorption along the digestive tract.

3. Erythropoietin (EPO) stimulates red blood cell production by the bone marrow.

4. *Renin* release leads to the formation of **angiotensin II,** the hormone that stimulates the adrenal production of aldosterone.

THE HEART *(p. 301)*

5. Specialized muscle cells in the heart produce **atrial natriuretic peptide (ANP),** which lowers blood pressure and/or blood volume.

THE DIGESTIVE SYSTEM *(p. 301)*

6. In addition to the pancreas, the lining of the digestive tract produces endocrine secretions that are essential to the normal breakdown and absorption of food.

THE ADIPOSE TISSUE *(p. 301)*

7. Adipose (fatty) tissue secretes *leptin,* a weight-control hormone.

HORMONES AND STRESS *(pp. 301–302)*

1. Stress is any condition within the body that threatens homeostasis. The body's general response to stress of different causes is similar. Stress causes the sympathetic division of the ANS to be activated, and then, if necessary, glucocorticoids to be released. If conditions of stress continue for long periods and corrective actions to restore homeostasis are not taken, organ systems fail and death results.

DISORDERS OF THE ENDOCRINE SYSTEM *(pp. 302–311)*

1. Endocrine disorders generally display symptoms of underproduction (inadequate hormonal effects) or overproduction (excessive hormonal effects) of hormones.

DIAGNOSIS OF ENDOCRINE DISORDERS *(p. 303)*

2. The study of endocrine disorders promotes an understanding of the normal endocrine system and homeostasis *(Figure 13-6).*

3. Many of the symptoms of specific endocrine disorders are predictable. *(Table 13-2)*

4. The correct diagnosis of endocrine disorders requires a physical examination and various tests. *(Table 13-3)*

THE PITUITARY GLAND *(p. 305)*

5. The overproduction of growth hormone causes **gigantism** if it occurs early in life, or **acromegaly** if it occurs later when the epiphyseal plates are closed and bone growth is over. An underproduction of GH in children causes **pituitary growth failure.** *(Figure 13-7)*

6. The underproduction of ADH causes **diabetes insipidus**, a condition characterized by **polyuria**, the loss of excessive amounts of water in urine. An overproduction of ADH results in increased water retention and edema.

7. *Prolactinoma* is a pituitary tumor that causes abnormal milk production in females and impotence in males.

THE THYROID GLAND *(p. 307)*

8. A **goiter** is an enlargement of the thyroid gland. **Hypothyroidism** in an infant causes **cretinism**, a condition with a very low metabolic rate. In an adult, hypothyroidism is called **myxedema**, with symptoms such as subcutaneous swelling, dry skin, hair loss, low body temperature, muscular weakness, and slowed reflexes. *(Figures 13-8, 13-9)*

9. Hyperthyroidism results from the overproduction of thyroid hormones. Symptoms include a high metabolic rate, blood pressure, and heart rate. In **Graves' disease**, excessive thyroid activity leads to goiter, the symptoms of hyperthyroidism, and **exophthalmos**, or protrusion of the eyes.

THE PARATHYROID GLANDS *(p. 308)*

10. Hypoparathyroidism, the underproduction of PTH, causes a decrease in body fluid calcium.

11. Hyperparathyroidism, the overproduction of PTH, causes an increase in body fluid calcium.

THE ADRENAL GLANDS *(p. 309)*

12. Addison's disease is caused by an inadequate production of glucocorticoids. *(Figure 13-10)*

13. Cushing's disease is caused by the overproduction of glucocorticoids, hormones involved in the response to stress. *(Figure 13-11)*

14. Hypoaldosteronism is caused by underproduction of aldosterone, which leads to the loss of water and sodium ions at the kidneys, and the gain of potassium ions.

15. Hyperaldosteronism, or **aldosteronism**, is caused by the overproduction of aldosterone.

16. Adrenogenital syndrome is a due to the overproduction of androgens (male hormones) in the adrenal cortex.

17. The growth of a **pheochromocytoma** tumor in the adrenal medullae, or in sympathetic ganglia, produces an excess of epinephrine.

THE PANCREAS *(p. 310)*

18. The most common endocrine disorder is diabetes mellitus. There are two forms; **IDDM (Type 1)** and **NIDDM (Type 2)**. It is characterized by **hyperglycemia** and **glycosuria**.

19. Insulin-Dependent Diabetes Mellitus (Type 1) is caused by inadequate levels of insulin. Also known as *juvenile diabetes*, it appears early in life.

20. Non-Insulin Dependent Diabetes Mellitus (Type 2), or *maturity-onset diabetes*, affects obese individuals over 40 years of age.

THE REPRODUCTIVE ORGANS *(p. 311)*

21. Hypogonadism is caused by low levels of FSH and LH from the anterior pituitary gland. Children with this condition will not develop sexually, and adults will not produce functional sperm or egg cells.

22. In males, the overproduction of androgens by the testes can cause **precocious** (premature) **puberty** as early as 5 or 6 years of age. In females, an overproduction of estrogens can cause precocious puberty at comparable ages.

Review Questions

MATCHING

Match each item in Column A with the most closely related item in Column B. Use letters for answers in the spaces provided.

	Column A	Column B
___	1. thyroid gland	a. islets of Langerhans
___	2. pineal gland	b. excessive urine production
___	3. adrenal medullae	c. atrial natriuretic peptide
___	4. parathyroid gland	d. increased cell growth and division
___	5. thymus gland	e. melatonin
___	6. adrenal cortex	f. hypophysis
___	7. heart	g. secretes epinephrine and norepinephrine
___	8. endocrine pancreas	h. thyroxine
___	9. gonadotropins	i. secretes regulatory hormones
___	10. hypothalamus	j. glucose in the urine
___	11. pituitary gland	k. secretes androgens, mineralocorticoids, and glucocorticoids
___	12. growth hormone	l. stimulated by low calcium levels
___	13. glycosuria	m. FSH and LH
___	14. polyuria	n. size reduced after puberty

MULTIPLE CHOICE

15. Cells of different tissues and organs respond to the same hormone if they have the same kind of _____ molecules.
(a) receptor   (b) target
(c) DNA   (d) protein

16. _____ lowers the level of glucose in the blood by aiding cells to take glucose in.
(a) calcitonin   (b) glucagon
(c) insulin   (d) growth hormone

17. Which of the following hormones causes ovulation in women? _____
(a) progesterone (b) estrogen
(c) FSH   (d) LH

18. Steroid hormones are produced from _____ .
(a) amino acids   (b) cholesterol molecules
(c) fatty acids   (d) proteins

19. The hormone that reduces the loss of sodium and water by the kidneys is _____ .
(a) parathyroid hormone (PTH)   (b) thyroxine
(c) aldosterone   (d) oxytocin

20. Milk production by the mammary glands is stimulated by _____ .
(a) prolactin   (b) estrogen
(c) progesterone (d) FSH

21. The hormone that increases the rate of cell growth and cell division is _____ .
(a) insulin   (b) testosterone
(c) glucagon   (d) growth hormone

22. One of a group of hormones that has an anti-inflammatory effect is _____ .
(a) aldosterone (b) cortisol
(c) epinephrine (d) thyroxine

23. Acromegaly is caused by _____ levels of growth hormone _____ in life.
(a) high; early (b) low; early
(c) high; late (d) low; late

24. _____ is caused by hyperthyroidism.
(a) Addison's disease (b) Graves' disease
(c) Cushing's disease (d) IDDM

TRUE/FALSE

____ 25. Atrial natriuretic peptide (ANP) is a hormone released by heart cells in response to low blood pressure.

____ 26. Prostaglandins are hormones built from amino acid molecules.

____ 27. The hormone-producing cells of the pancreas make up the islets of Langerhans.

____ 28. The anterior pituitary stores hormones produced by the hypothalamus.

____ 29. Blood glucose levels are increased by the action of glucagon.

____ 30. Secondary sex characteristics are determined by testosterone in males and by estrogen in females.

____ 31. A goiter is an enlarged pituitary gland.

____ 32. The removal of the thyroid gland results in hypoparathyroidism.

SHORT ESSAY

33. What is the primary difference in the way the nervous and endocrine systems communicate with their target cells?

34. How can a hormone modify the activities of its target cells?

35. What effects do calcitonin and parathyroid hormone have on blood calcium levels?

36. Julie is pregnant and is not receiving any prenatal care. She has a poor diet consisting mostly of fast food. She drinks no milk, preferring colas instead. How will this situation affect Julie's level of parathyroid hormone?

37. Describe the differences between Addison's disease and Cushing's disease.

38. Describe two symptoms diabetes mellitus and diabetes insipidus have in common. What tests could be used to distinguish between them?

APPLICATIONS

39. Russell wants to make the basketball team at his high school. He realizes that a few more inches in height would probably increase his chances. He asks his family physician for a prescription of growth hormone and the physician says the growth hormone won't help him grow taller. Why not?

40. Amanda has just been diagnosed as having diabetes mellitus. Her physician tells her to be sure to have her eyes examined regularly, and to check her feet regularly. Explain the reasons for this advice.

✔ Answers to Concept Check Questions

(p. 294) **1.** Hormones affect the workings of a target cell by changing *the identities*, *activities*, or *quantities* of its important enzymes and structural proteins. **2.** The regulatory hormones secreted by the hypothalamus regulate the secretions of endocrine cells in the anterior pituitary gland. One class of regulatory hormones, the releasing hormones (RH), stimulates the production of pituitary hormones; the other class, the inhibiting hormones (IH), prevents the synthesis and secretion of pituitary hormones.

(p. 296) **1.** The pituitary gland releases six hormones from its anterior portion and two from the posterior portion. Many of the pituitary's hormones stimulate or "turn on" other endocrine glands. As a result, the pituitary is called the "master" endocrine gland. **2.** Dehydration would cause an increase in the concentration of ions in the blood. In response, ADH would be released from the posterior pituitary gland. ADH reduces the amount of water lost at the kidneys and constricts blood vessels to increase blood pressure.

(p. 297) **1.** The manufacture of thyroxine requires the element iodine. **2.** Calcium concentration in the blood is regulated by the thyroid and parathyroid glands. Calcitonin produced by the thyroid reduces blood calcium levels, and parathyroid hormone produced by the parathyroid glands causes calcium levels to rise in the blood.

(p. 301) **1.** One of the functions of cortisol is to decrease the cellular use of glucose while increasing glucose availability by promoting the breakdown of glycogen and the conversion of amino acids to carbohydrates. The net result is an increase in blood glucose levels. **2.** Insulin increases the conversion of glucose into glycogen within skeletal muscle and liver cells. Glucagon stimulates the conversion of glycogen into glucose in the liver. Insulin reduces blood glucose levels, and glucagon increases blood glucose levels. **3.** Gonads of both sexes secrete steroid hormones. The testes secrete androgen hormones, such as testosterone, and the ovaries secrete estrogen and progesterone. These hormones are involved in the production of gametes and establishing secondary sex characteristics. **4.** The pineal gland receives nerve impulses from the optic tracts, and its secretion of melatonin is influenced by light–dark cycles. Increased amounts of light inhibit the production of and release of melatonin from the pineal gland.

(p. 301) **1.** EPO stimulates the production of red blood cells by the bone marrow. The increase in red blood cells improves the transport of oxygen to all body tissues. **2.** Atrial natriuretic peptide (ANP) is released by special cardiac muscle cells when they are excessively stretched. ANP lowers blood volume by increasing the amount of water and sodium lost at the kidneys and reducing the sensation of thirst.

(p. 302) **1.** Glucocorticoids act by maintaining adequate levels of glucose in the blood, primarily for use by the nervous tissue of the brain and spinal cord. At the same time, however, these secretions have side effects that decrease the immune response and increase the chances of infection.

(p. 311) **1.** The thyroid gland. Thyroid hormones stimulate cellular metabolism. **2.** Hyperparathyroidism is a disorder caused by the overproduction of parathyroid hormone (PTH). PTH acts to increase calcium levels in body fluids by stimulating osteoclasts to dissolve bone, releasing the calcium stored there. Excess PTH would cause extensive losses of calcium from bone.

CHAPTER
14

The Blood

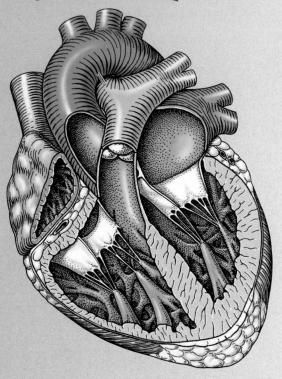

CHAPTER OUTLINE

CHAPTER OBJECTIVES

1 Describe the components of the cardiovascular system.
2 Describe the three major functions of blood.
3 Describe the important components of blood.
4 Discuss the composition and functions of plasma.
5 Discuss the characteristics and functions of red blood cells.
6 Describe the various kinds of white blood cells and their functions.
7 Describe the formation of the formed elements.
8 Describe the mechanisms that control blood loss after an injury.
9 Explain what determines blood type and why blood types are important.
10 Give an example of each of the primary blood disorders.
11 Distinguish between the different types of anemia.

We survive because of the constant exchange of chemicals between our bodies and the external environment. Eating, breathing, and waste excretion, for example, involve both absorbing and releasing chemicals. Absorbed nutrients are distributed to organs throughout the body by an internal transport network, the *cardiovascular system*, which also picks up waste products and delivers them to sites of excretion, such as the kidneys.

The cardiovascular system can be compared to the cooling system of a car. The basic components include a circulating fluid (blood versus water), a pump (the heart versus a water pump), and an assortment of conducting pipes (blood vessels versus rubber hoses). Although the cardiovascular system is far more complicated than a car's cooling system, both systems can experience sudden fluid losses, pump failures, or clogged hoses. This chapter focuses on the nature of the circulating blood. The next chapter (Chapter 15) examines the structure and function of the heart and the organization of blood vessels. Chapter 16 considers the *lymphatic system*, a body defense system closely interconnected with the cardiovascular system.

SYSTEM BRIEF

The **cardiovascular system** includes blood—a fluid connective tissue—the heart, and blood vessels. The cardiovascular system transports materials needed or excreted by the body's cells. These materials include nutrients, dissolved gases, hormones, and waste products. It also transports cells and molecules specialized for body defense.

FUNCTIONS OF BLOOD

The circulating fluid of the body is blood, a specialized connective tissue introduced in Chapter 4. ∞ *[p. 66]* Unlike the water in a car's cooling system, blood has a wider range of functions.

1. **Transportation.** Blood transports dissolved gases, nutrients, hormones, and metabolic wastes to and from the 75 trillion cells of the human body. Oxygen is carried from the lungs to the tissues, and carbon dioxide is carried from the tissues to the lungs. Nutrients absorbed in the digestive tract or released from storage in adipose tissue or the liver are distributed throughout the body. Hormones are carried from endocrine glands to their target tissues, and the wastes produced by body cells pass into the blood and are carried to the kidneys or other organs for excretion.

2. **Regulation.** Blood helps the body maintain homeostasis. It regulates the pH and electrolyte (ion) composition of interstitial fluids that bathe the cells of the body. For example, blood absorbs and neutralizes the acids generated by active tissues, such as the lactic acid produced by skeletal muscles. Blood also stabilizes body temperature by absorbing the heat generated by active skeletal muscles and redistributing it to other tissues. Finally, blood is able to restrict its own loss from the body. Blood contains enzymes and other substances that respond to breaks in blood vessel walls by initiating the process of *blood clotting*. The blood clot that develops acts as a temporary patch and prevents further blood loss.

3. **Protection.** Blood transports *white blood cells*, specialized cells that migrate into the body's tissues to fight infections or remove debris. It also delivers *antibodies*, special proteins that attack invading organisms or inactivate foreign toxins (poisons).

THE COMPOSITION OF BLOOD

Blood is usually collected from a superficial vein, such as the *median cubital vein* on the anterior surface of the elbow. If such a sample of blood is prevented from clotting and then spun in a centrifuge, it will separate into two layers. The lower layer, accounting for some 45 percent of the total volume, consists of **formed elements**, which include blood cells and cell fragments (Figure 14-1•). The upper layer, which is a yellowish liquid, makes up 55 percent of the volume of blood. It is called **plasma** (PLAZ-mah).

•**FIGURE 14-1**
The Composition of Blood.

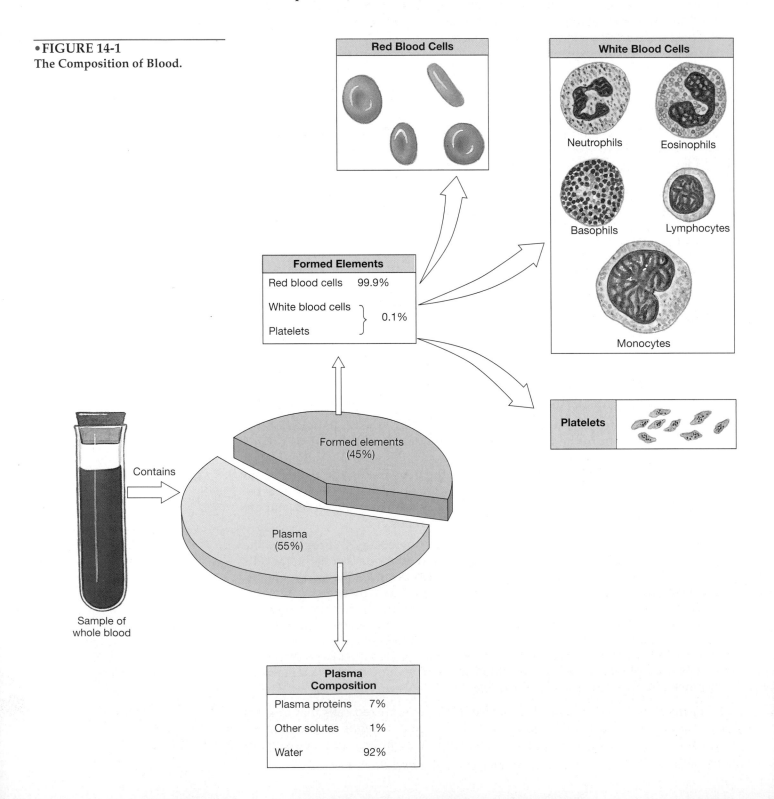

The **formed elements** are normally suspended in the plasma. **Red blood cells (RBCs)** transport oxygen and carbon dioxide. The less numerous **white blood cells (WBCs)** are components of the immune system. **Platelets** are small packets of cytoplasm that contain enzymes and other substances, called *clotting factors*, that are important to the blood clotting process.

The volume of blood in the body varies from 5 to 6 l in an adult man, and 4 to 5 l in an adult woman. The volume in a newborn is about 250 ml, or one-fourth of a liter.

Blood has certain basic physical characteristics. The temperature of the blood in the body is roughly 38°C (100.4°F), slightly higher than normal body temperature. Blood pH ranges from 7.35 to 7.45 and averages 7.4. It is therefore slightly alkaline. Blood is also sticky, cohesive, and somewhat resistant to flowing freely. Such characteristics determine the **viscosity** of a fluid. Blood is some five times "thicker," or more viscous, than water.

PLASMA

Plasma is 92 percent water. Proteins and a mixture of other solutes are dissolved in the watery plasma. There are about 7 g of plasma proteins and 1 g of other substances in each 100 ml of plasma. The large size of most blood proteins prevents them from crossing capillary walls, and they remain trapped within the cardiovascular system. This large concentration of plasma proteins marks the chief difference between plasma and interstitial fluid.

PLASMA PROTEINS

There are three primary classes of plasma proteins. **Albumins** (al-BŪ-mins), which constitute roughly 60 percent of the plasma proteins, are synthesized by the liver. Because albumins are the most abundant proteins, they are the major contributors to the high osmotic pressure of the plasma.

Globulins (GLOB-ū-lins), which make up 35 percent of the dissolved proteins, include forms with two important functions. The *immunoglobulins* (im-mū-nō-GLOB-ū-linz), or **antibodies,** attack foreign proteins and pathogens (disease-causing organisms). (The formation and function of antibodies are discussed in Chapter 16.) **Transport proteins**, which are globulins synthesized by the liver, bind small ions, hormones, or compounds that might otherwise be filtered out of the blood at the kidneys. One example is *thyroid-binding globulin*, which binds and transports the hormone thyroxine. Transport proteins are also important in carrying lipids that normally do not dissolve in water. When transport proteins bind lipids, the protein-lipid combination readily dissolves in plasma. Globulins involved in lipid transport are called *lipoproteins* (lī-pō-PRŌ-tēnz).

Fibrinogen (fī-BRIN-ō-jen), the third type of plasma protein, functions in blood clotting. It, too, is made by the liver. Under certain conditions, fibrinogen molecules interact to form large, insoluble strands of *fibrin* (FĪ-brin), the basic framework for a blood clot. If steps are not taken to prevent clotting in a blood sample, fibrinogen is converted to fibrin. The fluid left after clotting is known as **serum**. Clotting will be discussed later in this chapter.

fibra, fiber + *gennan*, to produce *fibrinogen*: a plasma protein that can interact to form insoluble strands of fibrin, the key ingredient of a blood clot

OTHER PLASMA SOLUTES

The remaining 1 percent of the plasma consists of solutes such as electrolytes (ions), organic nutrients, and organic wastes. The two major ions dissolved in plasma are sodium (Na^+) and chloride (Cl^-). Some of the other ions important for normal cellular activities include potassium (K^+), calcium (Ca^{2+}), and iodide (I^-). Organic nutrients include glucose, fatty acids, and amino acids. These molecules are used by cells for ATP energy production, growth, and maintenance. Organic wastes dissolved in the plasma are carried to sites of excretion. For example, *urea*, produced when the liver breaks down amino acids, is carried to the kidneys for excretion in urine.

FORMED ELEMENTS

The major cellular components of blood are red blood cells and white blood cells, or red and white **corpuscles** (KOR-pus-uls). In addition, blood contains *platelets*, small packets of cytoplasm that function in the clotting response (see Figure 14-1●).

The formed elements are produced through a process called **hemopoiesis** (hēm-ō-poy-Ē-sis). As we will see, all the red blood cells and platelets, and most of the white blood cells, are formed within bone marrow.

RED BLOOD CELLS

Red blood cells (RBCs) are also known as **erythrocytes** (e-RITH-rō-sītz), from *erythros* which means "red." The color of a RBC is due to *hemoglobin*, an iron-containing protein that binds and transports oxygen. The more oxygen that is bound to the hemoglobin, the brighter red it appears. Because RBCs are the most abundant blood cells (99.9 percent of all formed elements), they give whole blood its deep red color.

The number of RBCs in the blood of a normal individual staggers the imagination. For example, one cubic millimeter (mm^3) of blood contains roughly 4.8 to 5.4 million RBCs. In number, they make up roughly one third of the cells in the human body.

The **hematocrit** (hē-MAT-ō-krit) is the percentage of blood occupied by formed elements. In adult men it averages 46 (range: 40–54); in adult women, 42 (range: 37–47). Because blood contains roughly 1000 red blood cells for each white blood cell, the hematocrit closely approximates the volume of erythrocytes. The hematocrit is determined by centrifuging a blood sample until all the formed elements settle out from the plasma.

RBC Structure

RBCs have a number of specializations that increase their efficiency in transporting oxygen. One is their shape. Each cell is somewhat flattened, with a thin central region surrounded by a thick outer margin, a design that provides a relatively large surface area for the diffusion of gases. RBCs are also quite small; five or six could fit across the dot above this i. RBCs also lack nuclei and mitochondria, organelles found in most other cells. This leaves space for packing in additional molecules of hemoglobin, and it eliminates organelles that consume oxygen (mitochondria).

Hemoglobin

A mature red blood cell primarily consists of a cell membrane surrounding a mass of **hemoglobin** (HĒ-mō-glō-bin). Hemoglobin is responsible for the cell's ability to transport both oxygen and carbon dioxide. Each hemoglobin molecule contains iron ions that can reversibly bind oxygen molecules.

Where oxygen is abundant in the plasma (in the lungs), the iron ions of the hemoglobin molecules bind to oxygen. Where the plasma oxygen levels are low (in body tissues), the iron ions of the hemoglobin molecules readily release their oxygen. Where plasma oxygen concentrations are low, the carbon dioxide levels are usually high. Under such conditions, each hemoglobin molecule binds carbon dioxide molecules in a process that is just as reversible as the binding of oxygen to iron ions.

Much less reversible is the binding of *carbon monoxide (CO)* to the iron of the hemoglobin molecules. Carbon monoxide is a common gas produced by automobiles, fuel oil heaters, and cigarettes during combustion. CO binds much more strongly to hemoglobin than does oxygen (about 200 times more!). As a result, even very small amounts of CO can displace enough oxygen from hemoglobin that tissues can become oxygen-starved. In severe cases of carbon monoxide poisoning, the brain becomes unable to function, due to oxygen deprivation, and the result is unconsciousness and death.

---

aima, blood + *poiein*, to make
hemopoiesis: the production of formed elements

erythros, red + *kytos*, cell
erythrocytes: red blood cells, the most abundant formed elements in blood

aima, blood + *krinein*, to separate
hematocrit: a value that indicates the percentage of whole blood contributed by formed elements

This scanning electron micrograph of red blood cells reveals their three dimensional structure. (SEM × 1195)

Normal activity levels can be sustained only when tissue oxygen levels are kept within normal limits. If the hemoglobin content of the RBCs or the abundance of RBCs is reduced, the oxygen-carrying capacity of the blood is reduced, and a condition called **anemia** exists. Anemia causes a variety of symptoms, including premature muscle fatigue, weakness, and a general lack of energy.

RBC Life Span

A red blood cell has a relatively short life span of about 120 days. Because it lacks a nucleus and many other organelles it cannot repair itself the way other cells can. The continual elimination of red blood cells is usually unnoticed because new erythrocytes enter the circulation at a comparable rate. About 1 percent of the circulating erythrocytes are replaced each day, and in the process approximately 180 million new erythrocytes enter the circulation *each minute*.

Red Blood Cell Formation

Red blood cell formation, or **erythropoiesis** (e-rith-ry-poy-Ē-sis), occurs in the *red marrow* of bone. ∞ *[p. 143]* Blood cell production occurs in portions of the vertebrae, sternum, ribs, skull, scapulae, pelvis, and proximal limb bones.

For erythropoiesis to proceed normally, the red marrow tissues must receive adequate supplies of amino acids, iron, and B vitamins required for protein synthesis. RBC **stem cells** in the red marrow are stimulated directly by *erythropoietin* and indirectly by several hormones, including thyroxine, androgens, and growth hormone.

Erythropoietin, also called **EPO** or *erythropoiesis-stimulating hormone*, appears in the plasma when body tissues, especially the kidneys, are exposed to low oxygen concentrations. For example, EPO release occurs during anemia and when blood flow to the kidneys declines.

✔ What would be the effects of a decrease in the amount of plasma proteins?

✔ How would a decrease in the level of oxygen supplied to the kidneys affect the level of erythropoietin in the blood?

WHITE BLOOD CELLS

White blood cells are also known as WBCs or **leukocytes** (LOO-kō-sītz). WBCs are quite different from RBCs because WBCs have nuclei and lack hemoglobin. White blood cells help defend the body against invasion by pathogens and remove toxins, wastes, and abnormal or damaged cells in blood and other tissues.

A typical cubic millimeter of blood contains 6000–9000 leukocytes. Most of the white blood cells in the body, however, are found in body tissues, and those in the blood represent only a small fraction of the total population.

Circulating leukocytes use the bloodstream primarily for transportation to areas of invasion or injury. They also move by a process called *ameboid movement,* in which cytoplasm flows into cellular processes that are extended in front of the cell. This allows leukocytes to move along the walls of blood vessels and, when outside the bloodstream, through surrounding tissues. When problems are detected, these cells leave the circulation and enter the abnormal area. Leukocytes leave the bloodstream by squeezing between the cells forming the capillary wall. Once the leukocytes reach the problem area, they engulf pathogens, cell debris, or other materials by phagocytosis.

leukos, white + *kytos*, cell
leukocytes: white blood cells; nucleated blood cells that fight infection and remove abnormal or damaged cells in blood and other tissues

Types of Leukocytes

Traditionally, leukocytes have been divided into two groups based on their appearance after staining. On that basis leukocytes can be divided into *granulocytes* (with abundant stained granules) and *agranulocytes* (with few, if any, stained granules).

Typical leukocytes in the circulating blood are shown in Figure 14-2•. The granulocytes include *neutrophils, eosinophils,* and *basophils*; the two kinds of agranulocytes are *monocytes* and *lymphocytes.*

From 50 to 70 percent of the circulating white blood cells are **neutrophils** (NOO-trō-fils). Neutrophils are usually the first of the white blood cells to arrive at an injury site. They are very active phagocytes, specializing in attacking and digesting bacteria. Active neutrophils usually have a short life span outside the bloodstream, surviving for only about 12 hours.

ne, not + *uter*, either + *philein*, to love
neutrophil: the most abundant white blood cells; phagocytic cells that are drawn to injury sites, where they help protect against bacterial infection

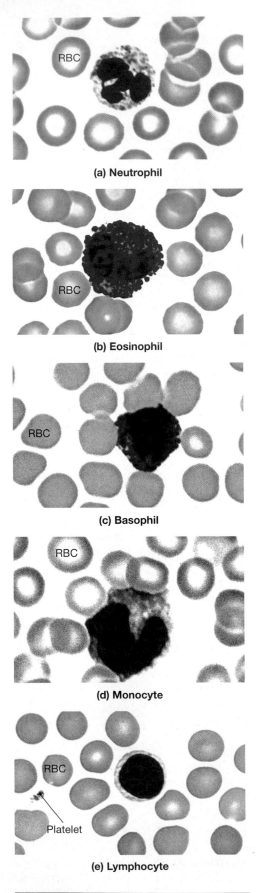

(a) Neutrophil

(b) Eosinophil

(c) Basophil

(d) Monocyte

(e) Lymphocyte

• **FIGURE 14-2**
White Blood Cells.
(a) Neutrophil. (b) Eosinophil. (c) Basophil. (d)
Monocyte. (e) Lymphocyte. (Magnified 1500X)

Eosinophils (ē-ō-SIN-ō-fils) are also phagocytic and usually represent 2 to 4 percent of the circulating white blood cells. Their numbers increase dramatically during an allergic reaction or infection by a parasite.

Basophils (BĀ-sō-fils) account for less than 1 percent of the leukocyte population. Basophils migrate to sites of injury, where they discharge their granules into the interstitial fluids. The granules contain *histamine* and *heparin*. Histamine enhances local inflammation, and heparin is a chemical that prevents blood clotting.

Monocytes (MON-ō-sītz) normally account for 2 to 8 percent of the circulating leukocytes. Outside the bloodstream in peripheral tissues they are called *macrophages*. Macrophages can engulf items as large as or larger than themselves by fusing together and forming giant cells. While doing so they release chemicals that attract and stimulate neutrophils, monocytes, and other phagocytic cells.

Lymphocytes (LIM-fō-sītz) account for 20 to 30 percent of the leukocyte population of the blood. Lymphocytes are the primary cells of the *lymphatic system.* These cells also act to protect the body and its tissues, but they do not rely on phagocytosis. The lymphatic system and its lymphocytes will be discussed in Chapter 15.

Changes in WBC Abundance

Changes in the numbers of WBCs in the blood can reveal a variety of disorders, including pathogenic infection, inflammation, and allergic reactions. By examining a stained blood smear on a microscope slide, a **differential count** can be obtained. A differential count ("diff count") indicates the number of each cell type in a random sample of 100 WBCs. Abnormal diff counts often accompany illness or injury. For example, neutrophils are more common than usual during bacterial infections and inflammation, eosinophils and basophils are more abundant during allergic reactions, and lymphocytes are more common during certain viral infections.

Such an examination also reveals whether there are too few or too many WBCs in the blood. The term **leukopenia** (loo-kō-PĒ-nē-ah) indicates inadequate numbers of white blood cells (*penia* means "poverty"). **Leukocytosis** (loo-kō-sī-TŌ-sis) refers to excessive numbers. Leukocytosis with white blood cell counts of 100,000 per cubic millimeter or more usually indicates the presence of some form of **leukemia** (loo-KĒ-mē-ah), a cancer of blood-forming tissues. Not all leukemias are characterized by leukocytosis; some forms result in the presence of abnormal or immature WBCs in the circulating blood. However, unless treated, all leukemias are fatal.

White Blood Cell Formation

Stem cells responsible for the production of white blood cells originate in the bone marrow. Neutrophils, eosinophils, and basophils complete their development in red bone marrow tissue; monocytes begin their differentiation in the bone marrow, enter the circulation, and complete their development after migrating into various body tissues, where they become macrophages. Stem cells responsible for the production of lymphocytes also originate in the bone marrow, but many of these migrate to *lymphoid tissues*, such as the thymus, spleen, and lymph nodes. As a result, lymphocytes are produced in these organs as well as in the bone marrow.

PLATELETS

Bone marrow contains a number of enormous cells with large nuclei. These cells are called **megakaryocytes** (meg-a-KAR-ē-ō-sītz; *mega* means "large"). Megakaryocytes continually shed small membrane-enclosed packets of cytoplasm into the circulation. These packets, called **platelets** (PLĀT-lets), start the clotting process and help close injured blood vessels. Platelets were once thought to be cells that had lost their nuclei, and they were then called *thrombocytes* (THROM-bō-sītz; clotting cells).

Platelets are continually replaced, and an individual platelet circulates for 10 to 12 days before being removed by phagocytes. On average, there are 350,000 per cubic millimeter of blood.

✔ What type of white blood cell would you expect to find in the greatest numbers in an infected cut?

✔ What cell type would you expect to find in elevated numbers in a person infected by parasitic worms?

✔ A sample of bone marrow has fewer-than-normal numbers of megakaryocytes. What body process would you expect to be impaired as a result?

HEMOSTASIS AND BLOOD CLOTTING

The process of stopping the flow of blood through the walls of damaged blood vessels is called **hemostasis**. The process involves three general responses that begin when cells lining a blood vessel are damaged. (1) Smooth muscles surrounding the vessel wall contract and decrease the diameter of the vessel. This slows or even stops the flow of blood if the vessel is small. (2) Within a few seconds of the injury, the membranes of cells lining the blood vessel become "sticky." Platelets begin to attach to the exposed sticky surfaces and, in turn, become sticky themselves. As more platelets arrive, they combine to form a mass of platelets that helps plug the break in the blood vessel lining. (3) Chemical clotting factors that begin the clotting process are released from the damaged tissues and from the clumping platelets.

aima, blood + *stasis*, standing
hemostasis: the response that stops blood loss after a blood vessel has been damaged

THE CLOTTING PROCESS

Blood clotting, or **coagulation** (kō-ag-ū-LĀ-shun), begins very quickly (about 15 seconds) after a blood vessel is damaged. Normal coagulation cannot occur unless the plasma contains the necessary **clotting factors**, which include calcium ions and 11 different plasma proteins. During the clotting process, the clotting proteins interact in sequence: One protein converts into an enzyme that activates a second protein, and so on, in a chain reaction.

When a blood vessel is damaged, clotting factors are released by both damaged tissue cells and the responding platelets. Along with calcium ions, these factors result in the conversion of a clotting protein called **prothrombin** into the enzyme **thrombin** (THROM-bin). Thrombin then completes the coagulation process by converting the dissolved **fibrinogen** to solid **fibrin**. As the fibrin network grows, blood cells and additional platelets are trapped within the fibrous tangle, forming a **blood clot** that seals off the damaged portion of the vessel. Figure 14 3 ● shows the sequence of events during the clotting process and a scanning electron micrograph of a blood clot.

Sometimes blood clots form in undamaged blood vessels. These clots often occur at roughened areas within the normally smooth inner walls of blood vessels. Such an area may cause platelets to stick to that area and form a clot called a **thrombus.** The thrombus gradually enlarges and reduces the diameter of the blood vessel. Eventually the vessel may be completely blocked, or a large chunk of the clot may break off. A drifting blood clot is called an **embolus** (EM-bo-lus), a term that means "plug." When an embolus becomes stuck in a blood vessel, it blocks circulation to the area downstream, killing the affected tissues. The blockage is called an **embolism.**

✔ What two sources release clotting factors after a cut in the skin?

✔ What is the difference between a thrombus and an embolus?

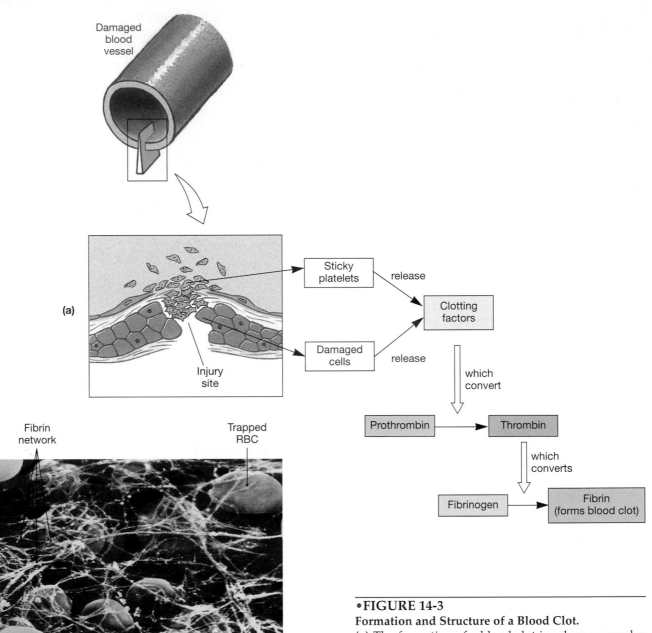

Damaged blood vessel

(a)

Injury site

Sticky platelets — release → Clotting factors

Damaged cells — release →

which convert ↓

Prothrombin → Thrombin

which converts ↓

Fibrinogen → Fibrin (forms blood clot)

Fibrin network

Trapped RBC

(b)

•FIGURE 14-3
Formation and Structure of a Blood Clot.
(a) The formation of a blood clot involves a complex chain of events. In the last few steps, prothrombin is converted to thrombin, and thrombin catalyzes the formation of fibrin threads from fibrinogen. **(b)** A scanning electron micrograph showing the network of fibrin that forms the framework of a clot. (SEM × 3561)

BLOOD TYPES

If you have ever donated blood or received a transfusion of blood you are already aware that not all blood is alike. There are different **blood types** that are determined by the presence or absence of specific chemical substances attached to the surfaces of red blood cells. These cell membrane molecules are called *antigens*. An **antigen** is any substance capable of stimulating the body to produce antibodies. The types of *surface antigens* on RBCs are genetically determined. There are at least 50 different kinds of antigens on the surfaces of RBCs. Three common antigens are known as **A**, **B**, and **Rh**.

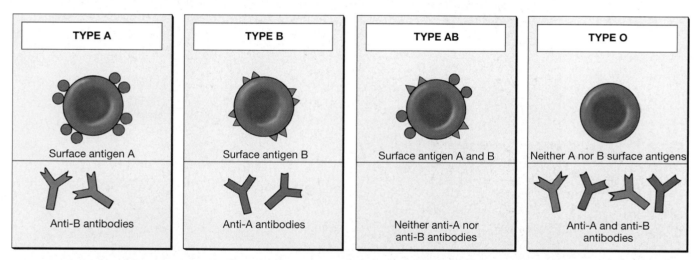

The red blood cells of a particular individual may have either antigen A or B on their surfaces, both A and B, or neither A nor B. For example, **Type A** blood has antigen A only, **Type B** has antigen B only, **Type AB** has both, and **Type O** has neither. Some blood types are more common than others. In the population of the United States, the distribution of blood types is Type O, 46 percent; Type A, 40 percent; Type B, 10 percent; and Type AB, 4 percent.

The presence of the Rh antigen, sometimes called the *Rh factor*, is indicated by the terms **Rh-positive** (present) or **Rh-negative** (absent). In recording the complete blood type, the term *Rh* is usually omitted, and the data are reported as O-negative (O⁻), A-positive (A⁺), and so forth. Some 85 percent of the U.S. population is Rh-positive.

ANTIBODIES AND CROSS-REACTIONS

The surface antigens on a person's own red blood cells are ignored by his or her own immune system. However, plasma contains antibodies, proteins that attack "foreign" antigens. An **antibody** is a protein that binds to specific antigens. For example, as Figure 14-4a• shows, the plasma of individuals with Type A blood

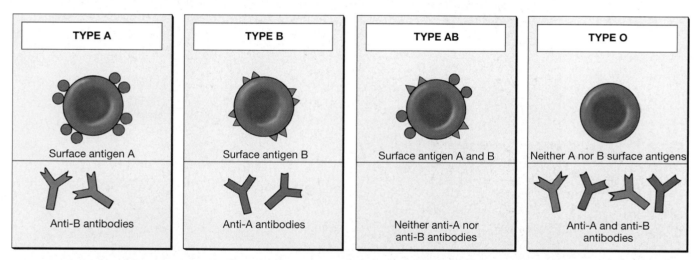

TYPE A	TYPE B	TYPE AB	TYPE O
Surface antigen A	Surface antigen B	Surface antigen A and B	Neither A nor B surface antigens
Anti-B antibodies	Anti-A antibodies	Neither anti-A nor anti-B antibodies	Anti-A and anti-B antibodies

(a)

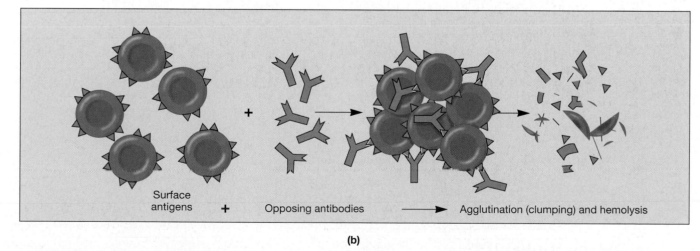

Surface antigens + Opposing antibodies ⟶ Agglutination (clumping) and hemolysis

(b)

•**FIGURE 14-4**
Blood Typing and Cross-Reactions.
(a) The blood type depends on the presence of antigens on RBC surfaces. The plasma contains antibodies that will react with foreign antigens. **(b)** A cross-reaction occurs when antibodies encounter their target antigens. The result is extensive agglutination, or clumping, of the affected RBCs, followed by cell breakdown, or hemolysis.

contains circulating anti-B antibodies that will attack Type B red blood cells. The plasma of Type B individuals contains anti-A antibodies. The red blood cells of an individual with Type O blood lack surface antigens altogether, so the plasma of such an individual contains both anti-A and anti-B antibodies. At the other extreme, Type AB individuals lack antibodies sensitive to either A or B antigens.

When an antibody meets its specific antigen, a **cross-reaction** occurs. Initially the red blood cells clump together, a process called **agglutination** (a-gloo-ti-NĀ-shun), and they may also **hemolyze**, or break down (Figure 14-4b•). The resulting clumps and fragments of red blood cells form drifting masses that can plug small vessels in the kidneys, lungs, heart, or brain, damaging or destroying tissues. To avoid such problems, care must be taken to ensure that the blood types of the donor and the recipient are **compatible**. In practice, this involves choosing a donor whose blood cells will not undergo cross-reaction with the plasma of the recipient (Figure 14-5•).

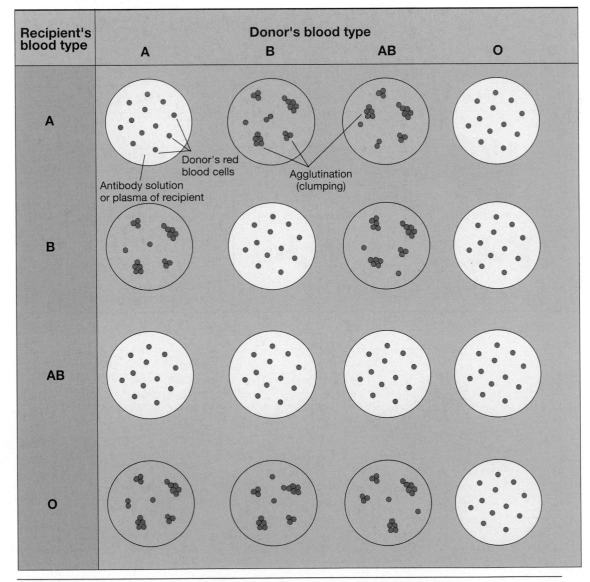

•**FIGURE 14-5**

Cross-Reactions Between Different ABO Blood Types.
In this test series, donor red blood cells are mixed with antibody solutions or plasma from potential recipients. Agglutination and hemolysis occur when the blood types are not compatible. Theoretically, Type O blood could be donated safely to all ABO blood groups, and a person with Type AB blood could receive blood from all ABO blood groups.

COMPATIBILITY TESTING

The standard test for blood type categorizes a blood sample in terms of the A, B, and Rh antigens, the three most likely to produce dangerous cross-reactions. The test involves mixing drops of blood with solutions containing anti-A, anti-B, and anti-Rh antibodies and noting any cross-reactions. For example, if the red blood cells clump together when exposed to anti-A *and* anti-B, the individual has Type AB blood. If no reactions occur, the person must be Type O. Figure 14-5• summarizes the range of cross-reactions possible between different ABO blood type donors and recipients. In practice, the presence or absence of the Rh antigen is also noted, and the individual is classified as Rh-positive or Rh-negative on that basis. In the most common type, Type O-positive, the red blood cells do not have antigens A and B, but they do have the Rh antigen.

Standard blood typing can be completed in a matter of minutes, and Type O blood can be safely administered in an emergency. Because their blood cells are unlikely to produce severe cross-reactions in a recipient, Type O individuals are sometimes called *universal donors*.

When time permits, additional tests are often performed, because with at least 48 other possible antigens on the cell surface, even Type O blood can occasionally trigger a dangerous cross-reaction. For this reason, **cross-match testing** is often performed to ensure complete compatibility. This test involves exposing the donor's red blood cells to a sample of the recipient's plasma under controlled conditions. This procedure reveals the presence of significant cross-reactions involving other antigens and antibodies.

Rh AND PREGNANCY

Unlike the situation with A and B antigens, the plasma of an Rh-negative individual does not normally contain anti-Rh antibodies. These antibodies appear only after an individual has been **sensitized** by previous exposure to Rh-positive red blood cells. Such exposure may occur accidentally during a transfusion, but it may also accompany a normal pregnancy involving an *Rh-negative mother* and an *Rh-positive fetus*. The sensitization that causes this condition usually occurs at delivery, when bleeding occurs at the placenta and uterus. Within 6 months after delivery, roughly 20 percent of Rh-negative mothers who carried Rh-positive children have become sensitized.

Because the Rh antibodies are not produced in significant amounts until *after* delivery, the first infant is not affected, but a sensitized mother will respond to a second Rh-positive fetus by producing massive amounts of anti-Rh antibodies. These can cross the placenta and break down the fetal red blood cells, producing a dangerous fetal anemia. This condition increases the fetal demand for blood cells, and the newly formed RBCs leave the fetal bone marrow and enter the fetal circulation before completing their development. These immature RBCs are called *erythroblasts*, and the condition is known as **erythroblastosis fetalis** (e-rith-rō-blas-TŌ-sis fē-TAL-is), or **hemolytic disease of the newborn (HDN)**. For the entire sequence of events, see Figure 14-6•. Without treatment, the fetus will usually die before delivery or shortly thereafter.

A newborn with severe HDN is anemic, and the breakdown products released by RBC destruction give the skin a characteristic yellow color. Because the mother's antibodies remain active for 1 to 2 months after delivery, the infant may need to have its entire blood volume replaced. This blood replacement removes most of the mother's antibodies as well as the degenerating red blood cells, reducing the chances for fatal complications.

To avoid the Rh problem completely, the mother's production of anti-Rh antibodies is prevented by administering anti-Rh antibodies (available under the name *RhoGam*) during pregnancy and following delivery. These "foreign" antibodies quickly destroy any fetal red blood cells that cross the placental barrier. Thus there are no Rh-positive antigens to stimulate the mother's immune system, sensitization does not occur, and Rh antibodies are not produced.

erythros, red + *blastos*, germ + *osis*, disease

erythroblastosis fetalis: a potentially fatal fetal disease caused by the destruction of fetal red blood cells by maternal antibodies

✔ What blood types can be transfused into a person with Type AB blood?

✔ Why can't a person with Type A blood receive blood from a person with Type B blood?

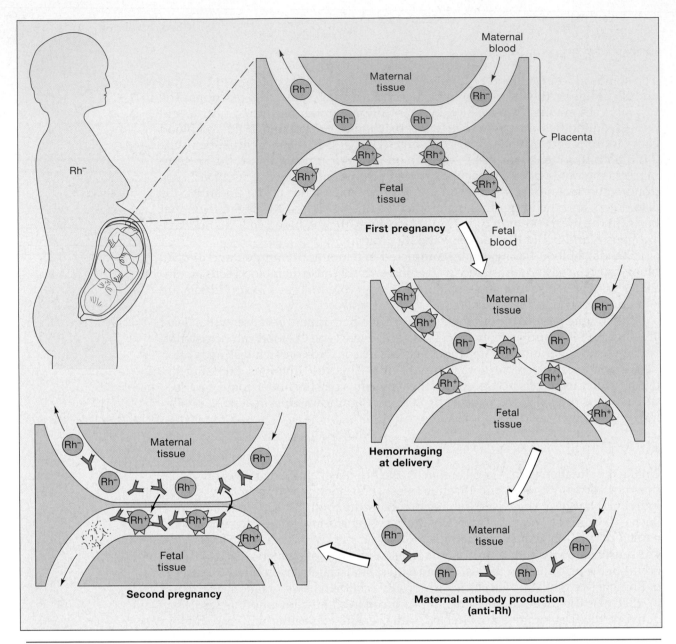

•**FIGURE 14-6**
Rh Factors and Pregnancy.
When an Rh-negative woman has her first Rh-positive child, mixing of fetal and maternal blood may occur at delivery
when the placental connection breaks down. The appearance of Rh-positive blood cells in the maternal circulation sen-
sitizes the mother, stimulating the production of anti-Rh antibodies. If another pregnancy occurs with an Rh-positive
fetus, the mother's antibodies can cross the placental barrier and attack fetal blood cells, producing symptoms of HDN
(hemolytic disease of the newborn).

BLOOD DISORDERS

Problems with blood can occur with any of its components—the proteins and
other solutes in the plasma, the red blood cells (RBCs), the white blood cells
(WBCs), and the platelets. A few examples of such problems were presented earlier. One of the most common problems of blood, *anemia*, is associated with low
numbers of RBCs and was introduced on p. 321. Variations in the number of WBCs
due to infections and leukemia were discussed on p. 322. Problems with clotting
may involve both platelets and plasma solutes leading to reduced coagulation
(excessive bleeding) or excessive coagulation (p. 323).

Important blood disorders are listed in Figure 14-7•. Blood disorders that
will be considered include those related to infection, congenital (including ge-

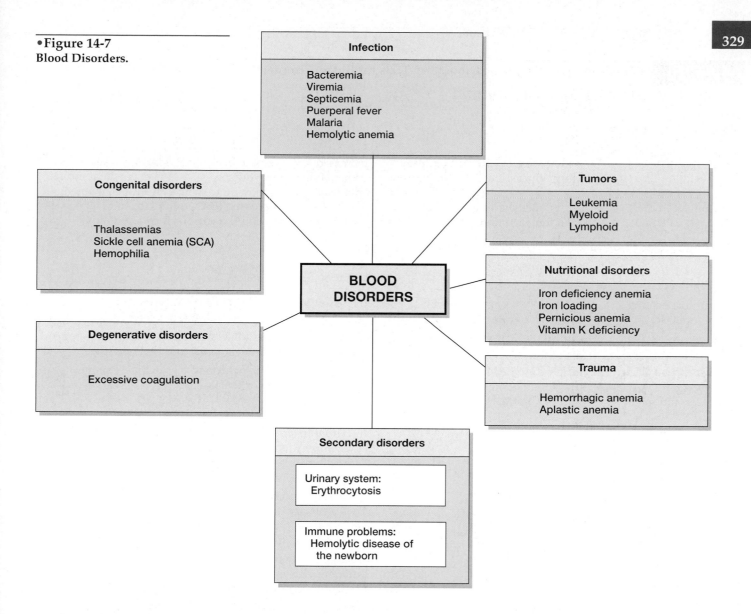

•Figure 14-7
Blood Disorders.

Infection

Bacteremia
Viremia
Septicemia
Puerperal fever
Malaria
Hemolytic anemia

Congenital disorders

Thalassemias
Sickle cell anemia (SCA)
Hemophilia

Tumors

Leukemia
Myeloid
Lymphoid

BLOOD DISORDERS

Nutritional disorders

Iron deficiency anemia
Iron loading
Pernicious anemia
Vitamin K deficiency

Degenerative disorders

Excessive coagulation

Trauma

Hemorrhagic anemia
Aplastic anemia

Secondary disorders

Urinary system:
Erythrocytosis

Immune problems:
Hemolytic disease of
the newborn

netic) problems, trauma, tumors, nutritional deficiencies, degenerative problems, and secondary conditions (blood disorders that result from non-cardiovascular systems). Blood disorders are commonly diagnosed by laboratory tests that examine the components of blood. Examples of common blood tests are presented in Table 14-1.

INFECTION

Blood is normally free of microorganisms. Bacteria and viruses may enter the blood through a wound or infection. **Bacteremia** is a condition in which bacteria circulate in blood, but do not multiply there. **Viremia** is a similar condition associated with viruses. The normal immune responses that destroy such invaders are discussed in Chapter 16. **Toxemia** is the presence of toxins, or poisons, in the blood, such as the botulinus toxin. ∞ *[p. 212]*

    Sepsis (SEP-sis), which means decay, is a poisoning due to infection in body tissues and body fluids. Sepsis of the blood, or **septicemia**, results if pathogens are present and multiplying in the blood. Septicemia, called *blood poisoning* in the past, results in pathogens and toxins being spread throughout the body.

Bacteria

One of the complicating factors of septicemia is **disseminated intravascular coagulation (DIC)**. Bacteria may cause DIC, although it may also result from trau-

Table 14-1   Representative Diagnostic Laboratory Tests for Blood Disorders

Laboratory Test	Normal Values in Blood Plasma or Serum	Significance of Abnormal Values
Complete blood count (CBC)		
RBC count	Adult males: 4.7–6.1 million/mm^3 Adult females: 4.0–5.0 million/mm^3	Increased RBC count, Hb, and Hct occur in *polycythemia vera* (a bone marrow disorder) and in events that induce hypoxia, such as moving to a high altitude or congenital heart disease. Decreased RBC count, Hb, and Hct occur with hemorrhage and the various forms of anemia
Hemoglobin (Hb)	Adult males: 14–18 g/dl Adult females: 12–16 g/dl	
Hematocrit (Hct)	Adult males: 42–52 Adult females: 37–47	
RBC Indices		
Mean corpuscular volume (MCV)	Adults: 80–98 μm^3 (measures average volume of single RBC)	Increased MCV and MCH occur in types of macrocytic anemia, including vitamin B$_{12}$ deficiency
Mean corpuscular hemoglobin (MCH)	Adults: 27–31 pg (measure of the average amount of Hb per RBC)	Decreased MCV and MCH in types of microcytic anemia, such as iron deficiency anemia and thalassemia
Mean corpuscular hemoglobin concentration (MCHC)	Adults: 32–36% (derived by dividing the total Hb concentration number by the Hct value)	Decreased levels (hypochromic erythrocytes) suggest iron deficiency anemia or thalassemia
WBC count	Adults: 5000–10,000/mm^3	Increased in chronic and acute infections; tissue death (MI, burns), leukemia, parasitic diseases and stress. Decreased in aplastic and pernicious anemias, overwhelming bacterial infection (sepsis), and viral infections
Differential WBC count	Neutrophils: 50–70%	Increased in acute bacterial infection, myelocytic leukemia, rheumatoid arthritis, and stress. Decreased in aplastic and pernicious anemia, viral infections, radiation treatment, and with some medications
	Lymphocytes: 20–30%	Increased in lymphocytic leukemia, infectious mononucleosis, and viral infections. Decreased in radiation treatment, AIDS, and corticosteroid therapy
	Monocytes: 2–8%	Increased in chronic inflammation, viral infections, and tuberculosis. Decreased in aplastic anemia and corticosteroid therapy
	Eosinophils: 2–4%	Increased in allergies, parasitic infections, and some autoimmune disorders. Decreased with steroid therapy
	Basophils: 0.5–1%	Increased in inflammatory processes and during healing. Decreased in hypersensitivity reactions
Platelet count	Adults: 150,000–400,000/mm^3	Increased count can cause vascular thrombosis; increased in polycythemia vera. Decrease can result in spontaneous bleeding; decreased in different types of anemia and in some leukemias
Factors assay (coagulation factors, I, II, V, VIII, IX, X, XI, XII)	Factors I–XI are measured for their activity and their minimal clotting activity	Decreased activity of the coagulation factors will result in defective clot formation. Deficiencies can be caused by liver disease or vitamin K deficiency
Plasma fibrinogen (Factor I)	200–400 mg/dl	Elevated in inflammatory conditions and acute infections and also with medications such as birth control pills. Decreased in liver disease, leukemia, and DIC (disseminated intravascular coagulation)

Table 14-1	Representative Diagnostic Laboratory Tests for Blood Disorders (continued)

Laboratory Test	Normal Values in Blood Plasma or Serum	Significance of Abnormal Values
Hemoglobin		
Hemoglobin A	Adults: within 95–98% of the total Hb	
Hemoglobin F (fetal form)	<2% of the total Hb after age 2 (newborn has 50–80% HbF)	Elevated levels after 6 months of age suggest thalassemia
Hemoglobin S	0% of the total Hb	Elevated levels occur in sickle cell anemia and the sickle cell trait
Serum bilirubin (hemoglobin breakdown product)	Adults: 0.1–1.2 mg/dl	Increased levels occur with severe liver disease, hepatitis, and liver cancer
Bone marrow aspiration biopsy	For the evaluation of hematopoiesis	Increased RBC precursors with polycythemia vera; increased WBC precursors with leukemia; radiation or chemotherapy therapy can cause a decrease in all cell populations

ma or other illnesses that release factors that stimulate blood clotting. In DIC, bacterial toxins activate several steps in the coagulation process, including thrombin, which then converts fibrinogen to fibrin within the circulating blood. Although much of the fibrin is removed by phagocytes or is dissolved by a fibrin-digesting enzyme (*plasmin*), small clots may block small vessels and damage nearby tissues. The liver is the source of circulating fibrinogen. If the liver cannot keep pace with the rate at which fibrinogen is being removed, through conversion to fibrin, clotting abilities decline and uncontrolled bleeding may occur.

Puerperal (pū-ER-per-al) **fever**, or *childbed fever*, is a bacterial infection that can lead to septicemia. It is caused by streptococci bacteria that spread from the birth canal to infect the uterus after childbirth, miscarriage, or abortion. This condition was a common cause of death before the use of antibiotics and sterile labor and delivery techniques. Symptoms include chills, fever, excessive bleeding, foul-smelling vaginal discharge, and abdominal and pelvic pain.

Parasites

Malaria is a parasitic disease caused by several species of the protozoan, *Plasmodium*. It is one of the most severe diseases in tropical countries, killing 1.5–3 million people per year, of which up to half are children under the age of 5. Malaria is transmitted by a mosquito from person to person. The parasite initially infects the liver, then enlarges and fragments into smaller forms that infect red blood cells (see Figure 6-2c•, p. 99). Periodically, at intervals of 2–3 days, all of the infected RBCs rupture simultaneously and release more parasites, which infect additional RBCs. The timing of the release and reinfection of RBCs corresponds to the cycles of fever and chills that characterize malaria. The release corresponds to the time of high fevers. Dead RBCs can block blood vessels leading to vital organs, such as the kidney and brain, leading to tissue death. Treatment rests on the use of a combination of drugs which either kill the parasites while in the blood plasma, the red blood cells, or the liver.

The loss of RBCs also reduces the oxygen capacity of the blood and results in weakness and fatigue, symptoms of anemia. This type of anemia, which is caused by the large-scale destruction of RBCs, is called **hemolytic anemia**. Other types of anemia are discussed below.

CONGENITAL DISORDERS

Common congenital blood disorders that are inherited include problems related to the hemoglobin molecules in RBCs (which can lead to hemolytic anemia) and to problems in the process of clotting (which can cause extensive bleeding).

Hemoglobin Disorders

Normal red blood cells are packed with hemoglobin molecules. As noted earlier, hemoglobin is responsible for the ability of RBCs to transport oxygen and carbon dioxide. ∞ *[p. 320]* Each hemoglobin molecule is made up of four protein molecules; a pair of "a," or *alpha* proteins, and a pair of "b," or *beta* proteins (Figure 14-8•). Each protein subunit contains an iron-containing group that binds oxygen. Abnormalities in the structure of its protein subunits leads to various hemoglobin disorders.

Thalassemias. The **thalassemias** (thal-ah-SĒ-mē-uhs) are a diverse group of inherited blood disorders caused by an inability to produce adequate amounts of normal protein subunits in their hemoglobin. The severity of different types of thalassemias depends on which and how many protein subunits are abnormal. The RBCs of individuals with **alpha-thalassemia** may be small and contain less than the normal quantity of hemoglobin, or be very small (*microcytic*) and relatively fragile. Individuals who are unable to make any normal alpha subunits cannot transport oxygen normally and die shortly after birth.

Two disorders are associated with the beta protein subunit. In **beta-thalassemia major**, or *Cooley's disease*, the individual inherits a severe form of hemolytic anemia. The symptoms of this condition include *microcytosis* (small RBCs), a low hematocrit (under 20), and enlargement of the spleen and liver (organs involved in recycling damaged RBCs). Treatment involves periodic transfusions—(the administration of blood components)—to keep adequate numbers of RBCs in the circulation. **Beta-thalassemia minor** seldom produces clinical symptoms. The rates of hemoglobin synthesis are depressed by roughly 15 percent, but this decrease does not affect their functional abilities, and no treatment is necessary.

Sickle cell anemia (SCA). **Sickle cell anemia** results from the production of an abnormal form of the beta subunits of hemoglobin. When the blood contains abundant oxygen, the hemoglobin molecules and the RBCs that carry them appear normal. But when the defective hemoglobin gives up enough of its bound oxygen, the cells change shape, becoming stiff and curved (Figure 14-9•). This "sickling" makes the RBCs fragile and easily damaged. Moreover, if this sickling occurs in a narrow capillary, the cell can become stuck and block the flow of other RBCs. Such circulatory blockages cause nearby tissues to become oxygen-starved. Symptoms of sickle cell anemia include pain and damage to a variety of organs and systems. In addition, the trapped red blood cells eventually die and break down, producing a type of hemolytic anemia.

Today, sickle cell anemia affects 60,000–80,000 African-Americans, or roughly 0.14 percent of the African-American population. To develop sickle cell anemia, an individual must have two copies of the sickling gene—one from each parent. If only one sickling gene is present, the individual has the sickling trait. In such cases, most of the hemoglobin is of the normal form, and the RBCs function normally. One African-American in 12 carries the sickling trait. But the presence of the abnormal hemoglobin gives the individual a better ability to tolerate the parasitic infection that causes malaria, a mosquito-borne illness (p. 331). The malaria parasites enter the bloodstream when an individual is bitten by an infected mosquito. The microorganisms then invade, and reproduce within, the RBCs. But when they enter the red blood cell of a person with the sickling trait, the cell responds by sickling. Such cells are engulfed by phagocytes, which then kill the parasites. As a result, the indi-

•**Figure 14-8**
The Structure of Hemoglobin.
Hemoglobin consists of four protein subunits. Each subunit contains an iron-containing group that binds oxygen.

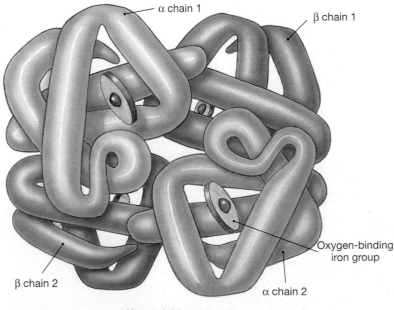

α chain 1

β chain 1

Oxygen-binding
iron group

β chain 2

α chain 2

Hemoglobin molecule

vidual remains less affected by the malaria, whereas normal individuals sicken and may die of that disease.

Symptoms of sickle cell anemia include pain and damage to a variety of organs and systems, depending on the location of the obstructions, and weakness and fatigue due to anemia. Transfusions of normal blood can temporarily prevent additional complications, and there are experimental drugs that can control or reduce sickling.

Clotting Disorders

Inherited problems in the clotting process result in excessive bleeding, or *hemophilia*.

Hemophilia. Hemophilia (hē-mō-FĒ-lē-a) is an inherited bleeding disorder. About 1 person in 10,000 is a hemophiliac, and of those, 80 to 90 percent are males. In most cases, hemophilia is caused by the reduced production of one clotting factor, *Factor VIII*. The severity of hemophilia varies, depending on how little clotting factor is produced. In severe cases, extensive bleeding occurs with relatively minor contact, and *hemorrhage* (blood loss) occurs at joints and around muscles.

The symptoms of hemophilia can be reduced or controlled by transfusions of the missing clotting factor. However, plasma from 2000 to 5000 individuals must be pooled to obtain enough clotting factor. This increases the cost of the treatment and increases the risk of infection with AIDS or hepatitis. Donated blood is now screened for these blood-carried diseases. New genetic engineering techniques that are now being used to produce Factor VIII should provide a safer and cheaper method of treatment.

TRAUMA

Hemorrhagic Anemia

Injuries may be accompanied by **hemorrhage** (HEM-or-ij), or blood loss. The loss of blood is considered a type of anemia, **hemorrhagic** (hem-o-RAJ-ik) **anemia,** because the oxygen-carrying capacity of the blood is reduced. The general symptoms of all types of anemia include premature muscle fatigue, weakness, lethargy, and a lack of energy.

Aplastic Anemia

In **aplastic** (ā-PLAS-tik) **anemia,** the bone marrow fails to produce new red blood cells (*aplastic* means "to not form"). This can occur following viral infection, exposure to toxins or radiation. For example, the 1986 nuclear accident in Chernobyl (U.S.S.R.) caused a number of cases of aplastic anemia. The condition is generally fatal unless surviving stem cells repopulate the marrow or a bone marrow transplant is performed. In aplastic anemia, the circulating red blood cells are normal in all respects, but because new RBCs are not being produced, the *reticulocyte* count is extremely low. A **reticulocyte** (re-TIK-ū-lō-sīt) is an immature red blood cell. Once they enter the blood from the bone marrow, they mature within 24 hours. In response to anemia, the body increases the rate of RBC production, and the reticulocyte count goes up.

TUMORS

Leukemias

The leukemias are cancers of blood-forming tissues. ∞ *[p. 74]* The cancerous cells of leukemia do not form a compact tumor, instead, they are spread throughout the body from their origin in the bone marrow. There are two types of leukemia, *myeloid* and *lymphoid*, and both are characterized by elevated levels of circulating WBCs. **Myeloid leukemia** is characterized by the presence of abnormal granulocytes (neutrophils, eosinophils, and basophils) or other cells of the bone marrow, such as the RBC stem cells, and megakaryocytes. **Lymphoid leukemia** involves lymphocytes and their stem cells. The first symptoms appear when immature and abnormal white blood cells appear in the circulation. As the number of white blood cells increases, they travel through the circulation, invading tissues and organs throughout the body.

•Figure 14-9
Sickling in Red Blood Cells.
(a) When fully oxygenated, the cells of an individual with the sickling trait appear relatively normal. **(b)** At lower oxygen concentrations, the RBCs change shape, becoming relatively rigid and sharply curved.

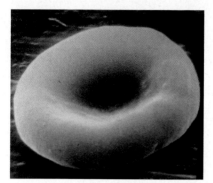

(a) Normal RBC

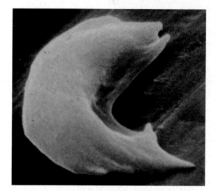

(b) Sickled RBC

These cells are extremely active, and they require abnormally large amounts of energy. As in other cancers, invading leukemic cells gradually replace the normal cells, especially in the bone marrow. Decreased numbers of RBCs and normal WBCs result in anemia and infection. A decline in platelet formation impairs blood clotting. An abnormally low platelet count is known as **thrombocytopenia** (throm-bō-sī-tō-PĒ-nē-ah). Other symptoms of thrombocytopenia include bleeding within the skin, along the digestive tract, and occasionally within the central nervous system.

Each type of leukemia is also classified as **acute** (short and severe) or **chronic** (prolonged). *Acute leukemias* may be linked to radiation exposure, hereditary susceptibility, viral infections, toxic chemical or drug exposure, or unknown causes. *Chronic leukemias* may be related to chromosomal abnormalities or immune system malfunctions. Survival in untreated acute leukemia averages about three months; individuals with chronic leukemia may survive for years.

Untreated leukemia is invariably fatal. Effective treatments exist for some forms of leukemia and not others. One option for treating acute leukemia is to perform a *bone marrow transplant*. If the bone marrow is extracted from another person, care must be taken to ensure that the blood types and tissue types are compatible. If they are not, the new lymphocytes may attack the patient's tissues, with potentially fatal results. Best results are obtained when the donor is a close relative. The bone marrow may also be removed from the patient, cleansed of cancer cells, and reintroduced after radiation or chemotherapy treatment eliminate cancer cells from the body. Although this method produces fewer complications, the preparation and cleansing of the marrow are technically difficult and time consuming.

NUTRITIONAL DISORDERS

Deficiencies in minerals and vitamins can cause various types of anemia. Excesses, especially in iron, can lead to other problems.

Iron Deficiency Anemia

In **iron deficiency anemia**, normal hemoglobin synthesis cannot occur, because iron reserves or the dietary intake of iron are inadequate. Developing red blood cells cannot synthesize functional hemoglobin; as a result, they are unusually small (*microcytic*). An estimated 60 million women worldwide have symptoms of iron deficiency anemia.

Women are especially dependent on a normal dietary supply of iron, because their iron reserves are smaller than those of men. A healthy male's body has about 3.5 g of iron. Of that amount, 2.5 g is bound to the hemoglobin of circulating red blood cells, and the rest is stored in the liver and bone marrow. In women, the total body iron content averages 2.4 g, with roughly 1.9 g incorporated into red blood cells. Thus, a woman's iron reserves consist of only 0.5 g, half that of a typical man.

Iron Loading

When the diet contains large amounts of iron the excess iron gets stored in peripheral tissues. This storage is called **iron loading.** Normally, body iron content is maintained at 4 g or less. As the iron deposits accumulate, cells eventually begin to malfunction. For example, iron deposits in pancreatic cells can lead to diabetes mellitus; deposits in cardiac muscle cells lead to abnormal heart contractions and heart failure. (There is evidence that iron deposits in the heart caused by the over-consumption of red meats may contribute to heart disease.) Liver cells become nonfunctional, and liver cancers may develop.

Comparable symptoms of iron loading may follow repeated transfusions of whole blood, because each unit of whole blood contains roughly 250 mg of iron. For example, as we noted previously, the various forms of thalassemia result from a genetic inability to produce adequate amounts of the protein subunits in hemoglobin. RBC production and survival are reduced, and so is the oxygen-carrying capacity of the blood. Most individuals with severe untreated thalassemia die in their twenties, but not because of the anemia. These patients are treated for se-

Clinical Note

What Is Blood Doping?

In the past, athletes involved with endurance sports such as cycling have turned to **blood doping** and its resulting erythrocytosis to give themselves an edge over their competitors. Blood doping involves removing whole blood from the athlete in the weeks before an event. The red blood cells are separated from the plasma and stored. By the time of the race, the competitor's bone marrow will have replaced the lost blood. Immediately before the event the red blood cells are reinfused, increasing the hematocrit. The objective is to elevate the oxy- gen-carrying capacity of the blood and thereby increase endurance. The consequence is that the athlete's heart is placed under a tremendous strain because it makes the blood thicker and harder to push around the circulatory system. The long-term effects are unknown, but the practice obviously carries a significant risk; it has recently been banned in amateur sports. Attempts to circumvent this rule in 1992–1993 by the use of administered EPO, a hormone that stimulates RBC production, resulted in the tragic deaths of 18 European cyclists.

vere anemia with frequent blood transfusions, which prolong life, but the excessive iron loading eventually leads to fatal heart problems.

Vitamin Deficiencies

Pernicious anemia. In **pernicious** (per-NISH-us) **anemia**, the RBC stem cells in the bone marrow do not receive enough vitamin B_{12} to continue their production of red blood cells. Erythrocyte production declines, and the red blood cells are abnormally large (*macrocytic*) and may develop a variety of bizarre shapes. Blood tests from a person with pernicious anemia indicate a low hematocrit, a low reticulocyte count, and an elevated MCV.

Vitamin K Deficiencies. Excessive bleeding, or poor coagulation, can result from deficiencies of vitamin K. This vitamin is required by the liver to synthesize four clotting factors, including prothrombin. Roughly half of the vitamin K is absorbed directly from the diet, and the rest is synthesized by intestinal bacteria.

DEGENERATIVE DISORDERS

Excessive Coagulation

The sensitivity of the clotting system increases with age, due to an increase in one of the protein factors involved in the clotting process. Abnormal, or excessive, clotting can sometimes occur, leading to the formation of a thrombus, or *thrombi* (*plural*), in undamaged blood vessels. ∞ *[p. 323]*

SECONDARY DISORDERS

Urinary System

Erythrocytosis. In **erythrocytosis** (e-RITH-rō-sī-TŌ-sis), the blood contains abnormally large numbers of red blood cells. Erythrocytosis generally results from the massive release of erythropoietin (EPO) by the kidneys when deprived of oxygen. ∞ *[p. 301]* After moving to high altitudes, people typically experience erythrocytosis, because the air there contains less oxygen than it does at sea level. The increased number of red blood cells compensates for the fact that each RBC is carrying less oxygen than it would at sea level. Mountaineers and those living at altitudes of 10,000–12,000 feet may have hematocrits as high as 65.

Immune Problems

Erythroblastosis fetalis, or hemolytic disease of the newborn (HDN), results when the mother has a Rh-negative and her fetus is Rh-positive. As discussed on p. 328, a second Rh-positive fetus is at risk because of the destruction of fetal RBCs by the mother's immune response.

✔ What inherited blood disorders are caused by defects in hemoglobin production?

✔ Inadequate levels of vitamin B_{12} and vitamin K are involved in what types of blood disorders?

CHAPTER REVIEW

Key Words

agglutination: Clumping of red blood cells due to interactions between surface antigens and plasma antibodies.

cardiovascular: Pertaining to the heart, blood, and blood vessels.

embolus (EM-bo-lus): A blood clot drifting in the circulation.

erythrocyte (e-RITH-rō-sīt): A red blood cell.

fibrin (FĪ-brin): Insoluble protein fibers that form the framework of a blood clot.

hematocrit (hē-MAT-ō-krit): Percentage of the volume of whole blood made up of cells.

hemoglobin (hē-mō-GLŌ-bin): The iron-containing protein found in red blood cells that gives them the ability to transport oxygen in the blood.

hemolysis: The breakdown of red blood cells.

hemopoiesis (hē-mō-poy-Ē-sis): Blood cell formation and differentiation.

hemostasis: The stoppage of bleeding.

leukocyte (LOO-kō-sīt): A white blood cell.

plasma (PLAZ-mah): The fluid that remains after the cells have been removed from a sample of whole blood.

platelets (PLĀT-lets): Small membrane-bound packets of cytoplasm that contain enzymes important in the clotting response.

serum: Blood plasma from which clotting factors have been removed.

Selected Clinical Terms

anemia: A condition in which the oxygen-carrying capacity of the blood is reduced because of a low hematocrit or low hemoglobin content.

aplastic anemia: Anemia caused by a lack of red blood cell production by the red bone marrow.

bacteremia: The presence of bacteria in the blood.

hemolytic anemia: Anemia caused by the breakdown of red blood cells.

hemorrhage (HEM-or-ij): Blood loss, or bleeding.

hemorrhagic anemia: Anemia caused by the loss of blood.

leukemia: Cancers of blood-forming tissues, characterized by extremely elevated numbers of circulating white blood cells.

pernicious (per-NISH-us) **anemia:** Anemia caused by an inadequate supply of vitamin B_{12} for the production of red blood cells.

septicemia: The presence and multiplication of bacteria in the blood.

thalassemia: A disorder resulting from production of an abnormal form of hemoglobin.

thrombus: A blood clot that forms at the inner surface of a blood vessel.

Study Outline

INTRODUCTION (p. 317)

1. The cardiovascular system provides a mechanism for the rapid transport of nutrients, waste products, and cells within the body.

SYSTEM BRIEF (p. 317)

1. The **cardiovascular system** is made up of the heart, blood vessels, and blood.

FUNCTIONS OF BLOOD (p. 317)

1. Blood is a specialized connective tissue. Its functions include (1) transporting dissolved gases, nutrients, hormones, and metabolic wastes; (2) the regulation of the pH and electrolyte composition of the interstitial fluids, the distribution of heat, and the restriction of fluid losses through damaged vessels; and (3) defense against pathogens and toxins.

THE COMPOSITION OF BLOOD (pp. 318–319)

1. Blood contains **plasma**, **red blood cells (RBCs)**, **white blood cells (WBCs)**, and **platelets**. The plasma and **formed elements** constitute **whole blood**, which can be separated for analytical or clinical purposes. (*Figure 14-1*)

PLASMA (p. 319)

1. Plasma accounts for about 55 percent of the volume of blood; roughly 92 percent of plasma is water. (*Figure 14-1*)

PLASMA PROTEINS (p. 319)

2. **Albumins** constitute about 60 percent of plasma proteins. **Globulins** constitute roughly 35 percent of plasma proteins; they include *immunoglobulins* (**antibodies**), which attack foreign proteins and pathogens, and **transport proteins**, which bind ions, hormones, and other compounds. In the clotting reactions, dissolved **fibrinogen** molecules are converted to solid **fibrin**. The plasma fluid remaining after clotting is called **serum.**

OTHER PLASMA SOLUTES (p. 319)

3. One percent of blood plasma includes ions, organic nutrients (food molecules), and organic wastes.

FORMED ELEMENTS (pp. 320–323)

1. The major cellular components of blood are red corpuscles, white corpuscles, and platelets.

RED BLOOD CELLS (p. 320)

2. Red blood cells, or **erythrocytes**, account for slightly less than half the blood volume and 99.9 percent of the formed elements. The **hematocrit** value indicates the percentage of whole blood occupied by cellular elements.

3. RBCs transport oxygen and carbon dioxide within the bloodstream. They are highly specialized cells with large surface-to-volume ratios. RBCs lack mitochondria and nuclei.

4. **Hemoglobin** is a globular protein that contains iron ions that bind to oxygen molecules.

5. Red blood cells usually degenerate after 120 days in the circulation.

6. **Erythropoiesis**, the formation of erythrocytes, occurs mainly within the red bone marrow in adults. RBC formation is increased by **erythropoietin** (**EPO**) stimulation, which occurs when body tissues are exposed to low oxygen concentrations.

WHITE BLOOD CELLS (p. 321)

7. White blood cells (**leukocytes**) defend the body against pathogens and remove toxins, wastes, and abnormal or damaged cells.

8. Leukocytes can move through the walls of small blood vessels.

9. Granular leukocytes include **neutrophils**, **eosinophils**, and **basophils**. From 50 to 70 percent of circulating WBCs are neutrophils, which are highly mobile phagocytes. The much less common eosinophils are also phagocytes. The relatively rare basophils migrate to damaged tissues and release *histamines*, aiding the inflammation response. *(Figure 14-2)*

10. Agranular leukocytes include monocytes and lymphocytes. **Monocytes** that migrate into peripheral tissues become macrophages. **Lymphocytes** are the primary cells of the *lymphatic system*. *(Figure 14-2)*

11. Changes in the number of WBCs reveal different disorders. Too few WBCs in the blood is called **leukopenia**, and excessive numbers of WBCs is called **leukocytosis**.

12. Stem cells responsible for the production of lymphocytes originate in the bone marrow, but many migrate to peripheral lymphoid tissues.

PLATELETS *(p. 323)*

13. **Megakaryocytes** in the bone marrow release packets of cytoplasm (**platelets**) into the circulating blood. Platelets are essential to the clotting process.

HEMOSTASIS AND BLOOD CLOTTING *(pp. 323–324)*

1. The process of **hemostasis** stops the loss of blood through the walls of damaged vessels.

THE CLOTTING PROCESS *(p. 323)*

2. Clot formation, or **coagulation**, occurs as **clotting factors** are released by damaged blood vessel cells and platelets. *(Figure 14-3)*

3. Excessive or inappropriate clotting can lead to the formation of an **embolus** (a drifting blood clot) or a **thrombus** (a clot attached to the wall of an intact vessel).

BLOOD TYPES *(pp. 324–328)*

1. **Blood type** is determined by the presence or absence of specific **antigens** on the RBC cell membranes; the antigens are **A**, **B**, and **Rh**.

ANTIBODIES AND CROSS-REACTIONS *(p. 325)*

2. **Antibodies** in the blood plasma react with RBCs bearing foreign antigens. *(Figure 14-4)*

COMPATIBILITY TESTING *(p. 327)*

3. In the ABO blood groups, persons with Type O blood are considered *universal donors*. *(Figure 14-5)*

Rh AND PREGNANCY *(p. 327)*

4. Rh antibodies are synthesized only after an Rh-negative individual becomes **sensitized** to the Rh antigen. *(Figure 14-6)*

BLOOD DISORDERS *(pp. 328–335)*

1. Blood disorders are due to changes in the components of blood and blood clotting. *(Figure 14-7)*

2. Laboratory tests that examine the components of blood are used in diagnosis of blood disorders. *(Table 14-1)*

INFECTION *(p. 329)*

3. In **bacteremia**, bacteria circulate in blood, but do not multiply; in **viremia**, viruses circulate in blood, but do not multiply. The presence of poisons in blood is **toxemia**. In **septicemia**, pathogens are present and multiplying in the blood.

4. Two examples of diseases caused by septicemia are **disseminated intravascular coagulation** (**DIC**) and **puerperal fever**, or *childbed fever*.

5. **Malaria** is transmitted by mosquitoes. It is caused by a protozoan that infects and destroys red blood cells. Their destruction leads to hemolytic anemia.

CONGENITAL DISORDERS *(p. 331)*

6. Inherited blood disorders involve abnormal hemoglobin molecules or problems in clotting *(Figure 14-8)*

7. **Thalassemias** are inherited blood disorders caused by abnormal hemoglobin. Some types of thalassemias are more severe than others. Symptoms of the **alpha-thalassemias** generally include small, fragile RBCs. **Beta-thalassemia major**, or *Cooley's disease*, is a severe form of hemolytic anemia that is treated with periodic transfusions. **Beta-thalassemia minor** seldom produces clinical symptoms.

8. **Sickle cell anemia (SCA)** results from an abnormal form of hemoglobin that causes the RBC to change its shape at low levels of oxygen. This can lead to the blockage of capillaries and oxygen-starved tissues, and when the RBCs break down, hemolytic anemia. *(Figure 14-9)*

9. **Hemophilia** is an inherited disorder characterized by reduced clotting ability that results in excessive bleeding.

TRAUMA *(p. 333)*

10. **Hemorrhagic anemia** is caused by a large loss of blood.

11. In **aplastic anemia**, no new red blood cells form in the bone marrow. A characteristic symptom is an extremely low count of immature RBCs, or **reticulocytes**.

TUMORS *(p. 333)*

12. The leukemias are cancers of blood-forming tissues. **Myeloid leukemia** is characterized by the presence of abnormal granulocytes or other cells of the bone marrow. **Lymphoid leukemias** involve lymphocytes and their stem cells.

NUTRITIONAL DISORDERS *(p. 334)*

13. In **iron deficiency anemia**, red blood cells are very small due to reduced amounts of hemoglobin.

14. **Iron loading** is the storage of excess iron. The stored iron can lead to complications in the pancreas, heart, and liver.

15. In **pernicious anemia**, deficiencies of vitamin B_{12} result in a decline in RBC production in the bone marrow.

16. **Vitamin K** deficiencies result in a reduced rate of synthesis of clotting factors, including prothrombin, by the liver. The result is excessive bleeding, or poor coagulation.

DEGENERATIVE DISORDERS *(p. 335)*

17. Blood coagulation potential increases with age, increasing the possible formation of a thrombus.

SECONDARY DISORDERS *(p. 335)*

18. The kidneys can stimulate the production of RBCs, resulting in a condition called **erythrocytosis**.

19. **Erythroblastosis fetalis**, or **hemolytic disease of the newborn (HDN)**, occurs when maternal antibodies attack red blood cells in the fetus. *(Figure 14-6)*

Review Questions

MATCHING

Match each item in Column A with the most closely related item in Column B. Use letters for answers in the spaces provided.

	Column A		Column B
___	1. EPO	a.	anti-A and anti-B
___	2. hemopoiesis	b.	high number of WBCs
___	3. embolus	c.	120 days
___	4. RBC life span	d.	neutrophil
___	5. RBC antigens	e.	found in red bone marrow
___	6. plasma antibodies	f.	hormone that stimulates RBC formation
___	7. RBC stem cells	g.	monocyte
___	8. leukopenia	h.	WBC of the lymphatic system
___	9. agranulocyte	i.	lowered oxygen capacity of the blood
___	10. leukocytosis	j.	A and B
___	11. granulocyte	k.	blood cell formation
___	12. platelet	l.	drifting blood clot
___	13. lymphocyte	m.	low number of platelets
___	14. thrombocytopenia	n.	low number of WBCs
___	15. anemia	o.	cell fragment involved in clotting

MULTIPLE CHOICE

16. The formed elements of blood include _____ .
(a) albumins, globulins, fibrinogen
(b) WBCs, RBCs, platelets
(c) plasma, fibrin, serum
(d) a, b, and c are correct

17. The amount of blood in the average person is in the range of _____ liters.
(a) 1 to 2   (b) 2 to 4
(c) 4 to 6   (d) 8 to 10

18. Plasma makes up about _____ percent of whole blood.
(a) 25   (b) 35
(c) 45   (d) 55

19. Which of the following plasma proteins is *not* made by the liver? _____
(a) albumins
(b) immunoglobulins
(c) transport proteins (such as lipoproteins)
(d) fibrinogen

20. The oxygen-carrying protein within an erythrocyte is _____ .
(a) hemoglobin   (b) fibrin
(c) fibrinogen   (d) plasmin

21. All red blood cells and most white blood cells are produced in the _____ .
(a) liver   (b) thymus
(c) spleen   (d) red bone marow

22. Dehydration would cause _____ .
(a) an increase in the hematocrit
(b) a decrease in the hematocrit
(c) no effect on the hematocrit
(d) an increase in plasma volume

23. _____ is characterized by large numbers of abnormal WBCs.
(a) erythrocytosis
(b) sickle cell anemia
(c) thalassemia
(d) leukemia

24. _____ is not associated with clotting disorders.
(a) hemophilia
(b) vitamin K deficiency
(c) septicemia
(d) thrombocytopenia

TRUE/FALSE

___ 25. The average pH of blood is 7.4.

___ 26. More than three-quarters of the U.S. population is Rh-negative.

___ 27. The most common blood type in the population of the United States is Type O.

___ 28. EPO appears in the plasma when body tissues are exposed to low levels of carbon dioxide.

___ 29. The most abundant WBCs in the blood of normal, healthy individuals are neutrophils.

___ 30. Leukocytes can move through the walls of small blood vessels.

___ 31. Malaria is a disease caused by roundworms and transmitted by mosquitoes.

SHORT ESSAY

32. What are the major types of functions performed by blood?

33. How do red blood cells differ from typical cells in the body?

34. What is hemostasis?

35. Why do patients suffering from advanced kidney disease frequently become anemic (have reduced RBC count or levels of hemoglobin)?

36. Describe each of the following types of anemia:

(a) hemolytic anemia

(b) aplastic anemia

(c) hemorrhagic anemia

(d) iron deficiency anemia

37. Why are sickled red blood cells harmful? Under what conditions could they be beneficial?

APPLICATIONS

38. How would you expect the extended use of antibiotics to affect blood clotting?

39. Linda was given RhoGam during and after the delivery of her baby. Why?

✔ Answers to Concept Check Questions

(p. 321) **1.** A decrease in the amount of plasma proteins in the blood may cause (1) a decrease in the blood osmotic pressure, (2) a decreased ability to fight infection, and (3) a decrease in the transport and binding of some ions, hormones, and other molecules. **2.** A decreased blood flow to the kidneys would result in a reduced oxygen supply, which would trigger the release of erythropoietin (EPO). The EPO would lead to an increase in *erythropoiesis* (red blood cell production).

(p. 323) **1.** In an infected cut we would expect to find a large number of neutrophils. Neutrophils are phagocytic white blood cells that are usually the first to arrive at the site of an injury and are specialized to deal with bacteria. **2.** An infection by a parasitic worm or other parasites would result in an increase in the number of phagocytic white blood cells called eosinophils. **3.** Megakaryocytes form cell fragments called platelets. Platelets play an important role in stopping blood flow and in the formation of blood clots. A decrease in the number of megakaryocytes would result in fewer platelets, which in turn would interfere with the ability to clot properly.

(p. 323) **1.** Clotting factors are released by damaged blood vessels and other body tissues, and platelets that have gathered at the cut. **2.** A *thrombus* is a blood clot that has formed within a blood vessel. A blood clot (thrombus) that drifts in the blood vessels is called an *embolus*.

(p. 327) **1.** A person with Type AB blood can accept Type A, Type B, Type AB, or Type O blood. **2.** If a person with Type A blood received a transfusion of Type B blood, the donated red blood cells would clump or agglutinate, potentially blocking blood flow to various organs and tissues.

(p. 335) **1.** The thalassemias and sickle cell anemia are both inherited diseases characterized by abnormal hemoglobin. **2.** Inadequate amounts of vitamin B_{12} results in pernicious anemia. Inadequate levels of vitamin K results in a decrease in the production of clotting factors that are necessary for normal blood coagulation.

CHAPTER
15

The Heart and Circulation

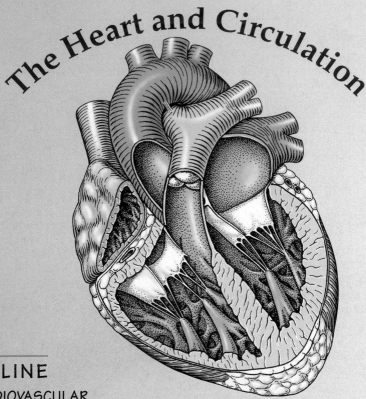

CHAPTER OUTLINE

CHAPTER OBJECTIVES

1 Describe the location and general features of
the heart.

2 Trace the flow of blood through the heart,
identifying the major blood vessels, chambers,
and heart valves.

3 Identify the layers of the heart wall.

4 Describe the events of a typical heartbeat, or
cardiac cycle.

5 Describe the components and functions of the
conducting system of the heart.

6 Describe the structure and function of arteries,
capillaries, and veins.

7 Describe how tissues and various organ systems interact to regulate blood flow and pressure in tissues.

8 Distinguish among the types of blood vessels
on the basis of their structure and function.

9 Identify the major arteries and veins and the
areas they serve.

10 Describe the age-related changes that occur in
the cardiovascular system.

11 Give examples of the major disorders of the
blood vessels.

12 Describe the major disorders of the heart and
distinguish between the various forms of
shock.

All the cells of the human body require a supply of oxygen and nutrients and a means of waste disposal. As we saw in the last chapter, blood meets both of these needs. However, blood must be kept moving. If blood remains stationary, its oxygen and nutrient supplies are quickly exhausted, its capacity to absorb wastes is soon reached, and neither hormones nor white blood cells can reach their intended targets.

All these vital activities thus depend on the pumping action of the heart. This muscular organ beats approximately 100,000 times each day, pumping roughly 8,000 l of blood—enough to fill forty 55-gal drums.

The blood pumped by the heart is carried within tubes of varying diameter (the blood vessels) to all parts of the body. Blood vessels carrying blood away from the heart, the *arteries*, gradually decrease in size; the vessels carrying blood back to the heart, the *veins*, gradually increase in size. The smallest of all blood vessels, the *capillaries*, connect the smallest arteries with the smallest veins. It is through the thin walls of the capillaries that gases, nutrients, and wastes are exchanged between the blood and the body's cells.

What makes the heart such a powerful organ? How are the blood vessels specialized for transporting blood? This chapter focuses on the cardiovascular structures involved in the pumping and transportation of blood. It also describes the major blood vessels of the body.

THE HEART AND CARDIOVASCULAR SYSTEM

Blood is pumped by the heart through a closed network of blood vessels that extend between the heart and all the tissues of the body. There are two general networks of blood vessels that make up the circulatory system. In the **pulmonary circulation** the blood completes a round-trip from the heart to the exchange surfaces of the lungs. In the **systemic circulation** the blood completes a round-trip from the heart to all portions of body *except* the exchange surfaces of the lungs. Each circuit begins and ends at the heart. **Arteries** carry blood away from the heart; **veins** return blood to the heart. **Capillaries** are small, thin-walled vessels that connect the smallest arteries to the smallest veins.

As indicated in Figure 15-1•, blood travels through the two circuits in sequence. For example, blood returning to the heart in the systemic veins must first go to the lungs and return to the heart before it can reenter the systemic arteries. The heart contains four muscular chambers. The **right atrium** (Ā-trē-um; hall) receives blood from the systemic circulation, and the **right ventricle** (VEN-tri-k'l) discharges it into the pulmonary circulation. The **left atrium** collects blood from the pulmonary circulation, and the **left ventricle** pushes it into the systemic circulation. When the heart beats, the two ventricles contract at the same time and eject equal volumes of blood. The heart, then, is really a double pump, with a ventricle and an atrium associated with each circuit.

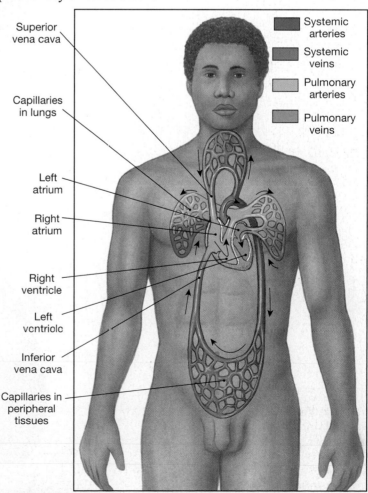

Superior vena cava

Capillaries in lungs

Left atrium

Right atrium

Right ventricle

Left ventricle

Inferior vena cava

Capillaries in peripheral tissues

Systemic arteries

Systemic veins

Pulmonary arteries

Pulmonary veins

•FIGURE 15-1
An Overview of Blood Flow.
Blood flows through separate pulmonary and systemic circulations, driven by the pumping of the heart. Each pathway begins and ends at the heart and contains arteries, capillaries, and veins.

STRUCTURE OF THE HEART

The heart is a small, cone-shaped organ roughly the size of a clenched fist. It lies near the anterior chest wall, directly behind the sternum, and sits at an angle with its blunt tip pointed downward and to the left side of the body. It is surrounded by the **pericardial** (per-i-KAR-dē-al) **cavity**, one of the three ventral body cavities introduced in Chapter 1. ∞ *[p. 13]* This cavity is lined by a serous membrane called the **pericardium**.

The pericardium is subdivided into a **visceral pericardium**, or **epicardium**, that covers the outer surface of the heart, and a **parietal pericardium** that lines the opposing inner surface of the fibrous sac surrounding the heart. The space be-

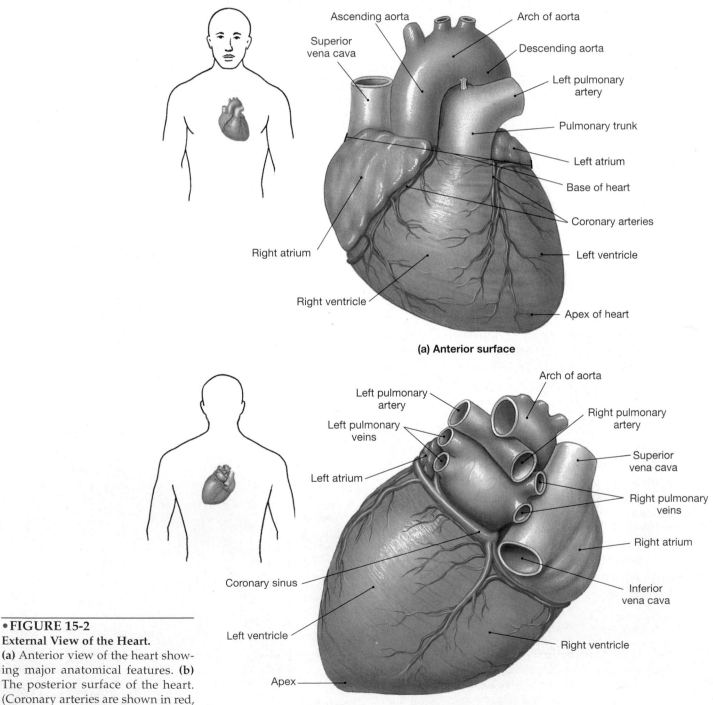

(a) Anterior surface

(b) Posterior surface

•**FIGURE 15-2**
External View of the Heart.
(a) Anterior view of the heart showing major anatomical features. **(b)** The posterior surface of the heart. (Coronary arteries are shown in red, coronary veins in blue.)

tween these surfaces contains a small amount of pericardial fluid that acts as a lubricant, reducing friction as the heart beats.

The great veins and arteries of the circulatory system are connected to the heart at the **base** (see Figure 15-2•). The pointed tip formed by the two ventricles is the **apex** (Ā-peks) of the heart. A typical heart measures approximately 12.5 cm (5 in.) from the attached base to the apex.

INTERNAL ANATOMY OF THE HEART

Figure 15-3• shows the internal appearance and structure of the four chambers of the heart. Muscular walls, or *septa*, separate the right atrium from the left atrium, and the right ventricle from the left ventricle. These dividing walls are called the **interatrial septum** and **interventricular septum**. The septa ensure that no mixing of blood occurs between the left and right chambers of the heart. On each side of the heart, blood flows from the atrium into the ventricle through an **atrioventricular (AV) valve**. The structure of this valve ensures that blood flows in one direction only, from the atrium into the ventricle.

Note that Figure 15-3• also shows the path of blood flow through both sides of the heart. The right atrium receives blood from the systemic circulation through two large veins, the **superior vena cava** (VĒ-na KĀ-va) and the **inferior vena cava**. The superior vena cava delivers blood from the head, neck, upper limbs, and chest. The inferior vena cava carries blood returning from the rest of the trunk, organs in the abdominopelvic cavity, and the lower limbs.

inter, between + *atrium*, room; or *ventriculus*, a little belly
septum, a wall
interatrial septum, interventricular septum: muscular walls that separate the two atria and the two ventricles, respectively

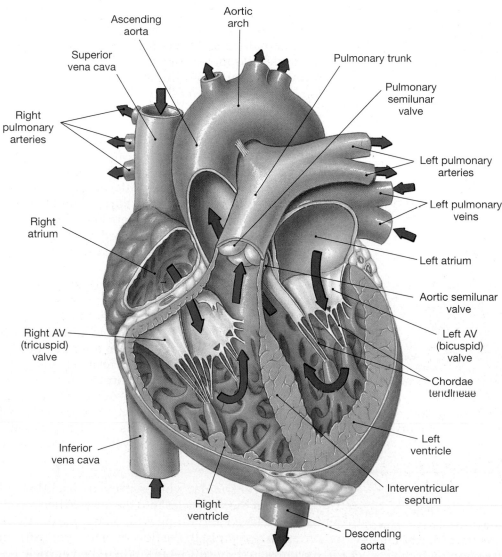

•FIGURE 15-3
Internal View of the Heart.
A diagrammatic frontal section through the heart showing major landmarks and the path of blood flow through the atria and ventricles.

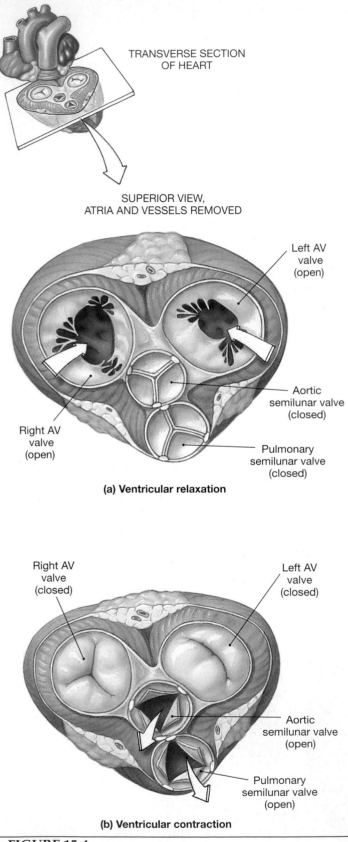

TRANSVERSE SECTION
OF HEART

SUPERIOR VIEW,
ATRIA AND VESSELS REMOVED

Left AV
valve
(open)

Right AV
valve
(open)

Aortic
semilunar valve
(closed)

Pulmonary
semilunar valve
(closed)

(a) Ventricular relaxation

Right AV
valve
(closed)

Left AV
valve
(closed)

Aortic
semilunar valve
(open)

Pulmonary
semilunar valve
(open)

(b) Ventricular contraction

• FIGURE 15-4
Valves of the Heart.
(a) Valve position during ventricular relaxation, when the AV valves are open and the semilunar valves are closed. **(b)** The appearance of the cardiac valves during ventricular contraction, when the AV valves are closed and the semilunar valves are open.

Blood travels from the right atrium into the right ventricle through a broad opening bounded by three flaps of fibrous tissue. These flaps, called *cusps*, are part of the **right atrioventricular (AV) valve**, also known as the **tricuspid** (trī-KUS-pid; *tri*, three) **valve**. Each cusp is braced by strong fibers, the *chordae tendineae* (KOR-dē TEN-di-nē-ē), that are connected to small muscles that project from the inner surface of the right ventricle.

Blood leaving the right ventricle flows into the large **pulmonary trunk** through the **pulmonary semilunar** (half-moon) **valve**. Once within the pulmonary trunk, blood flows into the **left** and **right pulmonary arteries**. These vessels branch repeatedly within the lungs, supplying the capillaries where gas exchange occurs. As blood leaves these respiratory capillaries it then passes through the **left** and **right pulmonary veins** before entering the left atrium.

As in the right atrium, the opening between the left atrium and the left ventricle has a valve—the **left atrioventricular (AV) valve**, or **bicuspid** (bī-KUS-pid; *bi*, two) **valve**—which prevents backflow from the ventricle into the atrium. As the name *bicuspid* implies, the left AV valve contains a pair of cusps rather than three. Because its shape resembles a bishop's hat, or *mitre*, this valve is also known as the **mitral** (MĪ-tral) **valve.**

Internally, the left ventricle resembles the right ventricle. Blood leaving the left ventricle passes through the **aortic semilunar valve** and into the systemic circuit by way of the *ascending aorta*.

The Ventricles

The close relationship between structure and function are clearly evident in the heart. For example, the function of an atrium is to collect blood returning to the heart and deliver it to the attached ventricle. Because the workloads placed on the right and left atria are very similar, the two chambers look almost identical. However, the demands placed on the right and left ventricles are very different, and there are structural differences between the two.

The lungs are close to the heart, and the pulmonary arteries and veins are relatively short and wide. As a result, the right ventricle normally does not need to push very hard to propel blood through the pulmonary circulation. As you can see in Figure 15-3•, the wall of the right ventricle is relatively thin compared with that of the left ventricle. Because the systemic circulation has a more extensive network of blood vessels, the left ventricle must exert much more force than the right ventricle. As might be expected, the left ventricle has an extremely thick muscular wall.

The Heart Valves

The atrioventricular valves prevent backflow of blood from the ventricles into the atria. When a ventricle is relaxed and filling with blood, the AV valve offers no resistance to the flow of blood from atrium to ventricle (Figure 15-4a•). When the ventricle begins to contract, blood moving back toward the atrium swings the cusps together, closing the valve (Figure 15-4b•). The chordae tendineae attached to the cusps prevent their ballooning back into the atrium and allowing the backflow of blood into the atrium when the ventricle contracts. However, a small amount of backflow often occurs, even in normal individuals. The swirling action creates a soft but distinctive sound, called a *heart murmur*.

The pulmonary and aortic semilunar valves prevent backflow of blood from the pulmonary trunk and aorta into the right and left ventricles. When these valves close, the three symmetrical cusps support one another like the legs of a tripod (Figure 15-4a•).

THE HEART WALL

The wall of the heart is made up of three distinct layers (Figure 15-5•). The **epicardium**, which covers the outer surface of the heart, is a serous membrane that consists of an exposed epithelium and an underlying layer of loose connective tissue. The **myocardium**, or muscular wall of the heart, contains cardiac muscle tissue and associated connective tissues, blood vessels, and nerves. The inner surfaces of the heart, including the valves, are covered by the **endocardium** (en-dō-KAR-dē-um), whose squamous epithelium is continuous with the epithelium, or *endothelium*, that lines the attached blood vessels.

Cardiac Muscle Cells

Cardiac muscle cells within the myocardium interconnect at specialized sites known as **intercalated** (in-TER-ka-lā-ted) **discs**. Specialized connections and tiny

•**FIGURE 15-5**
Layers of the Heart Wall.
A diagram of a section through the heart wall showing the epicardium, myocardium, and endocardium. The cardiac muscle cells of the myocardium branch and interconnect at sites called intercalated discs.

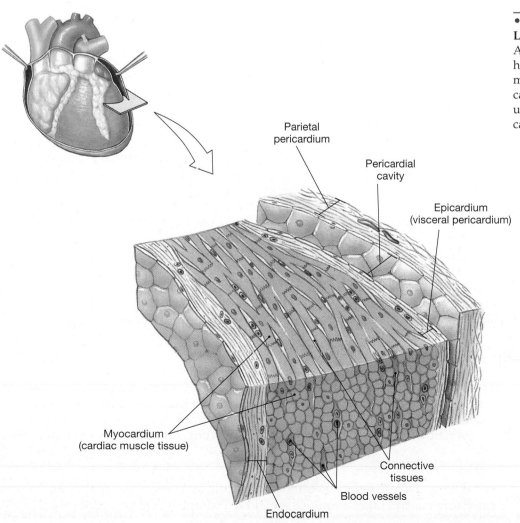

Parietal pericardium

Pericardial cavity

Epicardium (visceral pericardium)

Myocardium (cardiac muscle tissue)

Connective tissues

Blood vessels

Endocardium

Clinical Note

Coronary Circulation and Heart Attacks

Cardiac muscle cells need a constant supply of oxygen and nutrients, and any reduction in coronary circulation reduces the heart's performance. A reduction in the coronary blood supply, a process called a **coronary ischemia** (is-KĒ-mē-a), usually results from a partial or complete blockage of one or more coronary arteries. The usual cause is the formation of a fatty deposit, or *plaque*, in the wall of a coronary blood vessel.

A general term for these degenerative changes in the coronary circulation is **coronary artery disease (CAD)**. One of the first symptoms of CAD is **angina pectoris** (an-JĪ-na PEK-tō-ris) which means "a pain spasm in the chest." In the most common form of angina, pain radiates from the sternal area to the arms, back, and neck when the individual exerts himself or is under emotional stress. Angina develops with exertion because the partially blocked coronary vessels cannot meet the increased demand for oxygen and nutrients.

In a heart attack, or **myocardial** (mī-ō-KAR-dē-al) **infarction (MI)**, a portion of the coronary circulation becomes blocked and the cardiac muscle cells die from lack of oxygen. The affected tissue then degenerates, creating a nonfunctional area known as an *infarct*. Heart attacks most often result from severe coronary heart disease. The consequences of a heart attack depend on the site and duration of the circulatory blockage.

There are roughly 1.3 million MIs in the United States each year, and half of the victims die within a year of the incident. Factors that increase the risk of a heart attack include smoking, high blood pressure, high blood cholesterol levels, diabetes, male gender (below 70 years old), severe emotional stress, and obesity. Eliminating as many risk factors as possible will improve one's chances of preventing or surviving a heart attack.

---

pores at these sites strengthen and stabilize the cells and provide for the movement of ions, small molecules, and electrical impulses. In addition, the myofibrils of adjacent cells are attached to the disc. This increases the efficiency of the cells, as they can "pull together" whenever the heart contracts.

BLOOD SUPPLY TO THE HEART

The heart works continuously, and cardiac muscle cells require reliable supplies of oxygen and nutrients. During maximum exertion the oxygen demand rises considerably, and the blood flow to the heart may increase to nine times that of resting levels. Although the heart is filled with blood, the surrounding heart muscle relies on its own network of blood vessels. The vessels that supply blood to cells of the heart form the **coronary circulation**.

The coronary circulation begins with the left and right **coronary arteries**, which originate at the base of the ascending aorta (see Figure 15-2a•). Blood pressure here is the highest found anywhere in the systemic circulation, and this pressure ensures a continuous flow of blood to meet the demands of active cardiac muscle tissue. Each coronary artery splits in two, and smaller branches form a maze of interconnections called **anastomoses** (a-nas-to-MŌ-sēs). Because the arteries are interconnected in this way, the blood supply to the cardiac muscle remains relatively constant, regardless of pressure changes within the left and right coronary arteries. Two main **cardiac veins** carry blood away from the coronary capillaries. These veins drain into the **coronary sinus**, a large, thin-walled vein (see Figure 15-2b•). Blood from the coronary sinus flows into the right atrium near the base of the inferior vena cava.

✔ Damage to the semilunar valves on the right side of the heart would interfere with blood flow to what vessel?

✔ What prevents the AV valves from opening back into the atria?

✔ Why is the left ventricle more muscular than the right ventricle?

THE HEARTBEAT

The average heart beats about 100,000 times per day, or about 70 to 80 beats per minute. During 1 minute, the heart can pump 4 to 6 L (about 1 to 1.5 gal) of blood, that is, the entire volume of blood in the body. During strenuous activity, some five to seven times this volume of blood could be pumped in the same amount of time.

CARDIAC CYCLE

What happens during one heartbeat? One complete heartbeat makes up a single **cardiac cycle** and takes about 0.8 seconds to complete. During each cardiac cycle any one chamber undergoes alternating periods of contraction and relaxation. During contraction, or **systole** (SIS-to-lē), the chamber pushes blood into an adjacent chamber or into an arterial trunk (see Figure 15-6•). Systole is followed by the second phase, one of relaxation, or **diastole** (dī-AS-to-lē), during which the chamber fills with blood and prepares for the next heartbeat.

During the cardiac cycle, the pressure within each chamber rises in systole and falls in diastole. Because fluids will move from an area of high pressure to one of lower pressure, an increase in pressure in one chamber will cause the blood to flow to another chamber of lesser pressure. One-way valves between adjacent chambers help to ensure that blood flows in the desired direction.

The amount of blood ejected by a ventricle during a single heartbeat is known as the **stroke volume**. The amount of blood pumped by a ventricle in 1 minute is called the **cardiac output**. The cardiac output of an adult is about equal to the total volume of blood in the body.

Heart Sounds

When you listen to your own heart with a stethoscope, you hear the familiar "lubb-dupp" that accompanies each heartbeat. These sounds accompany the action of the heart valves. The first heart sound ("lubb") marks the start of ventricular systole, and the sound is produced as the AV valves close and the semilunar valves open. The second heart sound, "dupp," occurs at the beginning of ventricular diastole, when the semilunar valves close.

CONDUCTING SYSTEM OF THE HEART

The correct pressure relationships that occur in a normal heartbeat depend on the careful timing of contractions. In the normal pattern of blood flow, the atria contract first and then the ventricles. Each time the heart beats, the contractions of individual cardiac muscle cells are coordinated and harnessed to ensure that blood flows in the right direction at the proper time. Unlike skeletal muscle fibers, cardiac muscle cells can contract on their own without stimulation from hormones or nerves. This is a property called *automaticity*, or *autorhythmicity*. Normally, how-

---

•FIGURE 15-6
The Cardiac Cycle.
The atria and ventricles go through repeated cycles of systole and diastole. A cardiac cycle is made up of one period of systole and diastole.

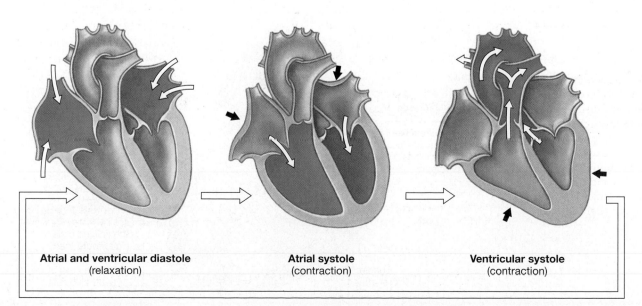

Atrial and ventricular diastole (relaxation)   **Atrial systole** (contraction)   **Ventricular systole** (contraction)

ever, cardiac muscle cells do not contract independently of one another; this would disrupt the cardiac cycle. Instead, the cardiac muscle cells wait for the arrival of a signal generated and distributed by smaller, specialized muscle cells that make up the *conducting system* of the heart. The conducting system includes **nodal cells** that establish the heartbeat rate and **conducting cells** that distribute the stimulus for contraction throughout the myocardium.

Nodal Cells

Nodal cells are unusual because they can generate electrical impulses, or *action potentials*, at regular intervals. ∞ *[p. 223]* Nodal cells are electrically connected to one another and to conducting cells, which are in turn connected to normal cardiac muscle cells. As a result, when an electrical impulse appears in a nodal cell, it sweeps through the conducting system, reaching all the cardiac muscle cells in less than 0.25 second. When the myocardial cells receive this stimulus, they contract. In this way nodal cells determine the heart rate.

There are two different groups of nodal cells, or **pacemaker cells**. The main pacemaker cells are found in the **cardiac pacemaker**, or **sinoatrial** (sī-nō-Ā-trē-al) **node (SA node)**, which is embedded in the wall of the right atrium near the entrance of the superior vena cava (Figure 15-7●). These pacemaker cells spontaneously generate 70 to 80 electrical impulses per minute.

The other group of nodal cells is located at the junction between the atria and the ventricles. They make up the **atrioventricular** (ā-trē-ō-ven-TRIK-ū-lar) **node (AV node)**. These cells are electrically connected to the SA nodal cells through conducting cells in the atrial walls. Isolated AV nodal cells generate only 40 to 60 electrical impulses per minute. In the normal heart, these cells are stimulated by impulses from the SA node before they can produce an action potential on their own. If the SA nodal cells are unable to function, then the cells of the AV node will become the new pacemakers of the heart, and the resting heart rate becomes unusually low.

Conducting Cells

The stimulus for a contraction must be distributed so that (1) the atria contract together, before the ventricles, and (2) the ventricles contract together, in a wave that begins at the apex and spreads toward the base. When the ventricles contract in this way, blood is pushed toward the base of the heart, into the aortic and pulmonary trunks.

The conducting network of the heart is shown in Figure 15-7●. The cells of the SA node are electrically connected to those of the AV node by conducting cells. After electrical impulses reach the AV node from the SA node, they travel to the

•FIGURE 15-7
The Conducting System of the Heart.
The stimulus for contraction is generated by pacemaker cells at the SA node. From there, impulses follow three different paths through the atrial walls to reach the AV node. After a brief delay, the impulses are conducted to the AV bundle (bundle of His), and then on to the left and right bundle branches, the Purkinje cells, and the ventricular myocardial cells.

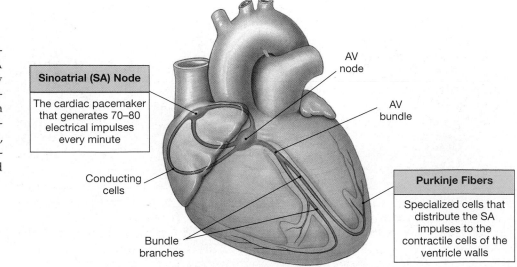

Sinoatrial (SA) Node

The cardiac pacemaker that generates 70–80 electrical impulses every minute

Conducting cells

Bundle branches

AV node

AV bundle

Purkinje Fibers

Specialized cells that distribute the SA impulses to the contractile cells of the ventricle walls

AV bundle, also known as the *bundle of His* (hiss). This bundle of conducting cells travels along the interventricular septum before dividing into **left** and **right bundle branches** that radiate across the inner surfaces of the left and right ventricles. At this point specialized **Purkinje** (pur-KIN-jē) **cells** (*Purkinje fibers*) carry the impulses to the contractile cells of the ventricles. This entire process takes place in slightly less than one-quarter of a second!

Normal pacemaker activity results in an average heart rate of 70 to 75 beats per minute (bpm). A number of clinical problems result from abnormal pacemaker activity. **Bradycardia** (brā-di-KAR-dē-a) means "slow heart" and is the term used to indicate a resting heart rate that is slower than normal (less than 60 bpm), whereas **tachycardia** (tak-i-KAR-dē-a), meaning "swift heart," indicates a faster than normal resting heart rate (100 or more bpm).

THE ELECTROCARDIOGRAM

The electrical impulses occurring in the heart can be detected by electrodes attached to the body surface. A recording of these electrical activities constitutes an **electrocardiogram** (ē-lek-trō-KAR-dē-ō-gram), also called an **ECG** or **EKG** (Figure 15-8●). Each time the heart beats, an electrical impulse radiates through the atria, reaches the AV node, travels down the interventricular septum, turns at the apex, and spreads through the muscle tissue of the ventricles toward the base.

Figure 15-8● shows the important features of a standard electrocardiogram. The small **P wave** accompanies the electrical impulse generated by the SA node. The atria begin contracting shortly after the start of the P wave. The **QRS complex** appears as the electrical impulse passes into the walls of the ventricles. This is a relatively strong electrical signal because the amount of ventricular muscle is much larger than that of the atria. The ventricles begin contracting shortly after the peak of the R wave. The smaller **T wave** indicates the return of the unstimulated condition in the ventricles as they prepare for another contraction. By this time, the atria have recovered and are preparing for their next contraction.

ECG analysis is especially useful in detecting and diagnosing **cardiac arrhythmias** (a-RITH-mē-as), abnormal patterns of cardiac activity. Momentary arrhythmias are not usually dangerous, and about 5 percent of the normal population experience a few abnormal heartbeats each day. Clinical problems appear when the arrhythmias reduce the pumping efficiency of the heart.

CONTROL OF HEART RATE

The amount of blood ejected by the ventricles (the cardiac output) is precisely regulated so that all the tissues of the body receive an adequate blood supply under a variety of conditions. Although the basic resting heart rate is established by the pacemaker cells of the SA node, this rate can be modified by the autonomic nervous system (ANS). Both the sympathetic and parasympathetic divisions of the ANS have direct connections at the SA and AV nodes and with the myocardium.

Neurotransmitters released by the sympathetic division cause an increase in the heart rate and in the force and degree of cardiac contraction. In contrast, neurotransmitter released by parasympathetic neurons (under control of the vagus nerves) slows the heart rate and decreases the force of cardiac contractions.

Both autonomic divisions are normally active at a steady background level, releasing neurotransmitters both at the nodes and into the myocardium. As a result, cutting the vagus nerves increases the heart rate, and administering drugs that prevent sympathetic stimulation slows the heart rate.

Other factors that can alter the heart rate include ions, hormones, and body temperature. For example, a lower than normal body temperature slows the production of electrical impulses at the SA node, reduces the strength of cardiac contractions, and lowers the heart rate. An elevated body temperature increases the heart rate and strength of cardiac contractions. This is one reason why your heart seems to be racing and pounding when you have a fever.

✔ Why is it important for the electrical impulses from the atria to be delayed at the AV node before passing into the ventricles?

✔ How does bradycardia differ from a normal heart rate?

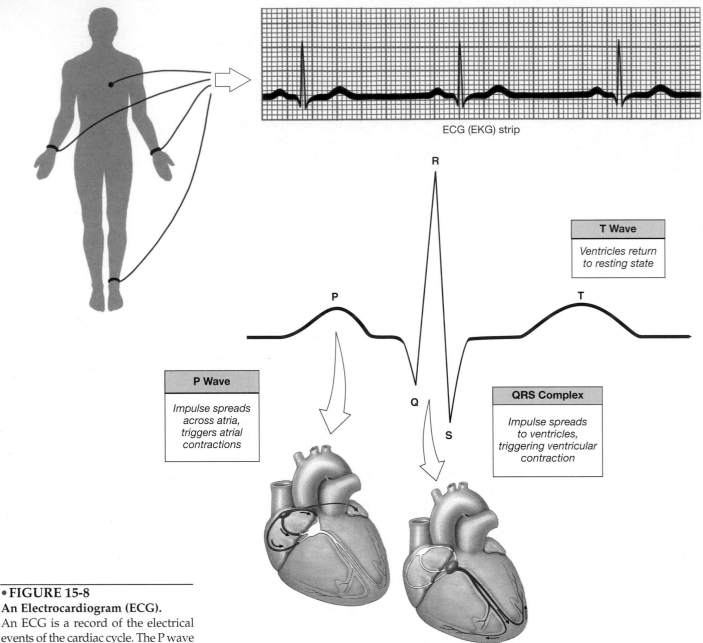

ECG (EKG) strip

R

T Wave

Ventricles return to resting state

P   **T**

P Wave

Impulse spreads across atria, triggers atrial contractions

Q

S

QRS Complex

Impulse spreads to ventricles, triggering ventricular contraction

•FIGURE 15-8
An Electrocardiogram (ECG).
An ECG is a record of the electrical events of the cardiac cycle. The P wave represents the signal from the SA node and the contraction of the atria. The QRS complex represents the impulse from the AV node and the contraction of the ventricles. During the T wave the muscles of the atria and ventricles are recovering and preparing for another set of contractions.

BLOOD VESSELS

Propelled by the heart, blood flows to and from the lungs and other body organs through tube-like arteries and veins. The large-diameter **arteries** that carry blood away from the heart branch repeatedly and gradually decrease in size until they become **arterioles** (ar-TĒ-rē-ōlz), the smallest vessels of the arterial system. From the arterioles blood enters the capillary networks that serve local tissues.

Barely larger in diameter than a single red blood cell, capillaries are the smallest blood vessels. It is at the capillaries that the vital functions of the cardiovascular system take place. All the exchanges of chemicals and gases between the blood and interstitial fluid take place across capillary walls.

Blood flowing out of the capillaries first enters the **venules** (VEN-ūlz), the smallest vessels of the venous system. These slender vessels subsequently merge with their neighbors to form small **veins**. Blood then passes through medium-sized and large veins before reaching the heart.

VESSEL STRUCTURE

The walls of arteries and veins contain three distinct layers, or *tunics*. The innermost layer of a blood vessel is made up of a squamous epithelium called an *endothelium*. The middle layer contains smooth muscle tissue in a framework of collagen and elastic fibers. When these smooth muscles contract, the vessel decreases in diameter (an action known as *vasoconstriction*), and when they relax, the diameter increases (an action termed *vasodilation*). The outer layer consists of a stabilizing sheath of connective tissue.

The multiple layers in their vessel walls give arteries and veins considerable strength, and the muscular and elastic components permit controlled alterations in diameter as blood pressure or blood volume changes.

Figure 15-9• shows the differences in structure of an artery, capillary, and vein. Arteries have the thickest walls and contain more smooth muscle and elastic fibers than veins. Note that the walls of capillaries are only one cell thick.

Arteries

The largest arteries are extremely resilient, elastic vessels with diameters of up to 2.5 cm (1 in.). Smaller arteries are less elastic and contain more smooth muscle tissue. The smallest arteries, the **arterioles**, have an average diameter of about 30 µm (0.03 mm). The smooth muscle cells help these vessels change their diameter. Such changes are important in altering the blood pressure and changing the rate of flow through body tissues.

Capillaries

Capillaries are the only blood vessels whose walls are thin enough to permit exchanges between the blood and the surrounding interstitial fluids. A typical capillary consists of a single layer of endothelial cells and an average diameter that is a mere 8 µm (0.008 mm), very close to that of a single red blood cell.

Capillaries work together as part of an interconnected network called a **capillary bed**. Upon reaching its target area, a single arteriole usually gives rise to a capillary bed made up of dozens of capillaries that, in turn, merge to form several **venules**, the smallest vessels of the venous system.

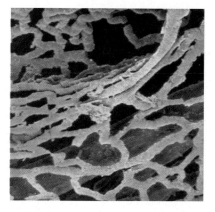

Scanning electron micrograph of a network of capillaries.

Veins

The walls of veins are thinner than those of arteries. Veins have relatively thin walls because they do not have to withstand much pressure. In the limbs, medium-sized veins contain **valves** that act like the valves in the heart, preventing the backflow of blood (see Figure 15-9b•).

BLOOD PRESSURE

The tissues and organs of the cardiovascular system (blood, heart, and blood vessels) work together to maintain an adequate flow of blood to the body's tissues. Adequate blood flow requires enough pressure to push the blood through smaller and smaller arteries, along the capillaries, and back to the heart through the venous system. Generally, blood flow is equal to cardiac output. The higher the cardiac output, the higher the blood flow and pressure, and the lower the cardiac output, the lower the blood flow and pressure.

Two factors affect blood flow: *pressure* and *resistance*. **Blood pressure** is the pressure, or force, that blood exerts against the walls of the blood vessels. Resistance to blood flow primarily results from the length and diameter of blood vessels and the viscosity, or "thickness," of blood. Blood pressure is highest in the larger arteries because they are closest to the heart. Resistance, however, is greatest within the smallest diameter vessels (arterioles and capillaries) because of the friction of blood rubbing against their walls. Because blood is some five times "thicker" than water, it is also harder to push through the blood vessels.

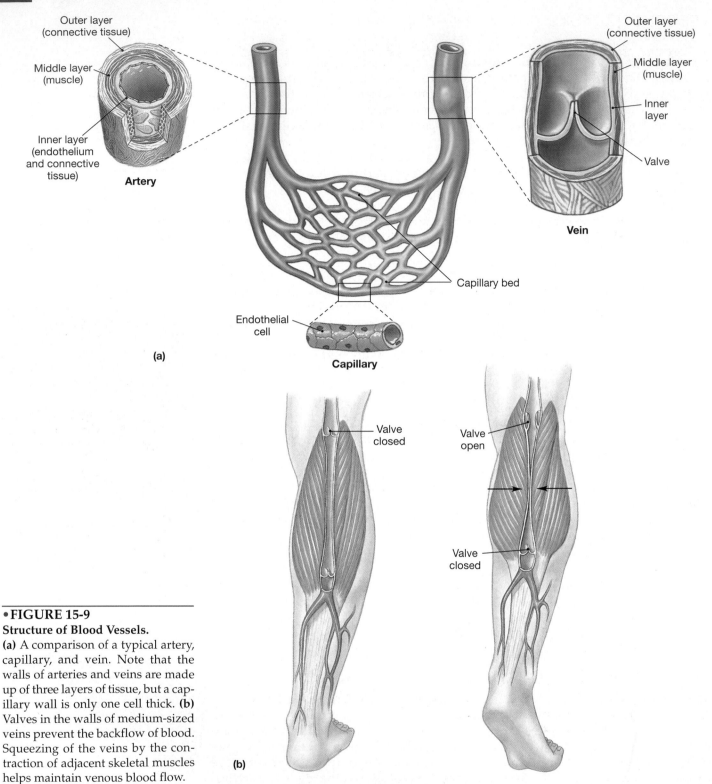

•FIGURE 15-9
Structure of Blood Vessels.
(a) A comparison of a typical artery, capillary, and vein. Note that the walls of arteries and veins are made up of three layers of tissue, but a capillary wall is only one cell thick. **(b)** Valves in the walls of medium-sized veins prevent the backflow of blood. Squeezing of the veins by the contraction of adjacent skeletal muscles helps maintain venous blood flow.

Blood pressure rises and falls in the larger arteries in response to the pumping activities of the heart. Blood pressure is measured in units of millimeters of mercury (mm Hg), a standardized unit of pressure. Blood pressure in the systemic circulation reaches a maximum "normal" level of 120 mm Hg in the largest arteries during ventricular systole. During ventricular diastole, blood pressure falls to about 80 mm Hg. Figure 15-10• shows the basic steps involved in measuring blood pressure.

The difference between the systolic and diastolic pressures is the **pulse pressure.** The pulse pressure becomes smaller as the distance from the heart increases.

The usual procedure for feeling the pulse involves squeezing an artery with the fingertips against a relatively solid mass, preferably a bone. When the vessel is compressed, the pulse is felt as a pressure against the fingertips. The inside of the wrist is often used because the *radial artery* can easily be pressed against the distal portion of the radius. The pulse is a measure of the heart rate and usually ranges between 70 and 75 beats per minute at rest.

✝ Clinical Note

How Is Blood Pressure Measured?

Blood pressure is determined with a *sphygmomanometer* (sfig-mō-ma-NOM-e-ter). Derived from *sphygmos*, meaning "pulse," and *manometer*, "a device for measuring pressure," this instrument is used to measure *arterial pressure* (the blood pressure in arteries). As shown in Figure 15-10 •, an inflatable cuff is placed around the arm in such a position that its inflation squeezes the brachial artery. A stethoscope is placed over the artery below the cuff, and the cuff is then inflated. A tube connects the cuff to a glass chamber containing liquid mercury, and as the pressure in the cuff rises, it pushes the mercury up into a vertical column. A scale along the column permits one to determine the cuff pressure in millimeters of mercury. Inflation continues until cuff pressure is roughly 30 mm Hg above the pressure sufficient to completely collapse the brachial artery, stop the flow of blood, and eliminate the sound of the pulse.

The investigator then slowly lets the air out of the cuff with the releasing valve. When the pressure in the cuff falls below systolic pressure, blood can again enter the artery. At first, blood enters only at peak systolic pressures, and the stethoscope picks up the sound of blood pulsing through the artery. As the pressure falls further, the sound changes because the vessel is remaining open for longer and longer periods. When the cuff pressure falls below diastolic pressure, blood flow becomes continuous and the sound of the pulse becomes muffled or disappears completely. Thus the pressure at which the pulse appears corresponds to the peak systolic pressure; when the pulse fades the pressure has reached diastolic levels.

When the blood pressure is recorded, systolic and diastolic pressures are usually separated by a slash mark, as in "120/80" ("one-twenty over eighty") or "110/75." These values are considered normal for young adults. High blood pressure, or *hypertension*, occurs when the systolic pressure is over 140, and the diastolic pressure is over 95 (140/95).

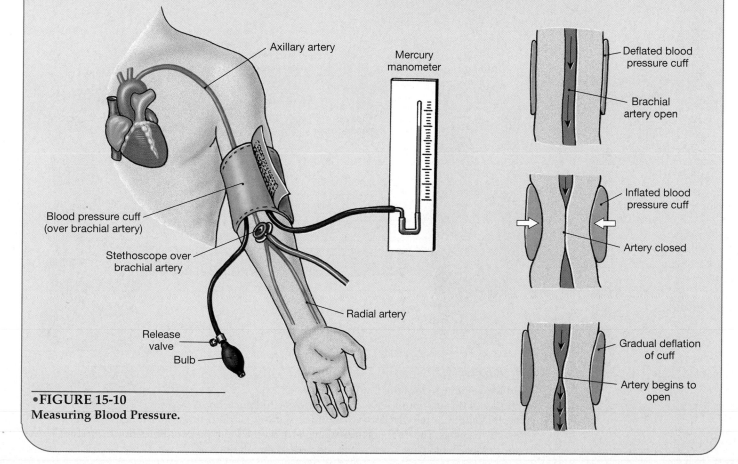

•**FIGURE 15-10**
Measuring Blood Pressure.

Capillaries and Fluid Movement

As blood passes through the capillaries, blood pressure drops even further, and the blood flow slows down. Unlike other blood vessels, however, capillaries are permeable to water, ions, and other small solutes (dissolved materials). Diffusion, blood pressure, and osmosis are the primary important forces that cause exchange across the walls of capillaries. As you will recall from Chapter 3, diffusion is the passive movement of dissolved materials from an area of relatively high concentration to an area of relatively low concentration. ∞ *[p. 43]* Diffusion across capillary walls tends to move oxygen and nutrients into the surrounding tissues, whereas carbon dioxide and waste products tend to diffuse from the tissues into the bloodstream.

Blood pressure tends to push water and dissolved substances out of the bloodstream and into the tissues. Osmosis refers to the movement of water between two solutions separated by a selectively permeable membrane, which in this case is the capillary wall. Osmosis always involves water movement into the solution that contains a higher concentration of solutes. ∞ *[p. 44]* As we noted earlier in the chapter, plasma proteins are the primary contributors to the osmotic concentration of the blood. The balance between blood pressure and osmosis tends to shift along a capillary from its start at an arteriole to its end at a venule (Figure 15-11•).

At the start of a capillary, blood pressure is relatively high. This tends to force water and other small dissolved materials out of the bloodstream and into the sur-

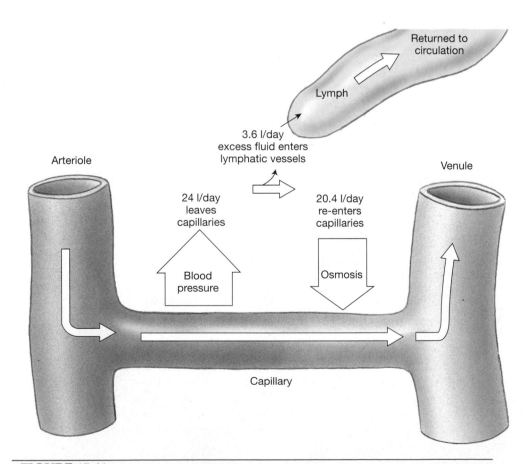

•**FIGURE 15-11**
Fluid Movement Across the Capillaries.
Fluid passes through capillary walls because of differences in osmosis and blood pressure along the length of the capillary. Most of the outflow of fluid from the capillaries is reabsorbed by osmosis. The fluid not immediately reabsorbed into the capillaries flows into lymphatic vessels, which then return it to the venous circulation.

Clinical Note

What Do Red Lips, Dark Circles Under the Eyes, and Blushing Have in Common?

Blood vessels! All these different features are due to blood vessels that lie near the surface of the skin.

For example, lips are reddish in color because of the color of their underlying capillaries. The color is visible through this part of the skin because the lips are covered only by a thin, almost transparent layer of epidermis.

Dark circles under the eyes are due to tiny blood vessels located in those areas. These vessels normally collect blood from the head. When an individual is ill or tired, these vessels be-

come wider, or dilated, and hold larger amounts of blood than normal. As with the lips, the skin in this region is also thin, and so the blood-filled vessels are more easily seen.

Blushing is a brief and sudden reddening of the face and sometimes the neck. It occurs when blood vessels close to the surface of the skin dilate, or widen, in an involuntary response to embarrassment. Blushing also occurs in some women during menopause in response to changes in the levels of sex hormones.

rounding tissue fluid. When this movement occurs, it increases the solute concentration in the blood, because plasma proteins cannot cross the capillary walls. At first, osmosis cannot eliminate this concentration gradient, because the blood pressure is too high—it is pushing water out of the bloodstream faster than osmosis can pull it back in. However, as blood flows along the capillary, the blood pressure gradually decreases. As the blood nears the end of the capillary, osmosis draws much of the released water and solutes back into the bloodstream. By the time the blood enters the venule, all but about one sixth of the total volume of fluid pushed into the tissues along the capillary has been reabsorbed. In the course of a typical day, 24 l of fluid are forced out of the capillaries into the interstitial fluids, and roughly 21.4 l are reabsorbed. The rest, 3.6 l, mixes with the interstitial fluid, flows through the tissues, and then enters the *lymphatic vessels*. These vessels, part of the lymphatic system, return 3.6 l of fluid, now called *lymph*, to the venous system each day.

REGULATION OF BLOOD FLOW

Body tissues require a constant blood flow to meet their demands for oxygen and nutrients. To meet these needs, homeostatic adjustments may occur in any of the three main variables that affect tissue blood flow: cardiac output, vessel resistance, and blood pressure.

Generally, short-term responses adjust cardiac output and blood vessel resistance to stabilize blood pressure and tissue blood flow. Long-term adjustments mainly involve changes in blood volume.

SHORT-TERM RESPONSES

Tissues can indirectly regulate their own blood flow through capillary networks. The smooth muscles of the arterioles that open into such networks contract and relax in response to varying levels of oxygen and carbon dioxide. When oxygen is abundant in a tissue, the smooth muscles contract and reduce blood flow. In contrast, when carbon dioxide concentration is high, the smooth muscles relax and increase blood flow. Such regulation by local factors provides a rapid response to immediate tissue needs.

The nervous system also provides a rapid response to maintain adequate blood flow. It does this through adjustments of cardiac output and blood vessel resistance. Changes in blood pressure and the levels of blood gases at specific sensory receptors stimulate these responses of the nervous system.

The medulla oblongata of the brain stem contains two important cardiovascular centers: the *cardiac* and *vasomotor centers*. The cardiac centers control cardiac output, and the vasomotor centers control blood vessel resistance. The cardiac centers include a *cardioacceleratory center* and a *cardioinhibitory center*. Sympathetic stimulation of the heart by the cardioacceleratory center increases cardiac output, and parasympathetic stimulation of the heart by the cardioinhibitory center decreases cardiac output. ∞ *[p. 232]*

The vasomotor center primarily controls the diameters of the arterioles. Inhibition of the vasomotor center leads to **vasodilation**, a dilation (widening) of arterioles that reduces vessel resistance. Stimulation of the vasomotor center causes **vasoconstriction**, a narrowing of the arterioles, which increases vessel resistance.

LONG-TERM RESPONSES

Hormones from the endocrine system are associated with both short-term and long-term responses. Epinephrine (E) and norepinephrine (NE) from the adrenal medullae act within seconds of their release to stimulate cardiac output and trigger the vasoconstriction of blood vessels.

Four other hormones—antidiuretic hormone (ADH), angiotensin II, erythropoietin (EPO), and atrial natriuretic peptide (ANP)—are concerned primarily with the long-term regulation of blood volume and blood pressure. ∞ *[pp. 296, 301]* All but ANP are stimulated by decreased blood pressure.

Angiotensin II appears in the blood following the release of renin by specialized kidney cells. Angiotensin II causes an extremely powerful vasoconstriction that elevates blood pressure almost at once. It also stimulates the secretion of ADH by the pituitary, and aldosterone by the adrenal cortex. These two hormones help increase blood pressure by increasing the volume of blood. ADH stimulates water conservation at the kidneys, and promotes thirst. Aldosterone stimulates the reabsorption of sodium ions and water at the kidneys, reducing their loss in the urine.

EPO is released at the kidneys in response to low blood pressure or low oxygen content of the blood. This hormone stimulates red blood cell production. Having more red blood cells increases the blood volume and improves the oxygen-carrying capacity of the blood.

ANP is produced by specialized cardiac muscle cells in the atrial walls when they are stretched by excessive amounts of venous blood. Unlike the three previous hormones, ANP reduces blood volume and blood pressure. It acts by increasing water losses at the kidneys, reducing thirst, and stimulating vasodilation. As blood volume and blood pressure decline, the stress on the atrial walls is removed, and ANP production decreases.

BLOOD VESSELS OF THE BODY

The circulatory system is divided into the pulmonary and systemic circulations. The pulmonary circulation, which transports blood between the heart and the lungs, begins at the right ventricle and ends at the left atrium. From the left ventricle, the arteries of the systemic circulation transport oxygenated blood to all the organs and tissues of the body except the lungs, ultimately returning deoxygenated blood to the right atrium. Figure 15-12• summarizes the primary circulatory routes within the pulmonary and systemic circulations.

PULMONARY CIRCULATION

Blood entering the right atrium is returning from a trip through capillary networks where oxygen was released and carbon dioxide absorbed. After passing through the right atrium and ventricle, blood enters the **pulmonary trunk**, the start of the pulmonary circulation. Along the pulmonary circulation a fresh sup-

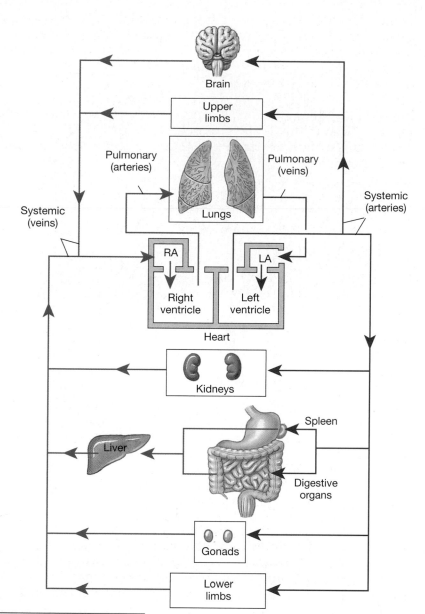

•**FIGURE 15-12**
An Overview of the Pattern of Circulation.

ply of oxygen enters the blood, carbon dioxide is excreted, and the oxygenated blood is returned to the heart for distribution in the systemic circulation.

Figure 15-13• shows the structure of the pulmonary circulation. The arteries of the pulmonary circulation differ from those of the systemic circulation in that they carry deoxygenated blood. (For this reason, color-coded diagrams usually show the pulmonary arteries in blue, the same color as systemic veins.) The pulmonary trunk gives rise to the **left** and **right pulmonary arteries**. These large arteries enter the lungs before branching repeatedly, giving rise to smaller and smaller arteries. The smallest arterioles provide blood to capillary networks that surround small air pockets or sacs, called *alveoli* (al-VĒ-ōl-ī). The walls of alveoli are thin enough for gas exchange to occur between the capillary blood and air, a process discussed in Chapter 17. As it leaves the alveolar capillaries, oxygenated blood enters venules that in turn unite to form larger vessels carrying blood to the **pulmonary veins**. These four veins, two from each lung, empty into the left atrium.

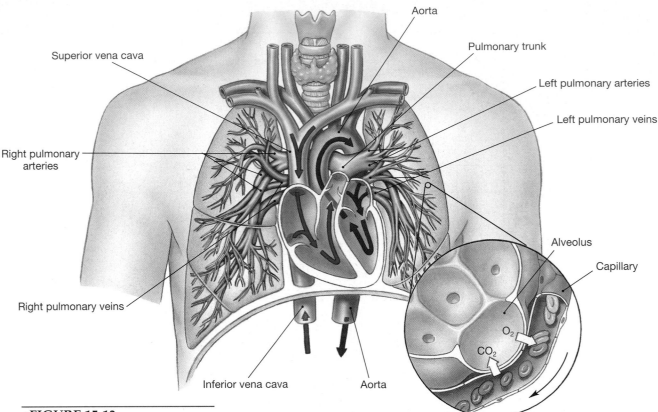

Aorta

Pulmonary trunk

Left pulmonary arteries

Left pulmonary veins

Superior vena cava

Right pulmonary
arteries

Right pulmonary veins

Inferior vena cava

Aorta

Alveolus

Capillary

O_2

CO_2

•FIGURE 15-13
The Pulmonary Circulation.

SYSTEMIC CIRCULATION: ARTERIES

The systemic circulation supplies the capillary beds in all other parts of the body. It begins at the left ventricle and ends at the right atrium. Figure 15-14• indicates the names and relative locations of the major arteries of the body.

The **ascending aorta** begins at the aortic semilunar valve of the left ventricle. The *left* and *right coronary arteries* originate near its base. The **aortic arch** curves over the top of the heart, connecting the ascending aorta with the **descending aorta**.

Arteries of the Aortic Arch

The three arteries that originate along the aortic arch are the **brachiocephalic** (brā-kē-ō-se-FAL-ik; *brachium*, arm + *cephalon*, head), the **left common carotid**, and the **left subclavian** (sub-KLĀ-vē-an; *sub*, below + *clavis*, a key). They deliver blood to the head, neck, shoulders, and upper limbs. The brachiocephalic artery branches to form the **right common carotid artery** and the **right subclavian artery**.

The Subclavian Arteries

The subclavian arteries supply blood to the upper limbs, chest wall, shoulders, back, and central nervous system. Before a subclavian artery leaves the thoracic cavity it gives rise to an artery that supplies the pericardium and anterior wall of the chest, and a **vertebral artery** that supplies blood to the brain and spinal cord.

After passing the first rib, the subclavian gets a new name, the **axillary artery**. The axillary artery crosses the axilla (armpit) to enter the arm, where its name changes again, becoming the **brachial artery**. The brachial artery provides blood to the arm before branching to create the **radial artery** and **ulnar artery** of the forearm. These arteries are interconnected by anastomoses at the palm, from which arteries to the fingers arise. In this way, blood can reach the fingers by various routes.

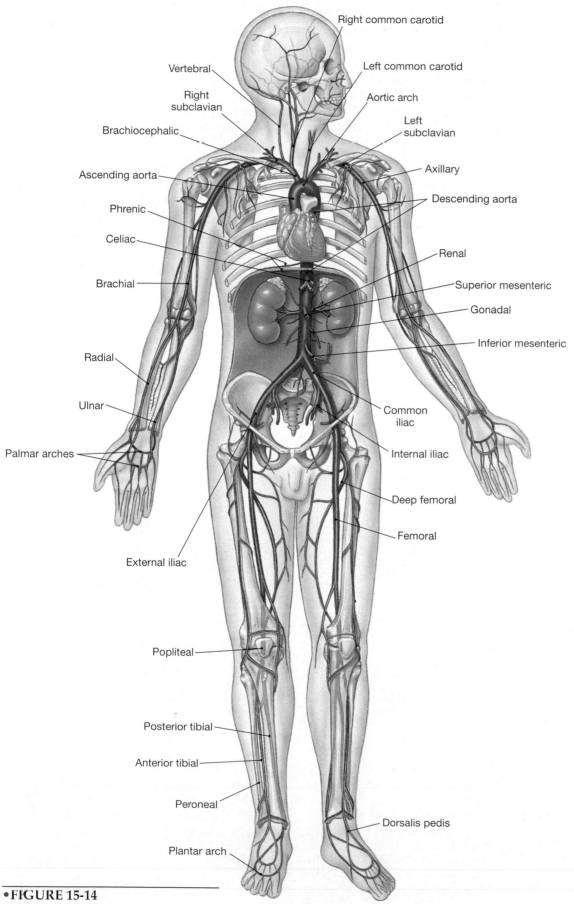

Right common carotid

Vertebral

Left common carotid

Right subclavian

Aortic arch

Brachiocephalic

Left subclavian

Ascending aorta

Axillary

Phrenic

Descending aorta

Celiac

Renal

Brachial

Superior mesenteric

Gonadal

Inferior mesenteric

Radial

Ulnar

Common iliac

Palmar arches

Internal iliac

Deep femoral

Femoral

External iliac

Popliteal

Posterior tibial

Anterior tibial

Peroneal

Dorsalis pedis

Plantar arch

•FIGURE 15-14
An Overview of the Arterial System.

Arteries Supplying the Brain

The common carotid arteries ascend deep in the tissues of the neck. A carotid artery can usually be located by pressing gently along either side of the trachea until a strong pulse is felt. Each common carotid artery divides into an **external carotid** and an **internal carotid** artery at an expanded chamber, the *carotid sinus*. (This sinus, which contains sensory receptors sensitive to blood pressure, will be discussed in Chapter 17.) The external carotids supply blood to the pharynx, larynx, and face. The internal carotids enter the skull to deliver blood to the brain.

The brain is extremely sensitive to changes in its circulatory supply. An interruption of circulation for several seconds will produce unconsciousness, and after 4 minutes there may be some permanent brain damage. Such circulatory problems are rare because blood reaches the brain through the vertebral arteries as well as by way of the internal carotids. Inside the cranium, the vertebral arteries fuse to form a large **basilar artery** that continues along the lower surface of the brain. This gives rise to the vessels indicated in Figure 15-15•.

Normally, the internal carotids supply the arteries of the front half of the cerebrum, and the rest of the brain receives blood from the vertebral arteries. But this circulatory pattern can easily change because the internal carotids and the basilar artery are interconnected in a ring-shaped anastomosis, the **cerebral arterial circle**, or *circle of Willis*, that encircles the stalk (infundibulum) of the pituitary gland. With this arrangement, the brain can receive blood from either the carotids or the vertebrals, and the chances for a serious interruption of circulation are reduced.

•FIGURE 15-15
Arterial Supply to the Brain.

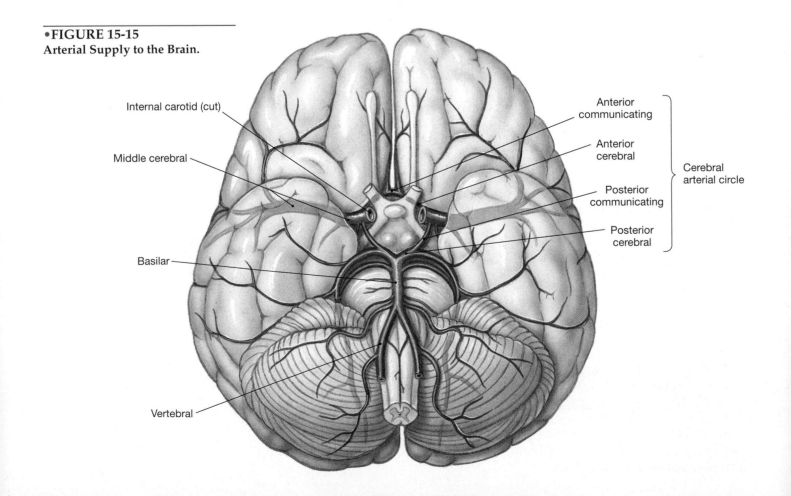

Internal carotid (cut)

Middle cerebral

Basilar

Vertebral

Anterior communicating

Anterior cerebral

Posterior communicating

Posterior cerebral

Cerebral arterial circle

Branches of the Thoracic Aorta

The portion of the descending aorta within the chest is called the *thoracic aorta*. It provides blood to the **intercostal** (*inter*, between + *costa*, rib) **arteries**, which carry blood to the spinal cord and the body wall. Other branches of this artery provide blood to the esophagus and pericardium. The **phrenic** (FREN-ik) **arteries** deliver blood to the muscular diaphragm that separates the thoracic and abdominopelvic cavities.

Branches of the Abdominal Aorta

The portion of the descending aorta inferior to the diaphragm is called the *abdominal aorta*. The abdominal aorta delivers blood to arteries that distribute blood to organs of the digestive system, and to the kidneys and adrenal glands. The **celiac** (SĒ-lē-ak), **superior mesenteric** (mez-en-TER-ik), and **inferior mesenteric arteries** provide blood to all the digestive organs in the abdominopelvic cavity. The celiac divides into three branches that deliver blood to the liver, spleen, and stomach. The superior mesenteric artery supplies the pancreas, small intestine, and most of the large intestine. The inferior mesenteric delivers blood to the last portion of the large intestine and the rectum.

Paired **gonadal** (gō-NAD-al) **arteries** originate between the superior and inferior mesenteric arteries; in males they are called *testicular arteries*; in females, *ovarian arteries*.

The **renal arteries** arise along the sides of the abdominal aorta and provide blood to the adrenal glands and kidneys.

Within the lumbar region, the abdominal aorta divides to form the **common iliac** (IL-ē-ak) **arteries** that carry blood to the pelvis and lower limbs. As it travels along the inner surface of the ilium, each common iliac divides to form an **internal iliac artery** that supplies smaller arteries of the pelvis and an **external iliac artery** that enters the lower limb.

Once in the thigh, the external iliac artery branches, forming the **femoral artery** and the **deep femoral artery**. When it reaches the leg, the femoral artery becomes the **popliteal artery**, which almost immediately branches to form the **anterior tibial**, **posterior tibial**, and **peroneal** arteries. These arteries are connected by two anastomoses, one on the top of the foot (the *dorsalis pedis*) and one on the bottom (the *plantar arch*).

SYSTEMIC CIRCULATION: VEINS

Figure 15-16● shows the major veins of the body. Arteries and veins supplying and draining the same organs and structures often run side by side, and in many cases they have comparable names. For example, the axillary arteries run alongside the axillary veins. In addition, arteries and veins often travel in the company of nerves that have the same names and innervate the same structures.

One significant exception to this pairing of arteries and veins occurs with the distribution of major veins in the neck and limbs. Arteries in these areas are not found at the body surface; instead, they are deep beneath the skin, protected by bones and surrounding soft tissues. In contrast, there are usually two sets of veins, one superficial and the other deep. The superficial veins are so close to the surface that they can be seen quite easily. This makes them easy targets for obtaining blood samples, and most blood tests are performed on venous blood collected from the superficial veins of the upper limb, usually a *median cubital vein* where it crosses the elbow.

This dual venous drainage helps control body temperature. When body temperature becomes abnormally low, the arterial blood supply to the skin is reduced and the superficial veins are bypassed. Blood entering the limbs then returns to the trunk in the deep veins. When overheating occurs, the blood supply to the skin increases and the superficial veins dilate. This is one reason why superficial veins in the limbs become so visible during periods of heavy exercise, or when sitting in a sauna, hot tub, or steam bath.

•FIGURE 15-16
An Overview of the Venous System.

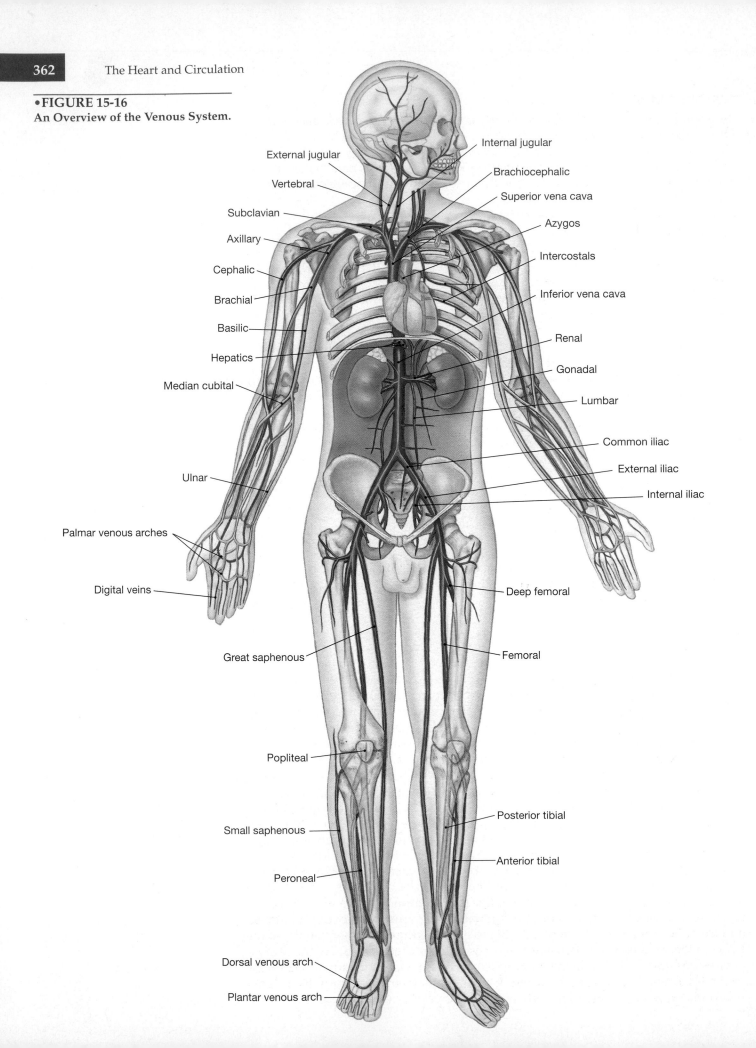

External jugular

Vertebral

Subclavian

Axillary

Cephalic

Brachial

Basilic

Hepatics

Median cubital

Ulnar

Palmar venous arches

Digital veins

Great saphenous

Popliteal

Small saphenous

Peroneal

Dorsal venous arch

Plantar venous arch

Internal jugular

Brachiocephalic

Superior vena cava

Azygos

Intercostals

Inferior vena cava

Renal

Gonadal

Lumbar

Common iliac

External iliac

Internal iliac

Deep femoral

Femoral

Posterior tibial

Anterior tibial

Superior Vena Cava

The **superior vena cava** (**SVC**) receives blood from the head and neck and the chest, shoulders, and upper limbs.

Head and neck. Small veins in the nervous tissue of the brain empty into a network of thin-walled channels, the **dural sinuses**. A sinus is a chamber or hollow in a tissue. Most of the blood leaving the brain passes through one of the dural sinuses and enters one of the **internal jugular veins** that descend in the deep tissues of the neck. The more superficial **external jugular veins** collect blood from the overlying structures of the head and neck. **Vertebral veins** drain the cervical spinal cord and the posterior surface of the skull.

Limbs and chest. A network of veins in the palms collects blood from each finger. These vessels drain into the **cephalic vein** and the **basilic vein**. The deeper veins of the forearm are the **radial** and **ulnar veins**. After crossing the elbow, these veins join to form the **brachial vein**. As the brachial vein continues toward the trunk it receives blood from the cephalic and basilic veins before entering the **axillary vein**.

The axillary vein then continues into the trunk, and at the level of the first rib it becomes the **subclavian vein**. After traveling a short distance inside the thoracic cavity, the subclavian meets and merges with the external and internal jugular veins of that side. This merger creates the large **brachiocephalic vein**. Near the heart, the two brachiocephalic veins (one from each side of the body) combine to create the superior vena cava. The superior vena cava receives blood from the thoracic body wall in the **azygos** (AZ-i-gos; unpaired) **vein** before arriving at the right atrium.

Inferior Vena Cava

The **inferior vena cava (IVC)** collects most of the venous blood from organs below the diaphragm.

Blood leaving the capillaries in the sole of each foot is collected by the *plantar venous arch,* which provides blood to the **anterior tibial vein**, the **posterior tibial vein**, and the **peroneal vein**, the deep veins of the leg. A *dorsal venous arch* drains blood from capillaries on the superior surface of the foot. This arch is drained by two superficial veins, the **great saphenous vein** (sa-FĒ-nus) and the **small saphenous vein**. The term *saphenous* means "prominent." (Surgeons use segments of the great saphenous vein, the largest superficial vein, as bypass vessels during coronary bypass surgery.) There are extensive interconnections between the plantar arch and the dorsal arch, and the path of blood flow can easily shift from superficial to deep veins.

At the knee, the small saphenous, tibial, and peroneal veins unite to form the **popliteal vein**, which, when it reaches the femur, becomes the **femoral vein**. Immediately before penetrating the abdominal wall, the femoral, great saphenous, and **deep femoral veins** unite. The large vein that results penetrates the body wall as the **external iliac vein**. The external iliac fuses with the **internal iliac vein**, which drains the pelvic organs. The resulting **common iliac vein** then meets its counterpart from the opposite side to form the inferior vena cava.

Like the aorta, the inferior vena cava lies posterior to the abdominopelvic cavity. As it ascends to the heart it collects blood from several **lumbar veins**, which may also empty into the common iliac vein. In addition, the IVC receives blood from the **gonadal**, **renal**, **suprarenal**, **phrenic**, and **hepatic veins** before reaching the right atrium.

Hepatic Portal System

You may have noticed that the veins just discussed did not include any veins from the digestive organs other than the liver. That is because blood from the capillaries serving the digestive organs does not enter veins that connect with the inferior vena cava. Instead, blood leaving the capillaries supplied by the celiac, superior, and inferior mesenteric arteries flows to the liver through the **hepatic portal system.** (A portal is a gate.) Blood in this system is quite different from that in other veins, because the hepatic portal vessels contain substances absorbed from the digestive tract. For example, levels of blood glucose, amino acids, fatty acids, and vitamins in the hepatic portal vein often exceed those found anywhere else in the cardiovascular system.

A portal system carries blood from one capillary bed to another and, in the process, prevents its contents from dilution by the entire bloodstream. The hepatic portal system begins at capillaries of the digestive tract and ends at capillaries in the liver. The liver regulates the concentrations of nutrients, such as glucose or amino acids, in the arriving blood before it continues into the inferior vena cava. When digestion is under way, the digestive tract absorbs high concentrations of nutrients, along with various wastes and an occasional toxin. The hepatic portal system delivers these compounds directly to the liver, where liver cells absorb them for storage, process them for immediate use, or excrete them. After passing through the liver capillaries, blood collects in the hepatic veins that empty into the inferior vena cava. The liver, then, is in an ideal position to regulate and maintain the composition of the blood in a relatively stable condition.

Figure 15-17• shows the vessels of the hepatic portal system. It begins in the capillaries of the digestive organs. Blood from capillaries along the lower portion of the large intestine enters the **inferior mesenteric vein**. As it nears the liver, veins from the spleen, the stomach, and the pancreas fuse with the inferior mesenteric, forming the **splenic vein**. The **superior mesenteric vein** collects blood from the entire small intestine, two thirds of the large intestine, and a portion of the stomach. The **hepatic portal vein**, which empties into the liver capillaries, forms through the fusion of the superior mesenteric and splenic veins.

✔ Blockage of which branch of the aortic arch would interfere with the blood flow to the left arm?

✔ Why would compression of the common carotid artery cause a person to lose consciousness?

✔ Grace is in an automobile accident and ruptures her celiac artery. What organs would be affected most directly by this injury?

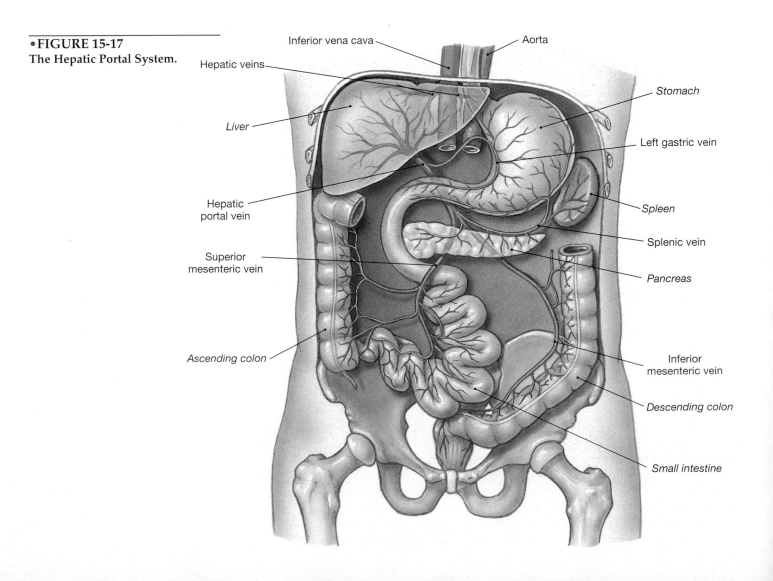

•FIGURE 15-17
The Hepatic Portal System.

FETAL CIRCULATION

There are significant differences in our cardiovascular systems before and after birth. The differences reflect differing sources of respiratory and nutritional support. For example, the lungs of the developing fetus are collapsed and nonfunctional, and the digestive tract has nothing to digest. Before birth, these needs are provided by the placenta.

Placental Blood Supply

Fetal circulation is shown in Figure 15-18a•. Blood flow to the placenta is provided by a pair of **umbilical arteries** that arise from the internal iliac arteries and enter the umbilical cord. Blood returns from the placenta in the **umbilical vein**, bringing oxygen and nutrients to the developing fetus. The umbilical vein delivers blood to capillaries within the developing liver and to the inferior vena cava by the **ductus venosus**. At birth, blood flow ceases along the umbilical vessels, and they soon degenerate.

Circulatory Changes at Birth

One of the most interesting aspects of circulatory development occurs at birth. Throughout embryonic and fetal life, the lungs are collapsed; yet following delivery, the newborn infant must be able to extract oxygen from inspired air rather than across the placenta.

•**FIGURE 15-18**
Fetal Circulation.
(a) Blood flow to and from the placenta. (b) Blood flow through the heart of a newborn baby.

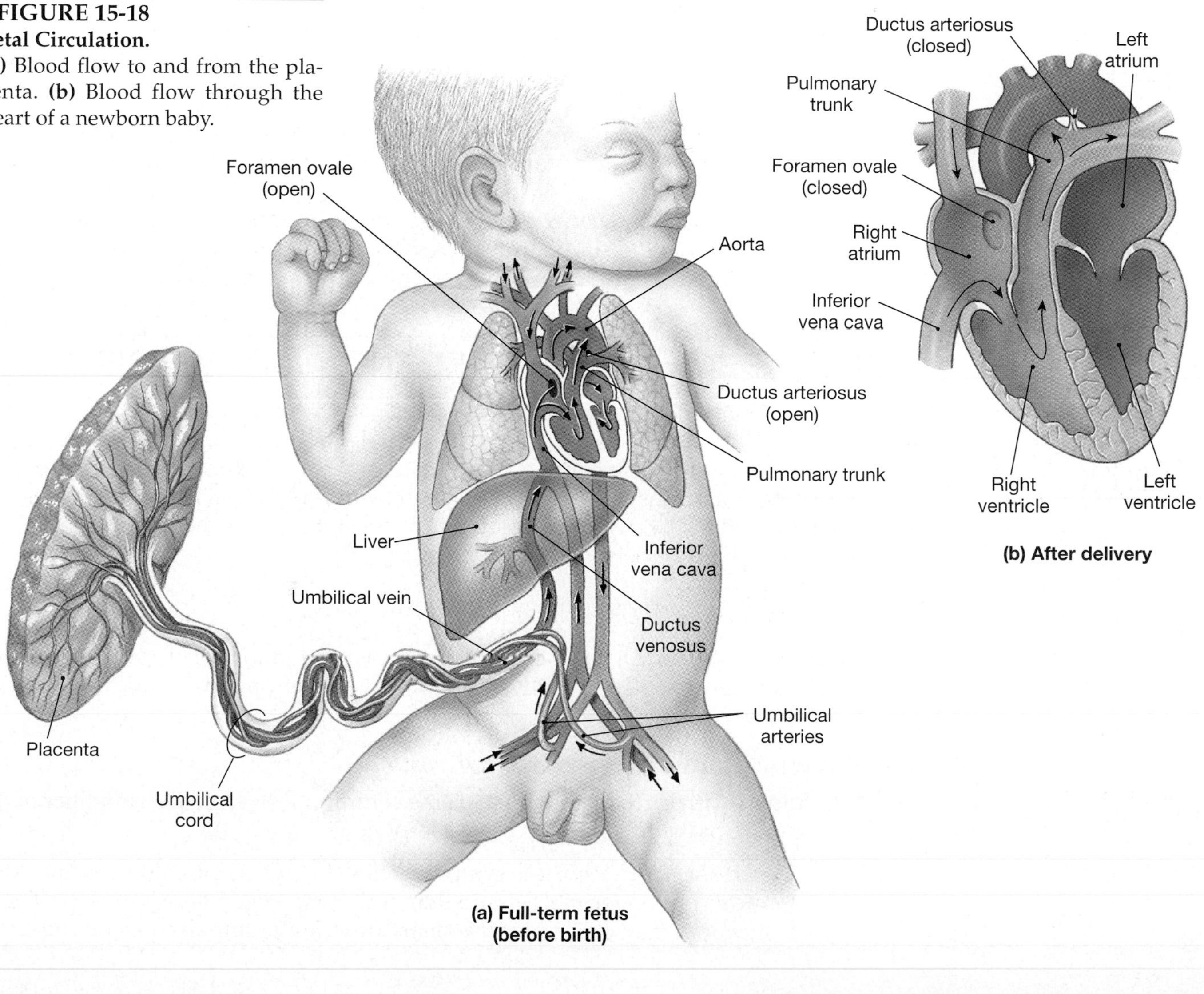

(a) Full-term fetus (before birth)

(b) After delivery

Although the septa between the chambers of the heart develop early in fetal life, the interatrial septum contains an opening up to the time of birth. This opening, called the **foramen ovale**, is covered by a flap that acts as a valve. Blood can flow freely from the right atrium to the left atrium, but any backflow will close the valve and isolate the two chambers. This allows blood to enter the heart at the right atrium and bypass the lungs. A second short-circuit exists between the pulmonary and aortic trunks. This connection, the **ductus arteriosus**, consists of a short, muscular vessel.

At birth dramatic changes occur in circulatory patterns (Figure 15-18b•). When an infant takes its first breath, the lungs expand, and so do the pulmonary vessels. Within a few seconds, the smooth muscles in the ductus arteriosus contract, isolating the pulmonary and aortic trunks, and blood begins flowing into and out of the lungs. As pressures rise in the left atrium, the valvular flap closes the foramen ovale and completes the circulatory remodeling.

AGING AND THE CARDIOVASCULAR SYSTEM

The capabilities of the cardiovascular system gradually decline with age. Age-related changes occur in the blood, heart, and blood vessels.

- **Blood.** Age-related changes in the blood may include (1) decreased hematocrit, (2) constriction or blockage of peripheral veins by formation of a thrombus (stationary blood clot), and (3) pooling of blood in the veins of the legs because the valves in these veins are no longer working effectively.

- **Heart.** Major age-related changes in the heart include (1) a reduction in the maximum cardiac output, (2) changes in the activities of the nodal and conducting cells, (3) replacement of damaged cardiac muscle cells by scar tissue, and (4) progressive blood vessel changes that restrict coronary circulation.

- **The Blood Vessels.** Age-related changes in blood vessels are often related to **arteriosclerosis** (ar-tē-rē-ō-skle-RŌ-sis), a thickening and toughening of arterial walls. **Atherosclerosis** (ath-er-ō-skle-RŌ-sis) is a type of arteriosclerosis characterized by changes in the endothelial lining of blood vessels. Atherosclerosis in coronary vessels restricts circulation through the myocardium and produces symptoms of *coronary artery disease*.

DISORDERS OF THE HEART AND BLOOD VESSELS

The cardiovascular system plays a key role in supporting all other body systems, and disorders of this system will affect virtually every cell in the body. In fact, in developed countries, the most common cause of death is due to heart disease. The most common forms of heart disease result from problems with the blood vessels that supply the heart with blood. Figure 15-19• summarizes major disorders of the heart and blood vessels.

THE PHYSICAL EXAMINATION

Cardiology (kar-dē-OL-o-jē) is the study of the heart and its functions. A **cardiologist** (kar-dē-OL-o-jist) is a physician skilled in the diagnosis and treatment of heart disease.

Symptoms and Signs

Individuals with cardiovascular problems commonly seek medical attention with one or more of the following as chief complaints:

- *Weakness and fatigue:* These symptoms develop when the cardiovascular system can no longer meet tissue demands for oxygen and nutrients. These symptoms may occur because heart function is impaired or because the

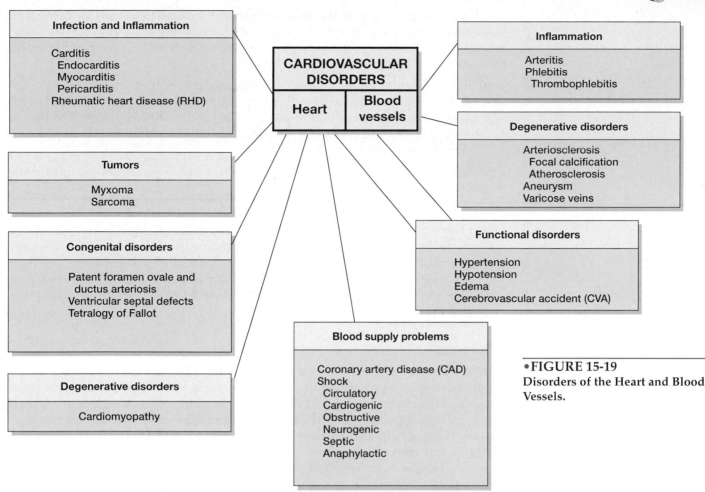

Infection and Inflammation

Carditis
  Endocarditis
  Myocarditis
  Pericarditis
Rheumatic heart disease (RHD)

Tumors

Myxoma
Sarcoma

Congenital disorders

Patent foramen ovale and
  ductus arteriosis
Ventricular septal defects
Tetralogy of Fallot

Degenerative disorders

Cardiomyopathy

CARDIOVASCULAR DISORDERS

Heart | Blood vessels

Inflammation

Arteritis
Phlebitis
  Thrombophlebitis

Degenerative disorders

Arteriosclerosis
  Focal calcification
  Atherosclerosis
Aneurysm
Varicose veins

Functional disorders

Hypertension
Hypotension
Edema
Cerebrovascular accident (CVA)

Blood supply problems

Coronary artery disease (CAD)
Shock
  Circulatory
  Cardiogenic
  Obstructive
  Neurogenic
  Septic
  Anaphylactic

•**FIGURE 15-19**
Disorders of the Heart and Blood Vessels.

blood is unable to carry normal amounts of oxygen, as in the various forms of anemia or the lungs cannot deliver enough oxygen to the blood (see Chapter 17). ∞ *[p. 329]*

- *Cardiac pain:* This is a deep pressure pain that is felt in the substernal region and typically radiates down the left arm or up into the shoulder and neck. Constant severe pain can result from inflammation of the pericardial membrane, a condition known as pericarditis. ∞ *[p. 69]* This pericardial pain may superficially resemble the pain experienced in a heart attack, or myocardial infarction (MI). Cardiac pain can also result from inadequate blood flow to the myocardium. This type of pain is called *myocardial ischemic pain*. Ischemic pain occurs in angina pectoris and in a myocardial infarction (see p. 346). Angina pectoris most commonly results from the narrowing of coronary blood vessels by atherosclerosis (p. 366). The associated pain appears during physical exertion, when myocardial oxygen demands increase. The pain associated with a myocardial infarction is often felt as a heavy weight or a constriction of the chest.

- *Palpitations:* Palpitations are a person's perception of an altered heart rate. The individual may complain of the heart "skipping a beat" or "racing." Palpitations result from an abnormal pattern of cardiac activity known as an arrhythmia (p. 349).

- *Pain on movement:* Individuals with advanced atherosclerosis of arteries to the limbs typically experience limb pain during exercise. The pain may become so severe that the person is unwilling or unable to walk or perform other common activities.

These are only a few of the many symptoms that can be caused by cardiovascular disorders. In addition, the individual may notice the appearance of characteristic signs of underlying cardiovascular problems. A partial listing of important cardiovascular signs includes the following:

- *Edema* is an increase of fluid in the tissues that occurs when (a) the pumping efficiency of the heart is decreased, (b) the plasma protein content of the blood is reduced, or (c) venous pressures are abnormally high. The tissues of the limbs are most commonly affected, and individuals experience swollen feet, ankles, and legs.

- Breathlessness, or *dyspnea* (DISP-nē-ah), occurs when cardiac output is unable to meet tissue oxygen demands.

- *Varicose veins* are swollen superficial veins that are visible at the skin surface.

- There may be characteristic and distinctive changes in skin coloration. For example, pallor is the lack of normal red or pinkish color in the skin of a Caucasian, or the conjunctiva and oral mucosa of darker-skinned people. Cyanosis is the bluish color of the skin that occurs with a deficiency of oxygen in the tissues. Cyanosis generally results from either cardiovascular or respiratory disorders.

Diagnostic Procedures

In many cases, the detection of a cardiovascular disorder occurs during the assessment stage of a physical examination. When the vital signs are taken, the pulse is checked for strength, rate, and rhythm. Weak or irregular heart beats will commonly be noticed at this time. The blood pressure is monitored with a stethoscope, blood pressure cuff, and sphygmomanometer (p. 353). Unusually high or low readings can alert the examiner to potential problems with cardiac or vascular function. However, a diagnosis of *hypotension* (low blood pressure) or *hypertension* (high blood pressure) is usually not made on the basis of a single reading but after several readings over a period of time.

The heart sounds are monitored by auscultation with a stethoscope. Cardiac rate and rhythm can be checked and an arrhythmia (variation in normal heart rhythm) detected. Abnormal heart sounds, or *murmurs*, may indicate problems with heart valves. Murmurs are noted in relation to their location in the heart, the time of occurrence in the cardiac cycle, and whether the sound is low or high pitched.

Abnormal functions of the heart and blood vessels can commonly be detected through physical assessment and the recognition of characteristic signs and symptoms discussed above. The structural basis of these problems is generally determined through the use of scans, X-rays, and the monitoring of electrical activity in the heart (Table 15-1). In addition to the blood tests listed in Table 14-1 (pp. 330–331), Table 15-1 also summarizes a series of other laboratory tests performed on blood samples that provide information specific for heart disorders. Taken together, these tables summarize the representative diagnostic procedures used to evaluate the health of the cardiovascular system.

DISORDERS OF THE BLOOD VESSELS

Arterial disorders are generally more life-threatening than venous disorders because blood is under higher pressure in arteries and the arteries must provide adequate blood flow to vital organs, such as the heart and brain, that are critical for survival. For these reasons, we will discuss disorders associated with blood vessels before those of the heart.

Inflammation

Blood vessel disorders occur in both arteries and veins. Inflammation in an artery is called **arteritis**, a condition which can lead to a narrowing of the vessel and re-

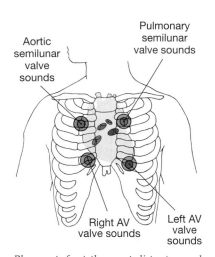

Aortic semilunar valve sounds

Pulmonary semilunar valve sounds

Right AV valve sounds

Left AV valve sounds

Placement of a stethoscope to listen to sounds generated by the action of individual valves. The positions of the valves are shown in matching colors.

Table 15-1	Representative Diagnostic Procedures and Laboratory Tests for Cardiovascular Disorders	
Diagnostic Procedure	**Method and Result**	**Representative Uses**
Electrocardiography (ECG)	Electrodes placed on the chest register electrical activity of cardiac muscle during a cardiac cycle; information is transmitted to a monitor for graphic recording	Heart rate can be determined through study of the ECG; abnormal wave patterns may occur with cardiac irregularities such as myocardial infarction, or arrhythmias
Echocardiography	Standard ultrasound examination of the heart	Detection of structural abnormalities of the heart while in motion
Exercise stress test	ECG, blood pressure, and heart rate are monitored during exercise on a treadmill	Coronary artery blockage is suspected if ECG patterns change and/or echocardiography results change
Chest X-ray	Film sheet with radiodense tissues in white on a negative image	Determination of abnormal shape or size of the heart and abnormalities of the aorta and other large blood vessels
Positron emission tomography (PET)	Radionuclides are injected into circulation; accumulation occurs at certain areas of the heart; a computer image for viewing is produced	Allows detection of damaged tissue areas; also useful in the determination of blood flow to the heart muscle
Pericardiocentesis	Needle aspiration of fluid from the pericardial sac for analysis	See pericardial fluid analysis below
Doppler ultrasound	Transducer is placed over the vessel to be examined, and the echoes are analyzed by computer to provide information on blood velocity and flow direction	Detection of venous occlusion; determination of cardiac efficiency; monitoring of fetal circulation and blood flow in umbilical vessels
Venography	Radiopaque dye is injected into peripheral vein of a limb, and an X-ray study is done to detect venous occlusions	Detection of venous blockage
Laboratory Test	**Normal Values in Blood Plasma or Serum**	**Significance of Abnormal Values**
Serum analysis		
Enzymes Creatine phosphokinase (CPK, CK)	0–36 U/l	Prolonged elevated levels of CK indicate myocardial damage; levels usually do not rise until 6–12 hours following a heart attack (myocardial infarct, or MI)
Lactate dehydrogenase (LDH)	Adults: 45–90 U/l	LDH enzyme converts pyruvic acid to lactic acid. Elevated in roughly 40% of patients with congestive heart failure
Electrolytes Sodium	Adults: 135–145 mEq/l	Decreased levels occur with heart failure, hypotension. Increased level occurs with essential hypertension
Potassium	Adults: 3.5–5.0 mEq/l	Decreased level when taking certain diuretics
Cholesterol	Adults: 150–240 mg/dl (desirable level <200 mg/dl)	High risk when >240 mg/dl for atherosclerosis and coronary artery disease; elevated levels occur in familial hypercholesterolemia
Triglycerides	Adults: 10–160 mg/dl (adults >50 years old may have higher values)	Elevated with acute MI and poorly controlled diabetes mellitus
Pericardial fluid analysis	Fluid is analyzed for appearance, number, and types of blood cells; protein and glucose levels; and cultured to detect infectious organisms	

duced blood flow. Inflammation of a vein is called **phlebitis** (fle-BĪ-tis). Phlebitis is often associated with blood clots, and then it is called *thrombophlebitis*.

Degenerative Disorders

Arteriosclerosis. *Arteriosclerosis* and its complications account for roughly half of all deaths in the United States. As described earlier (p. 366), arteries with arteriosclerosis become thick and tough, with a narrowed lumen (central passageway) that reduces blood flow. There are two major forms of arteriosclerosis, focal calcification and atherosclerosis. In *focal calcification* degenerating smooth muscle is replaced with calcium deposits. It occurs as part of the aging process, in association with atherosclerosis, and as a complication of diabetes mellitus.

Atherosclerosis is a type of arteriosclerosis that is associated with lipid deposits in the smooth muscle in the vessel wall and damage to the endothelial lining. A major factor involved in the development of atherosclerosis is high levels of lipids in the blood, especially cholesterol. Under such conditions, lipid deposits accumulate beneath the endothelium and produce a fatty mass of tissue called a **plaque** that projects into the lumen of the vessel (Figure 15-20●). If the levels of lipids in the blood remain high, changes occur in the endothelial lining that attract platelets and a blood clot, or *thrombus*, may form. The result is a narrowing of the vessel and reduced blood flow. If the blood clot loosens and drifts away, it is called an *embolus*. A circulatory blockage caused by an embolus is known as an *embolism*.

Aneurysm. An **aneurysm** (AN-ū-rizm) is a bulge in a weakened area of the wall of a blood vessel. Much like a bubble in the wall of a tire may cause a "blowout," so may an aneurysm in an artery or vein. The higher blood pressure in arteries makes an aneurysm there more serious than one in a vein. The most dangerous aneurysms are those involving arteries of the brain, where they cause strokes, and of the aorta, where a ruptured aneurysm will cause fatal bleeding in a matter of minutes. Unfortunately, because many aneurysms are painless until they actually rupture, they are likely to go undetected.

Aneurysms are normally caused by chronic high blood pressure, although any trauma or infection that weakens vessel walls can lead to an aneurysm. In addition, at least some aortic aneurysms have been linked to inherited disorders, such as Marfan's syndrome, that have weakened connective tissues in vessel walls.

Varicose Veins. **Varicose** (VAR-i-kōs) **veins** are sagging, swollen vessels that are often seen in the superficial veins of the legs. This condition generally arises in individuals who have an occupation requiring long hours of standing or sitting, accompanied by a lack of regular exercise. Because there is no muscular activity to help keep the blood moving, venous blood pools on the proximal (heart) side of each valve. As the venous walls are distorted, the valves close less tightly, and gravity can then pull blood back toward lower levels. This pulling further impedes normal blood flow, and the veins become grossly distended. Surgical procedures are sometimes used to remove or constrict the offending vessels.

Varicose veins may also arise in the network of veins in the walls of the anus. High pressures generated by the abdominal muscles can force blood into these veins, and repeated incidents leave them permanently distended. Such distended veins are known as **hemorrhoids** (HEM-o-roydz). Hemorrhoids are often associated with pregnancy, due to changes in circulation and abdominal pressures. Minor cases can be treated by the topical application of drugs that promote the contraction of smooth muscles within the venous walls. More severe cases may require the surgical removal or destruction of the distended veins.

Functional Disorders

High Blood Pressure. **Hypertension** is the presence of abnormally high blood pressure. Hypertension may result from a variety of different causes, including genetic factors, diet, stress levels, and circulating hormone levels. The usual criterion for hypertension is a resting blood pressure higher than 140/90 ("normal" is 120/80). (Temporary increases in blood pressure during exercise are not consid-

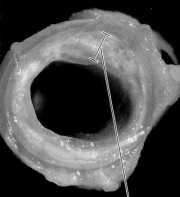

(a)

Plaque deposit in vessel wall

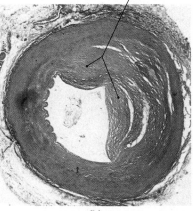

(b)

●**FIGURE 15-20**
A Plaque.
A plaque forms through the deposition of fat in the wall of a blood vessel. It narrows the vessel and can interfere with blood flow. **(a)** A coronary artery narrowed by plaque. **(b)** Sectional view of a large plaque.

ered as hypertension.) Hypertension significantly increases the workload on the heart, and the left ventricle gradually enlarges. More muscle mass means a greater oxygen demand. When the coronary circulation cannot meet the increased demand, symptoms of cardiac ischemia appear (p. 346).

Treatment consists of a combination of lifestyle changes and drug therapies. Quitting smoking, getting regular exercise, and restricting dietary intake of salt, fats, and calories will improve peripheral circulation, prevent increases in blood volume and total body weight, and reduce plasma cholesterol levels. These strategies may be sufficient to control hypertension if it has been detected before significant cardiovascular damage has occurred. Most therapies involve antihypertensive drugs, such as *calcium channel blockers*, *beta-blockers*, *diuretics*, and *vasodilators*, singly or in combination. Calcium blockers reduce heart rate by slowing the passage of calcium ions into cardiac muscle. They also relax smooth muscles in the walls of coronary blood vessels, and thereby improve blood flow to the heart. Beta-blockers reduce the effects of sympathetic stimulation on the heart, and the unopposed parasympathetic system lowers the resting heart rate and blood pressure. Diuretics promote the loss of water and sodium ions at the kidneys, lowering blood volume, and vasodilators further reduce blood pressure. **Hypotension**, or low blood pressure, is occasionally seen in patients who have received antihypertensive drugs.

Edema. Edema (e-DĒ-ma) is an abnormal accumulation of interstitial fluid. There are many different causes of edema, but the underlying problem in all types of edema is a disturbance in the normal balance of blood pressure and osmosis across the capillaries (p. 354).

In the U.S., serious cases of edema most commonly result from an increase in arterial blood pressure, venous pressure, or total circulatory pressure. The increase may result from heart problems, such as heart failure, venous blood clots that act to increase venous pressures, or kidney or liver problems. The net result is an increase in blood pressure that accelerates the movement of fluid into the tissues.

Cerebrovascular Disease. *Cerebrovascular disease* was introduced in Chapter 11 (see p. 260). Most symptoms of cerebrovascular disease appear when atherosclerosis reduces the circulatory supply to the brain. If the circulation to a portion of the brain is completely shut off, a cerebrovascular accident (CVA), or stroke, occurs. The most common causes of strokes include *cerebral thrombosis* (clot formation at a plaque), *cerebral embolism* (drifting blood clots, fatty masses, or air bubbles), and *cerebral hemorrhages* (rupture of a blood vessel, often at the site of an aneurysm). The symptoms and their severity vary with the vessel involved and the location of the blockage. For example, if the blood supply to the motor cortex is interrupted, there will be paralysis; if circulation to the speech center is cut off, speech problems occur.

DISORDERS OF THE HEART

Infection and Inflammation

Many different microorganisms may infect heart tissue, and inflammation can result from infection, trauma, or other factors that cause tissue damage.

Carditis. Carditis (kar-DĪ-tis) is a general term for inflammation of the heart. Clinical conditions resulting from cardiac infection are usually identified by the primary site of infection. For example, **endocarditis** is an inflammation of the endocardium. Endocarditis primarily damages the chordae tendineae and heart valves. The most severe complications result from the formation of blood clots on the damaged surfaces. These clots subsequently break free, entering the circulation as drifting emboli that may cause strokes, heart attacks, or kidney failure. Destruction of heart valves by infection may lead to valve leakage, heart failure, and death.

Inflammation or infection of the myocardium produces **myocarditis**. Some of the invading microorganisms include those responsible for diphtheria, syphilis, polio, and malaria. Initially, the membranes of infected heart muscle cells become easier to stimulate, and the heart rate rises dramatically. Over time, abnormal contractions may appear and the heart muscle weakens; these problems may eventually prove fatal.

Pericarditis, an inflammation of the pericardium, may restrict the expansion of the heart and reduce cardiac output. Pericarditis commonly results in an increased production of pericardial fluid, which pushes against the heart and reduces cardiac output. Such a condition is called **cardiac tamponade** (tam-pon-ĀD); cardiac tamponade can also result from trauma that results in bleeding into the pericardial cavity. If the reduced cardiac output is due to a loss of elasticity of the pericardium, then the condition is called *constrictive pericarditis*. Treatment involves draining the excess fluid or cutting the pericardial sac.

Rheumatic Fever. Rheumatic (roo-MA-tik) **fever** may follow an untreated throat infection by streptococcal bacteria. Rheumatic fever most commonly affects children of ages 5–15; symptoms include high fever, joint pain and stiffness, and a distinctive full-body rash. The damage to the heart probably results from a misdirected immune response that attacks the tissues of the heart valves. Rheumatic fever is seldom seen in the U.S., due to the widespread availability and use of antibiotics. However, inadequate antibiotic treatment (failure to complete the full 10 days of antibiotic therapy) increases the risk of rheumatic fever. In about one-half of patients, the carditis that does develop escapes detection, and scar tissue gradually forms in the myocardium and the heart valves. Valve condition deteriorates over time, although serious valve problems may not appear until 10–20 years after the initial infection.

The affected valves become thickened and may calcify. This thickening narrows the opening guarded by the valves, producing a condition called **valvular stenosis** (ste-NŌ-sis) (*stenos* means "narrow"). The resulting clinical disorder is known as **rheumatic heart disease (RHD)**. The thickened cusps stiffen in a partially closed position, but the valves do not completely block the circulation and much of the blood pumped out of the heart may flow back in. The most common forms of valvular heart disease are *mitral stenosis*, which affects the left AV valve (also called the bicuspid or mitral valve), and *aortic stenosis*, which affects the aortic semilunar valve. Because of a backflow of blood, both conditions decrease blood flow into the aortic trunk (and therefore into the systemic circulation) and cause an enlargement of the left ventricle. This combination can lead to heart failure.

Treatment for severe stenosis involves the replacement of the damaged valve with a prosthetic (artificial) valve. Figure 15-21• shows a stenotic heart valve and two possible replacements: a valve from a pig and a synthetic valve, one of a number of designs that have been employed. Valve replacement operations are quite successful, with about 95 percent of the surgical patients surviving for three years or more and 70 percent surviving more than five years. However, artificial valves require life-long treatment with anticoagulants to prevent clot formation, and pig valves wear out and must be periodically replaced.

Tumors

Primary tumors of the heart are rare, but, of those that do form, most are benign *myxomas*. They often form in the left atria, developing from the interatrial septum. Malignant cardiac *sarcomas* usually grow rapidly and death occurs within weeks to months after diagnosis. Secondary tumors of the heart are more common than primary tumors, with malignant tumors of the breast and lung usually serving as the source of metastasizing cells.

Congenital Disorders

Congenital circulatory problems are generally due to abnormal formations of the heart, or with interconnections between the heart and the pulmonary trunk or aorta. Three examples are shown in Figure 15-22•. All of these can be surgically corrected, although multiple surgeries may be required.

The incomplete closure of the foramen ovale or ductus arteriosus (Figure 15-22a•) result in serious functional problems. If the foramen ovale remains open, or **patent**, after birth, blood recirculates into the pulmonary circulation without entering the left ventricle. Oxygen is normal in the systemic circulation, but the right atrium and ventricle enlarge because they must work harder than usual, and pul-

(a)

(b)

(c)

•**FIGURE 15-21**
Artificial Heart Valves.
(a) A semilunar valve with stenosis (narrowing) **(b)** Whole Bioprosthetic™ heart valve, which uses the valve from a pig's heart. **(c)** Medtronic Hall™ prosthetic heart valve.

monary hypertension and pulmonary edema eventually result. If the ductus arteriosus remains open, the same basic problems develop. If, however, valve defects, narrowed pulmonary vessels, or other abnormalities occur as well, pulmonary pressures may increase enough to force blood into the systemic circulation through the ductus arteriosus. Because deoxygenated blood is now entering the systemic circulation, the blood is darker and the skin develops the blue tones of cyanosis, and the infant is known as a "blue baby."

Ventricular septal defects (Figure 15-22b•) are openings between the left and right ventricles, and are the most common congenital heart problems. When the more powerful left ventricle contracts, it pushes blood into the right ventricle and pulmonary circulation. The end results are the same as with patent foramen ovale and ductus arteriosus—pulmonary hypertension, pulmonary edema, and cardiac enlargement.

The *tetralogy of Fallot* (fa-LŌ) (Figure 15-22c•), is a complex group of four (*tetra-* means "four") heart and circulatory defects. In this condition the pulmonary trunk is narrowed, the interventricular septum is incomplete, the ductus arteriosus is open, and the right ventricle is enlarged.

Blood Supply Problems

Coronary Artery Disease. *Coronary artery disease (CAD)* refers to degenerative changes in the coronary circulation, a topic that was discussed on p. 346. As noted there, angina pectoris is a common symptom of coronary ischemia. Similar to the treatment for hypertension, angina may be controlled by a combination of changes in lifestyle and drugs. For example, lifestyle changes include limiting strenuous exercise and stressful situations, stopping smoking, and lowering fat consumption. Useful medications include calcium channel blockers, beta-blockers such as *propranolol* or *metoprolol*, vasodilators such as *nitroglycerin* (nī-trō-GLIS-er-in), and cholesterol-lowering drugs.

Angina can also be treated surgically. A long, slender, hollow **catheter** (KATH-e-ter) may be inserted into a large artery and be guided into a coronary artery to the blockage. The plaque may then be removed and sucked up in the catheter. In **balloon angioplasty** (AN-jē-ō-plas-tē), the catheter tip contains an inflatable balloon (Figure 15-23•). Once in position, the balloon is inflated, pressing the plaque against the vessel walls. Sometimes a *stent*, a small wire cylinder, is left to prevent further narrowing of the vessel.

In a **coronary bypass graft (CABG)**, a small section is removed from either a small artery or a peripheral vein and used to create a detour around the obstructed portion of a coronary artery. As many as four to six coronary arteries can be rerouted this way during a single operation. Coronary bypass surgery is often reserved for cases of severe angina that do not respond to other treatment.

Shock. Shock is a serious circulatory crisis marked by low blood pressure (hypotension) and inadequate blood flow to body tissues. Severe and potentially fatal symptoms develop as vital tissues become starved for oxygen and nutrients. Although there are different causes of shock, they all result from *heart failure*, an

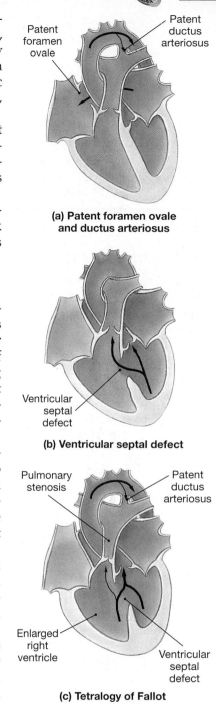

(a) Patent foramen ovale and ductus arteriosus

(b) Ventricular septal defect

(c) Tetralogy of Fallot

•**FIGURE 15-22**
Congenital Cardiovascular Problems.

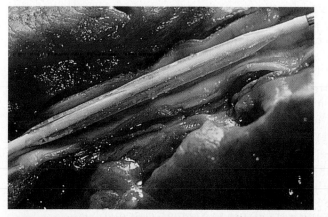

•**FIGURE 15-23**
Balloon Angioplasty.
A catheter with an inflatable cuff is inserted into a coronary artery. Inflation of the cuff flattens the plaque, and removes the obstruction, at least temporarily.

inability of the cardiac output to meet the circulatory demands of the body's tissues. Examples of different types of shock are:

- **Circulatory shock** or *hypovolemic shock*, is caused by a severe reduction in blood volume. The fluid loss may be caused by hemorrhaging or dehydration, such as after severe burns. All cases of circulatory shock share the same basic symptoms: (1) hypotension, (2) pale, cool, and moist ("clammy") skin, (3) confusion and disorientation, (4) a rise in heart rate and rapid, weak pulse, (5) an absence of urination, and (6) a drop in blood pH.

- **Cardiogenic** (kar-dē-ō-JEN-ik) **shock** occurs when the heart becomes unable to maintain a normal cardiac output. The most common cause is failure of the left ventricle as a result of a myocardial infarction. Cardiogenic shock may also be the result of arrhythmias, valvular heart disease, advanced coronary artery disease, cardiomyopathy, or ventricular arrhythmias.

- **Obstructive shock** occurs when the cardiac output is reduced because tissues or fluids are restricting the expansion and contraction of the heart. For example, fluid buildup in the pericardial cavity, called *cardiac tamponade*, can compress the heart and limit the filling of the ventricles.

- **Neurogenic** (noo-rō-JEN-ik) **shock** results from a widespread, uncontrolled vasodilation. It can be caused by general or spinal anesthesia and by trauma or inflammation of the brain stem. The underlying problem is damage to the vasomotor center or to the sympathetic nerves, leading to a loss of control of vasoconstriction of blood vessels.

- **Septic shock** also results from a widespread, uncontrolled vasodilation. It is caused by the massive release of bacterial poisons during a systemic infection. Symptoms of septic shock generally resemble those of other types of shock, but the skin is flushed, and the individual has a high fever. For this reason septic shock is also known as "warm shock."

- **Anaphylactic** (an-a-fi-LAK-tik) **shock** is a dangerous allergic reaction. It is also characterized by extensive peripheral vasodilation. This type of shock is discussed in chapter 16.

Degenerative Disorders

Cardiomyopathy. The **cardiomyopathies** (kar-dē-ō-mī-OP-a-thēz)) include an assortment of diseases with a common symptom: the progressive, irreversible degeneration of the myocardium. Cardiac muscle cells are damaged and replaced by fibrous tissue, and the muscular walls of the heart become thin and weak. As muscle tone declines, the ventricular chambers greatly enlarge. When the remaining cells cannot develop enough force to maintain cardiac output, symptoms of heart failure develop.

Chronic alcoholism and coronary artery disease are probably the most common causes of cardiomyopathy in the United States. Infectious agents, including viruses, bacteria, fungi, and protozoans, can also produce cardiomyopathies. Diseases affecting neuromuscular performance, such as muscular dystrophy, can also damage cardiac muscle cells, as can starvation or chronic variations in the extracellular concentrations of calcium or potassium ions.

Individuals suffering from severe cardiomyopathies may be considered as candidates for *heart transplant*s. This surgery involves the removal of the weakened heart and the replacement with a heart taken from a suitable donor. To survive the surgery, the recipient must be in otherwise satisfactory health. Because the number of suitable donors is limited, the available hearts are generally assigned to individuals younger than age 50. Out of the 8,000–10,000 U.S. patients each year who suffer from potentially fatal cardiomyopathies, only about 1,000 receive heart transplants. There is an 80–85 percent one-year survival rate and a 50–70 percent five-year survival rate after successful transplantation. This rate is quite good, considering that these patients would have died if the transplant had not been performed. However, the procedure remains controversial due to the high cost involved.

✔ What causes the development of varicose veins in the legs?

✔ How might an inflammation of the pericardium interfere with the output of the heart?

CHAPTER REVIEW

Key Words

anastomosis: The joining of two tubes, usually referring to a connection between two blood vessels without an intervening capillary bed.

arteriole: A small arterial branch that delivers blood to a capillary network.

atria: Thin-walled chambers of the heart that receive venous blood from the pulmonary or systemic circulation.

atrioventricular valve: One of the valves that prevent backflow into the atria during contraction of the ventricles (ventricular systole).

blood pressure: A force exerted against the vessel walls by the blood, as the result of the push exerted by cardiac contraction and the elasticity of the vessel walls.

capillary: Small blood vessel between arterioles and venules, whose thin wall permits the diffusion of gases, nutrients, and wastes between the blood plasma and interstitial fluids.

cardiac cycle: One complete heartbeat, including atrial and ventricular systole and diastole.

cardiac output: The amount of blood ejected by a ventricle each minute; normally about 5 l.

diastole: A period of relaxation within the cardiac cycle.

electrocardiogram (ECG, EKG): Graphic record of the electrical activities of the heart, as monitored at specific locations on the body surface.

endocardium: The simple squamous epithelium that lines the heart.

epicardium: Serous membrane covering the outer surface of the heart.

myocardium: The cardiac muscle tissue of the heart.

pericardium: The fibrous sac that surrounds the heart.

pulmonary circulation: Blood vessels between the pulmonary semilunar valve of the right ventricle and the entrance to the left atrium; the circulatory pathway between the heart and the lungs.

systemic circulation: Vessels between the aortic semilunar valve and the entrance to the right atrium; the blood vessels other than those of the pulmonary circulation.

systole: A period of contraction within the cardiac cycle.

vasoconstriction: A reduction in the diameter of arterioles due to the contraction of smooth muscles in the vessel wall.

vasodilation: An increase in the diameter of arterioles due to the relaxation of smooth muscles in the vessel wall.

venae cavae: The major veins delivering systemic blood to the right atrium.

ventricle: One of the large, muscular pumping chambers of the heart that discharges blood into the pulmonary or systemic circulations.

venule: Small, thin-walled vein that receives blood from capillaries.

Selected Clinical Terms

aneurysm: A bulge in the weakened wall of a blood vessel, generally an artery.

angina pectoris: A severe chest pain that occurs during exertion or stress; results from coronary ischemia when the heart's workload increases.

arteriosclerosis: A thickening and toughening of arterial walls.

arteritis: Inflammation of an artery.

atherosclerosis: A type of arteriosclerosis characterized by changes in the endothelial lining and the formation of plaque.

bradycardia: A heart rate slower than normal.

cardiomyopathies: A group of diseases characterized by the progressive, irreversible degeneration of the myocardium.

coronary ischemia: Restriction of the circulatory supply to the heart.

edema: An abnormal accumulation of fluid in peripheral tissues.

heart failure: A condition in which the heart weakens and peripheral tissues suffer from a lack of oxygen and nutrients.

hypertension: Abnormally high blood pressure.

hypotension: Blood pressure so low that circulation to vital organs may be impaired.

ischemia: Inadequate blood supply to a region of the body.

myocardial infarction (MI): A condition in which the coronary circulation becomes blocked and the cardiac muscle cells die from oxygen starvation; also called a heart attack.

phlebitis: Inflammation of a vein.

rheumatic heart disease (RHD): A disorder in which the heart valves become thickened and stiffen into a partially closed position, affecting the efficiency of the heart.

shock: An acute circulatory crisis marked by hypotension and inadequate blood flow to body tissues.

tachycardia: A heart rate that is faster than normal.

varicose veins: Superficial veins that are distended and swollen with blood.

Study Outline

INTRODUCTION *(p. 341)*

1. Contractions of the heart power the movement of blood through blood vessels to all parts of the body.

THE HEART AND CARDIOVASCULAR SYSTEM *(p. 341)*

1. The circulatory system can be subdivided into the **pulmonary circulation** (which carries blood to and from the lungs) and the **systemic circulation** (which transports blood to and from the rest of the body). **Arteries** carry blood away from the heart; **veins** return blood to the heart. **Capillaries** are tiny vessels that connect the smallest arteries and the smallest veins. *(Figure 15-1)*

2. The heart has four chambers: the **right atrium** and **right ventricle**, and the **left atrium** and **left ventricle**.

STRUCTURE OF THE HEART *(pp. 342–346)*

1. The heart is surrounded by the **pericardial cavity** (lined by the **pericardium**); the **visceral pericardium** (**epicardium**) covers the heart's outer surface, and the **parietal pericardium** lines the inner surface of the pericardial cavity.

2. The widest veins and arteries of the circulatory system are connected to the **base** of the heart; the pointed tip of the heart is the **apex**. *(Figure 15-2)*

INTERNAL ANATOMY OF THE HEART *(p. 343)*

3. The right atria and ventricle are separated from the left atria and left ventricle by *septa*. The right atrium receives blood from the systemic circuit through two large veins, the **superior vena cava** and **inferior vena cava**. *(Figure 15-3)*

4. Blood flows from the right atrium into the right ventricle via the **right atrioventricular (AV) valve (tricuspid valve)**.

5. Blood leaving the right ventricle enters the **pulmonary trunk** after passing through the **pulmonary semilunar valve**. The pulmonary trunk divides to form the **left** and **right pulmonary arteries**, which go to the lungs. The left and right pulmonary veins return blood from the lungs to the left atrium. Blood leaving the left atrium flows into the left ventricle via the **left atrioventricular (AV) valve (bicuspid valve** or **mitral valve)**. Blood leaving the left ventricle passes through the **aortic semilunar valve** and into the systemic circulation via the *ascending aorta*. *(Figures 15-3, 15-4)*

6. Valves normally permit blood flow in only one direction. *(Figure 15-4)*

THE HEART WALL *(p. 345)*

7. The bulk of the heart consists of the muscular **myocardium**. The **endocardium** lines the inner surfaces of the heart. *(Figure 15-5)*

8. Cardiac muscle cells are interconnected by **intercalated discs** that convey the force of contraction from cell to cell and conduct action potentials. *(Figure 15-5)*

BLOOD SUPPLY TO THE HEART *(p. 346)*

9. The **coronary circulation** meets the high oxygen and nutrient demands of cardiac muscle cells. The **coronary arteries** originate at the base of the ascending aorta. Interconnections between arteries called **anastomoses** ensure a constant blood supply. Two **cardiac veins** carry blood from the coronary capillaries to the **coronary sinus**.

THE HEARTBEAT *(pp. 346–349)*

1. The average heart beats about 100,000 times per day, or about 70 to 80 beats per minute.

CARDIAC CYCLE *(p. 346)*

2. The **cardiac cycle** consists of **systole** (contraction), followed by **diastole** (relaxation). Both sides of the heart contract at the same time, and they eject equal volumes of blood. *(Figure 15-6)*

3. The amount of blood ejected by a ventricle during a single beat is the **stroke volume**; the amount of blood pumped each minute is the **cardiac output**.

CONDUCTING SYSTEM OF THE HEART *(p. 347)*

4. The conducting system, composed of **nodal cells** and **conducting cells**, initiates and distributes electrical impulses within the heart. Nodal cells establish the rate of cardiac contraction, and conducting cells distribute the contractile stimulus to the general myocardium.

5. Unlike skeletal muscle, cardiac muscle contracts without neural or hormonal stimulation. **Pacemaker cells** found in the **cardiac pacemaker** (**sinoatrial**, or **SA**, **node**) normally establish the rate of contraction. From the SA node the stimulus travels to the **atrioventricular (AV) node**, then to the **AV bundle**, which divides into **bundle branches**. From here **Purkinje cells** convey the impulses to the ventricular myocardium. *(Figure 15-7)*

THE ELECTROCARDIOGRAM *(p. 349)*

6. A recording of electrical activities in the heart is an **electrocardiogram** (**ECG** or **EKG**). Important landmarks of an ECG include the **P wave** (atrial impulse and contraction), **QRS complex** (ventricular impulse and contraction), and **T wave** (ventricles recover). *(Figure 15-8)*

CONTROL OF HEART RATE *(p. 349)*

7. Although the basic heart rate is established by the pacemaker cells of the SA node, it can be modified by the autonomic nervous system (ANS).

BLOOD VESSELS *(pp. 350–355)*

1. Arteries, veins, and capillaries form a closed system of tubes that carry blood to all the cells of the body. **Arteries** branch repeatedly, decreasing in size until they become **arterioles**; from the arterioles blood enters the capillary networks. Blood flowing from the capillaries enters small **venules** before entering larger veins.

VESSEL STRUCTURE *(p. 351)*

2. In general, the walls of arteries are thicker than those of veins. *(Figure 15-9a)*

3. Capillaries are the only blood vessels whose walls permit exchange between blood and interstitial fluid.

4. Capillaries form interconnected networks called **capillary beds**.

5. Venules collect blood from the capillaries and merge into medium-sized veins and then large veins. The arterial system is a high-pressure system; blood pressure in veins is much lower. Valves in these vessels prevent the backflow of blood. *(Figure 15-9b)*

BLOOD PRESSURE *(p. 351)*

6. For blood to flow into peripheral capillaries, **blood pressure** (arterial pressure) must be greater than the resistance of the blood vessels.

7. Blood pressure is measured with a *sphygmomanometer*. *(Figure 15-10)*

8. Arterial pressure rises during ventricular systole and falls during ventricular diastole. The difference between these two pressures is **pulse pressure**.

9. Blood pressure forces fluid out of the capillaries, and osmosis "pulls" fluid back in. The net outflow is collected as lymph fluid and returned to the venous system. *(Figure 15-11)*

REGULATION OF BLOOD FLOW *(pp. 355–356)*

1. Homeostasis requires a continuous flow of blood so that tissues receive adequate levels of oxygen and nutrients.

2. Blood flow varies directly with cardiac output, blood vessel resistance, and blood pressure.

SHORT-TERM RESPONSES *(p. 355)*

3. Blood vessel resistance is adjusted at the tissues by local conditions, such as oxygen levels, that result in the widening (dilation) or narrowing (constriction) of arteriole walls.

4. Sympathetic activation leads to stimulation of the cardioacceleratory and vasomotor centers; parasympathetic activation stimulates the cardioinhibitory center. Epinephrine and norepinephrine stimulate cardiac output and peripheral vasoconstriction.

LONG-TERM RESPONSES *(p. 356)*

5. The endocrine system provides both short-term and long-term regulation. Epinephrine and norepinephrine from the adrenal medullae provide quick, short-term regulation of cardiac output and peripheral resistance. Hormones involved in long-term regulation of blood pressure and volume are antidiuretic hormone (ADH), angiotensin II, erythropoietin (EPO), and atrial natriuretic peptide (ANP).

BLOOD VESSELS OF THE BODY *(pp. 356–366)*

1. The distributions of arteries and veins are usually identical on both sides of the body, except near the heart. *(Figure 15-12)*

PULMONARY CIRCULATION *(p. 356)*

2. The pulmonary circulation includes the **pulmonary trunk**, the **left** and **right pulmonary arteries**, and the **pulmonary veins** that empty into the left atrium. *(Figure 15-13)*

SYSTEMIC CIRCULATION: ARTERIES *(p. 358)*

3. The **ascending aorta** gives rise to the coronary circulation. The **aortic arch** communicates with the **descending aorta.** *(Figures 15-14, 15-15)*

SYSTEMIC CIRCULATION: VEINS *(p. 361)*

4. Arteries in the neck and limbs are deep beneath the skin; in contrast, there are usually two sets of peripheral veins, one superficial and one deep. This dual venous drainage is important for controlling body temperature.

5. The **superior vena cava** receives blood from the head, neck, chest, shoulders, and arms. *(Figure 15-16)*

6. The **inferior vena cava** collects most of the venous blood from organs below the diaphragm. *(Figure 15-16)*

7. The **hepatic portal vein** collects blood from visceral organs in the abdominopelvic cavity. The hepatic portal vein delivers blood to capillary networks in the liver. *(Figure 15-17)*

FETAL CIRCULATION *(p. 365)*

8. Circulation changes at birth. Before birth, the developing fetus receives oxygen and nutrients through the **umbilical vein** from the placenta. *(Figure 15-18a)*

9. After birth, the lungs and pulmonary circuit become filled with blood as both the **foramen ovale** between the right and left atrium, and the **ductus arteriosus** between the pulmonary trunk and aortic arch are closed. *(Figure 15-18b)*

AGING AND THE CARDIOVASCULAR SYSTEM *(p. 366)*

1. Age-related changes occur in the blood, heart, and blood vessels.

2. Age-related changes in the blood can include (1) decreased hematocrit, (2) blockage of peripheral veins by a *thrombus* (stationary blood clot), and (3) pooling of blood in the veins of the legs.

3. Age-related changes in the heart include (1) a reduction in the maximum cardiac output, (2) changes in the activities of the nodal and conducting fibers, (3) replacement of damaged cardiac muscle fibers by scar tissue, and (4) blood vessel changes that can restrict coronary circulation.

4. Age-related changes in blood vessels are often related to **arteriosclerosis** and **atherosclerosis**.

DISORDERS OF THE HEART AND BLOOD VESSELS *(pp. 366–374)*

1. The leading cause of death in developed countries is heart disease. *(Figure 15-19)*

THE PHYSICAL EXAMINATION *(p. 366)*

2. A **cardiologist** is trained to diagnose and treat heart disease.

3. Common symptoms of cardiovascular disorders are *weakness* and *fatigue, cardiac pain, palpitations,* and *pain on exertion.*

4. Signs of cardiovascular disorders may include *edema* (an increase in tissue fluid), *dyspnea* (breathlessness), varicose veins, and changes in skin coloration (pallor or cyanosis).

5. Diagnosis of cardiovascular disorders is based on readings of pulse, blood pressure, and heart sounds taken during the physical examination; tests using scans, X-rays, and ECG recordings; and, laboratory tests of blood samples. *(Table 15-1)*

DISORDERS OF THE BLOOD VESSELS *(p. 368)*

6. Arteritis is the inflammation of an artery. **Phlebitis** is the inflammation of a vein.

7. Arteriosclerosis is a thickening and toughening of an artery. *Focal calcification* results in a hardening of the artery wall. Atherosclerosis is a process that leads to lipid deposits in the artery wall, and the formation of fatty **plaques**. A *thrombus* is a blood clot, and an *embolus* is a drifting blood clot. *(Figure 15-20)*

8. A bulge in a weakened section of an artery is an **aneurysm**.

9. Varicose veins are sagging and swollen veins caused by the pooling of blood in superficial veins due to problems with venous valves. **Hemorrhoids** are varicose veins that form in the anus.

10. Hypertension is abnormally high blood pressure. Antihypertensive drugs include *calcium channel blockers, beta-blockers, diuretics,* and *vasodilators.* **Hypotension** is abnormally low blood pressure.

11. Edema is an increased volume of interstitial fluid that results from an imbalance of blood pressure and osmotic forces across capillaries.

12. *Cerebrovascular disease* is caused by reduced blood supply to the brain. The blood flow may be disrupted by a *cerebrovascular accident (CVA),* or *stroke.* Common causes of strokes are *cerebral thrombosis, cerebral embolism,* and *cerebral hemorrhages.*

DISORDERS OF THE HEART *(p. 371)*

13. Carditis is an inflammation of the heart. **Endocarditis** and **myocarditis** are inflammations of different heart tissues. Cardiac output may be reduced as a result of **pericarditis**, an inflammation of the pericardium, through **cardiac tamponade** or *constrictive pericarditis.*

14. Rheumatic fever follows a streptococcal infection, usually in children. It may result in carditis that affects the heart valves and the myocardium. **Rheumatic heart disease (RHD)** is characterized by **valvular stenosis**, a narrowing and calcification of heart valves. Damaged valves are sometimes replaced with *prosthetic* (artificial) valves. *(Figure 15-21)*

15. Primary tumors of the heart are usually benign *myxomas;* a *sarcoma* is a malignant tumor.

16. Examples of congenital heart disorders include; **patent** (open) foramen ovale and patent ductus arteriosus, *ventricular septal defects,* and *tetralogy of Fallot.* All may be treated surgically. *(Figure 15-22)*

17. Coronary artery disease (CAD) is caused by degenerative changes in coronary circulation. **Coronary ischemia** is an inadequate blood flow to the heart. Chest pains, or *angina pectoris,* are the first symptoms of CAD. Treatment of angina includes changes in lifestyle, administration of drugs, or surgery to increase the blood flow. Surgical treatment includes **balloon angioplasty** and **coronary bypass graft (CBG)**. *(Figure 15-23)*

18. Shock is a circulatory crisis marked by low blood pressure and inadequate blood flow to body tissues. Shock may have different causes: **circulatory shock** (reduction in blood volume); **cardiogenic shock (inability of heart to maintain normal output); **obstructive shock** (heart expansion restricted); **neurogenic shock** (damage to vasomotor center or sympathetic nervous system); **septic shock** (bacterial infection); and **anaphylactic shock** (allergic reaction).

19. Cardiomyopathies are an assortment of diseases that lead to the degeneration of the myocardium. Treatment may involve a *heart transplant*.

Chapter Review

MATCHING

Match each item in Column A with the most closely related item in Column B. Use letters for answers in the spaces provided.

Column A	Column B
____ 1. epicardium	a. cardiac muscle cells
____ 2. right AV valve	b. cardiac pacemaker
____ 3. left AV valve	c. largest artery in body
____ 4. anastomoses	d. drains the liver
____ 5. myocardium	e. vasomotor center
____ 6. SA node	f. carries blood from digestive tract to the liver
____ 7. systole	g. mitral (bicuspid) valve
____ 8. diastole	h. interconnections between arteries
____ 9. hepatic vein	i. visceral pericardium
____ 10. hepatic portal vein	j. smallest arterial vessels
____ 11. aorta	k. inflammation of a vein
____ 12. arterioles	l. weakened arterial wall
____ 13. medulla oblongata	m. tricuspid valve
____ 14. saphenous vein	n. contractions of heart chambers
____ 15. phlebitis	o. largest superficial vein in body
____ 16. aneurysm	p. relaxation of heart chambers

MULTIPLE CHOICE

17. Blood supply to the muscles of the heart is provided by the _____ .
(a) systemic circulation
(b) pulmonary circulation
(c) coronary circulation
(d) hepatic portal system

18. Blood leaves the left ventricle by passing through the _____ .
(a) aortic semilunar valve
(b) pulmonary semilunar valve
(c) mitral valve
(d) tricuspid valve

19. The two-way exchange of substances between blood and body cells occurs only through _____ .
(a) veins   (b) capillaries
(c) arterioles   (d) arteries

20. The arteries of the pulmonary circulation differ from those of the systemic circulation in that they carry _____ .
(a) oxygen and nutrients
(b) deoxygenated blood
(c) oxygenated blood
(d) oxygen, carbon dioxide, and nutrients

21. The vein that collects most of the venous blood from below the diaphragm is the _____ .
(a) inferior vena cava
(b) superior vena cava
(c) azygos
(d) saphenous

22. In a blood pressure reading of 120/80, the 120 represents _____ and the 80 represents _____ .
(a) pulse pressure; mean arterial pressure
(b) systemic pressure; venous pressure
(c) diastolic pressure; systolic pressure
(d) systolic pressure; diastolic pressure

23. The fetus receives blood from the placenta through the _____ .
(a) umbilical arteries
(b) umbilical vein
(c) foramen ovale
(d) ductus arteriosus

24. Which one of the following hormones acts to reduce blood pressure and blood volume?
(a) EPO   (b) aldosterone
(c) ADH   (d) ANP

25. _____ is a heart disease that develops after a bacterial infection.
(a) thrombophlebitis
(b) edema
(c) coronary artery disease
(d) rheumatic fever

26. Circulatory shock may be caused by all of the following except _____ .
(a) myocardial infarction
(b) hemorrhaging
(c) dehydration
(d) severe burns

TRUE/FALSE

____ 27. The average heartbeat range for a healthy adult at rest is 40 to 50 beats per minute.

____ 28. Cardiac output is the amount of blood pumped by a ventricle in 1 minute.

____ 29. The vagus nerves carry nerve impulses that decrease heart rate.

____ 30. The normal heart sounds heard through a stethoscope are caused by the contractions of the ventricles.

____ 31. Blood is prevented from flowing backward in veins by valves.

____ 32. Nerve impulses control the contraction and relaxation of the arteriole walls at the entrance to capillary beds.

____ 33. Plaque is a fatty tissue that causes a narrowing of arteries in atherosclerosis.

____ 34. Ischemia is a condition of inadequate blood flow.

SHORT ESSAY

35. (a) What effect does sympathetic stimulation have on the heart? (b) What effect does parasympathetic stimulation have on the heart?

36. What circulatory changes occur at birth?

37. Why do capillaries permit the diffusion of materials, whereas arteries and veins do not?

38. Why is blood flow to the brain relatively continuous and constant?

39. List four common symptoms of cardiovascular disorders.

40. What is a "blue baby"?

APPLICATIONS

41. What is CAD? Explain why it can be so dangerous.

42. Two patients have aneurysms. In one, the aneurysm is located in the subclavian vein; in the other, it is in the right carotid artery. Which condition is more likely to be life-threatening? Why?

✔ Answers to Concept Check Questions

(p. 346) **1.** The semilunar valves on the right side of the heart guard the opening to the pulmonary artery. Damage to these valves would affect blood flow to the lungs. **2.** When the ventricles begin to contract, they force the AV valves to close, which pulls on the chordae tendineae, which then pull on the small muscles projecting from the ventricle wall. **3.** The wall of the left ventricle is more muscular than that of the right ventricle because the left ventricle has to generate enough force to propel the blood throughout all the body's systems except the lungs. The right ventricle has to generate only enough force to propel the blood a few centimeters to the lungs.

(p. 349) **1.** If the impulses from the atria were not delayed at the AV node, they would be conducted so quickly through the ventricles by the bundle branches and Purkinje fibers that the ventricles would begin contracting immediately, before the atria had finished their contraction. As a result, the ventricles would not be as full of blood as they could be and the pumping of the heart would not be efficient, especially during activity. **2.** Bradycardia refers to heart rates slower than normal, that is, less than 60 bpm compared with a normal heart rate of 70 to 75 bpm.

(p. 364) **1.** The left subclavian artery is the branch of the aorta that sends blood to the left shoulder and arm. **2.** The common carotid arteries carry blood to the head. A compression of one of the common carotid arteries would result in decreased blood flow to the brain and loss of consciousness or even death. **3.** Organs served by the celiac artery include the stomach, spleen, liver, and pancreas.

(p. 374) **1.** Muscular activity helps keep blood moving in the extremities. A lack of exercise and an occupation that requires an individual to sit or stand for long periods of time can lead to a pooling of blood in superficial veins of the limbs. The pooled blood reduces the effectiveness of the venous valves, leading to more pooling of blood, and swollen and distorted varicose veins. **2.** Cardiac output could be restricted or obstructed by pericarditis that leads to excess fluid in the pericardial sac (cardiac tamponade) or by an inelastic pericardium (constrictive pericarditis).

The Lymphatic System and Immunity

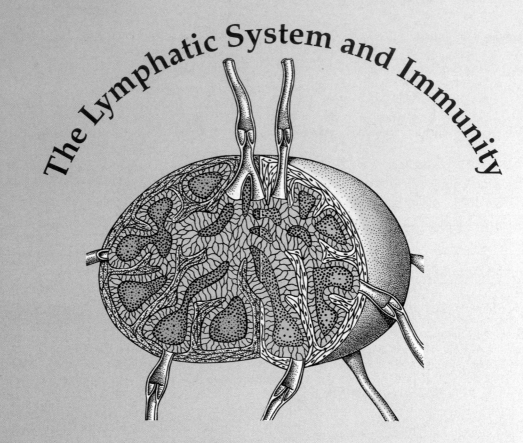

CHAPTER OBJECTIVES

1 Identify the major components of the lymphatic system and explain their functions.

2 Discuss the importance of lymphocytes and describe where they are found in the body.

3 List the body's nonspecific defenses and explain how each functions.

4 Describe the different categories of immunity.

5 Distinguish between cell-mediated immunity and antibody-mediated immunity.

6 Discuss the different types of T cells and the role played by each in the immune response.

7 Describe the primary and secondary immune responses to antigen exposure.

8 Relate allergic reactions and autoimmune disorders to immune mechanisms.

9 Describe examples of disorders of the lymphatic vessels, lymphoid tissue, lymphoid organs, and lymphocytes.

10 Distinguish among the three types of abnormal immune responses.

The **lymphatic system** works closely with the cardiovascular system. The vessels of this system return interstitial fluid to the venous system, from its origin at tissue capillaries. ∞ *[p. 354]* The lymphatic system also defends the body from infection and diseases, through the process of *immunity*. Immunity is a complex process that relies on specialized cells, vessels, and organs. This chapter describes the lymphatic system and examines the mechanisms responsible for its varied functions.

SYSTEM BRIEF

The primary functions of the lymphatic system include the following:

1. **The return of fluid from body tissues to the blood.** The return of 3.6 l (nearly 1 gal) of tissue fluids through the lymphatic system helps maintain the normal blood volume and prevent tissue swelling.

2. **The distribution of hormones, nutrients, and waste products from their tissues of origin to the general circulation.** Substances unable to enter the bloodstream directly may do so through the lymphatic vessels. For example, many lipids absorbed by the digestive tract form droplets too large to enter local capillaries. Instead, they bypass the capillaries and enter lymphatic vessels that empty into the venous system.

3. **The defense of the body from infection and disease.** In its protective role, the lymphatic system relies primarily on *lymphocytes*, a class of white blood cells introduced in Chapter 14. ∞ *[p. 321]* Lymphocytes respond to the presence of (1) disease-producing microorganisms (**pathogens**), such as bacteria or viruses; (2) abnormal body cells, such as virus-infected cells or cancer cells; and (3) foreign proteins, such as the toxins released by some bacteria. Lymphocytes use a combination of physical and chemical attack to overcome or eliminate these threats. The resistance to injuries and diseases caused by foreign chemical compounds and pathogens is called **immunity**.

STRUCTURE OF THE LYMPHATIC SYSTEM

The lymphatic system is made up of a network of *lymphatic vessels*, a fluid called *lymph*, and *lymphoid organs* that contain large numbers of lymphocytes. The lymph carried by the lymphatic vessels flows through the lymphatic organs before it is delivered to the venous system. Lymphocytes and other protective cells within the lymphatic organs filter foreign matter and pathogens from the lymph.

Figure 16-1• provides the names and general locations of the vessels and organs (lymph nodes, spleen, and thymus) of this system. It also shows the overall drainage pattern of lymph from the body into the right and left subclavian veins.

LYMPH AND THE LYMPHATIC VESSELS

The interstitial fluid forced out of the capillaries enters tiny lymphatic vessels that begin in all parts of the body (except the CNS). Once inside these vessels the fluid is called lymph. **Lymph** resembles blood plasma but contains a much lower concentration of proteins. The movement of lymph in lymphatic vessels is in one direction only, and ends when the lymph flows into the venous system.

The smallest lymphatic vessels begin as blind pockets and are called **lymphatic capillaries** (Figure 16-2•). Like blood capillaries, their walls consist of a single layer of cells. Interstitial fluid enters the lymphatic capillaries through small gaps between these endothelial cells. From the lymphatic capillaries, lymph flows into larger lymphatic vessels. Like veins, these larger lymphatics contain valves to prevent the backflow of fluid. Pressures within the lymphatic system are extremely low, and the valves are essential to maintaining normal lymph flow. Normal con-

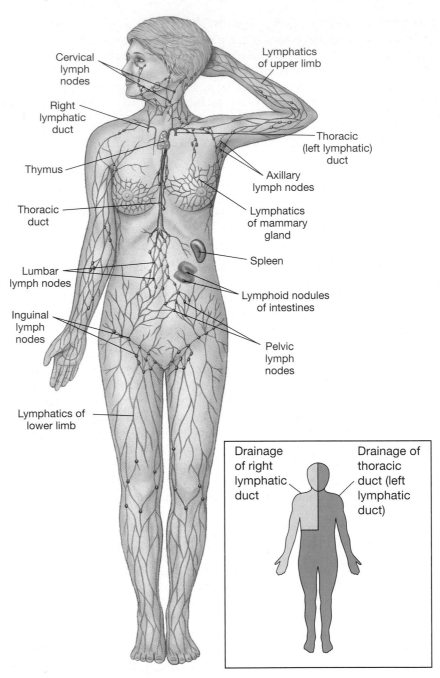

Cervical
lymph
nodes

Right
lymphatic
duct

Thymus

Thoracic
duct

Lumbar
lymph nodes

Inguinal
lymph
nodes

Lymphatics of
lower limb

Lymphatics
of upper limb

Thoracic
(left lymphatic)
duct

Axillary
lymph nodes

Lymphatics
of mammary
gland

Spleen

Lymphoid nodules
of intestines

Pelvic
lymph
nodes

Drainage
of right
lymphatic
duct

Drainage of
thoracic
duct (left
lymphatic
duct)

tractions of surrounding skeletal muscles squeeze the lymphatics and move the
lymph toward the venous system.

The lymphatic vessels empty into two large collecting ducts. As Figure 16-1•
shows, the **thoracic duct** collects lymph from the lower abdomen, pelvis, and
lower limbs, and from the left half of the head, neck, and chest. It empties its collected lymph into the venous system near the junction between the left internal
jugular vein and the left subclavian vein. The smaller **right lymphatic duct**, which
ends at the comparable location on the right side, delivers lymph collected from
the right side of the body superior to the diaphragm.

LYMPHOID TISSUES AND ORGANS

Lymphoid tissues are made up of connective tissue and lymphocytes. A **lymphoid
nodule** is one type of lymphoid tissue. Its lymphocytes divide within a central
zone called the *germinal center*. Nodules filter surrounding tissue fluid (interstitial

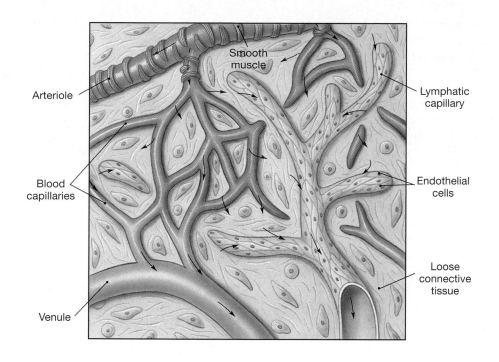

•**FIGURE 16-2**
Lymphatic Capillaries.
A three-dimensional view of the association of blood capillaries, body tissue, interstitial fluid, lymph, and lymphatic capillaries. Arrows show the direction of flow of interstitial fluid and lymph movement.

fluid) instead of lymph. About a millimeter in diameter, lymphoid nodules are found beneath the epithelia lining the respiratory, digestive, and urinary tracts. Because food is a common source of potentially dangerous material, the large number of lymphoid nodules associated with the small intestine provide an important defensive function for the body.

The **tonsils**, found within the walls of the pharynx, are familiar examples of lymphoid nodules. There are actually five tonsils: a single *pharyngeal tonsil*, or *adenoids*, a pair of *palatine tonsils*, and a pair of *lingual tonsils*. As Figure 16-3• shows, the tonsils guard the entrances to the digestive and respiratory tracts.

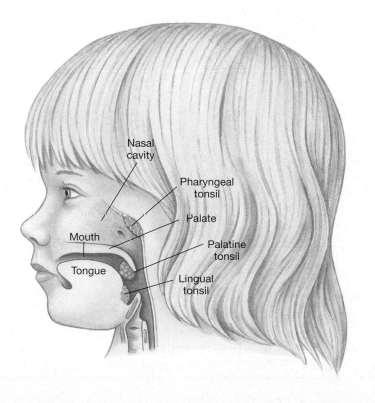

•**FIGURE 16-3**
The Arrangement of the Tonsils.
Diagrammatic view showing the position of the tonsils. The single pharyngeal tonsil (the adenoids) is located above the paired palatine and lingual tonsils. Notice the strategic position of the tonsils in relation to the internal opening to the nasal cavity and mouth.

The lymphocytes in a lymphoid nodule are not always able to destroy bacterial or viral invaders, and if pathogens become established in a lymphoid nodule, an infection develops. Two examples are probably familiar to you: *tonsillitis*, an infection of one of the tonsils (usually the pharyngeal tonsil), and *appendicitis*, an infection of lymphoid nodules in the *appendix*, an organ of the digestive tract.

Lymphoid organs, unlike lymphoid nodules, are separated from surrounding tissues by a fibrous capsule. Important lymphoid organs include the *lymph nodes*, the *thymus*, and the *spleen*.

Lymph Nodes

The kidney bean–shaped **lymph nodes** range in diameter from 1 to 25 mm. As Figure 16-4• shows, different lymphatic vessels carry lymph toward and away from a lymph node. The lymph node functions like a kitchen water filter: It filters and purifies the lymph before it reaches the venous system. "Purification" involves the removal of foreign molecules (antigens) and pathogens from the lymph by lymphocytes and phagocytic cells in the *cortex* and *medulla* of the lymph node. Lymph nodes are often called *lymph glands*, and "swollen glands" usually accompany tissue inflammation or infection. Such swelling occurs because the lymphocytes and phagocytic cells increase in number in order to deal with the infectious agents. Lymph nodes are located in regions where they can guard or trap harmful "intruders" before they reach vital organs of the body (see Figure 16-1•, p. 382).

•**FIGURE 16-4**
Structure of a Lymph Node.

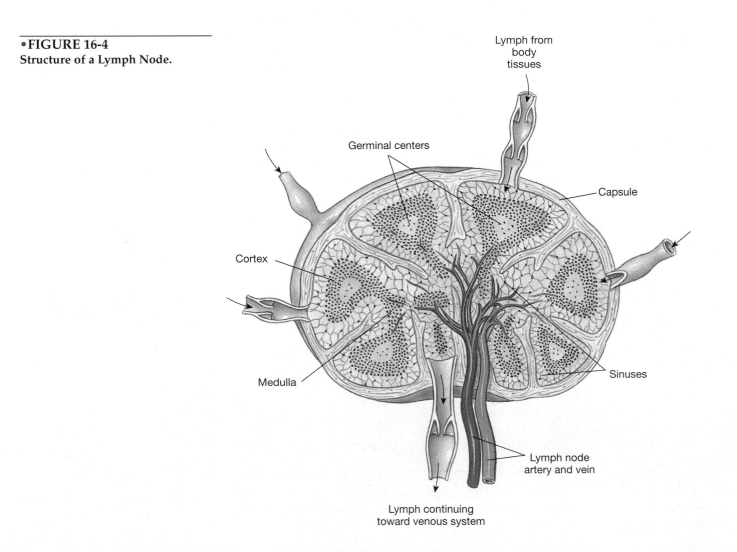

Lymph from body tissues

Germinal centers

Capsule

Cortex

Medulla

Sinuses

Lymph node artery and vein

Lymph continuing toward venous system

The Thymus

The **thymus** lies in the mediastinum posterior to the sternum. Within the thymus a class of lymphocytes called *T cells* become functionally mature. The thymus reaches its greatest size (relative to body size) in the first year or two after birth. It continues to enlarge up through puberty to about 35 g and then gradually decreases in size.

The thymus has two lobes, each divided into smaller compartments called *lobules*. Each lobule consists of a densely packed outer *cortex* and a central *medulla*. T cell lymphocytes divide in the cortex and as they mature, migrate into the medulla. From there they eventually enter a blood vessel for distribution throughout the body. Other cells within the lobules produce the thymic hormones (*thymosins*).

The Spleen

The adult **spleen** contains the largest collection of lymphoid tissue in the body. It is around 12 cm long and can weigh about 160 g. It lies wedged between the stomach, the left kidney, and the diaphragm. The spleen normally has a deep red color because of the blood it contains. The cellular components constitute the **pulp** of the spleen. Areas of *red pulp* contain large quantities of blood, whereas areas of *white pulp* resemble lymphoid nodules and contain lymphocytes.

The functions that the spleen performs for the blood are similar to those that the lymph nodes perform for lymph. The spleen (1) removes abnormal blood cells and components, (2) stores iron from recycled red blood cells, and (3) monitors and responds to pathogens and foreign antigens in the circulating blood.

As blood flows through the spleen, phagocytic cells engulf any damaged or infected cells. In addition, any microorganisms or other abnormal antigens will come to the attention of surrounding lymphocytes.

LYMPHOCYTES

Lymphocytes account for roughly one-fourth of the circulating white blood cell population. However, circulating lymphocytes are only a small fraction of the total lymphocyte population. The body contains around 10^{12} lymphocytes, with a combined weight of over a kilogram.

Types of Lymphocytes

There are three different classes of lymphocytes in the blood: **T cells**, **B cells**, and **NK cells** (natural killers). The letters T and B refer to the thymus and bone marrow. The T cells mature in the **thymus**, and B cells mature in **bone** marrow.

Three-fourths of the circulating lymphocytes are T cells. There are different groups of T cells, each with a different function. For example, one group directly attacks foreign cells or body cells infected by viruses. Others regulate the activities of other lymphocytes.

B cells, about one-eighth of the circulating lymphocytes, mature into **plasma cells**. Plasma cells, introduced in Chapter 14, are responsible for the production and secretion of **antibodies**, soluble proteins that are also known as **immunoglobulins**. ∞ *[p. 325]* These proteins react with specific chemical targets called **antigens**. Antigens are usually pathogens, parts or products of pathogens, or other foreign compounds. When an antigen-antibody complex forms, it starts a chain of events leading to the destruction of the target organism or compound.

The remaining one-eighth of circulating lymphocytes are **NK cells**. These lymphocytes readily attack foreign cells, normal cells infected with viruses, and cancer cells that appear in normal tissues. Their continual monitoring of body tissues is called *immunological surveillance*.

Origin and Circulation of Lymphocytes

Lymphocytes constantly move throughout the body; they wander through a tissue and then enter a blood vessel or lymphatic for transport to another site. In general, lymphocytes have relatively long life spans. Many survive for up to 4 years, and some last 20 years or more. Throughout life, normal lymphocyte populations are maintained through the divisions of stem cells in the bone marrow and lymphoid tissues.

Lymphocyte production involves the bone marrow and thymus (Figure 16-5•). During this process, each B cell and T cell gains the ability to respond to the presence of a specific antigen, and NK cells gain the ability to recognize abnormal cells.

Unspecialized cells in the bone marrow produce lymphoid stem cells with two distinct fates. One group remains in the bone marrow and generates NK cells and B cells that enter the circulation. The second group of lymphoid stem cells migrates to the thymus. Under the influence of thymic hormones, these cells divide repeatedly, producing large numbers of T cells that reenter the circulation. As these lymphocyte populations migrate through different body tissues, they retain the ability to divide and produce daughter cells of the same type. For example, division of a B cell produces other B cells, not T cells or NK cells. The ability to increase the number of lymphocytes of a specific type is important to the success of the immune response.

✔ How would blockage of the thoracic duct affect the circulation of lymph?

✔ If the thymus gland failed to produce the thymosins, what particular population of lymphocytes would be affected?

✔ Why do lymph nodes enlarge during some infections?

•**FIGURE 16-5**
The Origin of Lymphocytes.
Lymphocytic stem cells have two different fates. One group remains in the bone marrow, producing daughter cells that mature into B cells and NK cells. The second group migrates to the thymus, where their daughter cells mature into T cells. All three lymphocyte types circulate throughout the body in the bloodstream, leaving the circulation to take temporary residence in various body tissues.

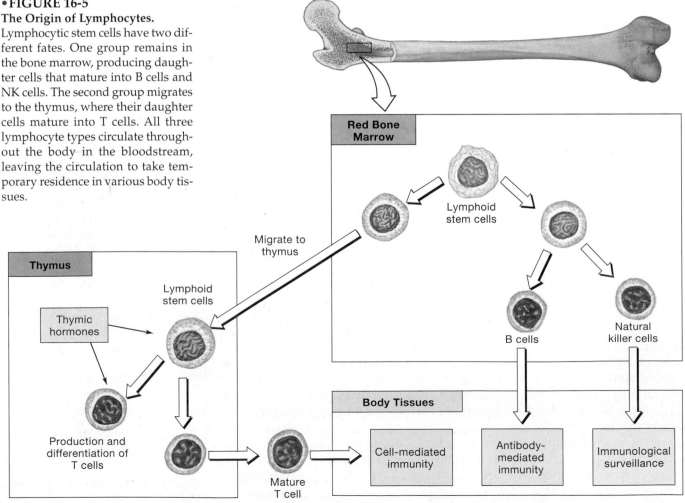

BODY DEFENSES

There are two defense strategies that act together to protect the human body, one general and one specific. General **nonspecific defenses** do not discriminate between one threat and another. These defenses, which are present at birth, include *physical barriers*, *phagocytic cells*, *immunological surveillance*, *interferon*, *complement*, *inflammation*, and *fever*. They provide the body with a defensive capability known as *nonspecific resistance*. **Specific defenses** provide protection against threats on an individual basis. For example, a specific defense may protect against infection by one type of bacteria but ignore other bacteria and viruses. Specific defenses are dependent on the activities of lymphocytes. Together, they produce a state of protection known as *specific resistance*, or immunity.

NONSPECIFIC DEFENSES

Nonspecific defenses deny entrance to, or limit the spread of, microorganisms or other environmental hazards.

Physical Barriers

To cause trouble, a foreign compound or pathogen must enter the body tissues, and that means crossing an epithelium. The epithelial covering of the skin, described in Chapter 7, has multiple layers, a "waterproof" keratin coating, and a network of tight seams that lock adjacent cells together. ∞ *[p. 118]* These create a very effective barrier that protects underlying tissues.

The exterior surface of the body is also protected by hairs that provide some protection against mechanical abrasion (especially on the scalp) and glandular secretions. Secretions from sebaceous and sweat glands flush the surface, washing away microorganisms and chemical agents. Some secretions also contain bacteria-killing chemicals, destructive enzymes (*lysozymes*), and antibodies.

The epithelia lining the digestive, respiratory, urinary, and reproductive tracts are more delicate, but they are equally well defended. Mucus bathes most surfaces of the digestive tract, and the stomach contains a powerful acid that can destroy many potential pathogens. Mucus moves across the lining of the respiratory tract, urine flushes the urinary passageways, and glandular secretions do the same for the reproductive tract.

Phagocytes

Phagocytes in body tissues remove cellular debris and pathogens by engulfing by phagocytosis, (cell eating). ∞ *[p. 45]* There are two general classes of phagocytic cells. **Microphages**, or "small eaters," are the neutrophils and eosinophils normally found in the circulating blood. These phagocytic cells leave the bloodstream and enter tissues subjected to injury or infection. **Macrophages**, or "large eaters," are large cells derived from white blood cells called monocytes. ∞ *[p. 322]* Almost every tissue in the body shelters resident or visiting macrophages. In some organs they have special names: For example, *microglia* are macrophages inside the CNS, and *Kupffer* (KOOP-fer) *cells* are found in and around blood channels in the liver. This relatively spread out collection of phagocytic cells in the body is called the monocyte-macrophage system.

Immunological Surveillance

The immune system generally ignores normal cells in the body's tissues, but abnormal cells are attacked and destroyed. The constant monitoring of normal tissues, called **immunological surveillance**, primarily involves the lymphocytes known as NK (natural killer) cells (see Figure 16-5•). These cells are sensitive to the presence of antigens characteristic of abnormal cell membranes. When they encounter these antigens on a cancer cell or a cell infected with viruses, NK cells secrete proteins that kill the abnormal cell by destroying its cell membrane.

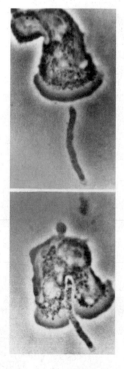

Phagocytosis of a bacterium by a human macrophage.

Unfortunately, some cancer cells avoid detection. They can then multiply and spread without interference by NK cells.

Interferons

Interferons are small proteins released by activated lymphocytes, macrophages, and by tissue cells infected with viruses. Interferons cause normal, uninfected cells to produce antiviral compounds that interfere with viral replication. They also stimulate macrophages and NK cells.

Complement

A group of *complement proteins* in the blood "complements," or supplements, the action of antibodies. These proteins interact with one another in chain reactions comparable to those of the clotting system. Their activation enhances phagocytosis, destroys cell membranes, and promotes *inflammation*.

Inflammation

Inflammation is a localized tissue response to injury. Introduced in Chapter 4, inflammation produces local sensations of swelling, redness, heat, and pain. ∞ *[p. 72]* Inflammation can be produced by any stimulus that kills cells or damages loose connective tissue. When stimulated by local tissue damage, special connective tissue cells called *mast cells* release histamine and heparin into the interstitial fluid. These chemicals initiate the process of inflammation.

Figure 16-6• follows the different actions that occur during inflammation of the skin. These actions serve to slow the spread of pathogens away from the injury, prevent the entry of additional pathogens, and set into motion a wide range of defenses that can overcome the pathogens and make permanent tissue repairs. The repair process is called *regeneration*.

Fever

A **fever** is a continued body temperature greater than 37.2°C (99°F). The hypothalamus contains groups of nerve cells that regulate body temperature and act as the body's thermostat. ∞ *[p. 6]* Macrophages, pathogens, bacterial toxins, and antigen-antibody complexes can reset the thermostat and cause a rise in body temperature.

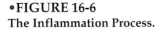

•FIGURE 16-6
The Inflammation Process.

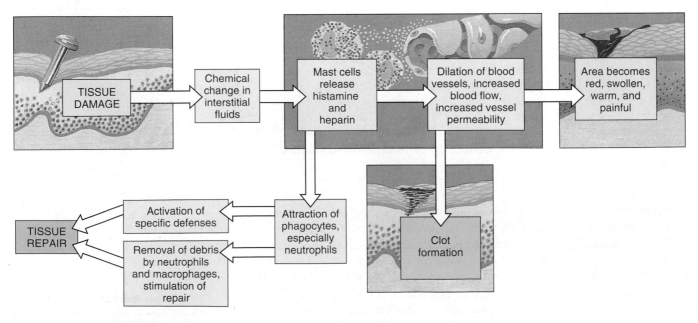

Within limits, a fever may be beneficial. For example, high body temperatures speed up the activities of the immune system. However, high fevers (over 40° C, or 104° F) can damage many different organ systems. For example, a high fever can cause CNS problems such as nausea, disorientation, hallucinations, or convulsions.

SPECIFIC DEFENSES AND IMMUNITY

Immunity is the resistance to injuries and diseases caused by *specific* foreign chemical compounds and pathogens. Such immunity, or *specific resistance*, may be inborn or may develop after birth. A number of different types of immunity are recognized.

Types of Immunity

The relationships between *innate* immunity and the different forms of *acquired* immunity are diagrammed in Figure 16-7•.

Innate immunity is inherited; it is present at birth and has no relation to previous exposure to the antigen involved. Such *inborn* immunity is what protects us from diseases that may infect farm animals or pets. For example, people are not subject to the same diseases as goldfish, and vice-versa.

Acquired immunity is an acquired resistance to infection. It may develop actively or passively. Active immunity may develop naturally, as a result of exposure to an antigen in the environment, or artificially, from deliberate exposure to an antigen. Naturally acquired immunity normally begins to develop after birth, and it is continually updated and broadened as the individual encounters new pathogens or other antigens. The purpose of artificially acquired immunity is to stimulate production of specific antibodies under controlled conditions so that

✔ What types of cells would be affected by a decrease in the monocyte-forming cells in the bone marrow?

✔ A rise in the level of interferon in the body would suggest what kind of infection?

•FIGURE 16-7
Types of Immunity.

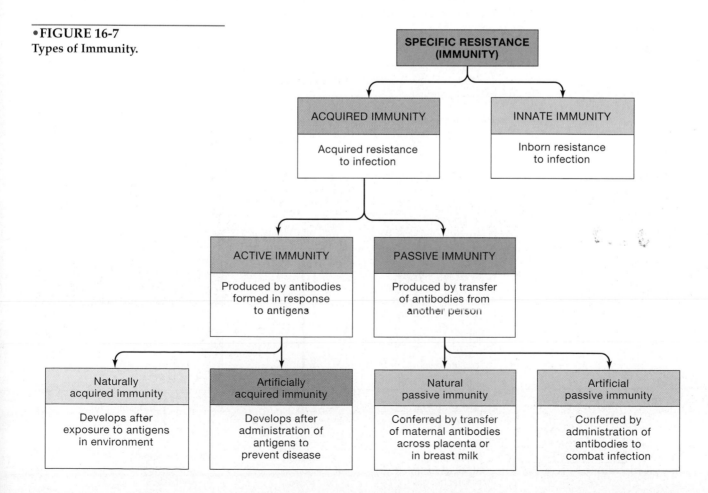

the individual will be able to overcome natural exposure to the pathogen at some future time. This is the basic principle behind *immunization* to prevent disease.

Immunization is accomplished by the administration of a **vaccine** (vak-SĒN), a preparation of antigens derived from a specific pathogen. The vaccine may be given orally or by injection. Most vaccines consist of the pathogenic organism, in whole or in part, living (but weakened) or dead. In some cases a vaccine contains one of the metabolic products of the pathogen.

Passive immunity is produced by the administration of antibodies produced by another individual. This occurs naturally when antibodies produced by the mother cross the placental barrier to provide protection against embryonic or fetal infections. Antibodies may also be administered to fight infection or prevent disease. Usually, antibodies are given to an individual who has already been exposed to a dangerous toxin or pathogen. In such cases, there is generally not enough time for the person to develop his or her own antibodies. Examples of antibody preparations given in such conditions include the antivenoms used to treat bites or stings from fishes, snakes, and spiders. Passive immunization provides only short-term resistance to infection, for the antibodies are gradually removed from circulation and are not replaced.

The Immune Response

The goal of the *immune response* is to destroy or inactivate pathogens, abnormal cells, and foreign molecules such as toxins. Active immunity appears after lymphocytes are exposed to an antigen. The resulting immune response depends on whether the activated lymphocyte is a T cell or a B cell. Activated T cells mount a direct attack on foreign or infected cells through a process called **cell-mediated immunity**. Activated B cells produce plasma cells, which produce antibodies that attack specific antigens. Because the blood transports the antibodies, B cells are said to be responsible for *humoral* ("liquid") *immunity*, or **antibody-mediated immunity**. Figure 16-8• compares these two different aspects of the immune response.

•**FIGURE 16-8**
An Overview of the Immune Response.

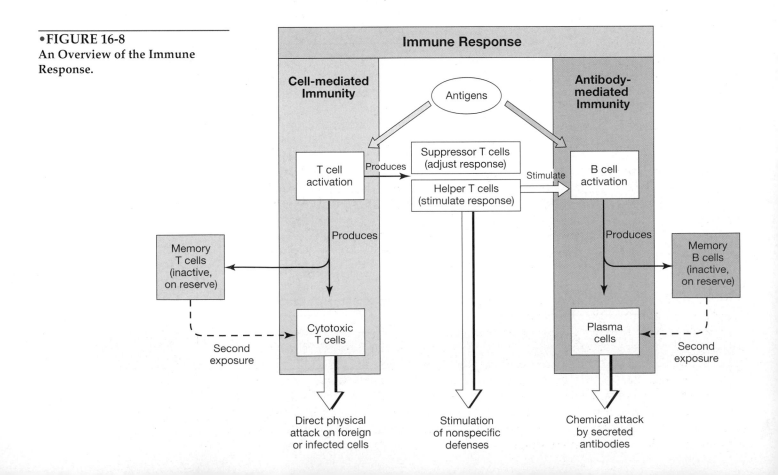

T Cell Immunity

Foreign antigens must usually be processed by macrophages and displayed on their cell membrane before they can activate T cells. T cells can also be activated by viral antigens displayed on the surface of virus-infected cells.

T cells activated by specific antigens divide and differentiate into four types of cells with different functions in the immune response. These cells are *cytotoxic T cells, memory T cells, suppressor T cells,* and *helper T cells.*

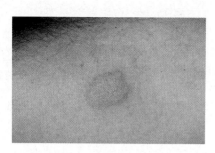

Inflammation on the forearm caused by cell-mediated immunity during a positive skin test for tuberculosis.

- **Cytotoxic T cells** are responsible for cell-mediated immunity. These cells track down and attack the bacteria, fungi, protozoa, or foreign tissues that contain the target antigen. For example, cytotoxic T cells are responsible for the rejection of skin grafts or organ transplants from other individuals. A cytotoxic T cell may kill its target cell by disrupting its metabolism with a poisonous compound, destroying its cell membrane, or activating genes that cause it to commit suicide.

- **Memory T cells** remain "in reserve." If the same antigen appears a second time, these cells will immediately differentiate into cytotoxic T cells, producing a more rapid and effective cellular response.

- **Suppressor T cells** depress the responses of other T cells and B cells. Because suppressor T cells take much longer to become activated than other types of T cells, they act *after* the initial immune response. In effect, these cells "put on the brakes" and limit the degree of the immune response to a single stimulus.

- **Helper T cells** release a variety of chemical compounds (cytokines) that (1) coordinate specific and nonspecific defenses and (2) stimulate the production of more T cells and the production of *antibodies* by plasma B cells. Helper T cells are the target cells of HIV, the virus that causes AIDS.

B Cell Immunity

B cell activation occurs in response to exposure to a specific antigen and to chemicals, called *interleukins,* secreted by helper T cells. Interleukins, which are also secreted by macrophages, are important to many aspects of the immune response. Once activated, a B cell divides several times, producing daughter cells that differentiate into **plasma cells** and **memory B cells**. Plasma cells begin synthesizing and secreting large numbers of antibodies with the same antigen target. Memory B cells perform the same role for antibody-mediated immunity that the memory T cells perform for cellular-mediated immunity. They remain in reserve to deal with subsequent exposure to the same antigens. At that time, they will respond and differentiate into antibody-secreting plasma cells.

Primary and Secondary Responses to Antigen Exposure

The initial response to antigen exposure is called the **primary response**. When an antigen appears a second time, it triggers a more extensive and prolonged **secondary response**. The magnitude of the secondary response is due to the presence of large numbers of memory cells that are already "primed" for the arrival of the antigen.

Because there is a lag between the activation of B cells, the production of plasma cells, and the secretion of antibodies, the primary response does not appear immediately. Instead, there is a gradual, sustained rise in the concentration of circulating antibodies in the blood that does not peak until several weeks after the initial exposure. Thereafter the antibody concentration declines, assuming that the individual is no longer exposed to the antigen.

Memory B cells do not differentiate into plasma cells unless they are exposed to the same antigen a second time. If and when that exposure occurs, these cells respond immediately, dividing and becoming plasma cells that secrete antibodies in massive quantities. This represents the secondary response to antigen exposure.

The secondary response produces an immediate rise in antibody concentrations to levels many times higher than those of the primary response. The secondary response appears even if the second exposure occurs years after the first, for memory cells are long-lived, potentially surviving for 20 years or more.

✚ Clinical Note

Why Do We Need Booster Shots?

Immunization against infection by various types of bacteria and viruses is commonly induced by using vaccines composed of weakened, or **attenuated** (a-TEN-ū-ā-ted), bacteria and viruses, or by **inactivated** or "killed" vaccines that consist of bacterial cell walls or the outer protein coat of viruses. Despite being weakened, live bacteria and viruses may produce mild symptoms comparable to those of the actual disease. Inactivated vaccines have the advantage of not producing even mild symptoms of the disease.

Unfortunately, inactivated vaccines do not stimulate as strong an immune response and so do not confer as long-last-ing an immunity as do vaccines using attenuated pathogens. In the years following vaccination, the antibody level in the blood declines, and the immune system eventually becomes unable to produce an adequate secondary response. As a result, the immune system must be "reminded" of the antigen periodically by the administration of *boosters*. Influenza (flu), typhoid, cholera, typhus, plague, and injected polio vaccines use inactivated viruses or bacteria. In some cases, the vaccine consists of fragments of the bacterial or viral walls, or their toxic products. A tetanus shot, for example, contains small quantities of the toxin released by these bacteria.

✔ A decrease in the number of cytotoxic T cells would affect what type of immunity?

✔ How would a lack of helper T cells affect the antibody-related immune response?

✔ Would the primary response or the secondary response be more affected by a lack of memory B cells for a particular antigen?

AGING AND THE IMMUNE RESPONSE

With advancing age the immune system becomes less effective at combating disease. T cells become less responsive to antigens, and so fewer cytotoxic T cells respond to an infection. Because the number of helper T cells is also reduced, B cells are less responsive, and antibody levels do not rise as quickly after antigen exposure. The net result is an increased susceptibility to viral and bacterial infection. For this reason, vaccinations for acute viral diseases, such as the flu (influenza), are strongly recommended for elderly individuals. The increased incidence of cancer in the elderly reflects the fact that immune surveillance declines, and tumor cells are not eliminated as effectively.

DISORDERS OF THE LYMPHATIC SYSTEM AND THE IMMUNE RESPONSE

As we've learned, the lymphatic system has three functions in the body: (1) to transport excess interstitial fluid to the bloodstream, (2) to distribute hormones, nutrients, and waste products, and (3) to protect the body through the immune response, or process of immunity. Disorders of the lymphatic system may involve problems with specific components (such as lymphocytes, vessels, tissues, and organs) or abnormal functioning of the immune response (Figure 16-9●).

THE PHYSICAL EXAMINATION AND THE LYMPHATIC SYSTEM

Individuals with lymphatic system disorders may experience a variety of symptoms. The observed symptoms depend on whether the problem affects the circulatory or immune functions of the lymphatic system. Important symptoms and signs include the following:

■ Enlarged lymph nodes are typically found during infection. They also develop in cancers of the lymphatic system or when primary tumors in other tissues have spread to nearby lymph nodes. The status of regional lymph nodes can therefore be important in the diagnosis and treatment of many different cancers. On palpation, cancerous nodes feel like dense masses rather than individual lymph nodes. In contrast, infected lymph nodes tend to be large, freely mobile, and very tender. Enlargement of the lymph nodes is called **lymphadenopathy** (lim-fad-e-NOP-ah-thē).

■ Red streaks on the skin may develop as a result of infection and inflammation of superficial lymph vessels.

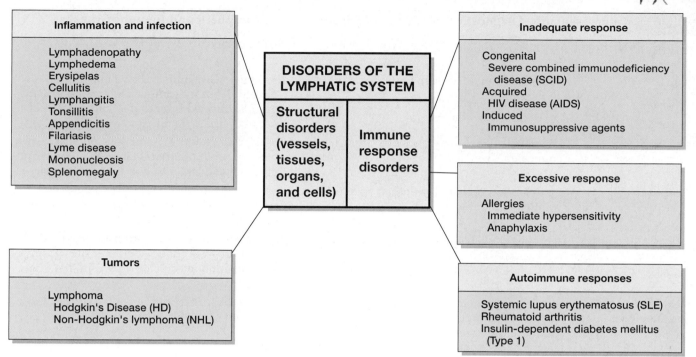

•FIGURE 16-9
Disorders of the Lymphatic System; Structures and Immune Response.

■ An enlargement of the spleen may result from acute infections. The spleen can be examined through palpation or percussion to detect splenic enlargement.

■ Weakness and fatigue typically accompany many different lymphatic disorders.

■ Skin lesions such as hives or urticaria can develop during allergic reactions. ∞ *[p. 124]* Immune responses to a variety of allergens, including animal hair, pollen, dust, medications, and some foods may cause such lesions.

■ Respiratory problems may accompany the allergic response to allergens such as pollen, animal hair, dust, and mildew.

■ When circulatory functions are impaired, the most common sign is **lymphedema**, a tissue swelling caused by the buildup of interstitial fluid. Lymphedema can result from a lymphatic blockage.

Diagnostic procedures and laboratory tests used to detect disorders of the lymphatic system are listed in Tables 16-1 and 16-2.

DISORDERS OF THE LYMPHATIC SYSTEM

Disorders of the components of the lymphatic system are primarily associated with inflammation and infection, or tumors. Disorders of the immune response are discussed in a later section.

Inflammation and Infection

Complications of Inflammation. In the period after an injury, tissue conditions generally become worse before they begin to improve. **Necrosis** (ne-KRŌ-sis) refers to the tissue degeneration that occurs after cells have been injured or destroyed. It is caused by the breakdown of lysosomes and the release of their digestive enzymes that first destroy the injured cells and then attack surrounding tissues. ∞ *[p. 48]*

As local inflammation continues, debris, fluid, dead and dying cells build up at the site of the injury. This thick fluid mixture is known as **pus**. An accumulation of pus in an enclosed tissue space is called an *abscess*. ∞ *[p. 124]* In the skin, an

Table 16-1   Representative Diagnostic Procedures for Disorders of the Lymphatic and Immune Systems

Diagnostic Procedure	Method	Results/Uses
Skin tests:		
Prick test	A small amount of antigen is applied as a prick to the skin	Erythema, hardening and swelling around puncture area indicates a positive test; usually the area affected is measured and must be of minimal size to qualify for a positive test
Intradermal test	Antigen is injected into the skin to form a 1–2 mm "bleb"	*Positive test:* Within 15 minutes, a reddened wheal is produced that is larger than 5 mm in diameter and larger than in the control group
Patch test	A patch is impregnated with the antigen and applied to the skin surface	*Positive test:* The patch provokes an allergic response, usually over several hours to several days
Tuberculin skin test	Injection of tuberculin protein into the skin	Red, hardened area >10 mm wide around injection site 48–72 hours later indicates positive test and antibodies to the organism that causes tuberculosis
Nuclear scan and CT scans of spleen	Scan images provided through energy from radiation emitted from radionuclides or through X-ray waves to reveal position, shape, and size of the spleen	To detect abscesses, tumors of the spleen, and an infarct of splenic tissue
Biopsy of lymphoid tissue	Surgical excision of suspicious lymphoid tissue for pathological examination in the laboratory	Determination of potential malignancy and staging of cancers
Lymphangiography	Dye is injected into lymphatic vessels in a limb and travels to nodes; X-rays are taken as the dye accumulates in the lymph nodes	Used in the diagnosis and monitoring of lymphomas and lymphedema

abscess can form as pus builds up inside the fibrin clot that surrounds the injury site. If the cellular defenses succeed in destroying the invaders, the pus will either be absorbed or surrounded by a fibrous capsule forming a type of *cyst*. (Cysts can also form in the absence of infection.)

Erysipelas (er-i-SIP-e-las), or "red skin," is a widespread inflammation of the dermis caused by bacterial infection. If the inflammation spreads into the subcutaneous layer and deeper tissues, the condition is called **cellulitis** (sel-ū-LĪ-tis). Erysipelas and cellulitis develop when bacterial invaders break through the fibrous wall of the cyst. These are serious conditions that require prompt administration of antibiotics.

Lymphangitis. **Lymphangitis** is an inflammation of superficial lymph vessels characterized by *erythematous* (red) streaks that originate at a site of infection. Before the linkage to the lymphatic system was known, this sign was called "blood poisoning." Lymphangitis commonly occurs in the limbs (Figure 16-10•).

Tonsillitis. **Tonsillitis** is a bacterial infection of the lymphoid nodules of the pharynx, the tonsils (see Figure 16-3•, p. 383). Symptoms include a sore throat, high fever, and *leukocytosis* (an abnormally high white blood cell count). The affected tonsil (usually the pharyngeal) becomes swollen and inflamed, sometimes enlarging enough to partially block the entrance to the trachea. Breathing and swallowing then becomes difficult, and in severe cases, impossible. As the infection proceeds, an abscess may develop within the tonsil. The bacteria may enter the bloodstream from the abscess by passing through the lymphatic capillaries and vessels to the venous system.

In the early stages, antibiotics may control the infection, but once an abscess has formed, the best treatment involves surgical drainage of the abscess. **Tonsillectomy**, the removal of a tonsil or tonsils, was once routinely performed to prevent recurring infections of the tonsils. Questions have arisen, however, concerning the benefits of such surgery to the individual; the tonsils represent a "first line" of defense against bacterial invasion of the pharyngeal walls.

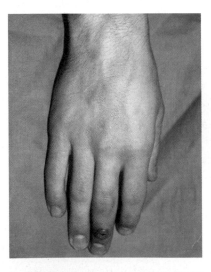

•FIGURE 16-10
Lymphangitis.
The red streaks mark the spread of the infection along inflamed lymphatic vessels.

Table 16-2	Examples of Laboratory Tests for Disorders of the Lymphatic and Immune Systems	
Laboratory Test	**Normal Values**	**Significance of Abnormal Values**
Complete blood count	See laboratory tests for blood disorders, Table 14-1 (p. 330)	
WBC count	Adults: 5,000–10,000/mm^3	Increased in chronic and acute infections, tissue death (MI, burns), leukemia, parasitic diseases, and stress. Decreased in aplastic and pernicious anemias and systemic lupus erythematosus (SLE) and with some medications.
Differential WBC count	Neutrophils: 50–70%	Increased in acute infection, myelocytic leukemia, and stress. Decreased in aplastic and pernicious anemias, viral infections, and radiation treatment and with some medications.
	Lymphocytes: 20–40%	Increased in chronic infections, lymphocytic leukemia, infectious mononucleosis, and viral infections. Decreased in radiation treatment, AIDS, and corticosteroid therapy.
	Monocytes: 2–8%	Increased in chronic inflammation, viral infections, and tuberculosis. Decreased in aplastic anemia and corticosteroid therapy.
	Eosinophils: 1–4%	Increased in allergies, parasitic infections, and some autoimmune disorders. Decreased in steroid drug therapy.
	Basophils: 0.5–1%	Increased in inflammatory processes and during healing. Decreased in hypersensitivity reactions and corticosteroid therapy.
Immunoglobulin electrophoresis (IgA, IgG, IgM)	Adults: IgA: 85–330 mg/dl IgM: 55–145 mg/dl IgG: 565–1,765 mg/dl IgD and IgE values should be minimal	Increased levels of IgG occur with infections; IgA levels increase with chronic infections and autoimmune disorders; IgE increases with allergic reactions and skin sensitivities; IgM levels are high in several infectious diseases, including liver disease
Total complement assay	Total complement: 41–90 hemolytic units	Total complement is decreased in SLE and glomerulonephritis and increased in rheumatic fever, rheumatoid arthritis, and certain types of malignancies
Rheumatoid factor test	Negative	Positive test indicates rheumatoid arthritis, but results may also be positive in SLE and myositis, other inflammatory conditions, and chronic infections
AIDS serology Enzyme-linked immunosorbent assay (ELISA)	Negative	Positive test indicates detection of antibodies against HIV. Tests given in the early stages of infection yield a negative result; positive results will not develop for several months. HIV-positive status is assigned after 2 different tests are positive for the antibodies and the Western blot is positive.
Western blot	Negative	Positive test indicates detection of antibodies to specific viral proteins

Appendicitis. **Appendicitis,** an inflammation of the appendix, generally follows a breakdown of the epithelial lining of the appendix. Bacteria that normally live within the large intestine then cross the epithelium and enter the underlying lymphoid tissues. Inflammation occurs, and the opening between the appendix and the rest of the intestinal tract may become constricted. Mucus secretion and pus formation accelerates, and the organ becomes increasingly distended. Eventually the swollen and inflamed appendix may rupture, or *perforate*. If this occurs, bacteria will be released into the abdominopelvic cavity, where they can cause a life-threatening *peritonitis*. The most effective treatment for appendicitis is the surgical removal of the organ, a procedure known as an **appendectomy.**

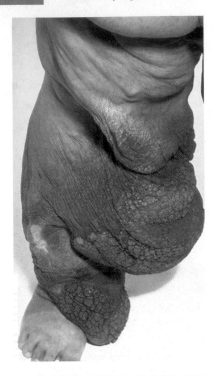

•FIGURE 16-11
Elephantiasis.
Elephantiasis of the leg is a chronic lymphadema caused by the parasitic infestation of lymph nodes and lymphatic vessels by roundworms.

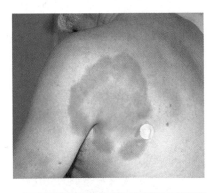

•FIGURE 16-12
Bull's Eye Rash of Lyme Disease.

Filariasis. Filariasis (fil-a-RĪ-a-sis) is a parasitic disease caused by roundworms that live within lymph nodes and lymphatic vessels. In this disease, larvae of a nematode (roundworm), usually *Wucheria bancrofti*, is transmitted from person to person by mosquitoes. Repeated scarring of the lymphatic passageways eventually blocks the movement of lymph and produces extreme lymphadema with the permanent distension of tissues. The limbs or external genitalia typically become grossly distended, a condition known as **elephantiasis** (el-e-fan-TĪ-a-sis) (Figure 16-11•). Early treatment usually prevents this complication.

Lyme Disease. **Lyme disease** is caused by *Borrelia burgdorferi*, a spirochete bacterium that normally lives in white-footed mice. The disease is transmitted to humans and other mammals, such as dogs and horses, through the bite of a tick that harbors that bacterium. Deer, which can carry infected adult ticks without becoming ill, have helped spread infected ticks through populated areas. The high rate of infection among children reflects the fact that they play outdoors during the summer, in fields where deer may also be found. A characteristic sign of Lyme disease is a skin rash that begins as a red bull's-eye centered on the bite (Figure 16-12•). Symptoms include fever, headache, enlarged lymph nodes, fatigue, and joint pain, swelling, and degeneration.

Many of the symptoms (fever, pain, skin rash) of Lyme disease develop in response to the release of *interleukin-1 (Il-1)* by activated macrophages. ∞ *[p. 391]* The cell walls of *B. burgdorferi* contain molecules that stimulate the secretion of Il-1 in large quantities. By stimulating the body's specific and nonspecific defense mechanisms, Il-1 exaggerates the inflammation, rash, fever, pain, and joint degeneration associated with the primary infection. Treatment for Lyme disease consists of the administration of antibiotics and anti-inflammatory drugs; a vaccine is now available.

Infectious Mononucleosis. **Infectious mononucleosis** is a chronic infection caused by a herpesvirus called the **Epstein-Barr virus (EBV)**. It is usually transmitted by oral contact and is also called "the kissing disease." The virus invades the lungs, bone marrow, and B cells in lymphoid organs. Symptoms include a fever, sore throat, fatigue, enlarged spleen, lymphadenopathy, or swelling of lymph nodes (especially those of the cervical region), and increased numbers of monocytes and lymphocytes in the blood.

The condition typically affects young adults (age 15 to 25) in the spring or fall. Treatment is symptomatic, as no drugs are effective against this virus. The most dangerous aspect of the disease is the risk of rupturing the enlarged spleen, which becomes fragile. Patients are therefore cautioned against heavy exercise and other activities that increase abdominal pressures. If the spleen does rupture, severe hemorrhaging may occur; death will follow unless transfusion and an immediate *splenectomy* are performed.

Splenomegaly. The spleen responds like a lymph node to infection, inflammation, or invasion by cancer cells. The enlargement that follows is called **splenomegaly** (splen-ō-MEG-a-lē), and *splenic rupture* may also occur under these conditions. An individual whose spleen is missing or nonfunctional has **hyposplenism** (hī-pō-SPLĒN-ism). Hyposplenism usually does not pose a serious problem. Hyposplenic individuals, however, are more prone to some bacterial infections, including infection by *Streptococcus pneumoniae*, than are individuals with normal spleens. Immunization against *S. pneumoniae* is therefore recommended. In **hypersplenism**, the spleen becomes overactive, and the increased phagocytic activities may lead to anemia (low RBC count), leukopenia (low WBC count), and thrombocytopenia (low platelet count). A splenectomy, surgical removal of the spleen, is the only current cure for hypersplenism.

Tumors

Lymphomas are malignant tumors consisting of cancerous lymphocytes found mainly within lymph nodes or the spleen. More than 30,000 cases of lymphoma are diagnosed in the United States each year, and that number has steadily been increasing. There are many types of lymphoma. One form, called **Hodgkin's disease (HD)**, accounts for roughly 40 percent of all lymphoma cases. Hodgkin's disease is more common in men than women and most commonly strikes individuals at ages 15–35 or those over age 50. The reason for this pattern of incidence is unknown. Although the cause of the disease is uncertain, an infectious agent (probably a virus) is suspected. Other types of lymphoma are usually grouped together under the heading of **non-Hodgkin's lymphoma (NHL)**. They are extremely diverse, and in most cases the primary cause remains a mystery.

The first symptom usually associated with any lymphoma is a painless enlargement of lymph nodes. The involved nodes have a firm, rubbery texture. Because the nodes are pain-free, the condition is typically overlooked until it has progressed and secondary symptoms appear. For example, patients seeking help for recurrent fevers, night sweats, gastrointestinal or respiratory problems, or weight loss may be unaware of any underlying lymph node changes. In the late stages of the disease, symptoms can include liver or spleen enlargement, central nervous system dysfunction, pneumonia, a variety of skin conditions, and anemia.

The most important factor influencing treatment selection is the stage of the disease. When diagnosed early, localized therapies may be effective. For example, the cancerous node(s) may be surgically removed and the region(s) irradiated to kill residual cancer cells. Success rates are very high when a lymphoma is detected in an early stage. Unfortunately, few lymphoma patients are diagnosed while in the early stages of the disease. The most effective treatment then may be some combination of surgery, local radiation, and chemotherapy.

DISORDERS OF THE IMMUNE RESPONSE

Disorders of the immune response can be sorted into three general categories, as were shown in Figure 16-9•:

- *Disorders resulting from an inadequate immune response.* Individuals with depressed immune defenses, such as occurs in *AIDS*, may develop life-threatening diseases caused by microorganisms that are harmless to other individuals.

- *Disorders resulting from an excessive immune response.* Conditions such as allergies can result from an immune response that is out of proportion with the size of the stimulus.

- *Disorders resulting from an inappropriate immune response.* **Autoimmune disorders** result when normal tissues are mistakenly attacked by T cells or the antibodies produced by activated B cells.

Inadequate Immune Response

An inadequate immune response is characteristic of *immunodeficiency diseases*. Such disorders are the result of congenital problems, acquired disorders such as viral infections, or induced disorders caused by treatments with immunosuppressive agents, such as radiation or drugs.

Congenital Disorders. Immunodeficiency diseases may result from problems with cell-mediated immunity and its T cells, with antibody-mediated immunity and its B cells, or both aspects of immunity. Individuals with **severe combined immunodeficiency disease (SCID)** fail to develop either cellular or antibody-mediated immunity due to a lack of functional T cells and B cells. Such individuals are unable to provide an immune defense, and even a mild infection can prove fatal. Infants with SCID usually die within the first year of life. However, isolation, bone marrow transplants, and gene therapy have helped some SCID patients.

Acquired Disorders. Acquired immune deficiency syndrome (AIDS), or late-stage HIV disease, is caused by a virus known as **human immunodeficiency virus (HIV)**. HIV is an example of a retrovirus, a virus that carries its genetic information in RNA, rather than DNA. Several types of immune cells, including

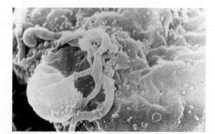

A scanning electron micrograph of a lymphocyte infected with HIV. The dots scattered across the surface of the cell are individual viruses.

macrophages can be infected by HIV, but it is the infection of helper T cells that leads to clinical problems. Cells infected with HIV are ultimately killed. The gradual destruction of helper T cells impairs the immune response, because these cells play a central role in coordinating immune responses. Circulating antibody levels decline, cellular immunity is reduced, and the body is left without defenses against a wide variety of microbial invaders. With immune function so reduced, ordinarily harmless pathogens can initiate lethal infections, known as opportunistic infections.

Infection with HIV occurs through intimate contact with the body fluids of infected individuals. Although all body fluids carry the virus, the major routes of transmission involve contact with blood, semen, or vaginal secretions. Most AIDS patients become infected through sexual contact with an HIV-infected person (who may not necessarily be suffering from the clinical symptoms of AIDS). The next largest group of patients consists of intravenous drug users who shared contaminated needles. A relatively small number of individuals have become infected with the virus after receiving a transfusion of contaminated blood or blood products. Finally, an increasing number of infants are born with the disease, having acquired it in the womb from infected mothers. At the current rate of increase there may be 40 million to 110 million infected individuals worldwide by the year 2000.

The best defense against AIDS consists of avoiding sexual contact with infected individuals. All forms of sexual intercourse carry the potential risk of viral transmission. The use of synthetic (latex) condoms greatly reduces the chance of infection (although it does not completely eliminate it). Condoms that are not made of synthetic materials are effective in preventing pregnancy but do not block the passage of viruses.

Despite intensive efforts, a vaccine has yet to be developed that will provide immunity from HIV infection. However, three different organizations (the World Health Organization, the U.S. Army, and the National Institute for Allergy and Infectious Diseases) have announced that they either have chosen or are choosing sites for vaccine trials. While efforts continue to prevent the spread of HIV, the survival rate for AIDS patients has been steadily increasing because new drugs are available that slow the progress of the disease, and improved antibiotic therapies help combat secondary infections. This combination is extending the life span of patients while the search for more effective treatment continues.

Induced Disorders. *Immunosuppressive agents* such as radiation and immunosuppressive drugs may destroy stem cells and lymphocytes, leading to complete loss of immune response. The treatment of late-stage lymphoma involves the use of radiation to destroy all the tumor cells in the body, including stem cells in the bone marrow. At that point, a **bone marrow transplantation** may be attempted. Barring infections and bleeding after surgery, the donor cells spread within the marrow and begin producing new RBCs, and WBCs within 2 weeks.

The transplantation of bone marrow and other organs, such as kidneys, liver, heart, lungs, and pancreas, requires that the tissues of the donor and host be as similar to each other as possible to avoid graft rejection. *Graft rejection* will not occur if the two individuals are identical, and so most grafts between identical twins are successful. To find those with compatible tissues, lymphocytes are collected and examined to determine the degree of similarity between the tissues of siblings and unrelated individuals. Because there will usually still be some differences in tissue types, immunosuppressive drugs must be administered to prevent graft rejection. New discoveries and improvements in such drugs are concerned with making their affects more specific, such as only inhibiting lymphocytes and not all aspects of the immune response.

Excessive Immune Response

Allergies are excessive immune responses to antigens. The sudden increase in T cell activity or antibody production can have unpleasant side effects. For example, neutrophils or cytotoxic T cells may destroy normal cells while attacking the antigen, or the antigen-antibody complex may trigger a massive inflammation. Antigens that trigger allergic reactions are called **allergens**.

There are several types of allergies. The most common is *immediate hypersensitivity*. One form, *allergic rhinitis*, includes "hay fever" and environmental allergies. **Immediate hypersensitivity** begins with **sensitization**. Sensitization is the initial exposure to an allergen that leads to the production of antibodies. Due to the lag time needed to activate B cells, produce plasma cells, and produce antibodies, the first exposure does not produce symptoms. These antibodies become attached to basophils and mast cells throughout the body. ∞ *[pp. 322, 388]* When exposed to the same allergen at a later date, these cells are stimulated to release histamine, heparin, prostaglandins, and other chemicals into the surrounding tissues. The result is a sudden, massive inflammation of the affected tissues.

The severity of the allergic reaction depends on the person's sensitivity and on the location involved. If allergen exposure occurs at the body surface, the response may be restricted to that area. If the allergen enters the systemic circulation, the response may be more widespread and perhaps lethal.

In **anaphylaxis** (a-na-fi-LAK-sis), a circulating allergen affects mast cells throughout the body. The entire range of symptoms can develop within minutes. Swellings and edema appear in the dermis, the respiratory passages contract, and breathing becomes difficult. In severe cases of anaphylaxis, an extensive peripheral vasodilation occurs, producing a fall in blood pressure that may lead to circulatory collapse. The overall response is called *anaphylactic shock*. ∞ *[p. 374]*

Many of the symptoms of immediate hypersensitivity can be prevented with **antihistamines** (an-tē-HIS-ta-mēnz), drugs that block the action of histamine. Benadryl® is a popular antihistamine available without a prescription. Treatment of severe anaphylaxis involves antihistamine, corticosteroid, and epinephrine injections.

Autoimmune Responses

Autoimmune disorders develop when the immune response mistakenly targets its own normal body cells and tissues. In most cases, the immune system recognizes and ignores the antigens normally found in the body. If the recognition process malfunctions, activated B cells begin producing antibodies, called **autoantibodies**, which attack the antigens normally found in healthy cells and tissues. The symptoms produced depend on the identity of the antigen attacked by these misguided antibodies. For example, *rheumatoid arthritis* occurs when autoantibodies form immune complexes within connective tissues, especially around synovial joints, and *insulin-dependent diabetes mellitus (IDDM)* is caused by autoantibodies that attack cells in the pancreatic islets. ∞ *[pp. 179, 310]*

Systemic lupus erythematosus (LOO-pus e-rith-ē-ma-TŌ-sis), or **SLE**, also appears to result from the production of autoantibodies. An individual with SLE manufactures autoantibodies against the body's own nucleic acids, ribosomes, clotting factors, blood cells, platelets, and lymphocytes. The immune complexes form deposits in peripheral tissues, producing anemia, kidney damage, arthritis, and blood vessel inflammation.

One possible sign of this condition is the presence of a butterfly-shaped discoloration of the face, centered over the bridge of the nose (Figure 16-13•). SLE affects women nine times as often as it affects men. There is no known cure, but almost 80 percent of SLE patients survive 5 years or more after diagnosis. Treatment consists of controlling the symptoms and depressing the immune response through the administration of specialized drugs or corticosteroids.

Many autoimmune disorders appear to be cases of mistaken identity. For example, proteins associated with measles, Epstein-Barr, influenza, and other viruses contain amino acid chains that are similar to those of myelin proteins. As a result, antibodies that target these viruses may also attack myelin sheaths. This may also cause some cases of *multiple sclerosis*. ∞ *[p. 233]*

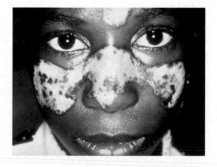

•FIGURE 16-13
Butterfly Rash of Systemic Lupus Erythematosus.

✔ What is lymphadenopathy?

✔ Hay fever is a common allergy. What causes an allergic reaction?

CHAPTER REVIEW

Key Words

antibody (AN-ti-bo-dē): A globular protein produced by plasma cells that binds to specific antigens and promotes their destruction or removal from the body.

antigen: A substance capable of inducing the production of antibodies.

B cells: Lymphocytes capable of differentiating into the plasma cells that produce antibodies.

immunity: Resistance to injuries and diseases caused by foreign compounds, toxins, and pathogens.

immunoglobulin (i-mū-nō-GLO-bū-lin): A circulating antibody.

lymph: Fluid contents of lymphatic vessels.

lymph nodes: Lymphatic organs that monitor the composition of lymph.

lymphocyte (LIM-fō-sīt): A cell of the lymphatic system that participates in the immune response; includes B cells and T cells.

macrophage: A large phagocytic cell derived from a monocyte.

T cells: Lymphocytes responsible for cellular immunity and for the coordination and regulation of the immune response; include helper and suppressor T cells and cytotoxic (killer) T cells.

Selected Clinical Terms

allergen: An antigen capable of triggering an allergic reaction.

allergy: An inappropriate or excessive immune response to antigens.

anaphylactic shock: A drop in blood pressure that may lead to circulatory collapse, resulting from a severe case of anaphylaxis.

anaphylaxis: A type of allergy in which a circulating allergen affects mast cells throughout the body, quickly causing widespread inflammation.

appendicitis: Infection and inflammation of the lymphoid nodules in the appendix.

autoantibodies: Antibodies that react with antigens on a person's own cells and tissues.

autoimmune disorder: A disorder that develops when the immune response mistakenly targets normal body cells and tissues.

immunosuppression: A reduction in the sensitivity of the immune system.

lymphadenopathy: Chronic or excessive enlargement of the lymph nodes.

lymphedema: The swelling of tissues due to an accumulation of interstitial fluid.

lymphoma: A malignant cancer consisting of abnormal lymphocytes found mainly within lymph nodes or the spleen.

necrosis: The death of cells or tissues from disease or injury.

splenomegaly: Enlargement of the spleen.

tonsillitis: Infection of one or more tonsils; symptoms include a sore throat, high fever, and a high white blood cell count (leukocytosis).

vaccine: A preparation of antigens derived from a specific pathogen.

Study Outline

INTRODUCTION (p. 381)

1. The **lymphatic system** absorbs the excess fluid that bathes body cells and returns it as lymph to the circulatory system. The cells, tissues, and organs of the lymphatic system also play a central role in the body's defense.

SYSTEM BRIEF (p. 381)

1. The lymphatic system helps maintain blood volume and eliminate local variations in the composition of the interstitial fluid.

2. The cells, tissues, and organs of the lymphatic system play a central role in the body's defenses against a variety of **pathogens** (disease-causing organisms).

3. Lymphocytes, the primary cells of the lymphatic system, provide an immune response to specific threats to the body. **Immunity** is the ability to resist infection and disease through the activation of specific physical and chemical defenses.

STRUCTURE OF THE LYMPHATIC SYSTEM (pp. 381–386)

1. The lymphatic system includes a network of *lymphatic vessels* that carry *lymph* and a series of *lymphoid organs* that are connected to the lymphatic vessels. *(Figure 16-1)*

LYMPH AND THE LYMPHATIC VESSELS (p. 381)

2. **Lymph** is a fluid similar to plasma but with a lower concentration of proteins. Lymph flows through a network of lymphatic vessels, or lymphatics, that originate as the **lymphatic capillaries**. *(Figure 16-2)*

3. The lymphatic vessels empty into the **thoracic duct** and the **right lymphatic duct**. *(Figure 16-1)*

LYMPHOID TISSUES AND ORGANS (p. 382)

4. A **lymphoid nodule** consists of loose connective tissue containing lymphocytes. **Tonsils** are examples of lymphoid nodules. *(Figure 16-3)*

5. Important **lymphoid organs** include the **lymph nodes**, the **thymus**, and the **spleen**. Lymphoid tissues and organs are distributed in areas especially vulnerable to injury or invasion.

6. Lymph nodes are encapsulated masses of lymphatic tissue containing lymphocytes. Lymph nodes monitor the lymph before it drains into the venous system. *(Figure 16-4)*

7. The **thymus** lies posterior to the sternum. *T cells* become mature in the thymus. *(Figure 16-1)*

8. The adult **spleen** contains the largest mass of lymphatic tissue in the body. The cellular components form the **pulp** of the spleen. *Red pulp* contains large numbers of red blood cells, and areas of *white pulp* contain lymphocytes. *(Figure 16-1)*

LYMPHOCYTES (p. 385)

9. There are three different classes of lymphocytes: **T cells** (thymus-dependent), **B cells** (bone marrow–derived), and **NK cells** (natural killer).

10. Different groups of T cells provide different defensive roles.

11. B cells can differentiate into **plasma cells**, which produce and secrete **antibodies** that react with specific chemical targets, or **antigens**. Antibodies in body fluids are **immunoglobulins**.

12. NK cells attack foreign cells, normal cells infected with viruses, and cancer cells. They provide *immunological surveillance*.

13. Lymphocytes continually migrate in and out of the blood through the lymphatic tissues and organs. Lymphocyte production involves the bone marrow, thymus, and peripheral lymphatic tissues. *(Figure 16-5)*

BODY DEFENSES *(pp. 387–392)*

1. The body's defenses fall into two categories. (1) **Nonspecific defenses** prevent the approach of, deny entrance to, or limit the spread of living or nonliving hazards. (2) **Specific defenses** provide protection against specific foreign compounds or pathogens.

NONSPECIFIC DEFENSES *(p. 387)*

2. **Physical barriers** include hair, epithelia, and various secretions of the integumentary and digestive systems.

3. *Microphages* and *macrophages* are two types of **phagocytic cells** that normally protect body tissues. **Microphages** are the neutrophils and eosinophils in circulating blood. **Macrophages** are derived from monocytes and reside in many body tissues.

4. **Immunological surveillance** involves constant monitoring of normal tissues by NK cells sensitive to abnormal antigens on the surfaces of otherwise normal cells. Cells selected to be killed include cancer cells and those infected with viruses.

5. **Interferons**, small proteins released by cells infected with viruses, trigger the production of antiviral proteins that interfere with viral replication inside the cell.

6. The **complement system** is made up of *complement proteins* that interact with each other in chain reactions to destroy target cell membranes, stimulate inflammation, attract phagocytes, and/or enhance phagocytosis.

7. **Inflammation** represents a coordinated nonspecific response to tissue injury. *(Figure 16-6)*

8. A **fever** (body temperature greater than 37.2°C or 99°F) can inhibit pathogens and accelerate metabolic processes.

SPECIFIC DEFENSES AND IMMUNITY *(p. 389)*

9. Specific immunity may involve **innate immunity** (genetically determined and present at birth) or **acquired immunity** (which appears following exposure to an antigen). Acquired immunity may occur naturally or be induced artificially. **Passive immunity** (the transfer of antibodies from one person to another) may be categorized as *natural* or *artificial*. *(Figure 16-7)*

10. The goal of the *immune response* is the destruction or inactivation of pathogens, abnormal cells, and foreign molecules. It is based on the activation of T or B cell lymphocytes by specific antigens. T cells provide **cell-mediated immunity**; B cells provide **antibody-mediated immunity**. *(Figure 16-8)*

11. T cells are activated by foreign antigens carried by macrophage cells or by viral antigens displayed on the surfaces of virus-infected cells.

12. Activated T cells may differentiate into *cytotoxic T cells*, *memory T cells*, *suppressor T cells*, and *helper T cells*.

13. Cell-mediated immunity results from the activation of **cytotoxic T cells**. Activated **memory T cells** remain on reserve to guard against future such attacks.

14. **Suppressor T cells** depress the responses of other T and B cells.

15. **Helper T cells** secrete chemicals that help coordinate specific and nonspecific defenses and regulate cellular and humoral immunity.

16. B cells responsible for antibody-mediated immunity are normally activated by helper T cells sensitive to the same antigen as they are.

17. An activated B cell may differentiate into a **plasma cell** or produce daughter cells that differentiate into plasma cells and **memory B cells**. Antibodies are produced by the plasma cells.

18. The antibodies first produced by plasma cells are the agents of the **primary response**. The maximum concentration of antibodies appears during the **secondary response** to antigen exposure.

AGING AND THE IMMUNE RESPONSE *(p. 392)*

1. With aging, the immune system becomes less effective at combating disease.

DISORDERS OF THE LYMPHATIC SYSTEM AND THE IMMUNE RESPONSE *(pp. 392–399)*

1. Disorders of the lymphatic system may be caused by problems with its components (lymphocytes, vessels, tissues, and organs) or with the immune response. *(Figure 16-9)*

THE PHYSICAL EXAMINATION AND THE LYMPHATIC SYSTEM *(p. 392)*

2. Signs and symptoms of lymphatic disorders include lymphadenopathy (enlarged lymph nodes), red streaks on the skin, spleen enlargement, weakness and fatigue, skin lesions, respiratory problems, and **lymphedema** (swollen tissues).

3. Diagnosis may require blood or other clinical tests. *(Tables 16-1, 16-2)*

DISORDERS OF THE LYMPHATIC SYSTEM *(p. 393)*

4. Inflammation, infection, and tumors are primarily responsible for problems with the components of the lymphatic system.

5. Complications of inflammation include **necrosis**, or tissue degeneration, the formation of **pus**, **erysipelas**, and **cellulitis**.

6. **Lymphangitis** is an inflammation of lymph vessels marked by red streaks in the skin from the point of infection. *(Figure 16-10)*

7. **Tonsillitis** is caused by bacterial infection of the tonsils, lymphoid tissue in the pharynx.

8. An inflammation of the appendix is called **appendicitis**.

9. **Filariasis**, an infection by a parasitic nematode, may scar and block lymphatic vessels causing **elephantiasis**. *(Figure 16-11)*

10. **Lyme disease** is a bacterial infection transmitted by the bite of a tick. *(Figure 16-12)*

11. **Infectious mononucleosis** is caused by the **Epstein-Barr virus (EBV)**. Also called the "kissing disease," it causes sore throat, fatigue, swelling of lymph nodes, and an enlarged spleen. Severe hemorrhaging may occur if the spleen ruptures.

12. **Splenomegaly** is an enlargement of the spleen. **Hyposplenism** is a condition characterized by a missing or nonfunctional spleen.

13. **Lymphomas** are malignant cancers consisting of cancerous lymphocytes found mainly within lymph nodes or the spleen.

DISORDERS OF THE IMMUNE RESPONSE *(p. 397)*

14. Disorders of the immune response may be categorized as *inadequate immune responses*, *excessive immune responses*, or *inappropriate* (**autoimmune**) *immune responses*.

15. Inadequate immune responses may be congenital, acquired, or induced. **Severe combined immunodeficiency disease (SCID)** is a congenital condition in which neither cell-mediated nor antibody-mediated immunity develops.

16. HIV disease is caused by a virus that infects helper-T cells. In the late stages, it is known as acquired immunodeficiency disease (AIDS).

17. *Immunosuppressive drugs* may be used to decrease the body's immune defenses after bone-marrow transplantations, and tissue and organ transplants.

18. Allergies are examples of excessive immune responses triggered by exposure to **allergens**, or antigens.

19. *Autoimmune disorders* develop when the body's immune system targets its own cells and tissues with **autoantibodies**. *(Figure 16-13)*

Review Questions

MATCHING

Match each item in Column A with the most closely related item in Column B. Use letters for answers in the spaces provided.

	Column A		Column B
___	1. humoral immunity	a.	produce antibodies
___	2. plasma cells	b.	system of circulating proteins
___	3. complement	c.	cancerous lymphocytes
___	4. microphages	d.	fluid in lymphatic vessels
___	5. macrophages	e.	largest lymphoid organ
___	6. microglia	f.	transfer of antibodies
___	7. interferon	g.	neutrophils, eosinophils
___	8. spleen	h.	secretion of antibodies
___	9. innate immunity	i.	type of lymphocyte
___	10. active immunity	j.	tissue swelling due to interstitial fluid
___	11. passive immunity	k.	monocytes
___	12. lymph	l.	exposure to antigen
___	13. T cell	m.	present at birth
___	14. lymphedema	n.	made by virus-infected cells
___	15. lymphoma	o.	CNS macrophages

MULTIPLE CHOICE

16. One of the functions of the lymphatic system is to return tissue fluid to the _____ .
(a) liver
(b) blood
(c) kidneys
(d) thymus

17. Lymph from the lower abdomen, pelvis, and lower limbs is received by the _____ .
(a) right lymphatic duct
(b) inguinal duct
(c) thoracic duct
(d) aorta

18. The _____ removes damaged or defective red blood cells from the circulation.
(a) thymus
(b) spleen
(c) lymph nodes
(d) tonsils

19. Inflammation _____ .
(a) aids in the temporary repair at an injury site
(b) slows the spread of pathogens
(c) produces swelling, redness, heat, and pain
(d) a, b, and c are correct

20. T cells and B cells can be activated only by _____ .
(a) exposure to a specific antigen
(b) pathogens
(c) cytokines
(d) cells infected with viruses

21. Memory B cells _____ .
(a) respond to a threat on the first exposure
(b) secrete antibodies
(c) respond to subsequent injuries or infections that involve the same antigens
(d) activate the complement system

22. Viruses from plants and most animals do not cause disease in people. This is an example of _____ .
(a) innate immunity
(b) naturally acquired immunity
(c) artificially acquired immunity
(d) passive immunity

23. _____ is an infection of the lymphoid nodules of the pharynx.
(a) lymphangitis
(b) tonsillitis
(c) appendicitis
(d) cellulitis

24. Which of the following disorders is due to an excessive immune response?
(a) anaphylaxis
(b) SCID
(c) AIDS
(d) SLE

TRUE/FALSE

___ 25. Tonsils are examples of lymph nodes.

___ 26. B cells are responsible for cell-mediated immunity.

___ 27. Plasma cells produce antibodies for one specific foreign antigen.

___ 28. Antibodies are composed of proteins.

___ 29. Valves prevent the backflow of lymph in larger lymphatic vessels.

___ 30. The thymus is most active in middle and old age.

___ 31. Anemia may be caused by hypersplenism.

___ 32. Autoantibodies attack antigens on the body's own cells and tissues.

SHORT ESSAY

33. What two large collecting vessels return lymph to the circulatory system? What areas of the body does each serve?

34. Give a function for each of the following:
(a) NK cells
(b) T cells
(c) B cells
(d) helper T cells
(e) plasma cells

35. How are the body's specific defenses different from its nonspecific defenses?

36. Distinguish between passively acquired natural and artificial immunity.

37. What is the "kissing disease"?

38. Elephantiasis is a condition marked by extreme lymphedema of the external genitalia and lower limbs. What causes this condition?

APPLICATIONS

39. An elderly man develops peritonitis following a ruptured appendix. Explain why this occurred.

40. Paula's grandfather is diagnosed as having lung cancer. His physician orders biopsies of several lymph nodes from adjacent regions of the body. Paula wonders why, as her grandfather's cancer is in the lungs. What would you tell her?

✔ Answers to Concept Check Questions

(p. 386) **1.** The thoracic duct drains lymph from the area beneath the diaphragm and the left side of the head and thorax. Most of the lymph enters the venous blood by way of this duct. A blockage of this duct would not only impair circulation of lymph through most of the body, it would also promote accumulation of fluid in the extremities. **2.** The thymosins from the thymus play a role in the maturation of T lymphocytes from lymphocyte stem cells. A lack of these hormones would result in a lack of T lymphocytes. **3.** During an infection, the lymphocytes and phagocytes in the lymph nodes in the affected region undergo cell division to better deal with the infectious agent. This increase in the number of cells in the nodes causes the nodes to become enlarged or swollen.

(p. 389) **1.** A decrease in the number of monocyte-forming cells in the bone marrow would result in a decreased number of macrophages in the body, since all the different macrophages are derived from monocytes. This would include the microglia of the CNS and *Kupffer cells* in the liver as well as others. **2.** A rise in interferon would indicate a viral infection. Interferon is released from cells that are infected with viruses. It does not help the infected cell but "interferes" with the virus's ability to infect other cells.

(p. 392) **1.** Cytotoxic T cells function in cell-mediated immunity. A decrease in their numbers would interfere with the body's ability to kill foreign cells and tissues as well as cells infected with viruses. **2.** Helper T cells promote B cell division, the maturation of plasma cells, and the production of antibody by the plasma cells. Without the helper T cells the humoral immune response would take much longer to occur and would not be as efficient. **3.** The secondary response would be affected by the lack of memory B cells for a specific antigen. The ability to produce a secondary response depends on the presence of memory B cells and T cells that are formed during the primary response to an antigen. These cells are not involved in the primary response but are held in reserve against future contact with the same antigen.

(p. 399) **1.** Lymphadenopathy refers to enlarged lymph nodes. It is generally caused by infection or cancer. **2.** Allergic reactions are due to excessive responses to allergens by the immune system. Hay fever is a type of reaction called immediate hypersensitivity. It results in a sudden, massive inflammation of the affected tissues.

The Respiratory System

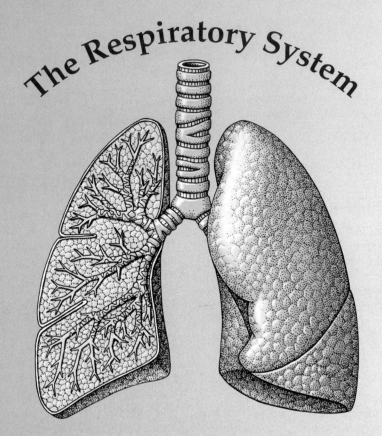

CHAPTER OBJECTIVES

1 Describe the primary functions of the respiratory system.

2 Explain how the delicate respiratory exchange surfaces are protected from pathogens, debris, and other hazards.

3 Relate respiratory functions to the structural specializations of the tissues and organs in the system.

4 Describe the process of breathing.

5 Describe the actions of respiratory muscles on respiratory movements.

6 Describe how oxygen and carbon dioxide are transported in the blood.

7 Describe the major factors that influence the rate of respiration.

8 Describe the changes that occur in the respiratory system with aging.

9 Describe characteristic symptoms and signs of respiratory system disorders.

10 Give examples of different disorders of the respiratory system and their causes.

Breathing is more essential to our lives than eating or drinking. In fact, 5 to 6 minutes without breathing will likely prove fatal. This is because the cells of our body must keep generating ATP to support their maintenance, growth, and replication. This cellular energy is obtained through chemical reactions that require oxygen, and we get that oxygen from the air around us. The same chemical reactions that consume oxygen generate carbon dioxide as a waste product. A constant supply of oxygen and a reliable method for removing the carbon dioxide are thus vital to our cells.

Every time we inhale we bring additional oxygen into our lungs, where it can be absorbed by the blood and distributed to our cells. Every time we exhale we eliminate carbon dioxide. The lungs provide the huge surface area needed for the exchange of gases between our blood and the external atmosphere. How huge is huge? In order to meet our oxygen demand, the total exchange surface of the lungs is at least 35 times larger than the total surface area of the body.

The *respiratory system* includes the lungs, the air-filled passageways leading to the lungs, and the skeletal muscles that move air into and out of the lungs. The following discussion of the organs of the respiratory system will follow the flow of air as it travels from the exterior toward the delicate exchange surfaces of the lungs. We will then consider how breathing occurs and how oxygen and carbon dioxide are transported and exchanged between the air and the blood, and between the blood and other body tissues.

SYSTEM BRIEF

Respiration is the exchange of gases between living cells and the environment. Respiration involves three different steps or activities:

1. **Pulmonary ventilation**: breathing, or the movement of air into and out of the lungs.

2. **Gas exchange**: the diffusion of oxygen and carbon dioxide between the lungs and blood (*external respiration*), and the diffusion of oxygen and carbon dioxide between blood and cells in the body tissues (*internal respiration*).

3. **Gas pickup and transport**: the transport of oxygen and carbon dioxide within in the blood.

The eventual use of oxygen within cells for the production of energy is called *cellular respiration*. The process of cellular respiration is discussed in Chapter 19.

The major function of the respiratory system is to move air to and from the lungs, where diffusion can occur between air and circulating blood. Additional functions include providing a defense against pathogenic invasion, producing sounds for speaking, assisting in the regulation of blood volume and pressure, and assisting in the control of body fluid pH.

STRUCTURE OF THE RESPIRATORY SYSTEM

The **respiratory system** includes the nose, nasal cavity, and sinuses; the pharynx (throat); the larynx (voice box); the trachea (windpipe); the bronchi and bronchioles (conducting passageways); and the lungs (Figure 17-1•). The bronchioles end at air sacs within the lungs called *alveoli* (al-VĒ-ō-lī); singular: *alveolus*. The walls of the alveoli are the gas-exchange surfaces of the lungs.

THE NOSE

Air normally enters the respiratory system through the paired **nostrils**, which open into the **nasal cavity**. Coarse hairs within the nostrils guard the nasal cavity from large airborne particles such as sand, dust, and insects.

The nasal cavity is divided into left and right compartments by the *nasal septum*. The bony nasal septum and the walls of the nasal cavity are formed by bones of the

•FIGURE 17-1
Components of the Respiratory System.

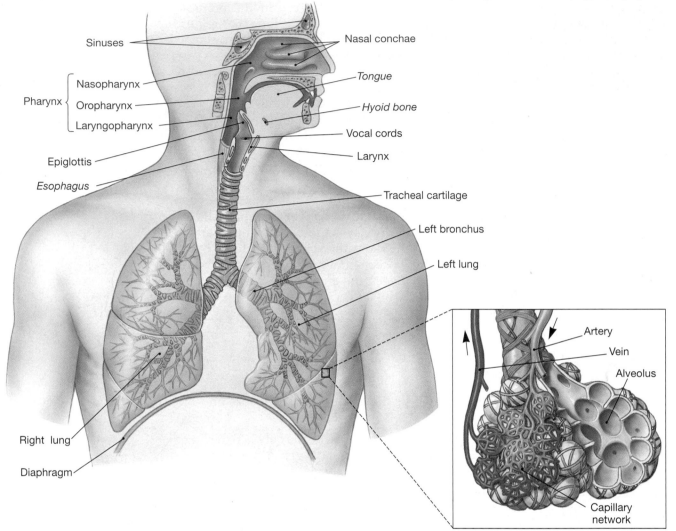

- Sinuses
- Nasal conchae
- Nasopharynx
- Oropharynx
- Laryngopharynx
- Pharynx
- Tongue
- Hyoid bone
- Vocal cords
- Larynx
- Epiglottis
- Esophagus
- Tracheal cartilage
- Left bronchus
- Left lung
- Artery
- Vein
- Alveolus
- Capillary network
- Right lung
- Diaphragm

•FIGURE 17-2
Ciliated Respiratory Epithelium.
Mucus is transported to the pharynx by ciliated respiratory epithelium within the nasal cavity and the lower air-conducting passageways.

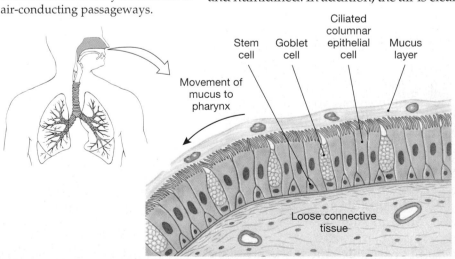

- Movement of mucus to pharynx
- Stem cell
- Goblet cell
- Ciliated columnar epithelial cell
- Mucus layer
- Loose connective tissue

cranium and face. The *superior, middle,* and *inferior nasal conchae* project toward the nasal septum from the sides of the nasal cavity. ⊂◯ *[p. 154]* The nasal conchae cause the incoming air to swirl, much like water flows turbulently over rapids. Because this movement slows down the passage of air, there is additional time for it to be warmed and humidified. In addition, the air is cleaned of small airborne particles as they come in contact with the mucus that coats the lining of the nasal cavity.

A bony **hard palate** forms the floor of the nasal cavity and separates it from the underlying oral cavity. A fleshy **soft palate** extends behind the hard palate. The area above the soft palate is the **nasopharynx** (nā-zō-FAR-inks), which lies between the nasal cavity and the portion of the pharynx connected to the oral cavity.

The nasal cavity and the lower air-conducting passageways are lined by a mucous membrane made up of a ciliated epithelium that contains many mucus-producing *goblet cells* (Figure 17-2•). The cilia sweep mucus and any trapped particles or microorganisms toward the pharynx, where they can be swallowed and exposed to the acids and enzymes of the stomach. The nasal cavity is also flushed by mucus produced in the *paranasal sinuses* (air-filled chambers in the facial bones) and by tears flowing through the *nasolacrimal duct*. Exposure to toxic vapors, large quantities of dust and debris, allergens, or pathogens usually causes a rapid increase in the rate of mucus production, and a "runny nose" develops.

THE PHARYNX

The **pharynx** is a chamber shared by the digestive and respiratory systems. Its three subdivisions are shown in Figure 17-1•. The upper portion, the **nasopharynx**, is lined by a ciliated epithelium. ∞ *[p. 63]* This region contains the pharyngeal tonsil and the entrances to the auditory tubes. The **oropharynx** extends from the level of the soft palate to the base of the tongue. The palatine tonsils lie in the lateral walls of the oropharynx. The narrow **laryngopharynx** (la-rin-gō-FAR-inks) extends from the oropharynx to the entrance to the esophagus. Food and liquids entering the digestive tract pass through both the oropharynx and laryngopharynx. These regions are lined by a stratified squamous epithelium that can resist mechanical abrasion, chemical attack, and pathogenic invasion. ∞ *[p. 63]*

THE LARYNX

Incoming air passes through the pharynx and then through a narrow opening, the **glottis** (GLOT-is), which is surrounded and protected by the **larynx** (LAR-inks), or voice box. The larynx contains nine cartilages that are stabilized by ligaments and/or skeletal muscles (Figure 17-3•). The two largest cartilages are the epiglottis and the thyroid cartilage. The elastic **epiglottis** (ep-i-GLOT-is) projects above the glottis. During swallowing the epiglottis folds back over the glottis, preventing the entry of liquids or solid food into the larynx and lower air passageways. The shield-shaped **thyroid cartilage** forms much of the anterior and lateral surfaces of the larynx. A prominent ridge on the anterior surface of this cartilage forms the "Adam's apple."

Two pairs of ligaments extend across the larynx, between the thyroid cartilage and other, smaller cartilages. The upper pair, known as the *false vocal cords*, help prevent foreign objects from entering the glottis and protect the more delicate lower folds, which are the true **vocal cords**.

The vocal cords vibrate when air passes through the glottis. These vibrations generate sound waves. As on a guitar or violin, short, thin strings vibrate rapid-

•FIGURE 17-3
Structure of the Larynx and Vocal Cords.
(a) Anterior view of the larynx. **(b)** Posterior view of the larynx. **(c)** Superior view of the larynx with the glottis open. **(d)** Superior view of the larynx with the glottis closed.

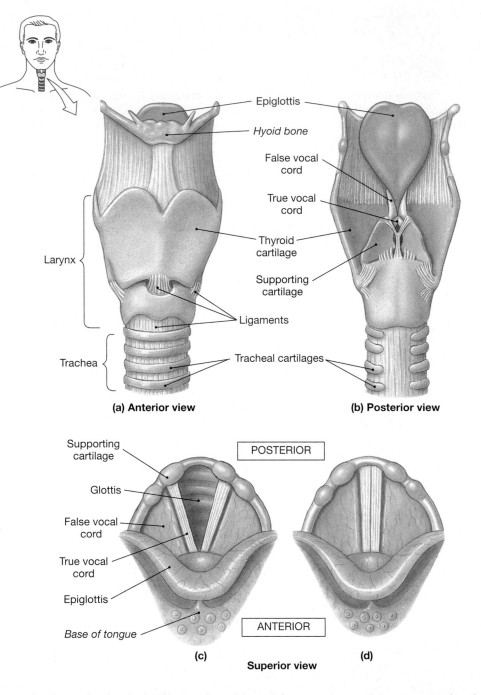

• FIGURE 17-3
Structure of the Larynx and Vocal Cords.
(a) Anterior view of the larynx. **(b)** Posterior view of the larynx. **(c)** Superior view of the larynx with the glottis open. **(d)** Superior view of the larynx with the glottis closed.

(a) Anterior view   (b) Posterior view

Superior view

ly, producing a high-pitched sound and large, long strings vibrate more slowly, producing a low-pitched tone. Because children of both sexes have slender, short vocal cords, their voices tend to be high-pitched. At puberty the larynx of a male enlarges more than that of a female. The true vocal cords of an adult male are thicker and longer and they produce lower tones than those of an adult female.

The pitch of the voice is regulated by the amount of tension in the vocal cords. The volume depends on the force of the air movement through the vocal cords. Further amplification and resonance occur within the pharynx, the oral cavity, the nasal cavity, and the paranasal sinuses. The final production of distinct words further depends on voluntary movements of the tongue, lips, and cheeks.

THE TRACHEA

The **trachea** (TRĀ-kē-a), or windpipe, is a tough, flexible tube about 2.5 cm (1 in.) in diameter and 11 cm (4.0 in.) long. The walls of the trachea are supported by about 20 **tracheal cartilages**. These C-shaped structures stiffen the tracheal walls and protect the airway. As you can see in Figure 17-3b•, the open portions of the tracheal cartilages face posteriorly, toward the esophagus. As a result, the posterior tracheal wall distorts easily, and this is important when large masses of food pass along the esophagus. The diameter of the trachea is adjusted by the ANS, which controls contractions of smooth muscle in its walls. Stimulation by the sympathetic nervous system increases the diameter of the trachea and makes it easier to move large volumes of air along the respiratory passageways.

THE BRONCHI

The trachea branches to form the **right** and **left primary bronchi** (BRONG-kī). The right primary bronchus is larger and more vertical than the left bronchus. As a result, accidentally inhaled objects, such as pieces of food, most often enter (and get stuck in) the right primary bronchus.

Each primary bronchus enters its lung at a groove or depression called the **hilus**. Pulmonary blood vessels, nerves, and lymphatic vessels also enter the lung at the hilus. Each primary bronchus branches as it enters the lung, giving rise to *secondary bronchi* that enter the lobes of the lung. In each lung, the secondary bronchi divide to form 10 *tertiary bronchi* in each lung, which then branch repeatedly. The primary bronchi and their branches form a network that is called the *respiratory tree*, or *bronchial tree*.

The cartilages surrounding the secondary bronchi are quite massive, but they decrease in size as the branches of the "tree" become smaller and smaller. When the diameter of the passageway has narrowed to around 1 mm, the cartilage rings disappear completely. This narrow passage represents a **bronchiole**.

All portions of the trachea and the various bronchi are lined by the same type of pseudostratified ciliated columnar epithelium found within the nasal cavity and nasopharynx. The beating of the cilia moves mucus and any foreign particles toward the pharynx for elimination by swallowing, coughing, or sneezing.

THE BRONCHIOLES

The walls of bronchioles contain more smooth muscle than the wider air-conducting passageways discussed above. Contraction and relaxation of the smooth muscle of the bronchioles is under autonomic control. Varying the diameter of the bronchioles regulates the amount of resistance to airflow and the distribution of air in the lungs. For example, impulses from the sympathetic nervous system can cause the smooth muscles in the walls of bronchioles to relax, thereby increasing the diameter of the respiratory passageways. This helps prepare us for intense activity by making it easier to move air quickly to and from the lungs. Extreme contraction of these smooth muscles can almost completely close the bronchioles, and this can make breathing difficult or even impossible. Significant reductions in the diameter of bronchioles accompany a severe asthma attack or an acute allergic reaction.

Bronchioles continue to divide until they form the **terminal bronchioles**, the end of the air-conducting portion of the respiratory system (Figure 17-4•). The terminal bronchioles branch to form passageways that open into clusters of *alveoli*.

THE ALVEOLI

Each lung contains approximately 150 million alveoli. The total gas-exchange area of all the alveoli in both lungs is about 140 m^2—roughly the size of a tennis court. Gas exchange at an alveolus is highly efficient because of its unusually thin epithelium. Figure 17-4b• shows the **alveolar macrophages** (*dust cells*) that patrol this epithelium, engulfing dust or debris that has reached the alveolar surfaces. **Surfactant** (sur-FAK-tant) **cells** produce an oily secretion, or **surfactant**, that forms a thin coating over the alveolar epithelium. Surfactant is important because it reduces surface tension within the alveolus. Surface tension results from the pull of

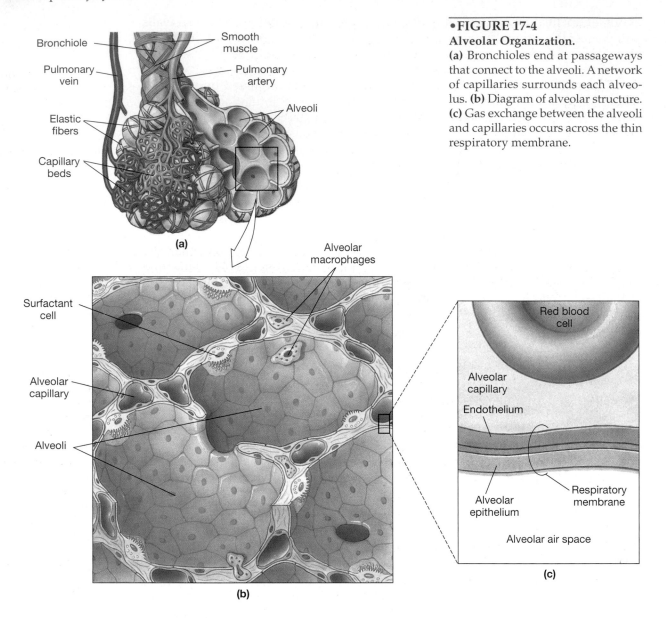

Bronchiole

Pulmonary vein

Elastic fibers

Capillary beds

Smooth muscle

Pulmonary artery

Alveoli

(a)

Alveolar macrophages

Surfactant cell

Alveolar capillary

Alveoli

(b)

Red blood cell

Alveolar capillary

Endothelium

Respiratory membrane

Alveolar epithelium

Alveolar air space

(c)

•FIGURE 17-4
Alveolar Organization.
(a) Bronchioles end at passageways that connect to the alveoli. A network of capillaries surrounds each alveolus. **(b)** Diagram of alveolar structure. **(c)** Gas exchange between the alveoli and capillaries occurs across the thin respiratory membrane.

attraction between water molecules at an air–water boundary. The alveolar walls are so delicate that without surfactant, they would collapse like deflating balloons.

The respiratory exchange surfaces receive blood from arteries of the pulmonary circulation. ∞ *[p. 356]* The pulmonary arteries enter the lungs and branch, following the bronchi to the alveoli. Each cluster of alveoli receives an arteriole and a venule, and a network of capillaries surrounds each alveolus (Figure 17-4a•). The walls of the alveoli, like the walls of capillaries, are made up of a single layer of epithelium. Together, these two epithelia make up the *respiratory membrane* that separates the alveolar air and the blood (Figure 17-4c•). Because it is so thin, as little as one ten-thousandth of a millimeter (0.1 μm), the exchange of respiratory gases by diffusion can occur very rapidly. After passing through the pulmonary venules, venous blood enters the pulmonary veins that deliver it to the left atrium.

THE LUNGS

The lungs are situated within the thoracic cavity, which has the shape of a broad cone. Its walls are the rib cage, and its floor is the muscular diaphragm. The mediastinum divides the thoracic cavity into two pleural cavities (see Figure 1-8•). ∞ *[p. 13]* Each lung occupies a single pleural cavity, lined by a serous membrane, or **pleura** (PLOO-ra). The *parietal pleura* covers the inner surface of the body wall and extends over the diaphragm and mediastinum. The *visceral pleura* covers the outer surfaces of the lungs. Because the parietal and visceral layers are in close contact with each other, the pleural cavity is not an open chamber. A thin layer of fluid between these layers lubricates and reduces friction between them as the lungs expand and contract.

The left and right lungs (Figure 17-1•, p. 406) occupy the left and right pleural cavities. Each lung has distinct **lobes** separated by deep grooves, or fissures. The right lung has three lobes and the left lung has two. Because most of the actual volume of each lung consists of air-filled passageways and alveoli, the lung has a light and spongy consistency. The many elastic fibers give the lungs the ability to tolerate large changes in volume.

✔ When the tension in the vocal cords increases, what happens to the pitch of the voice?

✔ Why are the cartilages that reinforce the trachea C-shaped instead of complete circles?

✔ What would happen to the alveoli if surfactant were not produced?

PHYSIOLOGY OF RESPIRATION

The process of respiration involves exchanges of gases between the air and the lungs, the lungs and the blood, the blood and body tissue fluids and cells, and the transport of gases by the blood. Problems that affect any of these steps will ultimately affect the gas concentrations of the tissue fluids and body cells.

PULMONARY VENTILATION

Pulmonary ventilation, or breathing, is the physical movement of air into and out of the lungs. A single breath, or *respiratory cycle*, consists of an **inhalation**, the movement of air into the lungs, and an **exhalation**, the movement of air out of the lungs. The movement of air into and out of the lungs during each breath is controlled by differences in air pressure between the lungs and the external environment.

Changes in the volume of the lungs are due to changes in the volume of their surrounding pleural cavities. Fluid in the pleural cavities causes the surface of each lung to stick to the inner wall of the chest and the superior surface of the diaphragm, much like the strong bond between the bottom of a wet glass and smooth table. As a result, when the diaphragm and chest wall move, the volume of the lungs changes.

When relaxed, the diaphragm has the shape of a dome and projects upward into the thoracic cavity, compressing the lungs. When the diaphragm contracts, it flattens and increases the volume of the thoracic cavity, expanding the lungs. Figure 17-5a• shows how the thoracic cavity enlarges when the diaphragm contracts.

•**FIGURE 17-5**
Breathing (Pulmonary Ventilation).
(a) Raising the curved bucket handle increases the amount of space between it and the bucket. Similarly, the volume of the thoracic cavity is increased when the diaphragm contracts. (b) Inhalation: As the thoracic cavity increases in volume, pressure within the lungs falls and air flows in. (c) Exhalation: When the rib cage returns to its resting position, the volume of the thoracic cavity decreases. Pressure rises, and air moves out of the lungs.

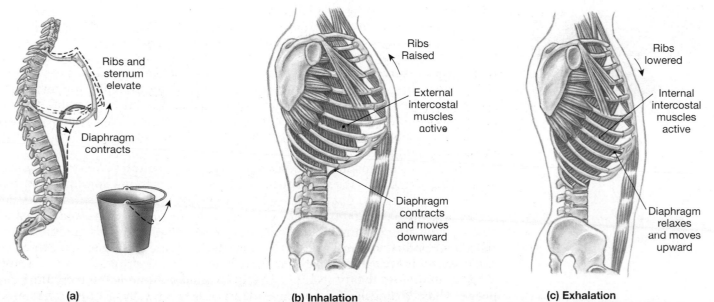

Ribs and sternum elevate
Diaphragm contracts

(a)

Ribs Raised
External intercostal muscles active
Diaphragm contracts and moves downward

(b) Inhalation

Ribs lowered
Internal intercostal muscles active
Diaphragm relaxes and moves upward

(c) Exhalation

Any change in volume has an effect on pressure. If you have ever watched a physician using a syringe, you are already aware of this. When the plunger is drawn back, the volume inside the syringe increases, and this lowers the internal pressure and pulls air or fluid into the syringe. When the plunger is pushed in, the volume inside decreases, and the air or fluid is pushed out of the syringe. The same principle is responsible for moving air in and out of the lungs.

Inhalation

At the start of a breath, pressures inside and outside the lungs are identical and there is no movement of air. When the thoracic cavity enlarges, the pressure inside the lungs drops, and air enters the respiratory passageways (Figure 17-5b●). Enlargement of the thoracic cavity involves the contractions of the diaphragm, aided by the external intercostal muscles.

The external intercostals elevate the rib cage. Because of the way the ribs and the vertebrae articulate, this movement increases the volume of the thoracic cavity. During heavy breathing, other muscles, such as the sternocleidomastoid, help the external intercostals elevate the ribs. This increases the amount of air moved into the lungs while reducing the time spent in inhalation.

Exhalation

Downward movement of the rib cage and upward movement of the relaxed diaphragm reverse the process and reduce the size of the lungs. Pressure inside the lungs now rises, and air moves out of the lungs (Figure 17-5c●).

During quiet breathing, as when sitting and reading, exhalation is a passive process. That is, when the respiratory muscles used during inhalation relax, gravity pulls the ribs downward, the stretched elastic fibers of the lungs recoil, and air is forced out of the lungs. When breathing heavily, contractions of the internal intercostal muscles and the abdominal muscles assist gravity in pulling the ribs downward and forcing the diaphragm into the thoracic cavity. These muscles increase the rate of exhalation as well as the amount of air moved out of the lungs.

Movement of Air

Sneezing forces irritating substances from the mouth and nose, sometimes at speeds up to 160 kph.

An inhalation, or *inspiration*, at rest does not completely fill the lungs with air. In fact, during a quiet respiratory cycle, about 500 ml of air is inhaled and 500 ml is exhaled. This amount of air movement is called the **tidal volume**. Only about two-thirds of the inspired air reaches the alveolar exchange surfaces within the lungs. The rest never gets farther than the conducting passageways, and thus does not participate in gas exchange with the blood. The volume of air in the conducting passages is known as the **dead space** of the lungs.

We can increase the tidal volume by inhaling more vigorously and exhaling more completely. Measurements of such increased volumes of air are used to evaluate how well an individual's respiratory system is functioning. For example, the **inspiratory reserve volume** is the amount of air that can be taken in over and above the resting tidal volume. Because lungs are larger in males, the inspiratory reserve volume averages 3,300 ml in males versus 1,900 ml in females. On the other hand, a forced exhalation, or *expiration*, at the end of a tidal cycle will push out an additional 1,000 ml of air in males (700 ml in females). This is the **expiratory reserve volume**.

The sum of the inspiratory reserve volume, the expiratory reserve volume, and the tidal volume is the **vital capacity**. This is the maximum amount of air that can be moved into and out of the respiratory system in a single respiratory cycle.

Roughly 1,100–1,200 ml of air remains in the respiratory passageways and alveoli, even after exhausting the expiratory reserve volume. This amount of air, called the **residual volume**, remains because the lungs are held against the thoracic wall, preventing them from contracting further.

The sum of the vital capacity and the residual volume is the **total lung capacity**. The average total lung capacity of males is about 6,000 ml, and 4,200 ml in

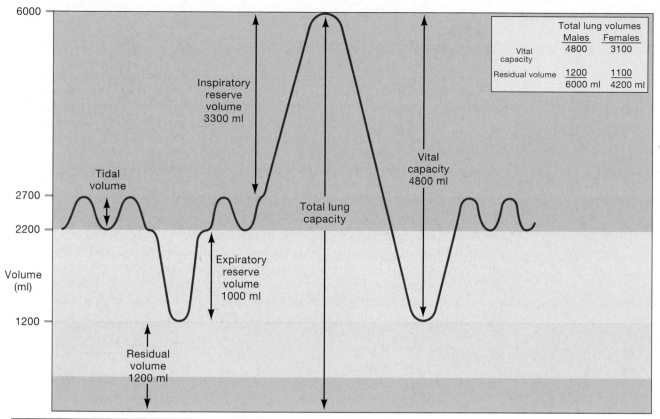

	Total lung volumes	
	Males	Females
Vital capacity	4800	3100
Residual volume	1200	1100
	6000 ml	4200 ml

•**FIGURE 17-6**

Respiratory Volumes and Capacities.

The graph diagrams the relationships among the respiratory volumes and capacities of an average male. The table compares the values for males and females.

females. Figure 17-6• shows the relationships among these different lung volumes and capacities in a male.

✔ How does contraction and relaxation of the diaphragm affect the volume of the thoracic cavity?

✔ What is meant by vital capacity?

Clinical Note

The Heimlich Maneuver

Choking on food that becomes lodged in the larynx or trachea is a common accident. In most cases, coughing expels the food, the airway clears, and no emergency measures need to be taken. Unfortunately, this is not always the case. In the United States, it is estimated that at least eight such choking events end in death every year.

If a person does get food stuck in the airway and cannot inhale or exhale, a procedure called the **Heimlich** (HĪM-lik) **maneuver**, or abdominal thrust, can save a life. This procedure relies on compressing the chest so forcefully that the remaining air in the lungs will blow the food mass out of the airway. In this procedure, a rescuer compresses the abdomen just beneath the diaphragm with a strong, upward squeeze. Figure 17-7• shows the placement of the rescuer's fist below the ribs and above the navel, and the direction of the upward thrust. Care must be taken to avoid damage to the ribs and underlying organs.

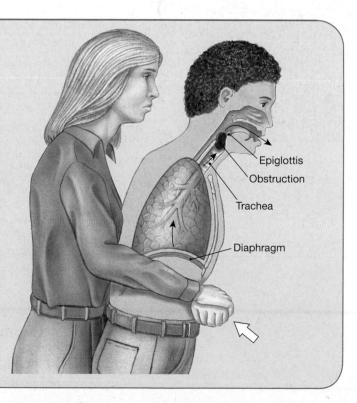

•**Figure 17-7**
The Heimlich Maneuver.

GAS EXCHANGE

The exchange of respiratory gases occurs across the membranes of living cells through the process of diffusion. ∞ *[p. 43]* External respiration is the diffusion of gases between the blood and alveolar air across the single cell layers (respiratory membrane) of the alveoli and lung capillaries. Internal respiration is the diffusion of gases between the blood and tissue fluid across the one-cell-thick capillary wall. The directions in which oxygen and carbon dioxide move by diffusion depend on their relative concentrations in air, blood, and tissue fluid.

External Respiration

The air we inhale contains 20.8 percent oxygen and 0.04 percent carbon dioxide, and exhaled air contains about 15 percent oxygen and 3.7 percent carbon dioxide. These differences are due to the uptake of oxygen by the blood and the removal of carbon dioxide from blood in the lung capillaries. The upper portion of Figure 17-8• shows the opposite directions in which oxygen and carbon dioxide diffuse between the alveoli and pulmonary capillaries in the lungs. Each gas diffuses independently from its area of relatively high concentration to its respective area of relatively low concentration.

Internal Respiration

Body cells are constantly using oxygen and generating carbon dioxide. As a result, oxygen concentrations are low within cells and carbon dioxide concentrations are high. The lower portion of Figure 17-8• shows the gas-exchange process during internal respiration; oxygen diffuses out of the capillaries, and carbon dioxide diffuses in until its concentration is the same as in the surrounding tissues.

GAS TRANSPORT

Oxygen and carbon dioxide do not readily dissolve in blood plasma. So how do body cells obtain adequate oxygen and eliminate enough carbon dioxide to survive? The answer lies in their red blood cells. The extra oxygen and carbon dioxide diffuse into the red blood cells, where the gas molecules are either tied up (in the case of oxygen) or used to make other chemical compounds (in the case of carbon dioxide). The important thing about these reactions is that they are completely reversible.

Oxygen Transport

Almost all the oxygen that diffuses into the pulmonary capillaries is bound to hemoglobin (Hb) molecules in red blood cells. The amount of oxygen held onto by hemoglobin depends primarily on the concentration of oxygen in the surrounding tissue. Therefore, the lower the oxygen content of a tissue, the more oxygen will be released by hemoglobin molecules as they circulate through the region.

Carbon Dioxide Transport

Carbon dioxide molecules enter the bloodstream and are transported to the lungs three different ways. As Figure 17-9• shows, most of the carbon dioxide enters red blood cells, and only a small amount (7 percent) actually dissolves in the blood plasma. Within the red blood cells, some 23 percent of the carbon dioxide binds directly to hemoglobin, and 70 percent is rapidly converted to *bicarbonate ions* (HCO_3^-).

The bicarbonate ions form in a two-step process. First, the carbon dioxide molecules combine with water to form carbonic acid. Second, the enzyme *carbonic anhydrase* catalyzes the rapid breakdown of carbonic acid molecules into bicarbonate ions and hydrogen ions. The newly formed bicarbonate ions move out of the red blood cells and into the blood plasma. The hydrogen ions produced by this reaction are bound to hemoglobin molecules inside the RBC.

•FIGURE 17-8
External and Internal Respiration.

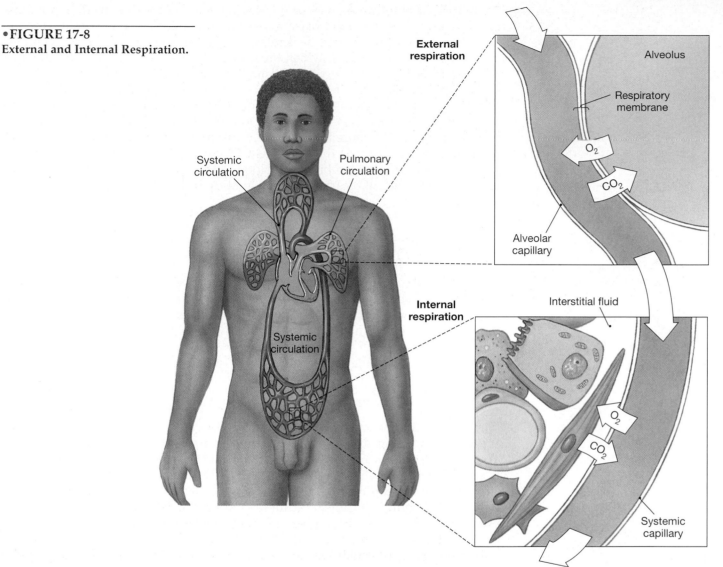

External respiration

Alveolus

Respiratory membrane

O_2

CO_2

Alveolar capillary

Systemic circulation

Pulmonary circulation

Systemic circulation

Internal respiration

Interstitial fluid

O_2

CO_2

Systemic capillary

•FIGURE 17-9
Carbon Dioxide Transport in the Blood.

Most of the carbon dioxide absorbed into the systemic capillaries enters red blood cells. Part of the carbon dioxide binds with hemoglobin, and the rest is converted to bicarbonate ions and released into the plasma. At the alveolar capillaries within the lungs, the carbon dioxide uptake processes are reversed, and carbon dioxide is released.

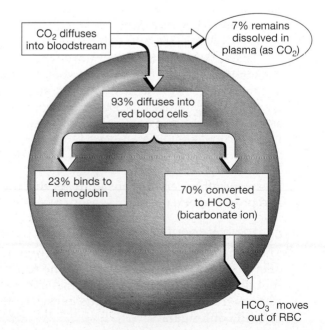

CO_2 diffuses into bloodstream

7% remains dissolved in plasma (as CO_2)

93% diffuses into red blood cells

23% binds to hemoglobin

70% converted to HCO_3^- (bicarbonate ion)

HCO_3^- moves out of RBC

On arrival at the pulmonary capillaries in the lungs, the dissolved carbon dioxide diffuses out of the blood plasma and into the alveoli. As this occurs, the hemoglobin-bound carbon dioxide is released, and bicarbonate and hydrogen ions combine into carbonic acid, which breaks apart into water and carbon dioxide. As this carbon dioxide enters the plasma it diffuses into the alveoli.

The pH of blood is affected by the amount of carbon dioxide that is transported. Some of the CO_2 dissolved in the blood plasma combines with water, forming carbonic acid that releases hydrogen ions and bicarbonate ions. The more CO_2 enters the blood, the more hydrogen ions are released, lowering the pH and increasing blood acidity. At the lungs, as CO_2 diffuses from the blood, hydrogen and bicarbonate ions recombine, and blood pH increases. As we will see in Chapter 20, the presence of bicarbonate ions in body fluids is essential for the control of body fluid pH.

✔ In what form is most of the carbon dioxide transported to the lungs?

✔ In what part of the body does external respiration take place?

CONTROL OF BREATHING

Under normal conditions the rates of gas exchange between the body cells and tissue capillaries, and between the alveoli and capillaries in the lungs are equal. If these rates become seriously unbalanced, problems soon develop. For example, if body cells are absorbing oxygen faster than it is being absorbed at the alveolar capillaries, tissues will soon become oxygen-starved. To prevent this the activities of the cardiovascular and respiratory systems must be adjusted.

RESPIRATORY CONTROL CENTERS

Normal breathing occurs automatically, without conscious control. Breathing is controlled by the **respiratory centers**, three groups of neurons in the brain stem. Two of these groups are located in the pons, and the other lies in the medulla oblongata. These respiratory centers regulate contraction of the respiratory muscles, control the **respiratory rate** (number of breaths per minute), and adjust the depth of breathing. The normal adult respiratory rate at rest ranges from 12 to 18 breaths per minute. Children breathe more rapidly, at around 18 to 20 breaths per minute.

The respiratory center within the medulla oblongata sets the basic pace for respiration by direct control of the respiratory muscles. The respiratory centers in the pons alter the basic rhythm established by the medulla oblongata. For example, the pons may adjust the respiratory rate and the depth of respiration in response to sensory stimuli, emotional states, or speech patterns. Figure 17-10● summarizes the factors involved in the involuntary control of respiration.

The activities of the respiratory centers can be affected by any factor that affects the metabolism of nervous tissues. For example, elevated body temperatures or central nervous system stimulants, such as amphetamines or even caffeine, increase the respiratory rate. Decreased body temperature or CNS depressants, such as barbiturates or opiates, reduce the respiratory rate.

SENSORY RECEPTORS AND RESPIRATION

The respiratory centers constantly receive messages from mechanical (stretching and pressure) and chemical sensory receptors that can alter the rate of respiration.

Stretch Receptors

The *inflation reflex* prevents the lungs from overexpanding as one breathes more deeply during strenuous activity. The receptors involved are stretch receptors that are stimulated when the lungs expand. Sensory fibers leaving these receptors travel within the vagus nerves to reach the respiratory centers in the medulla oblongata.

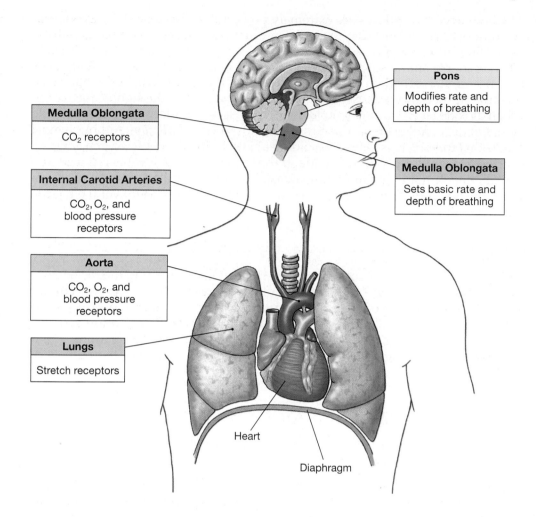

The Control of Breathing.
Various types of sensory receptors in the lungs, medulla oblongata, internal carotid arteries, and the ascending aorta can alter the rate and depth of breathing set by control centers in the medulla oblongata.

Pressure Receptors

Blood pressure receptors are located in the walls of the ascending aorta near the heart and at the bases of the internal carotid arteries of the neck. Nerve impulses from these sensory receptors affect the respiratory centers as well as the cardiac and vasomotor centers that control the heart and blood vessels. ∞ *[p. 355]* When blood pressure falls, the respiratory rate increases; when blood pressure rises, the respiratory rate declines. This adjustment results from stimulation or inhibition of the respiratory centers by sensory fibers in the glossopharyngeal (IX) and vagus (X) nerves.

Chemical Receptors

Chemoreceptors respond to chemical changes in the blood and cerebrospinal fluid. Receptors near the carotid arteries (close to the pressure receptors) and the aorta (near the aortic pressure receptors) are sensitive to the concentrations of carbon dioxide and oxygen in arterial blood; receptors in the medulla oblongata respond to the concentration of carbon dioxide in the surrounding cerebrospinal fluid (CSF).

Under normal conditions, carbon dioxide levels have a much more powerful effect on respiratory activity than does oxygen because the oxygen level in the arteries seldom declines enough to activate the oxygen receptors. But when oxygen levels do fall, the two types of receptors cooperate. Carbon dioxide is generated during oxygen consumption, so when oxygen concentrations are falling rapidly, carbon dioxide levels are usually increasing.

Chemoreceptor reflexes are extremely powerful respiratory stimulators, and they cannot be consciously suppressed. For example, you can hold your breath before diving into a swimming pool and thereby prevent the inhalation of water, but you cannot commit suicide by holding your breath "until you turn blue." Once the concentration of carbon dioxide rises to critical levels, you will be forced to take a breath. Inexperienced divers often provide an example of the significance of the carbon dioxide level. In order to hold their breath longer underwater they may take a series of deep, full breaths. This process is called *hyperventilation*. Although most believe that their increased breath-holding ability is due to "extra oxygen," it is really due to the loss of carbon dioxide. If the carbon dioxide level is reduced low enough, breath-holding ability may increase to the point that the individual becomes unconscious from oxygen starvation in the brain without ever feeling the urge to breathe.

VOLUNTARY AND INVOLUNTARY INFLUENCES

✔ Where are the automatic control centers for respiration located?

✔ Are the carotid and aortic chemoreceptors more sensitive to carbon dioxide levels or to oxygen levels?

Higher centers in the brain can influence respiration by affecting the respiratory centers of the pons, and by the direct control of respiratory muscles. For example, the contractions of respiratory muscles can be voluntarily suppressed or exaggerated; this must be done while talking or singing. Respiration can also be affected by activities of the limbic system. For example, activation of centers involved with rage, feeding, or sexual arousal will change the pattern of respiration.

AGING AND THE RESPIRATORY SYSTEM

Many factors interact to reduce the efficiency of the respiratory system in elderly individuals. For example,

1. With increasing age, elastic tissue throughout the body breaks down. This deterioration reduces the resilience of the lungs and lowers the vital capacity.

2. Movements of the chest cage become restricted because of arthritic changes in the rib joints and less flexible rib cartilages.

3. Some degree of *emphysema* is normally found in individuals age 50 to 70. Emphysema is a condition characterized by shortness of breath. It is caused by the destruction of the surfaces of alveoli, which, in turn, reduces the area available for gas exchange in the lungs. However, the degree of emphysema varies widely depending on the lifetime exposure to cigarette smoke and other respiratory irritants. Comparative studies of nonsmokers and those who have smoked for varying lengths of time clearly show the negative effect of smoking on respiratory performance.

DISORDERS OF THE RESPIRATORY SYSTEM

Disorders affecting the respiratory system occur through three basic mechanisms:

1. *Interfering with the movement of air along the respiratory passageways.* Internal or external factors may be involved. For example, within the respiratory tract, the constriction of small airways can reduce airflow to the lungs. External factors that interfere with air movement include the introduction of air or blood into the pleural cavity, which leads to lung collapse.

2. *Damaging or otherwise interfering with the diffusion of gases at the respiratory membrane.* The walls of the alveoli are part of the respiratory membrane, where gas exchange occurs. Any disease process that affects the alveolar walls will reduce the efficiency of gas exchange. For example, lung cancer and emphysema can destroy alveoli.

3. *Blocking or reducing the normal circulation of blood through the alveolar capillaries.* For example, blood flow to portions of the lungs may be prevented by a pulmonary embolism, a circulatory blockage of a pulmonary artery.

THE PHYSICAL EXAMINATION

The respiratory system is often divided into an upper respiratory tract and lower respiratory tract. The *upper respiratory tract* is made up of the nose, nasal cavity, paranasal sinuses, pharynx, and larynx. The *lower respiratory tract* includes the trachea, bronchi, bronchioles, and the lungs.

Symptoms

Many disorders of the upper respiratory tract are caused by viral and bacterial infections. General symptoms of such upper respiratory disorders include a runny nose, nasal congestion, sore throat, headache, and fever.

When they seek medical attention, individuals with lower respiratory disorders generally do so because they are experiencing chest pain and are having difficulty breathing. The chest pain associated with a respiratory disorder usually worsens when the person takes a deep breath or coughs. This pain is different from the chest pain experienced by individuals with angina (pain appears during exertion) or a myocardial infarction (pain is continuous, even at rest). Several disorders, such as those affecting the pleural membranes, cause chest pain that is localized to specific regions of the thorax. A person with such a condition will usually press against the sensitive area and avoid coughing or breathe slowly and shallowly in an attempt to reduce the pain.

Difficulty in breathing, or **dyspnea** (DISP-nē-ah), may be a symptom of pulmonary disorders, cardiovascular disorders, metabolic disorders, or environmental factors such as hypoxia (low blood oxygen levels) at high altitudes. It may be a chronic problem, or it may develop only during exertion, or only when the person is lying down.

Dyspnea due to respiratory problems generally indicates one of the following classes of disorders:

- *Obstructive disorders* result from increased resistance to air flow along the respiratory passageways. The individual usually struggles to breathe, even at rest, and expiration is more difficult than inspiration.

- *Restrictive disorders* result from factors that limit lung expansion or the diffusion of gases. Examples of such factors include scoliosis, paralysis of respiratory muscles, trauma, congenital disorders, and *pulmonary fibrosis*, where abnormal fibrous tissue in the alveolar wall slows oxygen diffusion into the bloodstream. Individuals with restrictive disorders usually experience dyspnea during exertion because pulmonary ventilation cannot increase enough to meet the respiratory demand.

Signs

Signs of respiratory disorders can be detected by the 4 techniques employed in a standard physical examination. ∞ *[p. 14]*

1. Inspection can reveal an abnormal clubbing of the fingers. ∞ *[p. 125]* A blue color of the skin and mucous membranes (cyanosis) usually indicates hypoxia.

2. Palpation of the bones and muscles of the thoracic cage can detect structural problems or abnormal musculature that might make breathing difficult.

3. Percussion on the surface of the thoracic cage over the lungs will yield sharp, resonant sounds. Dull or flat sounds may indicate structural changes in the lungs, such as those that accompany the collapse of part of a lung (*atelectasis*).

4. Auscultation of the lungs with a stethoscope yields the distinctive sounds of inspiration and expiration. These sounds vary in intensity, pitch, and duration. Examples of abnormal sounds include:

- *Rales* (rahls) are hissing, whistling, scraping, or rattling sounds associated with increased airway resistance. The sounds are created by turbulent airflow past accumulated pus or mucus or through narrowed airways.

- *Stridor* (STRĪ-dor) is a very loud, high-pitched sound that can be heard without a stethoscope. Stridor generally indicates acute airway obstruction, such as the partial blockage of the glottis by a foreign object.

- *Wheezing* is a whistling sound that can occur with inspiration or expiration. It generally indicates smaller airway obstruction.

- *Coughing* is a familiar sign of several respiratory disorders. Although primarily a reflex mechanism that clears the airway, coughing may also indicate irritation of the lining of the respiratory passageways. Coughing may eject **sputum** (SPŪ-tum), or *phlegm* (flem), a mucus secretion from the lungs, bronchi, or trachea. A *productive cough* ejects sputum; a *nonproductive cough* does not.

- A *friction rub* is a distinctive crackling sound produced by abrasion between abnormal serous membranes. A *pleural rub* accompanies respiratory movements and indicates problems with the pleural membranes. A *pericardial rub* accompanies the heartbeat and indicates inflammation of the pericardium.

During the assessment of vital signs, the **respiratory rate** (number of breaths per minute) is recorded, along with notations about the general rhythm and depth of respiration. The normal respiratory rate of an adult at rest ranges from 12 to 18 per minute (12–18/min). *Tachypnea* (tak-ip-NĒ-ah) is a respiratory rate faster than 20/min. in an adult; *bradypnea* (brad-e-NĒ-ah) is an adult respiratory rate below 12/min.

Table 17-1 introduces some of the important procedures and laboratory tests useful in diagnosing respiratory disorders.

RESPIRATORY DISORDERS

General causes of respiratory disorders in the upper and lower respiratory tract may be divided into inflammation and infection, trauma, and tumors, as well as congenital, degenerative, and secondary (immune and cardiovascular) disorders (Figure 17-11●).

Inflammation and Infection

Respiratory infections may be caused by viruses, bacteria, and fungi. Illnesses due to infections of the upper respiratory tract are some of the most common disorders of the human body, especially among children. They are commonly transmitted as droplets in the air, often emitted in a sneeze. Infections of the lower respiratory tract include some of most feared diseases in history, pneumonia and tuberculosis.

General terms for the inflammation and infection of various portions of the respiratory tract are:

- *Rhinitis*, affecting the mucous membranes of the nasal cavity,

- *Sinusitis*, affecting the paranasal sinuses,

- *Pharyngitis*, affecting the pharynx, or throat,

- *Epiglottitis*, affecting the epiglottis,

- *Laryngitis*, affecting the larynx,

- *Bronchitis*, affecting the bronchi,

- *Pneumonia*, affecting the lungs

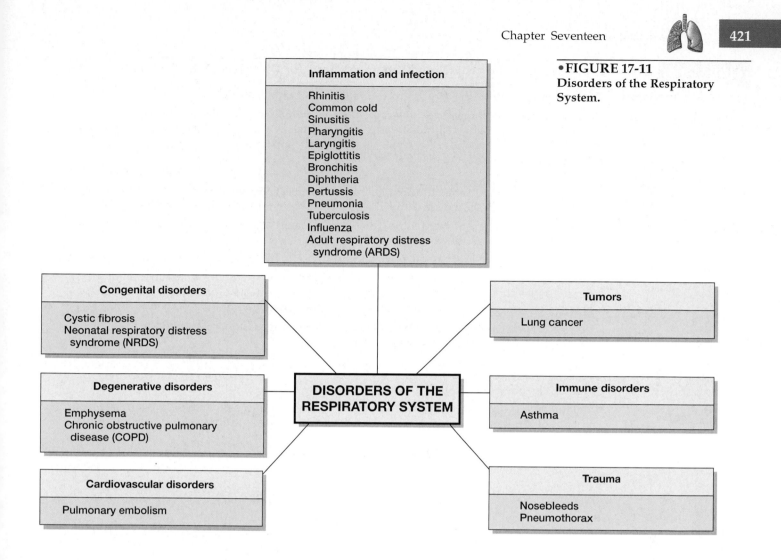

•FIGURE 17-11
Disorders of the Respiratory System.

Rhinitis. Rhinitis (ri-NĪ-tis), inflammation of the mucous membranes of the nose, is often accompanied by obstruction of the nasal passageway, nasal discharge, and sneezing. Common causes of rhinitis are viral infections and allergies. Allergies were discussed in Chapter 16 (p. 398).

The **common cold**, or **coryza** (ko-RĪ-zah), can be caused by any of more than 200 different viruses. Present all year round, different cold viruses predominate during different seasons. One group, the *rhinoviruses*, account for about half of all colds. They infect the epithelium of the upper respiratory tract. The *coronaviruses* are the second most common cause of colds. They replicate in the epithelium of the respiratory tract and digestive tract, where it can cause diarrhea and dehydration.

Colds are usually contracted by breathing in virus-containing droplets released by a person's sneeze, by handling contaminated materials, or in hand-to-hand contact. The symptoms of a cold appear 2 to 4 days after the initial infection. Most colds last about 1 week and are usually "head colds," restricted to the nose and throat. The severity of the symptoms, such as nasal discharge, fever, headache, sore throat, aching muscles, and chills, depend on the numbers of viruses released from the epithelial cells. The viral infection may predispose the person to secondary infections by bacteria, or activate herpes viruses to form cold sores. ∞ *[p. 128]*

Sinusitis. Flushing the nasal cavity epithelium with mucus produced in the paranasal sinuses often succeeds in removing mild irritants like dust particles (p. 407). **Sinusitis** (sī-nus-Ī-tis) most often arises from an infection that spreads from the nose along the drainage passageways. As the lining becomes irritated and swollen, the connecting passageways narrow. Drainage of mucus slows, congestion increases, and the individual experiences headaches and a feeling of

Table 17-1   Examples of Diagnostic Procedures and Laboratory Tests for Disorders of the Respiratory System

Diagnostic Procedure	Method and Result	Representative Uses
Pulmonary function studies	A spirometer or other measuring instrument is used to determine lung volumes and capacities. Forced vital capacity (FVC) is the amount of air forcibly expired after maximal inhalation	Useful in differentiating obstructive from restrictive lung diseases; also used to determine the extent of pulmonary disease
Bronchoscopy	Fiber-optic tubing is inserted through oral cavity and into trachea, larynx, and bronchus	Detects abnormalities; used to remove foreign objects, secretions, and tissue samples
Mediastinoscopy	A lighted instrument is inserted into an incision made superior to the sternum	Detection of abnormalities; biopsy removal
Lung biopsy	Lung tissue is removed for pathological analysis	Useful in differentiating pulmonary pathologies
Chest X-ray study	Standard X-ray produces film sheet with radiodense tissues shown in white on a negative image	Detection of tumors; inflammation of the lungs; rib or sternal fractures; pulmonary edema; pneumonia
Thoracentesis	A needle is inserted into the intrapleural space for removal of fluid for analysis or to relieve pressure	See pleural fluid analysis below
Pulmonary angiography	A catheter that releases a dye solution visible on X-ray is threaded into the pulmonary arteries	Detection of pulmonary embolism
Lung scan	Radionuclide is injected intravenously; radiation that is emitted is captured to create an image of the lungs	Determination of areas with decreased blood flow due to pulmonary embolism or pulmonary disease
Computerized tomography (CT) scan	Standard CT; contrast media are usually used	Detection of tumor, cyst, or other structural abnormality

Laboratory Test	Normal Values in Blood Serum or Plasma	Significance of Abnormal Values
Arterial blood gases and pH		
pH	7.35–7.45	<7.35 indicates acidosis >7.45 indicates alkalosis
CO_2 pressure	35–45 mm Hg	>45 mm Hg with pH <7.35 found in pulmonary disorders such as emphysema, COPD, and in CNS problems leading to irregular breathing. <35 mm Hg with a pH >7.45 accompanies prolonged hyperventilation; seen in emphysema
O_2 pressure	75–100 mm Hg	<75 mm Hg may occur in pneumonia and emphysema
HCO_3^-	22–28 mEq/l	>28 mEq/l (with elevated CO_2 and decreased pH) indicates chronic acidosis; <22 mEq/l (with decreased CO_2 and elevated pH) is characteristic of chronic alkalosis
Sputum studies		
Cytology	No malignant cells are present	Sloughed malignant cells indicate lung cancer
Culture and sensitivity (C&S)	Sputum sample is placed on growth medium	Identification of causative pathogenic organism and the organism's susceptibility to antibiotics
Tuberculin skin test	Skin wheal at 48–72 hours should be <10 mm in diameter	A skin lesion at injection site that measures >10 mm is a positive test for tuberculosis
Pleural fluid analysis		
Fluid color and clarity	No pus; fluid is clear	
Cells	WBC <1,000 per mm^3	WBC >1,000 per mm^3 indicates potential infectious or inflammatory process
Culture	Sample is placed on growth medium	Presence of bacteria indicates infection

painful pressure within the facial bones. Most cases of sinusitis are due to bacteria such as *Streptococcus pneumoniae* or *Haemophilus influenzae*, as a complication of the *common cold*. In addition to invading microorganisms, temporary sinus problems may accompany allergies or the exposure of the mucous epithelium to chemical irritants.

Chronic sinusitis may occur as the result of a **deviated (nasal) septum**. In this condition the nasal septum has a bend in it, generally at the junction between the bony and cartilaginous regions. Septal deviation often blocks drainage of one or more sinuses, producing chronic cycles of infection and inflammation. A deviated septum can result from developmental abnormalities or from injuries to the nose. The condition can usually be corrected or improved by surgery.

Pharyngitis. **Pharyngitis** (fair-in-JĪ-tis) that is caused by *Streptococcus pyogenes* is commonly called *strep throat*. Symptoms include a sore, inflamed throat, fever, and swollen tonsils and lymph nodes (Figure 17-12•). Like the common cold, there may be a cough or discharge from the nose. Antibiotics, generally penicillin, are used in treatment. Too long a delay in treatment is dangerous, because strep throat can lead to rheumatic fever.

Laryngitis. **Laryngitis** (lar-in-JĪ-tis) is characterized by a dry, sore throat and hoarse voice, or temporary loss of voice. Acute laryngitis may be caused by bacteria (*Streptococcus pneumoniae* and *Haemophilus influenzae*), viral infection, or an allergy to pollen, drugs, or other substances. Chronic laryngitis is usually due to other causes, such as overuse of the voice, or irritation due to smoking, alcohol, or chemical fumes. Treatment involves treating fever and pain, and keeping the throat moist with humidifiers. Persistent symptoms may require antibiotics or examination to check for signs of cancer.

Epiglottitis. **Epiglottitis** (ep-i-glot-TĪ-tis) is almost always caused by the bacterium, *Haemophilus influenzae.* Inflammation and swelling of the epiglottis can interfere with the flow of air making breathing difficult, or even block the air passageways, causing suffocation. Symptoms of this condition, most common in children age 2 to 6 years, is fever, noisy breathing (stridor), and drooling because swallowing saliva (or anything else) is difficult.

Bronchitis. **Bronchitis** (brong-KĪ-tis), an inflammation of the bronchial lining, is characterized by the overproduction of mucus and frequent coughing. An estimated 20 percent of adult males have had chronic bronchitis at some time. This condition is most commonly related to cigarette smoking, but it can also result from other environmental irritants, such as chemical vapors, air pollution, asbestos fibers, and coal dust. Over time the increased mucus production can block smaller airways and reduce respiratory efficiency. Inflammation may also be a secondary effect of a viral infection, or due to infection by bacteria such as *Streptococcus pneumoniae*, *Mycoplasma pneumoniae*, and others. Such infections may enter the alveoli and cause pneumonia.

Diphtheria. **Diphtheria** (dif-THĒ-re-ah) is a rare disease today in the U.S. A hundred years ago, however, it was a common cause of childhood deaths. Diphtheria is caused by a bacteria, *Corynebacterium diphtheriae* (kō-rī-nē-bak-TĒ-re-um dif-THĒ-rē-ē), which, in turn, carries viral DNA. The viral DNA contains the information for producing a toxin that interferes with protein synthesis, especially in the heart, kidneys, and nervous system. The bacteria commonly infects the pharynx, growing over the surface and forming a *pseudomembrane*, a fibrous network which leaves a bleeding wound when removed. If it is not removed, the pseudomembrane may cover the air passages and cause suffocation (Figure 17-13•).

Diphtheria is prevented with **DTP** (*diphtheria, tetanus, pertussis*) **vaccine**. It is administered in a series of vaccinations beginning at 2 months of age. Booster shots are also given throughout childhood and adulthood.

Pertussis. **Pertussis**, or *whooping cough*, is a highly contagious disease that usually occurs in infants and young children. Pertussis means "intensive cough." The symptoms of whooping cough are violent fits of coughing that end in a "whoop" sound as breath is drawn in.

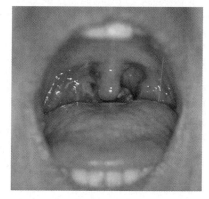

•**FIGURE 17-12**
Strep Throat.
The tonsils become swollen and red in this form of pharyngitis.

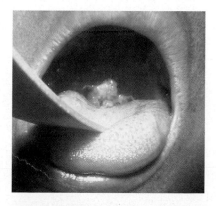

•**FIGURE 17-13**
Pseudomembrane of Diphtheria.
The tough, fibrous pseudomembrane produced in diphtheria can cause suffocation if it grows over the air passageway.

Pertussis is caused by bacteria, *Bordetella pertussis* (bor-de-TEL-la per-TUS-sis), that are spread in respiratory droplets and infect the upper respiratory tract. The bacteria release toxins into the blood and also destroy the ciliated respiratory epithelium, which allows the accumulation of mucus. Coughing is a reflex response that attempts to dislodge the mucus and clear the respiratory passage. The severe fits of coughing may continue to occur for up to 6 weeks, followed by milder coughing for several months. Treatment includes an antitoxin to neutralize the toxin and antibiotics to kill the bacteria. Pertussis has declined significantly where the pertussis vaccine is available. Part of the DTP vaccine, the pertussis vaccine is administered to infants at 2, 4, and 6 months of age, with a booster shot at 5 years of age.

Pneumonia. **Pneumonia** (noo-MŌ-nē-uh) is an inflammation of the lungs due to infection. As inflammation occurs, fluids leak into the alveoli and there is swelling and narrowing of the respiratory bronchioles. Pneumonia is the leading cause of death by infectious organisms in the U.S., and is a common complication of severe illnesses. It is most often caused by bacteria or viruses, but may also be due to fungi, protozoans, some allergies, chemicals, and radiation.

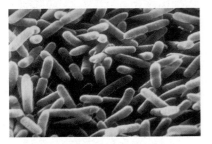

These pneumonia-causing bacteria use hollow, thread-like projections to attach to the ciliated respiratory epithelium.

Based on the site of infection, there are two types of pneumonia. In **lobar pneumonia**, the infection is in one of the five large divisions (*lobes*) of the lungs. **Bronchopneumonia** begins with an infection in the bronchi and bronchioles and then spreads to alveoli in the lungs. Lobar pneumonia is frequently caused by *Streptococcus pneumoniae* (noo-MŌN-nē-ī). The other types of bacteria that cause pneumonia are listed in Table 17-2. Bronchopneumonia may be caused by *S. pneumoniae*, as a secondary infection of other disorders, or as a result of breathing in (*aspiration*) fluids, such as vomit.

Symptoms of pneumonia appear suddenly a few days after the initial infection. They include chills, fever (up to 41° C or 106° F), chest pain, shortness of breath, cough, and sputum that contains mucus, pus, and blood. Treatment depends on the type of infectious organism. Different antibiotics are used for different bacteria. With proper treatment, the majority of individuals recover.

Tuberculosis. **Tuberculosis (TB)** is a major health problem throughout the world. With roughly 3 million deaths each year, it is the leading cause of death from infectious diseases. TB is caused by a bacterium called *Mycobacterium tuberculosis*. Unlike other deadly diseases, such as AIDS, TB can be transmitted through casual contact. Anyone who breathes is at risk of contracting this disease; all it takes is exposure to the causative bacterium. Coughing, sneezing, or speaking by an infected individual will spread the pathogen through the air in the form of tiny droplets that can be inhaled by others.

The tuberculin bacteria multiply slowly in WBCs and the alveolar macrophages within the lungs. When these cells die and rupture, the bacteria are taken in by other alveolar macrophages where they again multiply. Large amounts of fluids are released into the lungs because of the inflammatory response, which accumulates in the alveoli and produces symptoms similar to pneumonia. In many cases the immune system is able to kill the bacteria and fibrous scars form in the lung at the former sites of infection. If the immune system fails to kill the TB bacteria and no treatment is provided, the disease progresses in stages. At the site of infection, macrophages and fibroblasts proceed to wall off the area, forming an abscess, or *tubercle*. If the scar-tissue barricade fails, the bacteria move into the surrounding tissues, and the process repeats itself. The resulting masses of fibrous tissue distort the conducting passageways, increasing resistance and decreasing airflow. The bacteria also break down the alveoli. The combination severely reduces the area available for gas exchange (Figure 17-14•).

Treatment for TB is complex, because the bacteria can spread to many different tissues, such as the bones, lymphatic system, and meninges, and they can develop a resistance to standard antibiotics and require prolonged treatment.

As a result, several antibiotic drugs are administered in combination over a period of 6–9 months. The most effective drugs now available include *isoniazid*, which interferes with bacterial replication, and *rifampin*, which blocks bacterial protein synthesis.

The TB problem is much less severe in developed nations, such as the United States, than in developing nations. However, an increasing percentage of cases in the U.S. are caused by antibiotic-resistant strains. For example, in New York City, 33 percent of new cases are drug-resistant. The frightening part about this surge in resistant TB is that the fatality rate for infections resistant to two or more antibiotics is 50 percent. Fortunately, public health efforts that include isolation of cases and monitored compliance with treatment have thus far proven effective.

Influenza. Influenza, or "the flu," is a viral infection of the respiratory tract. The viruses are usually transmitted through the air in droplets produced by a sneeze or cough.

The **influenza virus** is an RNA virus in a group called the *orthomyxoviruses*. There are three different forms of the influenza virus: **types A, B,** and **C**. Once infected by type C virus, the immune system produces antibodies against type C and the individual has lifetime immunity. Antibodies are also produced against the A and B types, but because these forms mutate into new variations, infection with a new variant can occur. Type A forms new variations more frequently than type B. It is the type A strains that cause epidemics and worldwide pandemics, such as the Hong Kong flu in 1968.

Once the virus enter the body, they infect the ciliated respiratory epithelial cells, causing them to lose their cilia. This allows bacterial invasion and may lead to secondary infections, such as pneumonia. Symptoms begin to appear within 48 hours of infection. Typical flu symptoms include fever, headache, sore throat, runny nose, muscle weakness, fatigue, chest pain, and cough. The flu runs its course in a week or two, but a cough and general weakness may last a few more weeks.

Prevention involves anti-influenza vaccines, but, because types A and B change, the vaccines are not always successful in preventing infection. Vaccinations of new, updated vaccines need to be administered each year. They are recommended for older individuals and those with respiratory or circulatory disorders. Treatment is usually directed at relieving symptoms, such as analgesics for aches and pains. The drug *amantidine*, which blocks viral replication, may also be used in shortening the time and severity of an influenza A infection. New antiviral medicines may help with both A and B strains of influenza.

Table 17-2 summarizes the major infectious diseases of the respiratory system.

Trauma

Nosebleeds. The nasal cavity contains an extensive set of capillaries that help warm the incoming air before it enters the rest of the respiratory tract. These vessels and the relatively vulnerable position of the nose make a nosebleed, or **epistaxis** (ep-i-STAK-sis), a common event. Bleeding generally involves vessels of the mucous membrane covering the cartilaginous portion of the nasal septum. Nosebleeds may be caused by trauma, such as a punch in the nose, or drying, infections, allergies, and clotting disorders. Hypertension may also rupture the blood vessels.

Pneumothorax. An injury to the chest wall that penetrates the parietal pleura or damages the alveoli and the visceral pleura can allow air into the pleural cavity. This entry of air, called a **pneumothorax**, breaks the fluid bond between the pleural layers. This allows the elastic fibers of the lung to recoil and the result is a partially or completely collapsed lung. Treatment for a collapsed lung involves the removal of as much of the air as possible before the opening is sealed.

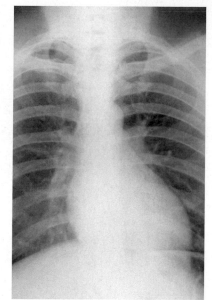

(a) Normal lungs

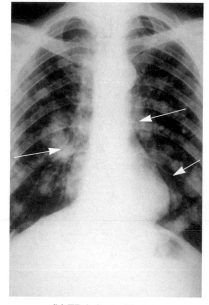

(b) TB-infected lungs

•FIGURE 17-14
X-rays of Tuberculosis.
(a) Chest X-ray showing healthy lungs. **(b)** Advanced case of tuberculosis. The white patches (arrows) indicate tubercles, where lung tissue has been permanently destroyed.

Table 17-2   Examples of Infectious Diseases of the Respiratory System

Disease	Organism(s)	Description
Bacterial Diseases		
Sinusitis	*Streptococcus pneumoniae* *Haemophilus influenzae*	Inflammation of the paranasal cavities; headaches, pain, and pressure in facial bones
Pharyngitis	*Streptococcus pyogenes*	Inflammation of the throat; strep throat; sore throat, fever, no cough or sputum
Laryngitis	*Streptococcus pneumoniae* *Haemophilus influenzae*	Inflammation of the larynx; dry, sore throat; hoarse voice or loss of voice
Epiglottitis	*Haemophilus influenzae*	Inflammation of the epiglottis; most common in children; fever, noisy breathing; swelling may block trachea and lead to suffocation
Bronchitis	*Streptococcus pneumoniae* *Mycoplasma pneumoniae*	Inflammation of the bronchial passageways; productive cough (with sputum); may enter lungs and cause pneumonia
Diphtheria	*Corynebacterium diphtheriae*	Inflammation of the pharynx; pseudomembrane in pharynx; bacterial toxin affects heart and other tissues
Pertussis (Whooping cough)	*Bordetella pertussis*	Highly contagious disease of children; mucus production; severe coughing ends in a "whoop" sound during inhalation
Pneumonia	*Streptococcus pneumoniae* *Staphylococcus aureus* *Klebsiella pneumoniae*	Inflammation of the lungs; alveoli fill with fluids; chills, high fever, cough, sputum of mucus, pus, and blood
Tuberculosis	*Mycobacterium tuberculosis*	Highly contagious infection of the lungs; may spread to other tissues; abscesses, or tubercles form in lungs; bacteria multiply in WBCs and alveolar macrophages and break down alveoli
Viral Diseases		
Common cold (coryza)	Rhinoviruses Coronaviruses	Nasal obstruction, nasal discharge, sneezing, and headache
Influenza	Orthomyxoviruses (Influenza virus A, B, C)	Frequent variations in Type A and B viruses may produce new epidemics; fever, headache, sore throat, nasal discharge, muscle weakness, fatigue, chest pain, and cough

Tumors

Lung cancer. **Lung cancer** is one of the most malignant cancers (see Clinical Note on p. 407). These cancers affect the epithelial cells that line the air conducting passageways, mucous glands, or alveoli. Symptoms do not generally appear until the condition has progressed to the point at which tumor masses are restricting airflow or compressing structures in the mediastinum.

Immune Disorders

Asthma. **Asthma** (AZ-ma) affects an estimated 3–6 percent of the U.S. population. There are several forms of asthma, but each is characterized by unusually sensitive and irritable conducting passageways. In many cases, the trigger appears to be an immediate hypersensitivity reaction to an allergen, such as pollen, in the inspired air. ∞ *[p. 399]* Drug reactions, air pollution, chronic respiratory infections, exercise, and/or emotional stress can also induce an asthmatic attack in sensitive individuals.

The most obvious symptoms are breathlessness and wheezing when exhaling. These symptoms are due to (1) *bronchoconstriction*, the narrowing of bronchioles (2) edema and swelling of the lining of the respiratory passageways, and (3) accelerated production of mucus. The combination makes breathing very difficult. Exhalation is affected more than inhalation; the narrowed passageways often collapse before exhalation is completed. Although mucus production increases, mucus transport slows, and fluids accumulate along the passageways. The bronchoconstriction and mucus production occur in a few minutes, in response to the release of histamine and other molecules by mast cells. The area then becomes inflamed, further reducing airflow and damaging respiratory tissues.

During a severe asthmatic attack the efficiency of gas exchange in the lungs is reduced, and peripheral tissues gradually become oxygen-starved, a condition that can prove fatal. Asthma fatalities have been increasing in recent years. The annual death rate from asthma in the United States is approximately 4 deaths per million population (for ages 5–34). Mortality among asthmatic African-Americans is twice that among Caucasian Americans.

Treatment of asthma involves the dilation of the respiratory passageways by administering **bronchodilators** (brong-kō-dī-LĀ-torz) and reducing inflammation and swelling of the respiratory mucosa. Important bronchodilators include *theophylline*, *epinephrine*, and *albuterol*; inhaled or injected corticosteroids are used to combat inflammation and swelling. Antihistamines may also help to some degree.

Congenital Disorders

Cystic Fibrosis. **Cystic fibrosis (CF)** is the most common lethal inherited disease that affects Caucasians of Northern European descent; it occurs at a frequency of 1 in 2,500 births. In the United States, 2,000 babies are born with CF each year, and there are roughly 30,000 persons with this condition. Individuals with classic CF seldom survive past age 30; death is generally the result of a massive bacterial infection of the lungs and associated heart failure.

The gene involved carries instructions for a transmembrane protein, responsible for ion transport, that is found in exocrine cells producing watery secretions. In persons with CF, the protein does not function normally. The exocrine cells cannot transport salts and water effectively, and the secretions produced are thick and gooey. Mucous glands of the respiratory tract and exocrine cells of the pancreas, salivary glands, and digestive tract are affected.

The most serious symptoms appear because the respiratory epithelium cannot transport such dense mucus, which then blocks the smaller respiratory passageways. This blockage reduces the diameter of the airways, and the inactivation of the normal respiratory defenses leads to frequent bacterial infections. Bacterial infections of the stagnant mucus further stimulates mucus production by epithelial cells.

Treatment has primarily been limited to supportive care and antibiotic therapy to control bacterial infections. In a few instances, lung transplants have provided relief. The normal and abnormal gene structure has now been determined, and the current goal is to correct the defect by inserting normal genes within the cells in critical areas of the body. In the meantime, it has been discovered that one of the factors contributing to the thickness of the mucus is the presence of DNA released from degenerating cells within areas of inflammation. Inhaling an aerosol spray containing an enzyme that breaks down DNA has proven to be effective in improving the performance of the respiratory system.

Respiratory Distress Syndrome. **Respiratory distress syndrome (RDS)** is a condition that affects infants and adults, and results in reduced levels of oxygen in the blood. It is caused by reduced levels of surfactant in the lungs.

Surfactant cells begin producing surfactants at the end of the sixth fetal month (p. 409). By the eighth month, surfactant production has risen to the level required for normal respiratory function. **Neonatal respiratory distress syndrome (NRDS)**, also known as **hyaline membrane disease (HMD)**, develops when surfactant production fails to reach normal levels. Although there are inherited forms of HMD, the condition most commonly accompanies premature delivery.

In the absence of surfactants, the alveoli tend to collapse during exhalation. Although the conducting passageways remain open, the newborn infant must then inhale with extra force to reopen the alveoli on the next breath. In effect, every breath must approach the power of the first, and the infant rapidly becomes exhausted. Respiratory movements become progressively weaker; eventually the alveoli fail to expand, and gas exchange ceases.

One method of treatment involves assisting the infant by administering air under pressure so that the alveoli are held open. This procedure, known as *positive end-expiratory pressure (PEEP)*, can keep the newborn alive until surfactant production increases to normal levels. Surfactant from other sources can also be provided; suitable surfactants can be extracted from cow lungs (Survanta), obtained from the liquid (amniotic fluid) that surrounds full-term infants, or synthesized by genetic engineering techniques (Exosurf). These preparations are usually administered in the form of a fine mist of surfactant droplets.

Surfactant abnormalities may also develop in adults as the result of severe respiratory infections or other sources of pulmonary injury. Alveolar collapse follows, producing a condition known as **adult respiratory distress syndrome (ARDS)**. PEEP is typically used in an attempt to maintain life until the underlying problem can be corrected, but at least 50–60 percent of ARDS cases result in fatalities.

Degenerative Disorders

Emphysema. **Emphysema** (em-fi-SĒ-ma) is a chronic, progressive condition characterized by shortness of breath and an inability to tolerate physical exertion. The underlying problem is the destruction of the respiratory exchange surfaces, the alveoli.

Emphysema has been linked to the inhalation of air that contains fine particulate matter or toxic vapors, such as those found in cigarette smoke. Early in the disease, the individual may not notice problems, even with strenuous activity. As the condition progresses, the reduction in exchange surface limits the ability to provide adequate oxygen. Obvious clinical symptoms, however, typically fail to appear until the damage is extensive.

Alpha-antitrypsin, an enzyme that is normally present in the lungs, helps prevent degenerative changes in lung tissue. Most people requiring treatment for emphysema are adult smokers; this group includes individuals with alpha-antitrypsin deficiency and those with normal tissue enzymes. At least 80 percent of nonsmokers with abnormal alpha-antitrypsin will develop emphysema, generally at ages 45–50. All smokers will develop at least some emphysema, typically by ages 35–40.

Unfortunately, the loss of alveoli and bronchioles in emphysema is permanent and irreversible. Further progression can be limited by cessation of smoking; the only effective treatment for severe cases is the administration of oxygen. Lung transplants have helped some patients. For persons with alpha-antitrypsin deficiency who are diagnosed early, attempts are underway to provide enzyme supplements by daily infusion or periodic injection.

Chronic Obstructive Pulmonary Disease. **Chronic obstructive pulmonary disease (COPD)** involves chronic airflow obstruction due to chronic bronchitis and/or emphysema. Some obstruction may be reversible with treatment of infection or with bronchodilators, but some degree of obstruction always remains. Individuals with COPD commonly expand their chests permanently in an effort to enlarge their lung capacities and make the best use of the remaining functional alveoli. This adaptation gives them a distinctive "barrel-chested" appearance. When severe, COPD sufferers may also have symptoms of heart failure, including widespread edema. Blood levels of oxygen are low, and the skin has the bluish coloration of cyanosis. The combination of widespread edema and bluish coloration has led to the descriptive term *blue bloaters* for individuals with this condition.

Cardiovascular Disorders

The most common secondary disorders affecting the respiratory system result from problems with the cardiovascular system. One important example is the condition known as a pulmonary embolism.

Pulmonary Embolism. Blood pressure in the pulmonary circulation is usually relatively low. As a result, pulmonary arteries can easily become blocked by small blood clots, fat masses, or air bubbles in the pulmonary arteries. A **pul-**

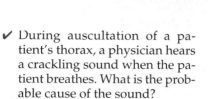

monary embolism results when a branch of a pulmonary artery is blocked and blood flow to the alveoli is stopped. A venous thrombosis (the formation of a blood clot in a vein) can promote the development of pulmonary embolism because there is a tendency for small blood clots to form, break off, and drift in the circulation. Large bone fractures may release globs of fat (from the bone marrow) into blood vessels, causing *fat emboli*. If a pulmonary embolism remains in place for several hours, the alveoli will permanently collapse. If the blockage occurs in a major pulmonary vessel, pulmonary resistance increases. This places more strain on the right ventricle, which may be unable to maintain cardiac output. Congestive heart failure may then result.

✔ During auscultation of a patient's thorax, a physician hears a crackling sound when the patient breathes. What is the probable cause of the sound?

✔ Why might an inflamed epiglottis be dangerous?

✔ What is the DTP vaccine?

CHAPTER REVIEW

Key Words

alveolus/alveoli (al-VĒ-o-lī): Air sacs within the lungs; the sites of gas exchange with the blood.

bronchiole: The finest subdivisions of the conducting passageways in the lungs.

diaphragm (DĪ-a-fram): The respiratory muscle that separates the thoracic cavity from the abdominopelvic cavity.

larynx (LAR-inks): A complex cartilaginous structure that surrounds and protects the glottis and vocal cords; the upper margin is bound to the hyoid bone, and the lower margin is bound to the trachea.

lungs: Paired organs of respiration that occupy the left and right pleural cavities.

nasal cavity: A chamber in the skull that extends between the nostrils and the nasopharynx.

paranasal sinuses: Air-filled chambers within bones of the skull; they are lined by a respiratory epithelium and they open into the nasal cavity.

pharynx: The throat; a muscular passageway shared by the digestive and respiratory tracts.

pleural cavities: Subdivisions of the thoracic cavity that contain the lungs.

respiration: Exchange of gases between living cells and the environment.

surfactant (sur-FAK-tant): Lipid secretion that coats alveolar surfaces and prevents their collapse.

trachea (TRĀ-kē-a): The windpipe; an airway extending from the larynx to the primary bronchi.

vital capacity: The maximum amount of air that can be moved in or out of the respiratory system; the sum of the inspiratory reserve, the expiratory reserve, and the tidal volume.

Selected Clinical Terms

asthma: An acute respiratory disorder caused by constricted airways due to an allergic response.

bronchitis: Inflammation of the bronchi and bronchioles.

cystic fibrosis: A lethal inherited disease that results in mucous secretions too thick to be transported easily, leading to respiratory problems.

dyspnea (DISP-nē-ah): Difficult, or labored, breathing.

emphysema: A chronic, progressive condition characterized by shortness of breath and inability to tolerate physical exertion.

epistaxis (ep-i-STAK-sis): A nosebleed.

Heimlich maneuver: Compression applied to the abdomen just beneath the diaphragm, to force air out of the lungs and clear a blocked trachea or larynx.

hypoxia: A condition of low oxygen concentrations in body tissues.

pneumonia: Inflammation of lungs due to infection; characterized by fluid leakage into the alveoli and/or swelling and constriction of the respiratory bronchioles.

pneumothorax: The entry of air into the pleural cavity.

respiratory distress syndrome: A condition resulting from inadequate surfactant production and associated collapse of the alveoli.

tuberculosis: A respiratory disorder caused by bacterial infection of the lungs.

Study Outline

INTRODUCTION *(p. 405)*

1. To continue functioning, body cells must obtain oxygen and eliminate carbon dioxide.

SYSTEM BRIEF *(p. 405)*

1. The process of **respiration** includes pulmonary ventilation (breathing), gas exchange, and gas transport. *Cellular respiration* refers to the cellular use of the oxygen obtained through respiration.

2. The functions of the respiratory system include providing for the gas exchange between air and circulating blood, defending the respiratory system from pathogens, permitting vocal communication, and aiding in the regulation of blood pressure and body fluid pH.

STRUCTURE OF THE RESPIRATORY SYSTEM *(pp. 405–411)*

1. The **respiratory system** includes the nose, nasal cavity and sinuses, pharynx, larynx, trachea, and conducting passageways leading to the gas-exchange surfaces of the lungs (*alveoli*). *(Figure 17-1)*

THE NOSE *(p. 405)*

2. Air normally enters the respiratory system through the **nostrils**, which open into the **nasal cavity**.

3. The **hard palate** separates the oral and nasal cavities. The **soft palate** separates the superior **nasopharynx** from the rest of the pharynx. *(Figure 17-1)*

4. Much of the respiratory epithelium is ciliated and produces mucus that traps incoming particles. *(Figure 17-2)*

THE PHARYNX *(p. 407)*

5. The **pharynx** is a chamber shared by the digestive and respiratory systems. The nasopharynx contains the pharyngeal tonsil and the entrances to the auditory tubes. The **oropharynx** is continuous with the oral cavity; the **laryngopharynx** includes the narrow zone between the oropharynx and the entrance to the esophagus. *(Figure 17-1)*

THE LARYNX *(p. 407)*

6. Inhaled air passes through the **glottis** on its way to the lungs; the **larynx** surrounds and protects the glottis. The **epiglottis** projects into the pharynx. *(Figure 17-3a,b)*

7. Two pairs of folds span the glottal opening: the relatively inelastic *false vocal cords* and the more delicate true **vocal cords**. Air passing through the glottis causes the true vocal cords to vibrate and produce sound. *(Figure 17-3c,d)*

THE TRACHEA *(p. 408)*

8. The **trachea** ("windpipe") is a tough, flexible tube. The tracheal wall contains C-shaped **tracheal cartilages** that protect the airway. The posterior tracheal wall can distort to permit passage of large masses of food along the esophagus.

THE BRONCHI *(p. 409)*

9. The trachea branches to form the **right** and **left primary bronchi**.

THE BRONCHIOLES *(p. 409)*

10. The primary bronchi and their branches are supported by cartilage rings or plates. The smallest bronchi are connected to smaller passageways called bronchioles. The bronchioles lack cartilage and have relatively larger amounts of smooth muscle than bronchi.

11. Each **terminal bronchiole** delivers air to clusters of alveoli. *(Figure 17-4)*

THE ALVEOLI *(p. 409)*

12. The alveoli are the respiratory exchange surfaces of the lungs. *(Figure 17-4)*

13. The alveolar lining consists of a simple squamous epithelium; **surfactant cells** scattered in it produce an oily secretion that keep the alveoli from collapsing. **Alveolar macrophages** patrol the epithelium and engulf foreign particles or pathogens. *(Figure 17-4)*

14. Each alveolus is surrounded by a network of capillaries, which make up the respiratory exchange region of the pulmonary circulation. Pulmonary veins collect and deliver the venous blood to the left atrium.

THE LUNGS *(p. 410)*

15. Each lung occupies a single pleural cavity lined by a **pleura** (a serous membrane).

PHYSIOLOGY OF RESPIRATION *(pp. 411–416)*

1. Respiratory physiology focuses on three integrated processes: *pulmonary ventilation*, or breathing (movement of air into and out of the lungs); gas exchange between the lungs, circulating blood, and body cells; and gas transport in the blood.

PULMONARY VENTILATION *(p. 411)*

2. A single breath, or *respiratory cycle*, consists of an **inhalation** (*inspiration*) and an **exhalation** (*expiration*).

3. Contractions of the diaphragm and the *external* and *internal intercostal* muscles of the rib cage alter the volume of the thoracic cavity and lungs. Increased volume lowers the air pressure within the lungs, and air enters the lungs; decreased volume increases air pressure within the lungs, causing air to be exhaled. *(Figure 17-5)*

4. The **vital capacity** is the **tidal volume** plus the **expiratory reserve** and the **inspiratory reserve volumes**. The air left in the lungs at the end of maximum expiration is the **residual volume**. *(Figure 17-6)*

GAS EXCHANGE *(p. 414)*

5. The directions in which oxygen and carbon dioxide diffuse across cell membranes depend on their concentrations in air, blood, and tissue fluid.

6. Inhaled and exhaled air differ in composition.

7. **External respiration** takes place in the lungs; oxygen diffuses into the blood, and carbon dioxide diffuses into the alveoli. **Internal respiration** takes place at tissue capillary beds; oxygen diffuses into interstitial fluid and body cells, and carbon dioxide diffuses into the blood. *(Figure 17-8)*

GAS TRANSPORT *(p. 414)*

8. Blood entering systemic capillaries delivers oxygen and absorbs carbon dioxide. The transport of oxygen and carbon dioxide in the blood involves reactions that are completely reversible.

9. Almost all the oxygen is transported attached to hemoglobin molecules within red blood cells.

10. Roughly 7 percent of the CO_2 transported in the blood is dissolved in the plasma, another 23 percent is bound to hemoglobin, and the rest is converted to carbonic acid, which breaks down into hydrogen ions and bicarbonate ions. *(Figure 17-9)*

CONTROL OF BREATHING *(pp. 416–418)*

1. Large-scale changes in the rates of gas exchange requires homeostatic adjustments by the cardiovascular and respiratory systems.

RESPIRATORY CONTROL CENTERS *(p. 416)*

2. The **respiratory centers** are located in the pons and medulla oblongata. These centers control the respiratory muscles and regulate the **respiratory rate** and depth of breathing. *(Figure 17-10)*

SENSORY RECEPTORS AND RESPIRATION *(p. 416)*

3. The *inflation reflex* prevents overexpansion of the lungs during forced breathing. Pressure receptors in blood vessel walls influence the respiratory rate. Chemoreceptors respond to changes in the concentrations of oxygen and carbon dioxide of the blood and cerebrospinal fluid. *(Figure 17-9)*

VOLUNTARY AND INVOLUNTARY INFLUENCES *(p. 418)*

4. Conscious and unconscious thought processes can affect breathing by affecting the respiratory centers.

AGING AND THE RESPIRATORY SYSTEM *(p. 418)*

1. The respiratory system is generally less efficient in the elderly because (1) elastic tissue deteriorates, lowering the vital capacity of the lungs; (2) movements of the chest cage are restricted by arthritic changes and decreased flexibility of costal cartilages; and (3) some degree of *emphysema*.

DISORDERS OF THE RESPIRATORY SYSTEM
(pp. 418–429)

1. Disorders of the respiratory system may: (1) interfere with the movement of air along the respiratory passages, (2) damage or impede the diffusion of gases at the respiratory membrane, and (3) block or reduce the normal flow of blood through the alveolar capillaries.

THE PHYSICAL EXAMINATION *(p. 419)*

2. Symptoms of upper respiratory problems include a runny nose, nasal congestion, sore throat, headache, and fever. Major symptoms of lower respiratory problems are chest pain and **dyspnea**, or difficulty in breathing.

3. Dyspnea is associated with *obstructive disorders* (increased resistance to air flow along the respiratory tract) and *restrictive disorders* (factors that limit lung expansion or gas diffusion).

4. Signs of respiratory disorders revealed in a physical examination include; (1) clubbing of the nails and cyanosis of the skin (inspection), (2) structural problems of the skeletal system or asymmetrical musculature (palpation), (3) collapsed lung (percussion), (4) sounds of breathing (auscultation). Examples of abnormal sounds are *rales*, *stridor*, *wheezing*, *coughing*, and *friction rub*.

5. The respiratory rate, or number of breaths per minute, lies between 12–18/minute in a resting adult. In *tachypnea*, the breathing rate is above this range, and in *bradypnea*, the breathing rate is below this range. *(Table 17-1)*

RESPIRATORY DISORDERS *(p. 420)*

6. General causes of respiratory disorders include inflammation and infection, trauma, tumors, immune problems, congenital, degenerative, and cardiovascular problems. *(Figure 17-11; Table 17-2)*

7. Rhinitis is an inflammation of the mucous membranes of the nose. Viral infections, such as the common cold, or **coryza**, and allergies cause rhinitis.

8. Sinusitis is an inflammation of the paranasal sinuses. Sinusitis may occur through infections or a **deviated (nasal) septum**.

9. Pharyngitis is an inflammation of the pharynx, or throat. *Strep throat* is an example of pharyngitis due to bacteria. *(Figure 17-12)*

10. Laryngitis is an inflammation of the larynx. It is marked by a sore throat and hoarse voice.

11. Epiglottitis, an inflammation of the epiglottis, is often due to infection. This condition restricts the passage of air and may cause suffocation.

12. Bronchitis is an inflammation of the bronchial lining that is often caused by smoking.

13. Diphtheria is a bacterial disease that results in the formation of a pseudomembrane in the pharynx that may block air passages. *(Figure 17-13)*.

14. Pertussis, or *whooping cough*, is a bacterial disease that causes the buildup of mucus in the respiratory passages of infants and young children.

15. Pneumonia is an infection of the lungs. During inflammation fluids fill the alveoli and respiratory bronchioles narrow, making breathing difficult.

16. Tuberculosis is a bacterial disease of the lungs that is a major worldwide problem. *(Figure 17-14)*

17. Influenza, or "the flu," is caused by a virus that exists in three variations; types A, B, and C.

18. Traumatic disorders of the respiratory system include **nosebleeds** and **pneumothorax**, where air enters the pleural cavity. The air causes a collapsed lung.

19. Lung cancer is a very malignant cancer responsible for one-third of all U.S. cancer deaths.

20. Asthma is characterized by breathlessness and wheezing. It is an immediate hypersensitivity reaction to an allergen in the air conducting passageways. Treatment often involves **bronchodilators**.

21. Cystic fibrosis is a lethal inherited disorder. Respiratory passageways fill with mucus, blocking the movement of air to the lungs. Bacterial infections then stimulate the production of more mucus.

22. Respiratory distress syndrome (RDS) is a condition caused by a low level of surfactant in the lungs. As a congenital condition in babies born prematurely, it is called **neonatal respiratory distress syndrome**, or **hyaline membrane disease (HMD)**.

23. Emphysema is a degenerative disorder in which the alveoli are progressively destroyed, leading to shortness of breath. Emphysema and chronic bronchitis are forms of chronic airways obstruction.

24. Pulmonary embolism occurs when arterial vessels of the pulmonary circulation become blocked. The result is collapsed alveoli, reduced oxygen in the blood, and a stressed right ventricle.

Review Questions

MATCHING

Match each item in Column A with the most closely related item in Column B. Use letters for answers in the spaces provided.

	Column A		Column B
___	1. nasopharynx	a.	capillary endothelium and alveolar epithelium
___	2. laryngopharynx	b.	macrophages in alveoli
___	3. thyroid cartilage	c.	lies above the soft palate
___	4. surfactant cells	d.	covers inner surface of thoracic wall
___	5. dust cells	e.	lines nasal cavity
___	6. parietal pleura	f.	difficulty in breathing
___	7. visceral pleura	g.	inspiration
___	8. ciliated epithelium	h.	inferior portion of pharynx
___	9. respiratory membrane	i.	Adam's apple
___	10. relaxed diaphragm	j.	covers outer surface of lungs
___	11. inhalation	k.	blocks entry of food into larynx
___	12. exhalation	l.	expands thoracic cavity volume
___	13. contracted diaphragm	m.	decreases thoracic cavity volume
___	14. epiglottis	n.	hoarse voice
___	15. dyspnea	o.	produce oily secretion
___	16. laryngitis	p.	expiration

MULTIPLE CHOICE

17. The _____ portion of the pharynx is a passageway for air only.
(a) nasopharynx
(b) oropharynx
(c) laryngopharynx
(d) trachea

18. The _____ lacks the support of cartilage.
(a) trachea
(b) primary bronchi
(c) secondary bronchi
(d) bronchioles

19. The movement of air into and out of the lungs is _____ .
(a) external respiration
(b) internal respiration
(c) pulmonary ventilation
(d) cellular respiration

20. The mucous membrane lining the nasal cavity performs all the following functions except _____ .
(a) moistens incoming air
(b) carries mucus, dust, and pathogens to the pharynx
(c) decreases carbon dioxide level of incoming air
(d) warms incoming air

21. The amount of air moved into or out of the lungs during a single respiratory cycle is the _____ .
(a) tidal volume
(b) residual volume
(c) vital capacity
(d) inspiratory volume

22. The respiratory control centers that set the basic rate of breathing are in the _____ .
(a) pons
(b) medulla oblongata
(c) lungs
(d) internal carotid arteries

23. Inflammation of the mucous membrane in the nasal cavity is _____.
(a) rhinitis
(b) sinusitis
(c) pharyngitis
(d) coryza

24. _____ is a lethal inherited disease of the respiratory system.
(a) pertussis
(b) tuberculosis
(c) asthma
(d) cystic fibrosis

TRUE/FALSE

___ 25. The voice box is the cartilaginous structure known as the larynx.

___ 26. Inspiration occurs when the air pressure within the lungs is greater than atmospheric pressure.

___ 27. Carbon dioxide is more important than oxygen as a chemical regulator of respiration.

___ 28. Most of the oxygen carried in the blood is dissolved in the blood plasma.

___ 29. The nasal conchae increase the area over which incoming air moves.

___ 30. The formation of bicarbonate ions from carbon dioxide in blood causes the blood to become more acidic.

___ 31. An epistaxis is a nosebleed.

___ 32. A certain amount of emphysema occurs normally with aging.

SHORT ESSAY

33. Why is breathing through the nasal cavity more desirable than breathing through the mouth?

34. How are surfactant cells involved with keeping the alveoli from collapsing?

35. What is the functional importance of the decrease in cartilage and increase in smooth muscle in the lower respiratory passageways?

36. You spend the night at a friend's house during the winter and his hot-air furnace lacks a humidifier. When you wake up in the morning you have a fair amount of nasal congestion and think you may be coming down with a cold. After a steamy shower and some juice at breakfast, the nasal congestion disappears. Explain.

37. Disorders of the respiratory system involve three basic mechanisms. Describe them and give an example of each.

38. Physicians recommend that older people get a flu shot in the fall before the winter flu season begins. Why do some of these individuals still get the flu?

APPLICATIONS

39. During an intramural football game, Joe is tackled so hard that he breaks a rib. On the way to the hospital he is having a difficult time breathing. What may be wrong with Joe?

40. Mrs. J. delivers her first child 8 weeks premature. The physician notices that the infant's inhalations are long and forced, and the infant seems to tire rapidly. What do you think the physician will recommend? Why?

✔ Answers to Concept Check Questions

(p. 411) **1.** Increased tension in the vocal cords will cause a higher pitch in the voice. **2.** The tracheal cartilages are C-shaped to allow room for expansion of the esophagus when large portions of food or liquid are swallowed. **3.** Without surfactant, surface tension in the thin layer of water that moistens their surfaces would cause the alveoli to collapse.

(p. 413) **1.** During contraction the diaphragm flattens and increases the volume of the thoracic cavity, which expands lung volume. During relaxation the diaphragm projects upward and decreases the volume of thoracic cavity and, in turn, lung volume. **2.** Vital capacity is the maximum amount of air that can be moved into and out of the respiratory system in a single respiratory cycle. It is the sum of the inspiratory reserve volume, the expiratory reserve volume, and the tidal volume.

(p. 416) **1.** Most of the carbon dioxide released by body tissues enters red blood cells, where it is converted into bicarbonate ions that then diffuse into the blood plasma. At the lungs the bicarbonate ions are converted back into carbon dioxide within the red blood cells and released into the alveoli. **2.** External respiration refers to the diffusion of gases between the blood in the lung capillaries and air in the alveoli of the lungs.

(p. 418) **1.** The respiratory control centers are located in the medulla oblongata and the pons. The respiratory center within the medulla oblongata sets the basic pace for respiration by sending involuntary nerve impulses to the respiratory muscles. The respiratory centers in the pons can alter the basic rhythm established by the medulla oblongata. **2.** Carbon dioxide levels have a much more powerful effect on the carotid and aortic chemoreceptors than does oxygen because, under normal conditions, the oxygen level in the arteries does not usually decline enough to activate the oxygen receptors.

(p. 429) **1.** The sound is a friction rub produced by abrasion between abnormal serous membranes. In this case, it is probably a pleural rub between the pleural membranes. **2.** Inflammation of the epiglottis causes it to swell. This swelling interferes with the passage of air. Swelling may be so severe that suffocation results. **3.** The DTP vaccine is a mixture of three vaccines: diphtheria, pertussis, and tetanus. It is commonly administered to infants at 2, 4, and 6 months of age. A booster is administered at the age of 5 years.

The Digestive System

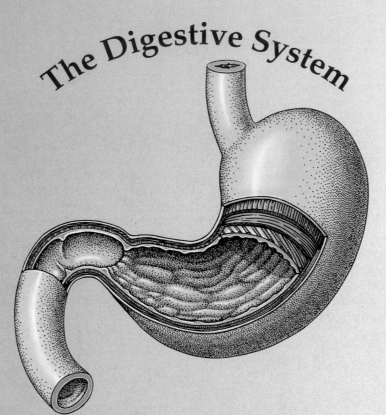

CHAPTER OUTLINE

CHAPTER OBJECTIVES

1 List the functions of the digestive system.
2 Identify the organs of the digestive tract and the accessory organs of digestion.
3 Explain how materials move along the digestive tract.
4 Describe how food is processed in the mouth and how it is swallowed.
5 Describe the stomach and its roles in digestion and absorption.
6 Describe digestion and absorption in the small intestine.
7 Describe the structure and functions of the pancreas, liver, and gallbladder.
8 Describe the structure and functions of the large intestine.
9 Describe the digestion and absorption of the substances in food.
10 Describe the changes in the digestive system that occur with aging.
11 Describe common symptoms and signs of digestive system disorders.
12 Describe examples of inflammation and infection, and other major disorders of the digestive system.

Few of us give any serious thought to the digestive system unless it begins misbehaving. Yet we devote a lot of time and effort to filling and emptying it. When something does go wrong with the digestive system, even something minor, most people seek help immediately. For this reason, television advertisements promote toothpaste and mouthwash, diet supplements, antacids, and laxatives on an hourly basis.

In its basic structure, the digestive system consists of a tube that passes through the body from the mouth to the anus. Various digestive glands secrete fluids into this tube. The digestive system as a whole supplies the energy and chemical building blocks for growth and maintenance from the food that moves down the digestive tube. It does this by first converting what we eat into a form that can be absorbed into the body. This involves breaking down the complex organic molecules in our food (proteins, starches, fats) into simpler molecules (amino acids, simple sugars, and fatty acids). This breakdown process is called *digestion*. *Absorption* is the movement of these breakdown products, along with other nutrients such as water, vitamins, or ions, across the wall of the digestive tube and into body tissues.

SYSTEM BRIEF

The digestive system consists of a muscular tube, the **digestive tract**, and various **accessory organs**, including the salivary glands, gallbladder, liver, and pancreas.

The functions of the digestive system involve a series of interconnected steps:

- *Ingestion* occurs when foods enter the digestive tract through the mouth.

- *Mechanical processing* is the physical manipulation of solid foods, first by the tongue and the teeth and then by swirling and mixing motions of the digestive tract.

- *Digestion* refers to the chemical breakdown of food into small molecules that can be absorbed by the digestive lining, or epithelium.

- *Secretion* aids digestion through the release of water, acids, enzymes, and buffers from the digestive tract lining and accessory organs.

- *Absorption* is the movement of small organic molecules, electrolytes, vitamins, and water across the digestive epithelium and into the interstitial fluid of the digestive tract.

- *Excretion* is the elimination of waste products from the body. Within the digestive tract, these waste products are compacted and discharged through a process known as *defecation* (def-e-KĀ-shun).

The lining of the digestive tract also plays a defensive role by protecting surrounding tissues against the (1) corrosive effects of digestive acids and enzymes and (2) pathogens that are either swallowed with food or residing inside the digestive tract.

STRUCTURE OF THE DIGESTIVE SYSTEM

Figure 18-1• shows the locations and functions of the accessory glands and different subdivisions of the digestive tract. The digestive tract begins with the oral cavity and continues through the pharynx, esophagus, stomach, small intestine, and large intestine before ending at the rectum and anus. The term *gastrointestinal tract*, or *GI tract*, is often used to refer to the lower portion of the digestive tract—the stomach and intestines.

THE WALL OF THE DIGESTIVE TRACT

The wall of the digestive tract is made up of four major layers. Along most portions of the digestive tract inside the peritoneal (abdominopelvic) cavity, the out-

•FIGURE 18-1
The Digestive System.

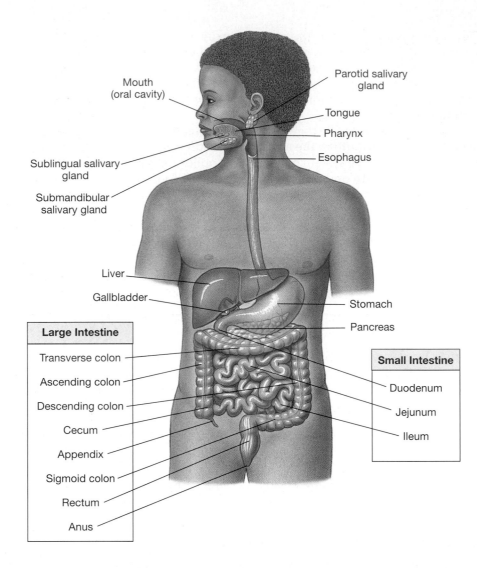

Mouth (oral cavity)

Parotid salivary gland

Tongue

Pharynx

Esophagus

Sublingual salivary gland

Submandibular salivary gland

Liver

Gallbladder

Stomach

Pancreas

Large Intestine

Transverse colon

Ascending colon

Descending colon

Cecum

Appendix

Sigmoid colon

Rectum

Anus

Small Intestine

Duodenum

Jejunum

Ileum

mesos, middle + *enteron*, intestine
mesentery: a double layer of serous membrane (peritoneum) that supports an organ within the abdominopelvic cavity

ermost layer is a *serous membrane* known as the **serosa**. This *visceral peritoneum* is continuous with the *parietal peritoneum* that lines the inner surfaces of the body wall. (These layers were introduced in Chapter 4.) ∞ *[p. 69]* In some areas the parietal and visceral peritoneum are connected by double sheets of serous membrane called **mesenteries** (MEZ-en-ter-ēz).

Figure 18-2a• shows the extensive mesenteries within the abdominopelvic cavity. Blood vessels, nerves, and lymphatics that service the digestive tract are sandwiched within the mesenteries (see Figure 18-2b•). These mesenteries also stabilize the position of attached organs, such as the stomach, and prevent the intestines from becoming knotted or kinked during digestive movements or sudden changes in body position.

The innermost layer, called the **mucosa**, is an example of a *mucous membrane*. ∞ *[p. 68]* It consists of an epithelial surface moistened by glandular secretions and an underlying loose connective tissue. Along most of the length of the digestive tract the mucosa is thrown into folds that (1) increase the surface area available for absorption and (2) permit expansion after a large meal. Figure 18-2c• shows an enlarged view of one such fold. Additionally, in the small intestine, the mucosa forms fingerlike projections called *villi* that further increase the area for absorption (see Figure 18-2d•). Most of the digestive tract is lined by a simple columnar epithelium that contains goblet cells and various other types of secretory cells.

The **submucosa** is a layer of loose connective tissue that surrounds the mucosa. It contains large blood vessels and lymphatics as well as a network of nerve fibers, sensory neurons, and parasympathetic motor neurons. This neural tissue helps

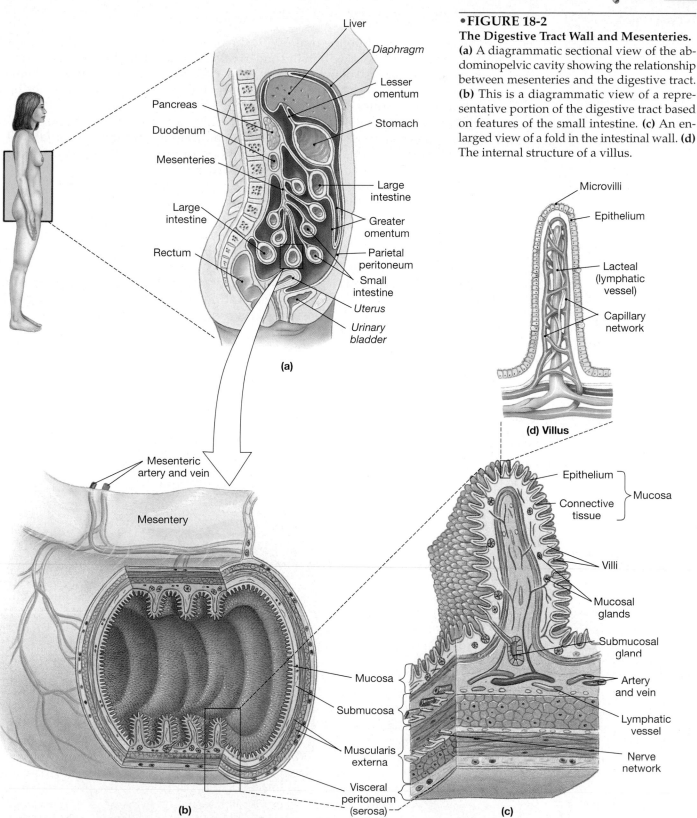

•FIGURE 18-2
The Digestive Tract Wall and Mesenteries.
(a) A diagrammatic sectional view of the abdominopelvic cavity showing the relationship between mesenteries and the digestive tract. **(b)** This is a diagrammatic view of a representative portion of the digestive tract based on features of the small intestine. **(c)** An enlarged view of a fold in the intestinal wall. **(d)** The internal structure of a villus.

control and coordinate the contractions of smooth muscle layers and the secretions of digestive glands.

The third layer, the **muscularis externa**, is composed of two sets of smooth muscle: an inner circular layer and an outer longitudinal layer. In a process called **peristalsis** (per-i-STAL-sis), these layers produce waves of contraction that push materials along the length of the digestive tract.

peri, around + *stalsis*, constriction
peristalsis: a wave of smooth muscle contraction that pushes materials along a tube such as the digestive tract

THE ORAL CAVITY

The mouth opens into the oral cavity, the part of the digestive tract that **ingests**, or receives, food. Food within the oral cavity is (1) tasted, (2) broken up into smaller pieces for easier swallowing (mechanical processing), (3) lubricated with mucus and salivary secretions, and (4) chemically digested as salivary enzymes break down carbohydrates (starches).

Figure 18-3a● shows the boundaries of the oral cavity. The cheeks form the sides of this chamber; anteriorly they are continuous with the lips, or **labia** (LĀ-bē-a; singular: *labium*). A pink ridge, the gums, or **gingivae** (JIN-ji-vē), surrounds the bases of the teeth of the upper and lower jaws.

The **hard** and **soft palates** provide a roof for the oral cavity, and the tongue forms its floor. The dividing line between the oral cavity and pharynx extends between the base of the tongue and the dangling **uvula** (Ū-vū-la), which is Latin for "little grape."

Tongue

The muscular tongue is used for manipulating food during chewing, preparing it for swallowing, and sensing its temperature and taste. Most of the tongue lies within the oral cavity, but the base of the tongue extends into the pharynx. A pair of *lingual tonsils*, lymphoid nodules that help resist infections, are located at the base of the tongue. ⚭ *[p. 383]*

•FIGURE 18-3
The Oral Cavity and Adult Teeth.
(a) An anterior view of the oral cavity, as seen through the open mouth. **(b)** The permanent teeth of an adult, with the age of eruption (first appearance) in years.

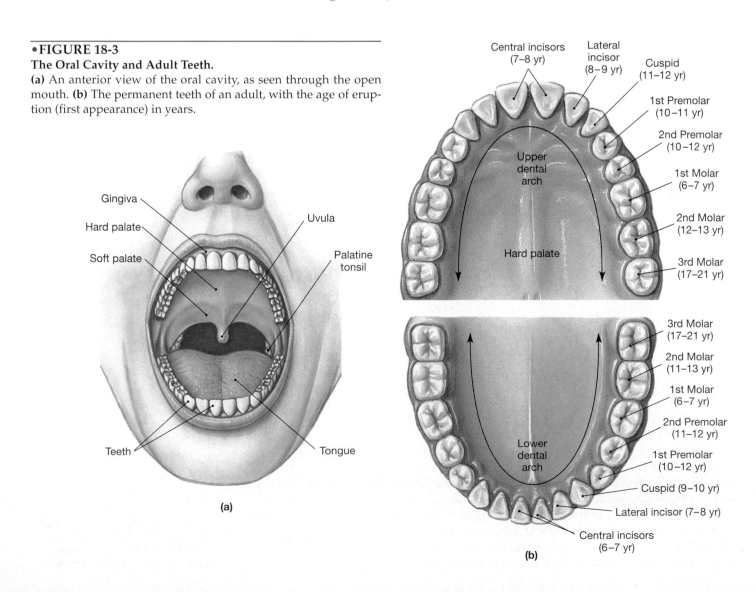

(a)

Gingiva
Hard palate
Soft palate
Uvula
Palatine tonsil
Teeth
Tongue

(b)

Central incisors (7–8 yr)
Lateral incisor (8–9 yr)
Cuspid (11–12 yr)
1st Premolar (10–11 yr)
2nd Premolar (10–12 yr)
1st Molar (6–7 yr)
2nd Molar (12–13 yr)
3rd Molar (17–21 yr)
Upper dental arch
Hard palate

3rd Molar (17–21 yr)
2nd Molar (11–13 yr)
1st Molar (6–7 yr)
2nd Premolar (11–12 yr)
1st Premolar (10–12 yr)
Cuspid (9–10 yr)
Lateral incisor (7–8 yr)
Central incisors (6–7 yr)
Lower dental arch

Salivary Glands

Figure 18-1•, p. 436, shows the locations and relative sizes of the three pairs of salivary glands. On each side the large **parotid salivary gland** lies below the zygomatic arch under the skin of the face. The **parotid duct empties** into the upper portion of the oral cavity. The ducts of the **sublingual salivary glands**, located beneath the mucous membrane of the floor of the mouth, and the **submandibular salivary glands**, found in the floor of the mouth along the inner surfaces of the mandible, open into the oral cavity beneath the tongue. Secretions from the salivary glands are controlled by the autonomic nervous system.

The salivary glands produce 1.0 to 1.5l of saliva each day. Saliva contains mostly water (99.4 percent), along with ions, buffers, waste products, antibodies, and enzymes. At mealtimes large quantities of saliva containing digestive enzymes are produced. One of these digestive enzymes, **salivary amylase**, breaks down carbohydrates, such as starches or glycogen, into smaller molecules that can be absorbed by the digestive tract. Between meals, small quantities of saliva are continually released onto the surfaces of the oral cavity. These secretions clean the oral surfaces and, through the action of a bacteria-destroying enzyme, reduce the numbers of bacteria.

Teeth

As food passes across the surfaces of the teeth, their opposing surfaces perform chewing of food. This action breaks down tough connective tissues and plant fibers, and helps saturate the materials with salivary lubricants and enzymes.

Adult teeth are shown in Figure 18-3b•. There are four different types of teeth, each with a different role. **Incisors** (in-SĪ-zerz), blade-shaped teeth found at the front of the mouth, are useful for clipping or cutting. The pointed **cuspids** (KUS-pidz), also called *canines* or "eyeteeth," are used for tearing or slashing. **Bicuspids** (bī-KUS-pidz), or *premolars*, and **molars** have flattened crowns with prominent ridges. They are used for crushing, mashing, and grinding.

During development, two sets of teeth begin to form. The first to appear are the **deciduous** (de-SID-ū-us; "falling off") **teeth**, also known as *baby teeth*. The first of these to appear are the lower, central incisors at about the age of 6 months. There are usually 20 deciduous teeth. Beginning at the age of 6 (and lasting until about age 21), these teeth are gradually replaced by the adult permanent teeth. Three additional teeth appear on each side of the upper and lower jaws as the person ages, extending the length of the tooth rows posteriorly and bringing the permanent tooth count to 32 (see Figure 18-3b•). The third molars, or *wisdom teeth*, may not erupt before age 21 because of abnormal positioning or inadequate space in the jaw. To avoid problems later in life, these teeth are often surgically removed.

A sectional view of a typical adult tooth is shown in Figure 18-4•. The bulk of each tooth consists of a mineralized material called **dentin** (DEN tin). It surrounds the connective tissue that makes up the central **pulp cavity**. The pulp cavity receives blood vessels and nerves through a narrow *root canal* at the base, or **root**, of the tooth. The root sits within a bony socket. The tooth is held in place by fibers of the **periodontal ligament** that extend from the dentin of the root to the surrounding bone. A layer of **cementum** (se-MEN-tum) covers the dentin of the root, providing protection and firmly anchoring the periodontal ligament. Cementum resembles bone, but it is softer. Where the

•FIGURE 18-4
A Typical Adult Tooth.

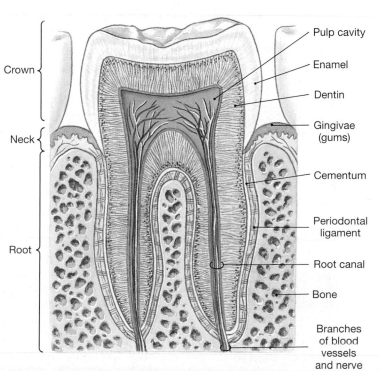

Crown

Neck

Root

Pulp cavity

Enamel

Dentin

Gingivae (gums)

Cementum

Periodontal ligament

Root canal

Bone

Branches of blood vessels and nerve

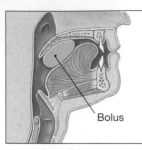

Bolus

(a)

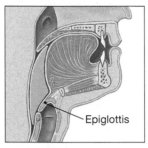

Epiglottis

(b)

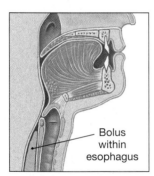

Bolus within esophagus

(c)

Esophagus

Lower esophageal sphincter

Diaphragm

(d)

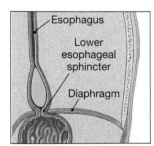

Thoracic cavity

Stomach

(e)

•**FIGURE 18-5**
The Swallowing Process.
The movement of materials from the mouth to the stomach.

tooth penetrates the gum surface, epithelial cells form tight attachments to the tooth and prevent bacterial access to the easily eroded cementum of the root. The **neck** of the tooth marks the boundary between the root and the **crown**. The dentin of the crown is covered by a layer of **enamel**, the hardest biologically manufactured substance.

THE PHARYNX

The pharynx, or throat, is a common passageway for solid food, liquids, and air. The pharyngeal muscles cooperate with muscles of the oral cavity and esophagus to initiate the process of swallowing. The muscular contractions during swallowing force the food along the esophagus and into the stomach.

THE ESOPHAGUS

The esophagus is a muscular tube that begins at the pharynx and ends at the stomach. It is approximately 25 cm long and 2 cm in diameter. The esophagus enters the peritoneal cavity through an opening in the diaphragm before emptying into the stomach.

The esophagus is lined by a stratified squamous epithelium that can resist abrasion, hot or cold temperatures, and chemical attack. The secretions of mucous glands lubricate this surface and prevent materials from sticking to the sides of the esophagus as swallowing occurs.

Swallowing

Swallowing involves both voluntary actions and involuntary reflexes that transport food from the pharynx to the stomach (Figure 18-5•). Before swallowing can occur, the food is compacted into a small mass, or **bolus**, of the proper texture and consistency.

Swallowing begins voluntarily, as the bolus is pressed up against the hard palate. The tongue then retracts, forcing the bolus into the pharynx and helping elevate the soft palate and uvula. These movements prevent the bolus from entering the nasal cavity (Figure 18-5a•). Involuntary movements then elevate the larynx and fold the epiglottis over the glottis to close off the trachea and direct the bolus into the esophagus (Figure 18-5b•). Once within the esophagus, the bolus is pushed toward the stomach by a series of peristaltic waves (Figure 18-5c•). The approach of the bolus triggers the opening of a circular muscle (the *lower esophageal sphincter*), and the bolus enters the stomach (Figures 18-5d,e•). Normally contracted, this sphincter prevents the movement of stomach acids into the esophagus. It is such leakage that causes "heartburn."

For a typical bolus the entire trip takes about 9 seconds to complete. Fluids may make the journey in a few seconds, arriving ahead of the peristaltic contractions; a relatively dry or bulky bolus travels much more slowly, and repeated peristaltic waves may be required to force it into the stomach.

THE STOMACH

The stomach is located under the diaphragm within the left upper quadrant of the abdominopelvic cavity. Its main functions include (1) the temporary storage of ingested food, (2) the mixing of ingested foods with gastric acids and enzymes, and (3) the breaking of chemical bonds (digestion) through the action of the acids and enzymes. The agitation of ingested materials with the gastric juices secreted by the glands of the stomach produces a thick, soupy mixture called **chyme** (kīm).

Figure 18-6• shows the stomach, a muscular organ with the shape of an expanded J. A short **lesser curvature** forms the inner-facing surface, and a longer **greater curvature** forms its lateral surface. Lying between these surfaces is the **body**, the largest region of the stomach. The curve of the J is the **pylorus** (pī-LOR-us), the part of the stomach connected to the small intestine. The muscular **pyloric sphincter** regulates the flow of chyme between the stomach and small intestine.

The dimensions of the stomach are extremely variable. When empty, the stomach resembles a muscular tube with a narrow cavity. When full, it can expand to contain 1 to 1.5l of material. This degree of expansion is possible because the stomach wall contains thick layers of smooth muscle, and the mucosa of the relaxed stomach contains numerous folds called **rugae** (ROO-gē). As the stomach expands, the smooth muscle stretches and the rugae gradually disappear.

The wall of the stomach contains three layers of smooth muscle. In addition to the circular and longitudinal layers found elsewhere along the digestive tract, there is an additional oblique (slanted) layer. This extra layer adds strength and assists in the mixing and churning activities essential to forming chyme.

The visceral peritoneum covering the outer surface of the stomach is continuous with a pair of mesenteries. The **greater omentum** (ō-MEN-tum), which means "fat skin," extends below the greater curvature and forms an enormous pouch that hangs over and protects the anterior surfaces of the abdominal organs (see Figure 18-2a●). The much smaller **lesser omentum** extends from the lesser curvature of the stomach to the liver.

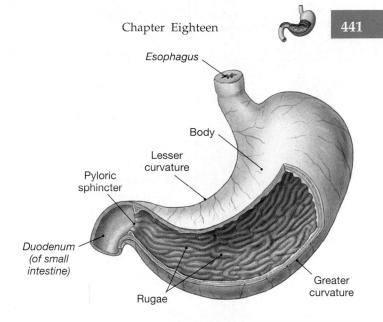

•FIGURE 18-6
The Stomach.
External and internal views of the stomach.

Gastric Juice

The stomach is lined by an epithelium dominated by mucous cells. The mucus produced helps protect the stomach lining from the acids, enzymes, and abrasive materials it contains. Shallow depressions called **gastric pits** open onto the internal surface of the stomach. Secretions from underlying gastric glands pass through the openings of the gastric pits. These glands produce about 1,500 ml of **gastric juice** each day.

Gastric juice contains **hydrochloric acid** (HCl) and **pepsin**, an enzyme that breaks down proteins. The hydrochloric acid lowers the pH of the gastric juice, kills microorganisms, and breaks down cell walls and the connective tissues in food.

✔ What layer of the wall of the digestive tract is responsible for the movement of food through peristalsis?

✔ Why is it essential that the glottis be covered during swallowing?

Control of Stomach Activity

The production of gastric juice by the stomach is controlled by the central nervous system (CNS) as well as by locally produced hormones. The secretion of gastric juice begins with the sight, smell, taste, or thought of food. These stimuli cause nerve impulses to be sent by the CNS along parasympathetic fibers of the vagus nerves to the cells of the gastric glands. Their secretions prepare the stomach to receive food.

The arrival of food in the stomach stimulates stretch receptors and chemical receptors in the stomach, and this triggers the release of a hormone, **gastrin,** into the circulatory system. Gastrin accelerates the production of gastric juice and this lowers the pH of the stomach contents. As this is occurring, gastric contractions begin to swirl and churn the stomach contents, mixing the ingested materials with the gastric secretions to form chyme. As digestion proceeds, the gastric contractions begin sweeping down the length of the stomach. Each time the pylorus contracts, a small quantity of chyme squirts through the pyloric sphincter into the small intestine.

The stomach performs preliminary digestion of proteins by pepsin. Protein digestion is not completed in the stomach, but there is usually enough time for pepsin to partially break down complex proteins before the chyme enters the small intestine. Little digestion of carbohydrates or fats occurs in the stomach.

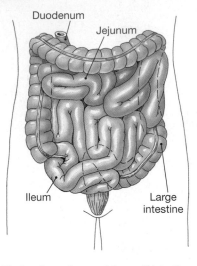

The location and parts of the small intestine.

THE SMALL INTESTINE

The small intestine is about 6 m long and ranges in diameter from 4 cm at the stomach to about 2.5 cm at the junction with the large intestine. Although longer than the large intestine, it is called the small intestine because its overall diameter is smaller.

The small intestine is made up of three regions. The **duodenum** (doo-ō-DĒ-num or doo-AH-de-num) is the 25 cm closest to the stomach. This segment receives chyme from the stomach and secretions from the pancreas and liver. The **jejunum** (je-JOO-num), the next region, is about 2.5 m long. Most chemical digestion and nutrient absorption occur in the jejunum. The third segment, the **ileum** (IL-ē-um), is the longest. It averages 3.5 m in length. The ileum ends at a sphincter muscle, the *ileocecal valve*, which controls the flow of chyme from the ileum into the large intestine. On average, it takes about 5 hours for chyme to pass from the duodenum to the end of the ileum, so the first of the materials to enter the duodenum after breakfast may leave the small intestine at lunch.

The small intestine fits in the relatively small peritoneal cavity because it is well packed, and the position of each of the segments is stabilized by mesenteries (see Figure 18-2a•).

Role of the Small Intestine

Most of the important digestive processes are completed in the small intestine, where the final products of digestion—simple sugars, fatty acids, and amino acids—are absorbed, along with most of the ions, vitamins, and water. Approximately 80 percent of all absorption takes place in the small intestine, with the rest divided between the stomach and the large intestine.

The Intestinal Wall

The intestinal lining bears a series of circular folds that are covered with a series of fingerlike projections called **villi** (see Figures 18-2b• and c•). These villi, in turn, are made up of a simple columnar epithelium whose exposed cell surfaces have many folds called microvilli. ∞ *[p. 47]* Together, the intestinal wall folds, villi, and microvilli tremendously expand the intestinal surface area available for absorbing digested food molecules. For example, if the small intestine were a simple tube with smooth walls, it would have a total absorptive area of around 3,300 cm², or roughly 3.6 ft². Instead, the arrangement of folds, villi, and microvilli increases the total area for absorption to approximately 2 million cm², or more than 2,200 ft²!

At the bases of the villi are openings of intestinal glands, which secrete a watery *intestinal juice*. In the duodenum, large intestinal glands secrete an alkaline mucus that helps neutralize the acids in the chyme arriving from the stomach. Intestinal glands also contain endocrine cells that produce several intestinal hormones.

Intestinal Secretions

Roughly 1.8l of watery **intestinal juice** enters the small intestine each day. Intestinal juice moistens the chyme, assists in neutralizing stomach acids, and dissolves both the digestive enzymes provided by the pancreas and the products of digestion.

Secretions of the larger digestive and accessory glands are controlled by hormones and the CNS. The focus of control activities is the duodenum, for it is there that the acids must be neutralized and the appropriate enzymes added. The submucosal glands (see Figure 18-2c•) protect the duodenal epithelium from gastric acids by secreting mucus and enzymes. As a result of parasympathetic (vagus nerve) stimulation, the submucosal glands begin secreting long before chyme reaches the pyloric sphincter. Sympathetic stimulation inhibits their secretions.

Table 18-1	Hormones of the Digestive Tract	
Hormone	**Source**	**Effects**
Gastrin	Stomach	Stimulates secretion of gastric juice by stomach; increases gastric contractions
Secretin	Duodenum	Stimulates secretion of buffers by pancreas; inhibits gastric juice secretion and gastric contractions
Cholecystokinin (CCK)	Duodenum	Stimulates secretion of digestive enzymes from pancreas and contraction of gallbladder; inhibits gastric juice secretion and gastric contractions
Gastric inhibitory peptide (GIP)	Duodenum	Stimulates release of insulin by endocrine portion of pancreas

The duodenum also contains endocrine cells that produce hormones that coordinate the secretions of the stomach, duodenum, pancreas, and liver. **Secretin** (sē-KRĒ-tin) is released when acids arrive in the duodenum. Its primary effect is to increase the secretion of water and buffers by the pancreas and liver. **Cholecystokinin** (kō-lē-sis-tō-KĪ-nin), or **CCK**, is secreted whenever chyme arrives in the duodenum, especially when it contains lipids and partially digested proteins. This hormone also targets the pancreas and liver. In the pancreas, CCK accelerates the production and secretion of all types of digestive enzymes. At the liver it causes the contraction of the gallbladder, which pushes bile into the duodenum. **Gastric inhibitory peptide**, or **GIP**, is released when chyme containing fats and glucose enters the small intestine. GIP causes the release of insulin from the pancreatic islets. Each of these hormones also reduces the contractions of the stomach and its secretion of gastric juice.

Functions of the major gastrointestinal hormones are summarized in Table 18-1.

chole, bile + *kystis*, bladder, + *kinein*, to move
cholecystokinin: a duodenal hormone that stimulates bile ejection and pancreatic enzyme secretion

✔ The digestion of which nutrient would be affected by damage to the parotid salivary glands?

✔ What type of tooth is best suited for chopping off bits of relatively rigid food such as a raw carrot?

✔ How is the small intestine adapted for the absorption of nutrients?

THE PANCREAS

The pancreas, shown in Figure 18-7•, lies behind the stomach, extending laterally from the duodenum toward the spleen. It is about 15 cm long. The pancreas is primarily an exocrine organ, producing digestive enzymes and buffers. Its endocrine cells, which secrete insulin and glucagon, account for only around 1 percent of the cells that make up the pancreas. The exocrine secretions, digestive enzymes and buffers, are carried to the duodenum by the **pancreatic duct**. The pancreatic duct passes through the duodenal wall along with the *common bile duct* from the liver and gallbladder.

Pancreatic enzymes are broadly classified according to their intended targets. **Lipases** (LĪ-pā-zez) attack lipids, **carbohydrases** (kar-bō-HĪ-drā-zez) digest sugars and starches, and **proteinases** break proteins apart.

Control of Pancreatic Secretion

The pancreas secretes a watery **pancreatic juice** in response to hormones from the duodenum. When acid chyme arrives in the small intestine, secretin is

•FIGURE 18-7
The Accessory Glands of Digestion.

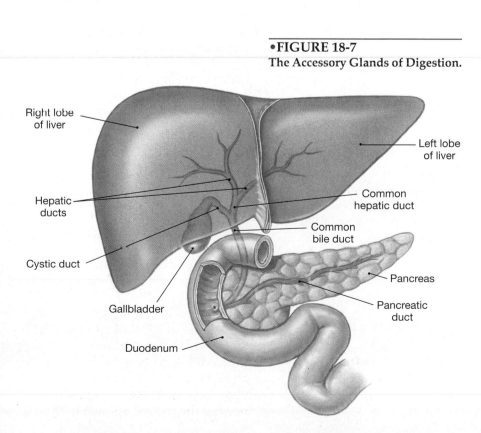

Right lobe of liver

Left lobe of liver

Hepatic ducts

Common hepatic duct

Common bile duct

Cystic duct

Pancreas

Gallbladder

Pancreatic duct

Duodenum

released, triggering the pancreatic production of an alkaline fluid with a pH of 7.5 to 8.8. Among its other components, this secretion contains neutralizing compounds, primarily *sodium bicarbonate*, that reduce the acidity of the chyme. A different intestinal hormone, cholecystokinin (CCK), controls the production and secretion of pancreatic enzymes. The specific enzymes involved are **pancreatic amylase**, similar to salivary amylase; **pancreatic lipase**, which breaks down fats; **nucleases** that break down nucleic acids; and several different **proteinases**.

Proteinases account for around 70 percent of the total pancreatic enzyme production. These enzymes are *trypsin* (TRIP-sin), *chymotrypsin* (kī-mō-TRIP-sin), and *carboxypeptidase* (kar-bok-sē-PEP-ti-dāz). Together these enzymes break up complex proteins into a mixture of single amino acids and small chains of amino acids.

THE LIVER

At 1.5 kg (3.3 lbs), the liver is the largest organ in the abdominopelvic cavity. In addition to its digestive role, this large, firm, reddish brown organ provides a wide range of other essential functions vital to life.

The liver lies directly under the diaphragm. As Figures 18-7• and 18-8• show, it is divided into the large **left** and **right lobes**, and two smaller lobes visible only from the posterior surface. The *gallbladder* is lodged within a recess in the right lobe of the liver. This muscular sac stores and concentrates bile prior to its excretion into the small intestine.

One of the main digestive functions of liver is the production of **bile**. Bile consists mostly of water, ions, cholesterol, and an assortment of lipids. The water and ions help dilute and neutralize acids in chyme as it enters the small intestine. The lipids are synthesized from cholesterol in the liver and are required for the normal digestion and absorption of fats.

Bile passes through passageways (*hepatic ducts*) within the lobes of liver until it eventually leaves the liver through the **common hepatic duct** (see Figure 18-7•). The bile within the common hepatic duct may either (1) flow into the duodenum through the **common bile duct** or (2) enter the *cystic duct* that leads to the gallbladder. Liver cells produce roughly 1l of bile each day, but a sphincter at the intestinal end of the common bile duct opens only at mealtimes. At other times, bile passes through the cystic duct for storage within the expandable gallbladder.

Liver Functions

The liver plays a central role in regulating the metabolism of the body. It can do this because all the blood leaving the absorptive areas of the digestive tract flows

---

•**Figure 18-8**
The Position of the Liver.
The position of the liver and its relationships to other visceral organs.

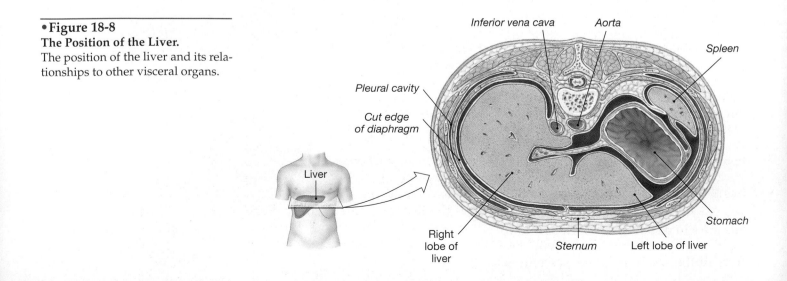

Inferior vena cava   Aorta

Spleen

Pleural cavity

Cut edge
of diaphragm

Liver

Right
lobe of
liver

Sternum   Left lobe of liver

Stomach

through the liver before reaching the general circulation. ∞ [p. 363] This enables the liver cells to extract absorbed nutrients or toxins from the blood before it reaches the rest of the body, and also to monitor and adjust the circulating levels of nutrients.

The liver also functions as the largest blood reservoir in the body. In addition to the blood arriving through the hepatic portal vein, the liver also receives about 25 percent of the cardiac output through the hepatic artery. Phagocytic cells in the liver constantly remove aged or damaged red blood cells, debris, and pathogens from the body's circulation. Equally important, liver cells synthesize the plasma proteins that determine the osmotic concentration of the blood, transport nutrients, and establish the clotting and complement systems.

Table 18-2 contains a partial listing of the over 200 different functions that have been assigned to the liver. As you might imagine, any condition that severely damages the liver represents a serious threat to life.

THE GALLBLADDER

The gallbladder (Figure 18-7•) is a pear-shaped muscular organ. The gallbladder stores excess bile formed by the liver. While bile remains in the gallbladder its composition gradually changes. Water is absorbed, and the bile salts and other components of bile become increasingly concentrated. If they become too concentrated, the salts may solidify, forming *gallstones* that can cause a variety of clinical problems.

The arrival of the intestinal hormone CCK in the circulating blood stimulates bile release. Cholecystokinin causes the walls of the gallbladder to contract and push bile through the cystic and common bile ducts into the small intestine. CCK is released whenever chyme enters the intestine, but the amount secreted increases if the chyme contains large amounts of fat.

THE LARGE INTESTINE

The large intestine begins at the ileum, the end of the small intestine, and ends at the anus. As can be seen in Figure 18-1•, the large intestine almost completely frames the small intestine. The main function of the large intestine is the reabsorption of water and the compaction of *feces* (solid wastes). It also absorbs important vitamins produced by bacteria that live in the large intestine.

Features of the Large Intestine

The large intestine, often called the *large bowel*, is approximately 1.5 m long and 7.5 cm in diameter. It is divided into three major regions: (1) the *cecum*, (2) the *colon*, and (3) the *rectum*.

Material arriving from the ileum first enters an expanded chamber called the **cecum** (SĒ-kum). A muscular sphincter, the **ileocecal** (il-ē-ō-SĒ-kal) **valve**, guards the connection between the ileum and the cecum. The cecum usually has the shape of a rounded sac, and the slender, hollow **vermiform appendix** (*vermis*, a worm) attaches to the cecum below the ileocecal valve. The average appendix is almost 9 cm long, and its walls contain lymphatic tissue. Infection of this tissue produces abdominal tenderness and pain, the symptoms of *appendicitis*.

Table 18-2	Major Functions of the Liver

Digestive and Metabolic Functions

Synthesis and secretion of bile

Storage of glycogen and lipid reserves

Maintenance of normal blood glucose, amino acid, and fatty acid concentrations

Synthesis and interconversion of nutrient types (such as the conversion of carbohydrates to lipids)

Synthesis and release of cholesterol bound to transport proteins

Inactivation of toxins

Storage of iron reserves

Storage of fat-soluble vitamins

Other Major Functions

Synthesis of plasma proteins

Synthesis of clotting factors

Phagocytosis of damaged red blood cells (by Kupffer cells)

Blood storage

Absorption and breakdown of circulating hormones (insulin, epinephrine) and immunoglobulins

Absorption and inactivation of lipid-soluble drugs

✔ How would a meal high in fat affect the level of cholecystokinin (CCK) in the blood?

✔ What secretions are produced by the exocrine portion of the pancreas?

✔ Where is excess bile produced by the liver stored?

The longest portion of the large intestine is the **colon**. The contractions of bands of longitudinal muscle produce the series of pouches characteristic of the colon. The **ascending colon** rises along the right side of the peritoneal cavity until it reaches the liver. It then turns horizontally, becoming the **transverse colon**. Near the spleen it turns down to form the **descending colon**. The descending colon continues along the left side until it curves and recurves as the S-shaped **sigmoid** (SIG-moid; the Greek letter Σ) **colon**. The sigmoid colon empties into the **rectum** (REK-tum), which opens to the exterior at the anus.

Functions of the Large Intestine

The reabsorption of water is an important function of the large intestine. Although roughly 1,500 ml of watery material arrives in the colon each day, some 1,300 ml of water is recovered from it, and only about 200 ml of feces is ejected.

In addition to water, the large intestine absorbs three vitamins produced by bacteria that live within the colon: *vitamin K*, which is needed by the liver to synthesize four clotting factors; *biotin*, which is important in glucose metabolism; and *vitamin B_5*, which is required in the manufacture of steroid hormones and some neurotransmitters.

Bacterial action breaks down protein fragments remaining in the feces into various compounds. Some of these compounds are reabsorbed, processed by the liver, released into the blood, and eventually excreted at the kidneys. Other compounds are responsible for the odor of feces, including hydrogen sulfide (H_2S), a gas that produces a "rotten egg" odor.

Some carbohydrates are not digested and arrive in the colon intact, where they provide a nutrient source for bacteria. The bacterial activities are responsible for the production of intestinal gas, or *flatus* (FLĀ-tus). Beans often trigger gas production because they contain a high concentration of indigestible polysaccharides.

Movement of materials from the cecum to the transverse colon occurs very slowly, allowing hours for the reabsorption of water. Movement from the transverse colon through the rest of the large intestine results from powerful peristaltic contractions, called *mass movements*, that occur a few times each day. The contractions force fecal materials into the rectum and produce the urge to defecate.

The rectum forms the end of the digestive tract. It is usually empty except when powerful peristaltic contractions force fecal materials out of the sigmoid colon. Stretching of the rectal wall then triggers the *defecation reflex*. **Defecation**, the excretion of solid wastes, occurs through the rectal opening, the **anus**.

DIGESTION AND ABSORPTION

A typical meal contains a mixture of carbohydrates, proteins, lipids, water, electrolytes (ions), and vitamins. Large organic molecules must be broken down through digestion before absorption can occur. Water, electrolytes, and vitamins can be absorbed directly.

Food contains large organic molecules, many of them insoluble. Because these large molecules cannot be absorbed into the body, the digestive system must first break them down into simpler molecules. Once absorbed, these molecules can be used by the body to generate ATP molecules and to synthesize complex carbohydrates, proteins, and lipids.

Organic molecules are usually complex chains of simpler molecules. In a typical dietary carbohydrate the basic molecules are simple sugars. In a protein, the building blocks are amino acids, and in lipids they are fatty acids and glycerol molecules. Digestive enzymes break, or split, the bonds between the different building-block molecules in a water-requiring process called **hydrolysis**, which means "to split with water." ∞ *[p. 25]*

Digestive enzymes differ according to their specific targets. *Carbohydrases* break the bonds between sugars, *proteinases* split the linkages between amino acids, and

Clinical Note

Why Do Many Adults Get Sick from Drinking Milk?

Lactose, also called milk sugar, is the primary carbohydrate in milk. Lactose is broken down into two simple sugars by lactase, an enzyme produced by intestinal cells. Lactase is quite important throughout infancy and early childhood. However, the intestinal cells often stop producing lactase during adolescence, and the individual becomes **lactose intolerant**. When such individuals drink a glass of milk or eat small amounts of other dairy products, they can suffer extreme discomfort and distress from gas production, cramps, and diarrhea.

How does lactose bring on such symptoms? Lactose does not merely pass through the digestive tract unused in those who are lactose intolerant. Undigested lactose provides a source of energy for the bacteria that live within the large intestine. Their activities and products are responsible for the noticeable symptoms.

There appears to be a genetic basis for lactose intolerance. In certain populations lactase production continues throughout adulthood. Only around 15 percent of Caucasians develop lactose intolerance, whereas estimates ranging from 80 to 90 percent have been suggested for the adult African-American and Oriental populations. These differences have an obvious effect on the foods these groups prefer.

lipases separate fatty acids from glycerols. Some enzymes within each class may be even more selective, breaking chemical bonds involving specific molecules. For example, a carbohydrase might ignore all bonds except those connecting two glucose molecules.

CARBOHYDRATES

Carbohydrate digestion begins in the mouth through the action of salivary amylase. Amylase breaks down complex carbohydrates into smaller fragments, producing a mixture composed primarily of two- and three-sugar molecules. Salivary amylase continues to digest the starches and glycogen in the meal for an hour or two before stomach acids render it inactive. In the duodenum, the remaining complex carbohydrates are broken down through the action of pancreatic amylase.

Further digestive activity occurs on the surfaces of the intestinal microvilli, where digestive enzymes reduce the two- and three-sugar molecules to simple single sugars. The simple sugars pass through the cells of the intestinal wall into the interstitial fluid. They then enter the intestinal capillaries for delivery to the liver by the hepatic portal vein.

PROTEINS

Proteins are relatively large and complex molecules, and different regions of the digestive tract cooperate to take proteins apart. In the mouth, the mechanical process of chewing increases the surface area of food exposed to gastric juices following ingestion. Stomach cells secrete acid and the protein-splitting enzyme pepsin. The strongly acidic environment of the stomach provides the proper pH for pepsin to begin the process of reducing the relatively huge protein molecules into smaller fragments.

After the chyme enters the duodenum and the pH has risen, pancreatic enzymes come into play. Working together, different proteinases from the pancreas break the protein fragments into a mixture of short amino acid chains and individual amino acids. Enzymes on the surfaces of the microvilli complete the process by breaking the short amino acid chains into single amino acids. In this form, they are absorbed by the intestinal cells. The amino acids then diffuse into intestinal capillaries and are carried to the liver.

LIPIDS

The most common dietary lipids are *triglycerides*. ∞ [p. 28] A triglyceride consists of three fatty acid molecules attached to a single molecule of glycerol.

Triglycerides and other dietary fats are relatively unaffected by conditions in the stomach, and they enter the duodenum in large lipid drops. Bile breaks the large drops apart in a process called **emulsification** (ē-mul-si-fi-KĀ-shun). The formation of tiny droplets increases the surface area available for pancreatic lipase to break the triglycerides apart. Although the resulting lipid fragments are absorbed by the intestinal cells, they do not then enter intestinal capillaries. Instead, they are attached to proteins that aid their transport within the water-based fluids of the body. Too large to enter the capillaries, these lipid-protein packages enter small lymphatic vessels called **lacteals** (LAK-tē-als), which underlie the intestinal villi (Figure 18-2d•). From the lacteals they proceed along the larger lymphatics and finally enter the venous circulation. The name lacteal means "milky" and refers to the pale, cloudy appearance of lymph containing large amounts of lipids.

WATER, IONS, AND VITAMINS

Each day 2 to 2.5l of water enters the digestive tract in the form of food and drink. The salivary, gastric, intestinal, and accessory gland secretions provide another 6 to 7l. Out of all these fluids, only about 150 ml of water is lost in the fecal wastes. This water conservation occurs through the process of osmosis. In osmosis, water always tends to flow into the solution containing the higher concentration of solutes (dissolved substances). The epithelial cells of the intestinal lining are continually absorbing solutes (dissolved nutrients and ions) from the contents of the intestinal tract. This results in water "following" the solutes into the surrounding tissues and then into the bloodstream.

Table salt, sodium chloride, dissolves in water and releases sodium ions and chloride ions. These are the most common electrolytes in the extracellular fluid. ∞ [p. 27] The absorption of sodium and chloride ions is the most important factor promoting water absorption through osmosis. Ions absorbed in smaller quantities include calcium ions, potassium ions, magnesium ions, iodine ions, bicarbonate ions, and iron ions.

Vitamins are organic compounds required in very small quantities for critical chemical reactions. There are two major groups of vitamins: **water-soluble vitamins** (vitamin C and the B vitamins) and **fat-soluble vitamins** (vitamins A, D, E, and K). All but one of the water-soluble vitamins (vitamin B_{12}) are easily absorbed by intestinal cells. The fat-soluble vitamins are absorbed from the intestine along with dietary lipids.

AGING AND THE DIGESTIVE SYSTEM

Essentially normal digestion and absorption occur in elderly individuals. However, there are many changes in the digestive system that parallel age-related changes already described for other systems.

1. **The rate of epithelial stem cell division declines**. The digestive epithelium becomes more easily damaged by abrasion, acids, or enzymes. In the mouth, esophagus, and anus the stratified epithelium becomes thinner and more fragile.

2. **Smooth muscle tone decreases**. General movements of the digestive tract decrease, and peristaltic contractions are weaker. This change slows the rate of movement along the digestive tract and promotes constipation. Problems are not restricted to the lower digestive tract. For example, weakening of muscular sphincters can lead to the leakage of stomach acids into the esophagus. The acid then attacks the lining of the esophagus, producing pain known as "heartburn."

3. **The effects of cumulative damage become apparent**. A familiar example is the gradual loss of teeth due to *dental caries* ("cavities") or *gingivitis* (in-

flammation of the gums). Cumulative damage can involve internal digestive organs as well. Toxins such as alcohol and other injurious chemicals that are absorbed by the digestive tract are transported to the liver for processing. The liver cells can be damaged by these compounds, and frequent exposure can lead to *cirrhosis* or other types of liver disease.

4. **Cancer rates increase.** As noted in Chapter 4, cancers are most common in organs where stem cells divide to maintain epithelial cell populations. Rates of colon cancer and stomach cancer rise in the elderly; oral and pharyngeal cancers are particularly common in elderly smokers.

5. **Changes in other systems have direct or indirect effects on the digestive system.** For example, the reduction in bone mass and calcium content in the skeleton is associated with erosion of the tooth sockets and eventual tooth loss. The decline in the senses of smell and taste can lead to dietary changes that affect the entire body.

✔ A narrowing of the ileocecal valve would interfere with the movement of materials between what two organs?

✔ Which section of the digestive tract functions in water reabsorption?

✔ What component of a meal would cause the lymph to become milky in appearance?

DISORDERS OF THE DIGESTIVE SYSTEM

Gastroenterology (gas-trō-en-ter-OL-o-jē) is the study of the digestive system and its diseases and disorders. Disorders of the digestive system may occur in any region of the digestive tract or its accessory organs. Problems in the local control and nervous system control of the activities of the digestive system may also lead to various disorders.

THE PHYSICAL EXAMINATION

Symptoms

Common symptoms of digestive disorders include the following:

1. Pain is a common symptom of digestive disorders. Pain in the oral cavity may be widespread, or focused, as in a tooth disorder. Abdominal pain is characteristic of a variety of digestive disorders. In most cases the pain is perceived as distressing but tolerable; if the pain is acute and severe, a surgical emergency may exist.

2. **Dyspepsia** (dis-PEP-sē-ah), or *indigestion*, is pain or discomfort in the upper abdomen that is often described as "burning pain."

3. **Nausea** (NAW-sē-ah) is a sensation that usually precedes or accompanies vomiting, or **emesis** (EM-e-sis). Nausea may result from digestive disorders or from disturbances of CNS function.

4. **Dysphagia** (dis-FĀ-jē-ah) is difficulty in swallowing. For example, the infections of tonsillitis, pharyngitis, and esophagitis may cause dysphagia.

Signs

A physical examination can provide information useful in the diagnosis of digestive system disorders. The abdominal region is particularly important, because most of the digestive system is located within the abdominopelvic cavity. As noted earlier, the four methods of a physical examination include *inspection, palpation, percussion,* and *auscultation.* ∞ *[p. 14]*

Inspection can provide a variety of useful diagnostic clues:

■ Bleeding of the gums, as in *gingivitis*, and characteristic oral lesions can be seen on inspection of the oral cavity. Examples of distinctive lesions include those of oral herpes simplex infections and *thrush*, lesions produced by infection of the mouth by the yeast, *Candida albicans*.

- Peristalsis in the stomach and intestines may be seen as waves passing across the abdominal wall in persons who do not have a thick layer of abdominal fat. The waves become very prominent during the initial stages of intestinal obstruction.

- A general yellow discoloration of the skin, a sign called *jaundice*, may indicate liver problems.

- Abdominal distention may be caused by conditions such as fluid accumulation in the peritoneal cavity (ascites), fluid or gas (*flatus*) within the digestive tract or obesity, tumors or enlargement of visceral organs, and pregnancy.

- *Striae* are multiple scars, 1–6 cm in length, that are visible through the epidermis. Striae develop in damaged dermal tissues after stretching and they are typically seen in the abdominal region after a pregnancy or other rapid weight gain.

Palpation of the abdomen may reveal specific details about the status of the digestive system, including:

- the presence of abnormal masses, such as tumors, within the peritoneal cavity.

- abdominal distention from either excess fluid within the digestive tract or peritoneal cavity, or gas within the digestive tract

- *hernia*, or protrusion of visceral organs into the *inguinal canal* or other weak spots in the abdominal wall.

- changes in the size, shape, or texture of visceral organs. For example, in several liver diseases, the liver becomes enlarged and firm, and these changes can be detected on palpation of the right upper quadrant.

- detection of voluntary or involuntary abdominal muscle contractions, usually in response to pain.

- identification of specific areas of tenderness and pain. For example, someone with *cholecystitis*, an inflammation of the gallbladder, generally experiences pain on palpation of the upper right quadrant. In contrast, a person with *appendicitis* generally experiences pain when the right lower quadrant is palpated.

Percussion of the abdomen reveals fewer signs than percussion of the chest, because the visceral organs do not contain extensive air spaces that would reflect the sounds conducted through surrounding tissues. However, the stomach usually contains a small air bubble, and percussion over this area produces a sharp, resonant sound. The sound becomes dull or disappears when the stomach fills, the spleen enlarges, or the peritoneal cavity contains abnormal quantities of peritoneal fluid, as in *ascites*.

Auscultation can detect gurgling abdominal sounds, or bowel sounds, produced by peristaltic activity along the digestive tract. Increased bowel sounds occur in persons with acute diarrhea, and bowel sounds may disappear in persons with advanced intestinal obstruction, peritonitis (an infection of the peritoneal lining), and spinal cord injuries that prevent normal innervation of the digestive tract.

A procedure called **endoscopy** (en-DOS-ko-pē) allows physicians to view internal body cavities and make more precise diagnoses. An **endoscope** consists of a bundle of flexible glass or plastic fibers that can transmit light and a viewing lens. Endoscopes have been developed for viewing the respiratory tract, urinary bladder, abdominal cavity, and the upper and lower digestive tracts. Information on representative diagnostic procedures and laboratory tests are given in Tables 18-3 and 18-4.

Table 18-3	Representative Diagnostic Procedures for Digestive System Disorders	
Diagnostic Procedure	Method and Result	Representative Uses
ORAL CAVITY		
Periapical (PA) X-rays	The periapical film is an X-ray of the crown and root area of several teeth	Detection of tooth decay, tooth impactions, fractures, progression of bone loss with periodontal disease, inflammation of the periodontal ligament
Bitewing X-rays of teeth	The bitewing film is an X-ray detailing areas where one tooth crown contacts another	Bitewing films reveal tooth decay between the teeth and early bone loss in periodontal disease
UPPER GI Esophagogastroduo-denoscopy, esophago-scopy, gastroscopy	A fiber-optic endoscope is inserted through the oral cavity into the esophagus, stomach, and duodenum (upper GI tract)	Detection of tumors, ulcerations, polyps, inflammation and obstructions; to perform tissue biopsy
Upper GI series	X-rays taken of the stomach and duodenum after swallowing barium sulfate to increase contrast	To determine cause of epigastric pain; to detect ulcers, polyps, gastritis, tumors and inflammation within the upper GI tract

DIGESTIVE SYSTEM DISORDERS

Due to the number and different kinds of digestive organs, there are many types of digestive disorders. We will restrict our discussion to representative disorders caused by inflammation and infection, tumors, congenital problems, and malabsorption disorders (Figure 18-9•).

Inflammation and Infection

The lining of the digestive tract is exposed to abrasive foods, strong chemicals, enzymes, and bacteria. It is therefore not surprising that inflammation or infection of the digestive tract are relatively common. In this discussion, specific examples are grouped by the affected region of the digestive tract.

The Oral Cavity. The mouth contains a resident population of more than 300 different types of bacteria. Under the right conditions, invasive bacterial infections can occur. These infections can be especially dangerous for patients with ineffective immune systems.

One of the most common problems with the oral cavity results from **dental caries**, or tooth decay. Dental caries (*cariosus* means "rottenness") is a condition in which the enamel and deeper layers of a tooth are eroded by bacteria. The first step is the appearance of **dental plaque**, a layer of bacteria and organic matter, on a tooth. Bacteria, such as *Streptococcus mutans*, break down sucrose into sources of energy (glucose and fructose molecules). They also link some of the resulting glucose molecules together to form the organic matter of plaque, which serves to bind the bacteria together. Bacterial action also produces acid which dissolves the underlying enamel, forming cavities in the surface of the tooth. If treatment is prompt, these cavities can be cleaned, sterilized, and filled with a patch, or "filling" of either synthetic material or a soft metal. If the erosion process continues, the bacteria will eventually reach the pulp cavity and infect it, producing painful **pulpitis** (pul-PĪ-tis). Treatment of pulpitis involves the removal of all pulp tissue, decayed areas, and sensory innervation; the pulp cavity is then filled and sealed with appropriate materials. This procedure is called a *root canal*. Reducing the amount of sugary foods, brushing the teeth regularly, and flossing help prevent dental caries by reducing the buildup of plaque. Fluoride reduces tooth decay by increasing the hardness of the enamel, making it more difficult to dissolve.

Table 18-4	Representative Laboratory Tests for the Diagnosis of Digestive System Disorders	
Laboratory Test	**Normal Values in Blood Plasma or Serum**	**Significance of Abnormal Values**
UPPER AND LOWER GI		
Serum electrolytes		
Potassium Sodium Magnesium	Adults: 3.5–5.0 mEq/l 135–145 mEq/l 1.5–2.5 mEq/l	Vomiting, diarrhea, and nasogastric intuba tion can cause potentially dangerous elec trolyte losses
Serum gastrin level	Adults: 45–200 pg/ml	Elevated with pernicious anemia, some gas tric ulcers and some pancreatic tumors
Lactose tolerance test	Fasting patient receives 100 mg of lactose. Plasma glucose should increase to >20 mg/dl within 2 hours	Blood glucose levels will not rise to normal levels in the absence of the enzyme lactase
Stool culture		
Culture and sensitivity (C&S) (involves culturing of stool sample)	Only normal intestinal flora such as *E. coli* should be isolated	Bacteria such as *Shigella*, *Campylobacter*, and *Salmonella* can cause acute diarrhea
Ova and parasites (O&P) (involves microscopic ex- amination of stool sample)	None found	Typical parasites are tapeworms, entamoe ba, and some protozoans, such as *Giardia*
Fecal analysis		
Occult blood	None found	Hidden (occult) blood in the feces can be due to inflammation, ulceration, or a tumor that is causing small amounts of bleeding
LIVER and GALLBLADDER		
Serum bilirubin (total) (Bilirubin, a hemoglobin breakdown product, is excreted in bile.)	Total serum bilirubin (adult): 0.1–1.2 mg/dl Total serum bilirubin (newborn): 1–12 mg/dl	Can be elevated with hepatitis and biliary obstruction; Bilirubin >15 mg/dl can result in serious neurological problems
Urine bilirubin	None	Presence of bilirubin may indicate obstruc tion of the bile duct, hepatitis, cirrhosis of the liver, or liver cancer
Liver Enzyme Tests (serum)		
Aspartate aminotransferase (AST or SGOT)	Adults: 0–35 U/l	Elevated levels occur with hepatitis, acute pancreatitis, and cirrhosis, although eleva tions are not specific to liver disease
Alanine aminotransferase (ALT or SGPT)	Adults: 0–35 IU/l	Elevated levels occur with hepatitis and other liver diseases; this test is more specific than the others for diseases of the liver
Alkaline phosphatase	Adults: 30–120 mU/ml	Elevated levels occur in biliary obstructions, hepatitis, and liver cancer
Serum proteins	Adults: Albumin 3.5–5.5 g/dl Globulin 2.0–3.0 g/dl	Decreased levels of albumin and increased levels of globulins occur in chronic liver disease
Ammonia (plasma)	Adult: 15–45 µg/dl	Elevated level occurs with liver failure
Hepatitis virus assays (serum)	None	Antibodies can be detected against hepatitis A (HAV), B (HBV), and C (HCV)
PANCREAS		
Amylase		
Urine	Adult: 35–260 U/l hour	Elevated levels occur for 7–10 days after pancreatic disease begins
Serum	Adult: 60–180 U/l	Elevated levels occur with pancreatic disease and with obstruction of the pancreatic duct
Lipase (serum)	Adult: 0–110 units/l	Elevated levels most commonly occur in acute pancreatitis

•FIGURE 18-9
Disorders of the Digestive System.

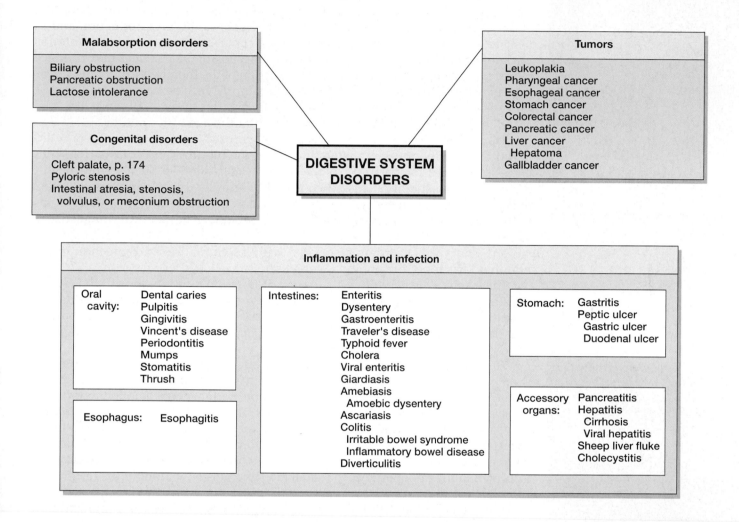

Malabsorption disorders

Biliary obstruction
Pancreatic obstruction
Lactose intolerance

Congenital disorders

Cleft palate, p. 174
Pyloric stenosis
Intestinal atresia, stenosis,
    volvulus, or meconium obstruction

DIGESTIVE SYSTEM DISORDERS

Tumors

Leukoplakia
Pharyngeal cancer
Esophageal cancer
Stomach cancer
Colorectal cancer
Pancreatic cancer
Liver cancer
  Hepatoma
Gallbladder cancer

Inflammation and infection

Oral cavity:	Dental caries
	Pulpitis
	Gingivitis
	Vincent's disease
	Periodontitis
	Mumps
	Stomatitis
	Thrush

| Esophagus: | Esophagitis |

Intestines:	Enteritis
	Dysentery
	Gastroenteritis
	Traveler's disease
	Typhoid fever
	Cholera
	Viral enteritis
	Giardiasis
	Amebiasis
	Amoebic dysentery
	Ascariasis
	Colitis
	Irritable bowel syndrome
	Inflammatory bowel disease
	Diverticulitis

Stomach:	Gastritis
	Peptic ulcer
	Gastric ulcer
	Duodenal ulcer

Accessory organs:	Pancreatitis
	Hepatitis
	Cirrhosis
	Viral hepatitis
	Sheep liver fluke
	Cholecystitis

Periodontal disease, the most common cause for the loss of teeth, occurs when dental plaque forms in the area between the gums and teeth. The bacterial activity may cause gum inflammation, tooth decay, and, eventually, breakdown of the periodontal ligament and surrounding bone. Its mildest form is called **gingivitis** (jin-ji-VĪ-tis), an inflammation of the gingivae, or gums (Figure 18-10a•). Severe bacterial infection and the formation of ulcers produces a painful condition called **acute necrotizing ulcerative gingivitis**, or *Vincent's disease*. Because this disease was common among soldiers fighting from trenches in WWI, it is also called *trench mouth*. In **periodontitis**, tissue around the teeth and supporting bone become infected, leading to an irreversible loss of bone and eventual loss of teeth (Figure 18-10b•).

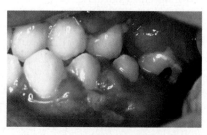
(a)

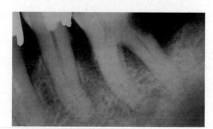
(b)

•FIGURE 18-10
Periodontal Disease.
(a) Gingivitis, an inflammation of the gums. (b) Periodontitis leads to the loss of bone around the teeth, which results in the loss of teeth.

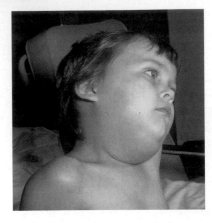

•FIGURE 18-11
Mumps.
Typical salivary gland swelling in mumps.

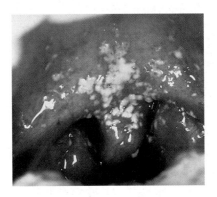

•FIGURE 18-12
Thrush.
Thrush is caused by a yeast and characterized by white patches covering the mucous membrane of the mouth.

The *mumps virus*, similar to the measles (rubeola) virus, causes an infection of the salivary glands called **mumps** (Figure 18-11•). The infection most often occurs in the parotid salivary gland, but it may also infect other salivary glands and other organs, including the gonads and the meninges. Infection typically occurs at 5 to 9 years of age. The first exposure stimulates antibody production and in most cases results in permanent immunity. In postadolescent males, the mumps virus infecting the testes may cause sterility. Infection of the pancreas may produce temporary or permanent diabetes, and infection of the meninges may cause deafness or other CNS problems. An effective mumps vaccine became available in 1967. The vaccine is usually combined with measles and rubella vaccines (the *MMR vaccine*), and administered to infants after the age of 15 months.

Stomatitis (sto-ma-TĪ-tis) is a general term used to refer to any form of inflammation of the mucous membrane lining the mouth. **Candidiasis** (KAN-di-DĪ-a-sis) is an infection caused by the yeast, *Candida albicans*. Such an infection of the oral mucous membrane is called **thrush** (Figure 18-12•). It appears as white, milky, removable patches within the mouth, and occurs in infants or individuals with diabetes, AIDS, or receiving long-term antibiotic treatment. Antifungal drugs are used to treat thrush.

The Pharynx. Disorders associated with inflammation and infection of the lymphoid tissues of the pharynx were discussed in Chapter 17. ∞ [p. 423]

The Esophagus. **Esophagitis** (ē-sof-a-JĪ-tis) is an inflammation of the esophagus. This painful condition usually results from the presence of stomach acids that leak through a weakened or permanently relaxed lower esophageal sphincter muscle. Such backflow, or *gastro-esophageal reflux*, is responsible for the symptoms of *heartburn* (p. 440).

The Stomach. Inflammation of the mucous membrane lining of the stomach is called **gastritis** (gas-TRĪ-tis). This condition may develop after swallowing drugs, including aspirin and alcohol. It may also appear after severe emotional or physical stress, bacterial infection of the gastric wall, or the ingestion of strong chemicals.

Gastritis may lead to ulcer formation. A **peptic ulcer** develops when digestive enzymes and acids erode through the stomach lining or duodenal lining. A **gastric ulcer**, a type of peptic ulcer, is located in the stomach and a **duodenal ulcer** is located in the duodenum of the small intestine. It is now thought that gastric infection by bacterium *Helicobacter pylori* (HE-lik-ō-bak-ter PĪ-lō-rē) is responsible for over 80 percent of peptic ulcers. Treatment for ulcers involves the administration of drugs, such as *cimetidine* (sī-MET-i-dēn) (*Tagamet*), to inhibit acid production by gastric glands, combined with antibiotics if *H. pylori* is present.

The Intestines. **Enteritis** is an inflammation of the intestine, usually applied to the small intestine. Enteritis usually causes watery bowel movements, or **diarrhea** (dī-a-RĒ-uh). **Dysentery** (dis-en-TER-ē) is the inflammation of the small and large intestine (colon). Frequently the diarrhea contains blood and mucus. **Gastroenteritis** is the inflammation of the stomach and the intestines. Gastroenteritis may be caused by infections by bacteria, viruses, protozoans, and parasitic worms. Important examples of each are discussed below, in addition to those listed in Table 18-5. Most of the these conditions are prevalent in areas of poor sanitation and low water quality.

International travelers frequently contract **traveler's disease**, a mild to severe diarrhea. It is most often caused by pathogenic strains of the bacterium, *Escherichia coli* (esh-er-IK-ē-a KŌ-lē). Symptoms include frequent, watery diarrhea, nausea, vomiting, fever, and abdominal pain. This bacteria normally lives in the intestinal tract of man and other animals. It is usually contracted by ingesting feces-contaminated food or water.

Typhoid fever is a serious epidemic disease of the digestive tract. It is caused by the bacterium, *Salmonella typhi* (sal-mō-NEL-a TĪ-fē). The bacteria are ingested in food or water and infect the upper regions of the small intestine. They spread to lymphoid tissues, where they are engulfed, but not killed, by phagocytes. After multiplying in the phagocytes, they spread to the gallbladder, grow in the bile,

Table 18-5   Examples of Infectious Diseases of the Digestive System

Disease	Organism(s)	Description
BACTERIAL DISEASES		
Dental caries	*Streptococcus mutans* and other oral bacteria	Tooth decay; bacteria within the dental plaque on teeth produce acids which dissolve tooth enamel leading to cavities
Pulpitis	As above	Infection of the pulp of the tooth
Gingivitis	As above	Infection of the gums
Vincent's disease	As above	Acute necrotizing ulcerative gingivitis, or trenchmouth; bacterial infection and ulcer formation
Periodontitis	As above	Infection of gums and bone; results in loosening and loss of teeth
Peptic ulcers	*Helicobacter pylori*	Ulcers in gastric lining
Traveler's disease	*Escherichia coli*	Mild to severe diarrhea, nausea, vomiting, abdominal pain, and general lack of energy
Typhoid fever	*Salmonella typhi*	Infection of the intestines and gallbladder; abdominal pain, abdominal distention, pain, low WBCs count, and enlarged spleen
Cholera	*Vibrio cholerae*	Intestinal infection; symptoms include nausea, vomiting, abdominal pain, and diarrhea; causes severe dehydration
VIRAL DISEASES		
Mumps	Mumps virus (paramyxovirus)	An infection of the salivary glands that may spread to the meninges or gonads
Viral enteritis	Rotaviruses	Intestinal infection; causes watery diarrhea, especially in young children
Viral hepatitis	Hepatitis A virus (HAV)	Infectious hepatitis; transmitted by fecal-contaminated water, food, milk, or shellfish
	Hepatitis B virus (HBV)	Serum hepatitis; transmitted by exchange of body fluid through blood transfusions, wounds, and sexual contact; pregnant carriers of the virus may pass it on to their children
	Hepatitis C virus (HCV)	Formerly non-A, non-B hepatitis; transmitted through blood and sexual contact
FUNGAL DISEASES		
Candidiasis (thrush)	*Candida albicans*	Thrush is a yeast infection of the oral mucosa that forms white, milky patches in the mouth
PARASITIC DISEASES		
Protozoa		
Giardiasis	*Giardia lamblia*	Intestinal infection; symptoms include diarrhea, dehydration, and weight loss; more common in children than adults
Amebiasis, or amoebic dysentery	*Entamoeba histolytica*	Infection of the large intestine; may produce ulcers and peritonitis; diarrhea contains blood
Helminths		
Ascariasis	*Ascaris lumbricoides*	Roundworm infestation of the intestines; larval movement to pharynx causes damage to intestinal wall, adults eat contents of the intestine
Sheep liver fluke	*Fasciola hepatica*	Flatworm infestation of the bile-conducting passageways in the liver, and gallbladder; cause inflammation of the liver; flukes consume blood

and reinfect the intestinal lining. Sepsis, or blood infection, may occur. Initial symptoms are fever, headache, and fatigue. Conditions worsen as the bacteria spread, causing abdominal distention and pain, and enlargement of the spleen. Unlike other digestive tract infections, there may be no diarrhea and the number of WBCs decline. If the individual survives, immunity develops within a month or so of the initial infection but carriers with poor hygiene may continue to spread the disease to others. A typhoid vaccine is available, and the antibiotics *chloramphenicol* or ampicillin are used in treatment.

Cholera (KOL-er-ah) is caused by a bacterium, *Vibrio cholerae* (VIB-rē-ō KOL-er-ī) present in fecal-contaminated water. *Vibrio* infects the intestinal lining after ingestion and produces a toxin that binds to cells of the small intestine. The toxin increases the permeability of the cell membranes, resulting in the loss of water and chloride ions, and an inhibition of sodium ion uptake. In addition, clumps of the intestinal lining flake off. Symptoms of cholera include nausea, vomiting, abdominal pain, and huge volumes of diarrhea. Cholera leads to severe dehydration, and a rapid fall in blood volume that is the cause of most deaths. Treatment primarily involves fluid and electrolyte replacement. Antibiotics do not rid the body of the bacteria, and an effective vaccine has yet to be developed. Individuals who recover only gain a temporary immunity.

Viral enteritis is caused by a group of viruses called *rotaviruses*. They infect infants and young children, and are a major cause of illness and death in developing countries where approximately 5 to 10 million children under age 5 die each year. Again, the viruses are transmitted through fecal contamination, and within 2 days of infecting the intestinal lining, they cause a watery diarrhea. Treatment involves restoring the lost fluids and electrolytes.

Giardiasis (jē-ar-DĪ-a-sis) is an infectious disease caused by the flagellated protozoan, *Giardia lamblia* (jē-AR-dē-a LAM-lē-a) (Figure 18-13•). *Giardia* is usually contracted as a *cyst* (a resting cell) from contaminated water or food. This protozoan infects the small intestine where it attaches to the epithelial lining and consumes mucus. It is more common in children than in adults. Symptoms of infection include diarrhea, abdominal pain, dehydration, and weight loss. It is thought that the weight loss is due to deficiencies in the absorption of fats and carbohydrates caused by large numbers of *Giardia* covering up the absorptive surfaces of the small intestine. It is transmitted by cysts that are released in the feces. Treatment involves the administration of antiprotozoan drugs.

Amebiasis (am-e-BĪ-a-sis), or **amoebic dysentery**, is an infection caused by the protozoan *Entamoeba histolytica* (en-ta-MĒ-ba his-tō-LI-ti-ka). It typically enters the body as a resting stage, or cyst in fecal-contaminated food or water. The cysts open in the colon, and the amoeba multiply and feed on the bacteria normally living in the large intestine. These amoeba then release more cysts during defecation. In acute cases, infection of the epithelial lining may result in the destruction of the mucosal cells and ulcer formation. Bacteria may then penetrate the wall of the large intestine and infect the body cavity, causing peritonitis, or the amoebas may enter blood vessels and travel throughout the body. Such intestinal infections cause many bloody diarrheal bowel movements a day, promoting dehydration. Antibiotics are available to kill the amoeba and to treat the secondary bacterial infections.

Ascariasis (as-kah-RĪ-a-sis) is an infection of the small intestine by the roundworm, *Ascaris lumbricoides* (AS-ka-ris lum-bri-KOY-dēz). This roundworm reaches lengths of 25–35 cm (10–14 in). It is acquired by consuming *Ascaris* eggs. The adult worms feed on the contents of the intestine, and can cause malnutrition. Treatment can eradicate the infection.

The Pancreas. Pancreatitis (pan-krē-a-TĪ-tis) is an inflammation of the pancreas. Factors that may cause this condition include blockage of the excretory ducts by gallstones (see below), viral infections, and toxic drugs, such as alcohol. Any of these stimuli may begin to injure exocrine cells in a portion of the organ. Lysosomes then activate digestive enzymes within the cell and the cells begin to break down. The released enzymes then digest surrounding, undamaged cells and ac-

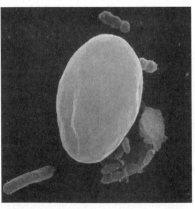

•FIGURE 18-13
Giardia.
(a) *Giardia* is a flagellated protozoan that infects the small intestine. **(b)** The cyst is a resting cell that is released in the feces.

tivate their enzymes, producing a chain reaction of destruction. In most cases, only a portion of the pancreas will be affected, and the condition subsides in a few days. In about one-eighth of the cases, the process does not stop, and the enzymes destroy the pancreas.

The Liver. Any condition that severely damages the liver represents a serious threat to life. **Hepatitis** is an inflammation of the liver. Major causes of hepatitis include alcohol consumption, drugs, or infection.

Excessive alcohol consumption causes fat accumulation in the liver and the gradual death of liver cells. Chronic alcohol abuse often leads to a condition called **cirrhosis** (sir-RŌ-sis), "a form of hepatitis," which is characterized by the degeneration of liver cells and their replacement by fibrous connective tissue (a process called *scarring*). The surviving liver cells divide, but the fibrous tissue prevents the re-establishment of normal tissue structure. As a result, liver function declines and a variety of other complications develop. The constriction of blood flow leads to high blood pressure in the hepatic portal vein, a condition known as *portal hypertension*. This abnormal elevation in pressure can push capillary fluid into the peritoneal cavity. This accumulation of fluid creates a characteristic abdominal swelling called **ascites** (a-SĪ-tēz).

Most cases of hepatitis and cirrhosis are caused by viral infections. There are many different forms of viral hepatitis: the most common are hepatitis A, B and C. **Hepatitis A**, or *infectious hepatitis*, typically results from the ingestion of food, water, milk, or shellfish contaminated by infected fecal wastes. It has a relatively short incubation period of 2–6 weeks. The disease generally subsides in a few months with no lasting problems, and progressive infection is rare unless the individual has pre-existing liver disease.

Hepatitis B, or *serum hepatitis*, is transmitted by the exchange of body fluids during intimate contact. For example, infection may occur through the transfusion of blood products, by exposure to infected body fluids through a break in the skin, or by sexual contact. Areas of infection include the liver, lymphoid tissues, and blood-forming tissues. The incubation period ranges from 1 to 6 months. Many people are carriers of the virus, and if a pregnant woman is a carrier, the newborn baby may become infected at birth. An effective, safe vaccine that prevents hepatitis B is now available.

Hepatitis C, originally designated *non-A, non-B hepatitis*, is most commonly transferred from individual to individual through the collection and transfusion of contaminated blood. Since 1990, screening procedures have been used to lower the incidence of transfusion-related hepatitis C. Hepatitis C (and hepatitis B) can also be transmitted among intravenous drug users through shared contaminated needles, and there is evidence that hepatitis C may also be sexually transmitted. Chronic hepatitis C infections produce significant liver damage in at least half the individuals infected with the virus.

The hepatitis viruses disrupt liver function by attacking and destroying liver cells. An infected individual may develop a high fever, and the liver may become inflamed and tender. As the disease progresses, enzymes normally confined to the cytoplasm of liver cells begin to appear in the bloodstream. Meanwhile, other signs appear as key liver functions are disrupted: blood protein levels decline, clotting time becomes extended, and the skin color often changes. Liver cells normally remove hemoglobin breakdown products (such as *bilirubin*) and excrete them in the bile. In a condition called **jaundice** (JAWN-dis), the skin and eyes develop a yellow color because these waste products are accumulating in body fluids. (Jaundice occurs in several other conditions that involve liver damage; cirrhosis is one example.)

Hepatitis may be acute (short-term) or chronic (long-term). *Acute hepatitis* is characteristic of hepatitis A. Almost everyone who contracts hepatitis A eventually recovers, although full recovery may take several months. Symptoms of acute hepatitis include severe fatigue and jaundice. In chronic hepatitis, fatigue is less pronounced and jaundice is rare. Chronic hepatitis is a progressive disorder that can lead to severe medical problems as liver function deteriorates and cirrhosis develops. Chronic hepatitis infections (especially B or C) may also lead to liver cancer. Immunization is available for both the hepatitis A and B viruses. There are no vaccines available to stimulate immunity to hepatitis C, or for the other forms of hepatitis.

Hepatitis may also be caused by helminth parasites. One example, the **sheep liver fluke**, *Fasciola hepatica* (fa-SĪ-ō-la he-PAT-I-ka), is found in humans in South America, Africa, and some parts of Europe (Figure 18-14•). The life cycle includes a snail as an intermediate host. The fluke larval stages live and develop in the snail, and are then released into the water. They then form resting cysts on water plants, such as watercress. When eaten by humans, the cysts hatch in the large intestine, burrow through the intestinal wall, and make their way to the liver where they feed on blood and block the bile-conducting passageways. Adult flukes are also found in the gallbladder. Infections can be treated with antihelminth drugs.

The Gallbladder. **Cholecystitis** (kō-lē-sis-TĪ-tis) is an inflammation of the gallbladder. It is usually caused by gallstones that block the cystic duct or common bile duct. **Gallstones** develop when the bile becomes too concentrated, and crystals of insoluble minerals and salts form. The condition of having gallstones is called *cholelithiasis* (kō-lē-li-THĪ-a-sis). Approximately, 1 million people develop acute symptoms of cholecystitis each year in the U.S. The gallbladder becomes swollen and inflamed, infections may develop, and signs of *obstructive jaundice* develop; skin color changes because the liver can no longer get rid of the hemoglobin breakdown products, and they accumulate in the bloodstream.

A blockage that does not work its way down the duct to the small intestine (duodenum) must be removed or destroyed. In most cases, surgery is required to remove large gallstones, and the entire gallbladder is removed to prevent recurrence. This procedure is called a *cholecystectomy*.

The Large Intestine. **Colitis** (kō-LĪ-tis) is a general term used to indicate an inflammation of the colon. Colitis often involves diarrhea or constipation. Diarrhea results when the lining of the colon becomes unable to maintain normal levels of water reabsorption or so much fluid enters the colon that its water reabsorption capacity is exceeded. **Constipation** is infrequent bowel movement, or defecation, generally involving dry, hard feces. It results when fecal material moves through the colon so slowly that excessive water reabsorption occurs.

Irritable bowel syndrome is characterized by diarrhea, constipation, or an alternation between the two. This condition may be called a *spastic colon*, or *spastic colitis*. It is believed that psychological factors contribute to irritable bowel syndrome, because the mucosa appears to be normal. Peristalsis is affected, however, and fecal materials are not moved normally. **Inflammatory bowel disease**, such as *ulcerative colitis*, involves chronic inflammation of the digestive tract, most commonly affecting the colon. The mucosa becomes inflamed and ulcerated with associated bleeding. Extensive areas of scar tissue develop. Acute bloody diarrhea with cramps, fever, weight loss, and anemia often develop. Treatment of inflammatory bowel disease normally involves anti-inflammatory drugs and corticosteroids that reduce inflammation. In severe cases, oral or intravenous fluid replacement is required.

Treatment of severe inflammatory bowel disease may also involve a **colectomy** (ko-LEK-to-mē), the removal of all or a portion of the colon. After a colectomy, the end of the intact digestive tube is attached to the abdominal wall if normal connection with the anus cannot be maintained. Wastes then accumulate in a plastic pouch or sac attached to the opening. If the attachment involves the colon, the procedure is a *colostomy* (ko-LOS-to-mē); if the ileum is involved, it is an *ileostomy* (il-ē-OS-to-mē). Surgical creation of an internal pouch made from the end of the remaining small intestine is another option.

In **diverticulosis** (dī-ver-tik-ū-LŌ-sis), pockets (*diverticula*) form in the mucosa of the colon, primarily the sigmoid colon. These pockets get forced outward, probably by the pressures generated during defecation. If they push through weak

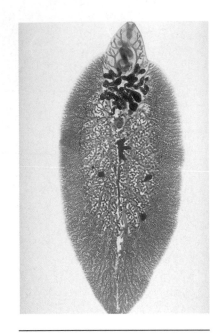

•FIGURE 18-14
Sheep Liver Fluke.
A whole, stained sheep liver fluke (*Fasciola hepatica*). (X 2.5)

points in the surrounding muscularis externa, they form somewhat isolated chambers that are subject to recurrent infection and inflammation. The infections cause pain, fever, and sometimes bleeding, a condition known as **diverticulitis** (dī-ver-tik-ū-LĪ-tis). Appendicitis, a comparable infection of a pocket (the appendix) at the start of the colon was described in Chapter 16, p. 395.

Tumors

The Oral Cavity. Malignant tumors, or *squamous carcinomas*, of the mouth are most common on the lower lip and the tongue. The main cause of lip and tongue cancers is tobacco smoking. The tumors of these cancers initially develop as whitish-gray patches called **leukoplakia** (loo-kō-PLĀ-ke-ah). Patches on the lip may form ulcers that bleed, and patches on the tongue may become hard, making eating, swallowing, and talking difficult. Tongue cancer is the most dangerous form, because it may spread rapidly to nearby lymph nodes. Treatment typically includes surgery and radiation.

The Pharynx. Malignant tumors of the pharynx are usually related to smoking and alcohol consumption. Symptoms of pharyngeal cancer include sore throat, difficulty swallowing (*dysphagia*), hoarseness, and sputum with blood. Treatment may include surgery, radiation therapy, and chemotherapy.

The Esophagus. Most tumors of the esophagus are malignant, only 10 percent are benign. They all usually occur in the middle to lower regions, and cause difficulty in swallowing. Respiratory infections are common complications because regurgitated food often enters the respiratory tract; this process is called *aspiration*. The most common causes of esophageal cancer are smoking and alcohol consumption. Generally, 5 year survival rates are low.

The Stomach. **Stomach cancer**, once a common fatal cancer has been declining in incidence for the last 50 years. Stomach, or gastric, cancers develop in the lining of the stomach. Diet and *H. pylori* infection are thought to play a role in causing stomach cancer, with the consumption of salted, pickled, and smoked foods apparently increasing the risk of developing tumors. Stomach cancer occurs more often in men and those over the age of 40. The initial symptoms of stomach cancer are vague and resemble those associated with indigestion, and, as a result, the condition may not be diagnosed in its early stages. The major method of treatment is surgery that removes part or the whole stomach, called **gastrectomy** (gas-TREK-to-mē). The loss of the stomach is not life-threatening but gastrectomy cannot cure the disease unless the surgery is performed while the cancer is at an early stage.

The Intestines. Malignant cancers are rare in the small intestine, but cancer of the colon and rectum is common. Each year roughly 152,000 new cases of **colorectal cancer** are diagnosed in the United States, and 57,000 deaths result from this condition. Most common among persons over 50 years of age, primary risk factors for colorectal cancer include (1) a diet rich in animal fats and low in fiber, (2) inflammatory bowel disease (p. 458), and (3) a number of inherited disorders that promote epithelial tumor formation along the intestines.

Successful treatment of colorectal cancer depends on early identification. It is believed that most colorectal cancers begin as small, localized tumors, or **polyps** (POL-ips), that grow from the mucosa lining the intestinal wall. The prognosis improves dramatically if cancerous polyps are removed before metastasis has occurred. If the tumor is restricted to the mucosa and submucosa, the 5-year survival rate is higher than 90 percent. If it extends into the serosa, the survival rate drops to 70–85 percent; after metastasis to other organs, the survival rate drops significantly.

One of the early signs of polyp formation is the appearance of blood in the feces. When blood is detected in the feces, X-ray techniques are commonly used as a first step in diagnosis. In the usual procedure, a large quantity of a liquid barium solution is introduced by *enema* (liquid injected into the rectum). Because this solution is radiopaque, the X-rays will reveal any intestinal masses, such as a large tumor, blockages, or structural abnormalities.

Early diagnosis and treatment of colorectal cancer is aided by a flexible **colonoscope** (ko-LON-o-skōp), which permits visual inspection of the lining of the large intestine. The colonoscope may be used to collect tissue samples, or remove polyps, and thereby avoid the potential complications of traditional surgery. **Sigmoidoscopy** is a procedure used to inspect the sigmoid colon and the rectum, and it utilizes an endoscope called a **sigmoidoscope**.

The Accessory Organs. Pancreatic cancer is a malignant tumor of the exocrine tissue of the pancreas. Although its cause is not known, it is thought to be connected to smoking, or diets high in fats or in the consumption of alcohol. Symptoms such as abdominal pain, indigestion, and weight loss do not appear until late in the development of the cancer. It also spreads and forms secondary tumors in the liver and lungs. If the diagnosis is made early, then the malignant tissue may be removed by surgery. The prognosis for late diagnosis is grim.

Liver cancer may be caused by either primary or secondary tumors in the liver. Most primary tumors develop from liver cells, and are called **hepatomas**. These cancers are relatively rare except in association with hepatitis B and C. The liver is a common site of secondary tumors, with most developing from malignant cells of the stomach, pancreas, or large intestine. Common symptoms of liver cancer are abdominal pain, weight loss, and loss of appetite. If cirrhosis is not a problem, treatment for hepatomas involves chemotherapy, surgical removal of the tumor, or a liver transplant. There is no cure for secondary tumors, but anticancer drugs may slow their progression.

Gallbladder cancer is a rare condition that mainly affects the elderly. It is associated with gallbladders containing gallstones. The cancer may cause jaundice or produce no symptoms at all. Treatment involves surgical removal of the cancer. If the cancer invades the liver, the prognosis is poor.

Congenital Disorders

Problems due to congenital disorders most often involve the oral cavity, stomach, pancreas, and intestines. Cleft palate, a major congenital disorder of the oral cavity, was discussed in Chapter 8 (p. 174).

The Stomach. Pyloric stenosis (ste-NŌ-sis) is a condition in which the pyloric sphincter muscle of the stomach is thicker than normal. The result is a blockage of food movement because of the narrowing of the passageway into the small intestine. Pyloric stenosis occurs in infants, and affects males more than females. Infants with this condition forcefully vomit their stomach contents over several feet (also called *projectile vomiting*). Treatment usually involves surgery.

The Pancreas. The pancreas of an individual born with cystic fibrosis may not produce adequate amounts of digestive enzymes. As a result, the individual suffers from *malabsorption* (inadequate absorption) of both fats and proteins. This causes an excess of fats to appear in the feces and poor growth ("failure to thrive") in the affected child.

The Intestines. Congenital disorders of the intestines generally cause obstruction problems. These problems may be due to atresia (a closed passageway), stenosis (a narrowing of a passageway), *volvulus* (a twisted or knotted intestine). In an *imperforate anus* the rectum is sealed off from the anus at the time of birth. This condition requires immediate corrective surgery.

Malabsorption Disorders

Difficulties in the absorption of all classes of organic compounds will result if the accessory glands or the intestinal lining malfunction. If the liver or pancreas are functioning normally but their secretions cannot reach the duodenum of the small intestine, the condition is called **biliary obstruction** (bile duct blockage) or **pancreatic obstruction** (pancreatic duct blockage). On the other hand, the ducts may remain open but the gland cells may be damaged and unable to secrete normally. Examples include pancreatitis and cirrhosis (pp. 456, 457).

Even if the normal enzymes are present in the intestine, absorption will not take place if the intestinal mucosa is not functioning properly. An inherited or acquired inability to manufacture specific enzymes will lead to malabsorption disorders; *lactose intolerance* is a good example (see p. 447). Damage to the intestinal lining due to ischemia (a deficiency in blood flow), radiation exposure, toxic compounds, or infection can all affect absorption and cause a depletion of nutrient and fluid reserves.

✔ What causes tooth decay?

✔ What makes mumps a more serious problem to males than females?

✔ What is an abdominal swelling caused by an accumulation of fluid in the peritoneum called?

✔ What do pancreatitis and cirrhosis of the liver have in common?

CHAPTER REVIEW

Key Words

absorption: The active or passive uptake of gases, fluids, or solutes.

bile: Exocrine secretion of the liver that is stored in the gallbladder and released into the duodenum, where it emulsifies lipids.

chyme (kīm): A semifluid mixture of ingested food and digestive secretions found in the stomach and proximal small intestine.

digestion: The chemical breakdown of ingested materials into simple molecules that can be absorbed by the cells of the digestive tract.

duodenum (doo-AH-de-num or doo-ō-DĒ-num): The first 25 cm of the small intestine.

gastric glands: Tubular glands of the stomach whose cells produce acid, enzymes, and hormones.

hydrolysis: The breakage of a chemical bond through the addition of water molecules.

ingestion: The introduction of materials into the digestive tract through the mouth.

lacteal (LAK-tē-al): A terminal lymphatic within an intestinal villus.

mesentery (MEZ-en-ter-ē): A double layer of serous membrane that supports and stabilizes the position of an organ in the abdominopelvic cavity and provides a passageway for the associated blood vessels, nerves, and lymphatics.

mucosa (mū-KŌ-sa): A mucous membrane; the epithelium plus the underlying loose connective tissue.

pancreatic juice: A mixture of buffers and digestive enzymes produced by the pancreas. It is discharged into the duodenum under the stimulation of the digestive enzymes secretin and cholecystokinin.

peristalsis (per-i-STAL-sis): A wave of smooth muscle contractions that propels materials along the axis of a tube such as the digestive tract.

villus: A slender fingerlike projection of the mucous membrane of the small intestine.

Selected Clinical Terms

ascites (a-SĪ-tēz): Fluid accumulation within the abdominal (peritoneal) cavity.

cholecystitis: Inflammation of the gallbladder due to blockage of the cystic or common bile duct by gallstones.

cholera: A bacterial infection of the digestive tract that causes massive fluid losses through diarrhea.

colitis: Inflammation of the colon.

diarrhea: Frequent watery bowel movements.

dyspepsia: Loss of the function of digestion; indigestion.

dysphagia: Difficulty or pain in swallowing.

emesis (EM-e-sis): Vomiting.

gastritis: Inflammation of the stomach lining.

hepatitis: Inflammation of the liver; usually caused by viral infection (hepatitis A, B, C) or toxins (alcohol).

peptic ulcer: Erosion of the gastric lining or duodenal lining by stomach acids and enzymes, or bacterial infection by *Helicobacter pylori*.

periodontal disease: A loosening of the teeth caused by breakdown of the periodontal ligaments by acids produced by bacterial activity.

Study Outline

INTRODUCTION (p. 435)

1. From the food that moves down the digestive tube, the digestive system supplies the energy and building materials needed for an individual's growth and maintenance.

SYSTEM BRIEF (p. 435)

1. The digestive system consists of the muscular **digestive tract** and various **accessory organs.**

2. Digestive functions include ingestion, mechanical processing, digestion, secretion, absorption, and excretion.

STRUCTURE OF THE DIGESTIVE SYSTEM (pp. 435–446)

1. The digestive tract includes the oral cavity, pharynx, esophagus, stomach, small intestine, large intestine, rectum, and anus. (*Figure 18-1*)

THE WALL OF THE DIGESTIVE TRACT *(p. 435)*

2. Double sheets of peritoneal membrane called **mesenteries** suspend the digestive tract within the abdominopelvic cavity. *(Figure 18-2a)*

3. The epithelium and underlying connective tissue form the inner **mucosa** (mucous membrane) of the digestive tract. Proceeding outward, the layers are the **submucosa** and the **muscularis externa**. Within the peritoneal cavity the muscularis externa is covered by a serous membrane called the **serosa**. *(Figure 18-2b)*

4. Folds in the intestinal wall and small villi increase the surface area of the digestive tract. *(Figures 18-2c,d)*

5. Contractions of the muscularis externa propel materials through the digestive tract by **peristalsis**.

THE ORAL CAVITY *(p. 438)*

6. The functions of the oral cavity include (1) tasting; (2) mechanical processing using the teeth, tongue, and oral surfaces; (3) lubrication by mixing with mucus and salivary secretions; and (4) digestion by salivary enzymes.

7. The cheeks form the sides of the oral cavity. The **hard** and **soft palates** form its roof, and the tongue forms its floor. *(Figure 18-3a)*

8. The primary functions of the tongue include manipulation of food to assist in chewing and swallowing, sensory analysis, and secretion of mucus.

9. The **parotid**, **sublingual**, and **submandibular salivary glands** discharge their secretions into the oral cavity. Saliva lubricates the mouth, dissolves chemicals, flushes the oral surfaces, and helps control bacteria. Its secretion is usually controlled by the autonomic nervous system. *(Figure 18-1)*

10. Chewing occurs through the contact of the opposing surfaces of the teeth. The 20 primary teeth, or **deciduous teeth**, are replaced by the 32 permanent teeth during development. *(Figure 18-3b)*

11. The **periodontal ligament** anchors the tooth in its bony socket. **Dentin** forms the basic structure of a tooth. The **crown** is coated with **enamel**, and the **root** with **cementum**. *(Figure 18-4)*

THE PHARYNX *(p. 440)*

12. The pharynx serves as a common passageway for solid food, liquids, and air. Pharyngeal muscles aid the swallowing process.

THE ESOPHAGUS *(p. 440)*

13. The esophagus carries solids and liquids from the pharynx to the stomach.

14. Swallowing begins with the compaction of a food mass, or **bolus**, and its movement into the pharynx, followed by the elevation of the larynx, folding of the epiglottis, and closure of the glottis. Peristalsis moves the bolus down the esophagus to the *lower esophageal sphincter*. *(Figure 18-5)*

THE STOMACH *(p. 440)*

15. The stomach has three major functions: (1) temporary bulk storage of ingested matter, (2) mixing of ingested foods, and (3) disruption of chemical bonds using acids and enzymes.

16. The **body** is the largest region of the stomach. The **pyloric sphincter** guards the exit from the stomach. In a relaxed state the stomach lining contains numerous **rugae** (ridges and folds). *(Figure 18-6)*

17. The **gastric glands** secrete **gastric juice**, which contains **hydrochloric acid** and the enzyme **pepsin**.

18. Gastric secretion is controlled by the central nervous system and digestive hormones.

THE SMALL INTESTINE *(p. 442)*

19. The small intestine includes the **duodenum**, the **jejunum**, and the **ileum**. A sphincter, the *ileocecal valve*, marks the transition between the small and large intestines.

20. The intestinal mucosa bears transverse folds and small projections called intestinal **villi**. These increase the surface area for absorption. *(Figures 18-2b,c)*

21. Intestinal glands secrete intestinal juice, mucus, and hormones. Intestinal juice moistens the chyme, helps buffer acids, and dissolves digestive enzymes and the products of digestion.

22. Intestinal hormones include **secretin, cholecystokinin (CCK), gastrin**, and **gastric inhibitory peptide (GIP)**. *(Table 18-1)*

23. Most of the important digestive and absorptive functions occur in the small intestine. Digestive enzymes and buffers are provided by the pancreas, liver, and gallbladder.

THE PANCREAS *(p. 443)*

24. The **pancreatic duct** penetrates the wall of the duodenum to deliver pancreatic secretions. *(Figure 18-7)*

25. The pancreas has two functions: endocrine (secreting insulin and glucagon into the blood) and exocrine (secreting water, ions, and digestive enzymes into the small intestine). Pancreatic enzymes include **lipases**, **carbohydrases**, and **proteinases**.

26. The pancreatic exocrine cells produce a watery **pancreatic juice** in response to hormones from the duodenum. When acid chyme arrives in the small intestine, (1) secretin triggers the pancreatic production of a fluid containing buffers, primarily sodium bicarbonate, that help neutralize the acid pH of the chyme, and (2) cholecystokinin is released.

27. CCK stimulates the pancreas to produce and secrete **pancreatic amylase**, **pancreatic lipase**, **nucleases**, and several protein-breaking enzymes (*trypsin*, *chymotrypsin*, and *carboxypeptidase*).

THE LIVER *(p. 444)*

28. The liver is the largest visceral organ in the body. *(Figure 18-7)*

29. Blood is supplied to the liver by the hepatic artery and hepatic portal vein.

30. The bile ducts from throughout the liver unite to form the **common hepatic duct**, which meets the *cystic duct* from the gallbladder to form the **common bile duct**, which empties into the duodenum.

31. In addition to producing **bile**, a fluid that aids the digestion of fats, the liver performs many other important functions. *(Table 18-2)*

THE GALLBLADDER *(p. 445)*

32. The gallbladder stores and concentrates bile.

THE LARGE INTESTINE *(p. 445)*

33. The main function of the large intestine is to reabsorb water and compact the feces. It also absorbs vitamins produced by bacteria.

34. The **cecum** collects and stores material from the ileum and begins the process of compaction. The **vermiform appendix** is attached to the cecum.

35. The **colon** has pouches and longitudinal bands of muscle.

36. The large intestine reabsorbs water and other substances such as vitamins. Bacteria are responsible for the production of intestinal gas, or *flatus*.

37. Distension of the stomach and duodenum stimulates peristalsis, or mass movements, of chyme from the colon into the **rectum**.

38. Muscular sphincters control the passage of fecal material to the **anus**. Distension of the rectal wall triggers the *defecation reflex*.

DIGESTION AND ABSORPTION *(pp. 446–448)*

1. The digestive system breaks down the physical structure of the ingested material and then disassembles the component molecules into smaller fragments through **hydrolysis**.

CARBOHYDRATES *(p. 447)*

2. Amylase breaks down complex carbohydrates into two- and three-sugar molecules. These are broken down into monosaccharides by enzymes at the epithelial surface and absorbed by the intestinal epithelium.

PROTEINS *(p. 447)*

3. Protein digestion involves the gastric enzyme pepsin and the various pancreatic proteinases. Single amino acids are absorbed and eventually delivered to the liver.

LIPIDS *(p. 447)*

4. During **emulsification**, bile reduces large drops of lipids into tiny droplets. Triglycerides within the droplets are broken down into fatty acids and monoglycerides by lipases. These products diffuse across the intestinal epithelium and enter lymphatic vessels called **lacteals**. *(Figure 18-2d)*

WATER, IONS, AND VITAMINS *(p. 448)*

5. About 2.0 to 2.5l of water is ingested each day, and digestive secretions provide 6 to 7l. Nearly all is reabsorbed by osmosis and only around 150 ml is lost in feces each day.

6. The absorption of sodium and chloride ions promote water absorption through osmosis.

7. The nine **water-soluble vitamins** are important to enzymatic reactions. **Fat-soluble vitamins** are enclosed within fat droplets and are absorbed with lipids.

AGING AND THE DIGESTIVE SYSTEM *(p. 448)*

1. Age-related changes include a thinner and more fragile epithelium due to a reduction in epithelial stem cell division, and weaker peristaltic contractions as smooth muscle tone decreases.

DISORDERS OF THE DIGESTIVE SYSTEM *(pp. 449–461)*

1. Gastroenterology is the study of the digestive system and its diseases and disorders.

THE PHYSICAL EXAMINATION *(p. 449)*

2. Symptoms of digestive system disorders include: pain, **dyspepsia** (indigestion), **nausea**, **emesis** (vomiting), and **dysphagia** (difficulty in swallowing).

3. Signs of digestive system disorders are revealed in a physical examination through *inspection*, *palpation*, *percussion*, and *auscultation*.

4. Inspection may reveal signs such as: bleeding gums, oral lesions, peristaltic waves across abdominal wall, yellowish skin, abdominal distention, or *striae*.

5. Signs revealed by palpation include: abnormal masses, abdominal distention, *hernia*, enlarged organs, abdominal muscle contractions, and tender and painful areas.

6. Percussion can distinguish abnormal fullness in the stomach, enlargement of some organs, or fluid in the abdominal cavity.

7. Auscultation can detect the presence of bowel sounds.

8. An **endoscope** is used to view internal cavities of the body, and improve diagnosis. *(Tables 18-3, 18-4)*

DIGESTIVE SYSTEM DISORDERS *(p. 451)*

9. Major digestive system disorders include inflammation and infection, tumors, congenital disorders, and malabsorption disorders. *(Figure 18-8)*

10. Dental caries, or tooth decay, is a common disease of the oral cavity that causes tooth enamel to dissolve. **Pulpitis**, infection of the pulp of the tooth, may occur if the bacteria penetrate the enamel.

11. Periodontal disease is associated with the loss of teeth. The mildest form is **gingivitis**, inflammation of the gums. *(Figure 18-10)*

12. Mumps is a viral disease, primarily of the salivary glands. *(Figure 18-11)*

13. Thrush is a type of **candidiasis**, or yeast infection, of the mouth. *(Figure 18-12)*

14. Esophagitis is an inflammation of the esophagus, often described as *heartburn*.

15. Gastritis is an inflammation of the stomach lining, or mucosa. A **peptic ulcer** that occurs in the stomach is called a **gastric ulcer**, and an ulcer in the duodenum is called a **duodenal ulcer**.

16. Enteritis, inflammation of the small intestine, often causes **diarrhea**, or watery, bowel movements. **Dysentery** is an inflammation of the large intestine, characterized by diarrhea containing blood and mucus. **Gastroenteritis** refers to inflammation of the stomach and intestines. *(Table 18-5)*

17. Traveler's disease is a bacterial infection that causes a mild to severe diarrhea.

18. Typhoid fever is a bacterial infection of the upper regions of the small intestine.

19. Cholera is caused by a bacterium, which produces a toxin that binds to cells of the small intestine.

20. Viral enteritis is a major cause of illness and death in young children.

21. Giardiasis is an infectious disease of the small intestine caused by a flagellated protozoan. *(Figure 18-13)*

22. Amebiasis, or **amoebic dysentery**, is a protozoan infection caused by an amoeba.

23. Ascariasis is an infection of the small intestine by a multicellular parasite, the roundworm *Ascaris*.

24. Pancreatitis is an inflammation of the pancreas that results in the loss of the exocrine cells which produce digestive enzymes.

25. Hepatitis, or inflammation of the liver, may be caused by alcohol, drugs, or infection. Inflammation leads to **cirrhosis**, a condition in which liver cells are destroyed and replaced by fibrous tissue.

26. Varieties of **viral hepatitis** include A, B, and C. **Hepatitis A** is known as **infectious hepatitis** and **hepatitis B** is called **serum hepatitis**.

27. Jaundice, yellow skin color, develops because of the loss of liver cells in hepatitis.

28. A multicellular parasite that causes hepatitis is the **sheep liver fluke**. *(Figure 18-14)*

29. Cholecystitis is an inflammation of the gallbladder that often develops because **gallstones** block the cystic duct or common bile duct.

30. Colitis, inflammation of the colon, may result in diarrhea or constipation.

31. In **diverticulosis**, pockets of the sigmoid colon become infected and inflamed. The pain and occasional bleeding from the infection causes a condition called **diverticulitis.**

32. The most common tumors of the oral cavity develop on the lip and tongue, usually from whitish-gray patches called **leukoplakia**.

33. Stomach cancer is one of the most common fatal cancers. Treatment involves **gastrectomy**, the surgical removal of part or all of the stomach.

34. Intestinal cancer is most common in the colon and rectum, and is known as **colorectal cancer**.

35. Malignant tumors may develop in the pancreas, liver, and gallbladder. Primary tumors in the liver are called **hepatomas**.

36. Congenital disorders of the mouth and stomach include cleft palate and **pyloric stenosis** (narrowed passageway between stomach and duodenum).

37. Congenital disorders of the intestines generally cause obstruction problems.

38. Malabsorption disorders may be due to accessory glands with blocked ducts (**biliary obstruction** or **pancreatic obstruction**), damaged glands, as in cirrhosis or pancreatitis, or abnormalities affecting the intestinal mucosa.

Review Questions

MATCHING

Match each item in Column A with the most closely related item in Column B. Use letters for answers in the spaces provided.

	Column A		Column B
___	1. pyloric sphincter	a.	serous membrane sheet
___	2. liver	b.	moves materials along digestive tract
___	3. mucosa	c.	regulates flow of chyme
___	4. mesentery	d.	increases muscular activity of digestive tract
___	5. gastric juice	e.	starch digestion
___	6. palate	f.	a narrowing of a passageway
___	7. lacteals	g.	inner lining of digestive tract
___	8. parasympathetic stimulation	h.	vomiting
___	9. sympathetic stimulation	i.	lymphatic capillaries
___	10. peristalsis	j.	inflammation of the gums
___	11. function of bile	k.	produces bile
___	12. salivary amylase	l.	emulsification of fats
___	13. stenosis	m.	inhibits muscular activity of digestive tract
___	14. gingivitis	n.	hydrochloric acid and pepsin
___	15. emesis	o.	roof of oral cavity

MULTIPLE CHOICE

16. The breakdown of large food molecules into their basic building blocks by enzymes is called _____ .
(a) absorption
(b) mechanical digestion
(c) secretion
(d) chemical digestion

17. The _____ is not considered an accessory organ of digestion.
(a) pharynx
(b) liver
(c) salivary glands
(d) pancreas

18. The part of the digestive tract that plays the primary role in the digestion and absorption of nutrients is the _____ .
(a) large intestine
(b) small intestine
(c) cecum and colon
(d) stomach

19. An adult who has a complete set of permanent teeth has _____ teeth.
(a) 20
(b) 28
(c) 32
(d) 36

20. Bile is stored in the _____ .
(a) pancreas
(b) common bile duct
(c) left lobe of the liver
(d) gallbladder

21. The function(s) of the large intestine is (are) _____ .
(a) reabsorption of water
(b) absorption of vitamins released by bacteria
(c) storage of fecal material before defecation
(d) a, b, and c are correct

22. The surface area of the small intestine is increased by all the following except _____ .
(a) rugae
(b) folds of the intestinal wall
(c) villi
(d) microvilli

23. The parotid salivary glands secrete _____ .
(a) gastrin
(b) saliva
(c) bile
(d) proteinases

24. _____ is an inflammation of the oral mucous membrane.
(a) enteritis   (b) stomatitis
(c) esophagitis   (d) diverticulitis

25. _____ is not caused by bacterial pathogens.
(a) mumps   (b) cholera
(c) typhoid fever   (d) traveler's disease

TRUE/FALSE

___ 26. Contraction of the lower esophageal sphincter prevents the leakage of stomach acid into the esophagus.

___ 27. The first part of the small intestine is the ileum.

____ 28. Bile contains no enzymes.

____ 29. Trypsin is a pancreatic enzyme that breaks down proteins.

____ 30. The intestinal villi contain capillaries and lacteals important for chemical digestion.

____ 31. Pancreatic juice contains sodium bicarbonate, which increases the acidity of the chyme released by the stomach into the duodenum.

____ 32. Dysphagia is a difficulty in swallowing.

____ 33. Ascariasis is an infection caused by protozoa.

____ 34. A person with a condition called cholelithiasis has gallstones.

SHORT ESSAY

35. What are the primary functions of the digestive system?

36. Name and describe the four layers of the digestive tract, beginning with the innermost layer.

37. How does the stomach promote and assist in the digestive process?

38. What five age-related changes occur in the digestive system?

39. Both leukoplakia and thrush affect the mucous membrane of the mouth. What is the difference between these two disorders?

40. Describe three causes of hepatitis.

APPLICATIONS

41. Patients diagnosed with liver cancer generally have a high mortality. Explain why.

42. Tony is a chronic alcoholic who suffers from cirrhosis of the liver. Explain how conditions such as a prolonged clotting time, jaundice, portal hypertension, and ascites can all be related to his cirrhosis.

✔ Answers to Concept Check Questions

(p. 441) **1.** Peristalsis is the result of alternate contractions of the circular and longitudinal layers of smooth muscle making up the muscularis externa. **2.** The covering of the glottis by the epiglottis during swallowing prevents solid foods and liquids from entering the trachea and lower air-conducting passageways.

(p. 443) **1.** Because the parotid salivary glands secrete salivary amylase, an enzyme that digests carbohydrates, damage to these glands would interfere with carbohydrate digestion. **2.** The *incisors* are the type of tooth best suited for chopping or cutting relatively rigid foods, such as raw vegetables. **3.** The small intestine has several adaptations that increase surface area to increase its capacity to absorb food. Its internal wall consists of folds that are covered by fingerlike projections of tissue called *villi*. The cells that cover the villi, in turn, have smaller fingerlike projections called *microvilli*. In addition, the small intestine has a rich blood and lymphatic supply to transport the absorbed nutrients.

(p. 445) **1.** The CCK level in the blood would increase. **2.** The exocrine pancreas secretes *pancreatic juice*, which neutralizes the acid chyme entering the small intestine, and different classes of digestive enzymes. Pancreatic amylase digests carbohydrates, lipases digest lipids, nucleases digest nucleic acids, and proteinases digest proteins. **3.** Bile is stored in the gallbladder until needed in the digestion of lipids.

(p. 449) **1.** A narrowing of the ileocecal valve would interfere with the flow of chyme from the small intestine to the large intestine. **2.** An important function of the large intestine is the reabsorption of water from intestinal contents. **3.** A meal high in fat would result in the uptake of combined protein–lipid fragments by the lymph and cause the lymph to become cloudy, or milky.

(p. 461) **1.** Tooth decay, or dental caries, is caused by bacteria that secrete acids which dissolve the enamel of the tooth. **2.** Mumps is primarily a viral infection of the salivary glands. However, the virus may also infect the testes, and may make postadolescent males sterile. **3.** Ascites. **4.** Gland cells are destroyed in both of these conditions. Exocrine cells that secrete digestive enzymes are lost in pancreatitis, and liver cells are lost in cirrhosis.

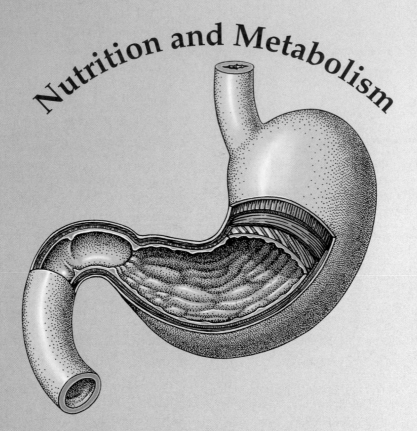

CHAPTER OUTLINE

CHAPTER OBJECTIVES

1   Define metabolism, and explain why cells need to build new structures.

2   Define catabolism and anabolism.

3   Describe the catabolism and anabolism of carbohydrates, lipids, and proteins.

4   Describe some of the roles of minerals and vitamins in the body.

5   Explain what constitutes a balanced diet and why it is important.

6   Contrast the energy content of carbohydrates, lipids, and proteins.

7   Define metabolic rate, and discuss the factors involved in determining an individual's metabolic rate.

8   Discuss the homeostatic mechanisms that maintain a constant body temperature.

9   Describe specific examples of eating, metabolic, and thermoregulatory disorders.

Energy is the ability to do work. You are able to sit and read these words because your cells can release and use energy. Your energy supply comes from the food you eat, in the form of carbohydrates, proteins, and fats (lipids). These molecules also form the basic raw materials needed to produce and maintain cells and tissues. Your diet also gives you other essentials, such as (1) water to replace the water lost in urination, perspiration, and so forth, (2) ions (electrolytes) to replace those lost with the water, and (3) vitamins, which are organic molecules required by some enzymatic reactions. The carbohydrates, proteins, lipids, water, electrolytes, and vitamins we get from our diet are collectively called **nutrients.** The absorption and use of nutrients is known as **nutrition.** Chapter 18 discussed the digestive breakdown and absorption of food through the lining of the digestive tract. This chapter considers what happens to nutrients after they are inside the cells of the body.

METABOLISM

Metabolism refers to all the chemical reactions of the body. The metabolic activities controlled by living cells are divided into two general classes of chemical reactions: *catabolism* and *anabolism*. Figure 19-1• shows the various ways that the major organic nutrients (simple sugars, amino acids, and lipids) are used by cells.

During **catabolism**, large organic molecules are broken down into smaller molecules, and energy is released that can be used to make ATP (adenosine triphosphate), the major high-energy compound of the cell. ∞ *[p. 31]* Catabolism usually occurs in two stages. The first stage takes place in the cytoplasm. For example, enzymes in the cytoplasm break carbohydrates into short carbon chains, triglycerides into fatty acids and glycerol, and proteins into individual amino acids.

Relatively little ATP is produced by these reactions in the cytoplasm. However, the simple molecules generated can then be absorbed and processed in a cell's mitochondria. This second, mitochondrial, stage of catabolism releases large

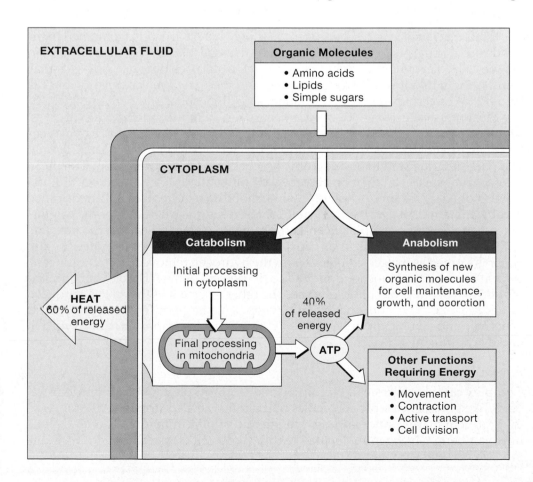

•**FIGURE 19-1**
Cellular Metabolism.
The cell obtains organic molecules from the extracellular fluid and breaks them down to release energy that is used to make ATP. Only about 40 percent of the energy released through catabolism is captured in ATP; the rest is radiated as heat. The ATP generated through catabolism provides energy for all vital cellular activities, including anabolism.

amounts of energy. About 40 percent of the released energy is captured and bound up in ATP. The rest escapes as heat that warms the cell and the surrounding tissues.

Anabolism, the building of new organic molecules, involves the formation of new chemical bonds. The ATP produced by mitochondria provides energy for anabolic reactions as well as for other cellular activities. These energy-requiring activities, such as muscle contraction, ciliary or cell movement, active transport, or cell division, vary from one cell to another. For example, skeletal muscle fibers need ATP to provide energy for contraction, and gland cells need ATP to synthesize and transport their secretions.

As you might expect, the cell tends to save the materials needed to build new compounds and breaks down the rest. The cell is continually replacing or repairing its membranes, organelles, enzymes, and structural proteins. This process requires more amino acids than lipids, and relatively few carbohydrates. Energy production (catabolism), however, tends to use these organic molecules in the reverse order. In general, a cell with excess carbohydrates, lipids, and amino acids will break down carbohydrates first. Lipids are a second choice, and amino acids are seldom broken down as long as carbohydrates or lipids are available.

CARBOHYDRATE METABOLISM

Carbohydrates, most familiar to us as sugars and starches, are the main sources of energy for cells.

Catabolism

Most cells produce ATP through the breakdown of carbohydrates, especially glucose, in a process called **cellular respiration**. The beginning and ending compounds involved in this process are shown in the following chemical equation:

$$C_6H_{12}O_6 \quad + \quad 6\,O_2 \quad \rightarrow \quad 6\,CO_2 \quad + \quad 6\,H_2O$$

| glucose | oxygen | carbon dioxide | water |

If this reaction occurred in one step, as shown here, the energy contained in the bonds holding the glucose together would be released in a burst of heat that would kill the cell. To avoid this problem, cells break the glucose down in a series of small steps. Although heat is still produced, several of these steps release energy that can be used to convert low-energy ADP (adenosine diphosphate) to high-energy ATP.

Most of the actual energy release occurs inside mitochondria. However, essential processing must first take place in the cytoplasm. In this process, a 6-carbon glucose molecule is split in half to form two 3-carbon molecules of *pyruvic acid*. The steps are said to be **anaerobic** because oxygen is not needed. The pyruvic acid molecules can then be absorbed by mitochondria and used for the production of ATP. ATP production in the mitochondria requires oxygen and is called **aerobic metabolism**. During these aerobic reactions, the fragments of the original glucose molecule are completely broken down into water and carbon dioxide, and ATP energy is produced. Carbon dioxide is split from the glucose molecule, and water is formed as oxygen combines with the hydrogen removed from the glucose.

On average, the mitochondria produce 95 percent of the ATP released from the breakdown of one molecule of glucose. The other 5 percent of the ATP is produced in the cytoplasm without the use of oxygen. This large difference between the anaerobic and aerobic production of ATP highlights the importance of oxygen in energy capture by our cells.

Anabolism

Complex carbohydrates can be constructed from simple monosaccharide molecules like glucose. Glucose molecules that are not needed immediately for energy production are converted to **glycogen** and stored in liver cells and skeletal muscle fibers. Glycogen molecules within these cells form compact, insoluble granules that take up very little space.

OTHER ENERGY-PRODUCING PATHWAYS

Aerobic metabolism is relatively efficient and capable of generating large amounts of ATP. It is the basis for normal cellular metabolism, but it has one obvious limitation: The cell must have adequate supplies of both oxygen and glucose. Cells simply cannot survive very long without oxygen. A lack of glucose, however, can be tolerated by most cells because they are able to break down other nutrients to give to their mitochondria (Figure 19-2•). Many cells can switch from one nutrient source to another as the need arises. For example, many cells can shift from glucose-based to lipid-based ATP production when glucose levels in the blood decline.

Cells break down proteins for energy only when lipids or carbohydrates are unavailable, primarily because proteins make up the enzymes and organelles that the cell needs to survive. Nucleic acids are present only in small amounts, and they are seldom used for energy, even when the cell is dying of acute starvation. This sparing of DNA makes sense because it is the DNA in the nucleus that determines all the structures and functional roles of the cell.

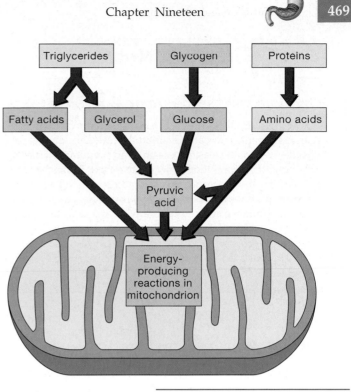

•**FIGURE 19-2**
Energy-Producing Pathways.

LIPID METABOLISM

Lipid molecules, like carbohydrates, contain carbon, hydrogen, and oxygen, but in different proportions. The most abundant lipids in the body are triglycerides.

Catabolism

The building blocks of triglycerides are glycerol and fatty acids. As Figure 19-2• shows, both of these molecules, especially fatty acids, can enter the mitochondria and produce ATP.

Cells with modest energy demands can shift over to lipid-based energy production when glucose supplies are limited. Skeletal muscle fibers normally cycle between lipid and carbohydrate metabolism. At rest, when energy demands are low, they break down fatty acids. When active and energy demands are high, skeletal muscle fibers shift over to metabolizing glucose.

Anabolism

Cells must often synthesize lipid molecules from their building blocks of fatty acids and glycerol. Unfortunately, our cells cannot *build* every fatty acid they can break down. *Linoleic acid*, an 18-carbon, unsaturated fatty acid, cannot be synthesized at all. Thus, a diet poor in linoleic acid slows growth and alters the appearance of the skin. *Arachidonic* and *linolenic acids* are other long-chain unsaturated fatty acids that the human body cannot synthesize. These three **essential fatty acids** must be included in the diet because they are needed to synthesize prostaglandins and the phospholipid components of cell membranes.

PROTEIN METABOLISM

There are roughly 100,000 different proteins in the human body. All contain varying combinations of the same 20 amino acids. Under normal conditions, there is a continual turnover of cellular proteins in the cytoplasm, and free amino acids are used to manufacture new proteins. If other energy sources are low, mitochondria can also break down amino acids to produce ATP (see Figure 19-2•).

Clinical Note

Why Do Athletes Binge on Carbohydrates Before an Event?

Although lipids and proteins can be broken down to provide organic molecules for catabolism in the mitochondria, carbohydrates require the least processing and preparation. It is not surprising, therefore, that athletes have tried to devise ways of exploiting these compounds as ready sources of energy.

Eating carbohydrates *just before* exercising does not improve performance and may actually decrease endurance by slowing the mobilization of existing energy reserves. Runners or swimmers preparing for lengthy endurance contests, such as a marathon or a 5-km swim, do not eat immediately before competing, and for 2 hours before the race their diet is limit-

ed to drinking water. However, these athletes often eat carbohydrate-rich meals for 3 days before the event. This **carbohydrate loading** increases the carbohydrate reserves of muscle tissue that will be called on during the competition.

Maximum effects can be obtained by exercising to exhaustion for 3 days before starting the high-carbohydrate diet; this practice is called *carbohydrate depletion/loading*. There are a number of potentially unpleasant side effects to carbohydrate depletion/loading, including muscle and kidney damage, and sports physiologists recommend that athletes use this routine fewer than three times per year.

Amino Acid Catabolism

Amino acid breakdown takes place primarily in liver cells. The first step in the breakdown of an amino acid is the removal of its nitrogen-containing amino group. Once the amino group is removed, another reaction converts it into an ammonia molecule. Ammonia molecules are highly toxic, even in low concentrations. Liver cells reduce the level of ammonia by converting it to **urea**, a relatively harmless compound. The urea is carried in the circulating blood to the kidneys, where it is excreted in the urine. The remaining carbon chain of the amino acid molecule can be used to produce ATP in the mitochondria.

Amino Acids and Protein Synthesis

The basic mechanism for protein synthesis was discussed in Chapter 3. ∞ [p. 49] The human body can synthesize roughly half the different amino acids needed to build proteins. These amino acids, called the **nonessential amino acids,** do not need to be ingested in food. There are 10 other amino acids, called **essential amino acids**, that cannot be synthesized and so must be supplied in the diet. Eight of them (*isoleucine, leucine, lysine, threonine, tryptophan, phenylalanine, valine,* and *methionine*) cannot be synthesized at all; the other two (*arginine* and *histidine*) can be synthesized in amounts that are adequate for adults but insufficient for growing children.

In amino acid anabolism, the amino group is removed from one amino acid and attached to another carbon chain, creating a "new" amino acid. This process enables a cell to synthesize many of the amino acids needed for protein synthesis. The liver, skeletal muscles, heart, lung, kidney, and brain have a high demand for amino acids because they synthesize large amounts of proteins.

✔ Which class of organic compounds serves as the major source of fuel for cells?

✔ What is the source of the urea excreted in urine?

✔ What makes one amino acid or fatty acid essential and another, nonessential?

DIET AND NUTRITION

Good health relies on the digestive tract to absorb fluids, carbohydrates, lipids, proteins, minerals, and vitamins on a regular basis and to keep pace with their use by the body's cells or their loss in urine, feces, or perspiration. The individual requirement for each of these nutrients varies from day to day and from person to person.

Nutritionists attempt to analyze a diet in terms of its ability to meet the needs of a specific individual. A *balanced diet* contains all the nutrients necessary to maintain homeostasis. A balanced diet prevents **malnutrition**, an unhealthy state resulting from inadequate or excessive intake of one or more nutrients.

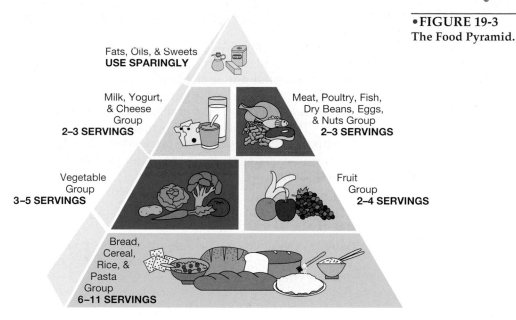

•FIGURE 19-3
The Food Pyramid.

A Guide to Daily Food Choices

Nutrient Group	Provides	Deficiencies
Fats, oils, sweets	Calories	Usually deficient in most minerals and vitamins
Milk, yogurt, cheese	Complete proteins, fats, carbohydrates; calcium, potassium, magnesium, sodium, phosphorus; vitamins A, B_{12}, pantothenic acid, thiamine, riboflavin	Dietary fiber, vitamin C
Meat, poultry, fish, dry beans, eggs, nuts	Complete proteins, fats; potassium, phosphorus, iron, zinc; vitamins E, thiamine, B_6	Carbohydrates, dietary fiber, several vitamins
Vegetables and fruits	Carbohydrates; vitamins A, C, E, folacin; potassium; and dietary fiber	Often low in fats, calories, and protein
Bread, cereal, rice, pasta	Carbohydrates; vitamins E, thiamine, niacin, folacin; calcium, phosphorus, iron, and sodium; and dietary fiber	Fats

THE SIX BASIC FOOD GROUPS

Nutritionists have devised a general guide to a balanced diet based on six different basic food groups. These groups are the *milk and dairy group*, the *meat group*, the *vegetable group*, the *fruit group*, the *bread and cereal group*, and the *fats, oils, and sweets group*. Each group differs from the others in their typical balance of proteins, carbohydrates, and lipids, as well as in the amount and identity of vitamins and minerals.

As Figure 19-3• shows, the six groups are arranged in a *food pyramid*, a shape based on their daily recommended number of servings. This arrangement emphasizes the need to restrict dietary fats, oils, and sugar and to increase consumption of breads and cereals, which are rich in complex carbohydrates (starches).

Although such groupings can help in planning a balanced diet, the wrong selections can lead to malnutrition even if all six groups are represented. What is important is to obtain nutrients in sufficient *quantity* (adequate to meet energy needs) and *quality* (including essential amino acids, essential fatty acids, vitamins, and minerals). The key is making intelligent choices about what you eat.

For example, consider the case of the essential amino acids. Some members of the meat and milk groups, such as beef, fish, poultry, eggs, and milk, contain all the essential amino acids in sufficient quantities. They are said to contain *complete proteins*. Many plants contain adequate *amounts* of protein, but they contain *incomplete proteins* that are deficient in one or more of the essential amino acids. True vegetarians, who restrict themselves to the fruit and vegetable groups (with or without the bread and cereal), must be careful that their meals include a combination of ingredients that will meet all their amino acid requirements.

MINERALS, VITAMINS, AND WATER

Minerals, vitamins, and water are essential components of the diet. The body cannot synthesize minerals, and our cells can generate only a small quantity of water and very few vitamins.

Minerals

Minerals are inorganic ions released through the breakdown of electrolytes, such as sodium chloride. Minerals play important roles in a wide variety of normal body functions. For example, minerals are essential for proper fluid balance, muscle contraction, conduction of nerve impulses, building teeth and bones, and transport of oxygen in the blood.

The major minerals and their functional roles are presented in Table 19-1. Large reserves of several important minerals in the body help reduce the effects of dietary variations in supply. Most mineral reserves are relatively small, however, and inadequate amounts in the diet can lead to various health problems. On the other hand, an excess of minerals in the diet can prove equally dangerous.

Vitamins

Vitamins are related to lipids and carbohydrates. Required in small amounts, vitamins form parts of enzymes and other compounds that control different aspects of cellular metabolism. They are assigned to either of two groups: *fat-soluble vitamins* and *water-soluble vitamins*. Because fat-soluble vitamins can dissolve in the many lipids in the body, they are easily stored. Our bodies thus contain reserves of these vitamins, and dietary deficiencies are usually not a problem. *Vitamin deficiency diseases* are more common with water-soluble vitamins because they are easily transported from one body fluid to another, and excesses are readily excreted in the urine. Current information concerning both groups of vitamins is summarized in Table 19-2.

Water

Our daily water requirement averages 2,500 ml (10 cups) of water per day. Although most of our water intake occurs through eating and drinking, a small amount (roughly 300 ml) is produced within the body. This *metabolic water* is a by-product of the aerobic reactions occurring within mitochondria.

The total amount of water required at any time varies with environmental conditions and metabolic activity. For example, exercise increases metabolic energy requirements, and it also accelerates water loss from evaporation and perspiration. The temperature rise accompanying a fever has a similar effect, and for each degree the temperature rises above normal, the daily water loss increases by 200 ml. Thus the advice "drink plenty of liquids" when one is sick has a definite physiological basis.

Table 19-1	Minerals		
Mineral	**Functions**	**Effects of Deficiency**	**Effects of Excess**
Bulk Minerals			
Sodium	Major cation (+) in body fluids; essential for normal muscle and nerve functions	Dehydration, muscle cramps	Tissue swelling, hypertension
Potassium	Major cation within cells; essential for normal muscle and nerve functions	Decline in heart rate; muscle weakness	Heartbeat weak and irregular
Chloride	Major anion (−) in body fluids; essential for acid-base balance	High blood pH; muscle cramps	Low blood pH; tissue swelling
Calcium	Essential for normal muscle and nerve functions; important mineral in bone	Muscles in extended state of contraction (tetany); rickets; osteoporosis; weak heartbeat	Calcium buildup in soft tissues (calcification); heart failure
Phosphorus	Essential for high-energy compounds (ATP), nucleic acids (DNA), and bone	Rickets	Unknown
Magnesium	Enzyme component; required for normal muscle and nerve functions	Muscles enter in extended state of contraction (tetany)	Unknown
Trace Minerals			
Iodine	Component of thyroid hormones; essential for normal metabolism	Enlarged thyroid gland (goiter)	Inhibits thyroid gland secretion
Iron	Component of hemoglobin and myoglobin; oxygen transport	Iron-deficiency anemia; weakness and fatigue	Deterioration of the liver
Zinc	Component of many enzymes	Abnormal cell growth	Fever, nausea
Copper	Required for synthesis of hemoglobin; part of some enzymes	Anemia	Liver damage
Manganese	Component of some enzymes	Retarded growth, bone abnormalities	Muscle weakness

DIET AND DISEASE

Diet has a profound influence on general health. Table 19-1 provides information on the uses of different minerals in the body, and Table 19-2 lists the health effects of consuming either excess or below-normal amounts of different vitamins. More subtle, long-term problems may be encountered when the diet includes the wrong proportions or combinations of nutrients. The average American diet contains too many calories, and too great a proportion of those calories are provided by fats. This diet increases the incidence of obesity, heart disease, atherosclerosis, hypertension, and diabetes in the U.S. population.

✔ In terms of the number of servings per day, which of the six food groups is most important?

✔ What is the difference between foods described as complete proteins and incomplete proteins?

Table 19-2   Vitamins

Vitamins	Importance	Source	Effects of Deficiency	Effects of Excess
Fat-soluble vitamins				
A (retinol)	Maintains epithelia; required for synthesis of visual pigments in the eye	Leafy green and yellow vegetables	Retarded growth, night blindness, breakdown of epithelial membranes	Liver damage, skin peeling, CNS effects (nausea, anorexia)
D (cholecalciferol)	Required for normal bone growth and calcium and phosphorus absorption	Synthesized in skin exposed to sunlight	Rickets, skeletal deformations	Calcium deposits in many tissues, disrupting functions
E (tocopherol)	Prevents breakdown of vitamin A and fatty acids	Meat, milk, vegetables	Anemia	None reported
K	Required for synthesis of blood clotting factors by the liver	Vegetables; produced by intestinal bacteria	Bleeding disorders	Liver dysfunction
Water-soluble vitamins				
B_1 (thiamine)	Part of enzyme involved in ATP formation in mitochondria	Milk, meat, bread	Muscle weakness, CNS and cardio-vascular problems (heart disease)	Low blood pressure (hypotension)
B_2 (riboflavin)	Same as above	Milk, meat	Epithelial and mucous membrane breakdown	Itching, tingling sensations
B_3 (niacin)	Same as above	Meat, bread, potatoes	CNS, GI, epithelial, and mucous membrane breakdown	Itching, burning sensations, dilation of blood vessels, death after large dose
B_5 (pantothenic acid)	Same as above	Milk, meat	Retarded growth	None reported
B_6 (pyridoxine)	Part of enzyme involved in amino acid and lipid metabolism	Meat	Retarded growth, anemia, epithelial changes	CNS alterations, sometimes fatal
Folacin (folic acid)	Part of enzyme involved in amino acid and nucleic acid metabolism	Vegetables, cereal, bread	Retarded growth, anemia, gastrointestinal disorders	Few noted except at massive doses
B_{12} (cobalamin)	Part of enzyme involved in red blood cell (RBC) production	Milk, meat	Reduced RBC production	Overproduction of RBCs
Biotin	Part of enzyme involved in various metabolic reactions	Eggs, meat, vegetables	Fatigue, muscular pain, nausea, dermatitis (inflammation of dermis in skin)	None reported
C (ascorbic acid)	Enzyme component for collagen synthesis	Citrus fruits	Epithelial and mucous membrane deterioration, loss of bone (scurvy)	Kidney stones

NUTRITIONAL AND METABOLIC DISORDERS

Nutritional and metabolic disorders may be due to overeating or undereating, or from consuming inadequate or excessive amounts of specific nutrients. Disorders may also arise if there are problems involving the absorption or metabolism of one or more nutrients. Disorders affecting temperature regulation, or *thermoregulation*, may disturb homeostasis throughout the body, with potentially fatal results. Figure 19-4• provides an overview, grouping nutritional and metabolic disorders into eating disorders, metabolic disorders, and thermoregulatory disorders.

EATING DISORDERS

Eating occurs in response to appetite or hunger. *Appetite* is a desire for food, and is associated with a pleasurable anticipation. *Hunger* is a craving, or need, for food. It occurs when the stomach is empty and the glucose level of the blood is low. These stimuli affect the hunger center in the hypothalamus. ∞ *[p. 248]* One result is contractions of the stomach, which cause hunger pains.

Eating disorders result in either inadequate or excessive food consumption. They may be caused by a variety of factors, acting alone or in combination and ranging from abnormal metabolic functions to psychological problems.

Inadequate Food Intake

Anorexia. Anorexia (an-o-REK-sē-ah) is the lack or loss of appetite. If prolonged, it is accompanied by a loss of weight. Anorexia may be caused by digestive disorders such as gastric cancer, pancreatitis, hepatitis, and diarrhea. It may also accompany disorders that involve other systems.

Anorexia nervosa is a form of self-induced starvation that appears to be the result of severe psychological problems. It is most common in females, with males accounting for only 5–10 percent of all cases. The condition is most typical in adolescent Caucasian women whose weight is roughly 30 percent below normal lev-

•**FIGURE 19-4**
Nutritional and Metabolic Disorders.

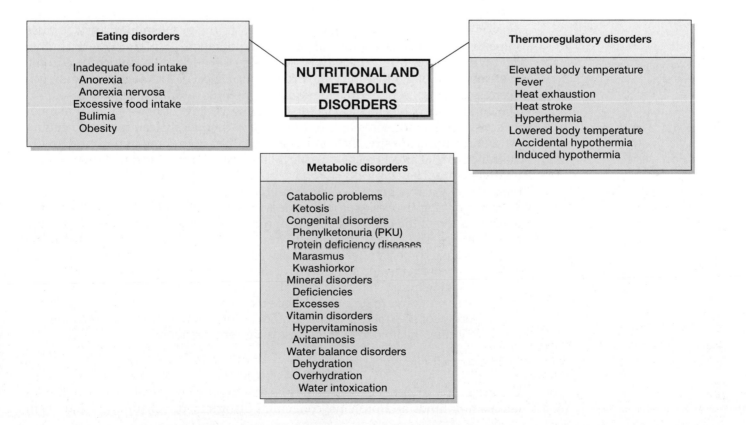

Eating disorders

Inadequate food intake
  Anorexia
  Anorexia nervosa
Excessive food intake
  Bulimia
  Obesity

NUTRITIONAL AND METABOLIC DISORDERS

Thermoregulatory disorders

Elevated body temperature
  Fever
  Heat exhaustion
  Heat stroke
  Hyperthermia
Lowered body temperature
  Accidental hypothermia
  Induced hypothermia

Metabolic disorders

Catabolic problems
  Ketosis
Congenital disorders
  Phenylketonuria (PKU)
Protein deficiency diseases
  Marasmus
  Kwashiorkor
Mineral disorders
  Deficiencies
  Excesses
Vitamin disorders
  Hypervitaminosis
  Avitaminosis
Water balance disorders
  Dehydration
  Overhydration
  Water intoxication

els. Although very obviously underweight, patients are convinced that they are too fat and refuse to eat normal amounts of food.

Young anorexic women may starve themselves down to a weight of 30–35 kg (66–77 lbs). Dry skin, peripheral edema (fluid accumulation), an abnormally low heart rate and blood pressure, a reduction in bone and muscle mass, and a lack of menstrual cycles are common symptoms. Some of the changes, such as decreased bone mass, are permanent. Treatment is difficult, and only about half who regain normal weight stay there for more than 5 years. Death rates from severe cases range from 10 to 15 percent.

Excessive Food Intake

Bulimia. Bulimia (bu-LĒM-ē-ah) is a disorder caused by an increase in the sensation of hunger. It is characterized by feeding binges, followed by vomiting, laxative use, or both. In this condition, the individual goes on an "eating binge" that may involve a meal that lasts 1–2 hours and may include 20,000 or more calories. The meal is followed by induced vomiting, usually along with the use of laxatives (to promote the movement of material through the digestive tract), and *diuretics* (drugs that promote fluid loss in the urine).

Bulimia is more common than anorexia nervosa, and generally involves women of the same age group. However, because individuals with bulimia may have normal body weight, many are not diagnosed until they are in their 30s–40s. The health risks of bulimia result from (1) cumulative damage to the stomach, esophagus, oral cavity, and teeth by repeated vomiting and exposure to stomach acids; (2) electrolyte imbalances resulting from the loss of sodium and potassium ions in the gastric juices, diarrhea, and urine; (3) edema; and (4) cardiac arrhythmias. Bulimia has been linked with depression.

Obesity. Obesity is defined as a condition of being 20 percent over ideal weight, because it is at this point that serious health risks appear. On that basis, some 20–30 percent of men and 30–40 percent of women in the U.S. can be considered obese. Basically, obese individuals take in more food energy than they are using.

There are two major categories of obesity: regulatory obesity and metabolic obesity. **Regulatory obesity**, the most common form, results from a failure to regulate food intake so that appetite, diet, and activity are in balance. In most instances, there is no obvious cause, although in rare cases the problem may arise because some disorder, such as a tumor, affects the hypothalamic centers in the brain that deal with hunger and satiety (the sensation of feeling full). Typically, chronic overeating is thought to result either from neurological or sociological factors, such as stress, neurosis, long-term habits, family or ethnic traditions, or from inactivity. The importance of biochemical factors is highlighted by the observation that medications such as antidepressants or *Depo-Provera* frequently lead to weight gain.

In **metabolic obesity**, the condition is secondary to some underlying bodily malfunction that affects cell and tissue metabolism. Cases of metabolic obesity are relatively rare. They typically involve chronic hypersecretion or hyposecretion of metabolically active hormones, such as glucocorticoids.

Categorizing an obesity problem is less important in a clinical setting than is the determination of the degree of obesity and the number and severity of the related complications. The affected individuals are at a high risk of developing diabetes, hypertension, and coronary artery disease as well as gallstones, thromboemboli, hernias, arthritis, varicose veins, and some forms of cancer. A variety of treatments may be considered, ranging from behavior modification, nutritional counseling, psychotherapy, and exercise programs to surgical methods that reduce the size of the stomach or bypass a portion of the small intestine. Unfortunately, more Americans are becoming obese, reflecting the limited effectiveness of current treatments.

METABOLIC DISORDERS

Disorders of nutrient metabolism may be due to problems associated with the catabolism of lipids and proteins, congenital disorders, or consuming insufficient

or excessive amounts of calories or nutrients (carbohydrates, lipids, proteins, minerals, vitamins, and water). We will discuss most of these disorders here; problems with water gains and losses will be covered in Chapter 20.

Catabolic Problems

Figure 19-5• shows the typical proportions of metabolic reserves, or major organic compounds, in the body in terms of their total energy content (Calories). The smallest reserves are the carbohydrates. During starvation, or a fast, lipids and proteins are broken down to provide the carbohydrates needed for aerobic metabolism (see Figure 19-2).

Ketosis. The breakdown, or catabolism, of the fatty acid portion of lipids and some of the amino acids that make up proteins results in the production of organic acids called **ketone bodies**. Although some of these ketone bodies are used for ATP production, a large fraction are not used and accumulate in body fluids, a condition called **ketosis** (kē-TŌ-sis). During prolonged starvation, the build up of ketone bodies in the blood reduces the pH of the blood. This acidification of the blood is called **ketoacidosis** (kē-tō-as-i-DŌ-sis). In severe cases, the pH may drop below 7.05, and cause *coma* (unconsciousness and unresponsiveness), cardiac arrhythmias, and death.

In Type 1 diabetes mellitus, most tissues cannot utilize glucose because of a lack of insulin. ∞ *[p. 310]* Most cells then survive by breaking down lipids and proteins, and, as a byproduct, produce large numbers of ketone bodies. The resulting drop in blood pH produces a condition called *diabetic ketoacidosis*, the most common and life-threatening form of ketoacidosis.

Congenital Disorders

Phenylketonuria. Phenylketonuria (fen-il-kē-tō-NOO-rē-a), or **PKU**, is one of about 130 disorders, called *inborn errors of metabolism*, that have been traced to the lack of specific enzymes. Individuals with PKU are missing the enzyme responsible for the conversion of the amino acid phenylalanine to another amino acid, tyrosine. This reaction is a necessary step in the synthesis of tyrosine, an important component of many proteins and the structural basis for melanin, epinephrine, and norepinephrine. If PKU is undetected and untreated, concentrations of phenylalanine in the blood may reach levels 6–7 times greater than normal. At such high levels central nervous system development is inhibited and severe brain damage, including mental retardation, results.

Fortunately, this condition is detectable shortly after birth, because it produces elevated levels of phenylalanine in the blood and *phenylketone*, a metabolic byproduct, in the blood and urine of the newborn infant. Treatment consists of controlling the amount of phenylalanine in the diet while plasma concentrations are monitored. This treatment is most important in infancy and childhood, when the nervous system is developing. Once the child has grown, dietary restriction of phenylalanine can be relaxed, except during pregnancy. A pregnant woman with PKU must protect the fetus from high levels of phenylalanine by following a strict diet that must actually begin before the pregnancy occurs.

Although the dietary restrictions are more relaxed for adults than for children, those with PKU must still monitor the ingredients used in the preparation of their meals. For example, one popular artificial sweetener, Nutrasweet®, consists of phenylalanine and aspartic acid. The consumption of food or beverages that contain this sweetener can therefore cause problems for PKU sufferers. Because tyrosine cannot be synthesized from dietary phenylalanine, the diet of these patients must also contain adequate amounts of tyrosine.

Protein Deficiency Diseases

Regardless of the energy content of the diet, if it is deficient in essential amino acids, the individual will be malnourished to some degree. In a **protein deficiency disease**, protein synthesis decreases throughout the body. As protein synthe-

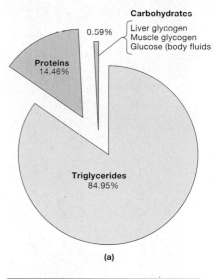

Carbohydrates
0.59%
Liver glycogen
Muscle glycogen
Glucose (body fluids)

Proteins
14.46%

Triglycerides
84.95%

(a)

•**FIGURE 19-5**
Metabolic Reserves.
Metabolic, or energy, reserves of carbohydrates, lipids, and proteins in a 70 kg (154 lb) individual.

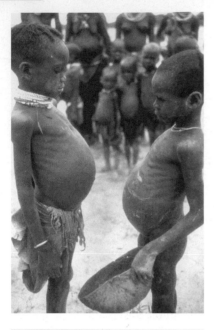

•FIGURE 19-6
Children with Kwashiorkor.

sis in the liver fails to keep pace with the breakdown of plasma proteins, the osmotic concentration of the blood falls. This reduction results in more water moving out of the capillaries and into interstitial spaces, the peritoneal cavity, or both. The longer the individual remains in this state, the more severe the ascites and edema that result. ∞ *[p. 457, 371]*

This clinical picture may occur in developing countries where dietary protein is often scarce or very expensive. Growing infants suffer from **marasmus** (ma-RAZ-mus) when deprived of adequate proteins and calories. **Kwashiorkor** (kwash-ē-OR-kor) occurs in children whose protein intake is inadequate, even if the caloric intake is acceptable (Figure 19-6•). In each case, additional complications include damage to the developing brain. It is estimated that more than 100 million children worldwide are suffering from protein deficiency diseases. War and civil unrest that disrupt local food production and distribution have been more important in producing recent famines, rather than a shortage of food itself.

Mineral Disorders

Mineral disorders due to deficiencies are generally more common than conditions caused by an excess of minerals. *Mineral deficiency disorders* may be caused by a deficiency of the mineral in the diet, or with problems in its uptake in the digestive tract. Table 19-1 lists important minerals, the effects of their deficiency (or excess), and the names of specific deficiency disorders. For example, a deficiency of calcium and phosphorus intensified by inadequate levels of vitamin D may lead to rickets, a condition characterized by soft bones (see Chapter 8, p. 175), iron deficiency anemia results from inadequate amounts of iron (see Chapter 14, p. 334), and a goiter, or enlarged thyroid gland, develops when iodine levels are low (see Chapter 13, p. 307).

Vitamin Disorders

Vitamin disorders due to excessive dietary intake are mostly characteristic of fat-soluble vitamins (D, A, K, and E). However, both fat-soluble and water-soluble vitamins may be associated with deficiency disorders. This difference primarily occurs because any excessive amounts of water-soluble vitamins are excreted in the urine, rather than stored. In contrast, excessive amounts of fat-soluble vitamins can be stored in the large lipid reserves of the body.

Hypervitaminosis (hī-per-vī-ta-min-Ō-sis) occurs when the amount ingested exceeds the body's abilities to store, use, or excrete a vitamin. Symptoms of *vitamin toxicity* can result when massive amounts (ten to thousands of times the recommended daily allowance) of fat-soluble vitamins are ingested. Vitamin A toxicity is the most common condition. In children, a single enormous overdose can produce nausea, vomiting, headache, dizziness, lethargy, and even death. Chronic overdose can lead to hair loss, joint pain, hypertension, weight loss, and liver enlargement.

Vitamin D toxicity can occur, but vitamins K and E are rarely if ever toxic. Inadequate amounts of dietary vitamins leads to **avitaminosis** (ā-vī-ta-min-Ō-sis), or **vitamin deficiency diseases**. Except for vitamins B₁₂ and C, which are stored in significant amounts in the body, this condition is more likely to occur with water-soluble vitamins (see Table 19-2). An inability to absorb a vitamin from the digestive tract, inadequate storage, or excessive demand may each produce avitaminosis. When reserves are depleted, scurvy (vitamin C deficiency) or pernicious anemia (vitamin B₁₂ deficiency) may eventually occur. With inadequate vitamin D, calcium absorption is low and osteoporosis more common.

Problems with Water Balance

Water gains and losses must be kept in balance to avoid problems with a variety of systems. **Dehydration**, a state of water depletion, results when water losses to the environment exceed water gains from drinking, eating, and metabolic water production. The loss of body water is associated with severe thirst, dryness and wrinkling of the skin, and a fall in blood volume and blood pressure. Eventually, circulatory shock develops, and the individual may die without treatment. Treatment involves providing fluids by mouth or intravenous infusions.

Overhydration, or *water excess*, causes a dilution of minerals in the body. Because the water content of the body cannot easily be determined, a condition of water excess is based on measurements of the concentration of sodium ions, one of the major ions in the body. Water excess may be caused by ingesting a large volume of fluid or infusion of intravenous fluids, or an inability to eliminate excess water due to various disorders. Overhydration can have affects on the function of the CNS. In its early stages, the individual behaves as if drunk on alcohol, a condition called *water intoxication*. If untreated, water intoxication can rapidly progress from confusion to hallucinations, convulsions, coma, and death. Treatment involves the use of *diuretics*, substances that increase the production of urine, and the infusion of a concentrated salt solution to increase the level of sodium ions to normal levels.

✔ Two eating disorders are anorexia nervosa and bulimia. What is the major difference between them?

✔ When carbohydrates are in short supply, lipids and proteins may be used as sources of energy. What byproduct is made when they are broken down, or catabolized?

✔ What do rickets, iron deficiency anemia, and goiter have in common?

ENERGY

When the chemical bonds holding glucose or other molecules together are broken, energy is released. The unit of energy measurement is the **calorie** (KAL-o-rē). Because heat energy is easily measured, a calorie is defined as the amount of energy required to raise the temperature of 1 gram of water 1 degree Celsius. This is such a small amount of energy in relation to that involved in keeping a 70-kg (154-lb) human alive, that a larger unit called the **kilocalorie** (KIL-o-kal-o-rē) (kc), or simply **Calorie** (with a capital C), is used instead. Each Calorie represents 1,000 calories or the amount of energy needed to raise the temperature of 1 *kilo*gram of water 1 degree Celsius. When you turn to the back of a dieting guide to check the caloric value of various foods, the numbers indicate Calories, not calories.

FOOD AND ENERGY

As we have seen, organic molecules are broken down, or catabolized, in living cells to produce carbon dioxide and water, and energy is released in the process. The catabolism of carbohydrates, proteins, and fats releases different amounts of energy. For example, fats release the most energy, roughly 9.46 Calories per gram (C/g). In contrast, carbohydrates release 4.18 C/g, and proteins release 4.32 C/g. Most foods are mixtures of fats, proteins, and carbohydrates, and the values in a "Calorie counter" vary as a result.

METABOLIC RATE

The **metabolic rate** of an individual is a measure of energy use. The metabolic rate changes according to the activity under way—the amounts of energy required for running and sleeping are obviously quite different. The **basal metabolic rate** (**BMR**) represents the minimum, or resting, energy use of an awake, alert person. An "average" individual has a BMR of 70 C per hour or about 1,680 C per day. Factors that can influence the BMR include age, sex, physical condition, and body weight.

The actual daily energy expenditures for individuals vary, depending on their activities. For example, a person leading a sedentary life may have near-basal energy demands, but a single hour of swimming can increase the daily caloric requirements by 500 C or more. If the daily energy intake exceeds the total energy demands, the excess will be stored, primarily as fats in adipose tissue. If the daily use of calories exceeds the amount in the diet, there will be a net reduction in the body's energy reserves and a corresponding loss in weight.

TEMPERATURE REGULATION

The BMR is an estimate of the rate of energy use. Our cells capture only a part of that energy as ATP, and the rest is "lost" as heat. However, heat loss serves an important homeostatic purpose; it raises the body's temperature to the narrow range within which our enzyme systems work best. (This range is about half a degree above and below a normal, average body temperature of 37° C, or about 1° above

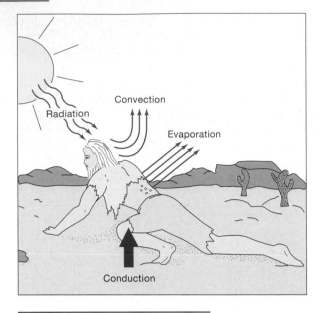

•FIGURE 19-7
Routes of Heat Gain and Loss.

and below 98.6° F.) Our bodies have various strategies to keep body temperatures within acceptable limits, regardless of the environmental conditions.

Mechanisms of Heat Transfer

Heat is exchanged with the environment in four different ways (Figure 19-7•).

1. **Radiation**. Warm objects lose heat as radiation. When we feel the heat from the sun, we are experiencing radiation. Our bodies lose heat the same way, and over half our heat loss occurs through radiation.

2. **Conduction**. Conduction is the direct transfer of energy through physical contact. When you sit down on a cold plastic chair in an air-conditioned room, you are immediately aware of this process. Conduction is usually not an effective mechanism for gaining or losing heat.

3. **Convection**. Convection is the result of conductive heat loss to the air that overlies the surface of the body. Warm air is lighter than cool air, and so it rises. As the body conducts heat to the air next to the skin, that air warms and rises; cooler air replaces it, and as it in turn becomes warmed, the cycle repeats.

4. **Evaporation**. Energy is absorbed when water evaporates (changes from a liquid to a vapor). The evaporation of water in the *perspiration* from sweat glands can remove large amounts of heat energy.

To maintain a constant body temperature, a person must lose heat as fast as heat is produced by metabolic reactions. For example, in a person at rest, over half the heat loss occurs through radiation, 20 percent through evaporation, 15 percent through convection, and the rest through conduction. To alter these rates, many different systems must be coordinated. These adjustments are made by the **heat-loss center** and the **heat-gain center** of the hypothalamus. The heat-loss center adjusts activity in the parasympathetic division of the autonomic nervous system, and the heat-gain center directs its responses through the sympathetic division. The overall effect is to control temperature by influencing two events: the rate of heat production and the rate of heat loss to the environment.

Promoting Heat Loss

When the temperature at the heat-loss center exceeds its normal setting, three major effects result: (1) Blood vessels dilate, and warm blood flows to the surface of the body. This causes the skin to take on a reddish color, to show a rise in temperature, and to increase heat loss through radiation and convection. (2) Sweat glands are stimulated, perspiration flows across the body surface, and evaporative heat losses accelerate. (3) The respiratory centers are stimulated, and breathing deepens. Often the individual begins breathing through the mouth, increasing evaporative losses through the lungs.

Restricting Heat Loss

The function of the heat-gain center of the brain is to prevent **hypothermia** (hī-pō-THER-mē-uh) or below-normal body temperature. When body temperature falls below acceptable levels, the heat-loss center is inhibited, and the heat-gain center is activated. Blood flow to the skin decreases, and with the circulation restricted, the skin may become pale or bluish. In addition, the pattern of blood flow changes. In warm weather, blood flows in a network of veins just under the skin. In cold weather, blood is diverted to a network of deep veins that lie beneath an insulating layer of subcutaneous fat.

Promoting Heat Production

In addition to conserving heat, the heat-gain center has two mechanisms for increasing the rate of heat production. During *shivering*, the rate of heat generation by skeletal muscles can increase by as much as 400 percent. Heat can also be produced by the release of hormones that increase the metabolic activity of all tissues. Both epinephrine from the adrenal gland and thyroxine from the thyroid gland have such effects.

THERMOREGULATORY DISORDERS

Failure to control body temperature effectively can result in a variety of physiological problems, as indicated in Figure 19-8•.

Elevated Body Temperatures

Fevers. A fever, or *febrile* condition, exists when the body temperature is maintained above 37.2°C (99°F). Fevers were introduced in Chapter 16 as one aspect of the nonspecific defenses (p. 388). A fever, known as **pyrexia** (pī-REK-sē-uh), often accompanies infections. Bacteria or other pathogens release or stimulate the production of substances called **pyrogens**, or "fever-producers," which reset the temperature control center in the brain and cause a rise in body temperature. A fever may also result from heat exhaustion or heat stroke.

Heat Exhaustion. In **heat exhaustion**, also known as *heat prostration*, the individual has problems maintaining blood volume under conditions of high environmental temperatures. The heat-loss center stimulates sweat glands, increasing their secretions and heat loss by evaporation. As fluid losses through perspiration increase, blood volume decreases. Normally, a decline in blood pressure would be countered by the constriction of peripheral blood vessels (vasoconstriction). Stimulation of the heat-loss center, however, causes a vasodilation (widening) of peripheral blood vessels. As blood flow to the brain declines, headache, nausea, and collapse follow. Treatment is simple: Provide fluids, salts, and a cooler environment.

Heat Stroke. **Heat stroke** is more serious and may follow an untreated case of heat exhaustion. The thermoregulatory center of the brain stops functioning, the sweat glands stop working, and the skin becomes hot and dry. Unless the situation is recognized in time, body temperature may climb to 41°–45°C (106°–113°F). Such excessive temperatures, called **hyperthermia**, will quickly disrupt body systems

The human body can maintain homeostasis near the equator as well as near the poles.

Condition	F°	C°	Thermoregulatory capabilities	Major physiological effects
Heat stroke	114		Severely impaired	Death
		44		Proteins denature, tissue damage accelerates
CNS damage	110	42		Convulsions
	106		Impaired	Cell damage
Disease-related fevers Severe exercise Active children	102	40	Effective	Disorientation
		38		Systems normal
Normal range (oral)	98	36		
Early mornings in cold weather Severe exposure	94	34	Impaired	Disorientation
	90	32		Loss of muscle control
	86	30	Severely impaired	Loss of consciousness
Hypothermia for open heart surgery	82	28		Cardiac arrest
	78	26		Skin turns blue
	74	24	Lost	Death

•**FIGURE 19-8**
Normal and Abnormal Variations in Body Temperature.

and destroy brain, liver, skeletal muscle, and kidney cells. Acute hyperthermia, as seen during heat stroke, in certain diseases, or in an occasional marathon runner, is life-threatening. Immediate treatment involves cooling the individual in an ice bath (to increase conduction) or giving alcohol rubs (to increase evaporation) in combination with the administration of *antipyretic drugs*, such as aspirin or acetaminophen.

Lowered Body Temperatures

Environmental Hypothermia. If body temperature declines significantly below normal levels, the thermoregulatory system begins to lose sensitivity and effectiveness. Cardiac output and respiratory rate decrease, and if body temperature falls below 28°C (82°F), cardiac arrest is likely. The individual then has no heartbeat, no respiratory rate, and no response to external stimuli, even painful ones. Body temperature continues to decline, and the skin appears blue or pale.

At this point, the individual looks dead. But because metabolic activities have decreased in all systems, the victim may occasionally survive, even after several hours have elapsed. Treatment consists of cardiopulmonary support and re-warming, both external and internal. The skin can be warmed up to 45°C (110°F) without damage; warm baths or blankets can be used. One effective method of raising internal temperatures involves the introduction of warm saline solution into the peritoneal cavity.

Hypothermia is a significant risk for those engaged in water sports, and its presence may complicate treatment of a drowning victim. Water absorbs heat roughly 27 times faster than does air, and the body's heat-gain mechanisms are unable to keep pace over long periods or when faced with a large temperature gradient. But hypothermia in cold water does have a positive side. On several occasions, small children who have drowned in cold water have been successfully revived after periods of up to 4 hours. Children lose body heat quickly, and their systems stop functioning very quickly as body temperature declines. This rapid drop in temperature prevents the oxygen-starvation and tissue damage that would otherwise occur when breathing stops.

Induced Hypothermia. Hypothermia may be intentionally produced during surgery to reduce the metabolic rate of a particular organ or of the entire body. In this process, the individual is first anesthetized to prevent the shivering that would otherwise slow the cooling. For example, during open-heart surgery the body is typically cooled to 25°–32°C (79°–89°F). This cooling reduces the metabolic needs throughout the body, which receives blood from an external pump and oxygenator (a heart/lung machine). Because the heart must be stopped and cannot be supplied with blood, it is maintained at a temperature below 15°C (60°F) during the operation. At this temperature, the cardiac muscle can survive several hours of ischemia (inadequate blood supply) without damage. Infants too small to use a heart/lung machine during heart surgery can survive for up to an hour without a heartbeat if they are kept hypothermic over that period.

✔ How would the BMR (basal metabolic rate) of a pregnant woman compare with her BMR in the nonpregnant state?

✔ Which glands are responsible for evaporative cooling of the body?

✔ What effect would constriction of blood vessels have on body temperature on a hot day?

CHAPTER REVIEW

Key Words

adenosine diphosphate (ADP): An adenosine molecule with two phosphate groups attached.

adenosine triphosphate (ATP): A high-energy compound consisting of an adenosine molecule with three phosphate groups attached; the third is attached by a high-energy bond.

aerobic: Requiring the presence of oxygen.

anabolism (a-NAB-ō-lizm): The synthesis of large and complex organic compounds from smaller, simpler compounds.

basal metabolic rate: The resting metabolic rate of a normal fasting subject under homeostatic conditions.

Calorie (C): The amount of heat required to raise the temperature of 1 kg of water 1°C; the unit commonly used in describing the energy values of foods.

catabolism (ka-TAB-ō-lizm): The breakdown of complex organic molecules into simpler compounds accompanied by the release of energy.

cellular respiration: The complete breakdown of glucose into carbon dioxide and water; a process that yields large amounts of ATP but requires mitochondria and oxygen.

glycogen (GLĪ-kō-jen): A polysaccharide that represents an important energy reserve; consists of a long chain of glucose molecules.

hypothermia (hī-pō-THER-mē-uh): Below-normal body temperature.

metabolism (me-TAB-ō-lizm): The total of all the biochemical processes under way within the human body at a given moment; includes anabolism and catabolism.

mineral: A solid, inorganic chemical element or compound naturally occurring in the earth.

nutrient: Any chemical component of food that supplies energy and/or materials for the body.

vitamin: An essential organic nutrient that functions as a component of enzymes in metabolism.

Selected Clinical Terms

avitaminosis (ā-vī-ta-min-Ō-sis): A vitamin deficiency disease.

bulimia (bu-LĒM-ē-ah): An abnormal increase in hunger characterized by binge eating that is followed by vomiting or laxative use.

dehydration: The loss of water from the body or a tissue.

heat exhaustion: A malfunction of the thermoregulatory system caused by excessive fluid loss in perspiration.

heat stroke: A condition in which the thermoregulatory center stops functioning and body temperature rises uncontrollably.

hypervitaminosis (hī-per-vī-ta-min-Ō-sis): A disorder caused by ingestion of excessive quantities of one or more vitamins.

hypothermia: A below-normal body temperature.

ketoacidodis (kē-tō-as-i-DŌ-sis): Acidification of the blood due to the presence of ketone bodies.

ketone bodies: Organic acids produced during the breakdown of lipids and some amino acids.

obesity: An increase in body weight, due to fat accumulation, to more than 20 percent above standard values.

phenylketonuria: An inherited metabolic disorder resulting from an inability to convert phenylalanine to tyrosine.

pyrexia (pī-REK-sē-uh): A fever.

Study Outline

INTRODUCTION (p. 467)

1. Cells in the human body are chemical factories that break down organic molecules from food to obtain energy.

2. In addition to energy, **nutrients**, the chemical components of food, are also used as building blocks for new cellular structures.

METABOLISM (pp. 467–470)

1. Metabolism refers to all of the chemical reactions of the body. These processes either break down organic compounds, the reactions of **catabolism**, or build them up, the reactions of **anabolism**.

2. In general, cells break down excess carbohydrates first, then lipids, and conserve amino acids. Only about 40 percent of the energy released through catabolism is captured in ATP; the rest is released as heat. *(Figure 19-1)*

3. Cells synthesize new compounds (1) to perform structural maintenance or repair, (2) to support growth, and (3) to produce secretions.

CARBOHYDRATE METABOLISM (p. 468)

4. Most cells generate ATP and other high-energy compounds by the breakdown of carbohydrates through **cellular respiration**.

5. Aerobic metabolism provides most of the ATP used by typical cells.

6. Oxygen-requiring reactions in the mitochondria produce much more ATP than the **anaerobic** reactions in the cytoplasm.

7. Through anabolic reactions, surplus glucose molecules are stored in liver and skeletal muscle cells as **glycogen**.

OTHER ENERGY-PRODUCING PATHWAYS (p. 469)

8. Cells can break down other nutrients to provide ATP if supplies of glucose are limited. *(Figure 19-2)*

LIPID METABOLISM (p. 469)

9. During lipid catabolism, lipids are broken down into simpler compounds that can either be converted into pyruvic acid or enter the mitochondria directly.

10. Cells can shift to lipid-based energy production when glucose reserves are limited.

11. In the synthesis of lipids almost any organic molecule can be used to form glycerol. **Essential fatty acids** cannot be synthesized and must be included in the diet.

PROTEIN METABOLISM (p. 469)

12. If other energy sources are inadequate, amino acids can be used. The amino group is removed in the liver, and the remaining carbon skeleton can enter the mitochondria to generate ATP.

13. Roughly half the amino acids needed to build proteins can be synthesized. However, there are 10 **essential amino acids** that need to be acquired through the diet.

DIET AND NUTRITION (pp. 470–474)

1. Nutrition is the absorption and use of nutrients from food. A *balanced diet* contains all of the ingredients necessary to maintain homeostasis; it prevents **malnutrition**.

THE SIX BASIC FOOD GROUPS (p. 471)

2. The six basic food groups are *milk and dairy* products; *meat, dry beans, eggs, and nuts; vegetables; fruits; bread and cereals;* and *fats, oils, and sweets.* (Figure 19-3)

MINERALS, VITAMINS, AND WATER (p. 472)

3. Minerals act as components of enzymes in various metabolic reactions. They also contribute to the balance of body fluids, and they play a role in nerve impulse conduction, neurotransmitter release, muscle contraction, skeleton building and maintenance, gas transport, and waste removal. *(Table 19-1)*

4. Vitamins are needed in very small amounts. Vitamins A, D, E, and K are the **fat-soluble vitamins**. Unlike fat-soluble vitamins, **water-soluble vitamins** are not stored in the body. *(Table 19-2)*

5. Daily water requirements average about 2,500 ml (10 cups).

DIET AND DISEASE (p. 473)

6. A balanced diet can improve general health. Most Americans consume too many calories, mostly in the form of lipids (fats).

NUTRITIONAL AND METABOLIC DISORDERS (pp. 475–479)

1. Nutritional and metabolic disorders are generally caused by the ingestion of too much or too few nutrients, or problems in metabolism of specific nutrients. *(Figure 19-4)*

EATING DISORDERS *(p. 475)*

2. *Appetite* is a desire for food; *hunger* is a need for food.

3. Anorexia is a loss of appetite, which results in a low food intake. **Anorexia nervosa** is a serious, self-induced starvation most common in adolescent females.

4. Excessive food intake occurs in **bulimia**, a condition of increased hunger followed by induced vomiting, the use of laxatives, and the use of diuretics.

5. Individuals 20 percent over ideal weight are considered to be obese. **Regulatory obesity** results from an inability to balance food intake with appetite, diet, and activity. Although rarer, **metabolic obesity** may occur when there is a hormone imbalance.

METABOLIC DISORDERS *(p. 476)*

6. Metabolic disorders may be associated with lipid and protein catabolism, congenital disorders, and a lack of calories or specific nutrients. *(Figure 19-5)*

7. Ketosis is a condition in which there is a high level of **ketone bodies** in body fluids.

8. Phenylketonuria (PKU) is an inborn error of metabolism. In this condition, high levels of the amino acid phenylalanine build up in the body and interfere with the development of the nervous system. If undiagnosed early in life, mental retardation results.

9. Protein deficiency diseases affect protein synthesis by the liver, resulting in a lack of blood proteins and the leakage of fluid into tissues (*edema*) and the abdominal cavity (*ascites*). *(Figure 19-6)*

10. *Mineral deficiency disorders* may be due to inadequate mineral levels in the diet or problems in their uptake in the digestive tract. *(Table 19-1)*

11. Hypervitaminosis is a condition in which the ingested amount of a vitamin is greater than the body's abilities to store, use, or excrete a vitamin. Inadequate vitamin levels result in **avitaminosis** and **vitamin deficiency diseases**.

ENERGY *(pp. 479–482)*

1. The energy content of food is usually expressed in **Calories** per gram (C/g).

FOOD AND ENERGY *(p. 479)*

2. The catabolism (breakdown) of lipids releases 9.46 C/g, about twice the amount as equivalent weights of carbohydrates and proteins.

METABOLIC RATE *(p. 479)*

3. The total of all the anabolic and catabolic processes under way in the body represents the **metabolic rate** of an individual. The **basal metabolic rate (BMR)** is the rate of energy usage at rest.

TEMPERATURE REGULATION *(p. 479)*

4. The homeostatic regulation of body temperature involves four processes of heat exchange with the environment: *radiation, conduction, convection,* and *evaporation. (Figure 19-7)*

5. The hypothalamus, which acts as the body's thermostat, contains the **heat-loss center** and the **heat-gain center**.

6. Physiological mechanisms for increasing heat loss include peripheral blood vessel dilation, increased perspiration, and increased respiration.

7. Body heat may be conserved by decreased blood flow to the skin. Heat may be generated by *shivering* or the release of hormones.

THERMOREGULATORY DISORDERS *(p. 481)*

8. Temperature disorders cause a variety of dangerous problems with physiological systems. *(Figure 19-8)*

9. A fever, or **pyrexia**, is caused by **pyrogens** ("fever producers") in the circulation, which reset the temperature control center in the brain.

10. Heat exhaustion, or *heat prostration*, occurs under conditions of high temperature, when the loss of fluids through perspiration results in a lowering of blood volume and vasodilation. A life-threatening condition called **heat stroke**, or **hyperthermia**, may follow heat exhaustion.

11. Hypothermia may occur accidentally or be induced.

Review Questions

MATCHING

Match each item in Column A with the most closely related item in Column B. Use letters for answers in the spaces provided.

	Column A		Column B
___	1. glycogen	a.	products of lipid catabolism
___	2. glycerol and fatty acids	b.	essential amino acid
___	3. anabolism	c.	below-normal body temperature
___	4. linoleic acid	d.	high-energy compound of cells
___	5. lysine	e.	fat-soluble vitamins
___	6. vitamins	f.	water-soluble vitamins
___	7. ammonia	g.	congenital disorder
___	8. urea	h.	nitrogenous waste
___	9. B complex and vitamin C	i.	protein deficiency disease
___	10. calorie	j.	production of protein from amino acids
___	11. A, D, E, K	k.	organic compounds needed in small amounts
___	12. hypothermia	l.	toxic nitrogen-containing compound
___	13. ATP	m.	unit of energy
___	14. phenylketonuria	n.	storage form of glucose
___	15. kwashiorkor	o.	essential fatty acid

MULTIPLE CHOICE

16. Cells build new organic compounds to _____ .
(a) maintain and repair cell structures
(b) support growth
(c) produce secretions   (d) a, b, and c are correct

17. The absorption of the proper amount and kind of nutrients from the diet is called _____ .
(a) digestion   (b) nutrition
(c) malnutrition   (d) catabolism

18. The phase of cellular respiration that does not require oxygen occurs in the _____ .
(a) nucleus   (b) cytoplasm
(c) mitochondria   (d) ribosome

19. _____ are the primary source of energy in cells.
(a) carbohydrates (b) lipids
(c) proteins (d) nucleic acids

20. The daily water ration is obtained by all the following except _____ .
(a) eating (b) drinking
(c) chemical reactions within the mitochondria
(d) exercise

21. Over half the heat loss from our bodies is due to _____ .
(a) radiation (b) convection
(c) conduction (d) evaporation

22. The basal metabolic rate (BMR) represents the _____ .
(a) maximum energy used when exercising
(b) minimum amount of energy used in light exercise
(c) minimum amount of energy used by an awake, resting individual
(d) total of a and b

23. Ketosis results from all of the following except _____ .
(a) the breakdown of lipids (b) the breakdown of proteins
(c) diabetes mellitus
(d) the breakdown of carbohydrates

24. Fat soluble vitamins _____ .
(a) cannot be stored in the body
(b) can cause hypervitaminosis
(c) include the B vitamins
(d) are rapidly excreted in the urine

TRUE/FALSE

____ 25. The hypothalamus contains the body's temperature-regulating centers.

____ 26. The major source of ATP for cells is glucose.

____ 27. More ATP is produced by the anaerobic phase of cellular respiration than by the aerobic phase.

____ 28. The energy yield of equal weights of carbohydrates, lipids, and proteins are about the same.

____ 29. Blood flow to the skin increases during cold weather.

____ 30. Unlike vitamins, minerals are inorganic substances.

____ 31. Our bodies cannot synthesize any minerals at all.

____ 32. High levels of ketone bodies cause a rise in blood pH.

SHORT ESSAY

33. Define the terms *metabolism*, *anabolism*, and *catabolism*.

34. Why are vitamins and minerals essential components of the diet?

35. How can the food pyramid be used to help an individual ensure that his or her diet contains the proper amounts and kinds of nutrients? Why are the dietary fats, oils, and sugars at the top of the pyramid and the bread and cereal at the bottom?

36. How is the brain involved in body temperature regulation?

37. Why do protein deficiency diseases result in the accumulation of fluid in body tissues and the abdominopelvic cavity?

38. It's a hot, sunny day and you are lying on a sandy beach. Describe the ways in which you are exchanging heat with the environment.

APPLICATIONS

39. Individuals suffering from anorexia nervosa typically have a relatively small heart, slow heart rate, and low blood pressure. These problems can eventually lead to death from heart failure. How does anorexia nervosa lead to these changes?

40. You are at a friend's outdoor barbecue on a hot summer day. One friend, who is drinking beer and sweating profusely, begins to complain of a headache and indigestion; suddenly he faints. What has happened? How could you determine if this was a life-threatening condition? Did the beer play a role in his collapse?

✔ Answers to Concept Check Questions

(p. 470) **1.** Carbohydrates are the major source of energy in cells. After carbohydrates, lipids are utilized next. Proteins and their amino acids are seldom broken down for energy unless there are inadequate supplies of carbohydrates and lipids. **2.** The use of amino acids as an energy source first requires the removal of the amino group. The free amino group is first converted into a toxic ammonia molecule, which is then converted into a nontoxic urea molecule. The urea is excreted in the urine. **3.** *Nonessential* molecules are those that can be synthesized by the cell; *essential* molecules cannot be synthesized and must be supplied in the diet.

(p. 473) **1.** In terms of recommended servings, between 6 and 11 per day, the bread, cereal, rice, and pasta group is the most important. **2.** Foods that contain all the essential amino acids in nutritionally required amounts are said to contain complete proteins. Foods that are deficient in one or more of the essential amino acids contain incomplete proteins.

(p. 479) **1.** Anorexia nervosa is a condition in which inadequate amounts of food are eaten. Bulimia is a condition in which excessive amounts of food are eaten, followed by vomiting or laxative use. **2.** The breakdown of lipids and proteins results in the formation of organic acids called ketone bodies. Increased levels of ketone bodies affect the pH of body fluids, making them more acidic. **3.** They are all caused by a deficiency of minerals. Rickets is due to a shortage of calcium and phosphorus, anemia due to a shortage of iron, and goiter due to a shortage of iodine.

(p. 482) **1.** The BMR of a pregnant woman should be higher than the BMR of the woman in a nonpregnant state because of increased metabolism associated with support of the fetus as well as the added effect of fetal metabolism. **2.** Merocrine sweat glands release perspiration, which absorbs large amounts of body heat as it evaporates. **3.** Constriction of peripheral blood vessels would decrease blood flow to the skin and decrease the amount of heat that the body could lose. As a result, body temperature would increase.

CHAPTER
20

The Urinary System and Body Fluids

CHAPTER OBJECTIVES

1　Identify the components of the urinary system and their functions.

2　Describe the structure of the kidneys.

3　Describe the structure of the nephron and the processes involved in urine formation.

4　List and describe the factors that influence filtration pressure and the rate of filtrate formation.

5　Describe the changes that occur in the filtrate as it moves along the nephron and exits as urine.

6　Describe how the kidneys respond to changing blood pressure.

7　Describe the structures and functions of the ureters, urinary bladder, and urethra.

8　Discuss the process of urination and how it is controlled.

9　Explain the basic concepts involved in the control of fluid and electrolyte regulation.

10　Explain the buffering systems that balance the pH of the intracellular and extracellular fluids.

11　Describe the effects of aging on the urinary system.

12　Describe the general symptoms and signs of disorders of the urinary system.

13　Give examples of the various disorders of the urinary system.

14　Give examples of conditions resulting from abnormal renal function.

The primary role of the urinary system is "pollution control"—it removes waste products that build up in the blood. These wastes are then discharged from the body. The elimination of dissolved and solid wastes from the body is called *excretion*. Excretion is performed by the urinary system, the digestive system, and the respiratory system. The cardiovascular system links these systems together.

The digestive tract absorbs nutrients from food and excretes solid wastes, and the liver adjusts the nutrient concentration of the circulating blood. The cardiovascular system delivers nutrients from the digestive system and oxygen from the respiratory system to the various tissues of the body. As blood leaves these tissues it carries carbon dioxide and other cellular waste products to sites of excretion. The carbon dioxide is eliminated at the lungs, as described in Chapter 17. Most of the organic waste products are removed from the blood at the kidneys, which produce a solution called *urine*. The urine is then stored by the urinary bladder until it can be excreted.

SYSTEM BRIEF

The urinary system's primary function is waste excretion. However, it also performs other functions that are often overlooked. It regulates blood volume and pressure and balances blood pH and the concentrations of sodium, potassium, chloride, and other ions in the blood.

All these activities are necessary if the composition of blood is to be kept within acceptable limits. A disruption of any one of these processes will have immediate and potentially fatal consequences. This chapter examines the work of the urinary system and the regulation of urine production.

STRUCTURE OF THE URINARY SYSTEM

The organs of the urinary system are shown in Figure 20-1•. The two **kidneys** produce **urine**, a liquid containing water, ions, and small soluble compounds. Urine leaving the kidneys travels along the left and right **ureters** (ū-RE-terz) to the **urinary bladder** for temporary storage. When **urination**, or **micturition** (mik-tū-RI-shun), occurs, contraction of the muscular bladder forces the urine through the **urethra** and out of the body.

THE KIDNEYS

The kidneys are located on either side of the vertebral column, at about the level of the eleventh and twelfth ribs. The right kidney usually sits slightly lower than the left (see Figure 20-1•), and both kidneys lie between the muscles of the body wall and the peritoneal (abdominal) lining. This position is called **retroperitoneal** (re-trō-per-i-tō-NĒ-al) because the organs are behind the peritoneum (*retro-* means "behind").

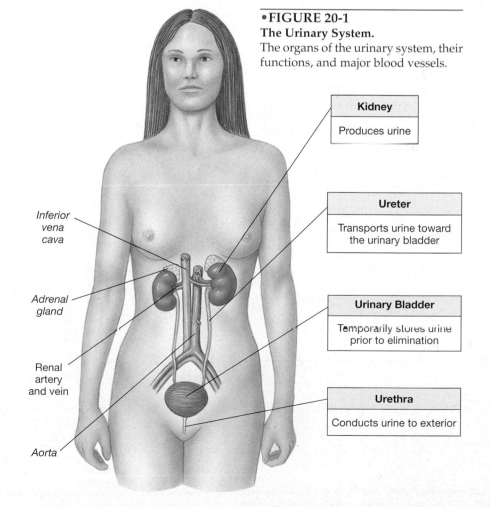

•FIGURE 20-1
The Urinary System.
The organs of the urinary system, their functions, and major blood vessels.

Inferior vena cava

Adrenal gland

Renal artery and vein

Aorta

Kidney
Produces urine

Ureter
Transports urine toward the urinary bladder

Urinary Bladder
Temporarily stores urine prior to elimination

Urethra
Conducts urine to exterior

The kidneys are held in place by the peritoneal lining and supporting connective tissues. Each kidney is also packed in a protective cushion of fatty tissue that helps absorb the jolts and shocks of daily existence. If a kidney is displaced, a condition called *floating kidney*, the ureters or renal blood vessels may become twisted or kinked, an extremely dangerous event.

KIDNEY STRUCTURE

Each kidney is shaped like a large kidney bean. A typical kidney measures 10 cm in length, 5.5 cm in width, and 3 cm in thickness. An indentation, called the **hilus**, is the point of entry for the renal artery and exit for the renal vein and ureter. (The adjective renal, comes from *renes*, which means "kidneys.") The surface of each kidney is covered by a capsule made of fibrous connective tissue.

Seen in section (Figure 20-2•), the kidney can be divided into an outer renal **cortex** and an inner renal **medulla**. The medulla contains 6 to 18 cone-shaped **renal pyramids**. Urine production occurs in the renal pyramids and overlying areas of renal cortex. Small ducts within the tips of each renal pyramid release urine into a cup-shaped drain, called a **minor calyx** (KA-liks; a cup, plural: *calyces*). Four or five minor calyces (KA-li-sēz) merge to form two to three **major calyces**, which combine to form a large, funnel-shaped chamber, the **renal pelvis**. The renal pelvis is connected to the ureter at the hilus of the kidney.

THE URINE-PRODUCING UNITS

The basic urine-producing units in the kidney are called **nephrons** (NEF-rons). There are about 1.25 million nephrons in each kidney. The location and structure of one nephron is shown in Figure 20-2•. Each nephron consists of a microscopic tube, or *renal tubule,* that is roughly 50 mm (2 in.) long. The tubule has two *convoluted* (coiled or twisted) segments separated by a simple U-shaped segment. The convoluted segments are in the cortex of the kidney, and the tube extends

---

•FIGURE 20-2
Structure of the Kidney and a Urine-Producing Nephron.

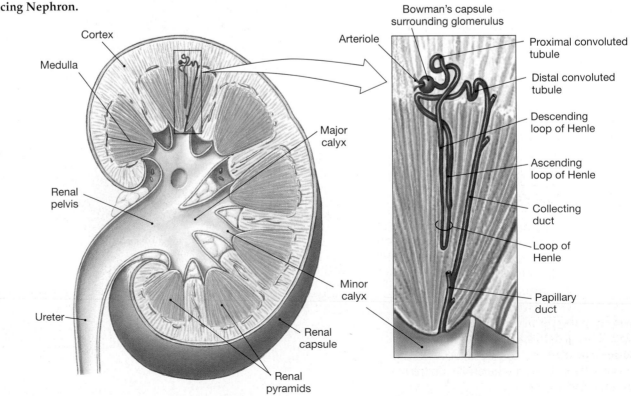

into the medulla. To illustrate the roles that different parts of a nephron play in producing urine, the nephron shown next, in Figure 20-3•, has been shortened and straightened out.

Structure of a Nephron

The nephron begins at a closed, expanded chamber that surrounds a knotted network of capillaries. The entire structure is called a **renal corpuscle** (KOR-pusl). The capillary knot is known as the *glomerulus* (glo-MER-ū-lus), and the enclosing chamber is the *Bowman's capsule*. Blood arrives at the glomerulus in the *afferent arteriole* and departs in the *efferent arteriole*. Filtration occurs in the renal corpuscle as blood pressure forces fluid and dissolved solutes out of the capillaries and into the Bowman's capsule. This produces a protein-free fluid known as a *filtrate*.

From the renal corpuscle the filtrate enters a long passageway that is subdivided into three regions: the *proximal convoluted tubule*, the *loop of Henle* (HEN-lē),

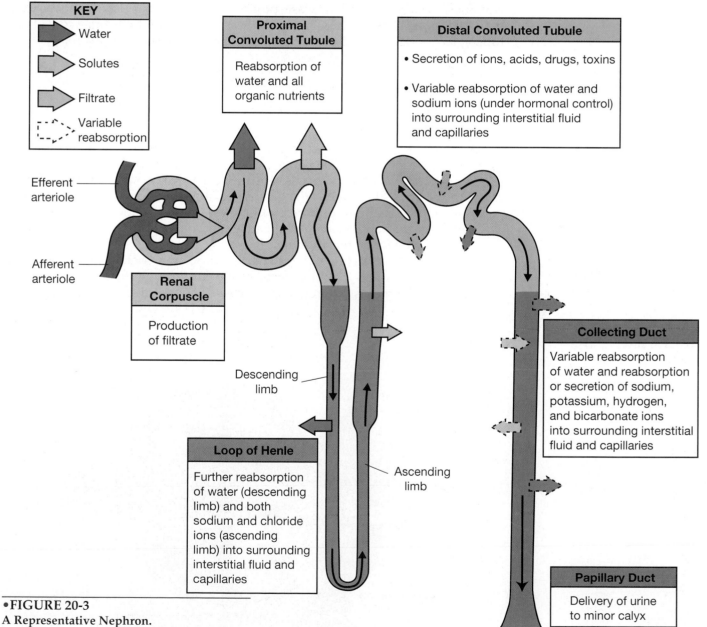

•FIGURE 20-3
A Representative Nephron.
Diagrammatic view showing major structures and functions of each segment of the nephron and collecting system.

and the *distal convoluted tubule*. As the filtrate, now called *tubular fluid*, travels along the tubule its composition is gradually changed. The changes that occur and the urine that results depend on the specialized activities underway in each segment of the nephron.

Each nephron empties into a **collecting duct**, the start of the *collecting system*. The collecting duct leaves the cortex and descends into the medulla, carrying tubular fluid from many nephrons toward a duct within the renal pyramid that delivers the fluid, now called *urine*, into the minor calyces.

Functions of a Nephron

Urine is very different from the filtrate produced at the renal corpuscle. The role of each segment of the nephron in converting filtrate to urine is described in Figure 20-3•. During filtration, not only do organic wastes pass through the glomerular capillary walls but so do water, ions, and organic nutrients such as glucose, fatty acids, and amino acids. These valuable substances must be reclaimed, along with most of the water, and the waste products must be excreted in a relatively concentrated solution. The three segments of the renal tubule (the proximal convoluted tubule, loop of Henle, and distal convoluted tubule) are responsible for

- reabsorbing all the useful organic molecules from the filtrate;
- reabsorbing over 90 percent of the water in the filtrate; and
- secreting into the tubular fluid any waste products that were missed by the filtration process.

Additional water and salts are removed in the collecting system before the urine is released into the renal pelvis.

The **glomerulus** and **Bowman's capsule** both have specialized features for filtering the blood. As Figure 20-4• shows, the relationship between the glomerulus and Bowman's capsule resembles that between the heart and the pericardium. The *capsular space*, which separates the inner and outer layers of Bowman's capsule, is connected to the renal tubule.

The endothelial cells lining the glomerular capillaries contain small pores. The pores are small enough that blood cells cannot fit through them, but water and most solutes (even plasma proteins) can easily move across the capillary wall.

•**FIGURE 20-4**
The Renal Corpuscle.

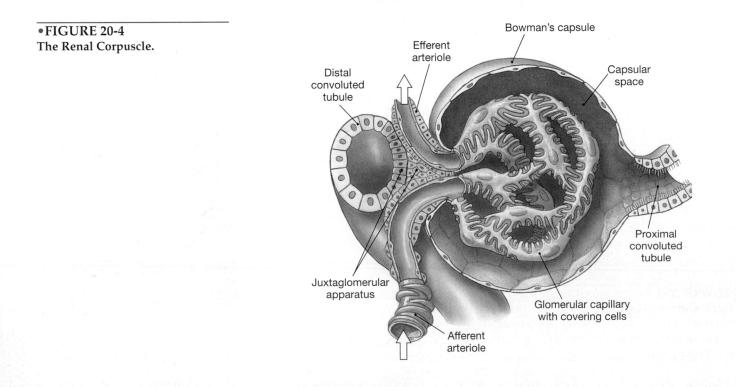

The inner layer of Bowman's capsule consists of specialized cells that cover the capillaries of the glomerulus. These cells have long processes that wrap around individual capillaries. The processes are very close together, separated only by narrow slits. The slits are much narrower than the capillary pores. As a result, most plasma proteins cannot enter the capsular space, but water, metabolic wastes, ions, glucose, fatty acids, amino acids, vitamins, and other small dissolved molecules can do so.

The filtrate next moves into the **proximal convoluted tubule (PCT)** where it is known as tubular fluid. While traveling along the PCT most of the valuable nutrients, such as glucose, amino acids, fatty acids, and various ions are reabsorbed from the tubular fluid. These materials are then released into the surrounding interstitial fluid. As solutes are transported from the tubular fluid into the interstitial fluid the solute concentration of the interstitial fluid increases, and that of the tubular fluid decreases. Water then flows out of the tubular fluid by osmosis. This eliminates the difference in solute concentration and in the process reduces the volume of the tubular fluid.

The PCT merges into the **loop of Henle,** which consists of a *descending limb* that travels toward the renal pelvis and an *ascending limb* that returns to the cortex. The ascending limb, which is impermeable to water and solutes, moves sodium and chloride ions out of the tubular fluid by active transport. As a result, the interstitial fluid of the medulla contains an unusually high solute concentration. The descending limb is permeable to water, and as it descends into the medulla, water moves out of the tubular fluid by osmosis.

The ascending limb of the loop of Henle ends where it bends and comes in close contact with the glomerulus and its vessels. At this point, the **distal convoluted tubule (DCT)** begins. The DCT passes between the afferent and efferent arterioles (Figure 20-4•). The cells of the DCT closest to the glomerulus and adjacent smooth muscle cells in the wall of the afferent arteriole form the **juxtaglomerular apparatus,** an endocrine structure that secretes an enzyme called *renin* and a hormone called *erythropoietin.* ∞ *[p. 301] Juxta-* means "near" and refers to the DCT and arteriole cells next to the glomerulus. Through its secretions, the juxtaglomerular apparatus helps regulate blood pressure and blood volume.

The DCT is an important site for the active secretion of ions, acids, drugs, and other materials into the tubular fluid and the selective reabsorption of sodium ions from the tubular fluid. Water may also be reabsorbed by osmosis in the final portions of the DCT, which helps concentrate the tubular fluid.

The last segment of the nephron is the **collecting duct**. Each collecting duct receives urine from many nephrons. Within the renal pyramid several collecting ducts merge and form larger *papillary ducts* that deliver urine to a minor calyx. In addition to transporting tubular fluid from the nephron to the renal pelvis, the collecting system can make final adjustments to the composition of the urine by reabsorbing water and by reabsorbing or secreting sodium, potassium, hydrogen, and bicarbonate ions.

THE BLOOD SUPPLY TO THE KIDNEYS

In normal individuals, about 1,200 ml of blood flows through the kidneys each minute, which amounts to 20 to 25 percent of the cardiac output. This is an amazing amount of blood for organs with a combined weight of less than 300 g (10.5 oz)! Figure 20-5a• diagrams the path of blood flow to each kidney. Each kidney receives blood from a *renal artery* that originates from the abdominal aorta. As the renal artery enters the kidney it divides into a series of smaller and smaller arteries on its way to the cortex (Figure 20-5b•).

The circulation pattern to one nephron is shown in Figure 20-5c•. Blood reaches each glomerulus through an **afferent arteriole** and leaves in an **efferent arteriole**. It then travels to the **peritubular capillaries** that surround the proximal and distal convoluted tubules. These capillaries pick up or deliver substances that are reabsorbed or secreted by these portions of the nephron.

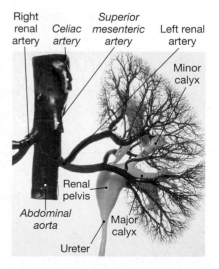
The blood supply to the kidneys.

✔ Which portions of the nephron are located in the renal cortex?

✔ Damage to which part of the nephron would interfere with the control of blood pressure?

✔ Why don't blood cells and large proteins in the plasma appear in the filtrate under normal circumstances?

The efferent arterioles and peritubular capillaries are further connected to a series of long, slender capillaries that accompany the loops of Henle into the medulla. These capillaries, known as the **vasa recta** because they are straight (*rectus*), absorb and transport solutes and water reabsorbed by the loops of Henle or collecting ducts.

Blood from the peritubular capillaries and vasa recta enters a network of venules and small veins that converge into larger veins that eventually empty into the **renal veins**.

•FIGURE 20-5

Blood Supply to the Kidneys.
(a) Sectional view, showing major arteries and veins; compare with Figure 20-2. (b) Circulation in the cortex. (c) Circulation to an individual nephron.

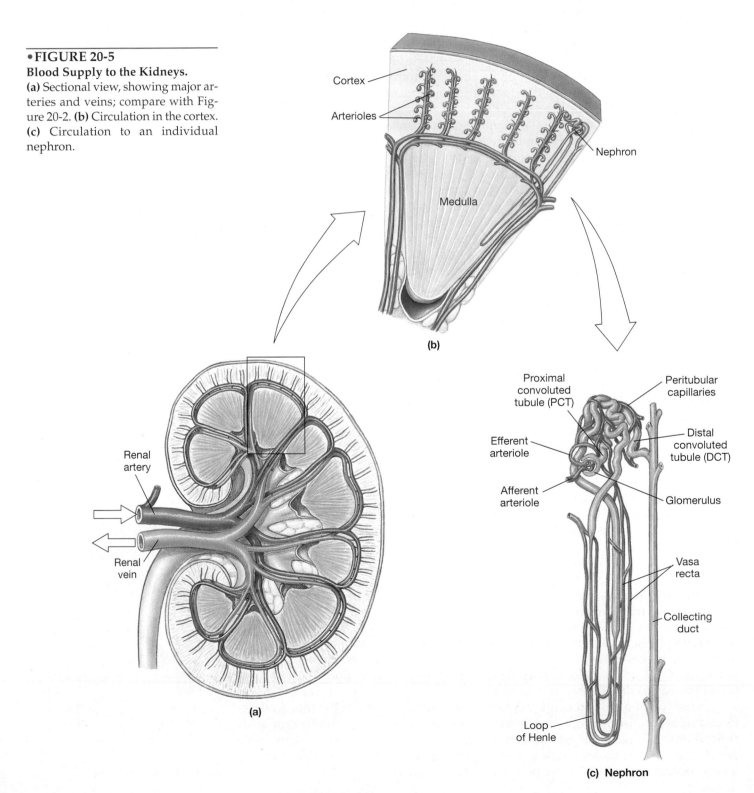

(b)

(a)

(c) **Nephron**

THE FORMATION OF URINE

The primary goal in urine production is to maintain the normal volume and composition of the blood. This involves the excretion of dissolved solutes and metabolic waste products, especially three nitrogen-containing compounds. The most abundant of these is **urea**, which is produced during the catabolism of amino acids. *Creatinine* is generated in skeletal muscle tissue through the breakdown of a high-energy compound similar to ATP that is involved in muscle contraction. Small amounts of **uric acid** are produced each day during the breakdown and recycling of RNA, a nucleic acid.

Since these nitrogen-containing wastes must be excreted in solution, water is also lost in the process. The kidneys can reduce water loss by producing urine that is four to five times more concentrated than normal body fluids. At the same time, the kidneys recover potentially useful compounds, such as sugars and amino acids, before they are lost in the excreted urine. The processes of *filtration*, *reabsorption*, and *secretion* are involved in kidney function. Together these processes create urine, a fluid that is very different from other body fluids. Table 20-1 shows the efficiency of the kidneys' work by comparing the composition of urine and blood plasma.

All segments of the nephron and collecting system participate in the process of urine formation. Most regions perform a combination of reabsorption and secretion, but the balance between the two shifts from one region to another. Referring to Figure 20-3•, p. 489, notice that filtration occurs exclusively in the renal corpuscle, nutrient reabsorption occurs primarily in the proximal convoluted tubule, active secretion occurs primarily in the distal convoluted tubule, and the loop of Henle and the collecting system interact to regulate the amount of water and the number of sodium and potassium ions lost in the urine.

The following sections trace the activities that occur in the nephron and collecting system as urine is produced.

FILTRATION

In **filtration**, blood pressure forces water across a filtration membrane. As noted earlier, blood pressure at the glomerulus forces water and solutes out of the bloodstream and into the capsular space. These substances are forced from the blood because blood pressure is slightly higher in the afferent arteriole than the efferent arteriole. This difference is due to the different diameters of these two arterioles. Because the diameter of the efferent arteriole is smaller, it offers a higher resistance to blood flow than the afferent arteriole. This results in "backing up" of blood in the afferent arteriole and an increase in blood pressure.

The filtrate-producing process is called *glomerular filtration*. The **glomerular filtration rate (GFR)** is the amount of filtrate produced in the kidneys each minute. Each kidney contains around 6 m² of filtration surface, and the GFR averages an astounding 125 ml per minute. This means that almost one fifth of the fluid delivered to the kidneys by the renal arteries leaves the bloodstream and enters the capsular spaces. In the course of a single day the glomeruli generate about 180l (50 gal) of filtrate, roughly 70 times the total volume of blood. But as the filtrate passes through the renal tubules, over 99 percent of it is reabsorbed.

Table 20-1	Significant Differences Between Urinary and Plasma Solute Concentrations	
Component	**Urine**	**Plasma**
Ions (mEq/l)		
Sodium (Na⁺)	147.5	138.4
Potassium (K⁺)	47.5	4.4
Chloride (Cl⁻)	153.3	106
Bicarbonate (HCO₃⁻)	1.9	27
Metabolites and Nutrients (mg/dl)		
Glucose	0.009	90
Lipids	0.002	600
Amino acids	0.188	4.2
Proteins	0.000	7.5 g/dl
Nitrogenous Wastes (mg/dl)		
Urea	1,800	305
Creatinine	150	8.6
Ammonia	60	0.2
Uric acid	40	3

Clinical Note

So That's What Gout Is!

The word *gout* is derived from a term that means "a drop," referring to a harmful substance falling drop after drop into a joint. **Gout** is actually a condition that results from the formation of solid, needle-shaped crystals of uric acid in body fluids. Such crystals form when uric acid concentrations build up in the blood, either because the body produces too much of it or the kidneys remove too little of it. The severity of the resulting swelling and pain depends on the amount and location of the crystal deposits. At first, uric acid crystals form in joints of the extremities, especially be-tween the metatarsal and first phalanges of the big toe. Also called *gouty arthritis*, it may last several days and then dis-appear for a period of days to years. Very high concentra-tions of uric acid can result in the formation of kidney stones and kidney failure.

Most sufferers of gout are males over the age of 50. Meats and fatty foods may initiate the onset of gout or make the con-dition worse. Because such foods usually cost more than car-bohydrates like bread and pastas, "rich foods" have often been associated with this condition.

Glomerular filtration is the vital first step essential to all kidney functions. Filtration depends on adequate circulation and blood pressure to maintain nor-mal filtration pressures at the glomerulus. If filtration pressure and the GFR de-cline, hormones are secreted by the kidney that restore normal glomerular filtration rates.

The *renin-angiotensin system* changes the glomerular filtration rate (GFR) by its effects on blood pressure and volume. When glomerular blood pressure de-clines, so does the GFR. Under these conditions the juxtaglomerular apparatus releases the enzyme *renin* into the blood. Renin starts a chain reaction that ulti-mately involves many different systems, bringing about increases in blood volume, blood pressure, and the GFR. This system is described in a later section on the control of kidney function.

REABSORPTION AND SECRETION

✔ How would a decrease in blood pressure affect the GFR?

✔ If the nephrons lacked a loop of Henle, how would this affect the volume and solute (osmotic) concentration of the urine they produced?

✔ Urea is a waste product that is neither secreted nor reabsorbed along the DCT. How does its concentration change in the DCT as water is reabsorbed?

Reabsorption is the return of water and solutes from the filtrate to the blood. The cells lining the proximal and distal convoluted tubules actively transport organ-ic nutrients and ions from the tubular fluid into the interstitial fluid. As the con-centration of these materials increases in the interstitial fluid, osmosis draws water out of the tubular fluid. Both the solutes and water then return to the circulation by diffusing into the peritubular capillaries.

Secretion is the active transport of materials from the interstitial fluid into the tubular fluid. Examples of secreted substances include various drugs, tox-ins, and acids. Throughout most of the DCT the tubular cells actively secrete potassium ions or hydrogen ions in exchange for sodium ions, which are re-absorbed. The rates of sodium ion reabsorption and potassium ion secretion are largely controlled by the hormone *aldosterone*. The higher the aldosterone levels, the more potassium ions are secreted, and the more sodium ions are re-absorbed.

A few compounds in the tubular fluid, including waste products such as urea and uric acid, are nei-ther reabsorbed nor secreted as fluid moves along the renal tubule. As water and other nutrients are re-moved, the concentration of these waste products gradually rises in the tubular fluid.

On average, 1.2l of urine is excreted daily. Nor-mal urine is a clear, bacteria-free solution with a yel-low color. The general characteristics of normal urine are listed in Table 20-2.

Table 20-2	General Characteristics of Normal Urine
pH	6.0 (range: 4.5–8)
Specific gravity	1.003–1.030
Water content	93–97 percent
Volume	1,200 ml/day
Color	Clear yellow
Odor	Varies depending on composition
Bacterial content	Sterile

THE CONTROL OF KIDNEY FUNCTION

Proper functioning of the urinary system depends on maintaining adequate oxygen levels and blood pressure at the kidneys. The regulation of normal kidney function involves the hormones erythropoietin, angiotensin II, aldosterone, ADH, and ANP. They should sound familiar because they have been discussed in earlier chapters. ∞ *[p. 356]*

- *Erythropoietin* is secreted by the kidneys when oxygen levels are low. This hormone stimulates the production of red blood cells in the bone marrow. As the numbers of red blood cells increase, so does the volume of blood and amount of oxygen delivered to tissues.

- *Angiotensin II* is part of the *renin-angiotensin system*, described later, that is activated by low renal blood pressures.

- *Aldosterone* stimulates the secretion of potassium ions and the reabsorption of sodium ions and water from tubular fluid. Aldosterone secretion can be stimulated by changes in circulating ion concentrations or as part of the renin-angiotensin system.

- *ADH* stimulates water reabsorption at the kidneys and promotes a sensation of thirst, which leads to increased water consumption. ADH secretion occurs when neurons in the hypothalamus are stimulated by a fall in blood pressure or an elevation in the solute concentration of the circulating blood. ADH secretion also occurs as part of the renin-angiotensin system.

- *Atrial natriuretic peptide* (ANP) promotes water loss at the kidneys and a general fall in blood pressure. It is produced by heart cells, in response to high blood pressure and excessive blood volume.

We will consider only two regulatory mechanisms in detail: (1) how the volume of urine is controlled and (2) the responses triggered by changes in blood pressure.

CONTROLLING URINE VOLUME

Urine composition is determined by the secretion and reabsorption of solutes along the renal tubule. Urine concentration depends on the volume of water in which those solutes are dissolved. By the time tubular fluid reaches the DCT, roughly 80 percent of the volume of the original filtrate has been reabsorbed. The final concentration of the urine is determined by events that occur along the DCT and collecting duct.

As we noted earlier, ion transport along the ascending limb of the loop of Henle creates a concentration gradient in the medulla. The highest solute concentration is found near the bend in the loop of Henle, deep in the medulla.

The amount of water reabsorbed along the DCT and collecting duct is controlled by circulating levels of antidiuretic hormone (ADH). In the absence of ADH, the distal convoluted tubule and collecting duct are impermeable to water. The higher the level of circulating ADH, the greater the water permeability and the more concentrated the urine.

If circulating ADH levels are low, little water reabsorption will occur, and virtually all the water reaching the DCT will be lost in the urine (Figure 20-6a•). If circulating ADH levels are high, as in Figure 20-6b•, the DCT and collecting duct will be very permeable to water. In this case the individual will produce a small quantity of urine with a solute concentration four to five times that of extracellular fluids.

RESPONSES TO CHANGES IN BLOOD PRESSURE

In response to low blood pressure at the kidneys, the juxtaglomerular apparatus releases renin into the circulation. Renin stimulates the conversion of an inactive

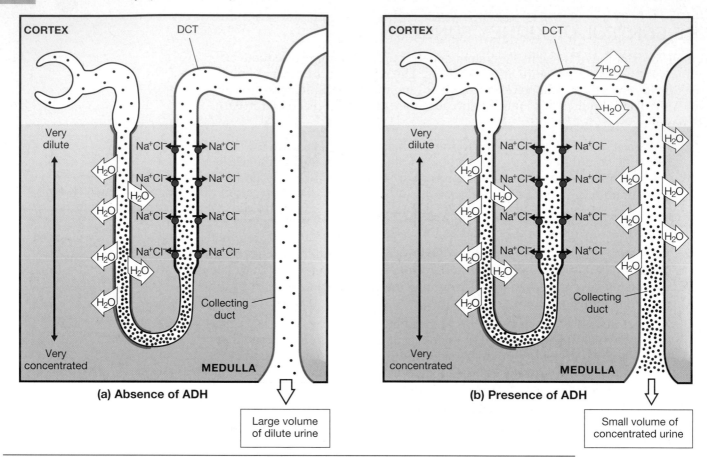

•FIGURE 20-6
The Loop of Henle and the Effects of ADH on the DCT and Collecting Duct.
(a) Permeabilities and urine production without ADH. **(b)** Permeabilities and urine production with ADH.

✔ ADH is a hormone that helps the body to retain water. On which portion(s) of the renal tubule does it act?

✔ What effects do the release of renin and erythropoietin have on blood pressure and blood volume?

blood protein into **angiotensin II**. Angiotensin II causes vasoconstriction, which increases blood pressure immediately. Angiotensin II also triggers the release of ADH and aldosterone. These hormones have long-term beneficial effects. ADH and aldosterone reduce water losses in the urine, and ADH promotes thirst, which leads to increased water consumption and a subsequent increase in blood volume and pressure.

The actions of atrial natriuretic peptide (ANP) oppose those of the renin-angiotensin system. The actions of ANP that affect the kidneys include (1) increased sodium ion loss in the urine, (2) increased glomerular filtration and an increase in the volume of urine, and (3) inactivation of the renin-angiotensin system through inhibition of renin, aldosterone, and ADH secretion. This combination lowers blood volume and blood pressure.

URINE TRANSPORT, STORAGE, AND EXCRETION

Urine production occurs in the kidneys. The other organs of the urinary system are responsible for the transport, storage, and elimination of the urine.

THE URETERS

The ureters (see Figure 20-1•, p. 487) are a pair of muscular tubes that begin at the renal pelvis of the kidneys and extend to the urinary bladder, a distance of about 30 cm. Like the kidneys, the ureters are retroperitoneal, and they penetrate the posterior wall of the bladder without entering the abdominal cavity.

The wall of each ureter contains an inner expandable epithelium, a middle layer of smooth involuntary muscle, and an outer connective tissue covering.

About every 30 seconds a peristaltic wave of contraction begins and travels along the ureter, squeezing urine from the renal pelvis to the urinary bladder.

Occasionally, solids composed of calcium deposits, magnesium salts, or crystals of uric acid develop within the collecting tubules, collecting ducts, or ureters. Such *kidney stones* in the ureters can be very painful. Kidney stones not only obstruct the flow of urine but may also reduce or eliminate filtration in the affected kidney.

THE URINARY BLADDER

The urinary bladder, shown in Figure 20-7•, is a hollow muscular organ that stores urine. Its dimensions vary with the volume of stored urine, but a full urinary bladder can contain a liter of urine.

The urinary bladder is held in place by ligaments that anchor it to the pelvic and pubic bones. A triangular area called the **trigone** (TRĪ-gōn) forms most of the floor of the urinary bladder. It is bounded by the openings of the ureters and the entrance to the urethra. The area surrounding the urethral entrance, called the *neck* of the urinary bladder, contains the **internal sphincter**, a muscle that provides involuntary control over the discharge of urine from the bladder.

The transitional epithelium that lines the renal pelvis and the ureter is continuous with the urinary bladder. This stratified epithelium can tolerate a considerable amount of stretching (see Figure 4-2•). ∞ *[p. 63]* The bladder wall also contains layers of smooth, involuntary muscle. Contraction of these muscles compresses the urinary bladder and expels its contents into the urethra.

THE URETHRA

The urethra extends from the urinary bladder to the exterior. Its length and functions differ in males and females. In the male, the urethra extends to the tip of the penis and is about 18 to 20 cm (7 to 8 in.) long. As Figure 20-7• shows, the initial portion of the male urethra is surrounded by the prostate gland. In addition to transporting urine, the male urethra carries male sex cells (sperm cells) and reproductive secretions from the prostate and other glands. In contrast, the female urethra is very short, extending 2.5 to 3.0 cm (about 1 in.), and transports only urine. In both sexes, the urethra contains a circular band of skeletal muscle that forms an **external sphincter**. ∞ *[p. 199]* The contractions of the external sphincter are under voluntary control.

URINE EXCRETION

The ejection of urine is called **urination**, or **micturition**. The urge to urinate usually appears when the bladder contains about 200 ml of urine. We become aware of this need through nerve impulses sent to the brain from stretch receptors in the wall of the urinary bladder. The stimulation of these receptors also results in involuntary contractions of the urinary bladder.

Such contractions increase the fluid pressure inside the bladder, but urination cannot occur unless both the internal and external sphincters are relaxed. We control the time and place of urination by voluntarily relaxing the external sphincter. When this sphincter relaxes, so does the internal sphincter. If the external sphincter does not relax, the internal sphincter remains closed, and the urinary bladder gradually relaxes. A further increase in bladder volume begins the cycle again, usually within an hour. Once the volume of the urinary bladder exceeds

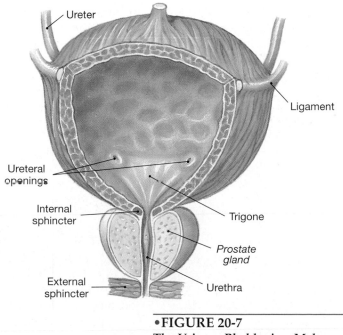

•**FIGURE 20-7**
The Urinary Bladder in a Male.

✔ What process is responsible for the movement of urine from the kidney to the urinary bladder?

✔ An obstruction of a ureter by a kidney stone would interfere with the flow of urine between which two points?

✔ The ability to control urination depends on one's ability to control which muscle?

500 ml, enough pressure is generated to force open the internal sphincter. This leads to an uncontrollable relaxation in the external sphincter, and urination occurs despite voluntary opposition or potential inconvenience. Normally, less than 10 ml of urine remains in the urinary bladder after urination.

BODY FLUIDS

Most of the weight of the human body, about 60 percent, is water. The cells of the body rely on water as a medium for the distribution of gases, nutrients, and waste products. If the water content of the body changes, cellular activities are jeopardized. For example, if the degree of *dehydration* (water loss) reaches about 10 percent of body weight, proteins denature, enzymes stop functioning, and cells die.

Figure 20-8• shows the three largest "pools" of water in the body. About 40 percent of the total body water content is found inside living cells, as the fluid medium of the **intracellular fluid (ICF)**, introduced in Chapter 3. ∞ *[p. 39]* The **extracellular fluid (ECF)** contains the rest of the body water. The two largest subdivisions of the ECF are the *interstitial fluid*, or *tissue fluid*, in body tissues (15 percent); and the *plasma* of the circulating blood (nearly 5 percent). Minor amounts of the ECF (less than 2 percent) include lymph, cerebrospinal fluid (CSF), synovial fluid, serous fluids (pleural, pericardial, and peritoneal fluids), aqueous humor, perilymph, and endolymph.

Exchange between the blood plasma and tissue fluid of the ECF occurs primarily across the linings of capillaries. Fluid may also travel from the interstitial spaces to the plasma through the lymphatic vessels.

The ICF and ECF are often called **fluid compartments** because their compositions are quite different from each other. For example, the principal ions in the ECF are sodium, chloride, and bicarbonate. The ICF contains an abundance of potassium, magnesium, and phosphate ions, plus large numbers of negatively charged proteins. Changes in the concentrations of ions and other dissolved substances can affect cell activities, cause changes in water movements between fluid compartments, and influence the acid-base balance of the body.

FLUID BALANCE

Water circulates freely within the extracellular fluid compartment. At capillary beds throughout the body, capillary blood pressure forces water out of the plasma and into the interstitial spaces. Some of that water is reabsorbed along the distal portion of the capillary bed, and the rest circulates into lymphatic vessels for transport to the venous circulation. ∞ *[p. 354]*

•FIGURE 20-8
Body Fluid Compartments.
The major body fluid compartments in the human body and their relative percentages of total body weight.

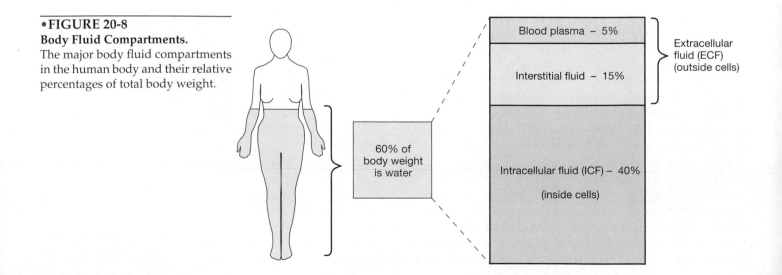

60% of body weight is water

Blood plasma – 5%

Interstitial fluid – 15%

Extracellular fluid (ECF) (outside cells)

Intracellular fluid (ICF) – 40%

(inside cells)

Water moves back and forth across the epithelial surfaces lining the abdominal, pleural, and pericardial cavities and through the synovial membranes lining joint capsules. Water also moves between the blood and the cerebrospinal fluid, the aqueous humor and vitreous body of the eye, and the perilymph and endolymph of the inner ear.

Water is also exchanged with the environment (see Table 20-3). Roughly 2,500 ml of water is lost each day through urine, feces, perspiration, and evaporation through the lungs. The losses due to perspiration vary, depending on the activities undertaken, but the additional deficits can be considerable, reaching well over a gallon (nearly 4 l) an hour. ∞ *[p. 122]* Most of the water losses are balanced by the gain of fluids through eating (48 percent) and drinking (40 percent). A smaller amount of water (12 percent) is gained primarily as the result of metabolic reactions in the mitochondria.

Table 20-3	Water Balance
Daily Input (ml)	
Water content of food	1,000
Water consumed as liquid	1,200
Metabolic water during catabolism	300
Total	2,500
Daily Output (ml)	
Urine	1,200
Evaporation at skin	750
Evaporation at lungs	400
Lost in feces	150
Total	2,500

ELECTROLYTE BALANCE

Electrolytes are compounds that break apart in water to form positive ions (cations) and negative ions (anions). Many ions in the human body are also called minerals; they were discussed in Chapter 19. ∞ *[p. 472]* An individual is in electrolyte balance when the rates of gain and loss are equal for each of the individual electrolytes in the body. Electrolyte balance is important for two reasons: (1) A gain or loss of electrolytes can cause a gain or loss in water, and (2) the concentrations of individual electrolytes affect a variety of cell functions.

Two minerals, sodium and potassium, are especially important in body fluids. Sodium is the most common positive ion within the extracellular fluid. Because it is so common, changes in sodium ion concentration have important effects on the osmotic movement of water from one fluid compartment to another. Potassium is the dominant positive ion in the intracellular fluid; extracellular potassium concentrations are normally low. Normal concentrations of both ions are necessary for transmitting nerve impulses.

Sodium

The amount of sodium in the ECF represents a balance between sodium ion absorption at the digestive tract and sodium ion excretion at the kidneys and other sites. The rate of uptake varies directly with the amount included in the diet. The kidneys are responsible for regulating sodium ion losses, in response to circulating levels of aldosterone (decreases sodium loss) and ANP (increases sodium loss).

Whenever the rate of sodium ion intake or loss changes, there is a corresponding gain or loss of water that tends to keep the sodium ion concentration constant. For example, eating a heavily salted meal does not raise the sodium ion concentration of body fluids because as sodium chloride crosses the digestive epithelium, osmosis brings additional water into the body. (This is why individuals with high blood pressure are told to restrict their salt intake; dietary salt is absorbed, and because "water follows salt," the blood volume and blood pressure increase.)

Potassium

Almost all (98 percent) of the potassium ions within the body lies within the ICF. The low potassium ion concentration of the ECF represents a balance between (1) the rate of entry across the digestive epithelium and (2) the rate of loss into the urine. The concentration of potassium ions in the extracellular fluid is controlled

✔ How would eating a meal high in salt affect the amount of fluid in the intracellular fluid compartment?

✔ What effect would being lost in the desert for a day without water have on the concentration of electrolytes in your blood?

primarily by the rate of its secretion along the distal convoluted tubule of the nephron. As with sodium ion excretion, the rate of potassium ion secretion is determined by aldosterone levels, but in an opposite manner. When aldosterone levels climb, potassium ions are secreted and sodium ions are reabsorbed; when aldosterone levels fall, potassium ions are conserved and sodium ions are secreted into the urine.

ACID–BASE BALANCE

The pH of body fluids represents a balance between the acids, bases, and salts in solution. This pH normally remains within relatively narrow limits, usually from 7.35 to 7.45.

When the pH falls below 7.35, a state of **acidosis** exists. **Alkalosis** exists if the pH increases above 7.45. These conditions affect virtually all systems, but the nervous system and cardiovascular system are particularly sensitive to variations in pH. For example, severe acidosis can be deadly because (1) CNS function deteriorates, and the individual becomes comatose; (2) cardiac contractions grow weak and irregular, and symptoms of heart failure develop; and (3) vasodilation of blood vessels produces a dramatic and potentially dangerous drop in blood pressure.

Carbon dioxide concentration is the most important factor affecting the pH in body tissues. In solution, carbon dioxide interacts with water to form molecules of carbonic acid (H_2CO_3). As noted in Chapter 17, the carbonic acid molecules then break apart to produce hydrogen ions (H^+) and bicarbonate (HCO_3^-) ions. ∞ [p. 414] The greater the number of hydrogen ions, the more acidic the solution becomes.

Because most of the carbon dioxide in solution is converted to carbonic acid, and most of the carbonic acid breaks apart and releases hydrogen ions, there is a direct relationship between the amount of CO_2 and the pH. When carbon dioxide concentrations rise, newly formed hydrogen ions cause the pH to go down. When carbon dioxide concentrations fall, pH rises because free hydrogen ions join with bicarbonate ions to form carbonic acid. (Remember that the smaller the pH value, the greater the acidity.)

Maintaining Acid–Base Balance

There are three main mechanisms that work together to maintain a normal balance of acids and bases in body fluids. These mechanisms rely on *buffers* and two organs, the lungs and the kidneys.

Buffer Systems. Buffers, introduced in Chapter 2, are dissolved compounds that can provide or remove hydrogen ions and thereby stabilize the pH of a solution. ∞ [p. 27] **Buffer systems** in the body include the protein, carbonic acid–bicarbonate, and phosphate systems.

- *Protein buffer systems* depend on the ability of amino acids to accept or release hydrogen ions. The plasma proteins and hemoglobin in red blood cells are important buffers in the blood.

- The *carbonic acid–bicarbonate buffer system* is based on the reversible conversions of carbon dioxide to carbonic acid, and carbonic acid into a hydrogen ion and a bicarbonate ion. The net effect is that $CO_2 + H_2O \rightleftharpoons H^+ + HCO_3^-$. If hydrogen ions are removed, they will be replaced through the combination of water and carbon dioxide; if hydrogen ions are added, most will be removed through the formation of carbon dioxide and water.

- The *phosphate buffer system* is quite important in buffering the pH of the intracellular fluids (ICF), where the concentration of phosphate ions is relatively high.

The Lungs and pH Regulation

The lungs contribute to pH regulation by their effects on the carbonic acid–bicarbonate buffer system through an increase or decrease in the rate of respiration. Increased respiration rates cause more CO_2 to be released at the lungs and plasma levels of CO_2 to decrease; decreased respiration rates cause less CO_2 to be released at the lungs and plasma levels of CO_2 to increase. Changes in respiratory rate affect pH because when the concentration of CO_2 in the plasma rises, the pH declines, and when the concentration of CO_2 in the plasma decreases, the pH increases.

The Kidneys and pH Regulation

Glomerular filtration puts hydrogen ions, carbon dioxide, and the other components of the carbonic acid–bicarbonate and phosphate buffer systems into the filtrate. The kidney tubules then modify the pH of the filtrate by secreting hydrogen ions or reabsorbing bicarbonate ions.

✔ How can carbon dioxide affect the pH of body fluids?

✔ What effect would a decrease in the pH of the body fluids have on the respiratory rate?

AGING AND THE URINARY SYSTEM

In general, aging is associated with an increased number of kidney problems. The major cause of such problems is a decline in the number of working nephrons and a reduction in the GFR. The total number of kidney nephrons drops by 30 to 40 percent between ages 25 and 85.

With age, the distal portions of the nephron and collecting system become less responsive to ADH. Less reabsorption of water and sodium ions occurs, and more potassium ions are lost in the urine.

Problems may also rise with urination. One is **incontinence** (in-KON-ti-nens), an inability to control urination voluntarily. Incontinence occurs as the sphincter muscles lose muscle tone and become less effective at voluntarily retaining urine. This condition may also follow a stroke, Alzheimer's disease, or other CNS problems affecting the cerebral cortex or hypothalamus.

In males, **urinary retention** may develop secondary to enlargement of the prostate gland. In this condition swelling and distortion of surrounding prostatic tissues compress the urethra, restricting or preventing the flow of urine.

DISORDERS OF THE URINARY SYSTEM

Disorders of the urinary system may occur in the kidneys, where urine production occurs, and the urine-conducting system, which includes the ureters, the urinary bladder, and the urethra. Although the kidneys perform the vital physiological functions of the urinary system, problems with the conducting system can have direct and immediate effects on kidney, or renal, function.

THE PHYSICAL EXAMINATION

Urology (u-ROL-o-jē) is the surgical branch of medicine concerned with all the components of the urinary system and its disorders. In addition to the urinary system, a *urologist* is also concerned with the male reproductive tract and its disorders. **Nephrology** (ne-FROL-o-jē) is the medical specialty concerned with the kidney and its disorders.

Symptoms and Signs

The primary symptoms of urinary system disorders are pain and changes in the frequency of urination. The nature and location of the pain can provide clues to the source of the problem. For example,

■ Pain superior to the pubic symphysis may be associated with urinary bladder disorders.

- Upper lumbar back pain radiating to the side and the right upper quadrant or left upper quadrant can be caused by kidney infections.

- Painful or difficult urination, called **dysuria** (dis-Ū-rē-ah), may occur with inflammation or infection of the urinary bladder, urethra, or obstructions to the passage of urine within the urinary tract.

Most individuals usually urinate 4–6 times per day, excreting a total amount of urine between 0.5 to 2 l (1 to 4 pints). Individuals with urinary system disorders may urinate more or less often than usual, and may produce normal or abnormal amounts of urine. The following symptoms may occur:

- Irritation of the lining of the urethra or urinary bladder can lead to the desire to urinate with increased frequency, although the total amount of urine produced each day remains normal. When these problems exist, the individual feels the urge to urinate when the urinary bladder volume is very small. The irritation may result from trauma, urinary bladder infection (*cystitis*) or tumors, or increased acidity of the urine.

- *Incontinence* may involve periodic involuntary urination or—a continual, slow trickle of urine from the urethra (p. 501). Incontinence may result from urinary bladder or urethral problems, damage or weakening of the muscles of the pelvic floor, or interference with normal sensory or motor innervation in the region. Kidney function and daily urinary volume are normal.

- In *urinary retention*, kidney function is normal, at least at first, but urination does not occur (p. 501). In members of both genders, urinary retention may result from obstruction of the outlet of the urinary bladder or from CNS damage, such as a stroke or damage to the spinal cord, which interferes with the reflex control of micturition.

- Changes in the volume of urine produced by a normally hydrated person indicate that there are problems either at the kidneys or with the control of renal function. **Polyuria**, the production of excessive amounts of urine (more than 2.5 l, or 5 pints), may result from hormonal or metabolic problems, such as those associated with diabetes (p. 310), or from damage to the glomeruli. **Oliguria** (a urine volume of 50–500 ml/day), and **anuria** (0–50 ml/day) are conditions that indicate serious kidney problems and potential *kidney*, or *renal*, *failure*. Renal failure can occur with heart failure, renal ischemia, circulatory shock (p. 374), burns (p. 134), and a variety of other disorders.

Important clinical signs of urinary system disorders include the following:

- **Hematuria** (hē-ma-TOO-rē-uh), the presence of red blood cells in the urine, indicates bleeding at the kidneys or conducting system. Hematuria producing dark red urine typically indicates bleeding in the kidney, and hematuria producing bright red urine indicates bleeding in the lower urinary tract. Hematuria most commonly occurs with trauma to the kidneys, kidney stones, tumors, or urinary tract infections.

- **Hemoglobinuria** is the presence of hemoglobin in the urine. Hemoglobinuria indicates increased RBC breakdown within the circulation due to cardiovascular or metabolic problems. Examples of conditions that result in hemoglobinuria include the thalassemias (p. 332), sickle cell anemia (p. 332), and hypersplenism (p. 396).

- Changes in urine color may accompany some kidney disorders. For example, the urine may become (1) cloudy due to the presence of bacteria, lipids, crystals, or epithelial cells; (2) red or brown from hemoglobin; (3) blue-green from a pigment formed during the breakdown of hemoglobin ; or (4) brown-black from excessive concentration. Not all color changes are abnormal, however. Some foods and several prescription drugs can cause changes in urine color. A serving of beets can give urine a reddish color, whereas eating rhubarb can give the urine an orange tint, and B vitamins give urine a vivid yellow color.

- Kidney disorders that result in the loss of proteins in the urine (*proteinuria*) may lead to a generalized edema in peripheral tissues. Facial swelling, especially around the eyes, is common.

- A fever commonly develops when the urinary system is infected by pathogens. Urinary bladder infections may cause a low-grade fever; kidney infections can produce very high fevers.

During the physical assessment, palpation can be used to check the state of the kidneys and urinary bladder. The kidneys lie in the region bounded by the lumbar region of the spine and the 12th ribs on either side. To detect tenderness due to kidney inflammation, the examiner gently thumps a fist over each side. This usually does not cause pain unless the underlying kidney is inflamed.

The urinary bladder can be palpated just superior to the pubic symphysis. On the basis of palpation alone, urinary bladder enlargement due to urine retention can be difficult to distinguish from the presence of an abdominal mass.

Many procedures and laboratory tests may be used in the diagnosis of urinary system disorders. The functional anatomy of the urinary system can be examined by using a variety of sophisticated procedures. For example, administering a radiopaque compound that will enter the urine permits the creation of an **intravenous pyelogram** (PĪ-el-ō-gram), or **IVP**, by taking an X-ray of the kidneys (Figure 20-9•). This procedure permits detection of unusual kidney, ureter, or urinary bladder structures and masses. Other diagnostic procedures and important laboratory tests are listed in Table 20-4.

URINARY SYSTEM DISORDERS

Major disorders of the urinary system include inflammation and infection, tumors, immune disorders, congenital disorders, and degenerative disorders. Figure 20-10• outlines the major classes of disorders of the urinary system and lists examples of each.

Inflammation and Infection

Urinary Tract Infections (UTIs) are second only to infections of the respiratory system in terms of clinical significance in the U.S. They usually result from the invasion of the urinary tract by bacteria. The intestinal bacterium *Escherichia coli* is most commonly involved. UTIs may originate in one area and spread to other regions of the urinary system. In the majority of cases, infections begin in the lowest reaches of the urethra, and may then ascend to the kidneys by way of the urinary bladder and ureters. Less frequently, infections descend from the kidneys to the urethra.

Kidney. Nephritis (nef-RĪ-tis) is an inflammation of the kidneys. Nephritis may result from bacterial infections, exposure to toxic or irritating drugs, or autonomic disorders. One of the major problems in nephritis is that the inflammation causes swelling, but the fibrous renal capsule prevents the kidney from increasing in size. The swelling compresses the nephrons and blood vessels, reducing the glomerular filtration rate (GFR).

Pyelitis is an inflammation of the renal pelvis region of the kidney. **Pyelonephritis** (pī-e-lō-ne-FRĪ-tis) is an inflammation of the kidney and the renal pelvis. This condition is usually caused by a bacterial infection that has ascended up to the kidney from

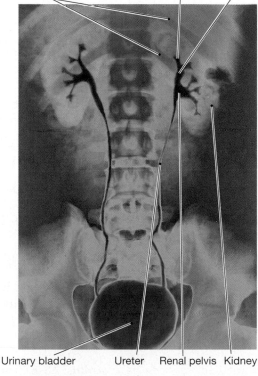

11th and 12th ribs · Minor calyx · Major calyx

Urinary bladder · Ureter · Renal pelvis · Kidney

•FIGURE 20-9
An X-ray of the Urinary System.
This pyelogram is an anterior-posterior X-ray that has been color-enhanced. Such techniques help in diagnosing kidney diseases, showing kidney stones, and other problems.

Table 20-4   Representative Diagnostic Procedures and Laboratory Tests for Disorders of the Urinary System

Diagnostic Procedure	Method and Result	Representative Uses
Cystoscopy	A small tube (cystoscope) is inserted along the urethra into the urinary bladder to inspect the lining of the urinary bladder	To obtain a biopsy specimen or to remove stones (calculi) and small tumors
Renal biopsy	Guided by ultrasound a biopsy needle is inserted into the kidney	To check for renal disease, rejection of transplanted kidney, or biopsy a tumor
Intravenous pyelography (IVP)	Dye injected intravenously is excreted into the urinary tract; a series of X-rays is then taken	To determine the presence of kidney disease, obstructions such as calculi or tumors, or anatomical abnormalities
Cystography	Dye is introduced into the urinary bladder through a catheter; a series of X-rays is then taken	Identification of tumors or rupture of the urinary bladder to check for backflow of urine into the ureter

Laboratory Test	Normal Values	Significance of Abnormal Values
URINALYSIS		
pH	4.6–8.0	Alkaline or acidic may indicate increased or decreased blood pH
Color	Pale yellow-amber	Color may change with certain drugs, foods, and renal infections; red or red brown color may indicate bleeding in the lower urinary tract or kidney, respectively
Appearance	Clear	Clouded urine may indicate bacterial infection or from certain foods
Odor	Aromatic	Acetone odor occurs in diabetic ketoacidosis
Specific gravity	1.005–1.030	Increased in dehydration, increased ADH production, heart failure, glycosuria, or proteinuria. Decreased in diabetes insipidus or renal failure
Protein	<100 mg/24 hr	Increased in kidney infections or inflammation and after strenuous exercise
Glucose	None	Appears in diabetes mellitus or Cushing's syndrome and after corticosteroid therapy
Ketones	None	Appear in poorly controlled diabetes mellitus, during dehydration, and in fasting
Sodium	40–220 mEq/day	Increased by dehydration. Decreased level with renal failure, liver disease, or congestive heart failure
Potassium	40–80 mEq/l/24 hr	Elevated by diuretics, dehydration, and starvation
SERUM ANALYSIS		
Sodium	Adults: 135–145 mEq/l	Increased by severe dehydration. Decreased with SIADH, renal failure, diuretic use
Potassium	Adults: 3.5–5.5 mEq/l	Increased by acute renal failure and acidosis. Decreased in some renal diseases and after diuretic use
Serum bicarbonate	Adults: 24–28 mEq/l	Elevation or reduction of bicarbonate levels is important in diagnosis of acid–base disorders
Urea (urea nitrogen)	6–21 g/day	Useful in estimation of GFR; values increase in liver disease and gout
Uric acid	Adults: 2.8–8.0 mg/dl	Elevated in renal failure and gout. Decreased in burns and some renal disorders

•**FIGURE 20-10**
Disorders of the Urinary System.

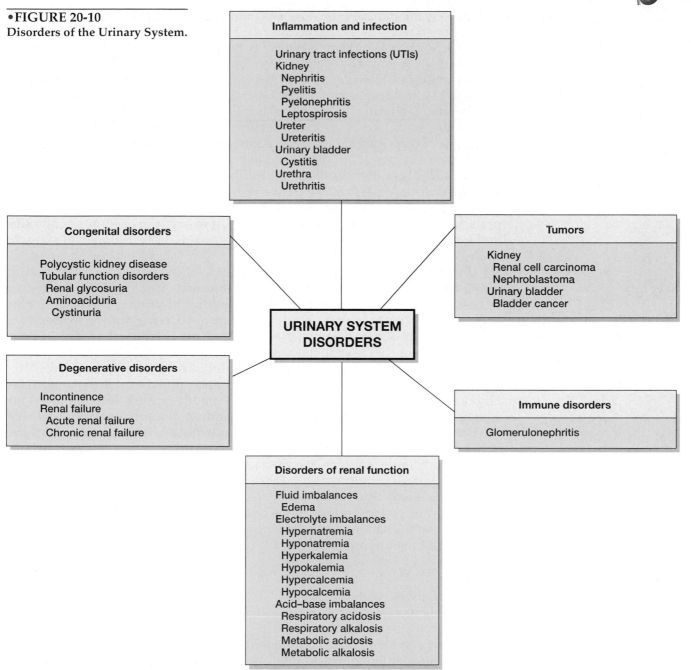

the bladder. *Acute pyelonephritis* occurs most often in women, and then, more commonly during pregnancy. Symptoms include high fever, vomiting, chills, and back pain on the affected side. Antibiotics are generally used in treating this condition. *Chronic pyelonephritis* begins in childhood in response to chronic infection, frequently caused by the reflux of urine into the kidney. It is caused by the abnormal development of the valve at the intersection of the ureter and urinary bladder. Repeated reflux of urine and subsequent infection may cause scarring, and lead to *renal insufficiency*, or inadequate kidney function. Treatment involves surgery to increase the flow of urine into the urinary bladder.

Leptospirosis is an infectious condition of the kidneys caused by the spirochete bacterium, *Leptospira interrogans* (lep-tō-SPĪ-ra in-TER-a-ganz). This is a relatively rare disease, with 50–150 cases reported each year in the United States. The bacteria normally infect the nephrons of the kidneys of dogs, cats, and wild mammals, such as rats and deer. They are released with urine from infected animals, and invade the human body through the eyes, nose, and mouth, or abraded skin. Humans often become infected when they clean up after their pets or go swimming with their dogs in a pool. Symptoms of fever generally occur between a week or two after exposure. Recovery is normally complete after 2 to 3 weeks. Leptospirosis may be treated with antibiotics. However, up to 30 percent of untreated cases may result in death. Immunity is only developed for specific strains of *Leptospira*, not all strains.

Ureters. Ureteritis (Ū-rē-ter-ī-tis) is inflammation of the ureters. The inflammation may be caused by an infection ascending from the bladder, or more commonly, blockage with a kidney stone. Kidney stones, or **calculi** (KAL-kū-lī), form within the urinary tract from calcium deposits, magnesium salts, or crystals of uric acid. This condition is called **nephrolithiasis** (nef-rō-li-THĪ-a-sis). The blockage of the urinary passage by a stone or by other factors, such as an enlarged prostate gland, results in **urinary obstruction**. Urinary obstruction is a serious problem because, in addition to causing pain, it will reduce or eliminate filtration in the affected kidney(s) by increasing the pressure in the filtrate and capsular space of the glomerulus.

Urinary Bladder. Inflammation of the lining of the bladder is called **cystitis** (sis-TĪ-tis). It is usually caused by a bacterial infection. Any conditions that reduce or obstruct urination, and that result in a buildup of urine in the bladder contribute to cystitis. Stagnant urine provides a good environment for bacterial growth. Cystitis is more common in women than men because the female urethra is shorter, and opens closer to the vagina and anus, potential sources of infectious organisms. Symptoms of cystitis include painful urination (dysuria), frequent urination with only small amounts being passed, and pus or blood in the urine. If antibiotics are used promptly, the condition is readily treated. If treatment is not sought early, the infection may ascend to the kidney and produce pyelonephritis.

Urethra. Urethritis (ū-re-THRĪ-tis) is an inflammation of the urethra. As noted above, this condition is more common in women than men. Urethritis may be caused by a number of different bacteria. For example, urethritis is a symptom of the sexually transmitted disease, *gonorrhea*.

Dysuria is a common symptom of urethritis. The urine may also contain blood (hematuria) or pus (especially in the case of gonorrhea). Urethritis may be followed by the formation of scar tissue, which may lead to *urethral stricture*, or a narrowing of the urethral passageway.

Tumors

Renal Tumors. Both benign and malignant tumors may form in the kidney. Benign tumors usually do not produce symptoms, unless they reach a large size, when they may be removed surgically.

The most common malignant tumor of the kidney is **renal cell carcinoma**. It occurs most often in adults over 40 years old, affecting twice as many men as women. Hematuria, or blood in the urine, is the most common symptom. Other symptoms include lumbar side (flank) pain, a lump in the abdomen, and weight loss. Metastasis may occur early, and spread to the bones, lungs, liver, and brain.

Nephroblastoma, also known as *Wilms' tumor*, accounts for one out of five cancers in children. It usually affects children under the age of 4, and is more common in boys. A lump in the abdomen and abdominal pain are characteristic symptoms. The tumor cells may spread to other organs just as in renal cell carcinoma. Early treatment increases the rate of survival to about 50 to 80 percent.

Urinary Bladder. Both benign and malignant tumors may form in the lining of the bladder. Occasionally, a benign tumor may become malignant. Malignant tumors may remain within the bladder, or spread through the bladder wall and into other nearby organs, such as the large intestine, uterus (in females) or prostate gland (in males). If the cancer cells enter lymphatic or blood vessels, they may metastasize throughout the body. The main sign of **bladder cancer** is hematuria. Symptoms include back pain, if the tumor is obstructing the ureter openings, or dysuria, if the tumor is blocking the urethral opening within the bladder. Bladder cancer is diagnosed by *cytoscopy* (the insertion of a viewing tube up the urethra) and biopsy.

Each year in the U.S., approximately 52,000 new cases of urinary bladder cancer are diagnosed, and 9,500 deaths occur. The incidence among males is three times that among females, and most patients are age 60–70. The rate of bladder cancer is highest among smokers and employees of chemical and rubber companies. The prognosis is reasonably good for superficial, localized cancers but is poor for persons with metastatic bladder cancer.

Immune Disorders

Glomerulonephritis. **Glomerulonephritis** (glo-MER-ū-lō-ne-FRĪ-tis) is an inflammation of the glomeruli caused by an immune response to infection. This condition may occur after an infection involving *Streptococcus* bacteria. The primary infection need not occur in or near the kidneys. As the immune system responds to the infection, the number of antigen–antibody complexes rapidly increases. These complexes clog the filtration system at the glomerulus (see Figure 20-4•, p. 490). This causes inflammation of the kidneys, and a drop in urine production. Any condition that leads to a massive immune response can cause glomerulonephritis, including viral infections and autoimmune disorders.

Mild forms of glomerulonephritis may not produce any symptoms, but severe inflammation can result in the loss of plasma proteins and red blood cells into the filtrate. Although small amounts of protein can be reabsorbed, when glomeruli are severely damaged the nephrons are unable to reabsorb all the plasma proteins that enter the filtrate. Plasma proteins then appear in the urine, a condition termed **proteinuria** (prō-tē-NŪ-rē-uh). This reduces the solute concentration of the blood, which can lead to widespread edema.

Congenital Disorders

Polycystic Kidney Disease. **Polycystic** (po-lē-SIS-tik) **kidney disease** is an inherited condition affecting the structure of the renal tubules. Swellings, or cysts develop along the length of the tubules, some growing large enough to squeeze adjacent nephrons and vessels. Kidney function deteriorates, and the nephrons may become nonfunctional, eventually leading to renal failure. The process is so gradual that serious problems seldom appear before the individual is 30–40 years old. Common symptoms include sharp pain in the sides, recurrent UTIs, and the presence of blood in the urine. Treatment is directed to the symptoms, focusing on the prevention of infection and reduction of pain with analgesics. In severe cases, *hemodialysis* or kidney transplantation may be required.

Tubular Function Disorders. The reabsorption of specific ions or compounds from the filtrate involves many different carrier proteins. Some individuals have an inherited inability to make one or more of these carrier proteins. For example, in **renal glycosuria** (glī-cō-SOO-rē-uh) a defective carrier protein makes it impossible for the PCT to reabsorb glucose from the filtrate. Although renal glucose levels are abnormally high, blood glucose is normal, which distinguishes this condition from diabetes mellitus. Affected individuals generally do not have any clinical problems except when demand for glucose is high, as in starvation, acute stress, or pregnancy.

Another example is **aminoaciduria** (a-mē-nō-as-i-DŪ-rē-uh), in which defective carrier proteins cannot reabsorb specific amino acids. **Cystinuria** is the most common disorder of amino acid transport, occurring in approximately 1 person in 12,500. Persons with this condition have difficulty reabsorbing *cystine* and similar amino acids. The most obvious and painful symptom is the formation of kidney and bladder stones that contain crystals of these amino acids. In addition to removal of these stones, treatment for cystinuria involves maintaining a high rate of urinary flow, so that amino acid concentrations do not rise high enough to promote stone formation, and a reduction of urinary acidity, because stone formation is enhanced by acidic conditions.

Degenerative Disorders

Incontinence. Incontinence is discussed on p. 501.

Renal Failure. Renal, or kidney, failure occurs when the kidneys become unable to perform the excretory functions to maintain homeostasis. When kidney filtration slows for any reason, urine production declines. As the decline continues, symptoms of renal failure appear because water, ions, and metabolic wastes are not excreted. Virtually all systems in the body are affected. For example, fluid balance, pH, muscular contraction, metabolism, and digestive function are disturbed. The individual's blood pressure increases (hypertension), anemia develops due to a decline in erythropoietin production, and CNS problems may lead to sleeplessness, seizures, delirium, and even coma.

Acute renal failure occurs when exposure to toxins, renal ischemia, urinary obstruction, or trauma causes filtration to slow suddenly or to stop altogether. With supportive treatment, the survival rate for individuals with acute renal failure is about 50 percent. In *chronic renal failure*, kidney function declines gradually, and the associated problems gradually develop over time. Management of this condition may involve restricting water, salt, and protein intake. These measures minimize the volume of urine produced and prevents the generation of large amounts of nitrogenous wastes. Acidosis, a common problem of patients with renal failure, can be neutralized by the ingestion of bicarbonate ions. The condition cannot be reversed, but progression can be delayed, although symptoms of acute renal failure eventually develop. Only *hemodialysis* or a kidney transplantation can prevent death from kidney failure.

In **hemodialysis** (hē-mō-dī-AL-i-sis), an artificial membrane is used to regulate the composition of the blood by means of a *dialysis machine* (Figure 20-11•). The basic principle involved in this process, called **dialysis**, involves the passive diffusion of molecules across a selectively permeable membrane. The patient's blood flows past one side of an artificial dialysis membrane that contains pores large enough to permit the diffusion of small ions and molecules but small enough to prevent the loss of the larger plasma proteins. A *dialysis fluid* flows on the other side. Various ions (phosphate, potassium, and sulfate), urea, creatinine, and uric acid diffuse into the dialysis fluid, and bicarbonate ions and glucose diffuse into the blood.

In practice, silastic tubes are inserted into a medium-sized artery and vein in one of the limbs. The two tubes are then connected by a *shunt* as shown in Figure 20-11b•. When connected to the dialysis machine, the individual sits quietly while blood circulates from the arterial shunt, through the machine, and back through the venous shunt. Inside the machine, the blood flows within a tube composed of dialysis membrane, and diffusion occurs between the blood and dialysis fluid. The original shunts were lined with Teflon to reduce problems with blood clotting. For chronic dialysis, surgery can create an artery-venous connection (a *fistula*) that replaces the artificial shunt.

DISORDERS OF RENAL FUNCTION

When gains are equal to losses, an individual is said to be in a state of balance. To balance every ml of water lost in the urine, one ml of water must be obtained; for every sodium ion excreted in sweat, a sodium ion must be absorbed from food; for every hydrogen ion released by a molecule of lactic acid, a hydrogen ion must be eliminated. Fluid balance, electrolyte balance, and acid–base balance have central roles in maintaining health. Treatment of any serious illness affecting the nervous, cardiovascular, respiratory, urinary, or digestive system must always include steps to restore normal fluid, electrolyte, and acid–base balance.

Fluid Imbalances

Fluid imbalances occur when the amount of water gain is not equal to water loss by the body.

•FIGURE 20-11
Hemodialysis.

(a) A patient connected to a dialysis machine, or "kidney machine." (b) A diagram of the dialysis procedure. Preparation for hemodialysis involves the insertion of a pair of shunts connected by a U-shaped loop (shunt) that permits normal blood flow when the patient is not connected to the machine.

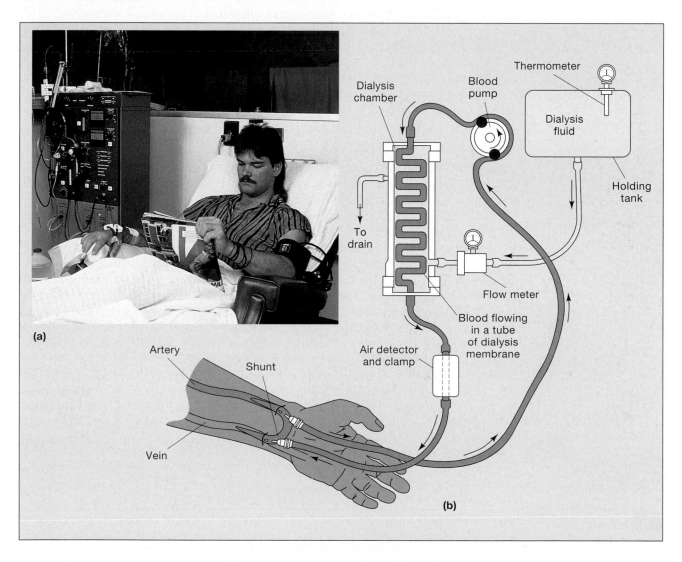

Water Losses Exceed Gains. Excess water loss soon produces dehydration (p. 498). Severe dehydration is dangerous because it reduces blood volume and blood pressure, and virtually all vital physiological processes occur in solution. Dehydration can occur from excessive perspiration (exercising in hot weather), inadequate water consumption (being lost in the desert), repeated vomiting, and diarrhea. These disorders promote water losses far in excess of electrolyte losses, so body fluids become increasingly concentrated (hypertonic). Normal responses that attempt to restore homeostasis include ADH and renin secretion, and increased thirst, which promotes an increase in fluid intake.

Water Gains Exceed Losses. When you drink a glass of pure water, such as distilled water, the water content of the body increases without a corresponding increase in solutes. Normally, this situation will be quickly corrected. The decreased solute concentration of the plasma depresses the secretion of ADH, reduces thirst, and increases water losses in the urine.

Diuretics (dī-ū-RET-iks) are drugs that promote the loss of water in the urine. Diuretics have many different mechanisms of action, but each affects transport activities and/or water reabsorption along the nephron and collecting system. Examples of some important diuretics in use today are shown in Table 20-5.

Edema. *Edema* is a condition in which there is an accumulation of fluid in the body tissues, one of the fluid compartments of the body. Edema has been discussed in earlier chapters as a symptom of disorders of various systems. ∞ *[p. 371]* Examples of edema include swelling associated with local inflammation, abdominal swelling (*ascites*), fluid filling the pleural cavity (*pleural effusion*), or as a general swelling of the body (*anasarca*, formerly known as *dropsy*). Swelling may not occur immediately, and edema may then only be recognized as a gain in weight. Examples of causes of edema include heart failure (p. 373), renal failure (p. 508), and cirrhosis of the liver (p. 457).

Electrolyte Imbalances

Electrolyte imbalances can result in deficiencies or excesses of minerals in body fluids. (Minerals in solution are also called electrolytes.) The general effects of such imbalances were listed in Table 19-1 (p. 473). This section provides a description of disorders associated with sodium and potassium, electrolytes that can produce major fluid shifts, and calcium, the most abundant mineral in the body.

Hypernatremia. **Hypernatremia** (hī-per-nah-TRĒ-mē-ah) is a condition of excess sodium in the blood plasma, or ECF. It is caused by dehydration, a topic discussed in Chapter 19 (p. 478). Symptoms of hypernatremia (and dehydration) include thirst, dryness and wrinkling of the skin, reduced blood volume and pressure, and an eventual circulatory collapse. Treatment is the ingestion of water or intravenous infusion of a hypotonic solution. *Normal saline*, a solution that contains a concentration of 0.9 percent sodium chloride, is isotonic with normal body fluids. It can therefore be infused to treat hypernatremia.

Hyponatremia. The condition of **hyponatremia** exists when the concentration of sodium in the ECF is lower than normal. It is caused by *overhydration*. ∞ *[p. 479]* Symptoms include disturbed CNS function (water intoxication), confusion, hallucinations, convulsions, coma, and, in severe cases, death. Treatment involves the use of diuretics and, rarely, the infusion of hypertonic salt solution.

Table 20-5	Representative Classes of Diuretic Drugs
Class (Example)	**Mechanism of Action**
Osmotic diuretics (Mannitol)	These are metabolically harmless substances that are filtered at the glomerulus and are not reabsorbed. Their presence in the urine increases its solute concentration and limits the amount of water reabsorption possible
Thiazides (Chlorothiazide)	These drugs reduce sodium and chloride transport in the proximal and distal tubules. They are often used to accelerate fluid losses in the treatment of hypertension and peripheral edema
Aldosterone blocking agents (Spironolactone)	By blocking the action of aldosterone, these drugs prevent the reabsorption of sodium along the DCT and collecting duct and thereby accelerate fluid losses
ACE (angiotensin-converting enzyme) inhibitors Capoten (Captopril)	These drugs prevent the activation of angiotensin II and thus inhibit aldosterone production. In the absence of aldosterone, water losses increase
Drugs with diuretic side effects (Caffeine, alcohol)	These drugs may work directly on the kidney tubules or indirectly, through effects on hormone production.

Hyperkalemia. Hyperkalemia (hī-per-kah-LĒ-mē-ah) occurs when there is a high concentration of potassium in the ECF. It may be caused by renal failure (which prevents normal potassium secretion), the use of some diuretics (which prevent sodium reabsorption and potassium secretion), and chronic acidosis (due to the movement of potassium ions out of cells and into the ECF). The most dangerous problem in hyperkalemia is severe cardiac arrhythmias. Treatment involves the infusion of hypotonic solution, selection of different diuretics, and the infusion of buffers to absorb hydrogen ions and combat the acidosis.

Hypokalemia. Hypokalemia is a condition of low potassium in the ECF. It may be caused by a low-potassium diet, diuretics, hypersecretion of aldosterone, or chronic alkalosis. Several diuretics can cause hypokalemia because they result in the production of large volumes of urine, and the large volume carries away significant amounts of sodium and potassium. Excess levels of aldosterone can also result in the retention of sodium ions and the loss of potassium ions. Alkalosis causes hypokalemia because cells respond by releasing hydrogen ions in exchange for potassium ions in the ECF. Treatment of hypokalemia includes increasing the amount of potassium in the diet, the ingestion of potassium tablets or solutions, and the intravenous infusion of a potassium-rich solution.

Hypercalcemia. Hypercalcemia (hi-per-kal-SĒ-me-ah) is a condition of a high calcium ion concentration in the ECF. The primary cause is hyperparathyroidism. Secondary causes include malignant cancers of the breast, lung, kidney, or bone marrow, and vitamin D toxicity. Symptoms include fatigue, confusion, muscle pain, cardiac arrhythmias, kidney stones, and calcification of soft tissues. Treatment involves surgery to remove the parathyroid glands, infusion of hypotonic fluid to lower calcium levels, and administration of calcitonin.

Hypocalcemia. Hypocalcemia, a condition of low levels of calcium in the blood plasma of the ECF, is less common than hypercalcemia. Hypoparathyroidism, vitamin D deficiency, or chronic renal failure are typically responsible for hypocalcemia. Symptoms include muscle spasms, convulsions, weak heartbeats, cardiac arrhythmias, and osteoporosis.

Acid–Base Imbalances

Normal acid–base balance is maintained by the various buffer systems, the lungs, and the kidneys, as was discussed earlier. ∞ *[pp. 500–501]* Together, these mechanisms are usually able to control pH so that the pH range of the extracellular fluids seldom varies outside the normal range of 7.35 to 7.45. When buffering mechanisms are severely stressed, the pH wanders outside of these limits, producing symptoms of alkalosis or acidosis.

There are two primary sources of acid–base, or pH, problems: respiratory disorders and metabolic disorders. **Respiratory acid–base disorders** result from abnormal carbon dioxide levels in the ECF. These conditions are related to an imbalance between the rate of carbon dioxide removal at the lungs and its generation in other tissues. **Metabolic acid–base disorders** are caused by the generation of acids during metabolism, or by conditions affecting the concentration of bicarbonate ions in the extracellular fluids.

Respiratory Acidosis. Respiratory acidosis develops when the respiratory system is unable to eliminate all of the CO_2 generated by body tissues. The primary symptom is low plasma pH due to *hypercapnia*, an elevated level of carbon dioxide in the plasma.

Respiratory acidosis represents the most frequent challenge to maintaining acid–base balance. The usual cause is *hypoventilation*, an abnormally low respiratory rate. Our tissues generate carbon dioxide at a rapid rate, and even a few minutes of hypoventilation can cause acidosis, reducing the pH of the ECF to as low as 7.0. Under normal circumstances, the chemoreceptors monitoring the CO_2 concentration of the plasma and cerebrospinal fluid (CSF) will eliminate the problem by stimulating an increase in the depth and rate of respiration.

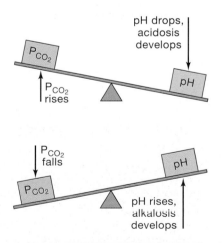

Changes in CO_2 levels can have a dramatic effect on blood pH.

Respiratory Alkalosis. Problems with **respiratory alkalosis** are relatively uncommon. This disorder develops when respiratory activity lowers plasma CO_2 concentration to below normal levels, a condition called *hypocapnia*. A temporary hypocapnia can be produced by hyperventilation, when increased respiratory activity leads to a reduction in the arterial CO_2 concentration. Continued hyperventilation can elevate the pH to levels as high as 7.8–8. This condition usually corrects itself, for the reduction in plasma CO_2 concentration removes the stimulation for the chemoreceptors, and the urge to breathe fades until carbon dioxide levels have returned to normal. Respiratory alkalosis caused by hyperventilation seldom persists long enough to cause a clinical emergency, although the associated tingling in the hands and feet and a light-headed feeling can be frightening. Breathing in and out of a paper bag for a few minutes increases the plasma CO_2 without lowering oxygen levels. This helps resolve the condition.

Metabolic Acidosis. **Metabolic acidosis** is the second most common type of acid–base imbalance. It is most frequently caused by the production of a large number of metabolic acids such as lactic acid or ketone bodies. It can also be caused by problems in excreting hydrogen ions at the kidneys. In fact, any condition accompanied by severe kidney damage can result in metabolic acidosis. Compensation for metabolic acidosis usually involves a combination of respiratory and renal mechanisms. Hydrogen ions interacting with bicarbonate ions form carbon dioxide molecules that are eliminated at the lungs. At the same time, the kidneys excrete additional hydrogen ions into the urine and generate bicarbonate ions that are released into the ECF. This combination elevates the pH and restores acid–base balance.

Metabolic Alkalosis. **Metabolic alkalosis** occurs when bicarbonate ion concentrations become elevated. The bicarbonate ions then interact with hydrogen ions in solution, forming carbonic acid, and the reduction in hydrogen ion concentrations then causes symptoms of alkalosis. Another cause of metabolic alkalosis is hydrochloric acid secretion by the stomach, because when H^+ is secreted into the stomach, bicarbonate ions are released into the extracellular fluid. The sudden surge in bicarbonate ion levels after a meal is known as the *alkaline tide*; it often results in an increase in blood pH and a mild alkalosis. When frequent vomiting occurs, stomach acid production is continuous; under these circumstances, acute metabolic alkalosis can develop. Cases of severe metabolic alkalosis are relatively rare. Compensation involves a reduction in breathing, coupled with the increased loss of bicarbonates in the urine.

✔ What is a UTI?

✔ Why is it important to urinate frequently?

✔ How do the lungs function to maintain acid–base balance during acidosis?

CHAPTER REVIEW

Key Words

acidosis (a-sid-Ō-sis): An abnormal condition characterized by a plasma pH below 7.35.

aldosterone: A steroid hormone produced by the adrenal cortex that stimulates sodium and water conservation at the kidneys; secreted in response to the presence of angiotensin II.

alkalosis (al-kah-LŌ-sis): Condition characterized by a plasma pH greater than 7.45.

angiotensin II: A hormone that causes an elevation in systemic blood pressure, stimulates secretion of aldosterone, promotes thirst, and causes the release of ADH.

buffer system: Interacting compounds that prevent increases or decreases in the pH of body fluids; includes the carbonic acid–bicarbonate buffer system, the phosphate buffer system, and the protein buffer system.

glomerulus: In the kidneys, a knot of capillaries that projects into the enlarged end of a nephron; the site of filtration, the first step in urine production.

juxtaglomerular apparatus: A complex responsible for the release of renin and erythropoietin; consists of the distal convoluted tubule and modified smooth muscle cells in the walls of the afferent and efferent arterioles adjacent to the glomerulus.

kidney: A component of the urinary system; an organ that regulates the composition of the blood, including excretion of wastes and maintenance of normal fluid and electrolyte balance.

micturition (mik-tū-RI-shun): Urination.

nephron (NEF-ron): The basic functional unit of the kidney.

peritubular capillaries: A network of capillaries that surrounds the proximal and distal convoluted tubules of the kidneys.

renin: Enzyme released by the juxtaglomerular cells when kidney blood pressure or oxygen levels in the blood decline; starts the conversion of an inactive protein into angiotensin II.

ureters: Muscular tubes that carry urine from the renal pelvis to the urinary bladder.

Selected Clinical Terms

calculi (KAL-kū-lī): Solid deposits (stones) that form within the urinary tract, usually from calcium, magnesium, or uric acid.

cystitis (sis-TĪ-tis): Inflammation of the lining of the urinary bladder.

dysuria (dis-Ū-rē-a): Painful urination.

glomerulonephritis (glo-MER-ū-lō-ne-FRĪ-tis): Inflammation of the glomeruli.

hematuria (hē-ma-TOO-rē-uh): Blood in the urine.

hemodialysis: A technique in which an artificial membrane is used to regulate the chemical composition of the blood.

hyperkalemia (hī-per-kah-LĒ-mē-ah): High potassium levels in the blood.

hypernatremia (hī-per-nah-TRĒ-mē-ah): High sodium levels in the blood.

incontinence: An inability to control urination voluntarily.

leptospirosis: A rare bacterial (spirochete) infection of the kidney.

nephritis (nef-RĪ-tis): Inflammation of the kidney.

nephrolithiasis (nef-rō-li-THĪ-a-sis): A condition resulting from the formation of kidney stones.

proteinuria (prō-tēn-OOR-ē-uh): Protein loss in the urine.

pyelonephritis (pī-e-lō-ne-FRĪ-tis): Inflammation of the kidney and renal pelvis.

renal failure: An inability of the kidneys to excrete water, ions, and metabolic wastes in sufficient quantities to maintain homeostasis.

respiratory acidosis: The most common form of acidosis, which results from inadequate respiratory activity (hypoventilation). It is characterized by elevated levels of carbon dioxide (hypercapnia) in body fluids.

urethritis (ū-re-THRĪ-tis): Inflammation of the urethra.

Study Outline

INTRODUCTION (p. 487)

1. Wastes generated by the cells of the body are excreted by the digestive, respiratory, and urinary systems.

SYSTEM BRIEF (p. 487)

1. The functions of the urinary system include (1) elimination of organic waste products, (2) regulation of blood volume and pressure, (3) regulation of plasma concentrations of ions, and (4) balance of blood pH.

STRUCTURE OF THE URINARY SYSTEM (p. 487)

1. The urinary system includes the kidneys, ureters, urinary bladder, and urethra. The kidneys produce urine (a fluid containing water, ions, and soluble compounds); during urination urine is forced out of the body. (Figure 20-1)

THE KIDNEYS (pp. 487–492)

1. The left kidney sits slightly higher than the right kidney. Both lie in a retroperitoneal position. (Figure 20-1)

KIDNEY STRUCTURE (p. 488)

2. The hilus provides entry for the renal artery and exit for the renal vein and ureter.

3. The ureter communicates with the renal pelvis. This chamber branches into two or three major calyces, each connected to four or five minor calyces. (Figure 20-2)

THE URINE-PRODUCING UNITS (p. 488)

4. The nephron (the basic functional unit in the kidney) includes the renal corpuscle and a twisted tubule that empties into the collecting system via a collecting duct. From the renal corpuscle the filtrate travels through the proximal convoluted tubule, the loop of Henle, and the distal convoluted tubule. (Figure 20-3)

5. Nephrons are responsible for (1) production of filtrate, (2) reabsorption of nutrients, and (3) reabsorption of water and ions.

6. The renal tubule begins at the renal corpuscle. It includes a knot of intertwined capillaries called the glomerulus surrounded by the Bowman's capsule. Blood arrives via the afferent arteriole and departs via the efferent arteriole. (Figure 20-4)

7. The proximal convoluted tubule (PCT) actively reabsorbs nutrients, plasma proteins, and electrolytes from the filtrate. They are then released into the surrounding interstitial fluid.

8. The loop of Henle includes a descending limb and an ascending limb. The descending limb is permeable to water; the ascending limb is impermeable to water and solutes.

9. The ascending limb delivers fluid to the distal convoluted tubule (DCT), which actively secretes ions and reabsorbs sodium ions from the urine.

10. The collecting ducts receive urine from nephrons and merge into a duct that delivers urine to a minor calyx. The collecting system makes final adjustments to the urine by reabsorbing water, or reabsorbing or secreting various ions.

THE BLOOD SUPPLY TO THE KIDNEYS (p. 491)

11. The renal arteries of the kidneys split up into a series of smaller arteries around the renal pyramids. Blood reaches the nephrons through afferent arterioles. Blood travels from the efferent arteriole to the peritubular capillaries and the vasa recta. It then returns through larger and larger veins to the renal vein. (Figure 20-5)

THE FORMATION OF URINE (pp. 493–494)

1. The primary goal in urine production is the excretion and elimination of dissolved solutes, principally metabolic waste products, such as urea, creatinine, and uric acid.

2. Urine formation involves filtration, reabsorption, and secretion. (Table 20-1)

FILTRATION (p. 493)

3. Glomerular filtration occurs as fluids move across the wall of the glomerulus into the capsular space, in response to blood pressure in the glomerular capillaries. The glomerular filtration rate (GFR) is the amount of filtrate produced in the kidneys each minute. Any factor that alters the filtration (blood) pressure will change the GFR and affect kidney function.

4. Dropping filtration pressures stimulate the juxtoglomerular apparatus to release renin. Renin increases blood volume and blood pressure.

REABSORPTION AND SECRETION *(p. 494)*

5. Reabsorption is the return of water, organic nutrients, and ions from the filtrate into the surrounding interstitial fluid and then into the blood.

6. Secretion is the active transport of materials from the interstitial fluid into the tubular fluid. Secreted materials include drugs, toxins, ions, and acids. Aldosterone levels affect the rates of sodium ion reabsorption and potassium ion secretion.

7. Some waste products such as urea and uric acid are neither reabsorbed nor secreted into the tubular fluid. Their concentration gradually rises as urine is formed. *(Table 20-2)*

THE CONTROL OF KIDNEY FUNCTION *(pp. 495–496)*

1. The kidneys act to maintain adequate oxygen levels and blood pressure in the body.

2. Hormones that regulate kidney function include erythropoietin, angiotensin II, aldosterone, ADH, and ANP.

CONTROLLING URINE VOLUME *(p. 495)*

3. ADH (*antidiuretic hormone*) regulates the volume of urine by controlling the amount of water reabsorbed along the DCT and collecting duct. If circulating ADH levels are low, urine volume is high; if circulating ADH levels are high, urine volume is low. *(Figure 20-6)*.

RESPONSES TO CHANGES IN BLOOD PRESSURE *(p. 495)*

4. Low blood pressure stimulates the secretion of erythropoietin, the *renin-angiotensin system*, and the secretion of angiotensin II, aldosterone, and ADH. Cardiac cells secrete ANP under conditions of high blood pressure. ANP inhibits the hormones of the renin-angiotensin system and reduces blood pressure.

URINE TRANSPORT, STORAGE, AND EXCRETION *(pp. 496–498)*

1. Urine produced by the kidneys is transported, stored, and eliminated by the rest of the urinary system. *(Figures 20-1 and 20-7)*

THE URETERS *(p. 496)*

2. The ureters extend from the renal pelvis to the urinary bladder. Peristaltic contractions by smooth muscles move the urine.

THE URINARY BLADDER *(p. 497)*

3. Internal features of the urinary bladder include the **trigone**, the *neck*, and the **internal sphincter**. Contractions of smooth muscle within the bladder wall compress the urinary bladder and expel the urine into the urethra. *(Figure 20-7)*

THE URETHRA *(p. 497)*

4. In both sexes, as the urethra passes through the pelvic floor a circular band of skeletal muscles forms the **external sphincter**, which is under voluntary control. *(Figure 20-7)*

URINE EXCRETION *(p. 497)*

5. The process of urination is initiated by stretch receptors in the urinary bladder wall. Voluntary urination involves the voluntary relaxation of the external sphincter, which allows the opening of the **internal sphincter**.

BODY FLUIDS *(pp. 498–501)*

1. The operations of body cells depend on water as a diffusion medium for dissolved gases, nutrients, and waste products. Maintenance of normal volume and composition in the extracellular and intracellular fluids is vital to life.

FLUID BALANCE *(p. 498)*

2. The **intracellular fluid (ICF)** contains nearly two-thirds of the total body water; the **extracellular fluid (ECF)** contains the rest. Exchange occurs between the ICF and ECF, but the two **fluid compartments** retain their distinctive characteristics. *(Figure 20-8)*

3. Water circulates freely within the ECF compartment. At capillary beds hydrostatic pressure forces water from the plasma into the interstitial spaces. Water moves back and forth across the epithelial lining of the peritoneal, pleural, and pericardial cavities; through synovial membranes lining joint capsules; and between the blood and cerebrospinal fluid, the aqueous humor and vitreous body of the eye, and the perilymph and endolymph of the inner ear.

4. Water losses are normally balanced by gains through eating, drinking, and metabolic generation. *(Table 20-3)*

ELECTROLYTE BALANCE *(p. 499)*

5. Electrolyte balance is important because total electrolyte concentrations affect water balance and because the levels of individual electrolytes can affect a variety of cell functions.

6. The rate of sodium uptake across the digestive epithelium is directly proportional to the amount of sodium in the diet. Sodium losses occur mainly in the urine and through perspiration.

7. Potassium ion concentrations in the ECF are very low. Potassium excretion increases as ECF concentrations rise, under aldosterone stimulation.

ACID–BASE BALANCE *(p. 500)*

8. The pH of normal body fluids ranges from 7.35 to 7.45; variations outside this relatively narrow range produces **acidosis** or **alkalosis**.

9. Carbonic acid is the most important factor affecting the pH of the ECF. In solution, CO_2 reacts with water to form carbonic acid; the dissociation of carbonic acid releases H^+ ions. An inverse relationship exists between the concentration of CO_2 and pH.

10. A **buffer system** prevents increases or decreases in the pH of body fluids. There are three major buffer systems: (1) *protein buffer systems* in the ECF and ICF; (2) the *carbonic acid–bicarbonate buffer system*, most important in the ECF; and (3) the *phosphate buffer system* in the intracellular fluids and urine.

11. The lungs help regulate pH by affecting the carbonic acid–bicarbonate buffer system; changing the respiratory rate can raise or lower the CO_2 concentration of body fluids, affecting the buffering capacity.

12. The kidneys vary their rates of hydrogen ion secretion and bicarbonate ion reabsorption depending on the pH of extracellular fluids.

AGING AND THE URINARY SYSTEM *(p. 501)*

1. Aging is usually associated with increased kidney problems. Age-related changes in the urinary system include declining numbers of functional nephrons and reduced GFR, reduced sensitivity to ADH, and problems with urination.

DISORDERS OF THE URINARY SYSTEM *(pp. 501–512)*

1. Disorders of the urinary system may occur primarily in the kidneys, or be caused by disorders in the urine transport and storage organs, the ureters, the urinary bladder, and the urethra.

THE PHYSICAL EXAMINATION *(p. 501)*

2. Urology is a surgical specialty concerned with the urinary system and its disorders. **Nephrology** (ne-FROL-o-jē) is restricted to the kidneys and their disorders.

3. Common symptoms of urinary system disorders are pain and changes in the frequency and amount of urination.

4. Urinary system disorders may cause individuals to urinate more or less often than usual, and produce normal or abnormal amounts of urine. **Polyuria** is the production of excessive amounts of urine. **Oliguria** is a below normal volume of urine production. **Anuria** is very low urine volume or none at all.

5. Signs of urinary system disorders include: **hematuria**, the presence of red blood cells in the urine; **hemoglobinuria**, the presence of hemoglobin in the urine; changes in urine color; edema; and, fever.

6. Many procedures and laboratory tests may be used in the diagnosis of urinary system disorders. *(Figure 20-9; Table 20-4)*

URINARY SYSTEM DISORDERS *(p. 503)*

7. Major disorders of the urinary system include inflammation and infection, tumors, immune disorders, congenital disorders, and degenerative disorders. *(Figure 20-10)*

8. *Urinary Tract Infections (UTIs)* occur more often in women than men.

9. **Nephritis**, inflammation of the kidneys, may result from bacterial infections or exposure to toxins or irritating drugs.

10. **Pyelitis** is an inflammation of the renal pelvis, and **pyelonephritis** is an inflammation of the renal pelvis and kidney.

11. **Leptospirosis** is a rare infection of the kidneys caused by the spirochete bacterium, *Leptospira interrogans*.

12. **Ureteritis** is inflammation of the ureters, that may be caused by an infection spreading from the bladder, or blockage with a kidney stone. In **nephrolithiasis**, kidney stones, or **calculi,** may form within the urinary tract from minerals or uric acid.

13. **Cystitis** is an inflammation of the lining of the urinary bladder, often due to bacterial infection.

14. **Urethritis** is an inflammation of the urethra, associated with bacterial infections.

15. **Renal cell carcinoma** is the most common malignant tumor of the kidney. **Nephroblastoma**, is a cancer that affects children under the age of 4, most commonly boys.

16. **Bladder cancer** may remain within the bladder, or spread through the bladder wall and into other nearby organs.

17. **Glomerulonephritis** is an inflammation of the glomeruli caused by an immune response to a streptococcal bacteria infection.

18. **Polycystic kidney disease** is an inherited condition that gradually causes cysts, or swellings, to form in the renal tubules of the nephrons.

19. **Renal glycosuria** results in high levels of glucose in the urine, despite normal blood glucose levels, and **aminoaciduria** causes high amino acid levels in the urine.

20. **Renal failure** occurs when the kidneys cannot perform the excretory functions to maintain homeostasis.

21. **Hemodialysis** is a technique used to regulate the composition of the blood artificially. *(Figure 20-11)*

DISORDERS OF RENAL FUNCTION *(p. 508)*

22. Fluid imbalances result in **fluid shifts**, water movements between different fluid compartments.

23. *Dehydration* from excessive perspiration, inadequate water consumption, repeated vomiting, and diarrhea cause body fluid solutes to become increasingly concentrated.

24. Water gain, or *overhydration*, may be treated with **diuretics**, drugs that promote the loss of water in the urine. *(Table 20-5)*

25. Electrolyte imbalances can result in mineral deficiency and excess disorders, and fluid imbalances.

26. Higher and lower than normal plasma levels of sodium are called **hypernatremia** and **hyponatremia**. Higher and lower than normal plasma levels of potassium are called **hyperkalemia** and **hypokalemia**. Higher and lower than normal plasma levels of calcium are called **hypercalcemia** and **hypocalcemia**.

27. Acid–base imbalances are commonly due to respiratory and metabolic disorders. They include **respiratory acidosis**, **respiratory alkalosis**, **metabolic acidosis**, and **metabolic alkalosis**.

Review Questions

MATCHING

Match each item in Column A with the most closely related item in Column B. Use letters for answers in the spaces provided.

	Column A		Column B
___	1. urination	a.	site of urine production
___	2. external sphincter	b.	capillaries around loop of Henle
___	3. hilus	c.	causes sensation of thirst
___	4. potassium	d.	micturition
___	5. nephrons	e.	voluntary control
___	6. renal corpuscle	f.	exit for ureter
___	7. sodium	g.	blood in urine
___	8. internal sphincter	h.	blood leaves glomerulus
___	9. efferent arteriole	i.	dominant cation in ICF
___	10. afferent arteriole	j.	dominant cation in ECF
___	11. vasa recta	k.	painful urination
___	12. ECF	l.	interstitial fluid
___	13. ADH	m.	contains glomerulus
___	14. dysuria	n.	blood to glomerulus
___	15. hematuria	o.	involuntary control

MULTIPLE CHOICE

16. The urinary system includes all the following except the

_____ .

(a) kidneys
(b) gallbladder
(c) ureters
(d) urinary bladder

17. Filtrate is formed within the glomerulus by the process of

_____ .

(a) diffusion
(b) osmosis
(c) filtration
(d) active transport

18. After the filtrate leaves the glomerulus it enters the

_____ .

(a) proximal convoluted loop
(b) distal convoluted loop
(c) collecting duct
(d) loop of Henle

19. The urinary system regulates blood volume and pressure by _____ .
(a) adjusting the volume of water lost in the urine
(b) releasing erythropoietin
(c) releasing renin
(d) all the above

20. The reabsorption of _____ ions and the resulting uptake of water by osmosis is stimulated by the hormone _____ .
(a) sodium; erythropoietin
(b) potassium; erythropoietin
(c) sodium; aldosterone
(d) potassium; aldosterone

21. The metabolism of amino acids results in the formation of a nitrogenous waste product called _____ .
(a) urea
(b) creatinine
(c) glycogen
(d) uric acid

22. Of the following, _____ do(es) not contribute to maintaining a relatively constant pH level in body fluids.
(a) kidneys
(b) buffer systems
(c) lungs
(d) liver

23. _____ is the production of excessive amounts of urine.
(a) oliguria
(b) polyuria
(c) anuria
(d) proteinuria

24. _____ is an inflammation of the lining of the urinary bladder.
(a) pyelitis
(b) urethritis
(c) ureteritis
(d) cystitis

25. _____ is the most frequent acid–base disorder.
(a) metabolic acidosis
(b) respiratory acidosis
(c) metabolic alkalosis
(d) respiratory alkalosis

TRUE/FALSE

___ 26. Water makes up about 30 percent of the human body.

___ 27. The kidneys lie behind the peritoneum.

___ 28. An increase in the CO_2 concentration in blood plasma decreases the pH of the plasma.

___ 29. Buffers help maintain a constant concentration of electrolytes within body fluids.

___ 30. The ureter carries urine from the urinary bladder to the outside.

___ 31. Interstitial fluid is an important part of intracellular fluid (ICF).

___ 32. Incontinence involves periodic involuntary urination.

___ 33. In cytoscopy, a viewing tube is inserted in the ureter to view the lining of the urinary bladder.

SHORT ESSAY

34. What is the primary function of the urinary system?

35. What structures are included as parts of the urinary system?

36. Why is it important to maintain a normal balance of electrolytes in body fluids?

37. Why do long-haul trailer truck drivers frequently experience kidney problems?

38. Mr. Casey is a healthy 67-year-old man. Explain how his urinary system would differ from that of his 10-year-old grandson.

39. The treatment of polycystic kidney disease and renal failure involve hemodialysis. What is hemodialysis?

40. Exercise physiologists recommend that adequate amounts of fluid be ingested before, during, and after exercise. Why is fluid replacement during extensive sweating important?

APPLICATIONS

41. For the past week, Patsy has felt a burning sensation in the urethral area when she urinates. She checks her temperature and finds she has a low-grade fever. What unusual substances are likely to be present in her urine? Explain why Patsy is more likely to suffer from this condition than her husband.

42. Laboratory tests show that Brandon, a normal 15-year-old, has a high glucose level in his urine and a normal glucose level in his blood. The normal level of blood glucose indicates he does not have diabetes mellitus. What condition would account for these glucose measurements?

✔ Answers to Concept Check Questions

(p. 492) **1.** The renal corpuscle, proximal convoluted tubule, distal convoluted tubule, and the proximal portions of the loop of Henle and collecting duct are all located within the renal cortex. **2.** Damage to the juxtaglomerular apparatus of the nephrons would interfere with the hormonal control of blood pressure. **3.** The openings between the cells that cover the glomerular capillaries are so fine that they will allow only substances smaller than plasma proteins to pass into the capsular space.

(p. 494) **1.** Decreases in blood pressure would reduce the blood hydrostatic pressure within the glomerulus and decrease the GFR. **2.** If the nephrons lacked a loop of Henle, the kidneys would not be able to form a concentrated urine. **3.** The reabsorption of water along the DCT results in an increase in the concentration of urea (and other waste products such as uric acid) in the tubular fluid.

(p. 496) **1.** ADH acts on the DCT and collecting duct by stimulating the reabsorption of water from the tubular fluid. **2.** Both renin and erythropoietin are released by kidney cells. Renin is released in response to low blood pressure, erythropoietin in response to low blood oxygen levels. Their release results in an increase in blood pressure and blood volume.

(p. 498) **1.** Under normal conditions, peristaltic contractions move urine along the minor and major calyces toward the renal pelvis, out of the renal pelvis, and along the ureter to the bladder. **2.** An obstruction of the ureters would interfere with the passage of urine from the renal pelvis to the urinary bladder. **3.** In order to control urination, one must be able to control the external sphincter, a ring of skeletal muscle that acts as a valve.

(p. 500) **1.** Consuming a meal high in salt would temporarily increase the solute concentration of the ECF. As a result some of the water in the ICF would shift into the ECF. **2.** Fluid loss through perspiration, urine formation, and respiration would increase the electrolyte concentration of body fluids.

(p. 501) **1.** Carbon dioxide affects pH because it combines with water to form carbonic acid. Carbonic acid molecules do not remain intact but break apart into hydrogen ions and bicarbonate ions. The greater the concentration of carbon dioxide, the more carbonic acid molecules form and break apart, releasing hydrogen ions. The greater the concentration of hydrogen ions, the lower the pH. **2.** A decrease in the pH of body fluids would have a stimulating effect on the respiratory center in the medulla oblongata. The result would be an increase in the rate of breathing. This would lead to the elimination of more carbon dioxide, which would tend to cause the pH to increase.

(p. 512) **1.** A UTI is a *Urinary Tract Infection*. UTIs are more common in women than in men. Most UTIs begin in the urethra and ascend through the urinary bladder and the ureters, before reaching the renal pelvis and kidneys. **2.** Stagnant urine provides an environment well suited for bacterial growth. An empty bladder has much less chance of becoming infected than one in which urine is present for long periods. **3.** A decrease in the pH of body fluids would result in an increase in the rate of breathing. This would lead to the elimination of more carbon dioxide, which would cause the pH to rise.

CHAPTER
21

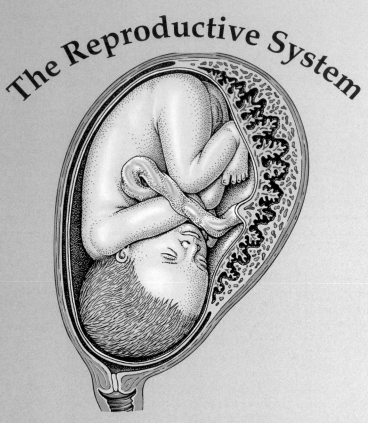

The Reproductive System

CHAPTER OUTLINE

Clinical Note: Birth Control Strategies

CHAPTER OBJECTIVES

1 Summarize the functions of the human reproductive system and its principal components.

2 Describe the components of the male reproductive system.

3 Describe the process of spermatogenesis.

4 Describe the roles of the male reproductive tract and accessory glands in the maturation and transport of spermatozoa.

5 Describe the hormones involved in sperm production.

6 Describe the components of the female reproductive system.

7 Describe the events of the ovarian and uterine cycles.

8 Describe the hormonal regulation of the ovarian and uterine cycles.

9 Describe the effects of age on the reproductive systems.

10 Describe the general symptoms and signs of male and female reproductive disorders.

11 Give examples of male and female reproductive disorders.

The reproductive system is the only one of our organ systems that is not essential for an individual's survival. It is indispensable, however, for the continuation of the human race.

Reproduction is the process that links one generation to the next in humans as well as in all other organisms. In simple, one-celled organisms reproduction occurs through the division of the organism itself. This type of reproduction is called *asexual*. Humans (and many familiar creatures) reproduce *sexually*, that is, they require two parents of opposite sexes for reproduction.

Human reproduction depends on the joining together of male and female reproductive cells, or **gametes** (GA-mēts), to form a new human being. The male and female gametes, a **sperm** from the father and an **ovum** (Ō-vum), or egg cell, from the mother, fuse in a process called **fertilization.** The resulting cell, called a **zygote** (ZĪ-gōt), divides repeatedly and gradually transforms into an individual with its own unique traits. This transformation process, called *development*, is the topic of the next chapter.

SYSTEM BRIEF

Human reproductive systems are specialized to produce, store, nourish, and transport male and female gametes. In both sexes, the gametes result from a special form of cell division called **meiosis** (mī-Ō-sis) (Figure 21-1•). The gametes produced through meiosis contain only half the number of chromosomes found in other cells. All other body cells contain 46 chromosomes, whereas gametes contain just 23. Thus the combination of a sperm and an egg produces a single cell with the normal complement of 46 chromosomes.

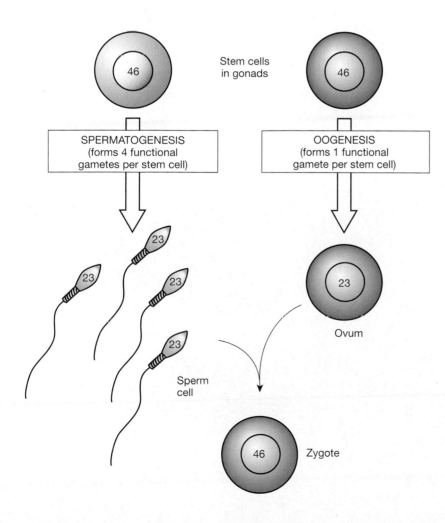

•**FIGURE 21-1**
Meiosis and the Formation of Gametes.
Special stem cells within male and female gonads undergo meiosis, which results in the formation of sex cells that contain half the normal number of chromosomes.

Both male and female reproductive systems include reproductive organs, or **gonads** (GŌ-nads), that produce gametes and hormones; ducts that receive and transport the gametes; accessory glands and organs that secrete fluids into various ducts; and structures associated with the reproductive system, collectively known as the **external genitalia** (jen-i-TĀ-lē-a).

The roles of the male and female reproductive systems are quite different. In the adult male the gonads, or **testes** (TES-tēz; singular: *testis*), secrete male sex hormones (*androgens*), principally testosterone, and produce a half-billion sperm each day. The sperm then travel along a lengthy duct system where they are mixed with the secretions of accessory glands, creating **semen** (SĒ-men). During **ejaculation** (e-jak-ū-LĀ-shun) the semen is expelled from the body.

The gonads, or **ovaries**, of adult females typically release only one ovum each month. This gamete travels along short **uterine tubes**, also called *fallopian tubes* or *oviducts*, that terminate in a muscular chamber, the **uterus** (Ū-te-rus). A short passageway, the **vagina** (va-JĪ-na), connects the uterus with the exterior. During intercourse the male ejaculation introduces semen into the vagina, and the sperm move farther along the female reproductive tract. If a single sperm fuses with an egg, fertilization occurs. The resulting cell divides repeatedly, and the process of development begins. The uterus provides protection and support as the cluster of cells becomes an *embryo* (months 1–2) and then a *fetus* (months 3–9), until the time of delivery.

STRUCTURE OF THE MALE REPRODUCTIVE SYSTEM

The principal structures of the male reproductive system are shown in Figure 21-2•. Proceeding from the testes, the sperm cells, or **spermatozoa** (sper-ma-tō-ZŌ-a), travel along the **epididymis** (ep-i-DID-i-mus), the **ductus deferens** (DUK-tus DEF-e-renz), the **ejaculatory** (ē-JAK-ū-la-tō-rē) **duct**, and the **urethra** before leaving the body. Accessory organs, notably the **seminal** (SEM-i-nal) **vesicles**, the **prostate** (PROS-tāt) **gland**, and the **bulbourethral** (bul-bō-ū-RĒ-thral) **glands**, empty their secretions into the ejaculatory ducts and urethra. Structures visible

•FIGURE 21-2
The Male Reproductive System.

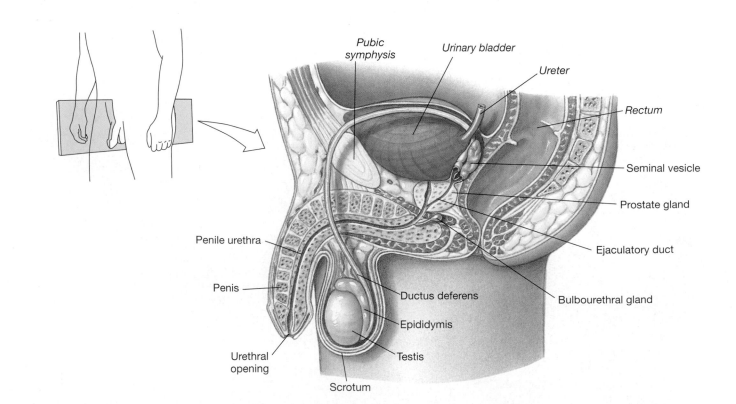

from the outside are the **scrotum** (SKRŌ-tum), which encloses the testes, and the **penis** (PĒ-nis). Together, the scrotum and penis constitute the external genitalia of the male.

THE TESTES

The *primary sex organs* of the male system are the testes. The testes hang within a fleshy pouch, the scrotum. The scrotum is subdivided into two chambers, each containing a testis. Each testis has the shape of an almond roughly 5 cm (2 in.) long and 2.5 cm (1 in.) wide.

The testes form in the abdominal cavity next to the kidneys. During fetal development the testes slowly descend from their original position and, in the seventh month, pass through openings (called *inguinal canals*) within the abdominal wall and into the scrotum. Each testis is attached to a **spermatic cord** made up of connective and muscle tissue surrounding a ductus deferens, blood vessels, lymphatic vessels, and nerves. The presence of the testes in the scrotum outside the body wall means that they are in an environment about 1.1° C (2° F) below normal body temperature. This cooler temperature is necessary for normal sperm development.

As Figure 21-3a • shows, each testis is subdivided into compartments called *lobules*. Within the lobules, sperm production occurs in tightly coiled **seminiferous** (se-mi-NIF-e-rus) **tubules**. Each tubule averages around 80 cm (31 in.) in length, and a typical testis contains nearly half a mile of seminiferous tubules.

Spermatozoa are produced by the process of **spermatogenesis** (sper-ma-tō-JEN-e-sis). This process begins with stem cells lying at the outermost layer of cells in the seminiferous tubules (Figure 21-3b •). The descendants of the stem cells un-

sperma, seed + *genesis*, origin
spermatogenesis: the process of sperm formation

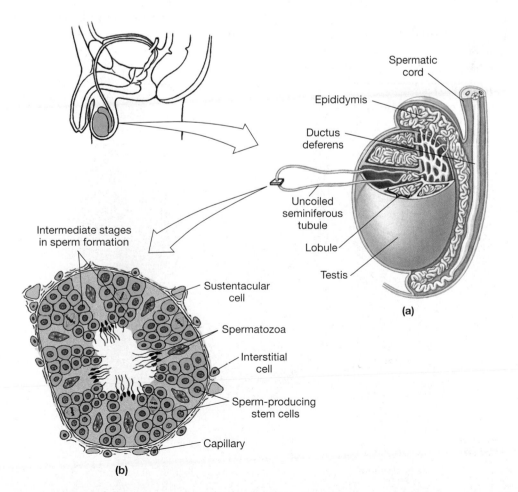

Spermatic cord

Epididymis

Ductus deferens

Uncoiled seminiferous tubule

Lobule

Testis

(a)

Intermediate stages in sperm formation

Sustentacular cell

Spermatozoa

Interstitial cell

Sperm-producing stem cells

Capillary

(b)

•**FIGURE 21-3**
Structure of the Testes.
(a) Diagrammatic sketch and structural relationships of the testes. **(b)** A section through a coiled seminiferous tubule showing its organization of cells.

dergo a halving of their chromosome number (meiosis) as they proceed to the central opening, or *lumen*.

In addition, each seminiferous tubule also contains **sustentacular** (sus-ten-TAK-ū-lar) **cells** (*Sertoli cells*). These large "nurse cells" provide nutrients and chemical signals that promote the development of spermatozoa. The entire process of spermatogenesis takes approximately 9 weeks.

The spaces between the tubules contain large **interstitial cells** that produce the male sex hormone **testosterone** (Figure 21-3b•). (Testosterone and other sex hormones were introduced in Chapter 13.) ∞ *[p. 300]*

The Sperm Cell

A sperm cell is quite small. For example, 1 ml of ejaculate can contain up to 100 million spermatozoa! Each sperm cell is made up of three regions: the *head*, the *middle piece*, and the *tail* (Figure 21-4•). The **head** is a flattened oval filled with densely packed chromosomes. The top of the head is covered by an **acrosomal** (ak-rō-SŌ-mal) **cap**, or **acrosome**, which contains enzymes essential for fertilization. A very short **neck** attaches the head to the **middle piece**, which is filled with mitochondria providing the energy for moving the flagellum that forms the **tail**.

THE MALE REPRODUCTIVE TRACT

The testes produce physically mature spermatozoa that are, as yet, incapable of fertilizing an ovum. The other portions of the male reproductive system, sometimes called the accessory structures, are concerned with the functional maturation, nourishment, storage, and transport of spermatozoa.

The Epididymis

Late in their development the spermatozoa enter the central opening of the seminiferous tubule. Although they have most of the physical characteristics of mature sperm cells, they are still functionally immature and incapable of locomotion.

•FIGURE 21-4
Structure of a Spermatozoon.

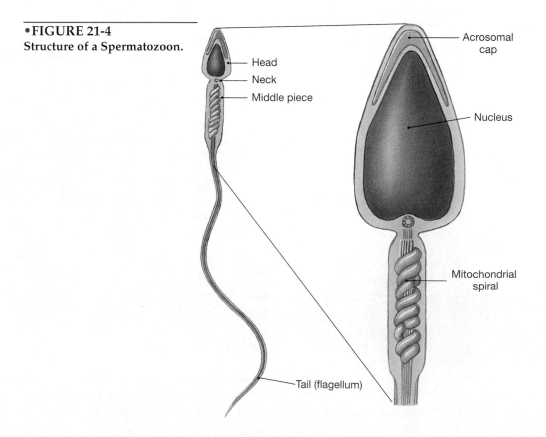

Head

Neck

Middle piece

Acrosomal cap

Nucleus

Mitochondrial spiral

Tail (flagellum)

At this point, fluid currents carry them into the **epididymis** (Figure 21-2•, p. 520). This tubule, almost 7 m (23 ft.) long, is so twisted and coiled that it actually takes up very little space. During the 2 weeks that a spermatozoon travels through the epididymis, it completes its physical maturation. The epididymis maintains suitable conditions for this process by absorbing cellular debris, providing organic nutrients, and recycling damaged or abnormal spermatozoa. Mature spermatozoa then arrive at the ductus deferens.

The Ductus Deferens

The **ductus deferens**, also known as the *vas deferens*, is 40 to 45 cm (16 to 18 in.) long. It extends toward the abdominal cavity within the spermatic cord. The ductus deferens curves downward alongside the urinary bladder toward the prostate gland (Figure 21-2•).

Peristaltic contractions in the muscular walls of the ductus deferens propel spermatozoa and fluid along the length of the duct. The ductus deferens can also store spermatozoa in an inactive state for up to several months.

The junction of the ductus deferens with the duct draining the seminal vesicle creates the **ejaculatory duct**, a relatively short (2 cm, or less than 1 in.) passageway that penetrates the muscular wall of the prostate and fuses with the ejaculatory duct from the other side before emptying into the urethra.

The Urethra

The urethra of the male extends from the urinary bladder to the tip of the penis, a distance of 15 to 20 cm (6 to 8 in.). The urethra in the male is a passageway used by both the urinary and reproductive systems.

THE ACCESSORY GLANDS

A typical ejaculation releases 2 to 5 ml of semen. This volume of fluid, called an **ejaculate**, is made up of spermatozoa, seminal fluid, and various enzymes. A normal **sperm count** ranges from 20 million to 100 million spermatozoa per milliliter. Because of their small size, sperm make up only about 1 percent of semen volume. **Seminal fluid**, the fluid component of semen, is a mixture of the combined secretions of the seminiferous tubules, the epididymis, and the accessory glands.

The fluids contributed by the seminiferous tubules and the epididymis account for only about 5 percent of the final volume of semen. The major fraction of seminal fluid is composed of secretions from the *seminal vesicles*, the *prostate gland*, and the *bulbourethral glands*. Major functions of these glandular organs include (1) activating the spermatozoa, (2) providing the nutrients that spermatozoa need for their own movement, (3) pushing spermatozoa and fluids along the reproductive tract by peristaltic contractions, and (4) producing buffers that counteract the acids found in the vagina.

The Seminal Vesicles

Each seminal vesicle is a tubular gland with a total length of around 15 cm (6 in.) that is compactly folded into a 5 cm x 2.5 cm (2 in. x 1 in.) mass.

The seminal vesicles contribute about 60 percent of the volume of semen. In particular, their secretions contain relatively high concentrations of fructose, a six-carbon sugar easily broken down by spermatozoa. The secretions are also slightly alkaline, and this alkalinity helps neutralize acids in the prostatic secretions and within the vagina. When mixed with the secretions of the seminal vesicles, previously inactive but mature spermatozoa begin beating their flagella and become highly mobile.

The Prostate Gland

The **prostate gland** is a small, muscular, rounded organ with a diameter of about 4 cm (1.6 in.) that surrounds the urethra as it leaves the bladder. The prostate produces an acidic secretion that contributes about 30 percent of the volume of semen.

The Bulbourethral Glands

The paired **bulbourethral glands**, or *Cowper's glands*, are round, with diameters approaching 10 mm (less than 0.5 in.) (Figure 21-2•). These glands secrete a thick, sticky, alkaline mucus that has lubricating properties.

THE PENIS

The **penis** is a tubular organ that surrounds the urethra (Figure 21-2•). It conducts urine to the exterior and introduces semen into the female vagina during sexual intercourse. The penis (Figure 21-5a•) is divided into three regions: (1) the **root**, the fixed portion that attaches the penis to the body wall; (2) the **body (shaft)**, the tubular portion that contains masses of *erectile tissue*; and (3) the **glans**, the expanded end that surrounds the external urethral opening.

The skin overlying the penis resembles that of the scrotum. A fold of skin, the **prepuce** (PRĒ-pūs), or *foreskin*, surrounds the tip of the penis. The prepuce attaches to the relatively narrow **neck** of the penis and continues over the glans.

•**FIGURE 21-5**
The Penis.
(a) Anterior and lateral view of the penis, showing positions of the erectile tissues. **(b)** Sectional view through the penis.

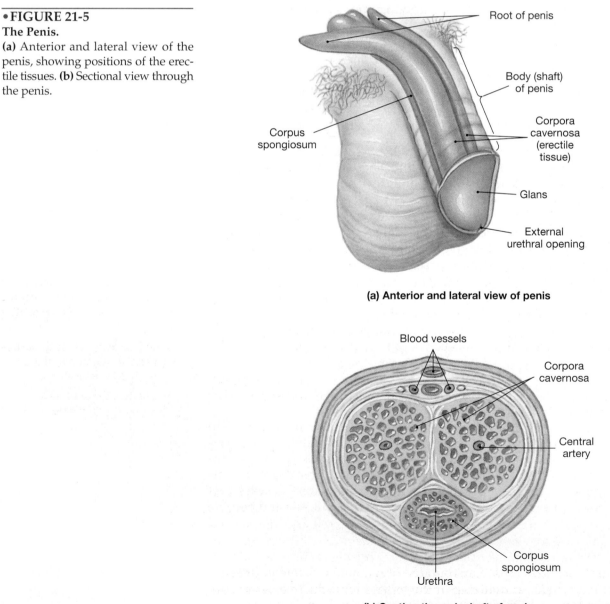

Root of penis

Body (shaft) of penis

Corpora cavernosa (erectile tissue)

Corpus spongiosum

Glans

External urethral opening

(a) Anterior and lateral view of penis

Blood vessels

Corpora cavernosa

Central artery

Corpus spongiosum

Urethra

(b) Section through shaft of penis

The surgical removal of the prepuce is called a *circumcision* (ser-kum-SIZH-un). In Western societies this procedure is usually performed soon after birth.

The body, or shaft, of the penis contains three columns, or bodies (singular: *corpus*; plural: *corpora*), of spongy, **erectile tissue** (Figure 21-5b●). Erectile tissue consists of a maze of blood channels incompletely divided by sheets of elastic connective tissue and smooth muscle. The two cylindrical **corpora cavernosa** (KOR-po-ra ka-ver-NŌ-sa) lie side by side and above the relatively slender **corpus spongiosum** (spon-jē-Ō-sum) that surrounds the urethra.

In the resting state, there is little blood flow into the erectile tissues because their arterial branches are constricted. In response to involuntary nerve impulses during *arousal*, the walls of the arterial blood vessels to the erectile tissue open, blood flow increases, the penis becomes engorged with blood, and **erection** occurs.

HORMONES AND MALE REPRODUCTIVE FUNCTION

The hormones that regulate the male reproductive system are produced by the pituitary gland, the hypothalamus, and the testes. The pituitary gland releases *follicle-stimulating hormone* (**FSH**) and *luteinizing hormone* (**LH**). ∞ *[p. 295]* The pituitary release of these hormones occurs in the presence of *gonadotropin-releasing hormone* (**GnRH**), a hormone released by the hypothalamus.

In the male, FSH stimulates the sustentacular cells of the seminiferous tubules. Under FSH stimulation, and in the presence of testosterone from the interstitial cells, sustentacular cells promote the production of sperm cells.

LH (formerly called *interstitial cell–stimulating hormone (ICSH)* in males) causes the secretion of testosterone by the interstitial cells of the testes. Testosterone levels and sperm production are both regulated by negative feedback. Low levels of testosterone stimulate the secretion of LH and FSH by the pituitary gland. The LH stimulates testosterone secretion, and testosterone and FSH stimulate sperm production. High levels of testosterone inhibit the secretion of LH and FSH, and the decline in LH inhibits testosterone secretion. The combination of reduced testosterone levels and reduced FSH levels slows the rate of sperm production.

Testosterone also has numerous functions in addition to promoting sperm production. ∞ *[p. 300]* It also determines the *secondary sexual characteristics* such as the distribution of facial hair, increased muscle mass and body size, and the quantity and location of fat deposits; stimulates protein synthesis and muscle growth; and influences brain development by stimulating sexual behaviors and sexual drive.

Testosterone production begins around the seventh week of embryonic development and reaches a peak after roughly 6 months of development. This early presence of testosterone stimulates the formation of the male duct system and accessory organs. Testosterone production then declines until it accelerates markedly at puberty, initiating sexual maturation and the appearance of secondary sexual characteristics.

✔ How do you think the lack of an acrosomal cap would affect the ability of a sperm cell to fertilize an egg cell?

✔ What happens when the arteries serving the penis dilate (increase in diameter)?

✔ What effect would low levels of FSH have on sperm production?

STRUCTURE OF THE FEMALE REPRODUCTIVE SYSTEM

A woman's reproductive system must produce sex hormones and gametes and also protect and support a developing embryo and nourish the newborn infant. The primary sex organs of the female reproductive system are the *ovaries*. The internal and external accessory organs include the *uterine tubes* (*fallopian tubes* or *oviducts*), the *uterus* (womb), the *vagina*, and the components of the external genitalia (Figure 21-6●). As in the male, various accessory glands secrete into the reproductive tract. Physicians specializing in the female reproductive system are called **gynecologists** (gī-ne-KOL-o-jists).

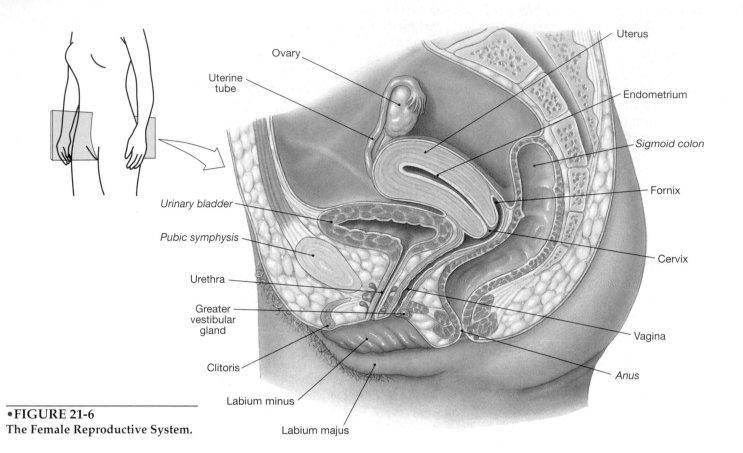

•FIGURE 21-6
The Female Reproductive System.

THE OVARIES

A typical ovary measures approximately 5 cm x 2.5 cm (2 in. x 1 in.). It has a pale white or yellowish coloration and a nodular consistency that resembles cottage cheese or lumpy oatmeal. The production of gametes occurs in the outer layer of the ovary. Arteries, veins, lymphatics, and nerves within its interior link the ovary with other body systems.

The ovaries are responsible for the production of female gametes, eggs or **ova** (singular: *ovum*), and the secretion of female sex hormones, including *estrogens* and *progestins*.

oon, egg + *genesis*, origin
oogenesis: the process of ovum formation

Through the process of meiosis, ovum production, or **oogenesis** (ō-ō-JEN-e-sis), produces gametes with half the number of normal chromosomes. Oogenesis begins before birth, accelerates at puberty, and ends at *menopause*. During the years between puberty and menopause, oogenesis occurs on a monthly basis as part of the *ovarian cycle*.

The Ovarian Cycle

Ovum production occurs within specialized structures called **ovarian follicles** (ō-VAR-ē-an FOL-i-klz). Before puberty, each ovary contains some 200,000 immature egg cells. Each of these egg cells is enclosed by a layer of follicle cells. The combination is known as a **primordial** (prī-MOR-dē-al) **follicle**. At puberty, rising levels of FSH begin to activate a different group of primordial follicles each month. This monthly process, shown in Figure 21-7•, is known as the **ovarian cycle**.

The cycle begins as the activated follicles develop into **primary follicles**. The follicle cells, which divide and form several layers around the immature ovum, or *oocyte*. The follicle cells provide nutrients to the oocyte.

Although many primordial follicles develop into primary follicles, usually only a few will take the next step. The deeper follicular cells begin secreting a fluid

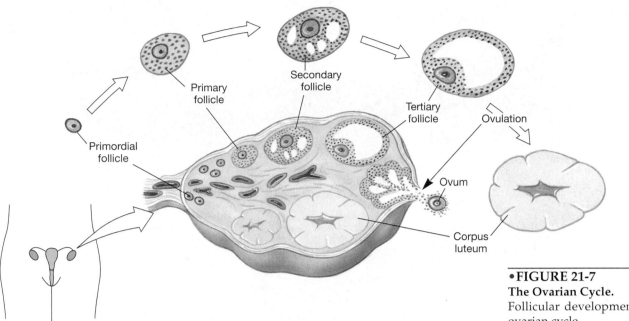

•FIGURE 21-7
The Ovarian Cycle.
Follicular development during the ovarian cycle.

that gradually accumulates and enlarges the entire follicle. At this stage the complex is known as a **secondary follicle**.

Eight to 10 days after the start of the ovarian cycle, the ovaries usually contain only a single secondary follicle. By days 10 to 14 of the cycle it has formed a mature **tertiary follicle**, or *Graafian* (GRAF-ē-an) *follicle*, roughly 15 mm (about 0.6 in.) in diameter. This complex is so large that it creates a prominent bulge in the surface of the ovary.

Ovulation, the release of the ovum, through the ovary wall, usually occurs at day 14 of a 28-day cycle. After ovulation, the ovum is drawn into the entrance of the uterine tube. The empty follicle collapses, and the remaining follicular cells multiply to create a hormone-producing structure known as the **corpus luteum** (LOO-tē-um). *Luteum* means "yellow," and the corpus luteum is also described as a "yellow body." Unless pregnancy occurs, after about 12 days the corpus luteum begins to degenerate. The disintegration marks the end of the ovarian cycle, but almost immediately the activation of another set of primordial follicles begins the next ovarian cycle.

THE UTERINE TUBES

Each **uterine tube** is about 13 cm (5 in.) long. The end closest to the ovary forms an expanded funnel, the *infundibulum*, with numerous fingerlike projections (Figure 21-8•). The projections, called *fimbriae* (FIM-brē-ē), and the inner surfaces of the uterine tube are carpeted with cilia that produce a current that moves the ovum into the broad entrance to the uterine tube. Once inside the uterine tube, the ciliary current and peristaltic contractions transport the ovum over 3 to 4 days to the uterine chamber. If fertilization is to occur, the ovum must encounter sperm during the first 12 to 24 hours of its passage down the uterine tube. Unfertilized ova disintegrate in the uterine tubes or uterus.

THE UTERUS

The **uterus** provides physical protection and nutritional support to the developing embryo and fetus (Figure 21-8•). The typical uterus is a small, pear-shaped organ about 7.5 cm (3 in.) long with a maximum diameter of 5 cm (2 in.). It weighs 30 to 40 g and is held in place by various ligaments.

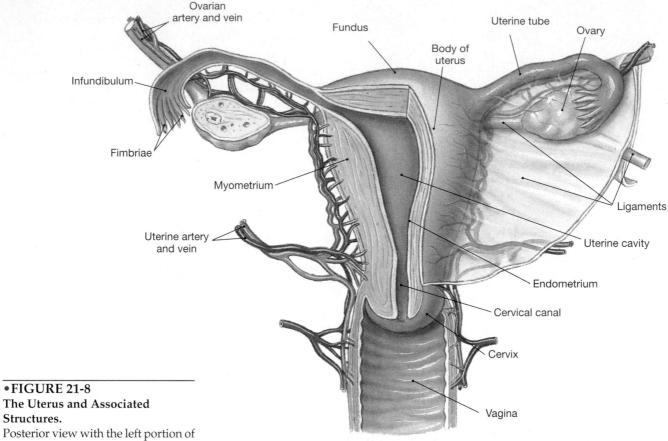

Posterior view

●**FIGURE 21-8**
The Uterus and Associated Structures.
Posterior view with the left portion of uterus, uterine tube, and left ovary shown in section.

endo, inside + *metria*, uterus
endometrium: the lining of the uterus; the region that supports a developing embryo and fetus

mys, muscle + *metria*, uterus
myometrium: the muscular walls of the uterus

The uterus can be divided into two regions: the *body* and the *cervix*. The **body** is the largest division of the uterus. The *fundus* is the rounded portion of the body superior to the attachment of the uterine tubes. The **cervix** (SER-viks) is the inferior portion of the uterus and projects a short distance into the vagina. The cervical opening leads into the *cervical canal*, a narrow passageway that opens into the **uterine cavity**.

In section, the thick uterine wall can be divided into an inner **endometrium** (en-dō-MĒ-trē-um) and a muscular **myometrium** (mī-ō-MĒ-trē-um) covered by a layer of visceral peritoneum. The myometrium is made up of interwoven smooth muscle cells capable of stretching during the growth of the embryo and fetus. The endometrium of the uterus undergoes regular, cyclical changes in response to changing levels of sexual hormones. These alterations produce the characteristic features of the *uterine cycle*.

The Uterine Cycle

The **uterine cycle**, or **menstrual** (MEN-stroo-al) **cycle,** is a repeating series of changes in the structure of the endometrium. This cycle averages 28 days in length, but it can range from 21 to 35 days in normal individuals. It can be divided into three stages: the *menstrual period (menstruation)*, the *proliferative phase*, and the *secretory phase* (Figure 21-9●).

The uterine cycle begins with **menstruation** (men-stroo-Ā-shun), a period of time marked by the breakdown and degeneration of the endometrium. The process is triggered by the decline in hormone concentrations as the corpus luteum disintegrates. Blood cells from broken blood vessels, and degenerating endometrial tissues and glands break away into the uterine chamber and then pass into the vagina. Menstruation usually lasts from 1 to 7 days, and over this period roughly 35 to 50 ml of blood is lost.

•**FIGURE 21-9**
**Phases of the Uterine (Menstrual)
Cycle.**

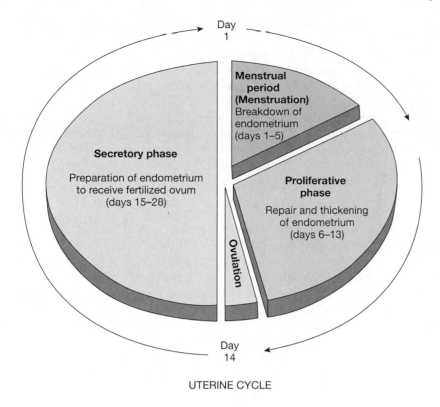

UTERINE CYCLE

The *proliferative phase* begins in the days following the completion of menstruation as the surviving epithelial cells multiply and spread across the surface of the endometrium. This repair process is stimulated by the rising hormone levels that accompany the growth of another set of ovarian follicles. By the time ovulation occurs, the repaired endometrium is filled with small arteries, and endometrial (uterine) glands are secreting a glycogen-rich mucus. The proliferative phase ends with ovulation.

During the *secretory phase*, which begins at ovulation, the endometrial glands enlarge, steadily increasing their rates of secretion. This activity is stimulated by the hormones from the corpus luteum. The secretory phase persists as long as the corpus luteum remains intact. Secretions from the corpus luteum peak about 12 days after ovulation. Over the next day or two the glandular activity declines, and the menstrual cycle comes to a close. A new cycle then begins with the onset of menstruation and the disintegration of the endometrium.

The first menstrual period, called **menarche** (me-NAR-kē), occurs at puberty, typically at age 11 to 12. Uterine cycles continue until age 45 to 50, when **menopause** (ME-nō-paws), the last uterine cycle, occurs. At that time, few, if any, follicles are left in the ovaries. Over the intervening years some 500 ova will have been ovulated.

THE VAGINA

The **vagina** is a muscular tube extending between the uterus and the external genitalia (Figures 21-6• and 21-8•). It has an average length of 7.5 to 9 cm (3 to 3.5 in.), but because the vagina is highly distensible, its length and width are quite variable. The cervix of the uterus projects into the **vaginal canal**. The vagina lies parallel to the rectum, and the two are in close contact. After leaving the urinary bladder the urethra turns and travels along the superior wall of the vagina.

The vagina (1) serves as a passageway for the elimination of menstrual fluids, (2) receives the penis during sexual intercourse, and (3) in childbirth forms the birth canal through which the fetus passes on its way to an independent existence.

The vagina is lined by a mucous membrane thrown into folds called *rugae*. Its walls contain a network of blood vessels and layers of smooth muscle. An elastic epithelial fold, the **hymen** (HĪ-men), may partially or completely block the entrance to the vagina.

The vagina normally contains resident bacteria supported by the nutrients found in the mucus secreted by the cervix. As a result of the bacteria's metabolic activities the normal pH of the vagina ranges between 3.5 and 4.5, and this acid environment restricts the growth of many pathogenic organisms.

THE EXTERNAL GENITALIA

The region enclosing the female external genitalia is the **vulva** (VUL-va) (see Figure 21-10•). The vagina opens into the **vestibule**, a central space bounded by the **labia minora** (LĀ-bē-a mi-NŌR-a; singular: *labium minus*). The labia minora are covered with a smooth, hairless skin. The urethra opens into the vestibule just anterior to the vaginal entrance. Anterior to the urethral opening, the **clitoris** (KLI-to-ris) projects into the vestibule. The clitoris is the female equivalent of the penis, derived from the same embryonic structures. Internally it contains erectile tissues that become engorged with blood during arousal.

During sexual arousal a pair of ducts discharges the lubricating secretions of the **greater vestibular glands** (refer to Figure 21-6•) into the vestibule near the vaginal entrance. These mucous glands resemble the bulbourethral glands of the male.

The outer limits of the vulva are established by the *mons pubis* and the *labia majora*. The prominent bulge of the **mons pubis** is created by fat tissue beneath the skin anterior to the pubic symphysis. The fleshy **labia majora** (singular: *labium majus*) encircle and partially conceal the labia minora and underlying vestibular structures.

•**FIGURE 21-10**
The Female External Genitalia.
An external view of the female perineum (outlined by dashed lines).

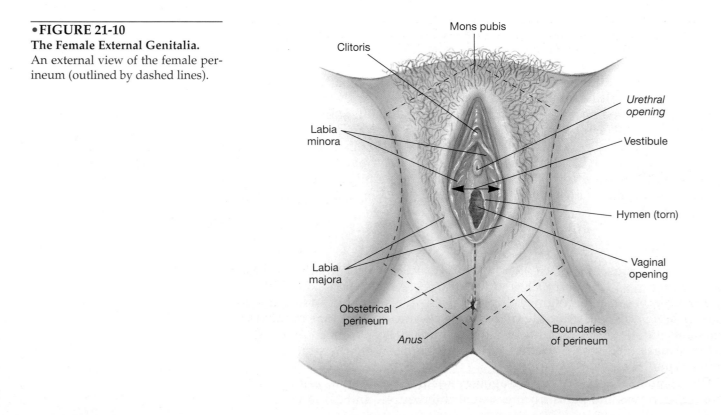

In both sexes, the **perineum** refers to the muscular pelvic floor that includes the external genitalia and anus. In the pregnant female, this term is also used more specifically for the region between the vaginal opening and the anus (the *obstetrical perineum*).

THE MAMMARY GLANDS

After its birth the newborn infant gains nourishment from the milk secreted by the mother's **mammary glands**. Milk production, or **lactation** (lak-TĀ-shun), occurs in the mammary glands of the breasts, specialized accessory organs of the female reproductive system (Figure 21-11•).

The mammary glands lie within fatty tissue beneath the skin of the chest. Each breast bears a small conical projection, the **nipple**, where the ducts of underlying mammary glands open onto the body surface. The skin surrounding each nipple has a reddish brown coloration, and this region is known as the **areola** (a-RĒ-ō-la).

The glandular tissue of the breast consists of a number of separate lobes, each containing several secretory lobules. Within each lobe the ducts leaving the lobules converge, giving rise to a single **lactiferous** (lak-TIF-e-rus) **duct**. Near the nipple, the lactiferous duct expands, forming an expanded chamber called a **lactiferous sinus**. There are usually 15 to 20 lactiferous sinuses opening onto the surface of each nipple. Bands of connective tissue, the *suspensory ligaments of the breast*, surround the duct system and help support the breasts. A layer of loose connective tissue separates the mammary complex from the underlying muscles, and the two can move relatively independently.

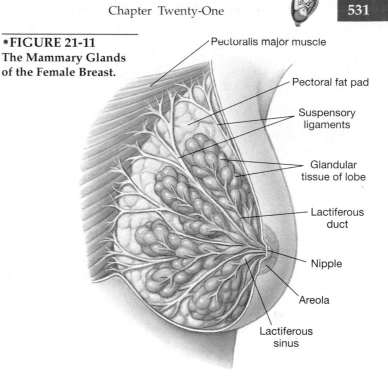

•**FIGURE 21-11**
The Mammary Glands of the Female Breast.

- Pectoralis major muscle
- Pectoral fat pad
- Suspensory ligaments
- Glandular tissue of lobe
- Lactiferous duct
- Nipple
- Areola
- Lactiferous sinus

✔ As the result of infections such as gonorrhea, scar tissue can block the opening of each uterine tube. How would this blockage affect a woman's ability to conceive?

✔ What tissue breaks away and is lost during menstruation?

✔ Would blockage of a single lactiferous sinus interfere with delivery of milk to the nipple? Explain.

HORMONES AND THE FEMALE REPRODUCTIVE CYCLE

As in the male, the activity of the female reproductive tract is controlled by both pituitary and gonadal hormones. But the regulatory pattern is much more complicated, for a woman's reproductive system does not just produce gametes; it must also coordinate the ovarian and uterine cycles. Circulating hormones, especially estrogen, regulate the **female reproductive cycle** to ensure proper reproductive function.

THE OVARIAN CYCLE

Hormonal regulation of the ovarian cycle differs between the *preovulatory period* and *postovulatory period*.

Hormones and the Preovulatory Period

Each month some of the follicles begin their development into primary follicles under the stimulation of FSH. As the follicular cells enlarge and multiply, they release steroid hormones collectively known as *estrogens*, the most important being **estradiol** (es-tra-DĪ-ol). Estrogens have multiple functions, including (1) stimulating bone and muscle growth, (2) maintaining female secondary sex characteristics such as body hair distribution and the location of adipose tissue deposits, (3) affecting CNS activity, including sex-related behaviors and drives, (4) maintaining functional accessory reproductive glands and organs, and (5) initiating repair and growth of the endometrium.

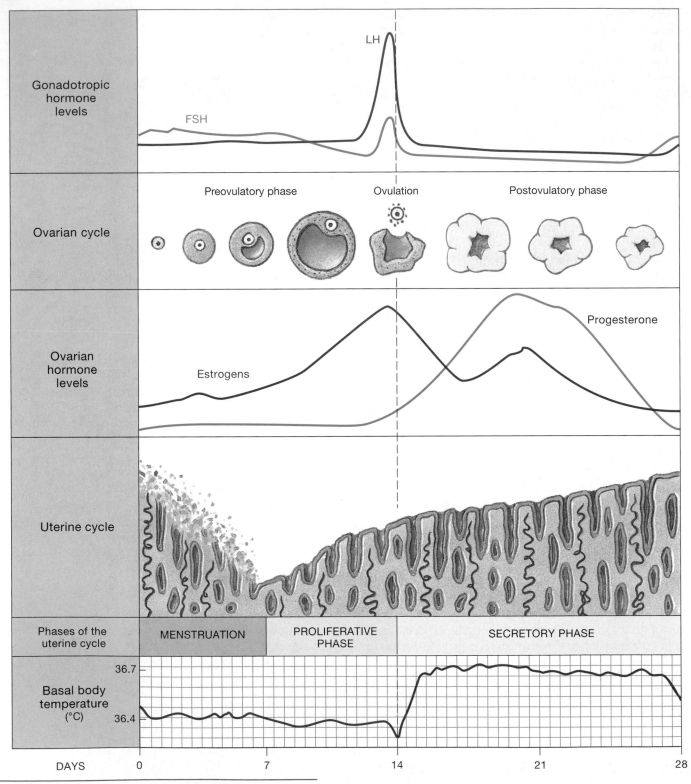

•FIGURE 21-12

Hormonal Regulation of the Female Reproductive Cycle.

The upper portion of Figure 21-12• summarizes the hormonal events associated with the ovarian cycle. As in the male, the primary hormones involved are produced by the anterior pituitary gland, under the control of the hypothalamus. As estrogen levels rise, they cause a decrease in the rate of FSH secretion. Estrogen also affects the rate of LH secretion—the rate of LH release increases as estrogen concentrations rise. Thus as the follicles develop and estrogen concentrations rise, the pituitary output of LH increases. The combination of estrogens, FSH, and LH continues to support follicular development and maturation.

Estrogen concentrations take a sharp upturn in the second week of the ovarian cycle, as one tertiary follicle enlarges in preparation for ovulation. At about day 14 estrogen levels peak, accompanying the maturation of that follicle. The high estrogen concentration then triggers a massive outpouring of LH from the anterior pituitary, which triggers ovulation.

Hormones and the Postovulatory Period

LH levels remain elevated for only 2 days, but that is long enough to cause the empty follicle to change into a functional corpus luteum. In addition to estrogen, the corpus luteum begins to manufacture steroid hormones known as **progestins** (prō-JES-tinz), in particular, **progesterone** (prō-JES-ter-ōn). Progesterone is the principal hormone of the postovulatory period. It prepares the uterus for pregnancy by stimulating the growth of the endometrium. It also stimulates metabolic activity, leading to a rise in basal body temperature.

Progesterone secretion continues at relatively high levels for the following week, but unless pregnancy occurs, the corpus luteum then begins to degenerate. Roughly 12 days after ovulation, the corpus luteum becomes nonfunctional, and progesterone and estrogen levels fall markedly. This decline leads to an increase in FSH and LH production in the anterior pituitary, and the entire cycle begins again.

THE UTERINE CYCLE

The hormonal changes that regulate the ovarian cycle also affect the uterus. The lower portion of Figure 21-12• follows changes in the endometrium during a single uterine cycle. The sudden declines in progesterone and estrogen levels that accompany the breakdown of the corpus luteum result in menstruation. The loss of endometrial tissue continues for several days until rising estrogen levels stimulate the regeneration of the endometrium. The preovulatory phase continues until rising progesterone levels mark the arrival of the postovulatory phase. The combination of estrogen and progesterone then causes a further thickening of the endometrium that prepares it for the arrival of an embryo.

HORMONES AND BODY TEMPERATURE

The changing levels of hormones also cause physiological changes that affect body temperature. During the preovulatory period, when estrogen is the dominant hormone, the resting, or "basal," body temperature measured upon awakening in the morning is about 0.3° C (or 0.5° F) lower than it is during the postovulatory period, when progesterone dominates. At the time of ovulation, basal temperature declines sharply, making the temperature rise over the following day even more noticeable (Figure 21-12•). By keeping records of body temperature over a few menstrual cycles a woman can often determine the precise day of ovulation. This information can be important for those wishing to avoid or promote a pregnancy, for this can occur only if an ovum becomes fertilized within a day of its ovulation.

✔ What changes would you expect to observe in the ovulatory cycle if the LH surge did not occur?

✔ What event occurs in the menstrual cycle when the levels of estrogen and progesterone decline?

AGING AND THE REPRODUCTIVE SYSTEM

The aging process affects the reproductive systems of men and women. The most striking age-related changes in the female reproductive system occur at menopause, whereas changes in the male reproductive system occur more gradually and over a longer period of time.

MENOPAUSE

Menopause is usually defined as the time that ovulation and menstruation cease. It typically occurs between the ages of 45 and 55, but in the years preceding it the ovarian and uterine cycles become irregular. A shortage of primordial follicles is the underlying cause of these developments; by age 50 there are often no primordial follicles left to respond to FSH.

Menopause is accompanied by a sharp and sustained rise in the production of FSH and LH and a decline in circulating concentrations of estrogen and progesterone. The decline in estrogen levels leads to reductions in the size of the uterus and breasts, accompanied by a thinning of the urethral and vaginal walls. The reduced estrogen concentrations have also been linked to the development of *osteoporosis*, presumably because bone deposition proceeds at a slower rate. A variety of nervous system effects are also reported, including "hot flashes," anxiety, and depression, but the hormonal mechanisms involved are not well understood. In addition, the risk of atherosclerosis and other forms of cardiovascular disease increase after menopause.

The symptoms accompanying and following menopause are sufficiently unpleasant that about 40 percent of menopausal women eventually seek medical assistance. Hormone replacement therapies involving a combination of estrogens and progestins can often prevent osteoporosis and the nervous and circulatory system changes associated with menopause.

THE MALE CLIMACTERIC

✔ Why does the level of FSH rise and remain high during menopause?

Changes in the male reproductive system occur more gradually, over a period known as the *male climacteric*. Circulating testosterone levels begin to decline between ages 50 and 60, coupled with increases in circulating levels of FSH and LH. Although sperm production continues (men well into their eighties can father children), there is a gradual reduction in sexual activity in older men.

☤ Clinical Note

Birth Control Strategies

Most adults practice some form of **contraception**, or birth control, during some part of their reproductive years. There are many different methods of preventing pregnancy. Fewer methods exist for the termination of a pregnancy.

CONCEPTION PREVENTION

Surgical Methods. *Sterilization* makes one unable to provide functional gametes for fertilization. Either sexual partner may be sterilized with the same net result. In a **vasectomy** (vaz-EK-to-mē) a segment of the ductus deferens is removed, making it impossible for spermatozoa to pass through the reproductive tract (see Figure 21-13a •). The surgery can be performed in a physician's office in a matter of minutes. With the section removed, the cut ends do not reconnect; in time, scar tissue forms a permanent seal. After a vasectomy the man experiences normal sexual function, for the epididymal and testicular secretions normally account for only around 5 percent of the volume of the semen. Spermatozoa continue to develop, but they remain within the epididymis until they degenerate. The failure rate for this procedure is 0.08 percent (a failure is defined as a resulting pregnancy).

In the female the uterine tubes can be blocked through a surgical procedure known as a tubal ligation (see Figure 21-13b •). Since the surgery involves entering the abdominopelvic cavity, complications are more likely than with vasectomy. The failure rate for this procedure is estimated at 0.45 percent.

As in a vasectomy, attempts may be made to restore fertility after a tubal ligation.

Hormonal Methods. Oral contraceptives manipulate the female hormonal cycle so that ovulation does not occur. Contraceptive pills contain a combination of progestins and estrogens. These hormones suppress pituitary production of GnRH, so FSH is not released and ovulation does not occur. The hormones are administered in a cyclic fashion, beginning 5 days after the start of menstruation and continuing for the next 3 weeks. Over the fourth week the woman takes placebo pills or no pills at all. There are now at least 20 different brands of combination oral contraceptives available, and over 200 million women are using them worldwide. In the United States, 25 percent of women under age 45 use the combination pill to prevent conception. The failure rate for the combination oral contraceptives, when used as prescribed, is 0.24 percent over a 2-year period. Birth control pills are not without their risks, however. For example, women with severe hypertension, diabetes mellitus, epilepsy, gallbladder disease, heart trouble, or acne may find that their problems worsen when taking the combination pills. Women taking oral contraceptives are also at increased risk for venous thrombosis, strokes, pulmonary embolism, and (for women over 35) heart disease.

Two progesterone-only forms of birth control are now available. *Depo-provera*™ is injected every 3 months. The silastic tubes of the Norplant™ system are saturated with progesterone and inserted under the skin. This method provides birth control for a period of approximately 5 years. Both Depo-provera™ and the Norplant™ system can interrupt or cause irregular menstruation, but they are easy to use and extremely convenient.

Barrier Methods. The **condom**, also called a *prophylactic* or "rubber," covers the body of the penis during intercourse and keeps spermatozoa from reaching the female reproductive tract. Condoms are also used to prevent transmission of sexually transmitted diseases, such as syphilis, gonorrhea, and AIDS. The condom failure rate has been estimated at over 6 percent. **Vaginal barriers** such as the *diaphragm, cervical cap,* or *vaginal sponge* rely on similar principles. A diaphragm, the most popular form of vaginal barrier in use at the moment, consists of a dome of latex rubber with a small metal hoop supporting the rim. Because vaginas vary in size, women choosing this method must be individually fitted. Before intercourse the diaphragm is inserted so that it covers the cervical opening. It is usually coated with a small amount of spermicidal jelly or cream, adding to the effectiveness of the barrier. The failure rate for a properly fitted diaphragm is estimated at 5 to 6 percent. The cervical cap is smaller and lacks the metal rim. It, too, must be fitted carefully, but unlike the diaphragm it may be left in place for several days. The failure rate (8 percent) is higher than that for a diaphragm. The vaginal sponge consists of a small synthetic sponge saturated with a *spermicide*—a sperm-killing foam or jelly. The failure rate for a contraceptive sponge is estimated at 6 to 10 percent.

Other Methods. An **intrauterine device (IUD)** consists of a small plastic loop or a T that can be inserted into the uterine chamber. The mechanism of action remains uncertain, but it is known that IUDs can change the intrauterine environment and lower the chances for fertilization and subsequent implantation of a fertilized ovum. IUDs are in limited use in the United States but they remain popular in many other countries. The failure rate is estimated at 5 to 6 percent.

The **rhythm method** involves abstaining from sexual activity on the days ovulation might be occurring. The timing is estimated based on previous patterns of menstruation and sometimes by following changes in basal body temperature. The failure rate for the rhythm method is very high, approaching 25 percent.

POST-CONCEPTION METHODS

If contraceptive methods fail, options exist to either prevent implantation or terminate the pregnancy.

Oral Methods. The "morning-after pills" contain estrogens or progestins. They may be taken within 72 hours of intercourse, and they appear to act by altering the transport of the zygote (fertilized egg) or preventing its attachment to the uterine wall. The drug known as *RU-486 (Mifepristone)* blocks the action of progesterone at the endometrial lining. The result is a normal menstrual period with degeneration of the endometrium whether or not fertilization has occurred.

Surgical Methods. **Abortion** refers to the termination of a pregnancy. Three classes of abortions are recognized. *Spontaneous abortions,* or *miscarriages,* occur naturally due to some developmental or physiological problem. *Therapeutic abortions* are performed when continuing the pregnancy represents a threat to the life and health of the mother. *Induced abortions* (elective abortions) are performed at the request of the individual. Induced abortions are currently legal during the first 3 months after conception, and many states permit abortions, sometimes with restrictions, until the fifth or sixth developmental month.

•**FIGURE 21-13**
Surgical Sterilization.
(a) In a male **vasectomy**, the removal of a 1-cm, (1/2 in.) section of the ductus deferens prevents the passage of sperm cells.
(b) In a female **tubal ligation**, the removal of a section of the oviduct prevents both the passage of sperm and the movement of the ovum or embryo into the uterus.

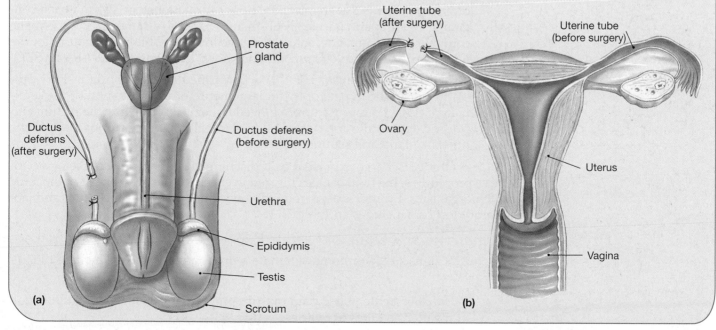

DISORDERS OF THE REPRODUCTIVE SYSTEM

This section examines disorders that affect the components of the male and female reproductive systems and that affect fertility. Disorders of pregnancy and development are discussed in Chapter 22.

THE PHYSICAL EXAMINATION

Symptoms and Signs of Male Reproductive System Disorders

An assessment of the male reproductive system begins with a physical examination. Common signs and symptoms of male reproductive disorders include the following:

- *Pain.* Testicular pain (of the testes) may result from infection, trauma, and testicular cancer. The pain may also originate elsewhere along the reproductive tract, such as along the ductus deferens or within the prostate gland.

- *Urethral discharge and dysuria.* These symptoms are commonly associated with sexually transmitted diseases (STDs). They may also accompany disorders of the epididymis or prostate gland that may be infectious or noninfectious.

- *Impotence.* **Impotence** is the inability to achieve or maintain an erection. It may occur as a result of psychological factors, such as fear or anxiety, medications, or alcohol abuse. It may also develop secondarily to neurological or cardiovascular problems that affect blood pressure or blood flow to the penile arteries.

- *Male infertility.* Infertility may be caused by a low sperm count, abnormally shaped sperm, or abnormal semen composition.

Examination of the male reproductive system normally involves the examination of the external genitalia and palpation of the prostate gland.

1. Inspection of the penis and scrotum for skin lesions, such as vesicles, *chancres*, and warts, which indicate the presence of sexually transmitted diseases. A **chancre** (SHANG-ker) is a painless, reddish ulceration associated with early-stage *syphilis*.

2. Palpation of each testis, epididymis, and ductus deferens to detect the presence of abnormal masses, swelling, or tumors. Possible abnormal findings include: (1) swelling of the scrotal cavity by serous fluid (a **hydrocele** (HĪ-dro-SĒL); (2) swelling of the testes due to infection, trauma, or *testicular cancer*; (3) swelling of the epididymis due to infection, tumor formation, or a spermatozoa-containing cyst (a **spermatocele** (SPER-mah-to-SĒL); (4) swelling of the spermatic cord due to inflammation, serous fluid accumulation, bleeding, trauma, or the formation of varicose veins, a condition known as a **varicocele** (VAR-i-ko-SĒL).

3. A *digital rectal examination (DRE)* is performed as a screening test for prostatitis, prostatic hypertrophy (enlargement), or inflammation of the seminal vesicles. In this procedure, a gloved finger is inserted into the rectum and pressed against the anterior rectal wall to palpate the posterior walls of the prostate gland and seminal vesicles.

If urethral discharge is present or if discharge occurs in the course of any of these procedures, the fluid can be cultured to check for the presence of pathogenic microorganisms. Examples of reproductive system diagnostic procedures and laboratory tests are included in Table 21-1.

Symptoms and Signs of Female Reproductive System Disorders

Important signs and symptoms of female reproductive disorders include the following:

- *Pelvic pain.* Acute pelvic pain may accompany inflammation and infection of the uterine tubes or ovaries.

Table 21-1   Representative Diagnostic Procedures and Laboratory Tests for Disorders of the Reproductive System

Diagnostic Procedure	Method and Result	Representative Uses
FEMALES		
Mammography	X-ray film is taken of the breast	Detection of cysts or tumors of the breast; effective in detecting early breast cancer
Laparoscopy	Fiber-optic tubing is inserted through incision in abdominal wall to view pelvic organs, remove tissue for biopsy, or for surgical procedures	Detection of cysts, adhesions, or endometriosis; diagnosis of pelvic inflammatory disease
Papanicolaou (PAP) smear	Cells from the cervix are removed for cytological analysis	Detection of cervical cancer; staging ranges from Class I (no abnormal cells) to Class V (definite malignancy)
Colposcopy	Special instrument is used to view cervical tissue microscopically	Detection of areas of dysplasia and malignancies of the cervix; follow-up to abnormal PAP smear
Cervical biopsy	Tissue is removed from the cervix for examination	Detection of dysplasia and malignancy
MALES		
Transrectal ultrasonography	Ultrasound transducer is inserted rectally and scan is performed	Detection of prostatic tumor and nodules, or abnormalities of seminal vesicles and surrounding structures; used to guide biopsy

Laboratory Test	Normal Values in Blood Plasma or Serum	Significance of Abnormal Values
FEMALES		
Serum analysis		
Estrogen	Early uterine cycle: 60–400 pg/ml Middle: 100–600 pg/ml Late: 150–350 pg/ml Postmenopausal: <30 pg/ml	Used to detect hypofunctioning ovaries and help determine timing of ovulation
Estradiol	Follicular phase: 20–150 pg/ml Ovulation: 100–500 pg/ml Luteal phase: 60–260 pg/ml	Decreased in ovarian dysfunction and in amenorrhea
FSH	Before and after ovulation: 4–20 mIU/ml Midcycle: 10–40 mIU/ml	Increased in the absence of estrogens; decreased in anorexia nervosa or hypopituitarism
LH	Follicular: 3–30 mIU/ml Midcycle: 30–150 mIU/ml	Elevated in ovarian hypofunction; useful in determining if ovulation has occurred
Progesterone	Before ovulation: <70 ng/dl Midcycle: 250–2,800 ng/dl	Useful in determining the timing of ovulation; levels are increased in early pregnancy
Prolactin	Nonlactating females: 0–23 ng/ml	Values >100 ng/ml in a nonlactating female indicate pituitary tumor
MALES		
Serum analysis		
Testosterone	Adult males: 0.3–1.0 µg/dl	Decreased in testicular disorders, alcoholism, or pituitary hypofunction
Prostate-specific antigen (PSA)	<4 ng/ml	Elevated in prostatic cancer, benign prostatic hypertrophy, and increasing age
Semen analysis Volume Sperm count Motility Sperm morphology	2–5.0 ml 60–150 million/ml 60–80% are motile 70–90% normal structure	Decreased sperm count causes infertility; infertility could result if >40% of sperm are nonmotile or >30% of sperm are abnormal

■ *Abnormal bleeding.* Bleeding between menstrual cycles can result from oral contraceptive use, hormonal fluctuations, infection, or a condition called *endometriosis*.

■ *Lack of menstruation.* This condition, called *amenorrhea*, may occur in women with anorexia nervosa (p. 475), women who overexercise and are underweight, in extremely obese women, in postmenopausal women, and in pregnancy.

■ *Abnormal vaginal discharge.* Such vaginal discharges may be the result of a bacterial, fungal, or protozoan infection.

■ *Dysuria.* Painful urination may accompany an infection of the reproductive system due to migration of pathogens to the urethral entrance. ∞ *[p. 502]*

■ *Infertility.* Infertility may be related to hormonal disturbances, a variety of ovarian disorders, or anatomical problems along the reproductive tract.

A physical examination generally includes the following steps:

1. Inspection of the external genitalia for skin lesions, trauma, or related abnormalities.

2. Inspection and/or palpation of the perineum, vaginal opening, labia, clitoris, urethral opening, and vestibule to detect lesions, abnormal masses, or discharge from the vagina or urethra. Samples of any discharge present can be cultured to detect and identify any pathogens involved.

3. Inspection of the vagina and cervix by using a *speculum*, an instrument that retracts (draws back) the vaginal walls to permit direct visual inspection. Changes in the color of the vaginal wall may be important diagnostic clues. For example, cyanosis of the vagina and cervix normally occurs during pregnancy, but it may also occur when a pelvic tumor exists or in persons with congestive heart failure. A reddening of the vaginal walls results from inflammation and infection.

   The cervix is inspected to detect lacerations, ulceration, abnormal growths, or cervical discharge. Cells are collected from around the cervical opening and examined in a procedure called a *Pap smear*. This test screens for the presence of cervical cancer.

4. *Bimanual examination* is a method for the palpation of the uterus, uterine tubes, and ovaries. The physician inserts two fingers vaginally and places the other hand against the lower abdomen to palpate the uterus and surrounding structures. The contour, shape, size and location of the uterus can be determined, and any swellings, masses, or areas of tenderness will be apparent. Abnormalities in the ovaries can also be detected in this way.

REPRESENTATIVE DISORDERS

Major reproductive system disorders are diagrammed in Figure 21-14•. They include inflammation and infection, trauma, tumors, congenital disorders, and uterine-associated disorders.

Inflammation and Infection

Male Reproductive System. An inflammation of the testis is called **orchitis** (or-KĪ-tis). It is characterized by swelling and pain. Orchitis is usually caused by infection with the mumps virus, but bacterial infections, such as tuberculosis and syphilis, may also be the cause. Orchitis may also lead to **oligospermia** (ol-i-go-SPER-mē-uh), a low number of spermatozoa in the semen. Oligospermia, which may be temporary or permanent, is a major cause of infertility.

Epididymitis (ep-i-did-i-MĪ-tis) is an acute inflammation of the epididymis that may indicate an infection of the reproductive or urinary tracts. This condition may also develop due to irritation caused by the backflow, or reflux, of urine into the ductus deferens.

•**FIGURE 21-14**
Disorders of the Reproductive
System.

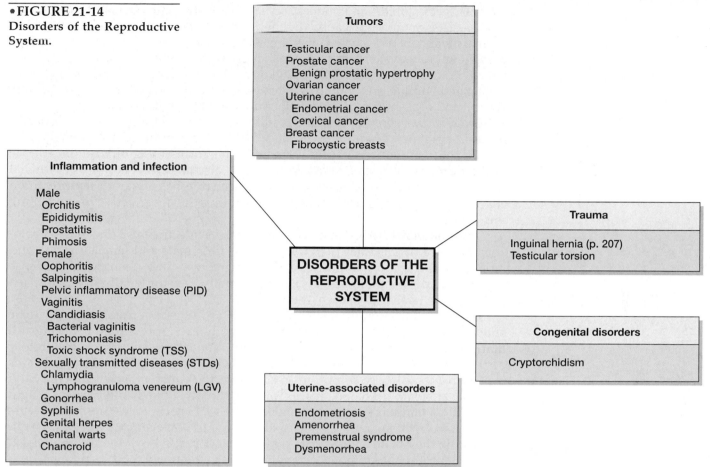

Inflammation of the prostate gland is called **prostatitis** (pros-ta-TĪ-tis). It can occur in males of any age, but most commonly afflicts older men. This condition may result from bacterial infections, or appear in the absence of any pathogens. Symptoms include pain in the lower back, perineum, or rectum, and sometimes painful urination with mucous discharge. Antibiotics are used to treat bacterial infections. Prostatitis symptoms should be taken seriously because they also resemble those of *prostatic cancer*.

Phimosis (fi-MŌ-sis) is an inability to retract the prepuce, or foreskin, from the tip of the penis in an uncircumcised adult male. This condition usually indicates inflammation of the prepuce and adjacent tissues.

Female Reproductive System. Inflammation of the ovaries is called **oophoritis** (ō-of-ō-RĪ-tis). It typically occurs with infection by the mumps virus. Oophoritis may also be associated with gonorrhea and *pelvic inflammatory disease*.

Salpingitis (sal-pin-GĪ-tis) is the inflammation of a tube within the body (*salpinx* means tube). In this case, it refers to inflammation of the uterine tubes. Salpingitis is associated with **pelvic inflammatory disease (PID)**, an infection of the uterine tubes and endometrium. PID is a major cause of infertility in women. Sexually transmitted pathogens are often involved, especially the bacteria that cause gonorrhea and *chlamydial* infections (discussed below). PID may also result from invasion of the region by bacteria normally found within the vagina. Symptoms of pelvic inflammatory disease include fever, lower abdominal pain, and elevated white blood cell counts. In severe cases, the infection may spread to other visceral organs or produce a generalized peritonitis. Sterility may occur because of damage and scarring of the uterine tubes, which prevent the passage of sperm to the oocyte or a zygote to the uterus.

Vaginitis (vaj-i-NĪ-tis) is an inflammation of the vagina. It is often the result of infections but may be associated with other factors, such as aging. There are several forms of vaginitis, and minor cases are relatively common. **Candidiasis** (kan-di-DĪ-a-sis) is an infection caused by the yeast, *Candida albicans*. This organism is a normal inhabitant of the vagina in many women. However, antibiotic usage, a suppressed immune system, stress, pregnancy, and other factors may change the vaginal environment and stimulate the overgrowth of the fungus. Symptoms include itching and burning, and a lumpy white discharge, called **leukorrhea** (lū-ko-RĒ-uh). Topical and oral antifungal medications are used to treat this condition.

Bacterial (nonspecific) vaginitis results from the combined action of several bacteria, primarily *Gardnerella vaginalis*. The bacteria involved are normally present in about one-third of adult women. In this form of vaginitis, an imbalance in the numbers of bacteria causes a vaginal discharge that contains epithelial cells and large numbers of bacteria. The discharge has a watery texture and a characteristic "fishy" odor. Topical or oral antibiotics are effective in controlling this condition.

Trichomoniasis (trik-ō-mō-NĪ-a-sis) involves infection by a flagellated protozoan parasite, *Trichomonas vaginalis*. It infects both male and female urinary and reproductive tracts, and produces a white or greenish-gray discharge and intense vaginal itching in females. Because it is a sexually transmitted disease, both partners must be treated to prevent reinfection.

A vaginal infection by *Staphylococcus aureus* bacteria is responsible for **toxic shock syndrome (TSS)**. Symptoms include high fever, sore throat, vomiting and diarrhea, and a generalized rash. As the condition progresses, shock, respiratory distress, and kidney or liver failure may develop, and 10–15 percent of all cases may prove fatal. These symptoms result from the entry of bacterial toxins, and even bacteria, into the bloodstream. TSS is apparently linked with the use of highly absorbent tampons, because the incidence of TSS declined as those items were removed from the market. TSS continues to occur, however, at a low but significant rate in those who use ordinary tampons, and in men or women after abrasion or burn injuries that promote bacterial infection. Treatment for TSS involves fluid administration, removal of the focus of infection (such as removal of a tampon or cleansing of a wound), and antibiotic therapy.

Sexually Transmitted Diseases. **Sexually transmitted diseases**, or **STDs**, are infectious diseases that are, in most cases, transferred from individual to individual by sexual contact. A variety of bacterial, viral, and fungal infections are included in this category. Although at least two dozen STDs are currently recognized, we will only discuss six: chlamydia, gonorrhea, syphilis, herpes, genital warts, and chancroid. HIV disease (AIDS), another important sexually transmitted disease, was considered in Chapter 16 (p. 397).

Infections by the bacterium *Chlamydia trachomatis*, which cause **chlamydia**, are currently the most prevalent STD. This organism causes up to one-half of all cases of PID, and is responsible for *nongonococcal urethritis*, **lymphogranuloma venereum (LGV)**, and conjunctivitis in the newborn (p. 282). In LGV, the lymph nodes in the groin become enlarged and inflamed. Abscesses and ulcers develop over the lymph nodes. Chlamydial infections may be cured with antibiotics (tetracycline and sulfa drugs), but all sexual partners must be treated to prevent reinfections.

The bacterium *Neisseria gonorrhoeae* (nī-SĒ-rē-a go-nor-RĒ-ī) is responsible for **gonorrhea**, a common STD in the U.S. Nearly 2 million cases are reported each year. These bacteria normally invade epithelial cells that line the male or female reproductive tract.

The symptoms of genital infection differ according to the gender of the infected individual. It has been estimated that up to 80 percent of women infected with gonorrhea experience no symptoms, or symptoms so minor that medical treatment is not sought. As a result, these women act as carriers, spreading the infection through their sexual contacts. An estimated 10–15 percent of women infected with gonorrhea experience more acute symptoms because the bacteria invade the epithelia of the uterine tubes, where they cause pelvic inflammatory disease (PID). As many as 80,000 women may become infertile each year as the result of scar tissue formation along the uterine tubes after gonorrheal and/or chlamydial infections.

Seventy to eighty percent of gonorrhea-infected males develop symptoms painful enough to make them seek antibiotic treatment. The urethral invasion is accompanied by pain on urination (dysuria) and typically by a viscous urethral discharge. A sample of the discharge can be cultured to permit positive identification of the organism involved.

Syphilis (SIF-i-lis) results from infection by the bacterium *Treponema pallidum* (trep-ō-NĒ-ma PAL-li-dum). Syphilis is a STD with a prolonged period of chronic illness. Its slow progression increases its likelihood of transmission to uninfected individuals. Untreated syphilis can cause serious cardiovascular and neurologic illness years after infection, or it can be spread to a fetus during pregnancy to produce congenital malformations.

Primary syphilis begins as the bacteria cross the mucous epithelium and enter the lymphatics and bloodstream. At the invasion site, the bacteria multiply; after an incubation period ranging up to 6 weeks, their activities produce a painless raised lesion, or *chancre* (Figure 21-15•). This lesion remains for several weeks before fading away, even without treatment. In heterosexual men, the chancre tends to appear on the penis; in women, it may develop on the labia, vagina, or cervix. Lymph nodes in the region often enlarge and remain swollen even after the chancre has disappeared.

Symptoms of *secondary syphilis* appear roughly 6 weeks later. Secondary syphilis generally involves a diffuse, reddish skin rash. Like the chancre, the rash fades over a period of 2–6 weeks. These symptoms may be accompanied by fever, headaches, and fatigue. The combination is so vague that unless a diagnostic blood test is ordered, the disease may easily be overlooked or diagnosed as something else.

The individual then enters the *latent phase*. The duration of the latent phase varies widely. Fifty to seventy percent of untreated individuals with latent syphilis fail to develop the symptoms of *tertiary syphilis*, or *late syphilis*, although the bacterial pathogens remain within their tissues. Those who do develop tertiary syphilis may do so 10 or more years after infection.

The most severe symptoms of tertiary syphilis involve the CNS and the cardiovascular system. *Neurosyphilis* may result from bacterial infection of the meninges or the tissues of the brain and/or spinal cord. In the cardiovascular system, the disease affects the major vessels, leading to aortic stenosis, aneurysms, or arterial calcification.

Syphilis may also be transmitted from mother to fetus across the placenta. These cases of *congenital syphilis* are marked by infections of the developing bones and cartilages of the skeleton and progressive damage to the spleen, liver, bone marrow, and kidneys. Treatment of syphilis involves the administration of penicillin or other antibiotics.

Genital herpes results from infection by herpes viruses. Two different viruses are involved. Up to 90 percent of genital herpes cases are caused by the virus known as HSV-2 (herpes simplex virus type 2), which is usually associated with the external genitalia. The remaining cases are caused by HSV-1, the virus that is also responsible for cold sores on the mouth. Typically within a week of the initial infection, the individual develops a number of vesicles on the external genitalia, which may develop into painful, ulcerated lesions (Figure 21-16•). In women, ulcerations may also appear on the cervix. These ulcerations gradually heal over the next 2–3 weeks. Recurring lesions are common, although subsequent incidents are less severe.

During delivery, infection of the newborn infant with herpes viruses present in the mother's vagina can lead to serious illness, because the newborn has few immunological defenses. Recent development of the antiviral agent *acyclovir* has helped treatment of initial infections.

A summary table of major infectious diseases of the reproductive system is presented in Table 21-2.

•**FIGURE 21-15**
A Syphilitic Chancre.

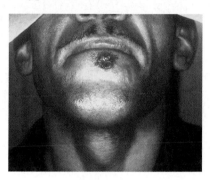

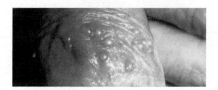
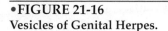
•**FIGURE 21-16**
Vesicles of Genital Herpes.

Table 21-2	Examples of Infectious Diseases of the Reproductive System	
Disease	**Organism(s)**	**Description**
Bacterial Diseases		
Bacterial vaginitis	Varied, but often *Gardnerella vaginalis*	Vaginitis caused by resident bacteria; symptoms include a watery discharge
Chancroid	*Haemophilus ducreyi*	A relatively rare STD; symptoms include soft chancres, which become ulcerated lesions, and enlarged lymph nodes in the groin
Chlamydia	*Chlamydia trachomatis*	Chlamydial infections cause PID, nongonococcal urethritis, and LGV
Gonorrhea	*Neisseria gonorrhoeae*	Infection of the epithelial cells of the male and female reproductive tracts; the majority of females show no symptoms, but others may develop PID; the majority of males develop painful urination (dysuria) and produce a viscous discharge
Lymphogranuloma venereum (LGV)	*Chlamydia trachomatis*	One type of chlamydia STD; symptoms include enlarged lymph nodes, which may abscess and form ulcers
Pelvic Inflammatory Disease (PID)	*Neisseria gonorrhoeae* *Chlamydia trachomatis*	PID is an infection of the uterine tubes (salpingitis); symptoms include fever, abdominal pain, and elevated WBC counts; may cause peritonitis in severe cases; sterility may result due to formation of scar tissue in the uterine tubes
Syphilis	*Treponema pallidum*	STD with a long period of chronic illness; symptoms of primary syphilis include chancres and enlarged lymph nodes; symptoms of secondary syphilis involve a reddish skin rash, fever, and headaches; tertiary syphilis affects the CNS and cardiovascular system
Toxic Shock Syndrome	*Staphylococcus aureus*	Vaginitis; symptoms include high fever, sore throat, vomiting and diarrhea, which may lead to shock, respiratory distress, and kidney or liver failure, and death
Viral Diseases		
Genital herpes	Herpes simplex viruses (HSV-1 and -2)	Most cases caused by HSV-2; ulcers develop on external genitalia, heal, and recur
Genital warts	Human papillomavirus	Genital warts appear on the external genitalia, perineum, anus, and on vagina and cervix of females
Fungal Diseases		
Candidiasis	*Candida albicans*	Yeast infection that causes vaginitis; symptoms include itching and burning, and a lumpy white discharge
Parasitic Diseases Protozoan		
Trichomoniasis	*Trichomonas vaginalis*	Flagellated parasite of both male and female urinary and reproductive tracts; produces a white or greenish-gray discharge and intense vaginal itching in females

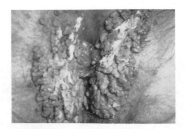

•FIGURE 21-17
Genital Warts.

Genital warts result from infection with one of a number of different strains of *human papillomavirus* (HPV) (Figure 21-17•). Warts appear on the external genitalia, perineum, anus, and, in females, on the vagina and cervix. Several of these strains are thought to be responsible for cases of cervical, anal, vaginal, and penile cancer. Roughly 1.2 million cases of genital warts are diagnosed each year in the U.S. There is no satisfactory treatment or cure for this problem. The traditional treatments have included *cryosurgery* (tissue destruction by extreme cold), erosion by caustic chemicals, surgical removal, and laser surgery to remove the warts. These treatments remove the visible signs of infection, but the virus remains within the epidermis.

Chancroid (SHANG-kroyd) is an STD caused by the bacterium *Haemophilus ducreyi* (hē-MOF-il-us doo-KRĀ-ē). Chancroid cases were rarely seen inside the U.S. before 1984, but since then the number of cases has risen dramatically, reaching 4,000–5,000 cases per year. The primary sign of this disease is the development of soft chancres, which develop into painful ulcers. The majority of chancroid patients also develop enlarged lymph nodes in the groin. Antibiotics are used to treat this condition.

Trauma

Two examples of trauma affecting the male reproductive system are *inguinal hernias* and *testicular torsion*. Inguinal hernias were described in Chapter 9 (p. 207).

Testicular torsion results if the spermatic cord twists such that the blood supply to a testis is obstructed. The result is pain and swelling of the testis. This condition occurs most often before or during puberty. The testis must be returned to its proper position within a short period of time after torsion occurs. Prompt surgery is needed to prevent permanent damage to the testis. If surgery occurs too late, then the nonfunctional testis is removed in a procedure called an **orchiectomy** (ōr-kē-EK-to-mē).

Tumors

Male Reproductive System. **Testicular cancer** is relatively rare, with about 7,200 new cases per year in the U.S. It is the most common cancer among males age 15–35. More than 95 percent of testicular cancers are the result of abnormal sperm-producing stem cells, rather than sustentacular or interstitial cells (p. 521). Treatment generally involves a combination of orchiectomy and chemotherapy. The survival rate is about 95 percent, primarily because of early diagnosis and effective treatment procedures.

In most cases, the prostate gland enlarges spontaneously in men over 50, a condition called **benign prostatic hypertrophy**. The increase in size may continue until the swelling constricts and blocks the urethra. If not corrected, the urinary obstruction can cause permanent kidney damage. Partial surgical removal is the most effective treatment at present. Drugs that relax the urethral muscles can often relieve the symptoms.

Prostate cancer is malignant and frequently metastatic. It is the second most common cancer in men and the second most common cause of cancer deaths in men. Most patients are elderly when diagnosed, over age 70.

Prostate cancer normally begins in one of the secretory glands. As the cancer progresses, a lump or swelling is produced on the surface of the gland. A digital rectal exam (DRE) is the easiest diagnostic procedure. A blood test for *prostate-specific antigen (PSA)* may also be used for screening purposes. If the condition is detected before metastasis has occurred, the usual treatment is localized radiation or a **prostatectomy** (pros-ta-TEK-to-mē), the surgical removal of the prostate gland. Side effects include urinary incontinence and loss of sexual function. If metastasis has occurred, secondary tumors may develop in the lymphatic system, lungs, bone marrow, liver, or the adrenal glands. Survival rates at this stage are relatively low.

Female Reproductive System. **Ovarian cancer** is the third most common reproductive cancer among women. However, it is the most lethal because ovarian cancer is seldom diagnosed in its early stages. The cancer is most common in women over age 50, and more common in those who have not had children. Lower rates of ovarian cancer occur in women who have taken birth-control pills.

The prognosis is relatively good for cancers that originate in the general ovarian tissues or from abnormal ova. These cancers respond well to some combination of chemotherapy, radiation, and surgery. However, more than three-fourths of ovarian cancers develop from epithelial cells, and sustained remission can be obtained in only about one-third of these patients. Early diagnosis would greatly improve the chances of successful treatment, but symptoms are vague or absent and as yet there are no reliable screening tests.

Treatment involves the surgical removal of malignant tissue. The surgical removal of an ovary is called an **oophorectomy** (ō-of-ō-REK-to-mē). Surgery for ovarian cancer may involve the removal of one ovary and uterine tube (a *unilateral salpingo-oophorectomy*), or both, in a *bilateral salpingo-oophorectomy*, or the removal of the uterus, called a **hysterectomy** (his-ter-EK-to-mē). Treatment of more-dangerous forms of early stage ovarian cancer includes radiation and chemotherapy in addition to surgery.

Uterine tumors are the most common tumors in women. It has been estimated that 40 percent of women over age 50 have benign uterine tumors involving smooth muscle and connective tissue cells. If small, these *leiomyomas* (lē-ō-mī-Ō-mas), or *fibroids*, usually cause no problems. If stimulated by estrogens, they can grow quite large, reaching weights as great as 13.6 kg (30 lb). Squeezing of the uterine tubes, distortion of adjacent organs, and compression of blood vessels may then lead to a variety of symptoms such as pelvic pain and abnormally heavy uterine bleeding. In symptomatic young women, observation or conservative treatment with drugs or restricted surgery to preserve fertility may be used. In older women, a decision may be made to remove the uterus in a hysterectomy.

Benign epithelial tumors in the uterus are called *endometrial polyps*. (A *polyp* is a protruding growth from an epithelial surface.) Roughly 10 percent of women probably have polyps, but because the polyps tend to be small and cause no symptoms, the condition passes unnoticed. If bleeding occurs, if the polyps become excessively enlarged, or if they protrude through the cervical opening, they can be surgically removed.

Uterine cancers are less common than uterine tumors. There are two types of uterine cancers, (1) endometrial and (2) cervical.

Endometrial cancer is an invasive cancer of the endometrial lining. The condition most commonly affects women age 50–70. Estrogen therapy, used to treat osteoporosis in postmenopausal women, increases the risk of endometrial cancer by 2–10 times. This increased risk can be eliminated by adding progesterone to the estrogen therapy.

There is no satisfactory screening test for endometrial cancer. The most common symptom is irregular bleeding, and diagnosis typically involves a biopsy of the endometrial lining. The prognosis varies with the extent of spread. Treatment of early-stage endometrial cancer involves a hysterectomy (frequently curative) perhaps followed by localized radiation therapy. In advanced stages, more aggressive radiation treatment is recommended. Chemotherapy has not proven to be very successful in treating endometrial cancers.

Cervical cancer is the most common reproductive system cancer in women age 15–34. Most women with cervical cancer fail to develop symptoms until late in the disease. At that stage, vaginal bleeding, especially after intercourse, pelvic pain, and vaginal discharge may appear. Early detection has dramatically reduced the mortality rate for cervical cancer. The standard screening test is the **Pap smear**, named for Dr. George Papanicolaou, an anatomist and cytologist who developed the technique in the late 1940s. The cervical epithelium normally sheds its superficial cells, and a sample of cells scraped or brushed from the epithelial surface can be examined for abnormal or cancerous cells. The American Cancer Society recommends yearly Pap tests at ages 20 and 21, followed by smears at 1-year to 3-year intervals until age 65.

Early treatment of abnormal, but not cancerous, lesions detected by mildly abnormal Pap smears may prevent progression to cancer formation. Treatment of localized cervical cancer involves the removal of the affected portion of the cervix. Treatment of more-advanced cancers typically involves a combination of radiation therapy, hysterectomy, lymph node removal, and chemotherapy.

Breast cancer is a cancer of the mammary gland. It is the leading cause of death for women between the ages of 35 and 45, but it is more common in women after age 50. An estimated 12 percent of women in the United States will develop breast cancer at some point in their lifetimes if they live to age 90. Notable risk factors include (1) a family history of breast cancer, (2) a pregnancy after age 30, and (3) early menarche (first menstrual period) or late menopause (last menstrual period). Breast cancers in males are very rare, but about 300 men die from breast cancer each year in the United States.

Early detection of breast cancer is the key to reducing mortalities. Most breast cancers are found through self-examination, but the use of clinical techniques, such as **mammography**, are also important. This technique uses low doses of X-rays to examine breast tissues.

Treatment of breast cancer begins with the removal of the tumor. Because in many cases the cancer cells begin to spread before diagnosis, a **mastectomy** (mas-TEK-to-mē), or surgical removal of part or all of the affected breast is required. For example, in a **segmental mastectomy**, or *lumpectomy*, only a portion of the breast is removed; in a **total mastectomy**, the entire breast is removed. The most common operation is a **modified radical mastectomy**, in which the breast and axillary lymph nodes are removed, but the muscular tissue remains. Chemotherapy, radiation treatments, and hormone treatments may be used to supplement the surgical procedures. Early detection and treatment has probably contributed to the reduced breast cancer death rate in the late 1990s.

Monthly stimulation of the mammary glands by hormones of the uterine cycle may cause the breasts to enlarge, become lumpy from swollen milk-forming lobules, and then return to normal. If lobules become inflamed, they may be walled off by fibrous scar tissue and form fibrous nodules and **cysts**. Clusters of such cysts can be felt as separate masses, a condition known as **fibrocystic breasts**. Because the symptoms are the same as breast cancer, biopsies may be needed to distinguish between this benign condition and breast cancer.

Congenital Disorders

Cryptorchidism. In **cryptorchidism** (kript-ŌR-ki-dizm), which means "hidden testis," one or both of the testis have not descended into the scrotum by the time of birth. Typically, the testes are lodged in the abdominal cavity or within the inguinal canal. This condition occurs in about 3 percent of full-term deliveries and in roughly 30 percent of premature births. In most instances, normal descent occurs a few weeks later, but the condition can be surgically corrected if it persists. Corrective measures are usually taken before puberty because testes in the abdomen will not produce sperm, and the individual will be sterile. If the testes cannot be moved into the scrotum, they will usually be removed, because about 10 percent of abdominal testes eventually develop testicular cancer. This surgical procedure is called an *orchiectomy*.

Uterine-Associated Disorders

There are a number of disorders associated with the uterus that do not fit within the major categories of disorders used in earlier chapters.

In **endometriosis** (en-dō-mē-trē-Ō-sis), an area of endometrial tissue begins to grow outside the uterus. The cause is unknown. One possibility is that pieces of the endometrium sloughed off during menstruation are forced through the uterine tubes into the peritoneal cavity, where they have reattached. The severity of the condition depends on the size of the abnormal masses and their location. Abdominal pain, bleeding, pressure on surrounding structures, and infertility are common symptoms.

Diagnosis is usually made using a *laparoscope* inserted through a small opening in the abdominal wall. Using this device, a physician can inspect the outer surfaces of the uterus and uterine tubes, the ovaries, and the lining of the pelvic cavity. Treatment may involve hormone administration or surgical removal of the tissue masses. If the condition is widespread, a hysterectomy or oophorectomy (removal of the ovaries) may be desirable.

Amenorrhea (ā-men-ō-RĒ-uh) is the absence or abnormal cessation of menstruation. *Primary amenorrhea* is the failure of menarche, the first menstrual period, to occur at puberty. This condition may be caused by nonfunctional ovaries, the absence of a uterus, or an endocrine or genetic disorder. *Secondary amenorrhea* is an interruption of the normal uterine cycle of an adult woman for 6 months or more. General causes of this condition are severe physical or emotional stresses. Examples include drastic weight-reduction programs, anorexia nervosa, and severe depression or grief. The condition also occurs in marathon runners and other women engaged in programs that require sustained high levels of exertion that reduce body lipid reserves. Pregnancy and menopause are normal causes of amenorrhea.

Premenstrual syndrome (PMS) is an array of symptoms that may develop in women 7–14 days before the start of menstruation. Fluid retention, breast enlargement, headaches, pelvic pain, and an uncomfortable feeling of bloating are common symptoms. These sensations may be associated with neurological changes producing irritability, anxiety, and depression. The actual mechanism responsible for PMS has yet to be determined. Exercise, reduction or avoidance of salt and caffeine, supplemental vitamins and calcium, and various hormone treatments reduce symptoms for some women.

Dysmenorrhea (dis-men-ō-RĒ-uh) is painful menstruation. It can result from uterine inflammation and contraction, or from conditions involving nearby pelvic structures. For example, endometriosis can cause severe dysmenorrhea.

✔ What problems may cause swelling of or in the scrotal cavity?

✔ What is the difference between an orchiectomy and an oophorectomy?

✔ The swelling of which gland in males leads to interference with urination?

CHAPTER REVIEW

Key Words

areola (a-RĒ-ō-la): Pigmented area that surrounds the nipple of the breast.

estrogens (ES-trō-jenz): Female sex hormones, notably estradiol; primary hormones regulating the female reproductive cycle.

genitalia: External organs of the reproductive system.

lactation (lak-TĀ-shun): The production of milk by the mammary glands.

meiosis (mī-Ō-sis): Cell division that produces gametes with half the normal chromosome number.

menarche: The first menstrual period that normally occurs at puberty.

menstruation (men-stroo-Ā-shun): The monthly flow of blood that signifies the start of the uterine cycle.

oogenesis (ō-ō-JEN-e-sis): Ovum production.

ovary: Female reproductive gland; site of gamete production and hormone secretion.

ovulation (ov-ū-LĀ-shun): The release of an ovum, following the rupture of the follicle wall.

progesterone (prō-JES-ter-ōn): The most important progestin secreted by the corpus luteum following ovulation; prepares the uterus for pregnancy.

semen (SĒ-men): Fluid ejaculate containing spermatozoa and the secretions of accessory glands of the male reproductive tract.

seminiferous tubules (se-mi-NIF-e-rus): Coiled tubules where sperm production occurs in the testis.

spermatogenesis: Sperm production.

spermatozoa (sper-ma-tō-ZŌ-a): Sperm cells; singular: spermatozoon.

testes (TES-tēz): The male gonads; sites of sperm production and hormone secretion.

testosterone (tes-TOS-te-rōn): The principal androgen produced by the interstitial cells of the testes.

Selected Clinical Terms

amenorrhea (ā-men-ō-RĒ-uh): The failure of menarche to appear before age 16, or a lack of menstruation for 6 months or more in an adult female of reproductive age.

breast cancer: Cancer of the mammary gland that is the primary cause of death for women age 35–45.

cervical cancer: Cancer of the cervix that is the most common reproductive cancer in women.

dysmenorrhea (dis-men-ō-RĒ-uh): Painful menstruation.

endometriosis (en-dō-mē-trē-Ō-sis): The growth of endometrial tissue outside the uterus.

gonorrhea: A sexually transmitted disease caused by bacterial infection.

impotence: Inability to achieve or maintain an erection.

mastectomy (mas-TEK-to-mē): Surgical removal of part or all of a cancerous mammary gland.

ovarian cancer: Cancer of the ovaries that is the most lethal reproductive cancer in women.

pelvic inflammatory disease (PID): An infection of the uterine tubes.

prostate cancer: Cancer of the prostate gland, the second most common cause of cancer deaths in males.

prostatectomy (pros-ta-TEK-to-mē): Surgical removal of the prostate gland.

sexually transmitted disease (STD): Diseases transferred from one individual to another primarily through sexual contact.

vaginitis (vaj-i-NĪ-tis): Inflammation of the vagina; often caused by fungal or bacterial infection.

Study Outline

INTRODUCTION (p. 519)

1. Human reproduction is *sexual*. It requires male and female **gametes** (reproductive cells). **Fertilization** is the fusion of a **sperm** from the father and an **ovum** from the mother to create a **zygote** (fertilized egg).

SYSTEM BRIEF (pp. 519–520)

1. The reproductive system includes **gonads**, ducts, accessory glands and organs, and the **external genitalia**.

2. Gametes are produced through **meiosis**. Meiosis produces reproductive cells with half the normal chromosome number. *(Figure 21-1)*

3. In the male the **testes** produce sperm, which are expelled from the body in **semen** during **ejaculation**. Each month, the **ovaries** (gonads) of a sexually mature female produce an egg that travels

along **uterine tubes** to reach the **uterus**. The **vagina** connects the uterus with the exterior.

STRUCTURE OF THE MALE REPRODUCTIVE SYSTEM
(pp. 520–525)

1. The **spermatozoa** travel along the **epididymis**, the **ductus deferens**, the **ejaculatory** duct, and the **urethra** before leaving the body. Accessory organs (notably the **seminal vesicles**, **prostate gland**, and **bulbourethral glands**) secrete fluids into the ejaculatory ducts and urethra. The **scrotum** encloses the testes, and the **penis** is an erectile organ. *(Figure 21-2)*

THE TESTES *(p. 521)*

2. Each testis is divided into a series of **lobules**. **Seminiferous tubules** within each lobule are the sites of sperm production. *(Figure 21-3a)*

3. Seminiferous tubules contain stem cells involved in **spermatogenesis**, and **sustentacular cells**, which nourish the developing spermatozoa. *(Figure 21-3b)*

4. Interstitial cells between the seminiferous tubules secrete sex hormones. *(Figure 21-3b)*

5. Each spermatozoon has a **head**, **middle piece**, and **tail**. *(Figure 21-4)*

THE MALE REPRODUCTIVE TRACT *(p. 522)*

6. From the testis the immature spermatozoa enter the **epididymis**, an elongate tubule. The epididymis aids their maturation and also serves as a recycling center for damaged spermatozoa. Spermatozoa leaving the epididymis are functionally mature, yet immobile.

7. The **ductus deferens**, or *vas deferens*, begins at the epididymis and passes through the inguinal canal as one component of the **spermatic cord**. The junction of the base of the seminal vesicle and the ductus deferens creates the **ejaculatory duct**, which empties into the urethra.

8. The urethra extends from the urinary bladder to the tip of the penis and carries products from both the urinary and reproductive systems.

THE ACCESSORY GLANDS *(p. 523)*

9. A typical ejaculation releases 2 to 5 ml of semen (an **ejaculate**), which contains 20 million to 100 million sperm per milliliter.

10. Each **seminal vesicle** is an active secretory gland that contributes about 60 percent of the volume of semen; its secretions contain fructose, which is easily metabolized by spermatozoa. These secretions also help neutralize the acids normally found in the urethra and vagina. The **prostate gland** secretes acidic fluids that make up about 30 percent of seminal fluid.

11. Alkaline mucus secreted by the **bulbourethral glands** has lubricating properties.

THE PENIS *(p. 524)*

12. The skin overlying the **penis** resembles that of the scrotum. Most of the body of the penis consists of three masses of **erectile tissue**. Beneath the superficial fascia there are two **corpora cavernosa** and a single **corpus spongiosum** that surrounds the urethra. Dilation of the erectile tissue with blood produces an **erection**. *(Figure 21-5)*

HORMONES AND MALE REPRODUCTIVE FUNCTION
(p. 525)

1. Important regulatory hormones include **FSH** (follicle-stimulating hormone), **LH** (luteinizing hormone, formerly called *ICSH* or interstitial cell–stimulating hormone). Testosterone is the most important androgen.

2. FSH, along with testosterone, promotes spermatogenesis.

3. LH causes the secretion of testosterone by the interstitial cells of the testes.

STRUCTURE OF THE FEMALE REPRODUCTIVE SYSTEM *(pp. 525–531)*

1. Principal organs of the female reproductive system include the *ovaries, uterine tubes, uterus, vagina,* and *external genitalia. (Figure 21-6)*

THE OVARIES *(p. 526)*

2. The ovaries are the site of ovum production, or **oogenesis**, which occurs monthly in **ovarian follicles** as part of the **ovarian cycle**.

3. As development proceeds **primordial**, **primary**, **secondary**, and **tertiary follicles** develop. At **ovulation**, an ovum is released at the ruptured wall of the ovary. *(Figure 21-7)*

THE UTERINE TUBES *(p. 527)*

4. Each **uterine tube** has an *infundibulum* with *fimbriae* (projections) that opens into the uterine cavity. For fertilization to occur, the ovum must encounter spermatozoa during the first 12 to 24 hours of its passage from the infundibulum to the uterus. *(Figure 21-8)*

THE UTERUS *(p. 527)*

5. The uterus provides protection and nutritional support to the developing embryo. Major anatomical landmarks of the uterus include the **body**, **cervix**, and **uterine cavity**. The uterine wall can be divided into an inner **endometrium** and a muscular **myometrium**. *(Figure 21-8)*

6. A typical 28-day **uterine cycle**, or **menstrual cycle**, begins with the onset of **menstruation** and the destruction of the endometrium. This process of menstruation continues from 1 to 7 days. *(Figure 21-9)*

7. After menstruation, the **proliferative phase** begins and the endometrium undergoes repair and thickens. Menstrual activity begins at **menarche** and continues until **menopause**. *(Figure 21-9)*

THE VAGINA *(p. 529)*

8. The **vagina** is a muscular tube extending between the uterus and external genitalia. A thin epithelial fold, the **hymen**, may partially block the entrance to the vagina.

THE EXTERNAL GENITALIA *(p. 530)*

9. The components of the **vulva** include the **vestibule**, **labia minora**, **clitoris**, **labia majora**, and the **lesser** and **greater vestibular glands**. *(Figure 21-10)*

THE MAMMARY GLANDS *(p. 531)*

10. At birth a newborn infant gains nourishment from milk secreted by maternal **mammary glands**. *(Figure 21-11)*

HORMONES AND THE FEMALE REPRODUCTIVE CYCLE
(pp. 531–533)

1. Hormonal regulation of the female reproductive system involves coordination of the ovarian and uterine cycles.

THE OVARIAN CYCLE *(p. 531)*

2. Estradiol, one of the *estrogens*, is the dominant hormone of the preovulatory period. Ovulation occurs in response to peak levels of estrogen and LH. *(Figure 21-12)*

3. Progesterone, one of the steroid hormones called *progestins*, is the principal hormone of the postovulatory period.

THE UTERINE CYCLE *(p. 533)*

4. Hormonal changes are responsible for the maintenance of the uterine cycle. *(Figure 21-12)*

HORMONES AND BODY TEMPERATURE *(p. 533)*

5. Body temperature is slightly lower during the preovulatory period than during the postovulatory period. *(Figure 21-12)*

AGING AND THE REPRODUCTIVE SYSTEM *(pp. 533–534)*

MENOPAUSE *(p. 533)*

1. Menopause (the time that ovulation and menstruation cease in women) typically occurs around age 50. Production of FSH and LH rises, while circulating concentrations of estrogen and progesterone decline.

THE MALE CLIMACTERIC *(p. 534)*

2. During the **male climacteric**, between ages 50 and 60, circulating testosterone levels decline, while levels of FSH and LH rise.

DISORDERS OF THE REPRODUCTIVE SYSTEM *(pp. 536–546)*

THE PHYSICAL EXAMINATION *(p. 536)*

1. Symptoms of male reproductive disorders include: pain, urethral discharge and dysuria, **impotence**, and male infertility.

2. Inspection may reveal skin lesions, such as vesicles, **chancres**, and warts, which may indicate sexually transmitted diseases.

3. Palpation may reveal abnormal masses, swelling, or tumors. **Hydrocele** is swelling of the scrotal cavity by serous fluid. **Spermatocele** is a spermatozoan-containing cyst. **Varicocele** is a varicose vein near the spermatic cord. *(Table 21-1)*

4. Symptoms of female reproductive disorders include: pelvic pain, abnormal bleeding, lack of menstruation, abnormal vaginal discharge, dysuria, and infertility.

5. Inspection of the external genitalia may reveal skin lesions, trauma, or other abnormalities.

REPRESENTATIVE DISORDERS *(p. 538)*

6. Major reproductive system disorders include inflammation and infection, trauma, tumors, congenital disorders, and uterine-associated disorders. *(Figure 21-14)*

7. Inflammation of the male reproductive system includes **orchitis**, **epididymitis**, and **prostatitis**.

8. Inflammation of the female reproductive system includes: **oophoritis**, **salpingitis**, **pelvic inflammatory disease (PID)**, **vaginitis**, and **toxic shock syndrome (TSS)**.

9. **Sexually transmitted diseases (STDs)** are infectious diseases that are transferred from individual to individual by sexual contact. STDs may include bacterial, viral, and protozoan infections.

10. Some STDs include **chlamydia** (the most frequent cause of all STDs), **lymphogranuloma venereum (LGV)**, **gonorrhea**, **syphilis**, **genital herpes**, **genital warts**, and **chancroid**. *(Figures 21-15, 21-16, 21-17)*

11. Two examples of trauma affecting the male reproductive system are *inguinal hernias* and **testicular torsion**.

12. Benign prostatic hypertrophy is most common in older males. Tumors in the male reproductive system cause **testicular cancer** and **prostate cancer**. **Prostatectomy** is the surgical removal of the prostate gland.

13. Tumors in the female reproductive system include **ovarian cancer** and **uterine tumors**. A **Pap smear** is a standard screening test for *cervical cancer*.

14. Breast cancer is a cancer of the mammary gland.

15. Congenital disorders include **cryptorchidism**, where one or both of the testis have not descended into the scrotum.

16. Uterine-associated disorders include **endometriosis, amenorrhea** (the absence or abnormal cessation of menstruation), **premenstrual syndrome** (the combination of symptoms that occur before menstruation), and **dysmenorrhea** (painful menstruation).

Review Questions

MATCHING

Match each item in Column A with the most closely related item in Column B. Use letters for answers in the spaces provided.

	Column A		Column B
___	1. gametes	a.	production of androgens
___	2. gonads	b.	muscular wall of uterus
___	3. interstitial cells	c.	high concentration of fructose
___	4. seminal vesicles	d.	painful menstruation
___	5. prostate gland	e.	secretes thick, sticky, alkaline mucus
___	6. bulbourethral glands	f.	inflammation of the testis
___	7. prepuce	g.	secretes progesterone
___	8. corpus luteum	h.	uterine lining
___	9. endometrium	i.	reproductive cells
___	10. myometrium	j.	ovum
___	11. female sex cell	k.	milk production
___	12. lactation	l.	surrounds urethra
___	13. clitoris	m.	foreskin of penis
___	14. dysmenorrhea	n.	reproductive organs
___	15. orchitis	o.	female erectile tissue

MULTIPLE CHOICE

16. Chromosomes are carried within the _____ of a sperm cell.
(a) acrosome
(b) head
(c) middle piece
(d) tail

17. Spermatogenesis (sperm production) occurs in the _____ .
(a) ductus deferens
(b) epididymis
(c) seminiferous tubules
(d) prostate gland

18. The role of the epididymis is to _____ .
(a) maintain an environment suitable for the maturation of spermatozoa
(b) produce spermatozoa
(c) transport spermatozoa to the urethra
(d) secrete testosterone

19. The hormone that stimulates the growth of ovarian follicles is _____ .
(a) estrogen   (b) FSH
(c) LH   (d) progesterone

20. The female structure that corresponds to the male penis is the _____ .
(a) vulva   (b) labia minora
(c) labia majora   (d) clitoris

21. At the time of ovulation, the basal body temperature _____ .
(a) is not affected
(b) increases noticeably
(c) declines sharply
(d) may increase or decrease a few degrees

22. A sudden surge in LH concentration causes _____ .
(a) the onset of menses
(b) the end of the uterine cycle
(c) menopause
(d) breakdown of the follicular wall and ovulation

23. _____ is a STD caused by a viral infection.
(a) syphilis
(b) gonorrhea
(c) chancroid
(d) genital warts

24. _____ is a congenital disorder.
(a) pelvic inflammatory disease
(b) endometriosis
(c) cryptorchidism
(d) oophoritis

TRUE/FALSE

___ 25. The scrotum is an external sac that holds the testes.

___ 26. The inner lining of the uterus is the myometrium.

___ 27. Fertilization is the union of the egg and sperm.

___ 28. Circumcision is the surgical removal of the prepuce.

___ 29. Sperm production in the testes is stimulated by testosterone secreted by the pituitary gland.

___ 30. The principal hormone of the postovulatory period is progesterone.

___ 31. Bacteria are normally present in the vagina.

___ 32. Salpingitis is the inflammation of the uterine tubes.

SHORT ESSAY

33. What accessory organs and glands contribute to the composition of semen? What are the functions of each?

34. Using an average duration of 28 days, describe each of the three phases of the uterine cycle.

35. What are the three major functions of the vagina?

36. Diane has an inflammation of the peritoneum (peritonitis), which her doctor says resulted from a urinary tract infection. Why can this situation occur in females but not in males?

37. How can pelvic inflammatory disease lead to sterility?

38. What is testicular torsion?

APPLICATIONS

39. Mrs. Sanchez, 38, comes to the doctor because she discovered a lump in her breast. What disorders could this indicate? What tests could you recommend to distinguish between these disorders?

40. Women body builders and women suffering from eating disorders such as anorexia nervosa commonly experience amenorrhea. What does this suggest about the relation between body fat and menstruation? What might be the benefit of amenorrhea under such circumstances?

✔ Answers to Concept Check Questions

(p. 525)   1. The acrosomal cap contains enzymes necessary for fertilization of the ovum. Without these enzymes, fertilization would not occur. 2. Dilation of the arteries serving the penis will result in erection. 3. FSH is needed for maintaining a high level of testosterone, which supports the formation of spermatozoa. Low levels of FSH would lead to low levels of testosterone in the seminiferous tubules and thus a lower rate of sperm production and low sperm count.

(p. 531)   1. Blockage of the uterine tube would cause sterility. 2. The outer layer of the endometrium is sloughed off during menstruation. 3. Blockage of a single lactiferous sinus would not interfere with the movement of milk to the nipple because each breast has between 15 and 20 lactiferous sinuses.

(p. 533)   1. If the LH surge did not occur during an ovulatory cycle, ovulation and corpus luteum formation would not occur. 2. A decline in the levels of estrogen and progesterone during the uterine, or menstrual cycle, signals the beginning of menstruation.

(p. 534)   1. At menopause, circulating estrogen levels begin to drop. Estrogen has an inhibitory effect on GnRH and FSH, and as the level of estrogen declines, the levels of FSH and GnRH rise and remain elevated.

(p. 546)   1. Swelling may be caused by serous fluid (hydrocele), by testicular cysts (spermatocele), or by varicose veins in the spermatic cord (varicocele). 2. An orchiectomy is the surgical removal of a testis and an oophorectomy is the surgical removal of an ovary. 3. The prostate gland.

Development and Inheritance

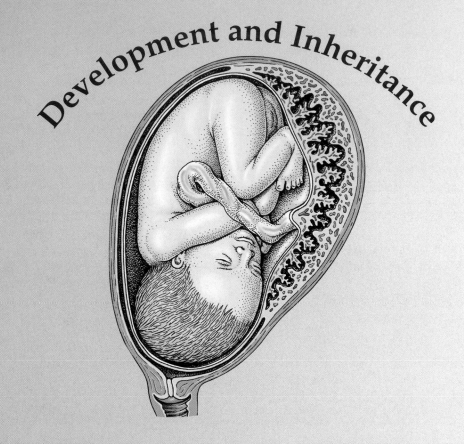

CHAPTER OUTLINE

CHAPTER OBJECTIVES

1   Describe the process of fertilization.
2   List the three prenatal periods, and describe the major events of each period.
3   List the three primary germ layers and their roles in forming major body systems.
4   Describe the roles of the different membranes of the embryo.
5   Describe the adjustments of the mother's organ systems in response to the presence of a developing embryo.
6   Discuss the events that occur during labor and delivery.
7   Describe the major stages of life after delivery.
8   Describe the basic patterns of inheritance of human traits.
9   Describe examples of disruptive factors and genetic disorders in development.
10   Describe examples of problems associated with pregnancy.

A human being develops in the womb for 9 months, grows to maturity in 15 to 20 years, and may live the better part of a century. During that whole time he or she will continue to change. Birth, growth, maturation, aging, and death are all parts of a single, continuous process. That process does not end with the individual, for human beings can pass at least some of their characteristics on to their offspring. Thus each generation gives rise to a new generation that will repeat the same cycle. In this chapter we examine **development**, the complex physical changes that occur from conception to maturity, and **inheritance**, the transfer of characteristics from parent to offspring.

STAGES OF DEVELOPMENT

Development begins at fertilization, or **conception**. The period of **prenatal development** extends from conception to birth. **Postnatal development** begins at birth and continues to maturity. Prenatal development involves two stages. Over the first 2 months, the developing individual is called an **embryo**. **Embryology** (em-brē-OL-ō-jē) is the study of embryonic development. After 2 months, the developing individual is called a **fetus**. Fetal development begins at the start of the ninth week and continues up to the time of birth.

FERTILIZATION

Fertilization involves the fusion of a sperm cell and an egg cell, each containing 23 chromosomes, to produce a single-celled *zygote* containing 46 chromosomes, the normal number of human chromosomes. As Figure 22-1a• shows, normal fertilization occurs in the upper third of the uterine tube, usually within a day of ovulation.

•FIGURE 22-1
Fertilization and Implantation.
(a) Cleavage and formation of a blastocyst occurs within the oviduct over the first 5 days after fertilization. (b) After implantation, the outer cells absorb nutrients from the enclosing endometrium before forming the placenta. The inner mass of cells, which becomes the body, is separated from the outer cells by the formation of an amniotic cavity and yolk sac.

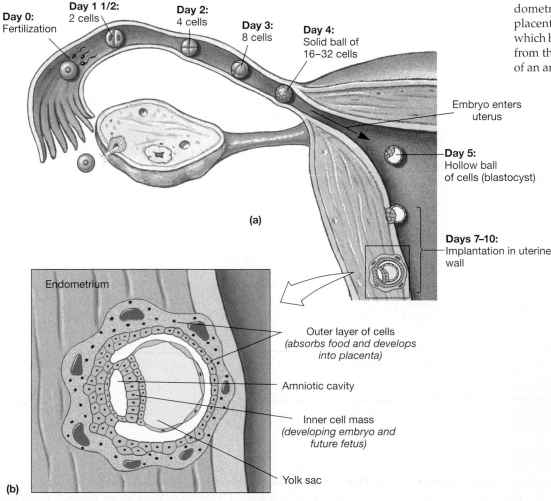

Day 0: Fertilization

Day 1 1/2: 2 cells

Day 2: 4 cells

Day 3: 8 cells

Day 4: Solid ball of 16–32 cells

Embryo enters uterus

Day 5: Hollow ball of cells (blastocyst)

(a)

Days 7–10: Implantation in uterine wall

Endometrium

Outer layer of cells (*absorbs food and develops into placenta*)

Amniotic cavity

Inner cell mass (*developing embryo and future fetus*)

Yolk sac

(b)

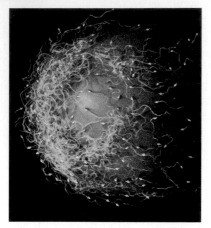

Sperm swarming around an egg, attempting to perform fertilization. Note the tremendous size difference between the sperm and egg.

Contractions of uterine muscles and ciliary currents in the uterine tubes aid the passage of sperm to the fertilization site. Of the 200 million spermatozoa introduced into the vagina in a typical ejaculate, only around 10,000 make it past the uterus, and fewer than 100 actually reach the egg. A male with a sperm count below 20 million per milliliter will usually be sterile because too few sperm survive to reach the egg.

Large numbers of sperm are needed for fertilization because when the egg leaves the ovary it is protected by a layer of follicle cells. Sperm cells release enzymes that separate these cells from one another to form an unobstructed passageway to the egg's cell membrane. One sperm cell cannot accomplish this—it takes the combined enzymes of many spermatozoa to clear the way to the egg. When one sperm does contact the egg surface, their cell membranes fuse, and the sperm nucleus is released into the cytoplasm of the egg. This event activates the egg and changes the cell membrane to prevent fertilization by other sperm.

GESTATION

The time spent in prenatal development is known as **gestation** (jes-TĀ-shun). An average gestation period, that is, the length of a normal pregnancy, is 38 weeks or 266 days. For convenience, the gestation period is usually considered as three **trimesters**, each 3 months long.

The **first trimester** is the period of embryonic and early fetal development. During this period the basic components of all of the major organ systems appear.

In the **second trimester** the organs and organ systems complete most of their development. The body proportions change, and by the end of the second trimester the fetus looks distinctively human.

The **third trimester** is characterized by rapid fetal growth. Early in the third trimester most of the major organ systems become fully functional, and an infant born 1 month or even 2 months prematurely has a reasonable chance of survival.

THE FIRST TRIMESTER

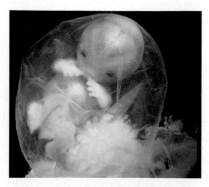

A 4-week old embryo has rudiments of all major organs and systems.

The first trimester is the most dangerous period in prenatal or postnatal life. Only about 40 percent of the eggs that are fertilized produce embryos that survive the first trimester, and an additional number of fetuses enter the second trimester already doomed or deformed by some developmental mistake. Because the developmental events of the first trimester are easily disrupted, pregnant women are usually warned to take great care to avoid drugs or other physical or chemical stresses during this period. Major highlights of the first trimester include (1) embryo formation and the *implantation* of the embryo within the uterine wall, (2) formation of tissues and embryonic membranes, (3) formation of the placenta, and (4) the beginning of organ formation.

EMBRYO FORMATION

After fertilization, the newly formed zygote undergoes a period of rapid cell divisions called **cleavage** (KLĒ-vij). During cleavage, new cells form so rapidly that they do not have time to grow. The first division, resulting in two cells, is completed about 1.5 days after fertilization. This early stage is about the size of the period at the end of this sentence. As Figure 22-1a• shows, by day 4, the embryo is a solid ball of cells, and it enters the chamber of the uterus. After 5 days of cleavage a hollow cavity appears, now forming an embryonic stage called a *blastocyst*. The cells making up this stage begin to form into outer and inner groups of cells. The outer layer of cells will be responsible for obtaining food for the *inner cell mass* that will form the developing embryo.

IMPLANTATION

Implantation begins on day 7 as the surface of the blastocyst touches and sticks to the uterine lining (Figure 22-1a•). Within the next few days, the blastocyst becomes completely embedded within the uterine wall and loses contact with the

A 12-week old embryo, which weighs just 26 grams.

uterine cavity; further development occurs entirely within the endometrium. Figure 22-1b• shows the implanted blastocyst and the inner cell mass that will develop into the body. The breakdown of uterine gland cells within the endometrium releases glycogen and other nutrients. These nutrients provide the energy needed to support the early stages of embryo formation.

Implantation requires a functional endometrium. As implantation is occurring, some of the cells of the blastocyst begin secreting a hormone called **human chorionic** (kō-rē-ON-ik) **gonadotropin (hCG)**. Because of the hCG, the corpus luteum does not degenerate, and it maintains its production of estrogens and progesterone. As a result, the endometrial lining remains perfectly functional, and menstruation does not occur and terminate the pregnancy. The production of hCG is taken over by the placenta as it develops. The presence of hCG in blood or urine samples provides a reliable indication of pregnancy, and kits sold for the early detection of pregnancy are sensitive for the presence of this hormone.

TISSUE AND MEMBRANE FORMATION

Germ Layers

After implantation, the inner cell mass of the blastocyst becomes organized into three layers of cells. Each of these **germ layers** will form different body tissues. The outer layer is the **ectoderm**, the middle layer is the **mesoderm**, and innermost layer is the **endoderm**. Together, these layers form the body's organs and organ systems.

Briefly, ectoderm gives rise to tissues of the skin and nervous system, mesoderm gives rise to connective and muscle tissues, and the endoderm gives rise to the inner epithelium lining the digestive system. Table 22-1 lists the specific contributions each germ layer makes to the body systems described in earlier chapters.

Table 22-1	Tissues and Organs Formed by the Primary Germ Layers
Primary Germ Layer	**Developmental Contributions to the Body**
Ectoderm	*Integumentary system:* epidermis, hair follicles and hairs, nails, and glands communicating with the skin (apocrine and merocrine sweat glands, mammary glands, and sebaceous glands) *Skeletal system:* pharyngeal cartilages of the embryo develop into portions of the sphenoid and hyoid bones, the auditory ossicles, and the styloid processes of the temporal bones *Nervous system:* all neural tissue, including brain and spinal cord *Endocrine system:* pituitary gland and the adrenal medullae *Respiratory system:* mucous epithelium of nasal passageways *Digestive system:* mucous epithelium of mouth and anus, salivary glands
Mesoderm	*Skeletal system:* all components except some ectodermal contributions *Muscular system:* all components *Endocrine system:* adrenal cortex, endocrine tissues of heart, kidneys, and gonads *Cardiovascular system:* all components, including bone marrow *Lymphatic system:* all components *Urinary system:* the kidneys, including the nephrons and the initial portions of the collecting system *Reproductive system:* the gonads and the adjacent portions of the duct systems *Miscellaneous:* the lining of the body cavities (pleural, pericardial, peritoneal) and the connective tissues that support all organ systems
Endoderm	*Endocrine system:* thymus, thyroid, and pancreas *Respiratory system:* respiratory epithelium (except nasal passageways) and associated mucous glands *Digestive system:* mucous epithelium (except mouth and anus), exocrine glands (except salivary glands), liver, and pancreas *Urinary system:* urinary bladder and distal portions of the duct system *Reproductive system:* distal portions of the duct system, stem cells that produce gametes

Extraembryonic Membranes of the Embryo

Four **extraembryonic membranes** start forming outside the embryo after implantation. These are the *yolk sac*, the *amnion*, the *allantois*, and the *chorion*. Figure 22-2• shows these membranes at weeks 4 and 10.

The first of the extraembryonic membranes to appear is the **yolk sac** (see also Figure 22-1b•). It aids in the transport of nutrients and is an important early site of blood cell formation. Its role is reduced after the first 6 weeks.

The **amnion** (AM-nē-on) encloses the *amniotic cavity*, which contains fluid that surrounds and cushions the developing embryo and fetus (see also Figure 22-1b•). The amnion increases in size during development.

The **allantois** (a-LAN-tō-is) extends away from the embryo. The allantois accumulates some of the small amount of urine produced by the kidneys during embryonic development. The base of the allantois will form the urinary bladder. The remaining portion of the allantois will form much of the umbilical cord.

The **chorion** (KOR-ē-on) forms outside the amnion. Blood vessels that develop within the chorion provide the nutrients and oxygen needed for continued growth and development. The chorion develops into the *placenta*.

THE PLACENTA

The **placenta** begins its development as projections from the chorion, called *chorionic villi*, first come in contact with maternal tissues (Figure 22-2a•). Embryonic

•FIGURE 22-2
Embryonic Membranes.
(a) The developing embryo and its external membranes at the fourth week of growth. **(b)** By week 10, the amnion has expanded greatly, filling the uterine cavity. The embryo, now making the transition to a fetus, is connected to the placenta (formed from the chorion) by the umbilical cord, which contains portions of the allantois and yolk sac membrane, and blood vessels.

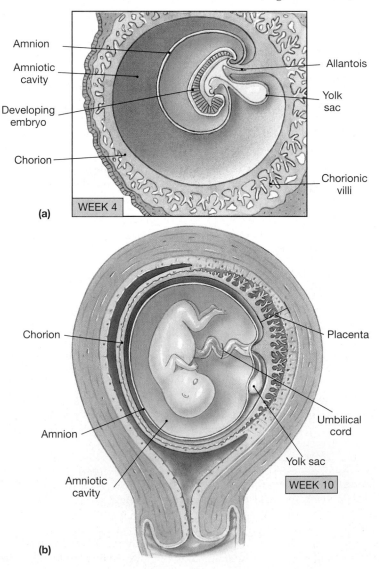

blood vessels develop within each of the villi, and circulation through these chorionic blood vessels begins early in the third week of development, when the heart starts beating. These villi continue to enlarge and branch, forming an intricate network within the endometrium. Blood vessels break down and maternal blood flows slowly through these newly opened spaces. Chorionic blood vessels pass close by, and exchange between the embryonic and maternal circulations occurs by diffusion across the chorion.

Placental Circulation

Figure 22-3• shows the fetal circulation at the placenta near the end of the first trimester. Blood flows to the placenta through the paired **umbilical arteries** and returns in a single **umbilical vein**. The chorionic villi provide the surface area for the exchange of gases, nutrients, and wastes between the fetal and maternal bloodstreams.

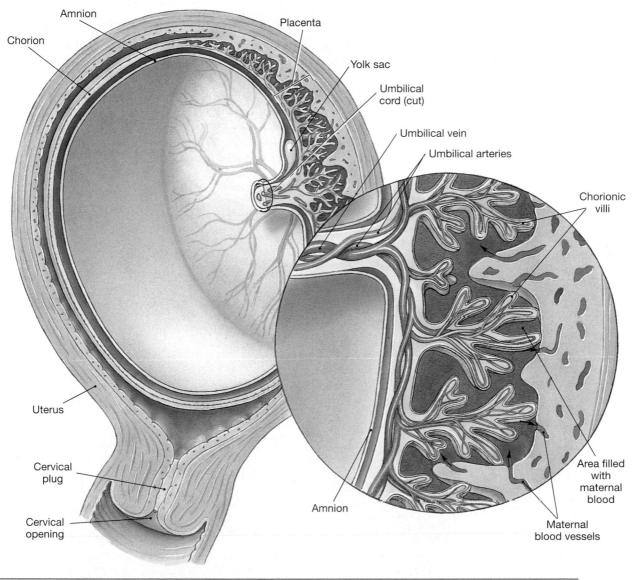

•FIGURE 22-3
A Three-Dimensional View of the Placenta.
For clarity the uterus is shown after the embryo has been removed and the umbilical cord cut. Blood flows into the placenta through broken maternal blood arteries. It then flows around fingerlike projections of the chorion (chorionic villi), which contain fetal blood vessels. Fetal blood arrives over paired umbilical arteries and leaves over a single umbilical vein. Maternal blood reenters the venous system of the mother through the broken walls of small uterine veins. Actual mixing of maternal and fetal blood does not occur.

✔ Why is it important that more than one spermatozoon contact the egg for fertilization to occur?

✔ What is the fate of the inner cell mass of the blastocyst?

✔ Which membrane encloses the fluid surrounding the developing embryo?

✔ What are two important functions of the placenta?

Placental Hormones

In addition to its role in the nutrition of the fetus, the placenta acts as an endocrine organ, releasing hormones. These hormones act to prevent menstruation during the pregnancy, prepare the mammary glands to produce milk, and prepare the body for labor and delivery.

Because of the human chorionic gonadotropin (hCG) secreted first by the blastocyst and then by the placenta, the corpus luteum persists for 3 to 4 months and maintains its production of progesterone, which helps keep the endometrium intact. The decline in the corpus luteum does not trigger menstruation because by the end of the first trimester the placenta is also secreting sufficient amounts of progesterone to maintain the endometrial lining and the pregnancy. As the end of the third trimester approaches, estrogen production by the placenta accelerates. The rising estrogen levels play a role in stimulating labor and delivery.

THE SECOND AND THIRD TRIMESTERS

By the start of the second trimester the basic frameworks of all the major organ systems have formed. Over the next 4 months, the fetus grows from a weight of 0.026 kg (about 1 oz) to a weight of around 0.64 kg (1.4 lb). The changes in body form that occur during the first and second trimesters are shown in Figure 22-4●.

During the third trimester, the basic structures of all the organ systems appear, and most become ready for their normal functions. The fetus also gains the most weight during this trimester. In 3 months the fetus puts on around 2.6 kg (5.7 lb), reaching a full-term weight of somewhere near 3.2 kg (7 lb).

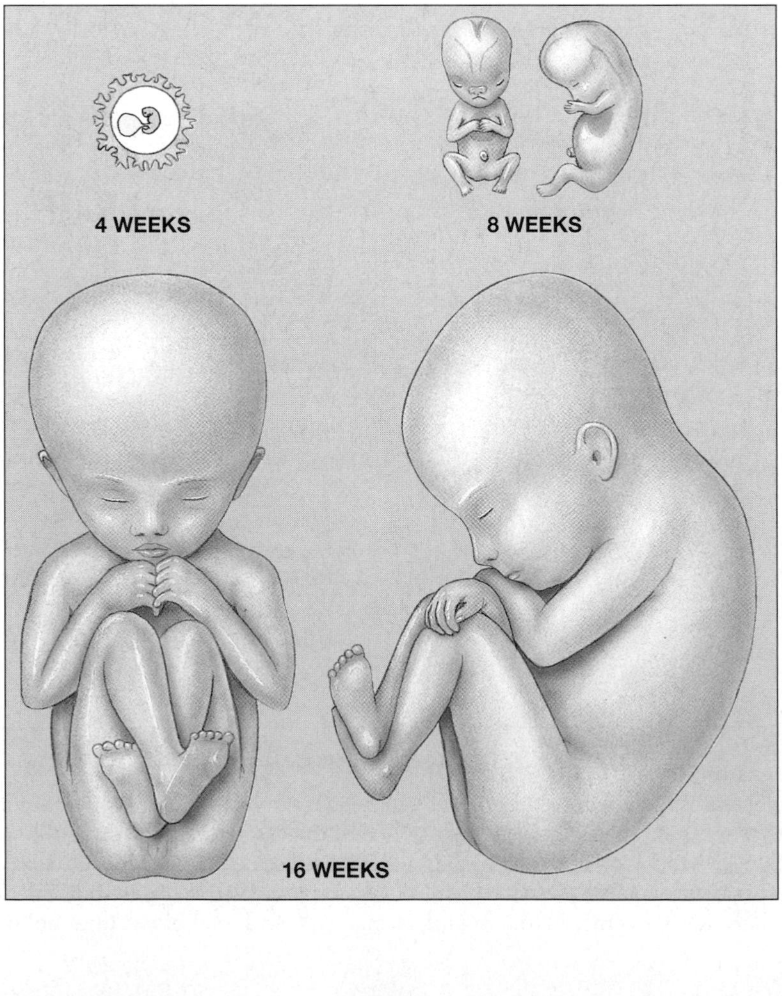

●FIGURE 22-4
Growth and Changes in Body Form. These views of the embryos (4 and 8 weeks) and the fetus (16 weeks) are shown at actual size.

4 WEEKS

8 WEEKS

16 WEEKS

PREGNANCY AND THE MOTHER'S SYSTEMS

The developing fetus is totally dependent on the mother's organ systems for food, oxygen, and waste removal. This means that the mother must absorb enough oxygen, nutrients, and vitamins for herself *and* her fetus, and she must eliminate all the generated wastes. In practical terms, the mother must breathe, eat, and excrete for two. As the fetus grows, the demands on the mother increase, and her body systems must make major adjustments.

Maternal changes during pregnancy include increases in respiratory rate, blood volume, and appetite. Because of increased blood volume, the kidneys must increase their rate of filtration. As a result of increased urine production and the weight of the uterus pressing down on the urinary bladder, pregnant women need to urinate frequently. Changes in the reproductive organs include a tremendous increase in size of the uterus, and by the end of the sixth month of pregnancy, the mammary glands have increased in size and have begun producing and storing their secretions.

CHANGES IN THE UTERUS

At the end of gestation a typical uterus will have grown from 7.5 cm (3 in.) in length and 60 g (2 oz) in weight to 30 cm (12 in.) in length and 1,100 g (2.4 lbs) in weight. It may then contain almost 5 l of fluid, giving the organ with its contained fetus a total weight of roughly 10 kg (22 lbs). This remarkable expansion occurs through the enlargement and elongation of existing cells (especially smooth muscle cells) in the uterus.

The tremendous stretching of the uterine wall is associated with a gradual increase in the rates of spontaneous smooth muscle contractions. In the early stages of pregnancy the contractions are short, weak, and painless. The progesterone released by the placenta has an inhibitory, or calming, effect on the uterine smooth muscle, preventing more extensive and powerful contractions.

Three major factors oppose the calming action of progesterone:

1. **Rising estrogen levels**. Estrogens, also produced by the placenta, increase the sensitivity of the uterine smooth muscles and make contractions more likely. Throughout pregnancy the effects of progesterone are dominant, but as the time of delivery approaches, estrogen production accelerates, and the uterine muscles become more sensitive to stimulation.

2. **Rising oxytocin levels**. Rising oxytocin levels stimulate an increase in the force and frequency of uterine contractions. Release of oxytocin by the pituitary gland is stimulated by high estrogen levels and by stretching of the uterine cervix.

3. **Prostaglandin production**. In addition to estrogens and oxytocin, uterine tissues late in pregnancy produce prostaglandins that stimulate smooth muscle contractions.

After 9 months of gestation, **labor contractions** begin in the muscles of the uterine wall. Once begun, the contractions continue until delivery has been completed.

LABOR AND DELIVERY

The goal of labor is the forcible expulsion of the fetus, a process known as **parturition** (par-tū-RISH-un), or **birth**. During labor, each contraction begins near the top of the uterus and sweeps in a wave toward the cervix. These contractions are strong and occur at regular intervals. As the birth nears, the contractions increase in force and frequency, changing the position of the fetus and moving it toward the cervical canal.

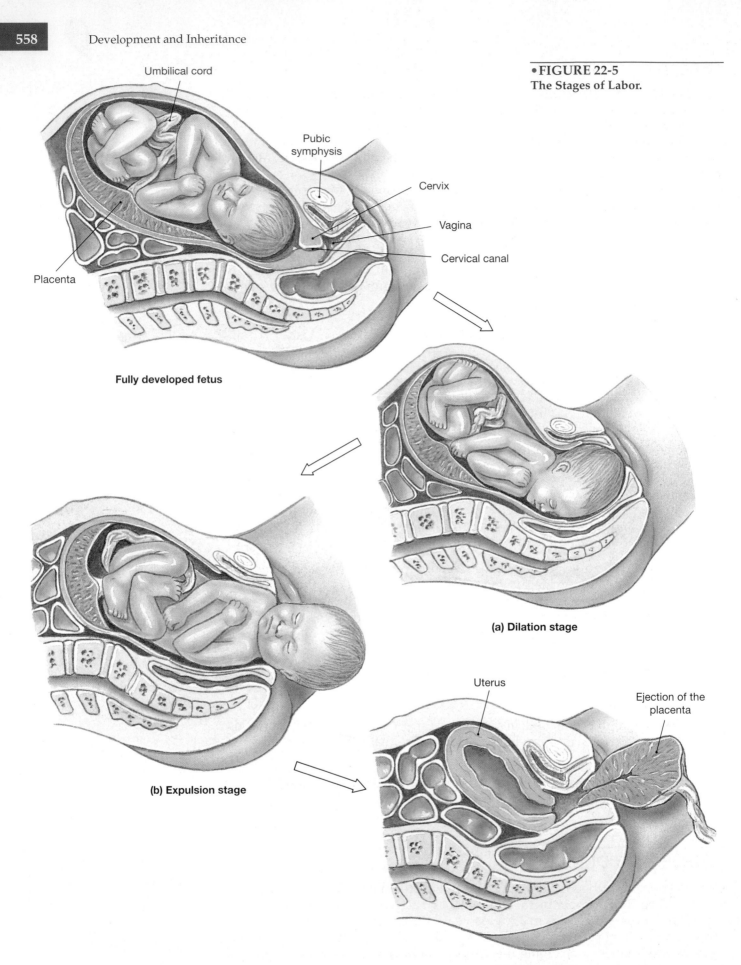

•FIGURE 22-5
The Stages of Labor.

Umbilical cord

Pubic symphysis

Cervix

Vagina

Cervical canal

Placenta

Fully developed fetus

(a) Dilation stage

(b) Expulsion stage

Uterus

Ejection of the placenta

(c) Placental stage

STAGES OF LABOR

Labor has traditionally been divided into three stages (Figure 22-5•), the *dilation stage*, the *expulsion stage*, and the *placental stage.*

The Dilation Stage

The **dilation stage** begins with the onset of labor, as the cervix dilates completely and the fetus begins to slide down the cervical canal. This stage may last 8 or more hours, but during this period the labor contractions occur at intervals of once every 10 to 30 minutes. Late in the process the amnion usually ruptures, releasing the amniotic fluid. This event is sometimes referred to as having the "water break."

The Expulsion Stage

The **expulsion stage** begins after the cervix dilates completely, pushed open by the approaching fetus. Expulsion continues until the fetus has totally emerged from the vagina, a period usually lasting less than 2 hours. The arrival of the newborn infant into the outside world represents the birth, or **delivery**.

If the vaginal entrance is too small to permit the passage of the fetus and there is acute danger of perineal tearing, the entryway may be temporarily enlarged by making an incision through the perineal musculature between the vagina and anus. ∞ *[p. 530]* After delivery, this **episiotomy** (e-pē-zē-O-to-mē) can be repaired with sutures, a much simpler procedure than dealing with a potentially extensive perineal tear. If unexpected complications arise during the dilation or expulsion stages, the infant may be removed by **cesarean section,** or "C-section." In such cases an incision is made through the abdominal wall, and the uterus is opened just enough to allow passage of the infant's head, which is the widest part of its body.

The Placental Stage

During the third, or **placental**, stage of labor the muscle tension builds in the walls of the partially empty uterus, and the organ gradually decreases in size. This uterine contraction tears the connections between the endometrium and the placenta. Usually within an hour after delivery the placental stage ends with the ejection of the placenta, or "*afterbirth.*" The disconnection of the placenta is accompanied by a loss of blood, perhaps as much as 500 to 600 ml, but because the maternal blood volume has increased during pregnancy the loss can be tolerated without difficulty.

MULTIPLE BIRTHS

Multiple births (twins, triplets, quadruplets, and so forth) may occur for several reasons. The ratio of twin to single births in the U.S. population is 1:89. About 70 percent of all twins are **fraternal**, and the other 30 percent are **identical**. Fraternal twins are produced when two eggs are fertilized at the same time, forming two separate zygotes. Fraternal twins may be of the same or different genders. Identical twins result when cells from one zygote separate during an early stage of development. Such individuals have the same genetic makeup and are always the same gender (either both male or both female). Triplets and larger multiples can result from the same processes that produce twins.

✔ Why does the rate of filtration at the mother's kidneys increase during pregnancy?

✔ By what process does the uterus increase in size during pregnancy?

✔ What effect would a decrease in progesterone have on the uterus during late pregnancy?

POSTNATAL DEVELOPMENT

Development does not stop at birth. In the course of postnatal development each individual passes through a number of **life stages**, that is, the neonatal period, *infancy, childhood, adolescence,* and *maturity.*

THE NEONATAL PERIOD, INFANCY, AND CHILDHOOD

The **neonatal period** extends from the moment of birth to 1 month thereafter. **Infancy** then continues to 2 years of age, and **childhood** lasts until puberty begins.

Two major events are under way during these developmental stages.

1. The major organ systems other than those associated with reproduction become fully operational and gradually acquire the functional characteristics of adult structures.

2. The individual grows rapidly, and there are significant changes in body proportions.

Pediatrics is the medical specialty that focuses on the period of life from birth through childhood and adolescence.

The Neonatal Period

A variety of structural and functional changes occur as the fetus becomes a newborn infant, or **neonate**. Before delivery, the transfer of dissolved gases, nutrients, waste products, hormones, and antibodies occurred across the placenta. At birth the newborn infant must become relatively self-sufficient, with the processes of respiration, digestion, and excretion performed by its own organs and organ systems.

A newborn infant has little ability to control its body temperature, particularly in the first few days after delivery. For this reason newborn infants are usually kept bundled up in warm coverings. As the infant grows larger, it adds on fatty tissue which acts as an insulating blanket, its metabolic rate also rises, and its thermoregulatory abilities become better developed.

The nutritional needs of an infant are normally best met by the milk produced by the mammary glands. By the end of the sixth month of pregnancy the mammary glands are fully developed, and the gland cells begin producing a secretion known as **colostrum** (ko-LOS-trum). Colostrum, which is provided to the infant during the first 2 or 3 days of life, contains relatively more proteins and far less fat than milk. Many of the proteins are antibodies that help the infant fight off infections until its own immune system becomes functional.

As colostrum production declines, milk production increases. Milk is a mixture of water, proteins, amino acids, lipids, sugars, and salts. It also contains large quantities of *lysozymes*, enzymes with antibiotic properties.

The actual secretion of the mammary glands is triggered when the newborn begins to suck on the nipple. Stimulation of touch receptors there leads to the release of oxytocin at the posterior pituitary. Oxytocin causes cells within the lactiferous ducts and sinuses to contract. This results in the ejection of milk, or *milk let-down* (Figure 22-6•). The milk let-down reflex continues to function until *weaning* occurs, typically 1 to 2 years after birth.

Infancy and Childhood

The most rapid growth occurs during prenatal development, and after delivery the relative rate of growth continues to decline. Postnatal growth during infancy and childhood occurs under the direction of circulating hormones, notably growth hormone from the pituitary, adrenal steroid hormones, and thyroid hormones. These hormones affect each tissue and organ in specific ways, depending on the sensitivities of the individual cells. As a result, growth does not occur uniformly, and as Figure 22-7• shows, the body proportions gradually change.

•**FIGURE 22-6**
The Milk Let-down Reflex.

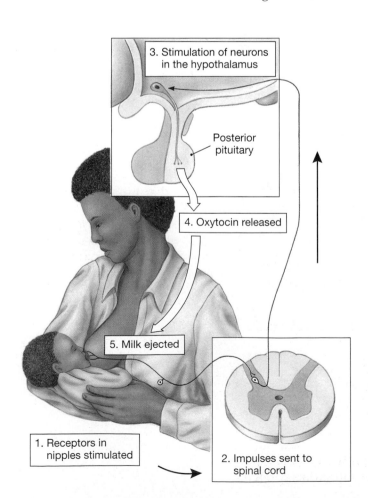

3. Stimulation of neurons in the hypothalamus

Posterior pituitary

4. Oxytocin released

5. Milk ejected

1. Receptors in nipples stimulated

2. Impulses sent to spinal cord

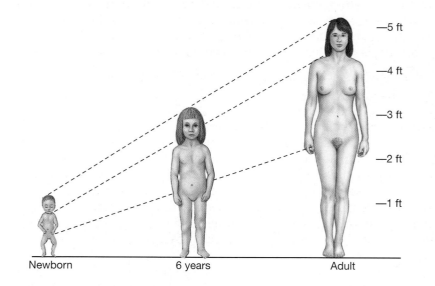

•FIGURE 22-7
Growth and Postnatal Changes in Body Form.
Notice the changes in body form and proportions as development proceeds. For example, the head, which contains the brain and sense organs, is relatively large at birth.

—5 ft
—4 ft
—3 ft
—2 ft
—1 ft

Newborn   6 years   Adult

ADOLESCENCE AND MATURITY

Adolescence begins at **puberty**, when three events interact to promote increased hormone production and sexual maturation:

1. The hypothalamus increases its production of *gonadotropin-releasing hormone* (*GnRH*).

2. The anterior pituitary becomes more sensitive to the presence of GnRH, and there is a rapid elevation in the circulating levels of FSH and LH.

3. Ovarian or testicular cells become more sensitive to FSH and LH. These changes initiate the production of male or female gametes and sex hormones that stimulate the appearance of secondary sexual characteristics and behaviors.

In the years that follow, the continual secretion of estrogens or testosterone maintains these sexual characteristics. In addition, the combination of sex hormones and growth hormone, adrenal steroids, and thyroxine leads to a sudden acceleration in the growth rate. The timing of the increase in size varies between the sexes, corresponding to different ages at the onset of puberty. In girls the growth rate is maximum between ages 10 and 13, whereas boys grow most rapidly between ages 12 and 15. Growth continues at a slower pace until ages 18 to 21.

Maturity is often associated with the end of growth in the late teens or early twenties. Although development ends at maturity, functional changes continue. These changes are part of the process of aging, or **senescence**. Aging reduces the efficiency and capabilities of the individual, and even in the absence of other factors will ultimately lead to death. We have discussed the effects of aging on the body systems in earlier chapters.

✔ How is colostrum different from normal human milk?

INHERITANCE AND GENES

"People-watching," one of the commonest activities at shopping malls, provides us with abundant examples of the range of human characteristics, or *traits*. Family groups are often obvious because they have similar traits, especially facial features such as eye shape and color. These are examples of inherited traits that are genetically passed on from one generation to the next. The study of *heredity*, or how different traits are inherited, is called **genetics**.

Except for sex cells (sperm and eggs), every cell in the body carries copies of the original 46 chromosomes present in the fertilized egg or zygote. Through development and differentiation, the instructions contained within the genes of the chromosomes are expressed in many different ways. No single living cell or tissue makes use of all its genetic information. For example, in muscle cells the genes important for contractile proteins are active, whereas a different set of genes is operating in neurons.

GENES AND CHROMOSOMES

Chromosome structure and the functions of genes were introduced in Chapter 3. ∞ [p. 49] Chromosomes contain DNA, and genes are segments of DNA. Each gene carries the information needed to direct the synthesis of a specific protein.

The 46 chromosomes of body cells occur in pairs, 23 pairs in all. One member of each pair was contributed by the sperm, and the other by the egg. Twenty-two of those pairs affect only somatic, or general body, characteristics such as hair color or skin pigmentation. The chromosomes of the 23rd pair are called the *sex chromosomes* because they determine whether a person will be male or female.

Except for the sex chromosomes (which are discussed later), both chromosomes in a pair have the same structure and carry genes that affect the same traits. If one member of the chromosome pair contains three genes in a row, with number 1 determining hair color, number 2 eye color, and number 3 skin pigmentation, the other chromosome will carry genes affecting the same traits, and in the same sequence.

Both chromosomes of a pair may carry the same form of a gene for a particular trait. For example, if a zygote receives a gene for curly hair from the sperm and one for curly hair from the egg, the individual will have curly hair. Because the chromosomes that make up a pair have different origins, one paternal and the other maternal, they do not have to carry the same forms of a gene. In that case the trait that is expressed in the individual will be determined by how the two genes interact.

A gene for a particular trait is said to be **dominant** if it is expressed regardless of the instructions carried by the other gene. A **recessive** gene is one that is expressed in the individual only if it is present on both chromosomes of a pair. For example, the albino skin condition is characterized by an inability to synthesize the yellow-brown pigment *melanin*. A single dominant gene determines normal skin coloration; two recessive genes must be present to produce an albino individual.

Not every gene for a trait can be neatly characterized as dominant or recessive. Some that can are included in Table 22-2. If you restrict attention to these genes it is possible to predict the characteristics of individuals based on those of their parents.

Polygenic inheritance involves interactions between two or more different genes. For example, the determination of eye color and skin color involves several genes. Polygenic inheritance is also involved in several important disorders, including hypertension (high blood pressure) and coronary artery disease. In these cases the particular genetic composition of the individual does not by itself determine the onset of the disease. Instead, the genes establish a susceptibility to particular environmental influences. Only if the individual is exposed to these influences will the disease develop. This means that not every individual with the genetic tendency for a disorder will actually get it. This makes it difficult to track polygenic conditions through successive generations and predict which family members might be affected. However, because many inherited polygenic conditions are *likely* but not *guaranteed* to occur, steps can often be taken to prevent the development of disease. For example, hypertension may be prevented or reduced by controlling diet and fluid volume, and coronary artery disease may be prevented by lowering the levels of cholesterol circulating in the blood.

Table 22-2   Different Patterns of Inheritance

Dominant Traits

Normal skin coloration

Brachydactyly (short fingers)

Free earlobes

Curly hair

Presence of Rh factor on red blood cell membranes

Presence of A and/or B antigens on red blood cell membranes
(in codominance inheritance, both dominant genes may be expressed)

Recessive Traits

Albinism

Blond hair

Red hair

Lack of A, B antigens on red blood cell membranes (Type O blood)

Inability to roll the tongue into a U-shape

Sex-Linked Traits

Color blindness

Polygenic Traits

Eye color

Hair color other than blond or red

Sex Chromosomes

The **sex chromosomes** determine the genetic sex of the individual. Unlike other chromosomal pairs, the sex chromosomes are not necessarily identical in appearance and in the number of genes they contain.

There are two different sex chromosomes, an **X chromosome** and a **Y chromosome**. The Y chromosome is much smaller than the X chromosome and contains fewer genes. The normal male chromosome pair is *XY*, and the female pair is *XX*. The eggs produced by a woman will always carry chromosome *X*, while a sperm can carry either chromosome *X* or *Y*. As a result, during fertilization, *Y*-carrying sperm will produce a male, and *X*-carrying sperm will produce a female. Figure 22-8• is a photograph of all the chromosomes of a normal male. Note the size difference between the Y and the X chromosomes.

The X chromosome also carries genes that affect body structures. The traits controlled by these genes are usually expressed because in most cases there are no corresponding genes on the Y chromosome. Such traits are known as **sex-linked** traits because the responsible genes are located on the sex chromosomes. The best known sex-linked characteristics are associated with noticeable diseases or disorders.

The inheritance of color blindness is sex-linked. A relatively common form of color blindness is associated with the presence of a dominant or recessive gene on the X chromosome. Normal color vision

•FIGURE 22-8
Chromosomes of a Normal Male.

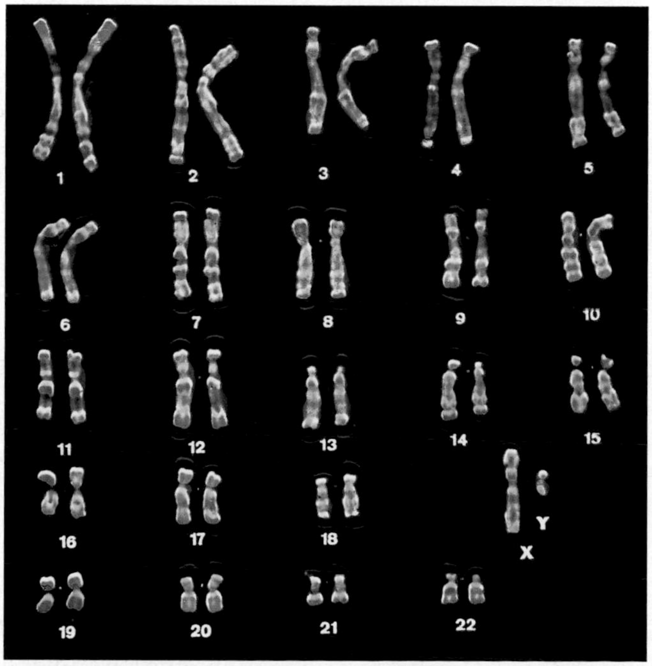

Clinical Note

The Human Genome Project

The study of human genetics has long been based on inherited disorders and how they are passed on in families. Because 20 to 25 years can often pass between generations, progress in learning how such disorders are inherited has been generally slow. However, increasing numbers of genes responsible for inherited disorders are now being identified and their locations pinpointed to specific chromosomes. Such accomplishments are due to the activities associated with the **Human Genome Project**. This project, funded by the National Institutes of Health and the Department of Energy, is attempting to identify the complete set of genes, or *genome*, contained in human chromosomes. The project began in October 1990 and

was expected to take 10 to 15 years. It is now expected to be completed by 2003.

What has been found so far? First, the locations of genes on eight chromosomes have been determined, or mapped, completely, and preliminary maps have been made for all other chromosomes. Over 38,000 genes have now been identified. Although a significant number, this is only a small fraction of the estimated 140,000 genes in the human genome. The specific genes responsible for more than 100 inherited disorders have also been identified. Because of these discoveries, *genetic screening*, a technique that can identify specific genes on chromosomes, is now available for many of these conditions.

✔ The ability to roll your tongue is a dominant trait. Would a person with a dominant and recessive gene for that trait be able to roll his tongue? Why, or why not?

✔ Joe has three daughters and complains that it's his wife's fault that he doesn't have any sons. What would you tell him?

is determined by the presence of a dominant gene, *C*, and color blindness results from the presence of the recessive gene *c*. A woman, with her two X chromosomes, can have two dominant genes, *CC*, or one dominant and one recessive, *Cc*, and still have normal color vision. Such an individual (*Cc*) with normal vision and a recessive gene is called a **carrier**. She will be color-blind only if she carries two recessive alleles, *cc*. But a male has only one X chromosome, so whatever that chromosome carries will determine whether he has normal color vision or is color-blind.

A number of other disorders are also X-linked traits, including certain forms of *hemophilia*, *diabetes insipidus*, and *muscular dystrophy*. In several instances, advances in molecular genetic techniques make it possible to locate specific genes on the X chromosome. This technique provides a relatively direct method of screening for the presence of a particular condition before the symptoms appear and even before birth.

DISORDERS OF DEVELOPMENT

Developmental disorders result in a variety of congenital problems affecting the anatomy or physiology of the newborn infant (Figure 22-9•). They often result from problems caused by exposure of the embryo to factors that disrupt normal developmental events or from genetic problems present at the time of fertilization. There may also be problems with the pregnancy itself, although development is proceeding normally. We will consider examples of each of these problems.

EXPOSURE TO DISRUPTIVE FACTORS

Teratogens

Teratogens (TER-a-to-jens) are factors that disrupt normal development by damaging cells, altering chromosome structure, or by changing the normal pattern of gene activation and expression. **Teratology** (ter-a-TOL-o-jē) is literally the "study of monsters." It is a branch of embryology that is concerned with abnormal development and congenital malformations, or departures from the pathways of normal development. Teratogens that affect the embryo in the first trimester will potentially disrupt multiple organ systems in fundamental ways. The embryonic survival rate will be low, and the survivors will usually have severe structural and functional defects. Teratogens introduced into the developmental process during the second and third trimesters will be more likely to affect specific organs or

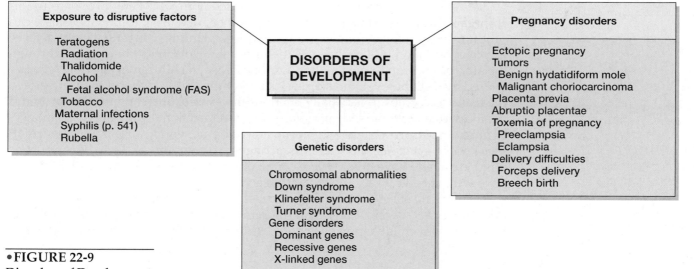

•FIGURE 22-9
Disorders of Development.

organ systems, because the major patterns of body organization are already established. Nevertheless, they may also reduce the chances for long-term survival.

Many powerful teratogens are encountered in everyday life. The location and severity of the resulting defects vary depending on the nature of the stimulus and the time of exposure. *Radiation* is a powerful teratogen that can affect all living cells. Even the X-rays used in diagnostic procedures can break chromosomes and produce developmental errors; thus procedures not based on radiation, such as ultrasound, are used to track embryonic and fetal development.

Virtually any unusual chemical that reaches an embryo has the potential for producing developmental abnormalities. For example, during the 1960s the European market was strong for *thalidomide*, a drug effective in promoting sleep and preventing nausea. Thalidomide was often prescribed for women in early pregnancy, with disastrous results. The drug crossed the placenta and entered the fetal circulation, where (among other effects) it interfered with the process of limb formation. As a result, many infants were born without limbs, or with drastically reduced ones. It should be noted that thalidomide had not been approved by the Federal Drug Administration, and it could not be sold legally in the United States. Although today the FDA is often criticized for the slow pace of its approval process, this was a case where the combination of rigorous testing standards and complex bureaucratic procedures protected the U.S. public.

The use of two powerful teratogens—alcohol and tobacco—is widespread. **Fetal alcohol syndrome (FAS)** occurs when maternal alcohol consumption produces developmental defects such as skeletal deformation, cardiovascular defects, and neurological disorders. Mortality rates can be as high as 17 percent, and the survivors are plagued by problems in later development. The most severe cases involve mothers who consume the alcohol content of at least 7 ounces of hard liquor, 10 beers, or several bottles of wine each day. But because the effects produced are directly related to the degree of exposure, there is probably no level of alcohol consumption that can be considered completely safe. Fetal alcohol syndrome is the number one cause of mental retardation in the United States today, affecting roughly 7,500 infants each year.

Smoking introduces potentially harmful chemicals, such as *nicotine*, lowers the oxygen content of maternal blood, and reduces the amount of oxygen arriving at the placenta. The fetus carried by a smoking mother will not grow as rapidly as one carried by a nonsmoker, and smoking increases the risks of spontaneous abortion, prematurity, and fetal death. There is also a higher rate of infant mortality after delivery, and postnatal development can be adversely affected.

Maternal infections may also affect fetal development. Fetal exposure to microorganisms responsible for syphilis (p. 541) or rubella ("German measles") can produce serious developmental abnormalities, including congenital heart defects, mental retardation, and deafness. **Rubella** is a viral infection that is spread through the air from person to person. It is a mild disease in children of age 6–12, characterized by a slight fever and skin rash that begins on the face and then spreads to the trunk and limbs. It is usually over within a week. Such an infection provides the individual with life-long immunity. Rubella infection during the first 4 months of pregnancy can interfere with normal development and cause tragic abnormalities known as *rubella syndrome*. Miscarriages are common, and newborns may be blind, deaf, and have other severe congenital problems. About one-fifth of babies born with rubella syndrome die within a year.

There is no specific treatment for rubella, but it can be prevented by vaccination. In the U.S., vaccination has almost eradicated rubella and rubella syndrome, but women of childbearing age should be aware of their immunization status.

GENETIC DISORDERS

The causes of genetic disorders, or genetic diseases, range from whole chromosomes to single genes on a chromosome. A major goal of the Human Genome Project involves identifying and locating the specific genes that cause genetic, or inherited, diseases.

Chromosomal Abnormalities

Some genetic disorders are caused by mistakes which occur during meiosis. ∞ *[p. 519]* These problems may result in the production of extra or missing copies of one or more chromosome. Fertilization by such an abnormal egg or sperm will produce a zygote with more or less than the normal complement of 46 chromosomes. Most zygotes and embryos with chromosomal abnormalities result in miscarriages, or spontaneous abortions. We will discuss three common genetic disorders caused by chromosomal abnormalities that do not kill the fetus before birth.

Down Syndrome. **Down syndrome** is the most common viable chromosomal abnormality. Individuals have an extra chromosome (number 21), giving them a total of 47 chromosomes. Estimates of the frequency of appearance range from 1.5 to 1.9 per 1,000 births for the U.S. population. The affected individual suffers from mental retardation and characteristic physical signs, including a fold of skin that covers the inner corners of the eyes, a flat nose, and relatively small hands with short fingers (Figure 22-10●). The degree of mental retardation ranges from moderate to severe, and few individuals with this condition lead independent lives. Anatomical problems affecting the cardiovascular system often prove fatal during childhood or early adulthood. Although some individuals survive to moderate old age, many develop Alzheimer's disease while still relatively young (before age 40).

For unknown reasons, there is a direct correlation between maternal age and the risk of having a child with Down syndrome. Below maternal age 25 the incidence approaches 1 in 2,000 births. For maternal ages 30–34, the odds increase to 1 in 900, and over the next decade they go from 1 in 290 to 1 in 46, or more than 2 percent for maternal age over 45. These statistics are becoming increasingly significant, for many women have delayed childbearing until their mid-30s or later.

Klinefelter Syndrome. In **Klinefelter syndrome**, the individual carries the sex chromosome pattern XXY. Such individuals have a total of 47 chromosomes. The individual is male, but the extra X chromosome causes reduced androgen production. As a result, the testes fail to mature, the individuals are sterile, and the breasts are slightly enlarged. The incidence of this condition among newborn males averages 1 in 750 births.

Turner Syndrome. Individuals with **Turner syndrome** have only a single female sex chromosome, resulting in a total of 45 chromosomes. Their sex chromosome complement is abbreviated XO. The incidence of this condition at delivery

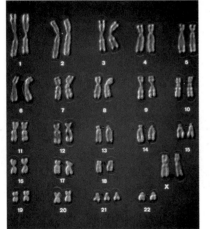

(a)

(b)

●**FIGURE 22-10**
Down Syndrome.
(a) The chromosomes of an individual with Down Syndrome. Note the extra copy of chromosome 21. **(b)** These girls have the characteristic features of individuals with Down Syndrome.

has been estimated as 1 in 10,000 live births. The condition may not be recognized at birth, for the phenotype is normal female, but maturational changes do not appear at puberty. The ovaries are nonfunctional, and estrogen production occurs at almost undetectable levels. The affected woman is short, infertile, and needs therapeutic hormones to develop adult sexual characteristics.

Gene Disorders

Inherited disorders may also be caused by single genes. The defective genes may be dominant or recessive. Some genetic disorders are sex-linked traits, and located on the X-chromosome (p. 563). Table 22-3 lists some of the relatively common disorders associated with defective genes.

Diagnosing Genetic Disorders

The detection of genetic disorders in a developing embryo or fetus relies on two basic techniques; (1) ultrasound, and (2) chromosomal analysis (see Table 22-4). Ultrasound is a noninvasive procedure that is good for revealing various physical abnormalities. The diagnosis of other disorders, such as chromosomal abnormalities and metabolic disorders, require a sample of the embryonic or fetal cells. *Amniocentesis* and *chorionic villi sampling* are used in such cases.

In **amniocentesis** (am-nē-ō-sen-TĒ-sis), a sample of amniotic fluid is removed, and the chromosomes of the fetal cells that it contains are analyzed. Chromosome abnormalities, such as Down syndrome, can be determined by constructing a **karyotype** (KAR-ē-ō-tīp), a picture of all the chromosomes within a cell (see Figure 22-10a•). The DNA within the cells may also be tested for specific, defective genes.

The needle inserted to obtain an amniotic fluid sample is guided into position using ultrasound. Unfortunately, amniocentesis has two major drawbacks:

1. Because the sampling procedure represents a potential threat to the health of the fetus and mother, amniocentesis is performed only when known risk factors are present. Examples of risk factors would include a family history of specific conditions, or in the case of Down syndrome, a maternal age over 35.

2. Sampling cannot safely be performed until the volume of amniotic fluid is large enough so that the fetus will not be injured during the sampling process. The usual time for amniocentesis is at a gestational age of 14–15 weeks. It may take several weeks to obtain results once samples have been collected, and by the time the results are received, the option of therapeutic abortion may no longer be available.

An alternative procedure, known as **chorionic villi sampling**, analyzes cells collected from the villi during the first trimester. Although it can be performed at an earlier gestational age, this technique is used less often because of an increased risk of spontaneous abortion. and abnormal limb development.

PREGNANCY DISORDERS

Pregnancy disorders may occur with implantation, the placenta, or other problems. Various tests are performed during pregnancy to monitor the health of the mother, and the developing embryo and fetus (Table 22-4). *Abortion* is the termination of a pregnancy. The three different classes of abortion were discussed in Chapter 21 (p. 535).

Table 22-3	Relatively Common Inherited Disorders
Disorder	**Page in Text**
Dominant	
Marfan's syndrome	p. 172
Huntington's disease	p. 259
Recessive	
Deafness	p. 285
Albinism	p. 119
Sickle-cell anemia	p. 332
Cystic fibrosis	p. 427
Phenylketonuria	p. 477
X-linked	
Duchenne's muscular dystrophy	p. 211
Hemophilia (one form)	p. 333
Color blindness	p. 283

Table 22-4	Representative Diagnostic Procedures and Laboratory Tests for Disorders of Development and Inheritance	

Diagnostic Procedure	Method and Result	Representative Uses
Amniocentesis	A needle collects amniotic fluid for analysis	Detection of chromosomal abnormalities, erythroblastosis fetalis, and birth defects such as spina bifida
Pelvic ultrasonography	Standard ultrasound	Detection of multiple fetuses, fetal abnormalities, and placenta previa; estimation of fetal age and growth
External fetal monitoring	Monitoring devices on the external abdominal surface measure fetal heart rate and force of uterine contraction	Detection of irregular heart rate or fetal stress
Internal fetal monitoring	Electrode is attached to fetal scalp to monitor heart rate; catheter is placed in uterus to monitor uterine contractions	As above
Chorionic villi biopsy	Test performed during weeks 8–10 of gestation; small pieces of chorionic villi are suctioned into a syringe	Detection of chromosomal abnormalities and biochemical disorders

Laboratory Test	Normal Values	Significance of Abnormal Values
Amniotic fluid analysis		
Karyotyping	Normal chromosomes	Detects chromosomal defects such as those in Down syndrome
Bilirubin	Traces only	Increased values indicate amount of hemolysis of fetal RBCs by mother's Rh antibodies
Meconium	Not present	Present in fetal distress
Maternal blood tests		
Serum Proteins Alpha-fetoprotein (AFP)	Week 16 of gestation: 5.7–31.5 ng/ml (lowers with increasing gestational age)	Increased values indicate possible neural tube defect such as spina bifida
Human chorionic gonadotropin (hCG)	Nonpregnant females: <0.005 IU/ml	Useful in determination of pregnancy; used in home pregnancy tests
Serum antibodies Toxoplasmosis Other (syphilis, group B beta-hemolytic strep and Varicella) Rubella Herpes simplex Type 2 virus	Pregnant females: negative for antibodies to these pathogens Neonates: negative for antibodies to these pathogens	Pathogens can cross the placenta and cause mild to severe problems
Blood type, Rh factor	Rh^+ or Rh^-	A sensitized Rh^- mother carrying an Rh^+ baby can result in erythroblastosis fetalis
Glucose Test	Pregnant females: Blood glucose levels should be 70–140 mg/dl 2 hr after glucose administration	Increased level indicates possible diabetic state
Neonatal blood tests		
Blood type, Rh factor	Rh^+ or Rh^-	Detection of maternal-fetal Rh incompatibility
Bilirubin	<12 mg/dl	Increased levels occur in jaundice due to immaturity of newborn's liver
Phenylalanine	1–3 mg/dl	>4 mg/dl occurs in phenylketonuria

Ectopic Pregnancies

Implantation usually occurs at the endometrial surface lining the uterine cavity. The precise location within the uterus varies, although most often implantation occurs in the body of the uterus. In an **ectopic** (ek-TOP-ik) **pregnancy**, however, implantation occurs somewhere other than within the uterus. (*Ectopic* means "displaced.")

The incidence of ectopic pregnancies is approximately 0.6 percent of all pregnancies. Conditions that appear to increase the chance of an ectopic pregnancy include douching and infection. Women douching regularly have a 4.4 times higher risk of experiencing an ectopic pregnancy, presumably because the flushing action pushes the zygote away from the uterus. If the uterine tube has been scarred by a previous episode of pelvic inflammatory disease, there is also an increased risk of an ectopic pregnancy.

Almost all (95 percent) ectopic pregnancies involve implantation within a uterine tube. Other implantation sites include the peritoneal cavity, the wall of the ovary, or in the cervix. The tube cannot expand enough to accommodate the developing embryo, and it usually ruptures during the first trimester. At this time the hemorrhaging that occurs in the peritoneal cavity may be severe enough to pose a threat to the woman's life.

In a few instances the ruptured uterine tube releases the embryo with an intact umbilical cord, and further development can occur. About 5 percent of these abdominal pregnancies actually complete full-term development; normal birth cannot occur, but the infant can be surgically removed from the abdominopelvic cavity. Because abdominal pregnancies are possible, it has been suggested that men as well as women could act as surrogate mothers if a zygote were surgically implanted in the peritoneal wall. It is not clear how the endocrine, cardiovascular, nervous, and other systems of a man would respond to the stresses of pregnancy. The procedure has been tried successfully in mice, however, and experiments continue.

Tumors

During normal development, the outer layer of cells of the embryo form the placenta. In about 0.1 percent of pregnancies, something goes wrong and a tumor forms in place of the placenta.

Hydatidiform Mole. A **hydatidiform** (hī-da-TID-i-form) **mole** is a tumor that develops from the outer layer of embryonic cells. Symptoms of such tumors include vaginal bleeding and severe morning sickness. Ultrasound scanning is useful in diagnosis, because no fetus develops. Urine and blood contain high levels of human chorionic gonadotropin (hCG), which is being produced by the tumor cells. Treatment is through surgical removal and chemotherapy. There is a chance of a malignant tumor developing, and, as a result, hCG levels are monitored after surgery.

Choriocarcinoma. A **choriocarcinoma** is a rare, malignant tumor that may develop from a hydatidiform mole in the uterus. It then spreads to the vagina, and possibly to other parts of the body such as the lungs, liver, brain, and bones. Persistent bleeding from the vagina is a characteristic symptom. Treatment involves anticancer drugs and surgery, such as a hysterectomy. Before advances in chemotherapy, only 10 percent of patients survived. Now, with appropriate care, deaths are rare.

Problems with the Placenta

Placenta Previa. In a **placenta previa** (PRĒ-vē-uh) implantation occurs in the lower portion of the uterus, in or near the cervical opening. This condition (*previa* means "in the way") causes problems as the growing placenta approaches the inner opening of the cervix. In a total placenta previa the placenta actually extends across this opening, while a partial placenta previa only partially blocks

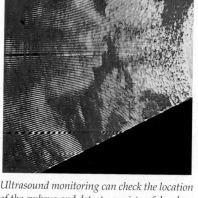

Ultrasound monitoring can check the location of the embryo and detect a variety of developmental problems.

the opening. The placenta contains a rich fetal blood supply and maternal blood vessels within the endometrium. Where the placenta passes across the internal opening the delicate complex hangs like an unsupported water balloon. As the pregnancy advances, even minor mechanical stresses can be enough to tear the placental tissues, leading to massive fetal and maternal hemorrhaging. Treatment in cases of total placenta previa usually involves bed rest for the mother until the fetus reaches a size at which cesarean delivery can be performed with a reasonable chance of fetal survival.

Abruptio Placentae. In an **abruptio placentae** (ab-RUP-shē-ō pla-SEN-tē) part or all of the placenta tears away from the uterine wall sometime after the fifth month of gestation. The bleeding into the uterine cavity and the pain that follows usually will be noted and reported, although in some cases the shifting placenta may block the passage of blood through the cervical canal. In severe cases the hemorrhaging leads to maternal anemia, shock, and kidney failure. Although maternal mortality is low, the fetal mortality rate from this condition ranges from 30 to 100 percent, depending on the severity of the hemorrhaging.

Problems with the Maintenance of a Pregnancy

Toxemia of Pregnancy. The rate of maternal complications during pregnancy is relatively high, but with prenatal monitoring providing early diagnosis and treatment of problems, the maternal death rate is currently less than 1 per 10,000 pregnancies. Pregnancy stresses maternal systems, and the stresses can overwhelm homeostatic mechanisms. The term **toxemia** (tok-SĒ-mē-uh) **of pregnancy** refers to disorders affecting the maternal cardiovascular system. It is thought that such disorders are caused by a substance released by the placenta, although it has yet to be identified. Some degree of toxemia occurs in 6–7 percent of late, or third-trimester pregnancies. Severe cases account for 20 percent of maternal deaths and contribute to an estimated 25,000 neonatal (newborn) deaths each year.

Toxemia of pregnancy includes **preeclampsia** (prē-ē-KLAMP-sē-uh) and **eclampsia** (ē-KLAMP-sē-uh). Preeclampsia most often occurs during a woman's first pregnancy. Blood pressure is elevated, with systolic and diastolic pressures reaching levels as high as 180/110 (120/80 is normal for young adults). Other symptoms include fluid retention and edema, along with CNS disturbances and alterations in kidney function. Roughly 4 percent of individuals with preeclampsia develop eclampsia.

Eclampsia may occur in late pregnancy, during labor, or even after delivery. Eclampsia, or *pregnancy-induced hypertension (PIH)*, is followed by severe convulsions and breathing difficulties lasting 1–2 minutes, followed by a variable period of coma. Other symptoms resemble those of preeclampsia, with additional evidence of liver and kidney damage. The mortality rate from eclampsia is approximately 5 percent; to save the mother the fetus must be delivered immediately. Once the fetus and placenta are removed from the uterus, symptoms of eclampsia disappear over a period of hours to days.

Common Problems with Labor and Delivery

There are many potential problems during labor and delivery. Two relatively common types of complications are *forceps deliveries* and *breech births*.

By the end of gestation the fetus has usually rotated within the uterus so that it will enter the birth canal head first, with the face turned toward the sacrum. In around 6 percent of deliveries the fetus faces the pubis rather than the sacrum. This slows or prevents descent through the birth canal, and is one cause of "failure to progress" through the stages of labor. Although these infants can eventually be delivered normally, risks to infant and mother increase the longer the fetus remains in the birth canal. Often the clinical response is the removal of the infant through a **forceps delivery**. The forceps used resemble a large, curved set of salad tongs that can be separated for insertion into the vaginal canal one side at a time.

Once in place they are reunited and used to grasp the head of the infant. An intermittent pull is applied, so that the forces on the head resemble those encountered during normal delivery. A cesarean section is another option for this and other problems that interfere with normal delivery.

In 3–4 percent of deliveries, the legs or buttocks of the fetus enter the vaginal canal first. Such deliveries are known as **breech births**. Risks to the infant are relatively higher in breech births because the umbilical cord may become constricted, and placental circulation cut off. Because the head is normally the widest part of the fetus, the cervix may dilate enough to pass the legs and body but not the head. Entrapment of the fetal head compresses the umbilical cord, prolongs delivery, and subjects the fetus to potential distress and severe damage. If the fetus cannot be repositioned manually, a cesarean section is usually performed.

MONITORING POSTNATAL DEVELOPMENT

Each newborn infant gets a close scrutiny after delivery. The maturity of the newborn may also be determined prior to delivery via ultrasound or amniocentesis (see Table 22-4).

The **Apgar rating** considers heart rate, respiratory rate, muscle tone, response to stimulation, and color at 1 and 5 minutes after birth. In each category the infant receives a score ranging from 0 (poor) to 2 (excellent), and the scores are then totaled. An infant's Apgar rating (0–10) has been shown to be an accurate predictor of newborn survival and the presence of neurological damage. For example, newborn infants with cerebral palsy (p. 259) usually have a low Apgar rating.

In the course of this examination, the breath sounds, the depth and rate of respirations, and the heart rate are noted. Both the respiratory rate and the heart rate are considerably higher in the infant than the adult (see Table 1-4, p. 15).

A physical examination of the newborn focuses on the status of vital systems and also on the following:

- The head of a newborn infant may be misshapen following vaginal delivery, but it usually assumes its normal shape over the next few days. The size of the head must be checked, however, to detect hydrocephalus (p. 259).

- The abdomen is palpated to detect abnormalities of internal organs.

- The external genitalia are inspected. The scrotum of a male infant is checked for the presence of descended testes.

- Cyanosis of the hands and feet is normal in the newborn, but the rest of the body should be pink. A generalized cyanosis may indicate congenital circulatory disorders, such as erythroblastosis fetalis (p. 327) or a patent foramen ovale (p. 372), or other problems.

Measurements of body length, head circumference and body weight are taken. A weight loss in the first 48 hours is normal, due to fluid shifts that occur as the infant adapts to the change from weightlessness (floating in amniotic fluid) to normal gravity. (Comparable fluid shifts occur in astronauts returning to earth after extended periods in space.)

The excretory systems of the newborn infant are assessed by examination of the urine and feces. The first bowel movement consists of a mixture of epithelial cells and mucus, and is greenish-black in color. This material is called the **meconium** (me-KŌ-ne-um).

The nervous and muscular systems are assessed for normal reflexes and muscle tone, which also help in identifying the presence of anatomical and physiological abnormalities. Examination of the newborn also provides baseline information useful in assessing postnatal development. In addition, newborn infants are often screened for genetic and/or metabolic disorders, such as phenylketonuria (PKU) (p. 477), congenital hypothyroidism (p. 308), and sickle cell anemia (p. 332).

✔ X-rays and alcohol are considered teratogens. What are teratogens?

✔ An extra sex chromosome is characteristic of what genetic disorder?

✔ Why is a breech birth more risky than a normal head-first birth?

CHAPTER REVIEW

Key Words

blastocyst (BLAS-tō-sist): Early stage in the developing embryo, consisting of an outer cell layer and an inner cell mass.

development: Growth and expansion from a lower to higher stage of complexity.

embryo (EM-brē-o): Developmental stage beginning at fertilization and ending at the start of the third developmental month.

fertilization: Fusion of egg and sperm to form a zygote.

fetus: Developmental stage lasting from the start of the third developmental month to delivery.

gestation (jes-TĀ-shun): The period of development within the uterus.

implantation (im-plan-TĀ-shun): The migration of a blastocyst into the uterine wall.

lactation: The production of milk by the mammary glands.

parturition (par-tū-RISH-un): Childbirth, delivery.

placenta: A complex structure in the uterine wall that permits diffusion between the fetal and maternal circulatory systems; forms the *afterbirth*.

pregnancy: The condition of having a developing embryo or fetus in the body.

puberty: Period of rapid growth, sexual maturation, and the appearance of secondary sexual characteristics; usually occurs between the ages of 10 and 15.

Selected Clinical Terms

abruptio placentae (ab-RUP-shē-ō pla-SEN-tē): Tearing of the placenta usually after the fifth gestational month.

amniocentesis (am-nē-ō-sen-TĒ-sis): An analysis of fetal cells taken from a sample of amniotic fluid.

Apgar rating: A method for evaluating newborn infants at 1 and 5 minutes post-delivery; a test for developmental problems and neurological damage.

breech birth: A delivery during which the legs or buttocks of the fetus enter the vaginal canal before the head.

choriocarcinoma: A malignant tumor that develops in the uterus from the placenta.

ectopic (ek-TOP-ik) **pregnancy:** A pregnancy in which the implantation occurs somewhere other than the uterus.

fetal alcohol syndrome: A neonatal condition resulting from maternal alcohol consumption; characterized by developmental defects typically involving the skeletal, nervous, and/or cardiovascular systems.

placenta previa (PRĒ-vē-uh): A condition in which the placenta covers the cervix and prevents normal birth; can cause fatal bleeding.

teratogens (TER-a-to-jens): Factors that disrupt normal development.

toxemia of pregnancy: Disorders affecting the maternal cardiovascular system; includes preeclampsia and eclampsia.

Study Outline

INTRODUCTION *(p. 551)*

1. Development is the gradual modification of physical characteristics from conception to maturity. **Inheritance** refers to the transfer of parental characteristics to the next generation.

STAGES OF DEVELOPMENT *(p. 551)*

1. Prenatal development occurs before birth and includes development of the **embryo** and **fetus**; **postnatal development** begins at birth and continues to maturity.

FERTILIZATION *(p. 551)*

1. Fertilization involves the fusion of two gametes, an egg (ovum) and sperm cell.

2. Fertilization normally occurs in the uterine tube within a day after ovulation. *(Figure 22-1a)*

3. The acrosomal caps of the spermatozoa release enzymes that separate follicular cells around the ovum, exposing its membrane. When a single spermatozoon contacts that membrane, fertilization follows.

GESTATION *(p. 552)*

1. The 9-month (266 days) **gestation period** can be divided into three **trimesters**.

THE FIRST TRIMESTER *(pp. 552–556)*

1. Major events of this period include (1) cell division and the formation of the embryo, (2) *implantation* of the embryo within the uterine wall, (3) formation of tissues and embryonic membranes, (4) formation of the placenta, and (5) the beginning of organ formation.

EMBRYO FORMATION *(p. 552)*

2. Cleavage subdivides the cytoplasm of the zygote in a series of mitotic divisions; the zygote becomes a **blastocyst**. Its inner cell mass develops into the individual. *(Figure 22-1b)*

IMPLANTATION *(p. 552)*

3. Implantation occurs about 7 days after fertilization as the blastocyst adheres and begins to burrow into the uterine lining. *(Figure 22-1b)*

4. Breakdown of the endometrium is prevented as outer cells of the blastocyst begin secreting a hormone called **human chorionic gonadotropin (hCG)**.

TISSUE AND MEMBRANE FORMATION *(p. 553)*

5. After implantation, the embryo is composed of three cell layers, an **endoderm**, **ectoderm**, and intervening **mesoderm**. It is from these **germ layers** that the body systems form. *(Table 22-1)*

6. These germ layers help form four **extraembryonic membranes**: the *yolk sac, amnion, allantois,* and *chorion. (Figure 22-2)*

7. The **yolk sac** is an important site of blood cell formation. The **amnion** encloses fluid that surrounds and cushions the developing embryo. The base of the **allantois** later gives rise to the urinary bladder. The **chorion** provides a means of nutrient uptake for the embryo.

THE PLACENTA *(p. 554)*

8. *Chorionic villi* extend outward into the maternal tissues, forming an intricate, branching network through which maternal blood flows. As development proceeds, the **umbilical cord** connects the fetus to the placenta. *(Figure 22-3)*

9. The placenta also synthesizes hormones such as hCG, estrogens, and progesterones.

THE SECOND AND THIRD TRIMESTERS *(pp. 556–557)*

1. The organ systems form during the second trimester and become functional during the third trimester. Additionally, the fetus enlarges and becomes more recognizably human. *(Figure 22-4)*

PREGNANCY AND THE MOTHER'S SYSTEMS *(p. 557)*

2. The developing fetus is totally dependent on maternal organs for nourishment, respiration, and waste removal. Maternal adaptations include increased blood volume, respiratory rate, nutrient intake, and kidney filtration.

CHANGES IN THE UTERUS *(p. 557)*

3. Progesterone produced by the placenta has an inhibitory effect on uterine muscles; its calming action is opposed by estrogens, oxytocin, and prostaglandins. At some point multiple factors interact to produce **labor contractions** in the uterine wall.

LABOR AND DELIVERY *(pp. 557–559)*

1. The goal of labor is **parturition** (forcible expulsion of the fetus).

STAGES OF LABOR *(p. 559)*

2. Labor can be divided into three stages: the **dilation stage**, **expulsion stage**, and **placental stage**. *(Figure 22-5)*

MULTIPLE BIRTHS *(p. 559)*

3. **Identical** twins are always the same gender. **Fraternal** twins may be the same or different genders.

POSTNATAL DEVELOPMENT *(pp. 559–561)*

1. Postnatal development involves a series of **life** stages, including the **neonatal period**, **infancy**, **childhood**, **adolescence**, and **maturity**. **Senescence** begins at maturity and ends in the death of the individual.

THE NEONATAL PERIOD, INFANCY, AND CHILDHOOD *(p. 559)*

2. The **neonatal period** extends from birth to 1 month of age. **Infancy** then continues to 2 years of age, and **childhood** lasts until puberty commences. During these stages major organ systems (other than reproductive) become operational and gradually acquire adult characteristics, and the individual grows rapidly.

3. In the transition from fetus to **neonate** the respiratory, circulatory, digestive, and urinary systems begin functioning independently. The newborn must also begin regulating its body temperature (thermoregulation).

4. Mammary glands produce protein-rich **colostrum** during the infant's first few days and then converts to milk production. These secretions are released as a result of the *milk let-down reflex*. *(Figure 22-6)*

5. Body proportions gradually change during postnatal development. *(Figure 22-7)*

ADOLESCENCE AND MATURITY *(p. 561)*

6. Adolescence begins at **puberty** when (1) the hypothalamus increases its production of GnRH, (2) circulating levels of FSH and LH rise rapidly, and (3) ovarian or testicular cells become more sensitive to FSH and LH. These changes initiate the production of gametes and sex hormones, and a sudden acceleration in growth rate.

INHERITANCE AND GENES *(pp. 561–564)*

1. Every body cell, except sex cells, carries copies of the original 46 chromosomes in the zygote. The chromosomes carry the genetic information.

GENES AND CHROMOSOMES *(p. 562)*

2. Every somatic human cell contains 23 pairs of chromosomes. Of these, 22 pairs affect only somatic, or general body, characteristics. The 23rd pair of chromosomes are the **sex chromosomes** because they determine the genetic sex of the individual.

3. Chromosomes contain DNA, and genes are functional segments of DNA. Both chromosomes of a pair may carry the same or different forms of a particular gene.

4. The different forms of a gene are considered dominant or recessive depending upon how their traits are expressed. *(Table 22-2)*

5. In **simple inheritance** an individual's traits are determined by interactions between a single pair of genes. **Polygenic inheritance** involves interactions among gene pairs on several chromosomes.

6. There are two different sex chromosomes, an **X chromosome** and a **Y chromosome.** The normal male sex chromosome complement is XY; that of females is XX. The X chromosome carries **sex-linked** genes that affect body structures but have no corresponding genes on the Y chromosome. *(Figure 22-8)*

DISORDERS OF DEVELOPMENT *(pp. 564–571)*

1. Developmental disorders result from exposure of the embryo to disruptive factors, genetic problems, and problems with the pregnancy. *(Figure 22-9)*

EXPOSURE TO DISRUPTIVE FACTORS *(p. 564)*

2. **Teratogens** are factors that disrupt normal development. **Teratology** is study of abnormal development.

3. Examples of teratogens include *radiation* (such as X-rays), drugs (such as *Thalidomide*), *alcohol* (may cause **fetal alcohol syndrome**), *tobacco*, and pathogens causing *maternal infections* such as **rubella**.

GENETIC DISORDERS *(p. 566)*

4. Genetic disorders may be the result of abnormal numbers of chromosomes or defective genes.

5. Genetic disorders caused by an extra chromosome or deletion of a chromosome include **Down syndrome**, **Klinefelter syndrome**, and **Turner syndrome**. *(Figure 22-10)*

6. Genetic diseases may be caused by dominant genes and recessive genes. They may also be inherited as sex-linked traits. *(Table 22-3)*

7. Genetic disorders may be diagnosed with ultrasound and chromosomal analysis. *(Table 22-4)*

8. Fetal cells for chromosomal analysis may be obtained through **amniocentesis** or **chorionic villi sampling**. A **karyotype** is a picture of an individual's chromosomes.

PREGNANCY DISORDERS *(p. 567)*

9. Pregnancy disorders may be due to problems with implantation, the placenta, or other factors. A variety of tests are performed during pregnancy to monitor maternal and fetal health. *(Table 22-4)*

10. **Ectopic pregnancies** result when implantation occurs in areas other than the uterus. Most occur in the uterine tubes.

11. Tumors may develop from the placenta-forming cells of the embryo or the placenta. A **hydatidiform mole** is a tumor within the uterus. A malignant tumor of the placenta is called a **chorio-carcinoma**.

12. **Placenta previa** is a condition in which implantation occurs near the cervix and the placenta grows over the cervical opening in the uterus. Problems arise because the placenta is not support-ed at the cervical opening and the placenta may tear, leading to he-morrhaging of both the mother and fetus.

13. **Abruptio placentae** is a condition in which part or all of the placenta tears away from the uterine wall sometime after the fifth month of gestation. Severe cases of hemorrhaging may lead to maternal anemia, shock, and kidney failure.

14. **Toxemia of pregnancy** refers to two disorders affecting the ma-ternal cardiovascular system. **Preeclampsia** occurs most often in a woman's first pregnancy. Symptoms are elevated blood pressure, fluid retention, edema, CNS disturbances, and changes in kidney function. **Eclampsia**, or *pregnancy-induced hypertension (PIH)*, may occur in late pregnancy, during labor, or after delivery, and may re-sult in seizures, coma, organ failure, and maternal death.

15. Complications of delivery may arise from the position of the fetus in the birth canal. A **forceps delivery** is sometimes under-taken when the head of the fetus is facing the mother's pubis and labor does not progress. In a **breech birth**, the fetus enters the birth canal with its legs or buttocks before its head.

MONITORING POSTNATAL DEVELOPMENT (p. 571)

16. A newborn is closely examined after delivery. A newborn's **Apgar rating** is a good predictor of its chances of survival. (Table 22-4)

Review Questions

MATCHING

Match each item in Column A with the most closely related item in Column B. Use letters for answers in the spaces provided.

	Column A		Column B
___	1. gestation	a.	blastocyst formation
___	2. cleavage	b.	ejection of placenta
___	3. ectoderm	c.	forms connective and muscle tissues
___	4. mesoderm	d.	indication of pregnancy
___	5. human chorionic gonadotropin	e.	embryo-maternal circulatory exchange
___	6. birth	f.	XY
___	7. episiotomy	g.	time of prenatal development
___	8. afterbirth	h.	X0
___	9. senescence	i.	perineal musculature incision
___	10. neonate	j.	newborn infant
___	11. male sex chromosomes	k.	process of aging
___	12. female sex chromosomes	l.	tumor
___	13. chorion	m.	gives rise to skin and nervous tissue
___	14. hydatidiform mole	n.	parturition
___	15. Turner syndrome	o.	XX

MULTIPLE CHOICE

16. A zygote forms in the _____ .
(a) vagina   (b) uterine (fallopian) tube
(c) ovary   (d) uterus

17. An embryo becomes a fetus after _____ weeks of development.
(a) 2   (b) 4   (c) 8   (d) 16

18. The gradual modification of body structures during the period from conception to maturity is _____ .
(a) development   (b) differentiation
(c) parturition   (d) senescence

19. The membrane that encloses the fluid that surrounds and cushions the developing embryo and fetus is the _____ .
(a) chorion   (b) amnion   (c) yolk sac   (d) allantois

20. Increased hormone production and sexual maturation during adolescence result from the activity of the _____ .
(a) hypothalamus   (b) anterior pituitary
(c) ovaries and testes   (d) all the above are correct

21. Milk let-down is associated with _____ .
(a) reflex action triggered by suckling
(b) the release of placental hormones
(c) implantation
(d) weaning

22. The normal number of chromosomes in our body cells is _____ .
(a) 23   (b) 24   (c) 46   (d) 48

23. _____ is caused by implantation near the cervix.
(a) abruptio placentae   (b) placenta previa
(c) fetal alcohol syndrome   (d) breech birth

24. Most ectopic pregnancies occur in the _____ .
(a) ovaries   (b) uterine tubes
(c) peritoneal cavity   (d) uterus

TRUE/FALSE

___ 25. Implantation refers to the formation of germ layers.

___ 26. The first blood cells of the embryo are formed by the yolk sac.

___ 27. An individual's gender is determined at birth.

___ 28. If both maternal and paternal copies of a gene must be present to express a trait, the genes are recessive.

___ 29. Delivery occurs during the third stage of labor.

___ 30. The hormone oxytocin stimulates strong contrac-tions of the uterus.

___ 31. Low blood pressure is a characteristic symptom of preeclampsia.

___ 32. Both amniocentesis and ultrasound can be used to reveal the sex of a fetus.

SHORT ESSAY

33. Describe the three life stages that occur between birth and ap-proximately age 10.

34. Identify the three stages of labor, and describe each of their characteristic events.

35. Discuss the changes that occur in the mother's systems during pregnancy.

36. Explain why more men than women are color-blind.

37. What is teratology?

38. What is the Apgar rating?

APPLICATIONS

39. A physician recommends to his pregnant patient, Marion, that she stay in bed during her pregnancy and that the fetus should be delivered by cesarean section, rather than normal vaginal birth. Why?

40. Jackie is two months pregnant and goes to her physician complaining of vaginal bleeding and severe morning sickness. The physician then orders an ultrasound, and urine and blood tests. These tests indicate an abnormal mass in the uterus, no embryo or fetus, and high levels of hCG in the urine and blood. What is your diagnosis of Jackie's condition?

✔ Answers to Concept Check Questions

(p. 556)   **1.** Multiple sperm are required for fertilization because one sperm does not contain enough enzyme to break down the connections between the follicle cells surrounding the egg cell. However, once the egg cell membrane is exposed, only one sperm cell will fertilize the egg. **2.** The inner cell mass of the blastocyst develops into the body. **3.** The amniotic membrane is the extraembryonic membrane that encloses the fluid and developing embryo. **4.** Placental functions include (1) supplying the developing fetus with a route for gas exchange, nutrient transfer, and waste product elimination, and (2) producing hormones that affect the mother's body systems.

(p. 559)   **1.** The glomerular filtration rate increases because of an increase in blood volume. The increased filtration rate speeds up the excretion of the additional metabolic wastes generated by the fetus. **2.** The uterus increases in size as smooth muscle cells enlarge and elongate. **3.** Progesterone decreases uterine contractions. A decrease in progesterone at any time during the pregnancy can lead to uterine contractions and, in late pregnancy, labor.

(p. 561)   **1.** Colostrum contains less fat and more proteins than normal human milk. Many of the proteins are antibodies that help the infant fight off infections prior to the functional development of its own immune system.

(p. 564)   **1.** A person with a dominant and a recessive gene would be able to roll their tongue. As long as a dominant gene is present it will be expressed. **2.** There are two different sex chromosomes, an X chromosome and a Y chromosome. The normal male chromosome pair is XY, and the female pair is XX. The eggs produced by Joe's wife will always carry an X, and the sperm produced during meiosis by Joe may carry X or Y. As a result, the sex of Joe's children depends on which type of sperm cell fertilizes the egg cell.

(p. 571)   **1.** Teratogens are agents or factors that disrupt the normal pattern of development. **2.** An extra sex chromosome is characteristic of Klinefelter syndrome. The individual has three sex chromosomes, XXY, and is a male. **3.** In a breech birth, the legs or buttocks emerge from the birth canal first. This orientation may result in the umbilical cord becoming constricted during the birth, and cutting off circulation to the fetus.

APPENDIX

I

Weights and Measures

Table 1   The U.S. System of Measurement

Physical Property	Unit	Relationship to Other U.S. Units	Relationship to Household Units
Length	inch (in.)	1 in. = 0.083 ft	
	foot (ft)	1 ft = 12 in.	
		= 0.33 yd	
	yard (yd)	1 yd = 36 in.	
		= 3 ft	
	mile (mi)	1 mi = 5280 ft	
		= 1760 yd	
Volume	fluid ounce (fl oz)	1 fl oz = 0.0625 pt	= 6 teaspoons (tsp)
			= 2 tablespoons (tbsp)
	pint (pt)	1 pt = 16 fl oz	= 32 tbsp
		0.5 qt	= 2 cups (c)
	quart (qt)	1 qt = 32 fl oz	= 4 c
		= 2 pt	
		0.25 gal	
	gallon (gal)	1 gal = 128 fl oz	
		= 8 pt	
		= 4 qt	
Mass	ounce (oz)	1 oz = 437.5 gr	
	pound (lb)	1 lb = 7000 gr	
		= 16 oz	
	ton (t)	1 t = 2000 lb	

Accurate descriptions of physical objects would be impossible without a precise method of reporting the pertinent data. Dimensions such as length and width are reported in standardized units of measurement, such as inches or centimeters. These values can be used to calculate the **volume** of an object, a measurement of the amount of space it fills. **Mass** is another important physical property. The mass of an object is determined by the amount of matter it contains; on earth the mass of an object determines its weight.

Most U.S. readers describe length and width in terms of inches, feet, or yards; volumes in pints, quarts, or gallons; and weights in ounces, pounds, or tons. These are units of the **U.S. system** of measurement. Table 1 summarizes familiar terms used in the U.S. system. For reference purposes, this table also includes a definition of the "household units," popular in recipes and cookbooks. The U.S. system can be very difficult to work with, because there is no logical relationship between the various units. For example, there are 12 inches in a foot, 3 feet in a yard, and 1,760 yards in a mile. Without a clear pattern of organization, converting feet to inches or miles to feet can be confusing and time-consuming. The relationships between ounces, pints, quarts, and gallons or ounces, pounds, and tons are no more logical.

In contrast, the **metric system** has a logical organization based on powers of 10, as indicated in Table 2. For example, a **meter (m)** represents the basic unit for the measurement of size. For measuring larger objects, data are commonly reported in terms of **kilometers (km**; *chilioi*, thousand); for smaller objects, data can be reported in **centimeters (cm** = 0.01 m; *centum*, hundred), **millimeters (mm** = 0.001 m; *mille*, thousand), and so forth. Notice that the same prefixes are used to report weights, based on the **gram (g)**, and volumes, based on the **liter (l)**. This text reports data in metric units, usually with U.S. equivalents. You should use this opportunity to become familiar with the metric system, because most technical sources report data only in metric units, and most of the rest of the world uses the metric system exclusively. Conversion factors are included in Table 2.

The U.S. and metric systems also differ in their methods of reporting temperatures; in the United States, temperatures are usually reported in degrees Fahrenheit (°F), whereas scientific literature and individuals in most other countries report temperatures in degrees centigrade or Celsius (°C). The relationship between temperatures in degrees Fahrenheit and those in degrees centigrade has been indicated at the bottom of Table 2.

APPENDIX

I

Weights and Measures

Table 1	The U.S. System of Measurement		
Physical Property	**Unit**	**Relationship to Other U.S. Units**	**Relationship to Household Units**
Length	inch (in.)	1 in. = 0.083 ft	
	foot (ft)	1 ft = 12 in.	
		= 0.33 yd	
	yard (yd)	1 yd = 36 in.	
		= 3 ft	
	mile (mi)	1 mi = 5280 ft	
		= 1760 yd	
Volume	fluid ounce (fl oz)	1 fl oz = 0.0625 pt	= 6 teaspoons (tsp)
			= 2 tablespoons (tbsp)
	pint (pt)	1 pt = 16 fl oz	= 32 tbsp
		0.5 qt	= 2 cups (c)
	quart (qt)	1 qt = 32 fl oz	= 4 c
		= 2 pt	
		0.25 gal	
	gallon (gal)	1 gal = 128 fl oz	
		= 8 pt	
		= 4 qt	
Mass	ounce (oz)	1 oz = 437.5 gr	
	pound (lb)	1 lb = 7000 gr	
		= 16 oz	
	ton (t)	1 t = 2000 lb	

Accurate descriptions of physical objects would be impossible without a precise method of reporting the pertinent data. Dimensions such as length and width are reported in standardized units of measurement, such as inches or centimeters. These values can be used to calculate the **volume** of an object, a measurement of the amount of space it fills. **Mass** is another important physical property. The mass of an object is determined by the amount of matter it contains; on earth the mass of an object determines its weight.

Most U.S. readers describe length and width in terms of inches, feet, or yards; volumes in pints, quarts, or gallons; and weights in ounces, pounds, or tons. These are units of the **U.S. system** of measurement. Table 1 summarizes familiar terms used in the U.S. system. For reference purposes, this table also includes a definition of the "household units," popular in recipes and cookbooks. The U.S. system can be very difficult to work with, because there is no logical relationship between the various units. For example, there are 12 inches in a foot, 3 feet in a yard, and 1,760 yards in a mile. Without a clear pattern of organization, converting feet to inches or miles to feet can be confusing and time-consuming. The relationships between ounces, pints, quarts, and gallons or ounces, pounds, and tons are no more logical.

In contrast, the **metric system** has a logical organization based on powers of 10, as indicated in Table 2. For example, a **meter (m)** represents the basic unit for the measurement of size. For measuring larger objects, data are commonly reported in terms of **kilometers (km;** *chilioi,* thousand); for smaller objects, data can be reported in **centimeters (cm** = 0.01 m; *centum,* hundred), **millimeters (mm =** 0.001 m; *mille,* thousand), and so forth. Notice that the same prefixes are used to report weights, based on the **gram (g),** and volumes, based on the **liter (l).** This text reports data in metric units, usually with U.S. equivalents. You should use this opportunity to become familiar with the metric system, because most technical sources report data only in metric units, and most of the rest of the world uses the metric system exclusively. Conversion factors are included in Table 2.

The U.S. and metric systems also differ in their methods of reporting temperatures; in the United States, temperatures are usually reported in degrees Fahrenheit (°F), whereas scientific literature and individuals in most other countries report temperatures in degrees centigrade or Celsius (°C). The relationship between temperatures in degrees Fahrenheit and those in degrees centigrade has been indicated at the bottom of Table 2.

Table 2 — The Metric System of Measurement

Physical Property	Unit	Relationship to Standard Metric Units	Conversion to U.S. Units	
Length	nanometer (nm)	1 nm = 0.000000001 m (10^{-9})	= 4 × 10^{-8} in.	25,000,000 nm = 1 in.
	micrometer (μm)	1 μm = 0.000001 m (10^{-6})	= 4 × 10^{-5} in.	25,000 mm = 1 in.
	millimeter (mm)	1 mm = 0.001 m (10^{-3})	= 0.0394 in.	25.4 mm = 1 in.
	centimeter (cm)	1 cm = 0.01 m (10^{-2})	= 0.394 in.	2.54 cm = 1 in.
	meter (m)	standard unit of length	= 39.4 in.	0.0254 m = 1 in.
			= 3.28 ft	0.3048 m = 1 ft
			= 1.09 yd	0.914 m = 1 yd
	kilometer (km)	1 km = 1000 m	= 3280 ft	
			= 1093 yd	
			= 0.62 mi	1.609 km = 1 mi
Volume	microliter (μl)	1 μl = 0.000001 l (10^{-6}) = 1 cubic millimeter (mm^3)		
	milliliter (ml)	1 ml = 0.001 l (10^{-3}) = 1 cubic centimeter (cm^3 or cc)	= 0.03 fl oz	5 ml = 1 tsp
				15 ml = 1 tbsp
				30 ml = 1 fl oz
	deciliter (dl)	1 dl = 0.1 l (10^{-1})	= 3.38 fl oz	0.29 dl = 1 fl oz
	liter (l)	standard unit of volume	= 33.8 fl oz	0.0295 l = 1 fl oz
			= 2.11 pt	0.473 l = 1 pt
			= 1.06 qt	0.946 l = 1 qt
Mass	picogram (pg)	1 pg = 0.000000000001 g (10^{-12})		
	nanogram (ng)	1 ng = 0.000000001 g (10^{-9})		
	microgram (μg)	1 μg = 0.000001 g (10^{-6})	= 0.000015 gr	66,666 mg = 1 gr
	milligram (mg)	1 mg = 0.001 g (10^{-3})	= 0.015 gr	66.7 mg = 1 gr
	gram (g)	standard unit of mass	= 0.035 oz	28.35 g = 1 oz
			= 0.0022 lb	453.6 g = 1 lb
	kilogram (kg)	1 kg = 1000 g	= 2.2 lb	0.453 kg = 1 lb
	metric ton (mt)	1 mt = 1000 kg	= 1.1 t	
			= 2205 lb	0.907 mt = 1 t

Temperature	*Centigrade*	*Fahrenheit*
Freezing point of pure water	0°	32°
Normal body temperature	36.8°	98.6°
Boiling point of pure water	100°	212°
Conversion	°C → °F: °F = (1.8 × °C) + 32	°F → °C: °C = (°F − 32) × 0.56

Figure A-1● spans the entire range of measurements that we will consider in this book. *Gross anatomy* traditionally deals with structures that can be seen with the naked eye or with a simple hand lens. A microscope can provide higher levels of magnification and reveal finer details. Before the 1950s, most information was provided by **light microscopes**. A photograph taken through a light microscope is called a *light micrograph (LM)*. Light microscopes can magnify cellular structures about 1,000 times and show details as fine as 0.25 μm. The symbol **μm** stands for **micrometer**; 1 μm = 0.001 mm, or 0.00004 inch. With a light microscope one can identify cell types, such as muscle cells or neurons, and see large structures within the cell. Because individual cells are relatively transparent, thin sections taken through a cell are treated with dyes that stain specific structures, making them easier to see.

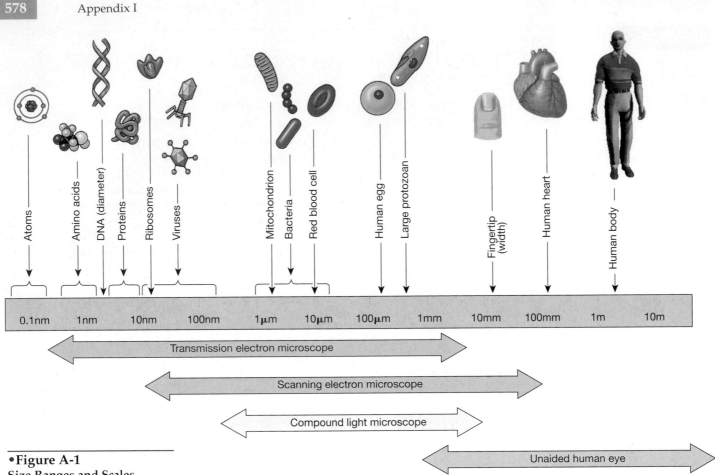

•**Figure A-1**
Size Ranges and Scales.

Although special staining techniques can show the general distribution of proteins, lipids, carbohydrates, and nucleic acids in the cell, many fine details of intracellular structure remained a mystery until investigators began using **electron microscopes**. These instruments use a focused beam of electrons, rather than a beam of light, to examine cell structure. In **transmission electron microscopes**, electrons pass through an ultrathin section to strike a photographic plate. The result is a *transmission electron micrograph (TEM).* Transmission electron microscopes show the fine structure of cell membranes and intracellular structures. In a **scanning electron microscope**, electrons bouncing off exposed surfaces create a *scanning electron micrograph (SEM).* Although a scanning microscope cannot achieve as much magnification as a transmission microscope, it provides a three-dimensional perspective on cell structure.

II

Normal Physiological Values

Tables 1 and 2 present normal averages or ranges for the chemical composition of body fluids. These values should be considered approximations rather than absolute values, as test results vary from laboratory to laboratory owing to differences in procedures, equipment, normal solutions, and so forth. Sources used in the preparation of these tables are indicated on p. 580. Blanks indicate tests for which data were not available.

Table 1	The Chemistry of Blood, Cerebrospinal Fluid, and Urine		
	Normal Ranges		
Test	**Blood[a]**	**CSF**	**Urine**
pH	S: 7.38–7.44	7.31–7.34	4.6–8.0
OSMOLARITY (mOsm/l)	S: 280–295	292–297	500–800
ELECTROLYTES	(mEq/l unless noted)		(urinary loss per 24-hour period[b])
Bicarbonate	P: 21–28	20–24	
Calcium	S: 4.5–5.3	2.1–3.0	6.5–16.5 mEq
Chloride	S: 100–108	116–122	120–240 mEq
Iron	S: 50–150 µg/l	23–52 µg/l	40–150 µg
Magnesium	S: 1.5–2.5	2–2.5	4.9–16.5 mEq
Phosphorus	S: 1.8–2.6	1.2–2.0	0.8–2 g
Potassium	P: 3.8–5.0	2.7–3.9	35–80 mEq
Sodium	P: 136–142	137–145	120–220 mEq
Sulfate	S: 0.2–1.3		1.07–1.3 g
METABOLITES	(mg/dl unless noted)		(urinary loss per 24-hour period[c])
Amino acids	P/S: 2.3–5.0	10.0–14.7	41–133 mg
Ammonia	P: 20–150 µg/dl	25–80 µg/dl	340–1,200 mg
Bilirubin	S: 0.5–1.2	<0.2	0.02–1.9 mg
Creatinine	P/S: 0.6–1.2	0.5–1.9	1.01–2.5
Glucose	P/S: 70–110	40–70	16–132 mg
Ketone bodies	S: 0.3–2.0	1.3–1.6	10–100 mg
Lactic acid	WB: 5–20[d]	10–20	100–600 mg
Lipids (total)	S: 400–1,000	0.8–1.7	0–31.8 mg
Cholesterol (total)	S: 150–300	0.2–0.8	1.2–3.8 mg
Triglycerides	S: 40–150	0–0.9	
Urea	P/S: 23–43	13.8–36.4	12.6–28.6
Uric acid	S: 2.0–7.0	0.2–0.3	80–976 mg
PROTEINS	(g/dl)	(mg/dl)	(urinary loss per 24-hour period[c])
Total	S: 6.0–7.8	20–4.5	47–76.2 mg
Albumin	S: 3.2–4.5	10.6–32.4	10–100 mg
Globulins (total)	S: 2.3–3.5	2.8–15.5	7.3 mg (average)
Immunoglobulins	S: 1.0–2.2	1.1–1.7	3.1 mg (average)
Fibrinogen	P: 0.2–0.4	0.65 (average)	

[a] S = serum, P = plasma, WB = whole blood
[b] Because urinary output averages just over 1 liter per day, these electrolyte values are comparable to mEq/l.
[c] Because urinary metabolite and protein data approximate mg/l or g/l, they must be divided by 10 for comparison with CSF or blood concentrations.
[d] Venous blood sample

| Table 2 | The Composition of Minor Body Fluids |

	Normal Averages or Ranges					
Test	Perilymph	Endolymph	Synovial Fluid	Sweat	Saliva	Semen
pH			7.4	4–6.8	6.4[a]	7.19
SPECIFIC GRAVITY			1.008–1.015	1.001–1.008	1.007	1.028
ELECTROLYTES (mEq/l)						
Potassium	5.5–6.3	140–160	4.0	4.3–14.2	21	31.3
Sodium	143–150	12–16	136.1	0–104	14[a]	117
Calcium	1.3–1.6	0.05	2.3–4.7	0.2–6	3	12.4
Magnesium	1.7	0.02		0.03–4	0.6	11.5
Bicarbonate	17.8–18.6	20.4–21.4	19.3–30.6		6[a]	24
Chloride	121.5	107.1	107.1	34.3	17	42.8
PROTEINS (mg/dl)						
Total	200	150	1.72 g/dl	7.7	386[b]	4.5 g/dl
METABOLITES (mg/dl)						
Amino acids				47.6	40	1.26 g/dl
Glucose	104		70–110	3.0	11	224 (fructose)
Urea				26–122	20	72
Lipids, total	12		20.9	0[d]	25–500[c]	188

[a] Increases under salivary stimulation

[b] Primarily alpha-amylase, with some lysosomes

[c] Cholesterol

[d] Not present in eccrine secretions

SOURCES

Ballenger, John Jacob. 1977. *Diseases of the Nose, Throat, and Ear.* Philadelphia: Lea and Febiger.

Davidsohn, Israel, and John Bernard Henry, eds. 1969. *Todd-Sanford Clinical Diagnosis by Laboratory Methods,* 14th ed. Philadelphia: W. B. Saunders.

Diem, K., and C. Lenter, eds. 1970. Scientific Tables, 7th ed. Basel, Switzerland: Ciba-Geigy.

Halsted, James A. 1976. *The Laboratory in Clinical Medicine: Interpretation and Application.* Philadelphia: W. B. Saunders.

Harper, Harold A. 1987. *Review of Physiological Chemistry.* Los Altos, Calif.: Lange Medical Publications.

Isselbacher, Kurt J., Eugene Braunwauld, Robert G. Petersdorf, Jean D. Wilson, Joseph B. Martin, Anthony S. Fauci, and Dennis L. Kaspar, eds. 1994. *Harrison's Principles of Internal Medicine,* 13th ed. New York: McGraw-Hill.

Lentner, Cornelius, ed. 1981. *Geigy Scientific Tables,* 8th ed. Basel, Switzerland: Ciba-Geigy.

Each entry starts with the commonly encountered form or forms of the prefix, suffix, or combining form followed by the word root (shown in italics) with its English translation. One example is given to illustrate the use of each entry form.

a-, *a-,* without: avascular
ab-, *ab,* from: abduct
-ac, *-akos,* pertaining to: cardiac
acr-, *akron,* extremity: acromion
ad-, *ad,* to, toward: adduct
aden-, adeno-, *adenos,* gland: adenoid
adip-, *adipos,* fat: adipose
aer-, *aeros,* air: aerobic metabolism
af-, *ad,* toward: afferent
-al, *-alis,* pertaining to: brachial
alb-, *albicans,* white: albino
-algia, *algos,* pain: neuralgia
ana-, *ana,* up, back: anaphase
andro-, *andros,* male: androgen
angio-, *angeion,* vessel: angiogram
ante-, *ante,* before: antebrachial
anti-, ant-, *anti,* against: antibiotic
apo-, *apo,* from: apocrine
arachn-, *arachne,* spider: arachnoid
arter-, *arteria,* artery: arterial
arthro-, *arthros,* joint: arthroscopy
-asis, -asia, state, condition: homeostasis
astro-, *aster,* star: astrocyte
aur-, *auris,* ear: auricle
auto-, *auto,* self: autonomic
baro-, *baros,* pressure: baroreceptor
bi-, *bi,* two: bifurcate
bio-, *bios,* life: biology
blast-, -blast, *blastos,* precursor: blastocyst
brachi-, *brachium,* arm: brachiocephalic
brachy-, *brachys,* short: brachydactyly
brady-, *bradys,* slow: bradycardia
bronch-, *bronchus,* windpipe, airway: bronchial
carcin-, *karkinos,* cancer: carcinoma
cardi-, cardio-, -cardia, *kardia,* heart: cardiac
cephal-, *cephalos,* head: brachiocephalic
cerebr-, *cerebrum,* brain: cerebral hemispheres
cerebro-, *cerebros,* brain: cerebrospinal fluid
cervic-, *cervicis,* neck: cervical vertebrae
chole-, *chole,* bile: cholecystitis
chondro-, *chondros,* cartilage: chondrocyte
chrom-, chromo-, *chroma,* color: chromatin
circum-, *circum,* around: circumduction
-clast, *klastos,* broken: osteoclast
coel-, -coel, *koila,* cavity: coelom
colo-, *kolon,* colon: colonoscopy
corp-, *corpus,* body: corpuscle
cortic-, *cortex,* rind or bark: corticosteroid
cost-, *costa,* rib: costal
cranio-, *cranium,* skull: craniosacral
cribr-, *cribrum,* sieve: cribriform
-crine, *krinein,* to separate: endocrine
cut-, *cutis,* skin: cutaneous
cyan-, *kyanos,* blue: cyanosis
cyst-, -cyst, *kystis,* sac: blastocyst
cyt-, cyto-, *kyton,* a hollow cell: cytology
de-, *de,* from, away: deactivation
dendr-, *dendron,* tree: dendrite
dent-, *dentes,* teeth: dentition
derm-, *derma,* skin: dermatome
desmo-, *desmos,* band: desmosome
di-, *dis,* twice: disaccharide
dia-, *dia,* through: diameter
digit-, *digit,* a finger or toe: digital
dis-, apart, away from: disability
diure-, *diourein,* to urinate: diuresis
ecto-, *ektos,* outside: ectoderm
-ectomy, *ektome,* excision: appendectomy
ef-, *ex,* away from: efferent
emmetro-, *emmetros,* in proper measure: emmetropia
encephalo-, *enkephalos,* brain: encephalitis
end-, endo-, *endos,* inside: endometrium
entero-, *enteron,* intestine: enteric

epi-, *epi,* on: epimysium
erythema-, *erythema,* flushed (skin): erythematosis
erythro-, *erythros,* red: erythrocyte
ex-, *ex,* out, away from: exocytosis
extra-, outside of, beyond, in addition: extracellular
fil-, *filum,* thread: filament
-form, *-formis,* shape: fusiform
gastr-, *gaster,* stomach: gastrointestinal
-gen, -genic, *gennan,* to produce: mutagen
genio-, *geneion,* chin: geniohyoid
gest-, *gesto,* to bear: gestation
glosso-, -glossus, *glossus,* tongue: hypoglossal
glyco-, *glykys,* sugar: glycogen
-gram, *gramma,* record: electrocardiogram
gran-, *granulum,* grain: granulocyte
gyne-, gyno-, *gynaikos,* woman: gynecologist
hem-, hemato-, *haima,* blood: hemopoiesis
hemi-, *hemi-,* half: hemisphere
hepato-, *hepaticus,* liver: hepatic vein
hetero-, *heteros,* other: heterosexual
histo-, *histos,* tissue: histology
holo-, *holos,* entire: holocrine
homeo-, homo-, *homos,* same: homeostasis
hyal-, hyalo-, *hyalos,* glass: hyaline
hydro-, *hydros,* water: hydrolysis
hyo-, *hyoeides,* U-shaped: hyoid
hyper-, *hyper,* above: hypertonic
hypo-, *hypo,* under: hypotonic
hyster-, *hystera,* uterus: hysterectomy
-ia, state or condition: insomnia
ile-, *ileum:* ileocecal valve
ili-, ilio-, *ilium:* iliac
in-, in, within, or denoting negative effect: inactivate
infra-, *infra,* beneath: infraorbital
inter-, *inter,* between: interventricular
intra-, *intra,* within: intracapsular
iso-, *isos,* equal: isotonic
-itis, *-itis,* inflammation: dermatitis
karyo-, *karyon,* body: megakaryocyte
kerato-, *keros,* horn: keratin
kino-, -kinin, *kinein,* to move: cholecystokinin
lact-, lacto-, -lactin, *lac,* milk: prolactin
leuko-, *leukos,* white: leukocyte
liga-, *ligare,* to bind together: ligament
lip-, lipo-, *lipos,* fat: lipid
lyso-, -lysis, -lyze, *lysis,* dissolution: hydrolysis
macr-, *makros,* large: macrophage
mal-, *mal,* abnormal: malabsorption
mamilla-, *mamilla,* little breast: mamillary
mast-, masto-, *mastos,* breast: mastoid
mega-, *megas,* big: megakaryocyte
melan-, *melas,* black: melanocyte
men-, *men,* month: menstrual
mero-, *meros,* part: merocrine
meso-, *mesos,* middle: mesoderm
meta-, *meta,* after, beyond: metaphase
micr-, *mikros,* small: microscope
mono-, *monos,* single: monocyte
morpho-, *morphe,* form: morphology
multi-, *multus,* much, many: multicellular
myo-, *mys,* muscle: myofibril
narc-, *narkoun,* to numb or deaden: narcotics
nas-, *nasus,* nose: nasolacrimal duct
natri-, *natrium,* sodium: natriuretic
necr-, *nekros,* corpse: necrosis
nephr-, *nephros,* kidney: nephron
neur-, neuro-, *neuron,* nerve: neuromuscular
oculo-, *oculus,* eye: oculomotor
odont-, *odontos,* tooth: odontoid process
-oid, *eidos,* form, resemblance: lymphoid
oligo-, *oligos,* little, few: oligodendrocyte
-ology, *logos,* the study of: physiology

-oma, *-oma,* swelling: carcinoma
onco-, *onkos,* mass, tumor: oncology
oo-, *oon,* egg: oocyte
ophthalm-, *ophthalmos,* eye: ophthalmic nerve
-opia, *ops,* eye: optic
orb-, *orbita,* a circle: orbicularis oris
orth-, *orthos,* correct, straight: orthopedist
osteon, osteo-, *os,* bone: osteocyte
oto-, *otikos,* ear: otolith
para-, *para,* beyond: parathyroid
patho-, -path, -pathy, *pathos,* disease: pathology
pedia-, *paidos,* child: pediatrician
per-, *per,* through, throughout: percutaneous
peri-, *peri,* around: periosteum
phag-, *phagein,* to eat: phagocyte
-phasia, *phasis,* speech: aphasia
-phil, -philia, *philus,* love: hydrophilic
-phobe, -phobia, *phobos,* fear: hydrophobic
phot-, *phos,* light: photoreceptor
-phylaxis, *phylax,* a guard: prophylaxis
physio-, *physis,* nature: physiology
platy-, *platys,* flat: platysma
-plegia, *plege,* a blow, paralysis: paraplegia
-plexy, *plessein,* to strike: apoplexy
pneum-, *pneuma,* air: pneumotaxic center
podo, *podon,* foot: podocyte
-poiesis, *poiesis,* making: hemopoiesis
poly-, *polys,* many: polysaccharide
post-, *post,* after: postnatal
pre-, *prae,* before: premolar
presby-, *presbys,* old: presbyopia
pro-, *pro,* before: prophase
proct-, *proktos,* anus: proctology
pulmo-, *pulmo,* lung: pulmonary
pyel-, *pyelos,* trough or pelvis: pyelitis
quadr-, *quadrans,* one quarter: quadriceps
re-, back, again: reinfection
retro-, *retro,* backward: retroperitoneal
rhin-, *rhis,* nose: rhinitis
-rrhage, *rhegnymi,* to burst forth: hemorrhage
-rrhea, *rhein,* flow, discharge: amenorrhea
sarco-, *sarkos,* flesh: sarcomere
scler-, sclero-, *skleros,* hard: sclera
-scope, *skopeo,* to view: colonoscope
-sect, *sectio,* to cut: transect
semi-, *semis,* half: semitendinosus
-septic, *septikos,* putrid: antiseptic
-sis, state or condition: metastasis
som-, -some, *soma,* body: somatic
spino-, *spina,* spine, vertebral column: spinalis
-stalsis, *staltikos,* contractile: peristalsis
-stomy, *stoma,* mouth, opening: colostomy
stylo-, *stylus,* stake, pole: styloid
sub-, *sub,* below: subcutaneous
super-, *super,* above or beyond: superficial
supra-, *supra,* on the upper side: suprarenal
syn-, *syn,* together: synthesis
tachy-, *tachys,* swift: tachycardia
telo-, *telos,* end: telophase
therm-, thermo-, *therme,* heat: thermoregulation
thromb-, *thrombos,* clot: thrombocyte
-tomy, *temnein,* to cut: appendectomy
tox-, *toxikon,* poison: toxin
trans-, *trans,* through: transport
tri-, *tres,* three: trimester
-tropic, *trope,* turning: adrenocorticotropic
tropho-, *trophe,* nutrition: trophoblast
-trophy, *trophikos,* nourishing: atrophy
uni-, *unus,* one: unicellular
uro-, -uria, *ouron,* urine: glycosuria
vas-, *vas,* vessel: vascular
zyg-, *zygotos,* yoked: zygote

Glossary/Index